Methods in Molecular Biology™

Series Editor
John M. Walker
School of Life Sciences
University of Hertfordshire
Hatfield, Hertfordshire, AL10 9AB, UK

For further volumes:
http://www.springer.com/series/7651

Transcriptional Regulation

Methods and Protocols

Edited by

Ales Vancura

Department of Biological Sciences, St. John's University, Queens, NY, USA

Humana Press

Editor
Ales Vancura, Ph.D
Department of Biological Sciences
St. John's University
Queens, NY, USA
vancuraa@stjohns.edu

ISSN 1064-3745 e-ISSN 1940-6029
ISBN 978-1-4939-5854-2 ISBN 978-1-61779-376-9 (eBook)
DOI 10.1007/978-1-61779-376-9
Springer New York Dordrecht Heidelberg London

Softcover re-print of the Hardcover 1st edition 2012

Printed on acid-free paper

Humana Press is part of Springer Science+Business Media (www.springer.com)

Preface

In the last 20 years, the methodologies used in the field of transcriptional regulation have undergone remarkable changes. Perhaps the most significant advancements are the development of genome-wide approaches for measuring gene expression, exemplified by gene chips (chip), and chromatin immunoprecipitation assays (ChIP) for measuring in vivo protein–DNA interactions at any genomic loci. The combination of genome-wide approaches and chromatin immunoprecipitation culminated in developing methods for measuring protein–DNA interactions at a genome-wide scale, such as ChIP–chip and ChIP–Seq. The regulatory role of chromatin and histone modifications also came to the forefront during the last 20 years, adding new layers of complexity to our understanding of transcriptional regulation. In addition, mass spectrometry methods identified an array of accessory proteins involved in all stages of RNA synthesis and its regulation.

This volume is divided in four parts: (1) Promoter elements, transcription factors, and preinitiation complex (PIC) assembly, (2) Chromatin structure, (3) Chromatin modifying complexes, and (4) RNA synthesis and regulation. Strict classification of the described methods into these four parts, however, is very difficult. For example, ChIP methods are used for studies of transcription factors binding to promoter elements as well as for the characterization of chromatin structure and histone modifications.

Most of the protocols presented in this volume have been developed using mammalian, yeast, or *Drosophila* cells and tissues, but most methods can be applied across the species and cell types. The methods in all four parts have been chosen to represent both classic and cutting-edge techniques to study transcriptional regulation. The reliability of all the protocols has been tested in laboratories around the world. Each chapter is appended by notes that navigate through the protocol and serves as a troubleshooting guide. It is our hope that this book will be useful not only to senior researchers and scientists experienced in transcriptional regulation, but also to graduate students and scientists who want to study transcriptional regulation for the first time.

I would like to thank all the authors for their enthusiastic help and support in assembling this volume; I fully realize that in the highly competitive environment of academic research, many scientists are reluctant to commit their time to writing book chapters and review articles. I also want to thank Dr. Ivana Vancurova for her significant help in editing this volume. Last, but not least, I would like to express my gratitude to the series editor, Dr. John Walker, and the outstanding staff of Humana Press for their support, help, and encouragement.

Queens, NY, USA *Ales Vancura*

Contents

Contributors

SHIGEKI ARAI • *Division of Gene Structure and Function, Research Center for Genomic Medicine, Saitama Medical University, Saitama-Ken, Japan*
DIDIER AUBOEUF • *INSERM, Lyon, France*
ALICIA BARROSO-DELJESUS • *Genomics Facility, Instituto de Parasitología y Biomedicina "López-Neyra", IPBLN-CSIC, Parque Tecnológico de Ciencias de la Salud, Av del Conocimiento s/n, Granada, Spain*
BLAINE BARTHOLOMEW • *Southern Illinois University School of Medicine, Carbondale, IL, USA*
BRENDAN BELL • *RNA Group, Département de microbiologie et d'infectiologie, Faculté de médecine et sciences de la santé, Université de Sherbrooke, Sherbrooke, QC, Canada*
ALFREDO BERZAL-HERRANZ • *Instituto de Parasitología y Biomedicina "López-Neyra", IPBLN-CSIC, Parque Tecnológico de Ciencias de la Salud, Av del Conocimiento s/n, Granada, Spain*
MARK D. BIGGIN • *Genomics Division, Lawrence Berkeley National Laboratory, Berkeley, CA, USA*
DANIELLE BITTENCOURT • *Department of Biochemistry and Molecular Biology, Keck School of Medicine, University of Southern California, Los Angeles, CA, USA*
ALLAN R. BRASIER • *Department of Internal Medicine, Institute for Translational Sciences, and Sealy Center for Molecular Medicine, University of Texas Medical Branch, Galveston, TX, USA*
ALEXANDER BREHM • *Institut für Molekularbiologie und Tumorforschung (IMT), Philipps-Universität Marburg, Marburg, Germany*
JENNIFER CHANG • *Department of Cell Biology, New York University School of Medicine, New York, NY, USA*
TZU-PEI CHANG • *Department of Biological Sciences, St. John's University, Queens, NY, USA*
ROMAIN CHARTON • *Département de Microbiologie et Infectiologie, Faculté de Médecine, Université de Sherbrooke, Sherbrooke, QC, Canada*
CHRIS CHEADLE • *Lowe Family Genomics Core, Division of Allergy and Clinical Immunology, Department of Medicine, The Johns Hopkins University School of Medicine, Baltimore, MD, USA*
YU-CHI CHEN • *Lowe Family Genomics Core, Division of Allergy and Clinical Immunology, Department of Medicine, The Johns Hopkins University School of Medicine, Baltimore, MD, USA*
YOUNG-WOOK CHO • *Nuclear Receptor Biology Section, NIDDK, NIH, Bethesda, MD, USA*
CHI-WING CHOW • *Department of Molecular Pharmacology, Albert Einstein College of Medicine, Bronx, NY, USA*

Antonio Conconi • *Département de Microbiologie et Infectiologie, Faculté de Médecine, Université de Sherbrooke, Sherbrooke, QC, Canada*
Chi V. Dang • *Division of Hematology, Departments of Medicine, Cell Biology, Oncology, and Pathology, The Johns Hopkins University School of Medicine, Baltimore, MD, USA*
Mekonnen Lemma Dechassa • *Southern Illinois University School of Medicine, Carbondale, IL, USA*
Mathew P. Dixon • *The Walter and Eliza Hall Institute of Medical Research, Melbourne, VIC, Australia*
Sherif Abou Elela • *RNA Group/Groupe ARN, Département de Microbiologie & Infectiologie, Faculté de médecine et des sciences de la santé, Université de Sherbrooke, Sherbrooke, QC, Canada*
Tamara Y. Erkina • *College of Pharmacy and Health Sciences, Butler University, Indianapolis, IN, USA*
Alexandre M. Erkine • *College of Pharmacy and Health Sciences, Butler University, Indianapolis, IN, USA*
Francesco Faiola • *Department of Biochemistry, University of California, Riverside, CA, USA*
Jinshui Fan • *Lowe Family Genomics Core, Division of Allergy and Clinical Immunology, Department of Medicine, The Johns Hopkins University School of Medicine, Baltimore, MD, USA*
Padmaja Gade • *Department of Microbiology & Immunology, Greenebaum Cancer Center, University of Maryland School of Medicine, Baltimore, MD, USA*
Jules Gagnon • *RNA Group/Groupe ARN, Département de Microbiologie & Infectiologie, Faculté de médecine et des sciences de la santé, Université de Sherbrooke, Sherbrooke, QC, Canada*
Luciano Galdieri • *Department of Biological Sciences, St. John's University, Queens, NY, USA*
Kai Ge • *Nuclear Receptor Biology Section, NIDDK, NIH, Bethesda, MD, USA*
Ghada Ghazal • *RNA Group/Groupe ARN, Département de Microbiologie & Infectiologie, Faculté de médecine et des sciences de la santé, Université de Sherbrooke, Sherbrooke, QC, Canada*
Myriam Gorospe • *Laboratory of Cellular and Molecular Biology, National Institute on Aging-Intramural Research Program, NIH, Baltimore, MD, USA*
Susanna F. Greer • *Division of Cellular and Molecular Biology and Physiology, Department of Biology, Georgia State University, Atlanta, GA, USA*
Joachim Griesenbeck • *Naturwissenschaftliche Fakultät III, Institut für Biochemie III, Universität Regensburg, Regensburg, Germany*
Ryoko Harada • *The Rosalind and Morris Goodman Cancer Research Centre, McGill University, Montreal, QC, Canada*
Anke Hoffmann • *Max Planck Institute for Psychiatry, Munich, Germany*
SunHwa Hong • *Nuclear Receptor Biology Section, NIDDK, NIH, Bethesda, MD, USA*
Swetansu K. Hota • *Southern Illinois University School of Medicine, Carbondale, IL, USA*

Jing Hu • *Center for Eukaryotic Gene Regulation, and Genetics Graduate Program, Pennsylvania State University, University Park, PA, USA*
Ashish Juvekar • *Department of Biological Sciences, St. John's University, Queens, NY, USA*
Dhan V. Kalvakolanu • *Department of Microbiology & Immunology, Greenebaum Cancer Center, University of Maryland School of Medicine, Baltimore, MD, USA*
Steven Kennedy • *Department of Biological Sciences, St. John's University, Queens, NY, USA*
Andrew J. Kueh • *The Walter and Eliza Hall Institute of Medical Research, Melbourne, VIC, Australia*
Riki Kurokawa • *Division of Gene Structure and Function, Research Center for Genomic Medicine, Saitama Medical University, Saitama-Ken, Japan*
Pingxin Li • *Center for Eukaryotic Gene Regulation, University Park, PA, USA; Department of Biochemistry and Molecular Biology, Pennsylvania State University, University Park, PA, USA; Genetics Graduate Program, Pennsylvania State University, University Park, PA, USA*
Xiao-Yong Li • *Genomics Division, Lawrence Berkeley National Laboratory, Berkeley, CA, USA*
Emilie Louvet • *Department of Cell and Molecular Biology, Karolinska Institutet, Stockholm, Sweden*
Subrata Manna • *Department of Biological Sciences, St. John's University, Queens, NY, USA*
Ernest Martinez • *Department of Biochemistry, University of California, Riverside, CA, USA*
Charles E. Massie • *CRUK Cambridge Research Institute, Cambridge, UK; Department of Haematology, Cambridge Institute for Medical Research, Cambridge, UK*
Tamara McLennan • *The Walter and Eliza Hall Institute of Medical Research, Melbourne, VIC, Australia*
Pablo Menendez • *Stem Cells, Development & Cancer Laboratory GENyO: Centre for Genomics and Oncology Pfizer-University of Granada-Andalusian Govermment, Parque Tecnológico de Ciencias de la Ilustración, Granada, Spain*
Eran Meshorer • *Department of Genetics, Institute of Life Sciences, The Hebrew University of Jerusalem, Jerusalem, Israel*
Ian G. Mills • *CRUK Cambridge Research Institute, Cambridge, UK; Centre for Molecular Medicine Norway, Nordic European Molecular Biology Laboratory Partnership, University of Oslo, Oslo, Norway*
Nuno P. Mira • *Institute for Biotechnology and Bioengineering, Technical University of Lisbon, Lisbon, Portugal*
John Moon • *Department of Biological Sciences, St. John's University, Queens, NY, USA*
Magdalena Murawska • *Institut für Molekularbiologie und Tumorforschung (IMT), Philipps-Universität Marburg, Marburg, Germany*

CHRIS MURGATROYD • *Max Planck Institute for Psychiatry, Munich, Germany*
ALAIN NEPVEU • *Departments of Biochemistry, Oncology, and Medicine, The Rosalind and Morris Goodman Cancer Research Centre, McGill University, Montreal, QC, Canada*
MICHAEL OHH • *Department of Laboratory Medicine and Pathobiology, University of Toronto, Toronto, ON, Canada*
TIMOTHY M. PALMER • *Institute for Cardiovascular and Medical Sciences, College of Medical, Veterinary and Life Sciences, University of Glasgow, Glasgow, Scotland, UK*
AMPARO PASCUAL-AHUIR • *Instituto de Biología Molecular y Celular de Plantas, CSIC-Universidad Politécnica de Valencia, Valencia, Spain*
NADINE PELLETIER • *Department of Medicine, The Rosalind and Morris Goodman Cancer Research Centre, McGill University Health Center, Montréal, Québec, Canada*
JULIE PELLOUX • *Département de Microbiologie et Infectiologie, Faculté de Médecine, Université de Sherbrooke, Sherbrooke, QC, Canada*
PIERGIORGIO PERCIPALLE • *Department of Cell and Molecular Biology, Karolinska Institutet, Stockholm, Sweden*
PUNIT PRASAD • *Southern Illinois University School of Medicine, Carbondale, IL, USA*
MARKUS PROFT • *Instituto de Biología Molecular y Celular de Plantas, CSIC-Universidad Politécnica de Valencia, Valencia, Spain*
SITHARAM RAMASWAMI • *Department of Biological Sciences, St. John's University, Queens, NY, USA*
OLGA ROCHE • *Department of Laboratory Medicine and Pathobiology, University of Toronto, Toronto, ON, Canada*
CRISTINA ROMERO-LÓPEZ • *Instituto de Parasitología y Biomedicina "López-Neyra", IPBLN-CSIC, Parque Tecnológico de Ciencias de la Salud, Av del Conocimiento s/n, Granada, Spain*
XIAOAN RUAN • *Genome Institute of Singapore, Singapore, Singapore*
YIJUN RUAN • *Genome Institute of Singapore, Singapore, Singapore*
ISABEL SÁ-CORREIA • *Institute for Biotechnology and Bioengineering, Technical University of Lisbon, Lisbon, Portugal*
BADI SRI SAILAJA • *Department of Genetics, Institute of Life Sciences, The Hebrew University of Jerusalem, Jerusalem, Israel*
WILLIAM A. SANDS • *Institute for Cardiovascular and Medical Sciences, College of Medical, Veterinary and Life Sciences, University of Glasgow, Glasgow, Scotland, UK*
DAVID ALAN SCHNEIDER • *Department of Biochemistry and Molecular Genetics, University of Alabama at Birmingham, Birmingham, AL, USA*
XIAOYUAN SONG • *Department of Medicine, University of California, San Diego School of Medicine, La Jolla, CA, USA*
DIETMAR SPENGLER • *Max Planck Institute for Psychiatry, Munich, Germany*
LEANNE STALKER • *Department of Biochemistry, McMaster University, Hamilton, ON, Canada*
CHRISTOPHER TAKACS • *RNA Group, Département de microbiologie et d'infectiologie, Faculté de médecine et sciences de la santé, Université de Sherbrooke, Sherbrooke, QC, Canada*

TAKUMI TAKIZAWA • *Graduate School of Biological Sciences, Nara Institute of Science and Technology, Nara, Japan*
MIGUEL C. TEIXEIRA • *Institute for Biotechnology and Bioengineering, Technical University of Lisbon, Lisbon, Portugal*
TIM THOMAS • *The Walter and Eliza Hall Institute of Medical Research, Melbourne, VIC, Australia*
BING TIAN • *Department of Internal Medicine, University of Texas Medical Branch, Galveston, TX, USA*
MAXIME TREMBLAY • *Département de Microbiologie et Infectiologie, Faculté de Médecine, Université de Sherbrooke, Sherbrooke, QC, Canada*
AGNIESZKA D. TRUAX • *Division of Cellular and Molecular Biology and Physiology, Department of Biology, Georgia State University, Atlanta, GA, USA*
ALES VANCURA • *Department of Biological Sciences, St. John's University, Queens, NY, USA*
IVANA VANCUROVA • *Department of Biological Sciences, St. John's University, Queens, NY, USA*
ANNE K. VOSS • *The Walter and Eliza Hall Institute of Medical Research, Melbourne, VIC, Australia*
XIANGTING WANG • *Department of Medicine, University of California, San Diego School of Medicine, La Jolla, CA, USA*
YANMING WANG • *Center for Eukaryotic Gene Regulation, University Park, PA, USA; Department of Biochemistry and Molecular Biology, Pennsylvania State University, University Park, PA, USA; Genetics Graduate Program, Pennsylvania State University, University Park, PA, USA*
YUAN-LIANG WANG • *Department of Biochemistry, University of California, Riverside, CA, USA*
TONYA WATKINS • *MS Lowe Family Genomics Core, Division of Allergy and Clinical Immunology, Department of Medicine, The Johns Hopkins University School of Medicine, Baltimore, MD, USA*
RAYMUND J. WELLINGER • *Département de Microbiologie et Infectiologie, Faculté de Médecine, Université de Sherbrooke, Sherbrooke, QC, Canada*
EMMANUELLE WILHELM • *RNA Group, Département de microbiologie et d'infectiologie, Faculté de médecine et sciences de la santé, Université de Sherbrooke, Sherbrooke, QC, Canada*
MANUEL WITTNER • *Naturwissenschaftliche Fakultät III, Institut für Biochemie III, Universität Regensburg, Regensburg, Germany*
HAYLEY D. WOOLSON • *Institute for Cardiovascular and Medical Sciences, College of Medical, Veterinary and Life Sciences, University of Glasgow, Glasgow, Scotland, UK*
CHAO-JUNG WU • *Department of Biochemistry, The Rosalind and Morris Goodman Cancer Research Centre, McGill University, Montréal, Québec, Canada; Department of Medicine, The Rosalind and Morris Goodman Cancer Research Centre, McGill University, Montréal, Québec, Canada*
CHRISTOPHER WYNDER • *Department of Biochemistry, Schulich School of Medicine and Dentistry, The University of Western Ontario, London, ON, Canada*
JOHN J. WYRICK • *School of Molecular Biosciences and Center for Reproductive Biology, Washington State University, Pullman, WA, USA*

KEZHI YAN • *Department of Biochemistry, The Rosalind and Morris Goodman Cancer Research Centre, McGill University, Montréal, Québec, Canada*
JUN YANG • *Department of Internal Medicine, University of Texas Medical Branch, Galveston, TX, USA*
TEDDY T. C. YANG • *Department of Molecular Pharmacology, Albert Einstein College of Medicine, Bronx, NY, USA*
XIANG-JIAO YANG • *Department of Biochemistry, The Rosalind and Morris Goodman Cancer Research Centre, McGill University, Montréal, Québec, Canada; Department of Medicine, The Rosalind and Morris Goodman Cancer Research Centre, McGill University, Montréal, Québec, Canada*
STEPHEN J. YARWOOD • *Institute for Molecular, Cell and Systems Biology, College of Medical, Veterinary and Life Sciences, University of Glasgow, Glasgow, Scotland, UK*
ADEEL ZUBAIR • *Department of Biological Sciences, St. John's University, Queens, NY, USA*

Part I

Promoter Elements, Transcription Factors, and Pre-initiation Complex Assembly

Chapter 1

Genome-Wide In Vivo Cross-linking of Sequence-Specific Transcription Factors

Xiao-Yong Li and Mark D. Biggin

Abstract

Immunoprecipitation of cross-linked chromatin in combination with microarrays (ChIP-chip) or ultra high-throughput sequencing (ChIP-seq) is widely used to map genome-wide in vivo transcription factor binding. Both methods employ initial steps of in vivo cross-linking, chromatin isolation, DNA fragmentation, and immunoprecipitation. For ChIP-chip, the immunoprecipitated DNA samples are then amplified, labeled, and hybridized to DNA microarrays. For ChIP-seq, the immunoprecipitated DNA is prepared for a sequencing library, and then the library DNA fragments are sequenced using ultra high-throughput sequencing platform. The protocols described here have been developed for ChIP-chip and ChIP-seq analysis of sequence-specific transcription factor binding in *Drosophila* embryos. A series of controls establish that these protocols have high sensitivity and reproducibility and provide a quantitative measure of relative transcription factor occupancy. The quantitative nature of the assay is important because regulatory transcription factors bind to highly overlapping sets of thousands of genomic regions and the unique regulatory specificity of each factor is determined by relative moderate differences in occupancy between factors at commonly bound regions.

Key words: In vivo cross-linking, Sequence-specific transcription factors, ChIP-chip, Chip-seq

1. Introduction

Sequence-specific transcription factors control the differential expression of genes in response to various physiological and pathological stimuli and during pattern formation and differentiation in animals and plants. To understand how these important biological events occur, it is critical to know where and at what level transcription factors bind throughout the genome (1). In vivo cross-linking followed by chromatin immunoprecipitation and either southern blot or PCR has long been used to detect direct binding

Ales Vancura (ed.), *Transcriptional Regulation: Methods and Protocols,* Methods in Molecular Biology, vol. 809,
DOI 10.1007/978-1-61779-376-9_1, © Springer Science+Business Media, LLC 2012

of transcription factors in vivo at a sample of genomic regions and has provided critical lessons in the distribution of factor binding (2–4). In the last decade, however, the advent of DNA microarray technology has allowed mapping of in vivo cross-linking to many more regions throughout the genome by the so-called ChIP-chip (5, 6), opening the way for computational/statistical analysis of the patterns of transcription factor binding and its relationship to other large-scale data sets for gene function and expression and chromatin structure. More recently, ultra high-throughput next generation DNA sequencing technologies, such as Roche 454, Illumina genome analyzer, the ABI SOLiD, and Helicos HeliScope (see ref. 7 for review), have become the predominant method for genome-wide mapping of in vivo cross-linking (8–10). This new ChIP-seq method has several advantages. It allows full genome coverage, regardless of the size of the genome and can be applied to any organism as long as the genome sequence is known, without the additional startup cost of building a microarray. It also has better dynamic range and higher resolution. For more very large genomes, ChIP-chip results tend to be noisy, presumably due to probe cross-hybridization, though this does not seem to be problem for genomes of the size of Drosophila (~180 Mb).

The ChIP-chip and ChIP-seq protocols used by different groups vary considerably. The protocols described here are designed for *Drosophila* embryos and have been optimized for high reproducibility and to provide a quantitative measure of relative levels of factor occupancy by seeking to minimize and control for any systematic experimental biases. The major steps described are shown in Fig. 1. The first difference between our protocol and others is specific to *Drosophila* embryos, which are surrounded by a hydrophobic vitelline membrane that must be permeabilized to allow entry of formaldehyde. In our protocol, this is accomplished by treating embryos with isopropanol followed by a hexane formaldehyde solution (11), whereas protocols for intact yeast or tissue culture cells readily accomplish fixation by treatment with aqueous formaldehyde. The second difference is more significant. Most groups isolate chromatin by sonicating intact nuclei to release short (~200–800 bp) length DNA fragments and then immunoprecipitate this unpurified chromatin (e.g. see refs. 5, 6, 8). The protocol presented here, instead, extracts long fragments of cross-linked chromatin (>20 kb) from nuclei using high concentrations of detergent and then purifies the cross-linked DNA by CsCl buoyant density ultracentrifugation, which dissociates non-covalently attached proteins from the DNA. The cross-linked DNA is then sonicated to smaller sizes and the immunoprecipitation is performed on this purified material. Recently, it has been shown that extracting DNA by sonicating intact nuclei, as many protocols do, may introduce an experimental bias by preferentially releasing DNA from accessible chromatin regions (12). In contrast, our protocol does not

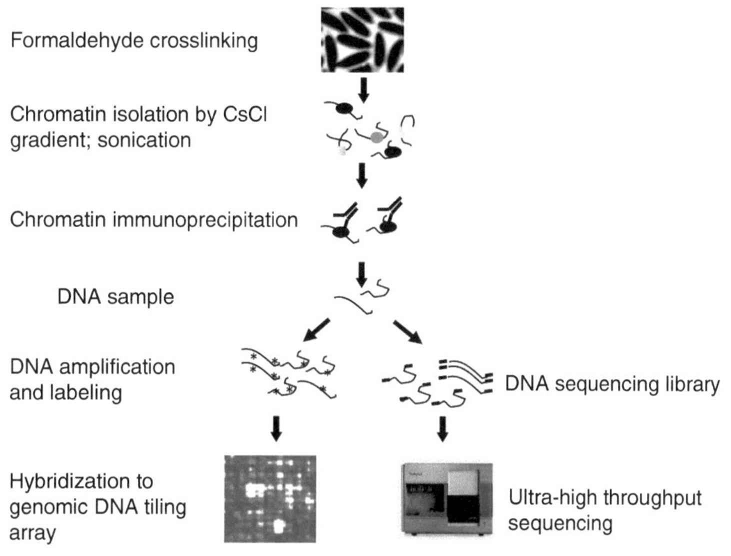

Fig. 1. Schematic representation of the ChIP-chip and ChIP-seq method.

introduce this bias (Fig. 2) (13, 14). Finally, another major difference between protocols is the approach used to amplify the immunoprecipitated DNA for ChIP-chip analysis. We have found that this step must be carefully optimized to reduce as much as possible stochastic experimental noise. Our amplification protocols have been optimized to give high reproducible between replicas and to faithfully preserve the relative levels of enrichment of different DNAs even when sub-nanogram amount of genomic DNA is used in either our ChIP-chip protocol (15) (Fig. 2).

For successful ChIP-chip or ChIP-seq experiments, it is critical to have first rate antibodies. In our experience, affinity-purified polyclonal antibodies are highly effective. We have been successful in obtaining ChIP quality antibodies for nearly every factor attempted, with good data being obtained for 21 proteins (16). Monoclonal antibodies or antibodies raised against a short protein sequence, in contrast, tend to be less effective; potential epitope masking may not only affect the general effectiveness, but may also cause bias in which genomic regions are detected. It is important that the portion(s) of the protein chosen for affinity purifying antibodies do not share homology with other proteins in the organism. Suitable nonhomologous regions can be identified using NCBI blast. Regions of approximately 100 amino acids in length are generally effective. To further ensure that the antibodies do not cross-react with other proteins, ideally, for each factor two ChIP experiments should be performed using affinity-purified antibodies against nonoverlapping parts of the same protein. We have found that the results between such antibody pairs are strikingly similar for the same factor, whereas data for different factors is, of course, different (15, 17) (Fig. 2).

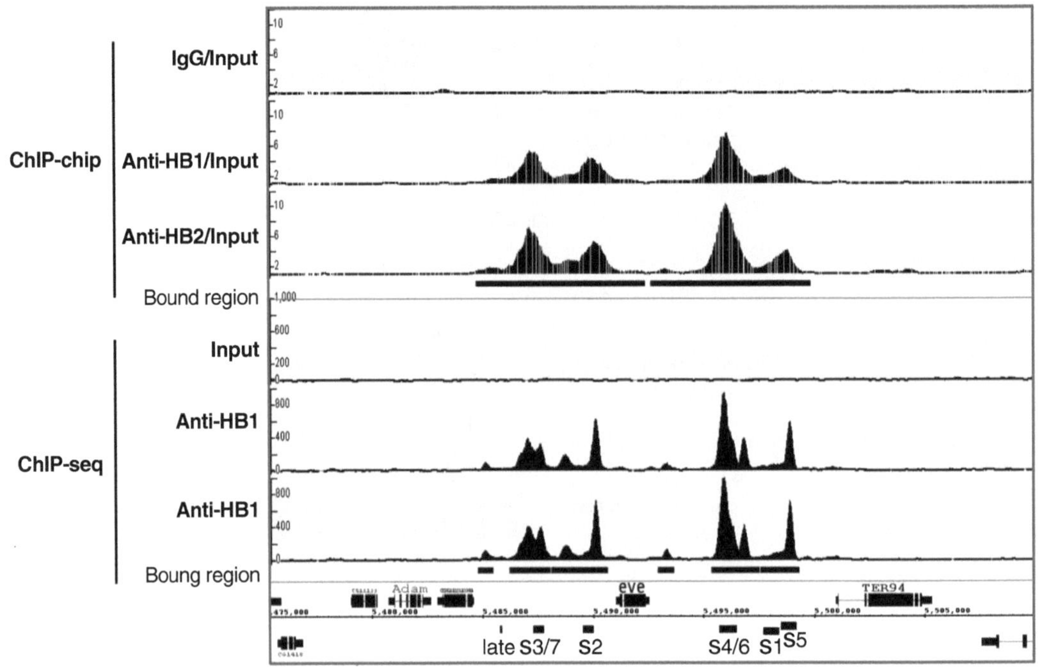

Fig. 2. The ChIP-chip score profile (*top half*) or ChIP-seq tag-density profile (*bottom half*) for anti-HB or normal rabbit IgG control chromatin immunoprecipitation. In each case, two ChIPs were performed using two different anti-HB antibodies against nonoverlapping epitopes (HB1 and HB2). The ChIP-chip score was calculated as $\log_2$ (mean Factor IP/mean input DNA) for the Factor IP track, and $\log_2$ (mean IgG control IP/mean input DNA) for the IgG control track, and a moving average with a window size of 675 bp was calculated. Shown in the figure are the unlogged window scores. The bound regions detected by TiMAT are marked underneath the binding profiles. For ChIP-seq, the short sequence tags aligned to the genome were extended by the average size of the fragments in the DNA library to generate the tag-density profiles, as shown. The signals shown for the different samples were scaled to the same total number of tags. The bound regions identified by MACS are shown underneath the profiles.

Our ChIP-chip studies have been carried out using the Affymetrix gene chip system, and ChIP-seq has been done using the Illumina Solexa genome analyzer. For our ChIP-chip and ChIP-seq experiments, besides the chromatin immunoprecipitation samples, two types of controls are always included: the input DNA and mock IP controls. All three types of samples are performed in duplicate for ChIP-chip. For ChIP-seq, to reduce costs the duplicates of each sample have been pooled prior to amplification and sequencing.

2. Materials

2.1. Fixation of Embryos with Formaldehyde

1. Large fly population cages (Genesee Scientific: #59-104).
2. Fine and coarse polyamide nylon mesh (Genesee Scientific: #:57-102 (fine) and #57-101 (coarse)).

3. 50% Clorox bleach (2.6% hypochlorite solution).
4. 10× PBS: 1.37 M NaCl, 27 mM KCl, 43 mM Na_2HPO_4, 14 mM KH_2PO_4 pH 7.3. Autoclave and store at room temperature.
5. Formaldehyde/hexane fixing solution: mix 210 ml hexane, 37 ml 37% formaldehyde (Sigma, #252549), and 27.5 ml 10× PBS, in a bottle with a magnetic stir bar. Vigorously stir the mixture for at least half an hour. Aliquot the top layer to a separate container. Make it the same day it is going to be used.
6. Embryo washing buffer: 1× PBS, 0.1% Triton X-100.

2.2. Purification of Chromatin

1. Motorized dounce homogenizer (Thomas Teflon Pestle Tissue Homogenizer #3431-E25).
2. Glass dounce (Bellco #1984-40040).
3. SW28 or SW41 rotor and centrifuge tubes.
4. SS34 rotor and centrifuge tubes.
5. A sonicator (e.g., Branson sonifier 450 or Bioruptor (Diagenode)).
6. Embryo homogenization buffer: 0.3 M sucrose, 15 mM NaCl, 5 mM $MgCl_2$, 15 mM Tris–Cl, pH 7.5, 60 mM KCl, 0.1 mM EDTA, 0.1 mM EGTA. Filter sterilize (0.22 μm) and store at 4°C.

   Add DTT to 0.5 mM and PMSF to 1 mM (from a 200 mM stock in ethanol) immediately before use.
7. Nuclear lysis buffer: 100 mM NaCl, 10 mM Tris–Cl, pH 8.0, 1 mM EDTA, 0.1%NP-40. Filter sterilize (0.22 μm) and store at 4°C. Add PMSF to 1 mM immediately before use.
8. 20% *N*-lauroylsarcosine (Sigma); 20% SDS (Sigma); 20% Triton X-100 (Sigma). Filter (0.22 μm) and store at room temperature.
9. CsCl buffer: 2% *N*-lauroylsarcosine, 1 mM EDTA. Filter (0.22 μm) and store at room temperature. Add 1 mM PMSF immediately before use.
10. CsCl solutions of densities at 1.5, 1.4, and 1.3 g/ml. For 10 ml solutions, dissolve 6.67 g, 5.34, and 4 g, in 8.33, 8.66, and 9 ml CsCl buffer, respectively. Adjust the volume to be prepared according to number of chromatin samples.
11. Dialysis tubing (Spectrum: Spec 2).
12. Dialysis buffer: 20 mM Tris–Cl, pH 8.0, 2 mM EDTA in sterile ddH_2O. Add PMSF to 1 mM immediately before use.

2.3. Chromatin Immunoprecipitation

All solutions and buffers for this step should be of high quality, especially for applications near end of the immunoprecipitation procedure and the subsequent processing of the samples for ChIP-chip and ChIP-seq. Preferably, the following solutions should be purchased from commercial resources: molecular grade water, 1 M Tris–Cl, pH 8.0, 0.5 M EDTA, 3 M sodium acetate.

1. 50 mg/ml BSA (Invitrogen, 15561-020), 10 mg/ml RNase A (Roche, #11119915001), 20 mg/ml proteinase K (Invitrogen, #AM2546).
2. 1 M Tris–Cl, pH 8.0 (Sigma, #93316), 0.5 M EDTA (Sigma, #E7889), 3 M NaOAc, pH 5.3 (Sigma, #71196).
3. 200 mM PMSF in 100% ethanol.
4. Protein A – sephacryl 1,000 beads (see Note 1).
5. IP buffer: 10 mM Tris–Cl pH 8.0, 1 mM EDTA, 150 mM NaCl, 0.5% Triton X-100, 0.1% sodium deoxycholate (Sigma, #5670), 0.5% *N*-lauroylsarcosine. Add PMSF to 1 mM immediately before use.
6. 5× IP buffer.
7. 0.5 M ChrIP wash buffer: same as ChrIP buffer, except the NaCl concentration is 0.5 M.
8. LiCl washing buffer: 10 mM Tris–Cl pH 8.0, 1 mM EDTA, 250 mM LiCl, 1% NP40 (or IGEPAL 630), 1% sodium deoxylcholate.
9. TE: 10 mM Tris–Cl pH 8.0, 2 mM EDTA.
10. TE (10:0.1): 10 mM Tris–Cl pH 8.0, 0.1 mM EDTA.
11. Antibodies: affinity purified.
12. Normal rabbit IgG (Santa Cruz Biotech, sc-2027).
13. Glycogen.
14. Phenol/chloroform.
15. Chloroform.
16. PCR primers flanking a known binding site(s) for the transcription factor to be analyzed. For ChIP-seq, where the DNA fragments are short, the amplicon should be as close to the true binding site as possible.

2.4. DNA Amplification and Labeling for ChIP-Chip

1. Sequenase 2.0 kit (USB corp, #70775Y).
2. Random prime primer A: GTTTCCCAGTCACGGTC(N)9. Custom synthesize, HPLC purify, dilute to 200 μM.
3. Random prime primer B: GTTTCCCAGTCACGGTC. Custom synthesize, HPLC purify. dilute to 100 μM.
4. 50 mg/ml BSA (Invitrogen, #15561-020), 0.1 M DTT, 25 mM dNTPs.
5. Microspin S-300 HR column (Amersham, #27-5130-01).
6. Taq2000 (Stratagene, #600196).
7. QIAquick PCR purification kit (Qiagen, #28104).
8. 1 U/μl DNase I (Epicentre, #D9902K).
9. One-Phor-All buffer (Roche).

10. Biotin-ddATP (Perkin Elmer, #NEL548).
11. TdTase (Roche, #3333566).
12. Thermocycler.

2.5. Array Hybridization and Scanning

1. 12× MES buffer: dissolve 70.4 g MES free acid monohydrate (Sigma, #69889), and 193.3 g MES-Na (Sigma, # M3058) in ddH_2O, adjust to 1 l. Filter sterilize and store at 4°C.
2. 3 nM B2 oligo (Affymetrix, #900301).
3. 2× MES-Triton solution: mix 1.66 ml 12× MES, 3.44 ml 5 M NaCl, 0.8 ml 0.5 M EDTA, 0.2 ml 1% Triton X-100, and 3.9 ml H_2O. Filer sterilize, store at 4°C, shield from light.
4. Hybridization cocktail: for each chip, prepare 200 μl cocktail by mixing 20.83 μl 12× MES, 150 μl 5 M TMAC (Sigma, #T3411), 3 μl 3 nM B2, 2 μl herring sperm DNA (10 mg/ml), and 18.67 μl H_2O.
5. Wash buffer A: for 1 l, mix 300 ml 20× SSPE, 1 ml 10% Tween-20, and 700 ml H_2O. Filter sterilize and store at room temperature.
6. Wash buffer B: for 500 ml, mix 41.6 ml 12× MES, 2.6 ml 5 M NaCl, 0.5 ml 10% Tween-20 (Pierce, #28320), and 455.2 ml H_2O. Filter sterilize and store at 4°C, shield from light.
7. 2× stain buffer: for 50 ml, mix 8.35 ml 12× MES, 18.5 ml 5 M NaCl, 0.5 ml 10% Tween-20, and 22.65 ml H_2O. Filter sterilize and store at 4°C.
8. R-phecoerythrin streptavidin (SAPE) solution: 1,200 μl each chip, mix 600 μl 2× MES stain buffer, 48 μl 50 mg/ml BSA, 12 μl 1 mg/ml SAPE (Molecular Probes, #S866), and 540 μl ddH_2O.
9. Ab solution: 600 μl each chip, 300 μl 2× MES stain buffer, 24 μl 50 mg/ml BSA (Invitrogen, #15561-020), 6 μl 10 mg/ml normal Goat IgG (Sigma, #I5256), 3.6 μl 0.5 mg/ml biotinylated anti-streptavidin (Vector Laboratories, #BA-0500), and 266.4 μl H_2O.
10. Affymetrix Fluidics Station 450.
11. GeneChip Scanner 3000 7G.

2.6. Solexa Sample Preparation

1. The Illumina ChIP-seq sample preparation kit (#IP-102-1001) provides all reagents needed. However, all reagents with the exception of adaptor and PCR primers for enrichment of adaptor ligated DNA fragments, both of which should be ordered as Genomic DNA Sample Prep Oligo Only Kit from Illumina (#FC-102-1003), can be purchased individually or as a kit, e.g., the NEBNext DNA Sample Prep Reagent Set 1 (#E600S or E6000L) from NEB, or from other sources (in particular, the ligase from enzymatics has been recommended based on a previous study (18)).

2. SYBR gold (Invitrogen, #S-11494).
3. Blue-light transilluminator (Invitrogen).
4. QIAquick PCR purification kit (Qiagen, #28104), MinElute PCR Purification Kit (Qiagen, #28004), MinElute Gel Extraction Kit (Qiagen, #28604).
5. Q_PCR primers: Illu-F, AGCAGAAGACGGCATACGAGCT CTTCCGATC, Illu-R, AATGATACGGCGACCACCGAGAT CTACACTC.
6. ABI Power SYBR green PCR master mix, (#4367659).

3. Method

3.1. Embryo Collection and Fixation

1. Maintain flies in large population cages in a chamber or incubator set at the proper temperature, humidity, and light cycle (see Note 2). For *Drosophila melanogaster*, we use 25°C and 50% humidity with a 14 h on, 10 h off light cycle.
2. On the day of embryo collection, the old molasses/yeast plates are replaced with fresh new plates covered with yeast, which is repeated twice for 1 h each to clear away the aged embryos retained by the mother flies.
3. For embryo collection, a molasses plate streaked lightly with yeast paste is placed in each cage for a certain time, usually 1 h.
4. Remove the plates with embryos and set them aside in the incubation chamber to let the embryos to age to the desired developmental stage.
5. Harvest the embryos on double Nitex mesh, with a coarse mesh on top to remove dead crushed adult flies, and a bottom fine mesh to retain embryos.
6. Wash away yeast, transfer embryos to a beaker containing 50% bleach, mix. Dechorionate the embryos for 2 min, then pour embryos back to the lower layer mesh, and rinse extensively.
7. Transfer embryos to a cup with mesh on the bottom, dry the embryos by placing paper towels beneath the mesh. Completely immerse the embryos in isopropanol in a beaker and disperse embryos well by using a narrow metal spatula. Remove cup and dry embryos quickly with paper towels beneath the mesh.
8. Transfer embryos to a 50-ml falcon tube and add 10 ml formaldehyde/hexane fixing solution per gram of embryo. Shake briefly, and rotate for 5 min at room temperature. Allow the embryos to settle, and pour off the fixative to a hazardous waste bottle.
9. Add embryo washing buffer to the embryos (use >10 ml per gram of embryos), shake vigorously, and then rotate at room

temp for 5 min. Pour onto a cup with mesh on the bottom to drain the buffer. Transfer the embryos to a new tube and repeat the wash with fresh buffer. Embryos should clump at first but then eventually become monodispersed.

10. Dry embryos and transfer them to plastic tubes. Flash freeze embryos in liquid nitrogen and store at −80°C.

3.2. Chromatin Purification and Sonication

3.2.1. Chromatin Purification

The amount of embryos used to set up each gradient, and the amount of chromatin obtained depend on the age of the embryos used. In the procedure described below, usually 5–7 g embryos at stage 5 are used for each gradient, and an SW28 rotor is used for ultracentrifugation (see Note 3). About 3 ml of purified chromatin at concentration of 75 μg/ml is obtained. For older embryos, adjust the amount of embryos according to their developmental stage (e.g., 4 g for stage 10 or 11, and 2.5 g for stage 14).

1. Thaw the frozen embryos, add 35 ml of cold embryo homogenization buffer + 0.5 mM DTT + 1 mM PMSF. Shake hard to break the clumps. Immediately transfer to a 40-ml glass dounce tube.
2. Dounce the embryos at 8,000 rpm for one stroke and two strokes at 7,000 rpm using a motor – drive homogenizer system (see Note 4).
3. Transfer to a 40 ml hand-held glass dounce and homogenize using five strokes with an A size pestle.
4. Pour into an SS34 tube and add 0.5 ml 20% Triton X-100 to a final concentration of 0.3%. Rock for 10 min at room temperature. This step removes remaining cell membranes, leaving largely nuclei suspended in cytosol and egg yolk. Spin down nuclei at 4,000 rpm (1,251 × *g*) for 15 min in an SS34 rotor at 4°C.
5. Pour off supernatant and add 5 ml nuclear lysis buffer. Resuspend pellet by pipetting up and down a few times.
6. Transfer to a small dounce and completely homogenize by ten strokes using a B size pestle.
7. Transfer homogenate to a 15-ml falcon tube, sonicate 20 s at low output setting using a Branson sonifier 450, or a Bioruptor, to partially fragment chromatin. This sonication step is not to reduce the chromatin fragment to sizes desired for ChIP, but rather to decrease viscosity of the sample. The DNA should be >20 kb is size after this step. Several samples may be combined at this step, with the sonication time increased accordingly.
8. Transfer to a clean SS34 tube. Add 1.8 ml 20% SDS (to final concentration of 3%) and quickly vortex.
9. Add 1.2 ml 20% *N*-lauroylsarcosine and 1.2 ml 20% Triton X-100. Incubate on a rotator at room temperature for 10 min.

10. Spin at 4,000 rpm (1,251 × *g*) for 10 min in an SS34 rotor. Transfer the cleared lysate to a new tube. Avoid the lipid layer on top that sometimes is present and the pellet, which may be loose.
11. Set up CsCl gradients by gently layering 8.5 ml of each of the three CsCl solution on top of one another in the order of 1.5, 1.4, and 1.3 g/ml. Finish by layering the nuclear lysate on top of the gradient.
12. Carry out ultracentrifugation in an SW28 rotor at 121,569 × *g* (26,000 rpm) for 40 h at 20°C.
13. After centrifugation, remove tubes and secure each with a clamp. Locate the chromatin, which usually does not form a uniform layer, but rather is fibrous and clumpy and differs from a band located about 0.5–1 cm above it that is more whitish. Gently insert an 18 1/2 gauge needle attached to a 5-ml syringe 0.5–1 cm below the chromatin band. Slowly pull the plunger and until all the chromatin is drawn into the syringe, yielding about 3 ml of chromatin for each gradient.
14. Dialyze the chromatin sample in a medium Spec 2 dialysis bag against dialysis buffer at a volume that is >100× the volume of the chromatin sample at 4°C for a total of 6 h, with two buffer changes during this time.
15. Recover the chromatin, which is partly insoluble at this stage but will go completely into solution after sonication. Flash freeze and store at −80°C.

3.2.2. Chromatin Sonication

Prior to being used for ChIP reactions, the chromatin needs to be fragmented to the desired size range by sonication. For our ChIP-chip experiment, we use a Branson sonifier 450. With this system, we usually sonicate 5 ml of chromatin at a time in a 15-ml falcon tube placed in an ice/water bath. With the power output set at 2.5, we carry out six cycles of sonication with 30 s on and 3 min off for each cycle, which produces chromatin fragments averaging about 500 bp, suitable for ChIP-chip. For ChIP-seq, we routinely use a Bioruptor, which can produce short fragments, ranging from 100 to 300 bp. Below is the detailed procedure for chromatin fragmentation using bioruptor.

1. First add *N*-lauroylsarcosine to 0.5%.
2. Aliquot 200–300 μl chromatin samples into 1.5-ml TPX hard plastic tubes (see Note 5).
3. Put samples in the Bioruptor. During sonication, the samples need to be kept cold. This is done by using a cold water circulation system attached to the bioruptor. Alternatively, it can be done by adding ice to the water bath, which needs to be replaced periodically, e.g., every 10 min. Adjust the amount of ice added so that at the end of each 10 min there is still small amount of ice remaining.

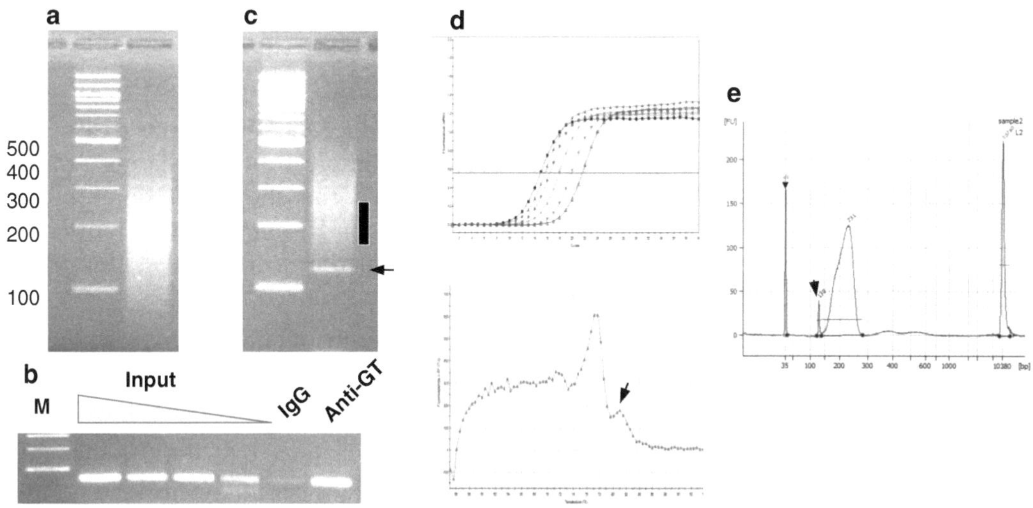

Fig. 3. Expected results at each step of the ChIP-seq procedure. (**a**) DNA isolated from sonicated chromatin. (**b**) PCR results for the ChIP and IgG control IP samples along with samples from serial dilution of input DNA. (**c**) DNA obtained following Solexa ChIP DNA sample preparation, *arrow* points to adaptor dimer by-products. (**d**) The amplification curve and dissociation curve from a Q_PCR analysis of a Solexa DNA library sample, a peak pointed to by *an arrow* corresponds to primer–dimer. (**e**) The electrophoregram of a library sample from an analysis on a Bioanalyzer, again, the *arrow* indicates the adaptor dimer.

4. Set Bioruptor intensity to H and the on/off cycle to 15 s ON/45 s OFF.
5. The process time should be determined empirically, but it is very reproducible for the same type of sample once a condition is found. For ChIP-seq, 100 min will produce fragments averaging 200–300 bp, or 140 min total processing time can produce fragments averaging less than 200 bp. Mix the samples after every two rounds (10 min) of sonication in the early stages of sonication.
6. To check the resulting DNA fragment size range and estimate chromatin DNA concentration: first take a 50 μl of sonicated chromatin, add 50 μl of TE and 1 μl 5 M NaCl, and incubate at 65°C overnight to reverse cross-links. Next morning, add 2 μl 0.5 mg/ml RNase and incubate at 37°C for 30 min. Then, add 2.5 μl 20% SDS and 2 μl of 20 mg/ml proteinase K and incubate at 55°C for 2 h. Purify the DNA using a Qiagen QIAquick PCR purification kit. Measure the DNA concentration by nanodrop, and run a 1.5% agarose gel to check the DNA fragment size distribution. Figure 3a shows the fragment size distribution of DNA purified from chromatin subjected to 14 × 10 min cycles of sonication.

3.3. Chromatin Immunoprecipitation

We use about 50 μg (based on DNA concentration determined as described above) chromatin for each immunoprecipitation reaction (ChIP) (see Note 6). Following ChIP, we use PCR to check enrichment of a DNA region known to be bound by the factor being studied, as described at the end of this section. If no target is known for certain, carry out a positive control IP using an antibody for another factor.

1. Transfer the chromatin needed for ChIP reactions to a 1.5-ml microcentrifuge tube (or multiple tubes if a larger volume of chromatin is needed). Spin the chromatin solution in a microcentrifuge at full speed for 15 min at 4°C. Transfer the chromatin solution to a new microcentrifuge tube. Discard the 30 μl or so of solution at the bottom of tube.
2. Add 5 × IP buffer at ¼ the volume of chromatin.
3. Add 7.5 μl of normal rabbit IgG per 50 μg chromatin. Incubate 30 min on ice.
4. During incubation, prepare protein-A sephacryl beads. Take 15 μl (packed beads volume) beads per 50 μg chromatin, wash beads twice with IP buffer, and after the washing is finished, remove buffer and leave just the beads in the tube.
5. Transfer the chromatin sample to a microcentrifuge tube containing protein-A sephacryl-1000 beads. Rotate 1 h at 4°C.
6. Spin in a microcentrifuge at 4,000 rpm (1,500 × *g*) for 2 min and then at full speed for 15 min. Transfer the precleared chromatin solution to a new microcentrifuge tube or to a 15-ml falcon tube if a relatively large volume of chromatin is used.
7. Dilute chromatin to 1.2 ml per 50 μg DNA with IP buffer, add 4 μl 50 mg/ml BSA for each 1 ml of diluted chromatin, and add PMSF to 1 mM. Save 24 μl of the chromatin solution as input sample (representing 2% of total input DNA).
8. For each factor IP or IgG control IP sample, use 1,200 μl of the diluted chromatin. For factor IP samples, add 1–3 μg of purified polyclonal antibody, or appropriate amount of monoclonal antibody. For IgG control IPs, add an equivalent amount of normal rabbit IgG. Incubate the samples on ice for 3 h or overnight at 4°C.
9. Transfer protein-A sephacryl-1000 beads to a microcentrifuge tube, 10 μl (packed beads volume) for each sample. Wash beads twice with 1 ml IP buffer + 200 μg/ml BSA. Resuspend beads in IP buffer with BSA at a volume of 100 μl for each 10 μl beads.
10. Spin the samples in a microcentrifuge at full speed for 15 min at 4°C. Transfer supernatant to tubes containing 100 μl of the protein A-sephacryl suspension.
11. Rotate 30 min at 4°C.
12. Pellet the beads by spinning in a microcentrifuge at 4,000 rpm for 1 min. Discard the supernatant.
13. Resuspend beads in 1.4 ml IP buffer, transfer to a new tube, rotate at room temperature for 15 min. Pellet beads, discard supernatant – this is the first wash.

14. Carry out the following consecutive washes at room temperature similar to step 13 (except no change of tube): one wash in IP buffer for 15 min, two washes in 0.5 M IP wash buffer for 20 min each, one wash in LiCl wash buffer for 20 min, and finally one wash in TE for 15 min.
15. Resuspend the pellet in 1 ml TE, transfer to a new tube, and spin down the beads. Discard the supernatant. Remove the remaining buffer in the beads with a 30 gauge ½ in. needle attached to a 1-ml syringe.
16. Add 100 μl TE + 2 μg RNase A and incubate at 37°C for 30 min. Resuspend the beads in middle of incubation once. For each input sample, add 80 μl TE and 2 μg RNase A.
17. Add 50 μl of 10 mM Tris–Cl pH 8.0, 2 mM EDTA, 300 mM NaCl and 1.5% SDS, and 40 μg of proteinase K.
18. Incubate all of the samples at 55°C, overnight.
19. Spin down and transfer the supernatant to a new tube.
20. Add 150 μl TE to the beads, resuspend, and spin again. Combine with the supernatant from the previous step.
21. Reverse cross-linking by incubating at 65°C for 6 h.
22. Add 30 μl 3 M sodium acetate and 20 μg glycogen to each sample.
23. Extract once with phenol: chloroform, once with chloroform.
24. Add 2.5 volume ethanol, mix. Leave at −80°C for 4 h or more.
25. Spin in a microcentrifuge at full speed for 20 min.
26. Wash with 75% ethanol twice.
27. Remove as much ethanol as possible after the last wash.
28. Air dry the DNA pellet at room temperature ~10 min.
29. Resuspend in TE (10:0.1): 20 μl for IP samples and 40 μl for input DNA samples. The samples are now ready to be analyzed by PCR or processed for chip or seq analysis.
30. Use PCR to check whether the IP worked. We set up PCR reactions that each contain 1 μl of 5×, 25×, 125×, or 625× diluted input DNA as standards, or 1 μl of the IP or IgG control IP samples. The number of PCR cycles should be adjusted so that the signals for the input samples are proportional to the dilutions used, instead of being saturated. An enrichment of a minimum 10 to more than a 100-fold, based on comparison of the signals between the IgG control IP and factor IP samples, is expected. Figure 3b shows an example of IP results.

3.4. DNA Amplification and Labeling for ChIP-Chip

For ChIP-chip, the DNA samples from the ChIP experiment (factor IP, IgG IP, and input) are amplified, labeled, and hybridized to an affymetrix tiling array. We have modified a commonly used random prime-based amplification method to improve the efficiency and reproducibility, particularly when small amounts of genomic DNA are used. The modifications include: (1) modification of the primer sequence to avoid primer–dimer formation, (2) 15× higher concentration of primer A, and a different purification procedure following random priming, (3) using higher sequenase concentration, (4) more random priming cycles (4 vs. 2), (5) changes in PCR conditions and other adjustments. With this protocol, highly reproducible amplification can be obtained even when as little as 0.5 ng of *Drosophila* whole genomic DNA is used, with correlation coefficients between duplicates ranging from 0.85 to 0.95 for all types of samples, including input, mock IP, and factor IP samples. Other commonly used methods include linker mediated PCR (5) and WGA (19).

The amount of input DNA used for amplification is about 20 ng, which is higher than those present in the ChIP or mock IP samples. This provides the best way to control for potentially intrinsic differences in the efficiency of amplification of different parts of the genome. We find, however, that the amplification of IgG control IP and factor IP samples are almost equally consistent.

3.4.1. ChIP-DNA Amplification (see Note 7)

1. Set up the first round reaction with a total volume of 18.5 μl as follows: 10.5 μl DNA (from factor IP, IgG control IP, or 2 ng/μl input) , 4 μl 5× sequenase buffer, 4 μl 200 μM primer A.
2. Prepare dNTP solution by mixing (for each reaction): 0.1 μl 20 mg/ml BSA, 1 μl 0.1 M DTT, 0.5 μl 25 mM dNTPs.
3. Dilute sequenase by mixing each 1 μl with 6 μl of sequenase dilution buffer.
4. Set up the following program that consists of four cycles each containing the following three segments on a thermocycler (typically a Perkin Elmer 9600 or 9700): (1) 95°C 4 min, (2) 10°C 5 min, (3) 37°C hold for 8 min. Set the ramp rate such that it takes about 10 min to ramp from 10°C from the previous step to 37°C.
5. Put the samples in the thermocycler. Start the reactions.
6. During each cycle, after the 4 min incubation at 94°C, pause the program, transfer the samples to ice, close the lid, and resume the program.
7. After the temperature reaches 10°C, transfer the samples back to the thermocycler, and add 1.6 μl dNTP mix (only for the first of the four random priming cycles) and 1 μl of 6× diluted sequenase (USB Corp). Make sure to add the solutions to all samples within the 5 min incubation time. Close thermocycler.
8. Repeat steps 6 and 7 till all four random-prime synthesis cycles are complete.
9. Purify the DNA as follows to remove the primers: take an Amersham microspin S-300 HR column; remove the buffer from the column by spinning at 3,000 rpm (735 ×*g*) for 1 min

in a microcentrifuge; add 20 μl of TE to each random-primed sample; load on top of the column; spin for 2 min; and save the flow through. Take a second Amersham microspin S-300 HR column; spin 1 min to remove buffer in column; reequilibrate the column by adding 300 μl of 10 mM Tris–Cl, pH 8.5, followed by a 1 min centrifugation; load the flow-through saved from the first column; spin 2 min. The purified DNA is ready for PCR amplification.

10. Set up PCR reaction with a total volume of 100 μl: 35 μl of the DNA sample from the first round (see Note 8), 10 μl 10× PCR buffer, 3 μl 25 mM $MgCl_2$, 1.5 μl 25 mM dNTPs, 4 μl 100 μM Primer B, 44.5 μl H_2O, 2 μl 5 U/μl Taq 2000.
11. Carry out PCR as follows: 15 cycles of: 95°C for 30 s, 45°C for 30 s, 55°C for 30 s, and 72°C for 1 min; then 15–20 cycles of 95°C for 30 s, 45°C for 30 s, 55°C for 30 s, and 72°C for 1 min.
12. Purify the amplified DNA using a Qiagen QIAquick PCR purification kit. Elute DNA using 35 μl H_2O. Check DNA concentration by measuring OD_{260}. Up to 12 μg of DNA can be obtained for each sample.
13. Run an aliquot of the amplified sample on a gel. The size range should be somewhat smaller than the starting material (use input DNA as a reference) to reflect efficient random-prime amplification.

3.4.2. DNA Fragmentation, Labeling

1. Dilute DNase I (1 U/μl; Epicentre) 15× into 1× One-Phor-All buffer (Roche).
2. Set up reactions as follows: mix 2.5 μg amplified DNA, 3.28 μl 10× One-Phor-All buffer, 2.2 μl diluted DNase I, and ddH_2O to total volume of 35 μl.
3. Incubate for 5 min at 37°C.
4. Inactivate DNase I at 99°C for 10 min, then place on ice.
5. Check 10% of the reaction by running it on a 1.5% agarose gel. The bulk of DNA should be 50–100 bp.

3.4.3. TdT Labeling

1. Set up a 50 μl reaction by mixing 31.2 μl DNase I treated DNA, 10 μl 5× TdT buffer, 5 μl 25 mM $CoCl_2$, 3.6 μl 1 mM biotin-ddATP, 0.18 μl TdTase (Roche).
2. Incubate at 37°C for 2 h.

3.5. Hybridization to Genomic DNA Tiling Array and Scanning

We use the Affymetrix Drosophila genomic DNA tiling array. All the hybridization to chip, chip washing and staining, as well as scanning are performed according to the manufacturer's recommendation. Consult the manufacturer if different versions of the instruments are used.

3.5.1. Hybridization

1. Dilute 2× MES-Triton to 1×. Inject 200 μl into each chip.
2. Prehybe for ~1 h in the Affymetrix hybridization oven at 45 rpm and 45°C.
3. Mix 200 μl hybridization cocktail with 50 μl of the sample from TdT reaction.
4. Boil sample for 10 min, transfer to a 45°C TempBlock, and incubate for 10 min.
5. Spin in a microcentrifuge at maximum speed for 3 min.
6. Inject 200 μl of the sample into the chip.
7. Carry out the hybridization in the hybridization oven at 45 rpm and 45°C for 18 h.

3.5.2. Washing and Staining

We use an Affymetrix GeneChIP Fluidics Station 450 to process the hybridized chips using the wash/stain-protocol EukGeWS2v4. For details of handling the Fluidics Station, follow the manufacturer's instruction.

1. Load buffer A, B, and ddH_2O and prime the station following the instructions.
2. Load the chip and the SAPE staining and antibody solution.
3. Carry out washes using the EUkGeWS2v4 protocol.

3.5.3. Scanning the Chip

Scan chips using Affymetrix GeneChip scanner, following the manufacturer's instruction.

3.6. ChIP-Seq Sample Preparation

The ChIP-seq sample preparation is carried out according to the "Illumina protocol for sample preparation for chip-seq" with some modifications. One major change is that the PCR amplification step is carried out before instead of after DNA size selection. This helps to decrease the loss of DNA complexity in the sample, which can be an issue since the amount of DNA in the ChIP sample is low. We have found that this modification leads to higher DNA yield. We also keep the number of PCR amplification cycles to a minimum to limit uneven amplification of different genomic regions, see Note 9.

3.6.1. Perform End Repair

1. Dilute Klenow DNA polymerase 1:5 with ddH_2O to a final Klenow concentration of 1 U/μl. An excess amount of enzyme may lead to DNA degradation when the amount of DNA sample is low.
2. Prepare the following reaction mix with a total volume of 50 μl: 10 μl DNA sample from ChIP, mock IP, or 0.1 ng/μl input DNA; 30 μl ddH_2O; 5 μl T4 DNA ligase buffer with 10 mM ATP; 2 μl dNTP mix; 1 μl T4 DNA polymerase; 1 μl diluted Klenow DNA polymerase; 1 μl T4 polynucleotide kinase.

3. Incubate in the thermal cycler for 30 min at 20°C.
4. Follow the instructions in the QIAquick PCR Purification Kit to purify samples each on one QIAquick column, eluting in 34 μl of EB.

3.6.2. Add "A" Bases to the 3′ End of the DNA Fragments

1. Prepare the following reaction mix with a total volume of 50 μl: 34 μl DNA from Subheading 3.6.1, 5 μl Klenow buffer, 10 μl dATP, 1 μl Klenow exo (3′–5′ exo minus).
2. Incubate for 30 min at 37°C.
3. Follow the instructions in the MinElute PCR Purification Kit to purify samples each on one MinElute column, eluting in 10 μl of EB.

3.6.3. Ligate Adapters to DNA Fragments

1. Dilute the adapter oligo mix 1:20 with water (see Note 9).
2. Prepare the following reaction mix with a total volume of 30 μl: 10 μl DNA from Subheading 3.6.2, 15 μl DNA ligase buffer, 1 μl diluted adapter oligo mix, 4 μl DNA ligase.
3. Incubate for 15 min at room temperature.
4. Follow the instructions in the MinElute PCR Purification Kit to purify samples, each on one MinElute column, eluting in 10 μl of EB.

3.6.4. Enrich the Adapter-Modified DNA Fragments by PCR

1. Set up the PCR reaction with total volume of 25 μl: 10 μl DNA, 7 μl ddH_2O 10 μl 5× Phusion buffer, 1.5 μl dNTP mix, 1 μl PCR primer 1.1, 1 μl PCR primer 2.1, 0.25 μl Phusion polymerase.
2. Amplify using the following PCR protocol: 30 s at 98°C; then 12–15 cycles (see Note 10) of 10 s at 98°C, 30 s at 65°C, and 30 s at 72°C; finally 5 min at 72°C.

3.6.5. Size Select the Library

1. Prepare a 1.5% agarose gel in 1× TAE or 0.5× TBE.
2. Load 500 ng of a 100 bp DNA ladder per lane of the gel.
3. Add loading buffer to the PCR product. Load the sample to the gel. Leaving at least one lane empty between the 100 bp ladders, load ladders on both sides of each sample lane. Be very careful to avoid sample escaping from the wells which will lead to cross-contamination. Minimize the number of samples (maximum three) purified on each gel.
4. Run gel at 100 V for 1.5 h.
5. Stain the gel with SYBR-gold for 10 min or longer until the DNA smear is clearly visible.
6. Place the gel over a blue-light transilluminator; the DNA ladder and the DNA smear in the sample lanes should be easily visible. In the sample lanes, besides the continuous DNA smear >150 bp, there is sometimes a band of about 140 bp, which corresponds

to adaptor ligation by-products. Figure 3c shows an example of the Solexa sample after run on a gel prior to size selection; the 140 bp band is indicated.

7. Cut the portion of the gel containing DNA of the desired size range, typically between 175 and 300 bp.
8. Take a photograph of the gel after the slice is excised.
9. Use a MinElute Gel Purification Kit to purify the DNA from the agarose gel slices and elute the DNA in 10 μl EB.

3.6.6. Quantifying the Solexa Library

Accurate determination of the DNA concentration in the sequencing library is very important. Otherwise, it may lead to either too little or too much of DNA from the library being loaded onto the flow cell. If too little DNA is used, few clusters (see below) will be generated and few reads will be obtained. On the other hand, if too much DNA is loaded, it will lead to overcrowding and overlap of images from neighboring clusters, which will also lead to too few high-quality reads.

Several methods can be used for quantification. One option is Nanodrop, but it can be inaccurate since the DNA concentration in a sequencing library is usually very low (due to limited amplification) and the measurement is further affected by residual contaminants from the column purification. More commonly used methods are Q_PCR and Bioanalyzer. We found that the concentrations measured by these two methods show a good correlation, but the absolute values can vary significantly. Thus, it is important to compare the concentrations measured to a library with known sequencing behavior that was quantified in a similar way. Here, we described the quantification using Q_PCR. For using the Bioanalyzer, follow the manufacturer's instruction.

1. Prepare library standard: dilute a Solexa library that has a known concentration and clustering behavior to the concentrations of 0.01, 0.0033, 0.0011, 0.00037, and 0.00012 ng/μl.
2. Dilute each of the experimental Solexa library samples by 200×, 1,000×, and 5,000× to ensure that at least one of the diluted samples has the amount of DNA within the standard range.
3. Set up reactions: to each well in a 96-well optical PCR plate, pipet 2 μl of library standard or diluted library samples, 2 μl Q_PCR primers (1 μM each), 6 μl ddH_2O, and 10 μl ABI Power SYBR Green PCR Master Mix.
4. Perform Q_PCR, on a Stratagene Mx3000p instrument, using the following parameters: 95°C for 10 min; 40 cycles of 95°C for 15 s, 65°C for 1 min, and 72°C for 30 s. Also, add a dissociation segment to the end of program.
5. Analyze the results using software associated with the Q_PCR machine to determine the concentration of the library samples.

6. Carry out dissociation curve analysis. There should be only one major peak in the middle of curve (Tm 77°C). Sometimes, a secondary curve with higher Tm 81°C is also observed. This secondary peak corresponds to the adaptor dimer. The adaptor dimer contamination is usually not an issue. If necessary, repurified the sample by agarose gel size selection.
7. Expected results: Fig. 3d shows the amplification curve for a Solexa sample along with the standards, and the bottom panels shows the dissociation curve, the arrow shows position of the adaptor dimer. In Fig. 3e, the electrophoregram of the sample after analyzed on a Bioanalyzer is also shown, and the minor contamination of the adaptor dimer is indicated by arrow. With 15 cycles of enrichment PCR, the DNA library is expected to have a concentration of up to 50 nM.

3.6.7. Sequencing of ChIP-Seq Samples

The sequencing step is usually carried out by a core facility. Consult with the person in charge of the sequencing about the amount sample to use. Based on standard sequencing protocols, 1 μl of a 20 nM library is sufficient. The details about the technology and work flow can be found from the Illumina company website. Briefly, sequencing a DNA library can be divided in two main steps. The first is cluster generation on a cluster station. At this step, clusters of DNA are generated on the surface of a flow cell through a clonal bridge amplification technology from each molecule in the library. The next sequencing step is performed on a Solexa Genome Analyzer. The sequencing process involves rounds of synthesis with the DNA molecules in the clusters as template and fluorophor-dNTPs as substrates. In each round of this sequence though synthesis process, the incorporation of each of the four bases coupled with a distinct fluorescence dye is captured by an imaging system, which is followed by the cleavage release of the fluorescence dye before a new round of sequencing is performed. The stacks of cluster images thus produced represent the raw data, which, after the sequencing process is finished, are processed by a suite of software, producing the actual sequence reads.

3.7. Data Analysis

3.7.1. ChIP-Chip Data Analysis

We use the program, TiMAT, developed by BDTNP, for our ChIP-chip data analysis (see ref. 15 for details). The data analysis includes the following main steps: (1) The complete set of six arrays from an experiment (Factor IP replicates, IgG control IP replicates, and input DNA replicates) are scaled to a common median value and then quantile normalized against each other; (2) Replicates are averaged and log (base 2) ratio scores are calculated for Factor IP and IgG control IP arrays: $\log_2$ (mean Factor IP/mean input DNA) and $\log_2$ (mean IgG control IP/mean input DNA). These scores are then smoothed using a sliding 675 bp window of trimmed means; (3) False discovery rate (FDR) estimates are calculated by two methods: a symmetric null distribution method, and IgG control

IP method; (4) Bound regions are then defined by collecting all windows with scores above the given FDR threshold into contiguous stretches of windows; (5) identify the primary peak for each bound region – the window having maximum intensity. Results from TiMAT, including oligonucleotide probe intensities, trimmed-mean window scores, bound region locations, peak magnitudes and locations, and nearby genes are reported in .sgr and .gff file formats as well as in TiMAT's own text-based report files.

The probe intensity profile or the window score profile for the whole genome or just the bound regions in .sgr format can be visualized using Affymetrix Integrated Genome Browser. It can also be converted to wig file for UCSC genome browser. Figure 2 shows the window score profiles at the *eve* gene locus for IPs using two different antibodies that recognize the transcription factor HB and also for the IgG control IP. The bound regions identified by TiMAT are marked underneath the profiles.

3.7.2. ChIP-Seq Data Analysis

The first step of ChIP-seq data analysis is to align the reads to the genome. This can be done using ELAND as part of the Illumina Analysis Pipeline which also includes cluster image processing and base calling, or using one of the several other programs, such as MAQ (20), Bowtie (21), etc. From the aligned reads, a tag-density profile can be generated by extending the short sequence tags by the average length of the DNA fragments in the library. The resulting files, in .sgr or wig format, can be viewed with a genome browser as described above. To identify the bound regions, a range of published software are available, among more recently released ones are CisGenome (22), Peakseq (23), SISSRs (24), MACS (25), GLITR (26), and Sole-Search (27). These programs usually use a local background model and at least one control (usually the input DNA) to calculate the statistical significance of each enriched peak, and/or determine FDRs.

In our studies (17), first we aligned the reads to the genome using bowtie. We used MACS (25) for peak finding. The output from MACS is a BED file containing a list of identified peak regions and an Excel file containing several types of information about the bound regions. The regions can be sorted based on the number of sequence tags in the identified regions, the fold enrichment, or FDR.

Figure 2 shows the tag-density profile around the eve gene for two IPs using different hunchback (HB) antibodies, and the input DNA sample. The enriched regions identified by MACS at 5% FDR are marked by horizontal bars underneath the tag-density tracks. A comparison with ChIP-chip results in Fig. 2 shows how ChIP-seq excels in resolution (and we note higher resolution can be achieved by selecting even smaller fragments for sequencing). Nevertheless, the bound regions identified by the two methods are highly correlated.

3.8. Expected Results

As shown in Fig. 2, the ChIP-chip and ChIP-seq results from our studies show strong enrichment at strongly bound regions and low background noise, which allows detection of not only the strong, but also fairly low levels of binding. The results are highly reproducible between biological replicate or between IPs using different antibodies. In our studies, the enrichment of known transcription factor targets in the ChIP step often exceeds 100-fold (see Note 11). For ChIP-chip, our optimization of the DNA amplification step significantly reduced the data noise, which leads to very high detection sensitivity of binding. As shown previously, we were able to detect >90% of the regions enriched by threefold at 1% FDR cutoff in a BAC spike-in experiment carried out based on our standard DNA amplification and hybridization procedure in the ChIP-chip protocol (15).

In our ChIP-chip and ChIP-seq studies, we found that the transcription factors often bind over a quantitative continuum to thousands of regions throughout the genome (15–17), in support of the widespread binding observed in our early studies (28). Such widespread binding has increasingly been seen by others and appears to be a common phenomenon for transcription factor binding in animal cells. In addition, we found that biochemically and functionally unrelated factors bind to highly overlapping sets of regions (15–17, 28). This widespread and highly overlapping binding is unlikely to be driven primarily by function. Instead, our analysis showed while the most highly bound regions are much more likely to be functional, many thousands of weakly bound regions are mostly likely nonfunctional (15–17, 28), and for each cis-regulatory element, where many factors bind, combinatorial regulatory output is determined by the relative levels of occupancy of each factor (16). Our more recent study indicates that widespread overlapping pattern of binding may be attributed to chromatin accessibility (13). In all, our studies demonstrate the importance of interpreting the genome – wide binding data in a quantitative manner.

4. Notes

1. We use protein A – sephacryl 1000 beads that we prepare ourselves (15) in our ChIP reactions. We found such beads are more efficient in IP reactions especially when chromatin fragments in the IP reactions are relatively large.
2. Maintaining flies health under proper conditions makes a big difference in minimizing egg retention and larvae contamination, and increases embryo yield. A cage containing about 100 ml flies can lay between 0.5 and 0.7 g embryos in 1 h.

3. For samples from fewer embryos (as little as 0.1 g embryo has been used successfully), use a SW41 rotor. In this case, reduce the amounts of all reagents, including buffers, detergents, and CsCl solution, as well as sonication time for the lysate, by three times compared to that described for an SW28 rotor. In addition, carry out the ultracentrifugation at 234,116 × g (37,000 rpm) for 24 h.
4. If no motor-drive available, go to next step, dounce the embryos a few more times while carrying out the homogenization using A pestle.
5. It is important to keep the volumes of different samples the same to achieve similar extent of sonication. In addition, sonication is more efficient and smaller fragment size can be achieved by carrying out the sonication in small volumes in a 1.5-ml tube than carrying out the sonication in a larger volume in a 15-ml plastic tube.
6. Good ChIP results have been obtained with ten times less chromatin (in this case, use low retention microcentrifuge tubes), but will likely produce noisier data in ChIP-chip and ChIP-seq.
7. The efficiency of random prime amplification methods is dependent on the average length of the DNA fragments in the sample. The method works well for ChIP-chip samples that contain DNA fragments averaging more than 500 bp. With DNA fragments <300 bp, the results become less reproducible. For the procedure described here, make sure to pipette accurately and use P2 pipettor for small volumes. If the amplification goes well, the amplification products will be smaller in size compared to starting material, as expected for frequent internal random-priming.
8. It is important not to use more than 35 μl for each 100 μl reaction; the residual amount of primer A remaining can inhibit the PCR reaction.
9. Diluting the adaptor is important to avoid or at least reduce an adaptor dimer ligation artifact. A 10–50× dilution is often recommended, but we have found that a dilution of 50× may lead to low yield of ligation products and lower reproducibility.
10. Based on the amount of chromatin we usually use (50 μg in each ChIP), 12 cycles produce just enough DNA to be sequenced if the standard Illumina presequencing sample preparation procedure is followed. Adjust the cycle number based on the amount of starting material or determine the cycle numbers empirically. The goal is to minimize the number of cycles as each cycle increases the amount of experimental noise.
11. The 100× enrichment was based on PCR, or the peak height in ChIP-seq. In ChIP-chip, the corresponding enrichment is about tenfold, which was seen for most factors in our ChIP-chip analysis (16).

References

1. Biggin MD, Tjian R (2001) Transcriptional regulation in Drosophila: the post-genome challenge. *Funct Integr Genomics.* **1**, 223–234.
2. Gilmour, D. S., Rougvie, A. E., and Lis, J. T. (1991) Protein-DNA cross-linking as a means to determine the distribution of proteins on DNA *in vivo*. *Meth. Cell Biol.* **35**, 369–381.
3. Carr, A. and Biggin, M.D. (1999) An in vivo UV crosslinking assay that detects DNA binding by sequence-specific transcription factors. *Methods Mol Biol.* **119**, 497–508.
4. Orlando, V. (2000) Mapping chromosomal proteins in vivo by formaldehyde-crosslinked-chromatin immunoprecipitation, *Trends Biochem. Sci.* **25**, 99–104.
5. Ren, B., Robert, F., Wyrick, J.J., Aparicio, O., Jennings, E.G., Simon, I., Zeitlinger, J., Schreiber, J., Hannett, N., Kanin, E., Volkert, T.L., Wilson, C.J., Bell, S.P., and Young, R.A. (2000) Genome-wide location and function of DNA binding proteins. *Science* **290**, 2306–9.
6. Iyer, V.R., Horak, C.E., Scafe, C.S., Botstein, D., Snyder, M., et al. (2001) Genomic binding sites of the yeast cell-cycle transcription factors SBF and MBF. *Nature* **409**, 533–538.
7. Metzker, M.L. (2010) Sequencing technologies — the next generation. *Nature Reviews Genetics* **11**, 31–46.
8. Johnson, D.S., Mortazavi, A., Myers, R.M.,and Wold, B. (2007) Genome-wide mapping of in vivo protein–DNA interactions. *Science* **316**, 1497–1502.
9. Barski, A., Cuddapah, S., Cui, K., Roh, T.Y., D.E. Schones *et al.*, (2007) High-resolution profiling of histone methylations in the human genome, *Cell* **129**, 823–837.
10. Robertson, G., Hirst, M., Bainbridge, M., Bilenky, M., Zhao Y. *et al.* (2007) Genome-wide profiles of STAT1 DNA association using chromatin immunoprecipitation and massively parallel sequencing, *Nat. Methods* **4**, 651–657.
11. Toth, J., and Biggin, M.D. (2000) The specificity of protein-DNA crosslinking by formaldehyde: in vitro and in drosophila embryos. *Nucleic Acids Res.* **15**, e4.
12. Auerbach, R.K., Euskirchen, G., Rozowsky, J., Lamarre-Vincent, N., Moqtaderi, Z., et al. (2009) Mapping accessible chromatin regions using Sono-Seq. *Proc Natl Acad Sci USA* **106**, 14926–14931.
13. Li, X.-Y., Thomas,S., Sabo, P.J., Eisen, M.B., Stamatoyannopoulos, J.A., and Biggin, M.D. (2011) The role of chromatin accessibility in directing the widespread, overlapping patterns of Drosophila transcription factor binding. *Genome Biology.* **12**, R34.
14. Kaplan, T., Li, X.-Y., Sabo, P.J., Thomas, S., Stamatoyannopoulos, J.A., Biggin, M.D. , and Eisen, M.B. (2011) Predicting the Landscape of Transcription Factor Binding During Early *Drosophila* Development. *PLoS Genet.* **7**, e1001290.
15. Li, X.-Y., Macarthur, S., Bourgon, R., Nix, D., Pollard, D.A., et al. (2008) Transcription Factors Bind Thousands of Active and Inactive Regions in the Drosophila Blastoderm. *Plos Biol.* **6**, e27.
16. Macarthur, S., Li, X.-Y., Li, J., Brown, J.B., Chu, H.C., et al. (2009) Developmental roles of 21 Drosophila transcription factors are determined by quantitative differences in binding to an overlapping set of thousands of genomic regions. *Genome Biol.* **10**, R80.
17. Bradley, R.K., Li, X.-Y., Trapnell, C., Davidson, S., Pachter, L., et al. (2010) Binding site turnover produces pervasive quantitative changes in transcription factor binding between closely related Drosophila species. *Plos Biol.* **8**, e1000343.
18. Quail, M.A., Kozarewa, I., Smith, F., Scally, A., Stephens, P.J. *et al.*, (2008) A large genome center's improvements to the Illumina sequencing system, *Nat. Methods* **5**, 1005–1010.
19. O'Geen, H., Nicolet, C.M., Blahnik, K., Green, R., Farnham, P.J. (2006) Comparison of sample preparation methods for ChIP-chip assays. *Biotechniques* **41**, 577–80.
20. Li, H., Ruan, J. & Durbin, R. (2008) Mapping short DNA sequencing reads and calling variants using mapping quality scores. *Genome Res.* **18**, 1851–1858.
21. Langmead B, Trapnell C, Pop M, Salzberg SL. (2009) Ultrafast and memory-efficient alignment of short DNA sequences to the human genome. *Genome Biol.* **10**, R25.
22. Ji, H., Jian, H., Ma., W., Johnson, D.S., Myers, R.M., and Wong, W,H. (2008) An integrated software system for analyzing CHIP-chip and ChIP-seq data. *Nat. Biotechnol.* **26**, 1293–1300.
23. Rozowsky, J., Euskirchen, G., Auerbach, R.K., Zhang, Z.D., Gibson, T., Bjornson, R., Carriero, N., Snyder, M., Gerstein, M.B. (2009) PeakSeq enables systematic scoring of ChIP-seq experiments relative to controls. *Nat Biotechnol.* **27**, 66–75.
24. Jothi, R., Cuddapah, S., Barski, A., Cui, K., Zhao, K. (2008) Genome-wide identification of in vivo protein-DNA binding sites from ChIP-Seq data. *Nucleic Acids Res.* **36**, 5221–31.

25. Zhang ,Y., Liu, T., Meyer, C.A., Eeckhoute, J., Johnson, D.S., Bernstein, B.E., Nussbaum, C., Myers, R.M., Brown, M., Li, W., et al. (2008) Model-based analysis of ChIP-Seq (MACS). *Genome Biol.* **9**, R137.

26. Tuteja, G., White, P., Schug, J., Kaestner, K.H.. (2009) Extracting transcription factor targets from ChIP-Seq data. *Nucleic Acids Res.* **37**, e113.

27. Blahnik, K.R., Dou, L., O'Geen, H., McPhillips, T., Xu, X., Cao, A.R., Iyengar, S., Nicolet, C.M., Ludäscher, B., Korf, I., and Farnham, P.J. (2010) Sole-Search: an integrated analysis program for peak detection and functional annotation using ChIP-seq data. *Nucleic Acids Res.* **38**, e13.

28. Carr. A., Biggin, M.D. (1999) A comparison of in vivo and in vitro DNA-binding specificities suggests a new model for homeoprotein DNA binding in Drosophila embryos. *EMBO J.* **18**,1598–608.

Chapter 2

Characterization of Complex Regulatory Networks and Identification of Promoter Regulatory Elements in Yeast: "*In Silico*" and "Wet-Lab" Approaches

Nuno P. Mira, Miguel C. Teixeira, and Isabel Sá-Correia

Abstract

Transcription is the first step in the flow of biological information from genome to proteome and its tight regulation is a crucial checkpoint in most biological processes occurring in all living organisms. In eukaryotes, one of the most important mechanisms of transcriptional regulation relies on the activity of transcription factors which, upon binding to specific nucleotide motifs (consensus) present in the promoter region of target genes, modulate the activity of RNA polymerase II activating and/or repressing gene transcription. The identification of binding sites for these transcription factors is crucial to the understanding of transcriptional regulation at the molecular level and to the prediction of putative target genes for each transcription factor. However, transcription regulation cannot simply be reduced to transcription factor-gene associations. Frequently, the transcript level of a given gene is determined by a multitude of activators and/or repressors resulting in intertwined and complex regulatory networks. Two case studies dedicated to the study of transcriptional regulation in the experimental model *Saccharomyces cerevisiae* are presented in this chapter. The computational tools available in YEASTRACT information system are explored in both studies, to identify the regulatory elements that serve as functional DNA-binding sites for a transcription factor (Rim101p), and to characterize the regulatory network underlying the transcriptional regulation of a given yeast gene (*FLR1*). A set of easily accessible experimental approaches that can be used to confirm the predictions of the bioinformatic analysis is also detailed.

Key words: Transcriptional regulation, Transcription factors, *cis*-Regulatory elements, Transcriptional regulatory network, Bioinformatics, *Saccharomyces cerevisiae*, YEASTRACT

Ales Vancura (ed.), *Transcriptional Regulation: Methods and Protocols*, Methods in Molecular Biology, vol. 809,
DOI 10.1007/978-1-61779-376-9_2,

1. Introduction

1.1. Identification of the Rim101p-Binding Site: In silico analysis of the promoter region of Rim101p-regulated genes

In this era of postgenomic research in the model eukaryote *Saccharomyces cerevisiae*, the identification of the genes regulated by a given transcription factor is being performed mostly at a genome-wide scale; for example, by comparing the transcriptomes of a parental strain and a deletion mutant lacking the transcription factor under study. Often, this approach leads to the identification of hundreds of putative, direct or indirect, target genes. With such a vast initial dataset it is difficult, if not impossible, to extract common regulatory elements within the promoter region of the co-regulated genes without the use of appropriate computational tools. Several algorithms, usually called "DNA motif finders", were developed to search the promoter region of co-regulated genes for overrepresented nucleotide sequences (e.g., AlignAce, http://www.psc.edu/general/software/packages/alignace/; MEME, http://meme.nbcr.net/meme4_4_0/intro.html – among many others) (1, 2). The success of these algorithms in identifying functional binding sites is, however, limited by the fact that the comparison of genome-wide expression in wild-type versus transcription factor deletion mutant cells provides direct and indirect target genes, which means that part of the genes used as an input dataset to the algorithm may not contain the binding site of interest in their promoter. Another problem is that transcription factors frequently do not bind to a single nucleotide sequence but, instead, they recognize a set of related (degenerated) sequences. The binding of a given transcription factor to its recognition sequence is also known to be largely influenced by environmental conditions; it is possible that a DNA-binding site that is functional under a specific condition is nonfunctional when the cells are under a different condition (3, 4).

The first example described in this chapter aims at demonstrating how the computational tools available in the YEASTRACT information system (5, 6), in particular the MUSA algorithm, can be used to analyze results from genomic expression analysis and extract a putative-binding site for the yeast transcription factor Rim101p. YEASTRACT is a database in which all the documented regulatory associations between yeast transcription factors and their target genes as well as between transcription factors and their DNA-binding sites have been gathered. This congregation of information turns YEASTRACT into a powerful and unique tool for the study of transcriptional regulation in *S. cerevisiae*.

Rim101p is the final effector of a signaling pathway well conserved among fungi whose physiological function was first associated with adaptation and resistance to alkaline pH conditions (7). In *S. cerevisiae*, Rim101p functions as a transcriptional repressor, binding to the promoter region of its target genes and recruiting the general transcriptional repressor machinery controlled by Tup1p (8). The DNA motif 5′-TGCCAAG-3′ was demonstrated to be the binding site used by Rim101p to repress the expression of alkaline pH-responsive genes (9). Recently, Rim101p and several members of the Rim101p pathway were identified as determinants of resistance to propionic acid in *S. cerevisiae* (10), extending the biological role of this regulatory system beyond alkaline pH response in this yeast species. Using a genome-wide transcriptomic analysis of cells cultivated in the presence or absence of propionic acid stress, a set of genes repressed by Rim101p was identified (10). Remarkably, a few of these Rim101p-repressed genes had not been identified before as targets of this transcription factor in alkaline pH-challenged yeast cells (9–11). Propionic acid is a weak acid largely used for the preservation of bakery and fresh dairy products. The molecular mechanisms behind yeast adaptation and resistance to this and other weak acids used as food preservatives have been studied in yeast (12) aiming to guide the design of more efficient preservation strategies. Although in this particular case, the DNA-binding site for the transcription factor under study is already known, the combined *in silico*/wet lab strategy described below works as a proof-of-principle for the characterization of other uncharacterized or poorly characterized yeast transcription factors.

1.2. Characterization of the Functional Transcription Regulatory Network Involved in FLR1 Up-regulation in Response to Mancozeb Stress

Usually, the transcript level of a gene depends on several activators and/or repressors. These transcription factors can also regulate themselves or be regulated by other transcription factors, resulting in hierarchical networks and adding additional layers of complexity to the problem of gene expression. In fact, according to the information gathered in YEASTRACT, each yeast gene has, on average, eight documented regulators. If we consider the binding sites present in the promoter region of the gene under analysis, this number can easily increase to dozens of possible regulators. Taking these numbers into consideration, the full characterization of a functional transcriptional regulatory network can be quite challenging.

In the second example described in this chapter, the regulatory network underlying the control of the transcription of the *FLR1* gene is characterized. *FLR1* encodes a plasma membrane drug:H^+-antiporter that was found to confer resistance to multiple drugs in yeast, in particular to the agricultural fungicide mancozeb, linked to Parkinson's disease and cancer development (13). Yeast has been used as a model organism to gain insights into the toxicity of this chemical and to understand the molecular mechanisms underlying mancozeb resistance in phytopathogenic fungi (14–16). One of the observed adaptive yeast responses to mancozeb is the dramatic up-regulation of *FLR1* transcription following sudden exposure to the drug (13). A strategy to uncover the transcription factors involved in mancozeb-induced up-regulation of *FLR1*, based on the use of the computational tools available in YEASTRACT, is described bellow as well as the methodology used to establish the regulatory associations between these transcription factors. This case study provides a useful example of how functional transcriptional regulatory networks can be characterized in yeast.

2. Materials

2.1. In Silico Analyses Using the YEASTRACT Database

1. YEASTRACT is of public access (www.yeastract.com) and a computer with Internet access is the only requirement to use the database.

2.2. Engineering of pNRG1 Promoter::lacZ Fusion

1. Restriction enzymes *BamHI*, *HIndII*, *XbaI*, *XhoI*, or *SalI* (Invitrogen).
2. Vector pNB404 (17) (available at Addgene repository) or other upstream activating sequence (UAS)-less vector.
3. Vector pAJ152 (18).
4. *E. coli* XL1-Blue competent cells (Invitrogen).
5. T4 ligase (Invitrogen).
6. LB growth medium (1% bacto-tryptone, 0.5% yeast extract, 1% NaCl, adjusted to pH 7.0 with 5 N NaOH). For solid medium, add 2% agar.
7. Ampicillin stock solution: 100 mg/mL. Prepare in water and filter sterilize prior to use.
8. Spin Miniprep Kit (Qiagen).

2.3. Site-Directed Mutagenesis

1. *Pfu* Turbo polymerase (Stratagene).
2. *DpnI* restriction enzyme (Stratagene).
3. Dimethyl sulfoxide (DMSO).

2.4. Comparison of lacZ Expression in S. cerevisiae Cells by Measuring β-Galactosidase Activity

1. Yeast strains *S. cerevisiae* BY4741, *S. cerevisiae* BY4741_Δ*pdr3*, *S. cerevisiae* BY4741_Δ*rpn4*, *S. cerevisiae* BY4741_Δ*yrr1* and *S. cerevisiae* BY4741_Δ*yap1* (available from the Euroscarf collection).
2. YPD growth medium (20 g/L glucose, 10 g/L yeast extract, 20 g/L peptone). For solid medium, add 2% agar.
3. Minimal growth medium MM4 (20 g/L glucose; 2.65 g/L ammonium sulfate; 1.7 g/L yeast nitrogen base without amino acids and ammonium, 20 mg/L histidine, 20 mg/L methionine, 60 mg/L leucine, 20 mg/L uracil). For solid medium, add 2% agar.

4. Lithium acetate/TE transformation solution (0.1 M lithium acetate, 0.01 M Tris–EDTA buffer). Prepare two stock solutions, in water, of 1 M lithium acetate (adjust to pH 7.5 with acetic acid) and of 0.1 M TE buffer (0.1 M Tris, pH 7.5, 0.01 M EDTA, pH 7.5). Autoclave both solutions and dilute them appropriately to achieve the above indicated composition of the lithium acetate/TE transformation solution.
5. PEG/lithium acetate transformation solution (32% PEG 4000, 0.1 M lithium acetate, 0.01 M TE buffer). Prepare a stock solution, in water, of 40% PEG (w/v) and autoclave. Use this solution and the solutions of 1 M lithium acetate and 0.1 M TE buffer to make the PEG/lithium acetate transformation solution.
6. Carrier DNA (2 mg/mL of salmon sperm DNA). Prepare the solution in sterile water and sonicate the DNA for shearing. Before each transformation, boil the DNA for 5 min and leave it on ice until further use.
7. 1 M propionic acid stock solution. Prepare the solution in water, adjust to pH 4.0 with 10 M NaOH and sterilize by filtration.
8. 1 mg/mL mancozeb stock solution. Prepare the stock solution of mancozeb by dissolving the powder (Sigma) in DMSO. Always use freshly prepared stock solutions.
9. Buffer Z: 60 mM $Na_2HPO_4 \cdot 7H_2O$, 40 mM $NaH_2PO_4 \cdot H_2O$, 10 mM KCl, 1 mM $MgSO_4 \cdot 7H_2O$. Sterilize by filtration. Supplement the buffer with 0.27% (v/v) β-mercaptoethanol before each utilization.
10. 0.05% (w/v) SDS solution.
11. Chloroform.
12. *o*-nitrophenil-β-D-galactosidase (ONPG) solution 4 mg/mL. Prepare the solution in 0.1 M phosphate buffer and sterilize by filtration.
13. 0.5 M Na_2CO_3.

2.5. Relative Quantification of FLR1, PDR3, YAP1, YRR1, and RPN4 Transcript Levels by Real-Time RT-PCR

1. Real-time PCR thermocycler (Applied Biosystems).
2. Reverse transcription kit (Applied Biosystems).
3. Power SYBR green mixture (Applied Biosystems).
4. 4 pmol/μL solution of primers specific for amplification of *FLR1*, *PDR3*, *YAP1*, *YRR1*, and *RPN4* genes. Primers are designed using PrimerExpress software (Applied Biosystems). Prepare the primer solutions by dissolving the dry oligonucleotides in molecular biology-grade water.
5. AE buffer solution (50 mM sodium acetate, 10 mM EDTA). Adjust pH to 5.3 with glacial acetic acid. Autoclave prior to use.

6. SDS 10% solution. Prepare the solution in water with 1% DEPC to reduce RNase activity.
7. Chloroform.
8. Equilibrated phenol solution (at pH 4.3) for RNA extraction (Sigma).
9. Chloroform: isoamylalcohol solution (24:1 proportion).
10. 3 M Sodium acetate solution (adjusted to pH 4.3 with glacial acetic acid).

3. Methods

3.1. Identification of the Rim101p-Binding Site: In silico analysis of the promoter region of Rim101p-regulated genes

The set of genes found to be regulated by Rim101p and used in this example includes: *DIT1*, *MNN4*, *DUR3*, *RIM8*, *YJR061w*, *YOR389w*, *HPF1*, *YLL053c*, *AQY2*, *SMP1*, *YPL277c*, *HSP26*, *NRG1*. These genes were identified based on a comparative microarray analysis of wild-type and Δ*rim101* mutant yeast cells under propionic acid stress (10). Since Rim101p is a transcriptional repressor (8), only the genes whose transcript levels increased in the Δ*rim101* mutant, compared to the wild-type strain, are used. This selection is based on the idea that these genes are more likely to be direct Rim101p targets and thereby more prone to harbor one (or more) Rim101p-binding site(s) in their promoters. According to our own experience, the performance of MUSA algorithm can be reduced by the existence of many genes in the dataset that are not directly regulated by the transcription factor under study. Nevertheless, the experimental strategy described below and summarized in Fig. 1 can be easily extrapolated to results obtained with other algorithms. For example, the freely accessible AlignAce algorithm (2) was found to be very useful in the identification of functional-binding sites using other yeast datasets (19).

3.2. In Silico Analysis of the Promoter Region of Rim101p-Regulated Genes Using the YEASTRACT Database

1. Access "DISCOVERER" on the upper left side ofYEASTRACT Web site (www.yeastract.com).
2. Enter the interface of "MUSA".
3. Introduce the gene list in the "ORFs/Gene list" box.
4. Fill the boxes corresponding to "recipient's mail" and "job description". Press "Run". The result of the analysis is shown in Fig. 2. Depending on the activity of the server, the answer may come in a period of time ranging from 20 min to several hours.
5. Select the first motif. Click "Match" to see whether this motif is described in the database as a binding site for a yeast transcription factor (Fig. 3).

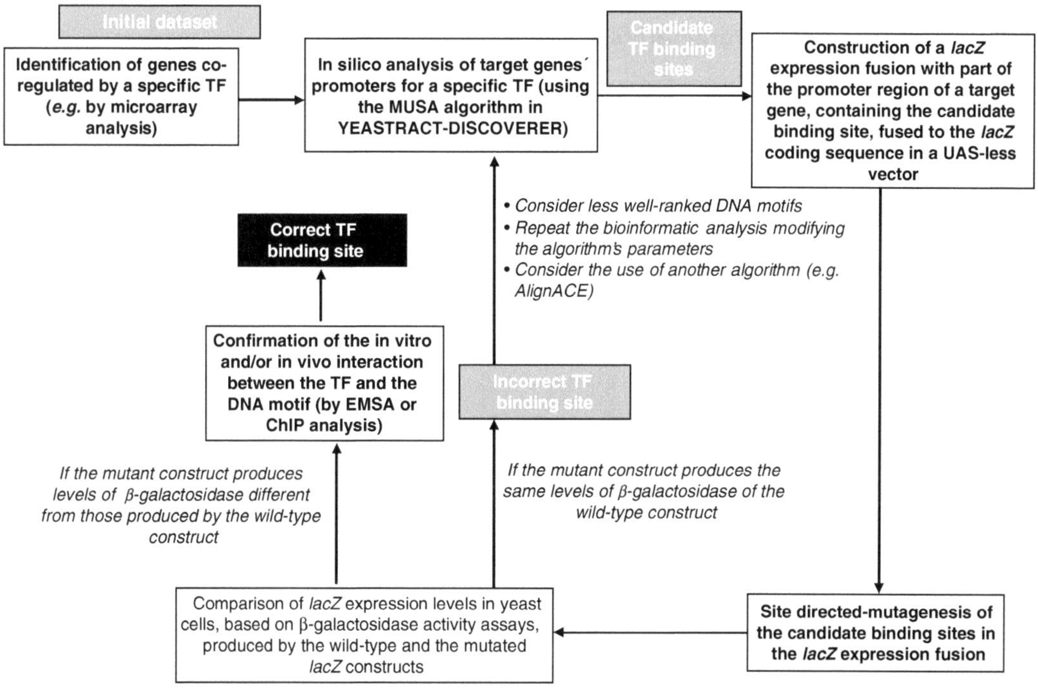

Fig. 1. Schematic representation of the main steps of the experimental methodology described here to identify functional-binding site(s) for a given yeast transcription factor (TF). This strategy is based on a preliminary *in silico* analysis of the promoter region of its target genes (direct and indirect), revealed for example by a comparative microarray analysis between the wild-type strain and a mutant with the gene encoding the transcription factor under study individually deleted.

6. The top two motifs identified by the algorithm as being enriched in the dataset (Fig. 2) correspond to the same binding site since the 5′-TGCCAAG-3′ motif, the second motif of the list, is the reverse complement of the first motif of the list, 5′-CTTGGCA-3′. Indeed, this DNA sequence corresponds to the binding site previously described for Rim101p (9), and YEASTRACT provides this information when this binding site sequence was matched with the information stored in the database (Fig. 3).

3.3. Experimental Approaches to Confirm the Bioinformatic Prediction of a Functional Rim101p-Binding Site

Since the presence of a DNA-binding site in the promoter region of a gene does not necessarily means that the TF binds to this nucleotide sequence, it is therefore necessary to demonstrate that the identified DNA sequence is required for Rim101p-dependent repression of the genes in the dataset under the conditions used for microarray analysis (10). A possible experimental strategy to demonstrate this is based on the use of *lacZ* expression fusions, as described below.

If the DNA motif 5′-TGCCAAG-3′ is indeed relevant for the repression by Rim101p of the genes in the dataset, then its alteration

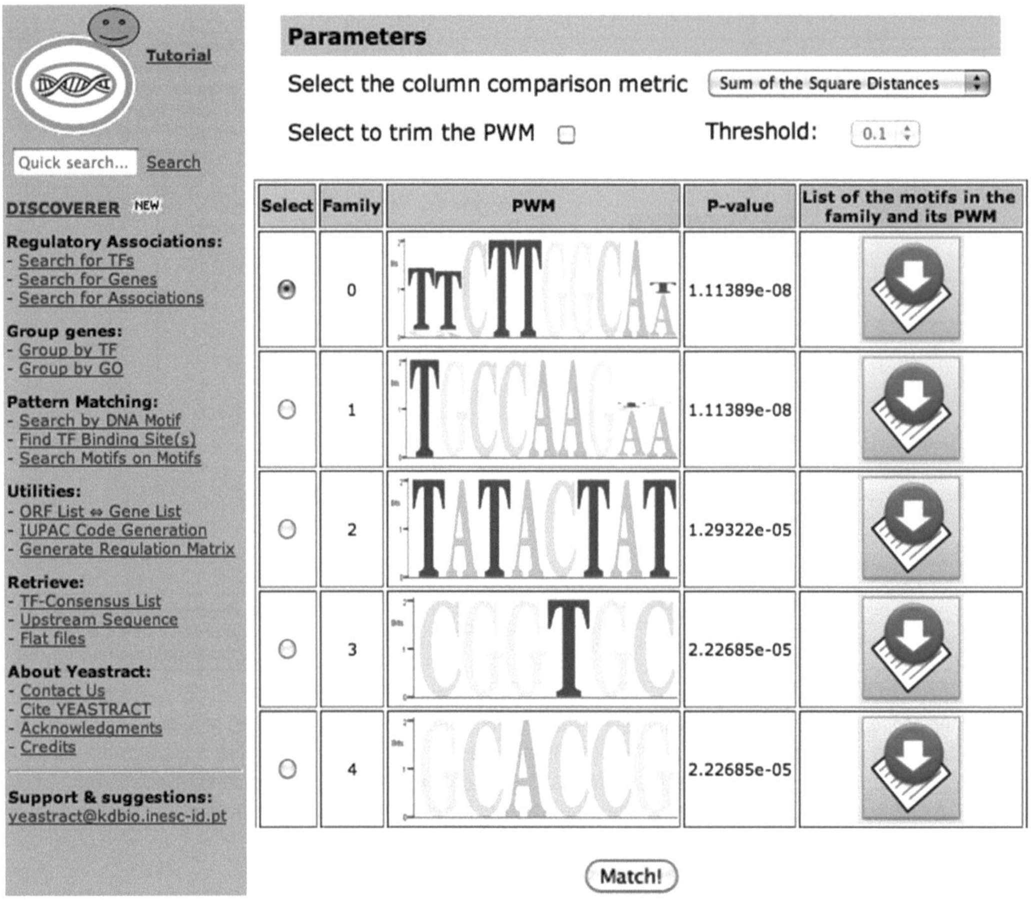

Fig. 2. DNA motifs enriched in the promoter region of the genes described to be repressed by Rim101p in the presence or absence of propionic acid stress (10), based on the *in silico* analysis of the DNA motif finder MUSA. The output provided by the algorithm includes a list of the enriched motifs, clustered in families and represented by a position weight matrix (PWM).

in the promoter region of one of the target genes is expected to abolish the repressive effect of the transcription factor. To monitor the transcription of one of the target genes, an expression fusion plasmid has to be constructed by fusing a portion of the promoter region of, for example, the *NRG1* gene, harboring the motif of interest in frame with the coding sequence of the *E. coli lacZ* gene, encoding a β-galactosidase. For the construction of the *lacZ* expression fusion, any other promoter region of the Rim101p-regulated genes harboring the candidate-binding motif TGCCAAG (or its reverse complement CTTGGCA) can be used in alternative to *NRG1*. The 5′-TGCCAAG-3′ DNA motif present in *NRG1* promoter has to be replaced, by site-directed mutagenesis, and the transcript levels of *lacZ* produced from both wild-type and mutated constructs are compared in the parental yeast cells cultivated in the presence or absence of propionic acid stress. Although this simple methodology provides a strong indication of the occurrence of an interaction between the Rim101p and the DNA sequences

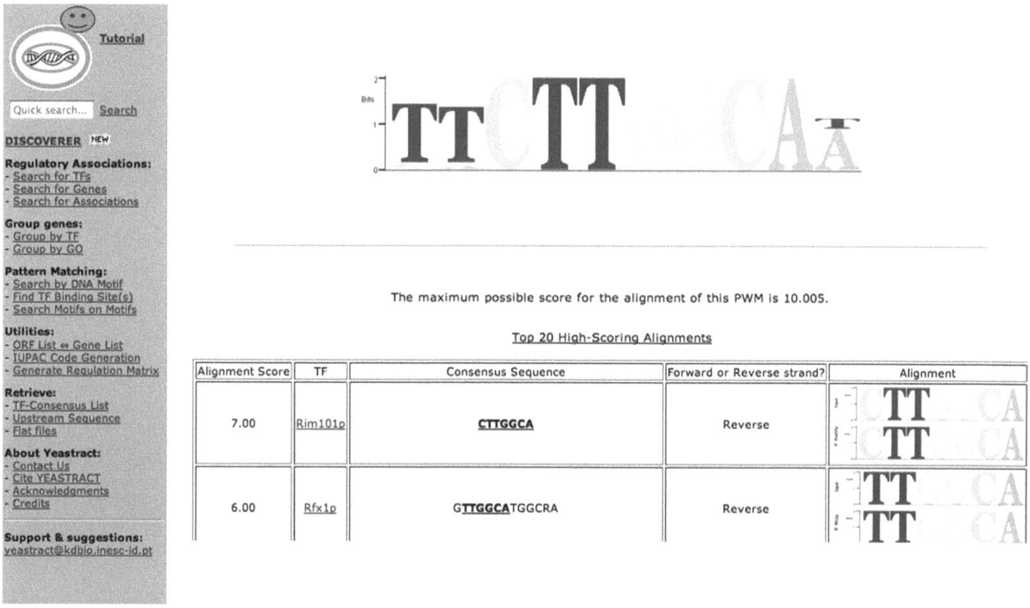

Fig. 3. Comparison of the CTTGGCA motif with the set of binding sites deposited in the YEASTRACT database. This motif, enriched in the promoter region of Rim101p-repressed genes, was demonstrated to be recognized in vitro by this transcription factor (9).

identified by the algorithm, the binding of Rim101p to the DNA sequence under study should be confirmed using, for example, electrophoretic mobility shift assays (EMSA) – the crucial parameters for the optimization of this technique are described in (20) – and/or chromatin immunoprecipitation (ChIP) – the experimental procedures and the information that can be extracted from ChIP analysis were recently reviewed (21), and is not described in this chapter.

3.4. Construction of pNRG1 Promoter::lacZ Fusion and Subsequent Site-Directed Mutagenesis

3.4.1. Design of PCR Primers for Synthesis of NRG1 Promoter (Insert Fragment)

1. Use the "Search by DNA motif" tool available in YEASTRACT to find the motif TGCCAAG in the *NRG1* promoter.
2. Copy the 75 nucleotides located upstream and downstream of the motif.
3. Design primers to amplify this 157 bp portion of *NRG1* promoter (see Note 1).
4. Use different restriction sites for each primer pair to support directional cloning (e.g., for cloning in the pNB404 vector, the enzymes *XbaI*, *XhoI*, *SalI*, and *EcoRI* can be used).
5. Purify the PCR product by electrophoresis in an agarose gel, followed by band excision.

3.4.2. Ligation of the NRG1 Promoter into the Vector pNB404 Vector (see Note 2)

1. Digest the insert fragment obtained in step 5, Subheading 3.4.1 at 37°C with appropriate restriction enzymes. If the buffer supplied for both restriction enzymes is the same, the two enzymes can be incubated simultaneously with the PCR product. Otherwise, perform the two restriction reactions

consecutively, precipitating the DNA with ethanol in between (see Note 3).

2. Digest, extract, and precipitate the pNB404 vector as it is described above for the insert DNA fragment.
3. Determine the concentration of the DNA vector and of the insert fragment by gel analysis using a molecular mass standard of known concentration or by spectrophotometry at 260 nm.
4. Perform a standard ligation reaction with T4 ligase using a 3:1 insert: vector ratio (see Note 4). Perform a ligation reaction using the vector DNA as a control.

3.4.3. Bacterial Transformation with the pNRG1 Promoter::lacZ Fusion

1. Add 80 μL of water to the ligation mixture.
2. Mix the ligation mixture with a batch of *E. coli* XL1-Blue competent cells.
3. Incubate the mixture on ice for 30 min.
4. Incubate the cells at 42°C for 45 s.
5. Add 1 mL of LB medium and incubate the cells in an orbital shaker (250 rpm) at 37°C for 60 min.
6. Spread an aliquot of the cell suspension (50–200 μL) onto LB agar plates supplemented with 150 mg/L of ampicillin. Incubate at 37°C overnight.
7. Select ten colonies from the plate and grow them overnight in 3–10 mL of liquid LB medium supplemented with 150 mg/L of ampicillin.
8. Pellet the cells by centrifugation at 4,000 × *g* for 5 min.
9. Extract the plasmid with the Qiagen Miniprep Kit.
10. Confirm the presence of the insert by digestion with the two restriction enzymes used for cloning and confirm the insertion of the portion of the *NRG1* promoter of interest in the plasmid by nucleotide sequencing.

3.4.4. Site-Directed Mutagenesis of the pNRG1 Promoter::lacZ Fusion

Design of Mutagenic Primers and PCR Reaction

1. Design two mutagenic primers to replace the candidate DNA motif TGCCAAG in the *pNRG1 promoter::lacZ* plasmid for AAAAAA taking into consideration that (see Note 5):
   - The two primers should be complementary to each other and homologous to the region where the mutation is to be introduced.
   - The site of desired mutation should be in the middle of the primer sequence.
   - The size of the primer should be close to 35 bp.
2. Set up a PCR reaction using 30 ng of the pNRB404-*NRG1* promoter plasmid (template DNA), 2 pmol of the engineered mutagenic primers, and 2.5 U of PfuTurbo polymerase (see Note 6).

3. Add 20 U of *DpnI* enzyme directly to the PCR mixture and incubate at 37°C for 1 h (see Note 7).
4. Transform *E. coli* cells with an aliquot of 1–15 μL of the digested PCR mixture (follow the steps described in Subheading 3.4, step 3).
5. Recover the plasmid DNA from the transformants using Qiagen Miniprep Kit and sequence to confirm the desired mutation.

3.5. Comparison of lacZ Expression Levels Produced by Wild-Type and Mutated pNRG1 Promoter-lacZ Constructs in S. cerevisiae Cells

3.5.1. Transformation of Yeast Cells with Wild-Type and Mutated pNRG1-Promoter::lacZ Plasmids

1. In a flask of 100 mL containing 30 mL of liquid YPD growth medium inoculate an isolated colony of wild-type *S. cerevisiae* BY4741.
2. Grow the cells at 30°C with orbital agitation (250 rpm) overnight.
3. Reinoculate the cells from both strains in 50 mL of fresh YPD liquid medium and let them duplicate at least twice (OD_{600nm}−0.2–0.4).
4. Harvest the cells by centrifugation (5 min, 4,000 × *g* at 4°C).
5. Wash the pellet with 30 mL of sterile water.
6. Resuspend the pellet in 250 μL of lithium acetate transformation solution.
7. Prepare the transformation mixtures by adding 1–2 μg of the wild-type and of the mutated plasmid p*NRG1* promoter::*lacZ*, 50 μL of cell suspension, 10 μL of single-stranded carrier DNA, and 300 μL of PEG-lithium acetate transformation solution.
8. Gently mix the suspension by inversion.
9. Incubate the mixture at 30°C with orbital agitation (250 rpm).
10. Incubate the suspensions for 20 min at 42°C.
11. Centrifuge 30 s at 4,000 × *g* and room temperature.
12. Discard the supernatant and resuspend the cells in 100 μL of water.
13. Plate the 100 μL aliquot in MM4 growth medium without uracil (see Note 8) and let the cells grow for 3 days at 30°C.

3.5.2. Cell Sampling in the Presence or Absence of Propionic Acid Stress

1. Inoculate three 100-mL sterile flasks, each containing 50 mL of MM4 growth medium without uracil (adjust the pH of the growth medium to 4.0 with HCl), with an isolated colony of the yeast cells transformed with the wild-type and mutated p*NRG1* promoter::*lacZ* plasmids.
2. Let the cells grow for a minimum period of 12 h at 30°C with orbital agitation (250 rpm).
3. Reinoculate the cells in 50 mL of fresh MM4 growth medium without uracil (at pH 4.0) and let them grow until mid-exponential

phase (OD_{600nm} of the culture between 0.5 and 0.8). Allow the cells to double at least four times.

4. Harvest the cells by filtration and reinoculate them in 50 mL of fresh MM4 growth medium without uracil (at pH 4.0, control), and in 50 mL of this same growth medium supplemented with 20 mM propionic acid. The OD_{600nm} of the culture should be close to 0.2.
5. Every 3 h harvest a sample of the cell suspension by filtration until the culture enters in stationary phase (see Note 9). A number of cells corresponding to a suspension with an OD_{600nm} of 1 in a volume of 5 mL provide a suitable number of cells to accurately assess *lacZ* expression.
6. Keep the filters at −20°C until further use.

3.5.3. Assessment of lacZ Expression Based on β-Galactosidase Activity (see Note 10)

1. Resuspend the filters containing the harvested cells in 5 mL of ice-cold buffer Z.
2. Measure the OD_{600nm} of the suspension.
3. Distribute 1 mL of the cell suspension in four tubes (labeled 1–4).
4. Add 50 μL of 0.05% SDS and 50 μL of chloroform to promote cell lysis.
5. Vortex vigorously for about 30 s.
6. Incubate the suspension for 5 min at 28°C.
7. Add 100 μL of ortho-nitrophenyl-β-galactoside (ONPG), the substrate of the β-galactosidase enzyme.
8. Stop the enzymatic reaction by adding 500 μL of Na_2CO_3 to each tube labeled as 1. Reactions occurring in tubes 2, 3, and 4 should be stopped at different times to get the enzyme activity.
9. Read the absorbance of each tube at 480 nm.
10. Calculate enzyme activity by estimating the increase in OD_{480nm} per unit of time of reaction. Normalize the enzymatic activity by dividing the result by the OD_{600nm} of the cell suspension (see Note 11).

Characterization of the Functional Transcriptional Regulatory Network Involved in *FLR1* Up-regulation in Response to Mancozeb Stress

The experimental strategy used to characterize the transcriptional regulatory network controlling *FLR1* up-regulation under mancozeb stress is summarized in Fig. 4. This methodology can be easily extended to any other yeast gene and, eventually, to groups of co-regulated genes. However, it should be emphasized that with the increase in the number of genes under analysis, the complexity of the problem can increase dramatically.

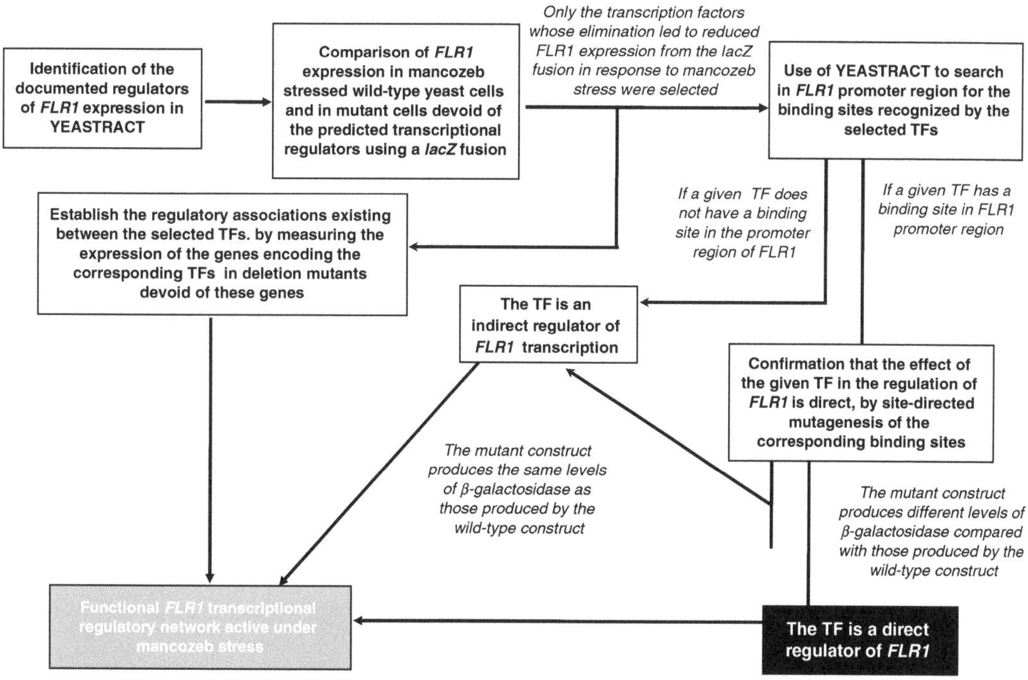

Fig. 4. Experimental strategy used to characterize the regulatory network underlying the activation of *FLR1* transcription under mancozeb stress.

3.6. Identification of Documented Transcriptional Regulators of FLR1 Transcription

The transcription factors documented to have an effect on *FLR1* expression can be easily extracted from the YEASTRACT database:

1. Access the YEASTRACT Web site http://www.yeastract.com.
2. Select the "Group by TF" tool. Fill the "ORFs/Gene" box with *FLR1* and leave the "Transcription factor" box blank.
3. Tick the "Check for all TFs" option *and* leave the option "Documented regulation" selected.
4. The database provides not only the documented regulators of *FLR1* gene, but also the documented regulatory associations established between them. Click on "Generate image" to see a prediction of the putative regulatory network underlying the control of *FLR1* transcription (Fig. 5).

3.7. Identification of the Transcription Factors Involved in the Up-regulation of FLR1 Under Mancozeb Stress

The involvement of the documented regulators of *FLR1* in the activation of this gene in response to mancozeb stress can be assessed by comparing the transcript levels of this gene in wild-type *S. cerevisiae* cells with the levels produced in single deletion mutants lacking each one of those transcriptional regulators. As this strategy implies the access to many strains, the use of a p*FLR1*::*lacZ* fusion plasmid is suggested as the easiest way to monitor *FLR1* expression in each case.

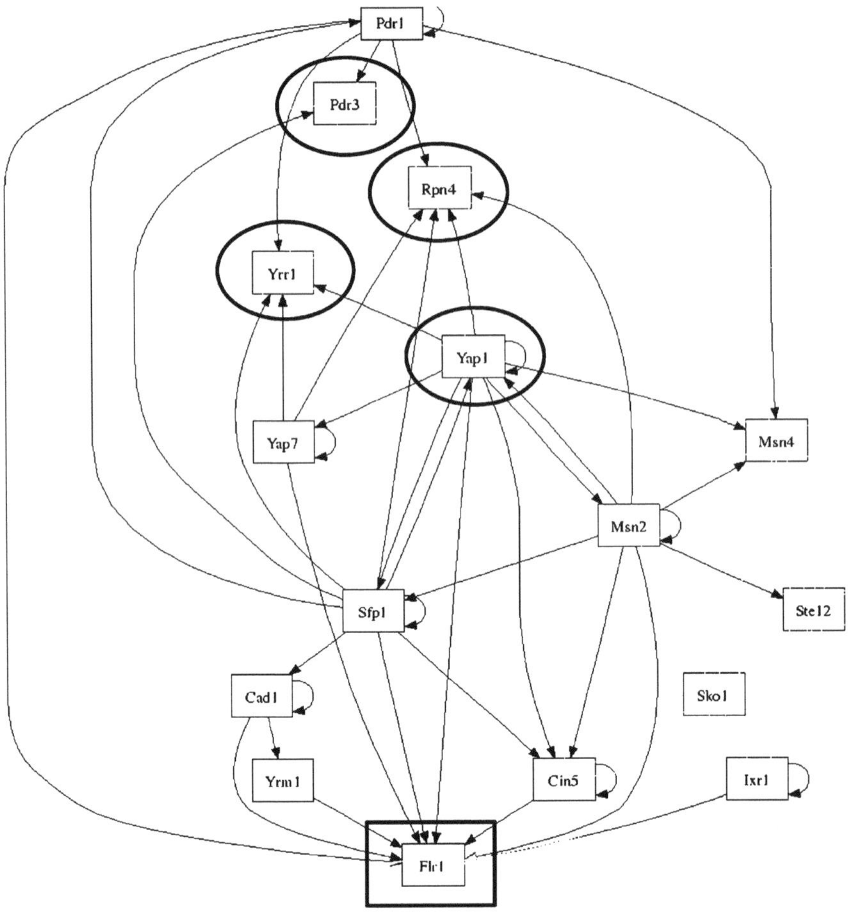

Fig. 5. Depiction of the regulatory network predicted to control *FLR1* gene expression, as obtained using the YEASTRACT database tools. Highlighted in the *square* is the target gene, *FLR1*, and in the *circles* are the transcription factors found to play a role in the control of *FLR1* up-regulation registered in mancozeb-challenged yeast cells (13).

3.7.1. Engineering the pFLR1::lacZ Fusion Plasmid

Design of PCR Primers for Synthesis of *FLR1* Promoter (Insert) by PCR

1. Design primers to amplify the promoter region of *FLR1* that is usually considered the 1000 bp upstream of the ATG start codon (see Note 12). Use different restriction sites for each primer pair to support directional cloning (e.g., for cloning in the pAJ152 vector, the enzymes *HIndIII* and *BamHI* may be used).
2. Purify the PCR product by electrophoresis in an agarose gel, followed by band excision.

Ligation of Insert into the pAJ152 Vector and Propagation in *E. coli* XL1-Blue Cells

1. Use the protocol described in Subheading 3.4.2, to digest the PCR-amplified *FLR1* promoter with appropriate restriction enzymes (e.g., with *BamHI* and *HIndIII*).
2. Digest the pAJ152 vector with the same restriction enzymes (steps described in Subheading 3.4.2).

3. Ligate the insert DNA fragment to pAJ152 vector as described in Subheading 3.4.2.
4. Transform *E. coli* XL1-Blue competent cells with the engineered plasmid (steps described in Subheading 3.4.3).
5. Confirm the presence of the correct insert of DNA fragment first by digestion with the two restriction enzymes used for cloning followed by nucleotide sequencing.

3.7.2. Comparison of lacZ Expression Levels Produced from pFLR1::lacZ Fusion Plasmid in S. cerevisiae Wild-Type and in the Mutants Lacking the Documented Regulators of FLR1 Transcription

1. Transform wild-type *S. cerevisiae* BY4741 cells and the mutants deleted for the 15 transcription factors indicated in Fig. 4 with the engineered p*FLR1*::*lacZ* plasmid (steps described in Subheading 3.5.1).
2. Inoculate 50 mL of MMB growth medium without uracil with an isolated colony of the wild-type cells transformed with p*FLR1*::*lacZ*. Repeat the procedure for the 15 mutants transformed with the same expression fusion.
3. Let the cells grow for a minimum period of 12 h at 30°C with orbital agitation (250 rpm).
4. Reinoculate the cells in 50 mL of fresh MM4 growth medium without uracil and let them grow until mid-exponential phase (OD_{600nm} of the culture between 0.5 and 0.8). Allow the cells to double at least four times.
5. Harvest the cells by filtration and reinoculate them in 50 mL of fresh MM4 growth medium without uracil, and in 50 mL of this same growth medium supplemented with 2 mg/L mancozeb so that the OD_{600nm} of the culture is 0.2.
6. Let the cells grow for 2 h in the presence or absence of the drug. After that time, harvest a volume of culture in order to have a suspension with an OD_{600nm} of 1 in a volume of 5 mL.
7. Use the steps described in Subheading 3.5.3 to compare β-galactosidase activity in the different strains harboring the p*FLR1*::*lacZ* fusion.
8. Select the mutants in which β-galactosidase activity, corresponding to *FLR1* expression, is lower than the values attained in the parental strain under mancozeb stress.

3.8. Effect of Pdr3p, Yap1p, Yrr1p, and Rpn4p in the Regulation of FLR1 Transcription Under Mancozeb Stress

Screening with the *lacZ* fusion plasmid allows the identification of the regulators contributing to the up-regulation of *FLR1* under mancozeb stress. The obtained results show that among the 15 transcription factors tested only the elimination of *PDR3*, *YAP1*, *YRR1*, or *RPN4* reduced the expression of *lacZ* in cells cultivated in the presence of mancozeb (13). These transcription factors are highlighted in the hypothesized transcriptional regulatory network shown in Fig. 5. The positive effect of Pdr3p, Yap1p, Yrr1p, and Rpn4p in *FLR1* expression can result from their direct

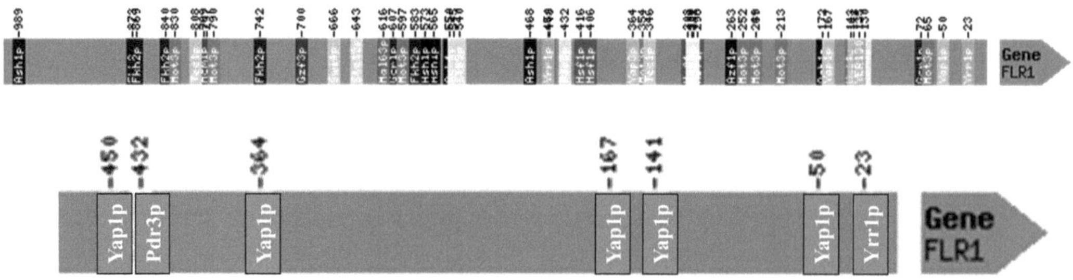

Fig. 6. Distribution of all the potential transcription factor-binding sites found to occur in the *FLR1* promoter region, using the YEASTRACT database tools. A detailed view of the functional region of the *FLR1* promoter is shown below, highlighting the binding sites recognized by the transcription factors found to play a role in the control of *FLR1* up-regulation under mancozeb stress are displayed.

binding to the promoter region of this gene or it can be an indirect effect. Indeed, the examination of *FLR1* promoter (see Note 13) only identified the presence of binding sites for Pdr3p, Yap1p, and Yrr1p (Fig. 6), suggesting that the effect of these transcription factors is direct. On the other hand, Rpn4p is likely to be an indirect regulator of *FLR1* transcription. The in vivo binding of Pdr3p, Yap1p, and Yrr1p to *FLR1* promoter can be confirmed using ChIP but with the increase in the number of transcription factors in the regulatory network this approach may become too laborious. A simpler experimental strategy, based on the site directed mutagenesis of the binding sites of Pdr3p, Yap1p, and Yrr1p in *FLR1* promoter in the p*FLR1*::*lacZ* fusion, is here described to get an idea on the involvement of those sequences in the up-regulation of *FLR1* under mancozeb stress. The results obtained confirm that these sequences are indeed critical for maximal transcription of *FLR1* in mancozeb-stressed cells (22), suggesting that Pdr3p, Yap1p, and Yrr1p are direct regulators of *FLR1* gene under mancozeb stress. Assuming that the effect of Rpn4p over *FLR1* transcription is indirect, the role of this transcription factor upon *FLR1* up-regulation is hypothesized to occur through its action in the regulation of *PDR3*, *YAP1*, and/or *YRR1* expression. To check this premise, the expression of each of these genes in the wild-type strain and in the Δ*rpn4* deletion mutant has to be assessed. Moreover, to establish how this regulatory network acts in the control of *FLR1* transcription under mancozeb stress, it is also necessary to study the transcriptional profile of each of the players in the network and to understand how each of the transcription factors Yap1p, Pdr3p, Yrr1p, and Rpn4p influences their own expression and one of the remaining genes. Overall, the transcript levels of each of the five players in the network, *YAP1*, *YRR1*, *PDR3*, *RPN4*, and *FLR1* must be measured, for example, by real-time RT-PCR, in cells of the parental strain challenged with mancozeb and compared to the levels registered in Δ*yap1p*, Δ*pdr3*,

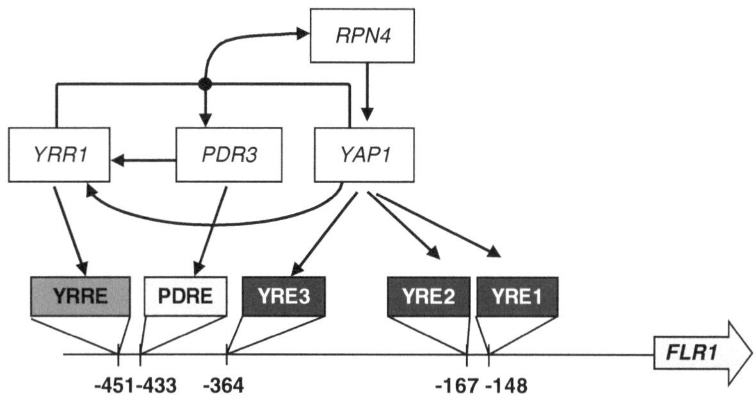

Fig. 7. Regulatory network structure proposed to control *FLR1* up-regulation in mancozeb-challenged cells (ref. 22; and our unpublished results).

Δ*yrr1*, and Δ*rpn4* cultivated in the same conditions. The combination of the data gathered through the method described below allows us to propose the transcriptional regulatory network structure depicted in Fig. 7.

3.8.1. Site-Directed Mutagenesis of pFLR1::lacZ Fusion

1. Design mutagenic primers to replace the guanine and cytosine nucleotides of the Pdr3p-, Yap1p- and Yrr1p-binding sites present in *FLR1* promoter by adenine or thymine nucleotides (see Note 14).
2. Transform wild-type *S. cerevisiae* cells with the original p*FLR1*::*lacZ* fusion and with the engineered mutant constructs (see Note 15).
3. Compare the levels of β-galactosidase activity produced from the different constructs in cells exposed for 2 h to 2 mg/L of mancozeb (see Note 16).

3.8.2. Quantification of YAP1, YRR1, RPN4, FLR1, and PDR3 in Mancozeb-Stressed Yeast Cells by Real-Time RT-PCR

Cell Sampling in the Presence or Absence of Mancozeb Stress

1. Inoculate 50 mL of MMB growth medium with an isolated colony of wild-type *S. cerevisiae* BY4741.
2. Let the cells grow for a minimum period of 12 h at 30°C with orbital agitation (250 rpm).
3. Reinoculate the cells in 50 mL of fresh MM4 growth medium and let them grow until mid-exponential phase (OD_{600nm} of the culture between 0.5 and 0.8). Allow the cells to double at least four times.
4. Harvest the cells by filtration and reinoculate them in 100 mL of fresh MM4 growth medium, and in 100 mL of this same growth medium supplemented with 2 mg/L mancozeb so that the culture OD_{600nm} is 0.2.
5. Let the cells grow for 1 h. After that time, centrifuge the culture at 4,000 ×*g* for 5 min (at 4°C).

6. Discard the supernatant and freeze the cell pellet rapidly in liquid nitrogen. Keep the pellets at –80°C until further use.
7. Repeat the procedure for Δ*rpn4*, Δ*pdr3*, Δ*yrr1*, and Δ*yap1* mutants.

RNA Extraction (see Note 17)

1. Resuspend the cell pellets in 450 μL of ice-cold AE buffer. Transfer the mixture for a 2-mL Eppendorf tube.
2. Add 90 μL of 10% SDS and vortex 5 s.
3. Add 800 μL of phenol solution and vortex 5 s.
4. Incubate the mixture for 4 min at 65°C.
5. Place tubes on ice until the formation of crystals is observed (see Note 18).
6. Centrifuge the mixture at 14,000 × *g* for 5 min (at 4°C).
7. Transfer the top phase to a fresh Eppendorf tube.
8. Add 450 μL of phenol and 450 μL of the chloroform:isoamyl alcohol solution.
9. Vortex for 5 s, and centrifuge 5 min at 14,000 × *g* at 4°C.
10. Transfer the top phase to a fresh tube and repeat steps 8 and 9.
11. Transfer the top-phase to a fresh tube containing 900 μL of chloroform. Vortex 5 s and centrifuge at 14,000 × *g* for 5 min.
12. Take the top-phase to a new tube. Add 50 μL of sodium acetate.
13. Add 1 mL of ice-cold ethanol. Vortex for 5 s and leave the mixture at –80°C for 20 min (see Note 19).
14. Centrifuge at 14,000 × *g* for 20 min (at 4°C).
15. Wash the pellet with 750 μL of 70% ethanol. Centrifuge at 14,000 × *g* for 10 min.
16. Dry the pellet for 5 min in a speed-vacuum.
17. Resuspend the pellet in 50 μL of RNase free water.
18. Determine RNA concentration by spectrophotometry at 260 nm.

Relative Quantification of *YAP1, RPN4, PDR3,* and *YRR1* Transcript Levels by Real-Time RT-PCR

Conversion of Total RNA to cDNA

1. Prepare a stock solution of the RNA samples (500 ng/μL concentration).
2. Prepare the following reaction mixtures in plastic tubes (see Note 20):
   - 2 μL of 500 ng/μL RNA solution.
   - 1 μL of 10× reverse transcriptase buffer.
   - 2 μL of dNTPs stock solution.
   - 0.5 μL of random primers.
   - 1.85 μL of RNAse free water.
   - 0.25 μL of reverse transcriptase.
3. Set up the PCR program in the thermocycler: 10 min at 25°C, 30 min at 48°C, and 5 min at 95°C.

Relative Quantification of Transcript Levels by Real-Time PCR

1. After the PCR program is complete, dilute the cDNA with 30 μL of RNAse-free water (see Note 21).
2. The cDNA from the different strains is used to quantify the transcript levels of the gene of interest (in this case of *YAP1*, *YRR1*, *PDR3*, and *RPN4*) and also of the *ACT1* gene, used as an endogenous control (see Note 22).
3. To quantify *YAP1* expression in the different strains prepare, in real-time PCR optical tubes, the following mixture:
   - 2.5 μL of *YAP1* forward primer solution.
   - 2.5 μL of *YAP1* reverse primer solution.
   - 2.5 μL of cDNA diluted solution.
   - 5 μL of water.

   12.5 μL of power SYBR green master mix.
4. Make similar mixtures using the primers specific for the amplification of *FLR1*, *YRR1*, *PDR3*, *RPN4*, and *ACT1* genes.
5. Set up the PCR program in the thermocycler: 2 min at 50°C, 10 min at 95°C, 40 cycles of 15 s at 95°C (denaturation), and 1 min at 60°C (amplification). Collect the fluorescence data during the amplification step.
6. Use the software provided with the real-time thermocycler to calculate the relative expression of the genes in all the strains cultivated under mancozeb stress (see Note 23).

4. Notes

1. Smaller DNA fragments (up to 50 bp) spanning the candidate DNA motif can also be used but fragments of such size are frequently more difficult to clone.
2. For the study of *cis*-acting regulatory elements, it is highly recommended to use *lacZ*-based plasmids free of UAS; pNB404 (commercially available at Addgene) is suggested. In this vector, *lacZ* expression was put under the control of the minimal promoter *CYC1* which only contains a transcription and a translation initiation site (17).
3. DNA precipitation with ethanol is performed by adding 4 μL of 3 M sodium acetate and 100 μL of 100% ethanol to the restriction reaction mixture (considering that the reaction is carried out in a volume of 40 μL). Leave the mixture for 30 min at −20°C. Pellet the DNA by centrifugation at maximum speed for 30 min at 4°C. Wash the pellet with 500 μL of 70% ethanol. Dry the pellet on air or using a speed-vacuum.
4. To an autoclaved 1-mL microcentrifuge tube, add 4 μL of 5× ligase reaction buffer, 15–60 fmol of vector DNA, 45–180 fmol

of insert DNA (0.1–1.0 μg total DNA), and 2.5 U of T4 DNA ligase (Invitrogen). Bring the total volume to 20 μL, mix gently and centrifuge briefly. Incubate the mixture at 25°C for 1 h. If no transformants are obtained different proportions of insert:vector can be tried or the ligation can be carried out for a longer period of time (maximum overnight) at 16°C.

5. Free softwares are available in the Internet to design the mutagenic primers, an example being PrimerX (http://www.bioinformatics.org/primerx/).
6. The amount of primer to be used in a mutagenic PCR is variable and depends mostly on the size of the sequence; 2 pmol is, however, a good starting point. The amount of template DNA to be used is also variable, being recommended to start with 30 ng and increasing, in steps of 15 ng, to a maximum of 500 ng, until transformants are obtained. An annealing temperature of 60°C and a nucleotide incorporation rate by the *Pfu* Turbo enzyme of 1 kb/min is recommended. In general, only 18 cycles of amplification are required. 1 to 5 μL of DMSO should be used in the PCR mixture to improve polymerase activity, this being particularly important if plasmids larger than 4 kb are used as templates.
7. Increase the duration of the digestion step with *DpnI* enzyme up to 2 h if 200–500 ng of template DNA are used.
8. Uracil is the selection mark of pNB404 plasmid.
9. In this growth medium, stationary phase is achieved at OD_{600nm} 2.5–3.
10. As described in Miller method (23).
11. Normalization can be performed based on total protein concentration, using a Lowry or Bronsted assay; however, normalization based on the culture OD_{600nm} is simpler and produces reliable results.
12. The nucleotide sequence of *FLR1* promoter can be obtained either in YEASTRACT (use the "Quick search" tool to access the page dedicated to *FLR1* gene) or in the SGD Web site (www.yeastgenome.org).
13. Use the "Search for TF" tool to map the binding sites of Pdr3p, Yap1p, and Yrr1p in *FLR1* promoter. Fill the "ORF/Gene" box with *FLR1*. Select the "Potential" option in the Search type menu and press search. Select the transcription factors Rpn4p, Pdr3p, Yap1p, and Yrr1p and press "Redisplay" (Fig. 6).
14. Follow the recommendations described in Subheading 3.4.1.
15. Follow the experimental procedure described in Subheading 3.5.1.
16. The experimental procedure of cell sampling in the presence of mancozeb is described in Subheading 3.8.2 and the

determination of β-galactosidase activity is described in Subheading 3.5.3.

17. Other alternatives to the hot-phenol method that is described here can also be used to extract RNA from yeast cells, including the use of specific kits such as those available at Qiagen.
18. Typically, crystals begin to form after 10 min. A dry-ice bath can be used instead of ice to speed up the process.
19. The mixture can be left at −20°C, but in this case it is convenient to increase incubation time up to 30 min.
20. The conversion of total RNA into cDNA can be performed directly in the thermocycler used for real-time PCR or, alternatively, it can be performed in a regular thermocycler. If a regular thermocycler is used the mixtures can be made in regular PCR tubes.
21. For less abundant genes, it can be convenient to perform lower dilutions of cDNA samples.
22. Other genes can be used as house-keeping genes, such as *PDA1* and *IPP1*; however, according to our experience *ACT1* is reliable under the conditions described in this work.
23. The calculation of each gene transcript level is performed automatically using the software provided with Applied Biosystems real-time thermocycler. These values can be calculated manually based on the instructions provided by the manufacturer.

Acknowledgments

Work on gene and genome-wide transcription regulation in yeast in our laboratory is financially supported by FEDER and Fundação para a Ciência e Tecnologia (FCT) (contracts PTDC/AGR-ALI/102608/2008, PTDC/EBB-EBI/108517/2008 and PTDC/BIO/72063/2006). N.P.M. acknowledges a postdoctoral grant (SFRH/BPD/46982/2008) from FCT.

References

1. Bailey, T. L., Williams, N., Misleh, C., and Li, W. W. (2006) MEME: discovering and analyzing DNA and protein sequence motifs, *Nucleic Acids Res 34*, W369–373.
2. Hughes, J. D., Estep, P. W., Tavazoie, S., and Church, G. M. (2000) Computational identification of cis-regulatory elements associated with groups of functionally related genes in Saccharomyces cerevisiae, *J Mol Biol 296*, 1205–1214.
3. Harbison, C. T., Gordon, D. B., Lee, T. I., Rinaldi, N. J., Macisaac, K. D., Danford, T. W., and al., e. (2004) Transcriptional regulatory code of a eukaryotic genome, *Nature 431*, 99–104.
4. Nguyen, D. T., Alarco, A. M., and Raymond, M. (2001) Multiple Yap1p-binding sites mediate induction of the yeast major facilitator FLR1 gene in response to drugs, oxidants, and alkylating agents, *J Biol Chem 276*, 1138–1145.

5. Teixeira, M. C., Monteiro, P., Jain, P., Tenreiro, S., Fernandes, A. R., Mira, N. P., Alenquer, M., Oliveira, A., Freitas, A. T., and Sá-Correia, I. (2006) The YEASTRACT database: a tool for the analysis of transcriptional regulatory associations in *Saccharomyces cerevisiae*, *Nucleic Acids Res. 34*, D446–D451.
6. Monteiro, P., Mendes, N., Teixeira, M. C., d'Orey, S., Tenreiro, S., Mira, N. P., Pais, H., Francisco, A., Carvalho, A., Lourenço, A. B., Sá-Correia, I., Oliveira, A., and Freitas, A. T. (2008) YEASTRACT-DISCOVERER: new tools to improve the analysis of transcriptional regulatory associations in *Saccharomyces cerevisiae*, *Nucleic Acids Res 36(Database issue)*, D132–136.
7. Penalva, M. A., Tilburn, J., Bignell, E., and Arst, H. N., Jr. (2008) Ambient pH gene regulation in fungi: making connections, *Trends Microbiol 16*, 291–300.
8. Lamb, T. M., and Mitchell, A. P. (2003) The transcription factor Rim101p governs ion tolerance and cell differentiation by direct repression of the regulatory genes *NRG1* and *SMP1* in Saccharomyces cerevisiae, *Mol Cell Biol 23*, 677–686.
9. Lamb, T., and Mitchell, A. P. (2003) The transcription factor Rim101p governs ion tolerance and cell differentiation by direct repression of the regulatory genes *NRG1* and *SMP1* in *Saccharomyces cerevisiae*, *Mol Cell Biol 23*, 677–686.
10. Mira, N. P., Lourenço, A. B., Fernandes, A. R., and Sá-Correia, I. (2009) The *RIM101* pathway has a role in yeast adaptation and resistance to propionic acid and other weak acids, *FEMS Yeast Res.. 9*, 202–216.
11. Lamb, T. M., Xu, W., Diamond, A., and Mitchell, A. P. (2001) Alkaline response genes of *Saccharomyces cerevisiae* and their relationship to the *RIM101* pathway, *J Biol Chem 276*, 1850–1856.
12. Mira, N. P., Teixeira, M. C., and Sá-Correia, I. (2010) Adaptation and tolerance to weak acid stress in *Saccharomyces cerevisiae*: a genome-wide view, *OMICS: A Journal of Integrative Biology 14*, 525–540.
13. Teixeira, M. C., Dias, P. J., Simoes, T., and Sa-Correia, I. (2008) Yeast adaptation to mancozeb involves the up-regulation of *FLR1* under the coordinate control of Yap1, Rpn4, Pdr3, and Yrr1, *Biochem Biophys Res Commun 367*, 249–255.
14. Cabrito, T. R., Teixeira, M. C., and Sá-Correia, I. (2009) Global adaptive response and resistance to agricultural fungicides: lessons from yeast and phytopathogenic fungi, In *Fungicides: Chemistry, Environmental Impact and Health Effects* (Columbus, F., Ed.), Nova Science Publishers, New York, USA.
15. Dias, P. J., Teixeira, M. C., Telo, J. P., and Sá-Correia, I. (2010) Insights into the mechanisms of toxicity and tolerance to the agricultural fungicide mancozeb in yeast, as suggested by a chemogenomic approach, *OMICS: A Journal of Integrative Biology 14*, 211–227.
16. Santos, P. M., Simões, T., and Sá-Correia, I. (2009) Insights into yeast adaptive response to the agricultural fungicide mancozeb: A toxicoproteomics approach, *Proteomics 3*, 657–670.
17. Bachhawat, N., Ouyang, Q., and Henry, S. A. (1995) Functional Characterization of an Inositol-sensitive Upstream Activation Sequence in Yeast, *J. Biol. Chem. 270*, 25087–25095.
18. Andre, B., Hein, C., Grenson, M., and Jauniaux, J. C. (1993) Cloning and expression of the *UGA4* gene coding for the inducible GABA-specific transport protein of Saccharomyces cerevisiae, *Mol Gen Genet 237*, 17–25.
19. Mira, N. P., Henriques, S. F., Keller, G., Matos, R., Arraiano, C., Teixeira, M. C., Winge, D. R., and Sá-Correia, I. (2011) Identification of a DNA-binding site for the transcription factor Haa1, required for Saccharomyces cerevisiae response to acetic acid stress, Nucleic Acids Research, 16, 6896–6907.
20. Hellman, L. M., and Fried, M. G. (2007) Electrophoretic mobility shift assay (EMSA) for detecting protein-nucleic acid interactions, *Nat Protoc 2*, 1849–1861.
21. Grably, M., and Engelber, D. (2010) A Detailed Protocol for Chromatin Immunoprecipitation in the Yeast *Saccharomyces cerevisiae*, In *Methods in Molecular Biology*, pp 211–224.
22. Teixeira, M. C., Dias, P. J., Monteiro, P., Sala, A., Oliveira, A., Freitas, A. T., and Sá-Correia, I. (2010) Refining current knowledge on the yeast *FLR1* regulatory network by combined experimental and computational approaches, *Molecular Biosystems 6*, 247–281.
23. Miller, J. (1972) *Experiments in Molecular Genetics*, Cold Spring Harbor Laboratory Press, New York.

Chapter 3

Electrophoretic Mobility Shift Assay Analysis of NFκB Transcriptional Regulation by Nuclear IκBα

Ashish Juvekar, Sitharam Ramaswami, Subrata Manna, Tzu-Pei Chang, Adeel Zubair, and Ivana Vancurova

Abstract

Transcription factor NFκB is a key regulator of genes involved in immune and inflammatory responses, as well as genes regulating cell proliferation and survival. In addition to many inflammatory disorders, NFκB is constitutively activated in a variety of human cancers and leukemia. Thus, inhibition of NFκB DNA binding activity represents an important therapeutic approach for disorders characterized by high levels of constitutive NFκB activity. We have previously shown that NFκB DNA binding activity is suppressed by the nuclear translocation and accumulation of IκBα, which is induced by inhibition of the 26S proteasome. In this chapter, we describe a protocol that uses small inhibitory RNA (*si* RNA) interference followed by electrophoretic mobility shift assay (EMSA) to analyze the regulation of NFκB DNA binding by nuclear IκBα induced by the proteasome inhibitor MG132. Using this protocol, we show that in human leukemia Hut-78 cells that exhibit high levels of NFκB DNA binding activity, MG132 induces nuclear translocation and accumulation of IκBα, which then specifically inhibits NFκB DNA binding. This protocol uses human leukemia Hut-78 cells; however, it can be easily adapted for other cells exhibiting high levels of constitutive NFκB DNA binding.

Key words: Electrophoretic mobility shift assay, MG132, NFκB, Nuclear IκBα, Nuclear translocation, Proteasome inhibition, si RNA, Transcriptional regulation

1. Introduction

Transcription factor NFκB comprises a family of proteins that serve as crucial regulators of genes involved in host immune and inflammatory responses, cell survival, proliferation, and differentiation (1–6). In addition to many pro-inflammatory genes, NFκB induces transcription of anti-apoptotic genes, such as genes of the Bcl-2 family (7–11). Since increased activation of NFκB has been associated

Ales Vancura (ed.), *Transcriptional Regulation: Methods and Protocols*, Methods in Molecular Biology, vol. 809,
DOI 10.1007/978-1-61779-376-9_3, © Springer Science+Business Media, LLC 2012

with the pathogenesis of numerous inflammatory disorders as well as with many types of human cancer and leukemia (7–13), inhibition of NFκB activity represents an important therapeutic target.

In most resting cells, the NFκB proteins (p50 NFκB1, p65/Rel A, c-Rel, NFκB2/p52, and Rel B) form dimers that are retained in the cytoplasm through their association with the inhibitory protein IκBα. Following cell stimulation with inflammatory signals, IκBα is phosphorylated on Ser-32 and Ser-36 by the enzymes of IκB kinase complex (IKK), ubiquitinated and selectively degraded by the 26S proteasome (14). The cytoplasmic degradation of IκBα by the proteasome then results in the release of NFκB dimers from the inhibitory complex, followed by their translocation to the nucleus, where they bind to promoter regions of NFκB-dependent anti-apoptotic and pro-inflammatory genes and stimulate their transcription. Thus, the 26S proteasome inhibitors, known to inhibit NFκB activity by inhibiting the inducible degradation of phosphorylated IκBα in stimulated cells, have been used as therapeutic targets in many inflammatory disorders as well as in malignancies characterized by high levels of NFκB-dependent transcription (15, 16).

Cutaneous T cell lymphoma (CTCL) are aggressive tumors that are characterized by high levels of constitutive NFκB DNA binding activity and high resistance to apoptosis (17–20). We have previously shown that in metastatic prostate cancer cells exhibiting high levels of NFκB DNA binding activity, the 26S proteasome inhibitor MG132 blocks NFκB DNA binding by a novel and previously unrecognized mechanism that consists of inducing IκBα translocation from the cytoplasm to the nucleus, where it inhibits NFκB-dependent transcription (21).

Here, we describe a protocol that uses the proteasome inhibitor MG132 to induce the nuclear translocation and accumulation of IκBα in CTCL Hut-78 cells. MG132 increases the nuclear levels of IκBα in Hut-78 cells, thus inhibiting NFκB DNA binding activity. To confirm that the inhibition of NFκB is directly caused by the nuclear IκBα, we silenced IκBα expression by IκBα-specific *si* RNA, and then treated cells with MG132. The decreased nuclear expression of IκBα in cells transfected with IκBα *si* RNA, compared to cells transfected with control RNA, resulted in a substantially increased NFκB DNA binding activity, both in untreated Hut-78 cells, and in cells treated with increasing concentrations of MG132. The main points of this protocol are (1) cell culture and transfection with *si* IκBα or control RNA; (2) incubation of cells with MG132; (3) isolation of cytoplasmic and nuclear extracts; (4) analysis of nuclear and cytoplasmic proteins by western blotting; and (5) analysis of NFκB DNA binding activity by electrophoretic mobility shift assay (EMSA).

2. Materials

2.1. Cell Culture

1. Hut-78 cells obtained from the American Type Culture Collection (Manassas, VA).
2. RPMI medium (Gibco, Bethesda, MD) supplemented with 10% fetal bovine serum (FBS), 2 mM glutamine, 1 mM sodium pyruvate, and 1% penicillin–streptomycin solution.
3. 75-cm^2 culture flasks.
4. Standard 12-well plates with clear flat bottom.
5. 2-ml microcentrifuge tubes.

2.2. Cell si RNA Transfection

1. Human IκBα (Santa Cruz Biotechnology, Santa Cruz, CA; sc-29360) and control (sc-37007) small interfering RNAs (*si* RNAs), *si* RNA transfection medium (sc-36868), and *si* RNA transfection reagent (sc-29528) (see Note 1).
2. Standard 12-well plates with clear flat bottom.
3. Antibiotic free RPMI medium (Gibco, Bethesda, MD) supplemented with 10% FBS, 2 mM glutamine, 10 mM HEPES, pH 7.4, and 1 mM sodium pyruvate.
4. 1.5-ml microcentrifuge tubes.

2.3. Induction of Nuclear IκBα Accumulation by MG132

1. Sterile DMSO (American Type Culture Collection, Manassas, VA).
2. MG132 (EMD Chemicals, San Diego, CA) is dissolved in sterile DMSO at a concentration 100 mM and stored at −80°C (see Note 2).

2.4. Preparation of Cytoplasmic and Nuclear Extracts

1. Relaxation buffer (RB): 10 mM HEPES, pH 7.5, 10 mM KCl, 3 mM NaCl, 3 mM $MgCl_2$, 1 mM EDTA, 1 mM EGTA (see Note 3).
2. 10% (v/v) Igepal CA-630 (Sigma): In a sterile 15-ml centrifuge tube, dilute 1 ml of 100% Igepal CA-630 in 9 ml of sterile water. Mix well by inverting the tube several times. Avoid vortexing, which could cause foaming. Store at room temperature. If turbidity or sediment develop during storage, a clear liquid can be obtained by heating to 40°C (see Note 4).
3. Nuclear extract buffer (NEB): 20 mM HEPES, pH 7.5, 25% (v/v) glycerol, 500 mM KCl, 1 mM $MgCl_2$, 1% (v/v) Igepal CA-630, 1 mM EDTA (see Note 5).
4. 1 M dithiothreitol (DTT): Dissolve 154 mg of DTT (Cleland's Reagent) in 1 ml of sterile water. Aliquot and store at −20°C (see Note 6).
5. 100 mM phenylmethylsulfonyl fluoride (PMSF): Dissolve 17 mg of PMSF in 1 ml of absolute ethanol. Store at −20°C (it will not freeze) (see Note 7).

6. Protease inhibitor cocktail for mammalian cell extracts containing pepstatin A, bestatin, leupeptin, aprotinin, 4-(2-aminoethyl]-benzenesulfonyl fluoride (AEBSF] and trans-epoxysuccinyl-L-leucylamido(4-guanidino]butane (E-64), obtained from Sigma (P-8340) (see Note 8).
7. 5× sample buffer (5×SB): 62.5 M Tris–HCl, pH 6.8, 10% (v/v) glycerol, 2% (w/v) SDS, 0.05% (w/v) bromophenol blue, and 5% (v/v) 2-mercaptoethanol (added just before use).
8. 2× sample buffer (2×SB): 0.4 ml of 5×SB, 0.55 ml of deionized water, and 50 μl of 2-mercaptoethanol (added just before use).

2.5. SDS-Polyacrylamide Gel Electrophoresis (SDS-PAGE)

1. 90% ethanol
2. Acrylamide/bis (30%): Dissolve 87.6 g of acrylamide and 2.4 g of *N′N′-bis*-methylene-acrylamide in 200 ml of deionized water. Adjust volume to 300 ml. Filter and store at 4°C in the dark (see Note 9).
3. 10% SDS: Dissolve 10 g of SDS in 90 ml of water with gentle stirring. Bring volume to 100 ml and filter. Store at room temperature.
4. 1.5 M Tris–HCl, pH 8.8: Dissolve 27.23 g of Tris base in 80 ml of deionized water with gentle stirring. Adjust pH to 8.8 with 6N HCl. Bring total volume to 150 ml, filter and store at 4°C.
5. 0.5 M Tris–HCl, pH 6.8: Dissolve 6 g of Tris base in 60 ml of deionized water. Adjust pH to 6.8 with 6 N HCl. Bring total volume to 100 ml, filter, and store at 4°C.
6. 10% Ammonium persulfate (APS): Dissolve 100 mg of APS in 1 ml water. Prepare fresh daily.
7. TEMED (Bio-Rad, Hercules, CA).
8. 10× running buffer (10×RB): Dissolve 30.3 g of Tris base, 144 g of glycine, and 10 g of SDS in 800 ml of deionized water. Do not adjust pH with acid or base. Bring the total volume to 1,000 ml and store at 4°C.
9. Prestained molecular weight markers (Bio-Rad, Hercules, CA).

2.6. Western Blotting

1. Nitrocellulose membrane Hybond C (Amersham, Piscataway, NJ).
2. Filter paper.
3. Transfer buffer: 25 mM Tris–HCl, 192 mM glycine, 20% methanol, pH 8.3.
4. 10×TBS buffer: 100 mM Tris–HCl, pH 7.5, 1.4 M NaCl, 15 mM $MgCl_2$.
5. TBST: TBS buffer containing 0.1% (v/v) Tween-20.
6. TBSTM blocking buffer: TBST containing 5% nonfat dry milk (Carnation) (see Note 10).

7. IκBα rabbit polyclonal antibody (Santa Cruz Biotechnology, sc-371) is diluted 1:250 in TBSTM.
8. Actin rabbit polyclonal antibody (Sigma, A5060) is diluted 1:2,000 in TBSTM.
9. Horse-radish labeled anti-rabbit IgG secondary antibody (Amersham, Piscataway, NJ) is diluted 1:2,000 in TBSTM.
10. ECL PLUS Western Blotting Analysis kit (Amersham, RPN2132).
11. Stripping buffer: 62.5 mM Tris–HCl, pH 6.7, 2% (w/v) SDS, 100 mM 2-mercaptoethanol.

2.7. Electrophoretic Mobility Shift Assay

1. NFκB double-stranded DNA oligonucleotide 5′-TTGTT ACAAGGGGACT-TTCCGCTGGGGACTTTCCAGG GAGGC-3′containing two tandemly repeated NF-κB binding sites (underlined) (see Note 11).
2. T4 polynucleotide kinase (Promega, M4101).
3. γ-(32P]ATP, specific activity 6,000 Ci (222TBq)/mMole (Perkin-Elmer, Waltham, MA) (see Note 12).
4. 1% SDS/100 mM EDTA STOP solution.
5. MicroSpin G-25 columns (Amersham, Piscataway, NJ).
6. 10× binding buffer (10×BB): 200 mM HEPES, pH 7.5, 5 mM EDTA, 5 mM DTT, 10% (v/v) Igepal CA-630, 50% (v/v) glycerol.
7. Acetylated BSA (Sigma): Dissolve content of the vial (10 mg) in 0.5 ml of sterile water to obtain stock concentration of 20 mg/ml. Aliquot and store at −20°C.
8. Poly-dI.dC (Pharmacia): Dissolve content of the vial (0.5 mg) in 0.5 ml of sterile water to obtain stock concentration of 1 mg/ml. Aliquot and store at −20°C.
9. 5×TGE buffer: Dissolve 30.3 g of Tris base and 150 g of glycine in 900 ml of deionized water. Add 20 ml of 0.5 M EDTA, pH 8. Measure pH to make sure that it is 8.4–8.5. Adjust volume to 1,000 ml, filter, and store at room temperature.
10. Acrylamide:Bis (30%): Dissolve 29.2 g of acrylamide and 0.8 g of *N′ N′-bis*-methylene-acrylamide in 70 ml of deionized water. Adjust volume to 100 ml, filter, and store at 4°C in the dark.
11. 10% APS: Dissolve 100 mg of APS in 1 ml of water. Prepare fresh daily.
12. TEMED (Bio-Rad).
13. 6× DNA loading dye: Dissolve 25 mg of bromophenol blue in 7 ml of sterile water in 15-ml tube. Add 3 ml of 100% glycerol, mix well, aliquot in microcentrifuge tubes, and store at 4°C.
14. Whatman DE-81 chromatography paper.

3. Methods

This section describes the protocol for analysis of NFκB transcriptional regulation by nuclear IκBα. Even though this protocol uses human CTCL Hut-78 cells that have high levels of constitutive NFκB DNA binding activity, it should be readily adaptable for other cells with high levels of NFκB DNA binding as well. First, cells are transfected with IκBα (or control) *si* RNA to reduce the total cellular levels of IκBα. The nuclear translocation and accumulation of IκBα is then induced by inhibition of the 26 proteasome using MG132 (or other 26S proteasome inhibitor). To monitor the cellular levels and nuclear translocation of IκBα, the nuclear and cytoplasmic fractions are isolated and analyzed by western blotting (22, 23). Finally, the NFκB DNA binding activity in the absence and presence of nuclear IκBα is analyzed by the EMSA (24). Figure 1 shows the cellular levels of IκBα in cells transfected with control (top panel) or IκBα (bottom panel) *si* RNA. As expected, transfection with IκBα *si* RNA results in a considerably decreased expression of IκBα. Cell incubation with increasing concentrations of MG132 for 24 h induces translocation of IκBα from the cytoplasm to the nucleus. Figure 2a illustrates NFκB DNA binding activity measured by EMSA in control and IκBα *si* RNA transfected cells treated 24 h with increasing concentrations of MG132. Figure 2b represents a densitometric evaluation of the NFκB DNA binding complexes shown in Fig. 2a, analyzed by Phosphoimager. Together, the results illustrated in Figs. 1 and 2 demonstrate that increased nuclear accumulation of IκBα induced by the proteasome inhibition, results in the suppression of NFκB DNA binding activity. The following steps can be accomplished in 5–7 days.

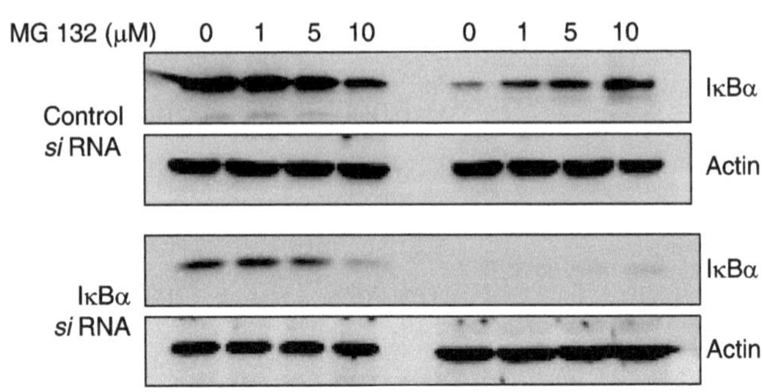

Fig. 1. Western analysis of IκBα in Hut-78 cells transfected with IκBα and control *si* RNA. Representative western blotting of CE and NE prepared from Hut-78 cells transfected with control (*top panel*) or IκBα (*bottom panel*) *si* RNA, followed by treatment with increasing concentrations of MG132 for 24 h. The membranes were analyzed by using IκBα and actin antibodies. Each lane corresponds approximately to 4×10^4 cells.

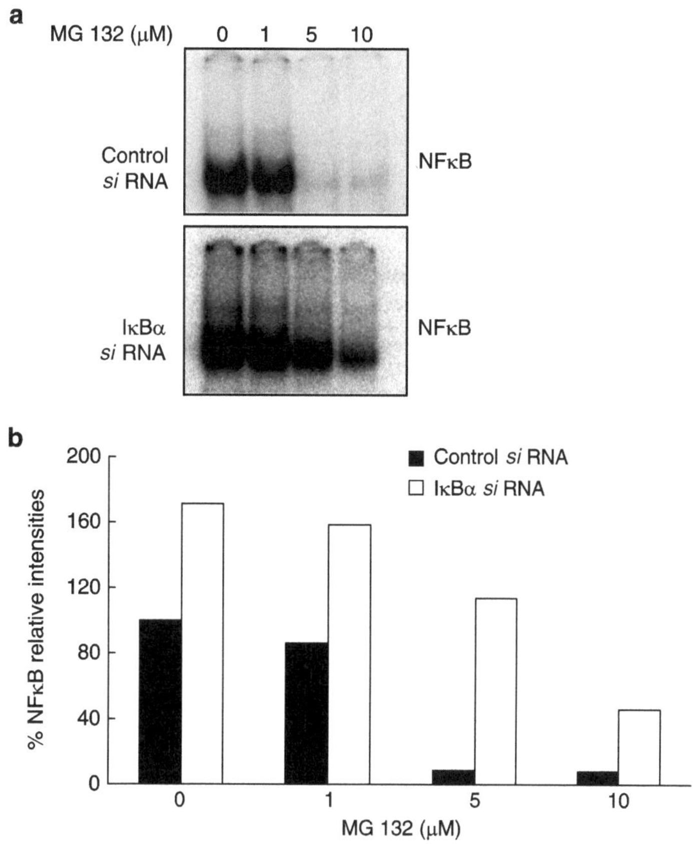

Fig. 2. The MG132-induced nuclear translocation of IκBα results in the suppression of NFκB DNA binding activity in Hut-78 cells. (**a**) Representative EMSA assay of NFκB DNA binding activity measured in Hut-78 cells transfected with control (*top panel*) or IκBα (*bottom panel*) *si* RNA, followed by incubations with increasing concentrations of MG132 for 24 h. (**b**) Densitometric evaluation of the EMSA assays of NFκB DNA binding activity measured in Hut-78 cells transfected with control (*full columns*) or IκBα (*empty columns*) *si* RNA as described above in (**a**).

3.1. Cell Culture

1. Grow Hut-78 cells in RPMI medium supplemented with 10% FBS, 2 mM glutamine, 10 mM HEPES, 1 mM sodium pyruvate and penicillin–streptomycin solution to a concentration 1×10^6 cells/ml.
2. Transfer Hut-78 cells at the concentration of 2×10^5 cells/ml into a 12-well plate and incubate in a humidified 5% CO_2 atmosphere at 37°C in an antibiotic-free RPMI medium supplemented with 10% FBS, 2 mM glutamine, 10 mM HEPES, and 1 mM sodium pyruvate solution for 24 h to 80% confluence.

3.2. Cell si RNA Transfection

1. For each transfection, use 60 μmol of either control *si* RNA-A or IκBα *si* RNA (Santa Cruz Biotechnology, CA).
2. Transfect the cells for 6 h in transfection medium with *si* RNA transfection reagent according to manufacturer's instructions (Santa Cruz Biotechnology, Santa Cruz, CA).

3. After 6 h, add fresh RPMI medium containing 2 mM glutamine, 10 mM HEPES, 1 mM sodium pyruvate, 20% FBS and 2% penicillin–streptomycin solution and incubate overnight.
4. After transfection, add fresh RPMI medium supplemented with 10% FBS, 2 mM glutamine, 10 mM HEPES, 1 mM sodium pyruvate, and 1% penicillin–streptomycin solution.

3.3. Induction of Nuclear IκBα Accumulation by MG132

1. Incubate the transfected cells (1×10^6 cells/ml) with increasing concentrations of MG132 (0, 1, 5, and 10 μM) in a standard 24-well plate for 24 h in tissue culture incubator (37°C, 5% CO_2 humidified atmosphere), to allow the nuclear IκBα translocation and accumulation, induced by MG132.
2. After 24 h, transfer the cells from the plate to 2-ml prechilled microcentrifuge tubes and follow the protocol for preparation of cytoplasmic and nuclear extracts (see Subheading 3.4).

3.4. Preparation of Cytoplasmic and Nuclear Extracts

All steps must be performed at 4°C or on ice, unless stated otherwise.

1. At the specified times, transfer the cells from the cell culture plates to 2-ml microcentrifuge tubes.
2. Centrifuge the tubes containing the cell suspension in a refrigerated centrifuge at 1,700 × *g* for 5 min.
3. Carefully aspirate and discard the supernatants, resuspend the cells in 1.5 ml of ice-cold PBS, and centrifuge at 1,700 × *g* for 5 min.
4. Carefully aspirate and discard the supernatants.
5. Add 0.15 ml of ice-cold relaxation buffer (RB) containing 2 mM DTT, 2 mM PMSF and the protease inhibitor cocktail (Sigma) used in a concentration 15 μl/10^6 cells. Mix well but gently by pippetting up and down, and by tapping on the bottom of the centrifuge tube. The relaxation buffer is hypotonic; the cells will swell in this buffer.
6. Incubate the cell suspensions on ice for 15 min.
7. Solubilize the plasma membrane by adding 0.05 volumes (=7 μl per 0.15 ml of cell suspension) of 10% Igepal CA-630. Vortex vigorously for 10 s to lyse the cells. Do not leave the cells in RB containing Igepal CA-630 for longer than 15 min, since this might also lyse the nuclear membranes.
8. Centrifuge for 5 min at 200 × *g* (see Note 13).
9. Carefully aspirate the supernatant (=cytoplasmic extract), add 50 μl of 5×SB, and boil immediately for 7 min on a boiling water bath. Label as cytoplasmic extract (CE) and store at −70°C (see Note 14).
10. Once the CE is removed, the nuclear pellets will become more visible. Wash the nuclear pellets by adding 0.2 ml of RB

containing the Sigma protease inhibitors, 2 mM PMSF and 2 mM DTT.

11. Centrifuge as above (step 8), remove and discard the supernatant.
12. For analysis by western blotting, resuspend each nuclear pellet in 50 μl of 2×SB, and boil immediately for 7 min on a boiling water bath. Label as nuclear extracts (NE), aliquot into several tubes, and store at −70°C (see Note 15).
13. For analysis of NFκB DNA binding activity, resuspend the nuclear pellets in 50 μl of ice-cold NEB containing 2 mM DTT, 2 mM PMSF, and the protease inhibitor cocktail (Sigma) used in a concentration 15 μl/10^6 cells. Vortex, incubate on ice 20 min, vortex again and centrifuge 10 min at 14,000×*g*. Transfer the supernatants (=nuclear extracts) to prechilled tubes, remove small amount (5 μl) for determining protein concentration, aliquot, and store at −70°C.

3.5. SDS-Polyacrylamide Gel Electrophoresis

To run SDS-PAGE, we use the Bio-Rad Mini PROTEAN-3 System. For separation of IκBα, 12% gels are used.

1. Clean the glass plates with 90% ethanol and assemble according to the Bio-Rad Mini-PROTEAN 3 Instructions. Make sure the plates are flush, to prevent gel leakage. Make a mark 2 cm off the top of the short plate.
2. Prepare resolving gel solution (10 ml) by mixing 3.4 ml of deionized water, 4.0 ml of acrylamide/bis solution (30%), 2.5 ml of 1.5 M Tris–HCl, pH 8.8 buffer, and 0.1 ml of 10% SDS.
3. Immediately before pouring the gels, add 50 μl of 10% APS and 5 μl of TEMED. Swirl gently and pour the gels immediately.
4. Overlay the gels with water.
5. Allow the gels to polymerize for about 45 min.
6. Prepare 4% gel solutions for the stacking gels. Mix 6.1 ml of deionized water, 1.3 ml of acrylamide/bis solution, 2.5 ml of 0.5 ml Tris–HCl, pH 6.8, and 0.1 ml of 10% SDS.
7. Before casting the stacking gels, pour off water from the top of the resolving gels.
8. Add 50 μl of 10% APS and 10 μl of TEMED to the stacking gel solution. Swirl gently, and pour the solution between the glass plates on the top of the resolving gel. Insert combs immediately.
9. Allow the gels to polymerize for about 1 h.
10. Gently remove the combs and rinse the wells thoroughly with 1× running buffer (1×RB) that is prepared by diluting 50 ml of 10×RB with 450 ml of deionized water.

11. Assemble plates in the tank according to the Bio-Rad Mini-PROTEAN 3 Cell Instruction Manual. Add about 125 ml of 1×RB to the inner chamber, and about 200 ml of 1×RB to the tank (lower chamber).
12. Load the samples into the wells with a pipette using gel loading tips.
13. Run the gels at constant 120 V for about 50 min, or until the blue dye from the samples reaches the bottom of the gel.
14. Turn off the power, disassemble the gel apparatus, and carefully remove the gel from the glass plates. Follow the protocol for western blotting (Subheading 3.6).

3.6. Western Blotting

To analyze the cytoplasmic and nuclear levels of IκBα, we first probe the membranes with IκBα antibody, which is then followed by actin antibody to confirm equal protein loading. The procedure described below uses the Bio-Rad semidry transfer apparatus.

1. Soak two thick filter papers, nitrocellulose membrane (cut to the size of the filter paper) and the gel in transfer buffer for 15 min.
2. Assemble the transfer sandwich (from bottom to top): Thick filter paper, membrane, gel, thick filter paper. Make sure there are no bubbles between membrane and the gel.
3. Once the lid is put on the transfer sandwich, activate the power. For one 12% gel, the transfer conditions are 18 V (constant), 35 min. For two 12% gels, the conditions are 20 V (constant), 50 min.
4. After the transfer is complete, turn off the power, and dissemble the transfer sandwich. The gel can be now discarded, and the colored prestained markers should be clearly visible on the membrane.
5. Transfer the membrane into blocking solution (TBSTM), and incubate 1 h with gentle rocking at room temperature (see Note 16).
6. Pour off the blocking solution, and incubate the membrane with IκBα primary antibody diluted 1:250 in TBSTM for 1 h at room temperature.
7. Pour off the primary antibody and wash the membrane five times in TBST (5 min each wash).
8. After washing, incubate the membrane for 1 h at room temperature with HRP-labeled secondary anti-rabbit antibody diluted 1:2,000 in TBSTM.
9. Wash the membrane five times in TBST.

10. Pour off TBST from the membrane, place the membrane on a sheet of saran wrap, and develop the signal with ECL PLUS system.
11. Detect the signal with digital imager or autoradiographic film. The image shown in Fig. 1 illustrates IκBα nuclear levels detected by digital imager *Syngene* (Frederick, MD).
12. To confirm equivalent amounts of loaded proteins, strip the membrane for 30 min at 50°C with stripping buffer, and incubate it for 1 h at room temperature with actin antibody diluted 1:2,000 in TBSTM.
13. Develop the signal with HRP-labeled secondary antibody and ECL PLUS system as described above.

3.7. Electrophoretic Mobility Shift Assay

The NFκB oligonucleotide used as a probe for EMSA is a 42-bp double-stranded construct (5′-TTGTTACAAGGGGACTTTCC GCTGGGGACTTTCCAGGGAGGC-3′) containing two tandemly repeated NFκB binding sites (underlined) (see Note 11).

1. End label annealed double-stranded NFκB oligonucleotide by using T4 kinase. To accomplish this, mix the following components in a microcentrifuge tube on ice: 7 μl of water, 2 μl of 10× kinase buffer, 4 pmol of annealed double-stranded NFκB oligonucleotide, 2 μl (15 U) of T4 kinase, and 50 μCi (γ-^{32}P) ATP (see Note 12).
2. Incubate for 30 min at 37°C in a water bath.
3. Stop the reaction by adding 5 μl of 1% SDS/100 mM EDTA STOP solution, vortex, and spin briefly.
4. Purify the labeled oligonucleotide on a MicroSpin G-25 column.
5. Prepare 7.5% gel in TGE buffer by mixing 41 ml of water, 15 ml of 5× TGE buffer, 18.8 ml of acrylamide:bis (30%) solution, 0.6 ml of 10% APS and 60 μl of TEMED. Pour the gel and let it polymerize for 45 min. Rinse the wells with 1× TGE buffer using a 10-ml hypodermic syringe.
6. Prerun the gel without samples in 1× TGE buffer at 150 V (see Note 17).
7. Prepare binding reactions by mixing the following components in microcentrifuge tubes on ice: 8 μl of water, 2 μl of 10× binding buffer, 2 μl of poly-dI.dC stock solution, 1 μl of BSA stock solution, 5 μl of the nuclear extract containing approximately 5 μg protein, and 1 μl of ^{32}P-labeled NFκB oligonucleotide (see Note 18).
8. Mix gently by tapping on the bottom of the tubes, pulse spin, and incubate 15 min at room temperature.

9. Load the samples on the prerun gel.
10. Run the gel for 2.5 h at 180 V (see Note 17).
11. After electrophoresis, transfer the gel to Whatman DE-81 paper, cover it with Saran wrap, and dry it.
12. Expose the dried gel to PhosphorImager (Molecular Dynamics). The image shown in Fig. 2a illustrates NFκB DNA complexes detected in the nuclear extracts of Hut-78 cells, while Fig. 2b represents a densitometric evaluation of the images shown in Fig. 2a.

4. Notes

1. In this study, we used IκBα *si* RNA obtained from Santa Cruz Biotechnology. However, a similar product can be obtained from Cell Signaling Technology, Inc.
2. MG132 (MW 475.6) can be purchased from EMD Chemicals, Sigma, or Biomol. MG132 is a potent, reversible, and cell-permeable inhibitor of the 26S proteasome. For the stock concentration of 100 mM, dissolve the content of the vial (5 mg) in 100 μl of sterile DMSO, aliquot, and store at −80°C.
3. Relaxation buffer can be prepared in 100 ml amount and stored at 4°C. Just before use, take out the amount required for the day, and add DTT (2 mM final concentration), PMSF (2 mM final concentration), and the Sigma protease inhibitor cocktail mix (15 μl/10^6 cells); keep on ice.
4. Igepal CA-630 is a nonionic, nondenaturing detergent that is chemically indistinguishable from Nonidet P-40. It solubilizes plasma membranes to release the nuclei.
5. NEB can be prepared in 10-ml amount, and stored at 4°C. Just before use, take out the amount required for the day, and add DTT, PMSF, and the Sigma protease inhibitor cocktail mix; keep on ice.
6. DTT is a strong reducing agent. It is used to prevent formation of intra- and intermolecular disulfide bonds of proteins. DTT should be added to RB and NEB in the final working concentration 2 mM just before use.
7. PMSF is unstable in aqueous environment. It is essential that it is dissolved in absolute alcohol (ethanol, methanol, or isopropanol), and is added to the RB and NEB in the final working concentration 2 mM just before use.
8. This protease inhibitor cocktail contains protease inhibitors with broad specificity for the inhibition of serine, cysteine, aspartic proteases, and aminopeptidases. It should be stored

at −20°C, and added to the RB and NEB just before use, in the final concentration 15 μl/10^6 cells.

9. Unpolymerized acrylamide is extremely toxic when ingested, and absorbs easily into the skin, so use care and wear gloves while handling solutions that contain it. Once acrylamide is polymerized, it is no longer absorbable, but care still should be taken when handling the gel.
10. Although blocking with 5% Carnation nonfat dry milk in TBST works well in most cases, alternatively 1% protease-free BSA in TBST can be used.
11. This oligonucleotide was designed to detect NFκB activity in human leukocytes (24). Alternatively, commercially available double-stranded NFκB oligonucleotides can be used (such as from Santa Cruz Biotechnology).
12. Perform the ^{32}P-labeling steps in an area designated for use of radioactive materials. Follow the safety guidelines for work with radioactivity.
13. It is helpful to always centrifuge the tubes in one position (e.g., the cap snaps facing toward the center of the rotor); this way you can always expect the pellets to be at the same place.
14. Once the cytoplasmic extracts are collected, keep them on ice, and boil them immediately with 5×SB to prevent protein degradation. Once boiled, the samples are more stable and can be stored at −70°C for several weeks. To prevent repeated thawing of the samples, store them in smaller aliquots.
15. Similarly as with the CE, boil the NE with 2×SB immediately to prevent protein degradation. Once boiled, the samples are more stable and can be stored at −70°C for several weeks. Avoid repeated thawing by storing the samples in smaller aliquots.
16. The western blotting can be interrupted at this point, and the membranes can be incubated in TBSTM overnight at 4°C.
17. The EMSA gel can be run either at room temperature with cooling using running water, or at 4°C.
18. It is convenient to prepare a common binding mix by mixing all components (for desired amount of reactions) except for the nuclear extracts and labeled oligonucleotide.

Acknowledgements

This work was supported by NIH grants GM079581 and AI085497 to I.V.

References

1. Baeuerle, P, A. and Baltimore, D. (1996) NFκB: ten years after. *Cell* **87**, 13–20.
2. Ghosh, S. and Karin, M. (2002) Missing pieces in the NFκB puzzle. *Cell* 109, S81–96.
3. Barnes, P. J. and Karin, M. (1997) NFκB – a pivotal transcription factor in chronic inflammatory diseases. *New Engl. J. Med.* **336**, 1066–1071.
4. Baldwin, A. S. (2001) The transcription factor NFκB and human disease. *J. Clin. Invest.* 107, 3–6.
5. Yamamoto, Y. and Gaynor, R. B. (2001) Therapeutic potential of inhibition of the NFκB pathway in the treatment of inflammation and cancer. *J. Clin Invest.* **107**, 135–142.
6. Hayden, M. S. and Ghosh, S. (2008) Shared principles in NFκB signaling. *Cell* **132**, 344–362.
7. Prasad, S., Ravindran, J. and Aggarwal, B. B. (2010) NFκB and cancer: how intimate is this relationship. *Mol. Cell. Biochem.* **336**, 25–37.
8. Bharti, A. C. and Aggarwal, B. B. (2002) NFκB and cancer: its role in prevention and therapy. *Biochem. Pharmacol.* **64**, 883–888.
9. Dutta, J., Fan, Y., Gupta, N., Fan, G. and Gelinas, C. (2006) Current insights into the regulation of programmed cell death by NFκB. *Oncogene* **25**, 6800–6816.
10. Karin, M. (2006) NFκB in cancer development and progression. *Nature* 441, 431–436.
11. Wan, F. and Lenardo, M. J. (2010) The nuclear signaling of NFκB: current knowledge, new insights, and future perspectives. *Cell. Res.* **20**, 24–33.
12. Lenz, G. and Staudt, L. M. (2010) Aggressive lymphomas. *N. Engl. J. Med.* **362,** 1417–1429.
13. Mancino, A. and Lawrence, T. (2010) NFκB and tumor-associated macrophages. *Clin. Cancer Res.* **16,** 784–789.
14. Karin, M. and Neriah, Y. B. (2000) Phosphorylation meets ubiquitination: The control of NF-κB activity. *Annu. Rev. Immunol.* **18**, 621–663.
15. McConkey, D.J. and Zhu, K. (2008) Mechanisms of proteasome inhibitor action and resistance in cancer. *Drug Resist. Updat.* **11**, 164–179.
16. Shah, J. J. and Orlowski, R. Z. (2009) Proteasome inhibitors in the treatment of multiple myeloma. *Leukemia* **23**, 1964–1979.
17. Izban, K. F., Ergin, M., Qin, J. Z., Martinez, R. L., Pooley, R. J., Saeed, S. and Alkan, S. (2000) Constitutive expression of NFκB is a characteristic feature of mycosis fungoides: implications for apoptosis resistance and pathogenesis. *Hum. Pathol.* **31**, 1482–1490.
18. Sors, A., Jean-Louis, F., Pellet, C., Laroche, L., Dubertret, L., Courtois, G., Bachelez, H. and Michel, L. (2006) Down-regulating constitutive activation of the NFκB canonical pathway overcomes the resistance of cutaneous T-cell lymphoma to apoptosis. *Blood* **107**, 2354–2363.
19. Sors, A., Jean-Louis, F., Bégué, E., Parmentier, L., Dubertret, L., Dreano, M., Courtois, G., Bachelez, H. and Michel, L. (2008) Inhibition of IκB kinase subunit 2 in cutaneous T-cell lymphoma down-regulates NFκB constitutive activation, induces cell death, and potentiates the apoptotic response to antineoplastic chemotherapeutic agents. *Clin. Cancer Res.* **14**, 901–911.
20. Kiessling, M. K., Klemke, C. D., Kaminski, M. M., Galani, I. E., Krammer, P. H. and Gulow, K. (2009) Inhibition of constitutively activated NFκB induces reactive oxygen species and iron-dependent cell death in cutaneous T-cell lymphoma. *Cancer Res.* 69, 2365–2374.
21. Vu, H.Y., Juvekar, A., Ghosh, C., Ramaswami S., Le, D. H. and Vancurova, I. (2008) Proteasome inhibitors induce apoptosis of prostate cancer cells by inducing nuclear translocation of IκBα. *Arch. Biochem. Biophys.* **475**, 156–163.
22. Castro-Alcaraz, S., Miskolci, V., Kalasapudi, B., Davidson, D. and Vancurova, I. (2002) NFκB regulation in human neutrophils by nuclear IκBα: correlation to apoptosis. *J. Immunol.* **169**, 3947–3953.
23. Ghosh, C. C., Ramaswami, S., Juvekar, A., Vu, H. Y., Galdieri, L., Davidson, D. and Vancurova, I. (2010) Gene specific repression of proinflammatory cytokines in stimulated human macrophages by nuclear IκBα. *J. Immunol.* **185**, 3685–3693.
24. Vancurova, I., Miskolci, V. and Davidson, D. (2001) NFκB activation in TNFα-stimulated neutrophils is mediated by protein kinase Cδ. Correlation to nuclear IκBα. *J. Biol. Chem.* **276**, 19746–19752.

Chapter 4

Probing Endogenous RNA Polymerase II Pre-initiation Complexes by Electrophoretic Mobility Shift Assay

Emmanuelle Wilhelm, Christopher Takacs, and Brendan Bell

Abstract

RNA polymerase II (Pol II) plays a crucial role in eukaryotic biology since it is necessary for the expression of all protein-coding genes as well as most microRNAs and several small nuclear RNAs. Pol II is specifically recruited to core promoter DNA via its association with general transcription factors (GTFs) that possess DNA binding activity such as TFIID, TFIIA, and TFIIB. The large multi-protein assemblies of Pol II together with the GTFs required for productive transcription are termed pre-initiation complexes (PICs). To date, studies of the interaction of PICs with promoter DNA have relied on the use of purified or recombinant GTFs. Recent findings have demonstrated an astonishing diversity in the function of core promoters as well as in the protein composition of PICs. The currently known subset of GTFs alone cannot account for observed PIC and core promoter diversity. In order to identify the full complement of factors that impart PIC specificity, techniques to analyze the DNA binding of endogenous PIC are essential. Analysis of endogenous PIC formation has remained out of reach due to technical hurdles presumably including the large size of endogenous PIC, their highly dynamic association with core promoters, and the complex topology of DNA bound to PIC. We have optimized electrophoretic mobility shift assays (EMSAs) to achieve the detection of endogenous Pol II PIC from nuclear extracts of human cells. Here, we provide a robust and sensitive EMSA method for the analysis of endogenous Pol II PICs.

Key words: Pre-initiation complex, RNA polymerase II, TFIID, TATA-binding protein, Electrophoretic mobility shift assay, TBP-associated factor, General transcription factor

1. Introduction

In eukaryotes, RNA polymerase II (Pol II) plays a pivotal role in gene expression since it is required for the transcription of all protein-coding messenger RNAs (1), most microRNAs (2), and certain small nuclear RNAs (snRNAs) (3). Given the fact that the regulation of gene expression by Pol II impacts on virtually all

Ales Vancura (ed.), *Transcriptional Regulation: Methods and Protocols*, Methods in Molecular Biology, vol. 809,
DOI 10.1007/978-1-61779-376-9_4, © Springer Science+Business Media, LLC 2012

aspects of normal growth and development as well as infection and disease, the association of Pol II with promoters has been intensely investigated. Extensive work from numerous laboratories has established that accurate promoter-dependent transcription by Pol II requires the assembly of pre-initiation complexes (PICs) that include the general transcription factors (GTFs) TFIID, TFIIA, TFIIB, TFIIF, and TFIIH that recruit Pol II to the promoter (4). The electrophoretic mobility shift assay (EMSA) has been an instrumental experimental tool in dissecting the assembly pathway of Pol II PICs upon the core promoter (5).

Work to date on PIC formation has employed purified and/or recombinant GTFs. In most cases, model core promoters have been studied, particularly the Adenovirus major late promoter. While special EMSA conditions have been developed for the detection of affinity-tag purified TFIID (6), detection of more complex endogenous PICs has not been achieved. Factors hampering endogenous PIC detection by EMSA are thought to include the large size of the complexes. For example, TFIID alone contains approximately 16 proteins and has a molecular mass of more than 700 kilodaltons (kDa) (7). Moreover, the Mediator complex that is included in complete PIC contains 26 subunits and has a molecular weight of 1.2 megadaltons (MDa) (8). PICs contain more than 70 polypeptides even considering only Pol II plus the essential GTFs (4, 8). The highly dynamic nature of PIC can also pose technical problems as PICs are known to cycle on and off promoter DNA (9–12). Finally, the complexity of topological winding of promoter DNA around PIC (13) could also confound EMSA analysis.

Although PIC formation upon core promoters was initially envisaged as generic process occurring in a similar fashion at different genes, recent advances suggest a very different scenario. The last decade of research has revealed that core promoter sequence elements and their functions in transcription display a high degree of diversity (14, 15). Moreover, the PICs that recruit Pol II to these diverse core promoters are also highly variable in protein composition and function (16, 17). Major challenges in the field of gene regulation are therefore to match the diversity of core promoters to their cognate PIC, and to identify cellular factors that impart specificity to core promoter recognition. Thus, the use of purified and/or recombinant proteins in EMSA limits the analysis of complexity of core promoter to known GTFs. These limitations highlight the need for new approaches to identify and dissect endogenous PICs and their as yet unidentified core promoter-specific components.

To facilitate the analysis of PIC by EMSA, we have taken a reductionist approach using relatively short synthetic oligonucleotides coupled with careful optimization of binding conditions to detect endogenous PIC. We have focused on TATA box-containing core promoters since these correlate with highly regulated genes (18). Here, we provide a detailed method that reproducibly detects bona fide endogenous PICs by EMSA using the HIV core promoter as a model. The technique provides new avenues for studying PIC and core promoter specificity, as well as for identifying novel core promoter DNA motifs required for PIC binding.

2. Materials

2.1. Annealing of Oligonucleotides

1. T4 DNase Ligase buffer 10× (New England Biolabs).
2. NEB2 buffer 10× (New England Biolabs).
3. Oligonucleotides (Integrated DNA Technologies) (see Note 1), resuspended in distilled deionized water at 200 μM:
   (a) HIV promoter sense (HIV wt): CTAGATGCTGCATA TAAGCAGCTGCTTTTTGCATG.
   (b) HIV promoter antisense: CAAAAAGCAGCTGCTTAT ATGCAGCAT.
   (c) Non-specific oligo sense (non-specific): GATCCGGAGTAC TTCAAGAACG.
   (d) Non-specific oligo antisense: GATCCGTTCTTGAAGT ACTCCG.
   (e) Mutant HIV promoter sense (mutHIV): CTAGATGATCAGAGCTCGGATATCCTTTT TGCATG.
   (f) Mutant HIV promoter antisense: CAAAAAGCAGCTG CTTATATGCAGCAT.
   (g) TATAKO HIV promoter sense (TATAKO): CTAGATG CTGAGAGCTCGCAGCTGCTTTTTGCATG.
   (h) TATAKO HIV promoter antisense: CAAAAAGCAGCTGC GAGCTCTCAGCAT.

2.2. Radioactive Labelling of Oligonucleotides

1. Deoxynucleotides: 20 mM dATP, dGTP, dTTP, and dCTP.
2. Radioactively labelled deoxynucleotides: 3 μM (α-^{32}P) dATP (BLU/512H) and 3 μM (α-^{32}P) dCTP (BLU/513H), both from Perkin Elmer stored at +4°C.
3. Klenow (large fragment) (see Note 2).
4. 0.5 M EDTA.
5. Microspin™ G-25 column (GE Healthcare).

2.3. Polyacrylamide Gel Electrophoresis

1. TBE 5×: 0.45 M Tris–borate, 10 mM EDTA.
2. Acryl-bisacryl (30:0.8) prepared in house with 30 g of acrylamide, 0.8 g of bisacrylamide, and completed to 100 ml with distilled water.
3. 50% Glycerol.

4. 10% Ammonium persulphate (APS) (see Note 3).
5. TEMED.

2.4. EMSA Sample Preparation

1. Binding buffer 5×: 100 mM HEPES, 25 mM $MgCl_2$, 40% glycerol, 500 mM KCl in distilled water.
2. 10 mg/ml Acetylated bovine serum albumin.
3. 10 mg/ml Poly(dI-dC)·Poly(dI-dC).
4. HeLa cells (ATCC # CCL-2) nuclear extracts were prepared as detailed in the protocol of Dignam et al. (19) (see Note 4).
5. Antibodies for supershift: anti-HIP116/SMARCA3 (#A300-230A Bethyl laboratories, Montgomery, TX); anti-Pol II clone 8WG16 (#MMS-126R Covance, Emeryville, CA); anti-TBP 1/mAb2C1 (20); anti-TBP 2/mAb4C2; anti-TBP 3/mAb3G3 (21); and anti-TLS/FUS/mAb1D2 (22).

2.5. Gel Drying

1. Whatman paper (3 M), cut slightly larger than the size of the gel.
2. Plastic wrap (Saran Wrap).

3. Methods

A critical parameter for the detection of endogenous PICs is the amount of competitor nucleic acid. The optimization of Poly(dI-dC)·Poly(dI-dC) concentration must be undertaken as illustrated in Fig. 1. Here, we have used the HIV-1 core promoter as an experimental model. This method has also been successfully used with several TATA-containing cellular (e.g., hsp70) or viral (e.g., Adenovirus major late) core promoters. Despite advances in non-radioactive detection systems, we find that the most sensitive and most cost-effective methodology remains radioactive labelling of test core promoters. We therefore remind readers that all manipulations dealing with radioactive material should be done under proper radioprotection conditions. This is particularly true for gel handling after gel migration. All radioactive waste must be disposed of according to institutional guidelines.

3.1. Annealing of Oligonucleotides

1. In a microfuge tube, mix 22.5 μl of sense with 22.5 μl of antisense single-stranded oligonucleotides together with 5 μl of T4 DNA ligase buffer. Heat the mixture for 2 min at 95°C, and immediately put the microtube (held by a float) into approximately 1 L of water that has just been brought to boil in a beaker and let the water and microfuge tube cool down gradually (see Note 5). When room temperature has been reached, freeze at –20°C.

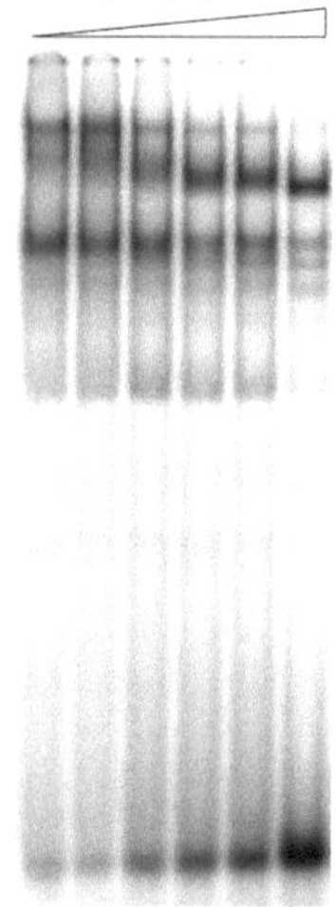

Fig. 1. Optimization of non-specific competitor concentration is critical for EMSA detection of endogenous PICs. Increasing amounts (200, 300, 400, 500, 600, and 800 ng from *lane 1* to *lane 6*) of poly(dI-dC) were titrated into the binding reaction in the absence of other non-specific competitors. The results show that sharp PIC bands are visualized only in the presence of optimal amounts of poly(dI-dC). Insufficient poly(dI-dC) results in poor resolution of the complex bands and in complete utilization of the probe in non-specific binding as shown by free probe level. The concentration of nuclear extract should likewise be optimized for best results (data not shown).

3.2. Radioactive Labelling of Oligonucleotides with α-^{32}P dATP and α-^{32}P dCTP

1. Thaw all material required on ice: NEB2 buffer, double-stranded oligonucleotide, and deoxynucleotides.
2. In a sterile Eppendorf tube, first combine 1.5 μl of sterile water and 2 μl of NEB2 buffer.
3. Add 5 μl of 1 μM double-stranded oligonucleotide (5 pmol) to be labelled, and 0.25 μl of both 20 mM dTTP and 20 mM dGTP (5 nmol).
4. Behind a protective plexiglass shield and using proper radio-protection precautions, add 5 μl of both 3 μM α-^{32}P dATP and 3 μM α-^{32}P dCTP (15 pmol) to the Eppendorf tube.
5. Add 1 μl of the Klenow fragment to the tube and mix contents by slight flicking or vortexing, then incubate the sample at room temperature for 15 min.
6. Chase the Klenow polymerase end-filling reaction of the probe by adding 0.25 μl of both dATP and dCTP (20 mM, non-radioactive). Then incubate the sample at room temperature for another 5 min (see Note 6).
7. Add 1 μl of 0.5 M EDTA to stop the reaction (see Note 7).
8. Add sterile H_2O to complete the volume to 50 μl, giving a final concentration of 0.1 pmol of oligonucleotide per 1 μl.

9. Pipette the full volume of the sample onto the Microspin™ G-25 purification column matrix, making sure to deliver the volume onto the centre of the matrix, halfway up the slope. Centrifuge at 735 × *g* for 2 min (see Note 8). Check for incorporation of radioactivity by comparing the counts of the filtrate to the counts of the column after centrifugation (easily accomplished with a Geiger counter). Roughly one fourth of the radioactivity is typically integrated into the probe. Dispose of the used column in proper radioactive waste container, and conserve the labelled oligonucleotide at –20°C until use.

3.3. PAGE Preparation

1. The following instructions assume the use of a Bio-Rad PROTEAN® II xi Cell gel system. All glass plates are cleaned with a detergent, rinsed with distilled water before use, and finally cleaned with ethanol.
2. Prepare a native 4.5% polyacrylamide gel by first mixing 40 ml of distilled H_2O with 5.5 ml of TBE 5×, 8.25 ml of acryl-bisacryl (30:0.8), and 1.1 ml of 50% glycerol. Degas the solution by applying a vacuum (see Note 9).
3. Add 200 μl of 10% APS and 55 μl of TEMED and gently mix the solution by swirling to initiate the polymerization of the acrylamide. Immediately pour the solution into the gel sandwich from an upper corner of the gel, avoiding any bubbling and carefully insert the comb and let polymerize for 30–45 min.
4. After complete polymerization, remove the comb, attach the gel sandwich to the cooling core, and place this assembly into the buffer chamber.
5. Pour running buffer composed of TBE 0.5× and 1% glycerol into the setup and rinse the wells using a syringe to remove any residual acrylamide from the wells (see Note 10).
6. A pre-run is performed for a minimum of 1.5 h at 200 V, after which the running buffer is removed (see Note 11) and replaced with a fresh buffer for migration.

3.4. EMSA Sample Preparation

1. Basic protocol:
   (a) All the solutions required are thawed and kept on ice.
   (b) Prepare a mastermix for the whole series by sequential addition of the following components: for one lane, add sterile H_2O required for a final total volume of 20 μl, 4 μl of Binding buffer 5×, 0.4 μl of acetylated BSA 10 mg/ml, 0.2 μl of Poly(dI-dC)·Poly(dI-dC) 10 mg/ml, and 0.2 μl of non-specific oligonucleotides (from a stock of concentration 90 μM) (see Fig. 1 and Note 12).
   (c) Add 20 μg of nuclear extract (2–4 μl, depending on the concentration). The nuclear extract will bring to the reaction

the endogenous proteins and factors essential to form PICs on the radiolabelled promoter used as a probe.

(d) Incubate this mixture on ice for 5 min.

(e) Add 1 μl of labelled oligonucleotide (following radioprotection protocols) and mix with light flicking, then incubate at room temperature for 15 min.

(f) Load the sample gently into the well, being careful to load it as close to the bottom as possible while avoiding the introduction of bubbles.

(g) The gel is run for approximately 4 h at 150 V with a constant flow of cold water running through the cooling core (see Note 13).

2. EMSA Sample Preparation with Competitor (Fig. 2):

Follow the same steps as for a normal sample preparation, adding the unlabelled competition oligonucleotide in the initial mix, at the same time than the unspecific competitor (step 1b) at

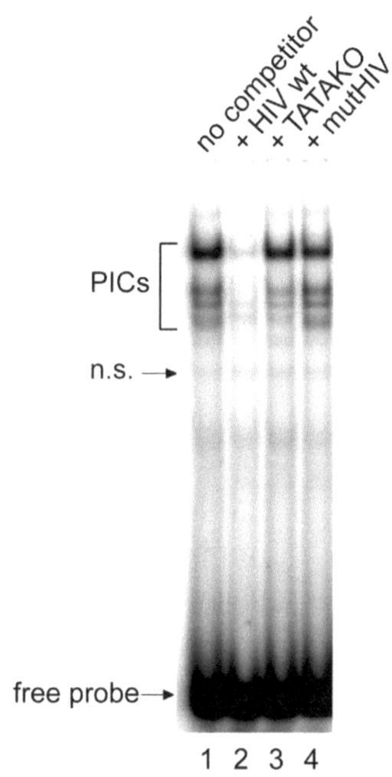

Fig. 2. Competition experiments define the specificity of the complexes. A 200-fold molar excess of unlabeled double-stranded oligonucleotide competitors were added to the binding reaction. An excess of unlabeled wt HIV promoter to the reaction prevents the formation of most complexes, except the non-specific ones (n.s.) (*lane 2* vs. *lane 1*), indicating these complexes bind specifically to the wild-type promoter. In contrast, a mutated competitor of scrambled sequence leads to no changes in complex formation, consistent with the conclusion that PIC formation on the wt HIV core promoter is sequence-specific (*lane 4* vs. *lane 1*). Competition with an unlabeled HIV core promoter bearing a mutation in the TATA box is no longer able to efficiently compete for PIC formation with the wt HIV core promoter, consistent with the requirement of an intact TATA box for PIC binding (*lane 3* vs. *lane 1*).

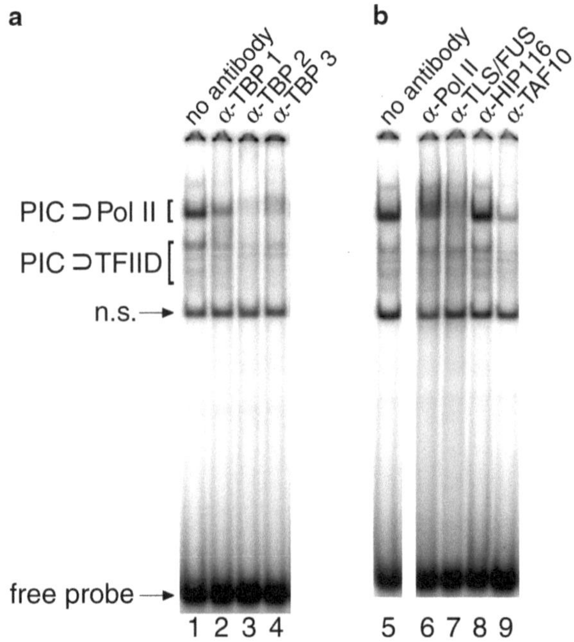

Fig. 3. Use of supershift analysis to identify proteins within pre-initiation complexes (PICs). (**a**) The addition of three independent monoclonal antibodies directed against three distinct epitopes of TATA-binding protein (TBP) reveals the dependence on TBP for all PIC (*lanes 2*, *3*, and *4* vs. *lane 1*). (**b**) Endogenous PIC contain the GTFs RNA polymerase II (Pol II) and TFIID as revealed by supershift analysis. Complexes supershifted by antibodies directed against RNA polymerase II (*lane 6*) are indicated (PIC Pol II). PICs that are supershifted or blocked by antibodies against TFIID subunits (*lanes 7* and *9*) are indicated (PIC TFIID). Antibodies directed against HIP116 have no influence on the formation of any PICs (*lane 8*).

the desired concentration (see Fig. 2 and Note 14). Keep in mind the final total volume of 20 μl for each sample.

3. EMSA Sample Preparation with Supershift with Specific Antibody (Fig. 3):

   Follow the same steps as for a normal sample preparation, except after the 15-min incubation period with the labelled oligonucleotide (step 1d) add 0.2–5 μl of the antibody (dependent on the concentration and affinity of the antibody) and incubate on ice for 45 min to 1.5 h, flicking the tube every 15 min to mix the contents. Keep in mind the final total volume of 20 μl for each sample. Once this incubation period is over, the sample is ready to be loaded into the well.

3.5. Gel Removal and Drying

1. Turn the power source off and take the cooling core carrying the gel out of the chamber (see Note 15). Place it in the sink to allow the water to drain from the core.
2. Carefully remove the glass plate sandwich from the cooling core, and free them from the clamps.

3. Place the glass plates on a flat surface with the long plate down, the top of the gel oriented towards you.
4. Push each of the spacers at the top of the gel out towards the side so that enough spacer is exposed for you to fit your forefinger underneath it.
5. While holding the long plate down with your thumbs, insert your forefingers underneath the spacers and slowly and gently push up on them, separating the two glass plates (see Note 16).
6. Place a piece of 3 M paper on the gel, flip the sandwich formed by the glass plate and the paper, and slowly and gently lift the plate while maintaining the paper with the gel sticking on it. Remove the plate completely and cover the gel with saran wrap.
7. For homogeneous drying, put a second 3 M paper underneath the first one before drying the gel at 80°C for 45 min to 1 h under vacuum.
8. After drying, discard the lower paper sheet and overnight expose the gel on a phosphoimager screen or a radiography film before revealing.

4. Notes

1. When annealed, the double-stranded oligonucleotides are flanked by cohesive ends allowing Klenow filling for radioactive labelling. It is convenient to include restriction enzymes sites for subsequent cloning into promoter-reporter constructs for further cell-based or in vitro transcription assays.
2. The large fragment of Klenow enzyme was purchased from New England Biolabs (M0210S) or prepared according to Joyce and Derbyshire's method (23) with modifications according to Turner et al. (24).
3. 10% APS solution is better prepared fresh, or can be prepared as a stock solution and frozen in small single use aliquots.
4. Typically, we have employed HeLa cell nuclear extracts prepared by the classical method of Dignam et al. (19). The method described herein has also performed well using nuclear extracts from different cell lines (e.g., U937, Jurkat, HEK293, HL-60), as well as nuclear extracts from freshly isolated human peripheral blood mononuclear cells. Furthermore, nuclear extracts prepared by alternative protocols for small-scale preparations from suspension cells (25) or adherent cells (26) have also been successfully used.

5. Evaporation can occur in the microfuge tube during the first hour of this step. To avoid concentration of the reaction components, it is recommended to periodically flick microfuge tube a downward motion to force liquid back to the bottom of the tube.
6. This ensures that all the oligonucleotides are polymerized to completion.
7. EDTA chelates the Mg^{2+}, thus inhibiting the enzymatic activity of the Klenow fragment.
8. Microspin™ G-25 column is prepared by vortexing lightly, breaking off the tip, centrifuging at 735 × *g* for 1 min in a microtube, and finally transferring to a new microtube. The matrix in the column purifies via size exclusion; therefore, the smaller molecules (i.e., the dNTPs) are withheld in the column and the filtrate contains the labelled oligonucleotide.
9. This removes the dissolved air from the solution, which would otherwise interfere with the polymerization of the gel.
10. First pour running buffer into the upper chamber and check for any leak that would drop at the bottom of the setup into the lower chamber. If leaks occur, remove the sandwich from the cooling core and put it properly in place again.
11. The pre-run is crucial to remove APS from the gel that could otherwise interfere with migration.
12. The three last components act as unspecific competitors. Their relative amounts have been carefully optimized to obtain sharp bands of the specific complexes. An example is shown in Fig. 1.
13. It is inadvisable to add loading buffer to the samples, as the glycerol contained in the binding buffer is sufficient to cause the sample to rest at the bottom of the well. Instead, one well can be loaded with a easily visible amount of standard DNA loading buffer containing bromophenol and xylene cyanol to track the migration of the samples. The bromophenol blue dye should typically migrate approximately three fourth of the way down the gel.
14. Unlabelled competitor oligonucleotide must be titrated for each core promoter used. Typically, a 200-fold molar excess of competitor oligonucleotide provides useful information for specificity studies. A range of competitor concentrations should be used to compare more subtle differences in affinities for core promoters bearing point mutations.
15. At all times when handling the gel, a protective plexiglass screen must be placed between the gel and anyone in the vicinity.

16. Ideally, the gel remains attached to the long plate and slowly detaches from the short plate as you continue to gradually lift it up, although it may happen that it sticks to the short plate. In this case, the best approach is to flip the whole sandwich upside down to have the gel sticking to the lower plate for easier handling. It is best to separate plates slowly to ensure uniform detachment of the gel from the plate.

Acknowledgements

We thank Drs. Arndt Benecke and Antonio Conconi for critical comments on the manuscript. We are grateful to Dr. Laszlo Tora for the generous gift of monoclonal antibodies directed against TFIID subunits.

References

1. Young R. A. (1991) RNA polymerase II. *Annu Rev Biochem* **60**: 689–715
2. Lee Y., Kim M., Han J., Yeom K. H., Lee S., et al. (2004) MicroRNA genes are transcribed by RNA polymerase II. *EMBO J* **23**: 4051–60
3. Jawdekar G. W., Henry R. W. (2008) Transcriptional regulation of human small nuclear RNA genes. *Biochim Biophys Acta* **1779**: 295–305
4. Thomas M. C., Chiang C. M. (2006) The general transcription machinery and general cofactors. *Crit Rev Biochem Mol Biol* **41**: 105–78
5. Buratowski S., Hahn S., Guarente L., Sharp P. A. (1989) Five intermediate complexes in transcription initiation by RNA polymerase II. *Cell* **56**: 549–61
6. Zerby D., Lieberman P. M. (1997) Functional analysis of TFIID-activator interaction by magnesium-agarose gel electrophoresis. *Methods* **12**: 217–23
7. Pugh B. F., Tjian R. (1992) Diverse transcriptional functions of the multisubunit eukaryotic TFIID complex. *J Biol Chem* **267**: 679–82
8. Taatjes D. J. (2010) The human Mediator complex: a versatile, genome-wide regulator of transcription. *Trends Biochem Sci* **35**: 315–22
9. Sprouse R. O., Karpova T. S., Mueller F., Dasgupta A., McNally J. G., Auble D. T. (2008) Regulation of TATA-binding protein dynamics in living yeast cells. *Proc Natl Acad Sci USA* **105**: 13304–8
10. van Werven F. J., van Teeffelen H. A., Holstege F. C., Timmers H. T. (2009) Distinct promoter dynamics of the basal transcription factor TBP across the yeast genome. *Nat Struct Mol Biol* **16**: 1043–8
11. Metivier R., Penot G., Hubner M. R., Reid G., Brand H., Kos M., Gannon F. (2003) Estrogen receptor-alpha directs ordered, cyclical, and combinatorial recruitment of cofactors on a natural target promoter. *Cell* **115**: 751–63
12. de Graaf P., Mousson F., Geverts B., Scheer E., Tora L., Houtsmuller A.B., Timmers H.T. (2010) Chromatin interaction of TATA-binding protein is dynamically regulated in human cells. *J Cell Sci* **123**: 2663–2671
13. Coulombe B., Burton Z. F. (1999) DNA bending and wrapping around RNA polymerase: a "revolutionary" model describing transcriptional mechanisms. *Microbiol Mol Biol Rev* **63**: 457–78
14. Smale S. T., Kadonaga J. T. (2003) The RNA polymerase II core promoter. *Annu Rev Biochem* **72**: 449–79
15. Juven-Gershon T., Kadonaga J. T. (2010) Regulation of gene expression via the core promoter and the basal transcriptional machinery. *Dev Biol* **339**: 225–9
16. Bell B., Tora L. (1999) Regulation of gene expression by multiple forms of TFIID and other novel TAFII-containing complexes. *Exp Cell Res* **246**: 11–9
17. Mousson F., Kolkman A., Pijnappel W. W., Timmers H. T., Heck A. J. (2008) Quantitative proteomics reveals regulation of dynamic components within TATA-binding protein (TBP) transcription complexes. *Mol Cell Proteomics* **7**: 845–52

18. Tora L., Timmers H. T. (2010) The TATA box regulates TATA-binding protein (TBP) dynamics in vivo. *Trends Biochem Sci* **35**: 309–14
19. Dignam J. D., Lebovitz R. M., Roeder R. G. (1983) Accurate transcription initiation by RNA polymerase II in a soluble extract from isolated mammalian nuclei. *Nucleic Acids Res* **11**: 1475–89
20. Bell B., Scheer E., Tora L. (2001) Identification of hTAF(II)80 delta links apoptotic signaling pathways to transcription factor TFIID function. *Mol Cell* 8: 591–600
21. Lescure, A., Lutz, Y., Eberhard, D., Jacq, X., Krol, A., Grummt, I., Davidson, I., Chambon, P., and Tora, L. (1994) The N-terminal domain of the human TATA-binding protein plays a role in transcription from TATA-containing RNA polymerase II and III promoters. *EMBO J* **13**: 1166–75
22. Bertolotti A., Lutz Y., Heard D. J., Chambon P., Tora L. (1996) hTAF(II)68, a novel RNA/ssDNA-binding protein with homology to the pro- oncoproteins TLS/FUS and EWS is associated with both TFIID and RNA polymerase II. *EMBO J* **15**: 5022–31
23. Joyce C. M., Derbyshire V. (1995) Purification of *Escherichia coli* DNA polymerase I and Klenow fragment. *Methods Enzymol* **262**: 3–13
24. Turner R. M., Jr., Grindley N. D., Joyce C. M. (2003) Interaction of DNA polymerase I (Klenow fragment) with the single-stranded template beyond the site of synthesis. *Biochemistry* **42**: 2373–85
25. Li Y. C., Ross J., Scheppler J. A., Franza B. R., Jr. (1991) An in vitro transcription analysis of early responses of the human immunodeficiency virus type 1 long terminal repeat to different transcriptional activators. *Mol Cell Biol* **11**: 1883–93
26. Zerivitz K., Akusjarvi G. (1989) An improved nuclear extract preparation method. *Gene Anal Tech* **6**: 101–9

Chapter 5

Elucidating Protein: DNA Complex by Oligonucleotide DNA Affinity Purification

Teddy T.C. Yang and Chi-Wing Chow

Abstract

Transcription factors recruit a wide variety of associated co-factors to regulate gene expression. These co-factors include protein kinases, phosphatases, deacetylases, methylases, and ubiquitin ligases, etc. To identify novel protein kinases associated with transcription factor NFAT, we took advantage of the increased ability of DNA binding and used an oligonucleotide affinity-binding approach. Coupling with in-gel kinase assays to detect phosphotransferase activity, we were able to identify p90 ribosomal S6 kinase (RSK) and p70 S6 kinase (S6K) that are present in the NFAT:DNA complex. We further demonstrated that RSK and S6K binds to and physically interacts with NFATc4. Similar oligonucleotide affinity-binding approach can be coupled with other enzymatic reactions, such as dephosphorylation, deacetylation, methylation, ubiquitination, etc. Mass spectrometry can also be carried out to systemically identify these transcription co-factors in the protein:DNA complex. Lastly, gene-specific enhancer elements can also be devised based on their respective sequence to identify distinctive protein:DNA complexes.

Key words: Transcription activation complex, Oligonucleotide DNA affinity purification, Transcription factor NFAT, Transcription co-factors, Protein kinases

1. Introduction

The transcription factor NFAT family was first identified as an important regulator for IL-2 gene expression (1). Among the five distinct NFAT genes, four NFAT members (NFATc1 to NFATc4) exhibit high sequence homology and similar regulatory mechanisms by the calcium-mediated calcineurin phosphatase (reviewed in refs. 2, 3). Current models suggest that two signaling pathways are converged onto NFAT to mediate optimal NFAT activation. A calcium-mediated signaling pathway is involved to activate the calcineurin phosphatase, which binds to the conserved NH_2-terminal NFAT homology domain and dephosphorylates

Ales Vancura (ed.), *Transcriptional Regulation: Methods and Protocols*, Methods in Molecular Biology, vol. 809,
DOI 10.1007/978-1-61779-376-9_5, © Springer Science+Business Media, LLC 2012

NFAT. Dephosphorylation promotes nuclear accumulation of NFAT. In addition to the calcium-mediated NFAT dephosphorylation, a second signaling pathway – the Ras-mediated ERK MAP kinase – is involved in phosphorylation and activation of NFAT and NFAT partners.

Our initial goal was to identify transcription co-factors that are present in the NFAT transcription complex. Given the distinct subcellular pools of NFAT at resting state and upon activation, cytosolic NFAT is likely associated with different co-factors as compared to NFAT resides in nucleus. Furthermore, phosphorylation status of NFAT is markedly different in cytosol and nucleus and such modification could further regulate the extent and stability of interaction with NFAT co-factors. Once in the nucleus, NFAT interacts with NFAT partners to form intimate composite complexes, which are sequence dependent. Different co-factors may interact with the diverse NFAT composite elements depending on the architecture of NFAT target genes. Common co-immunoprecipitations, therefore, may not allow differentiation of distinct pools of NFAT in cytosol, nucleus and specific target gene loci. In addition, antibody immunoprecipitation is largely dependent on specificity and efficiency on the antibody, which could pull down co-factors independent of NFAT. The interaction of antibody may also displace binding of certain co-factors that are critical for NFAT regulation.

To capture nuclear resided NFAT and to identify NFAT-associated co-factors, we exploited the increased DNA binding of dephosphorylated NFAT and utilized oligonucleotide DNA affinity purification (4–6). Unlike antibody immunoprecipitation, specificity of affinity DNA precipitation can readily be tested by competition and using mutated NFAT binding oligonucleotides (Figs. 1–4). In the present review, we demonstrate that by coupling oligonucleotide affinity binding to in-gel kinase assay, we have identified novel NFAT kinases (Figs. 1–3) (see Note 1).

2. Materials

2.1. Preparation of Cell Lysates

1. 3T3/L1 and COS-7 cells are cultured in Dulbecco's modified Eagle's medium (Sigma, St. Louis, MO) supplemented with 10% fetal calf serum, 2 mM L-glutamine, penicillin (100 U/ml), and streptomycin (100 μg/ml) (Invitrogen).
2. Ionomycin (Ion, Sigma) is dissolved at 2 mM in dimethyl sulfoxide (DMSO, Sigma) and stored in single use aliquots at –20°C.
3. Phorbol 12-myristate 13-acetate (PMA, Sigma) is dissolved at 100 μM in DMSO and stored in single use aliquots at –20°C.

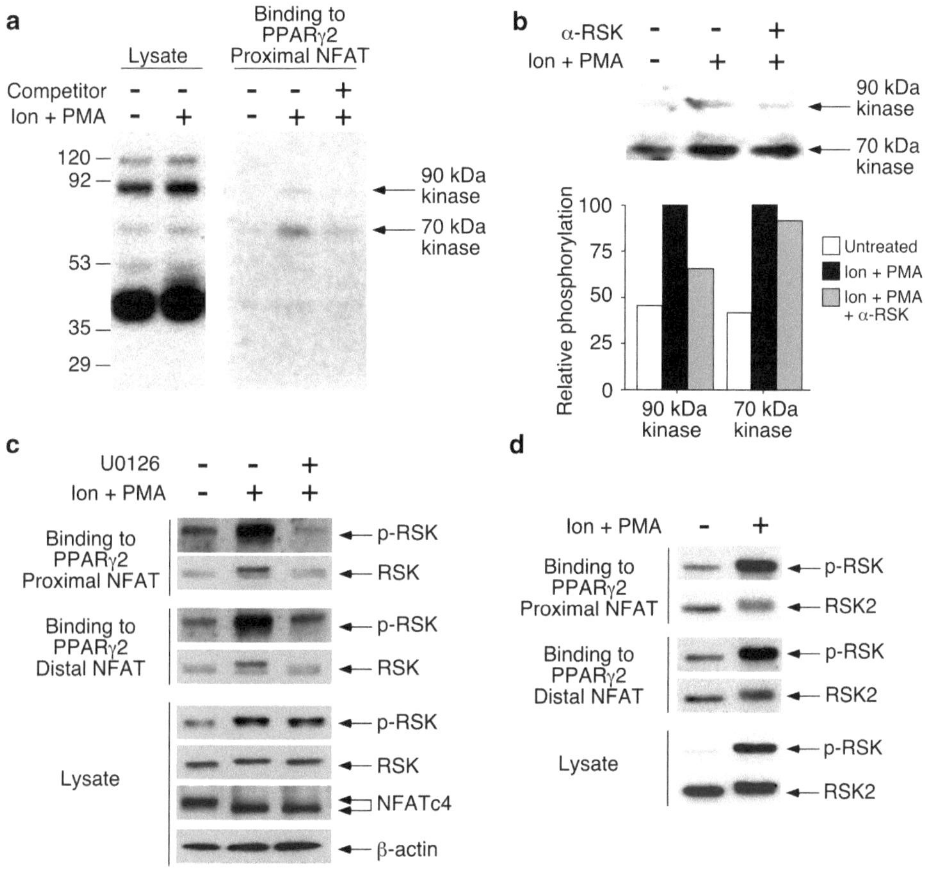

Fig. 1. Ribosomal p90 S6 kinase associates with activated NFAT:DNA complex. (**a**) Identification of protein kinases present in the activated NFAT:DNA complex. Biotinylated PPARγ2 proximal NFAT DNA binding element was incubated with untreated (–) or ionomycin (Ion, 2 μM) and phorbol ester (PMA, 100 nM) treated (+, Ion + PMA) 3T3/L1 cell extracts. Protein kinases in the extracts and bound to the immobilized NFAT DNA element were detected by in-gel kinase assays and visualized by autoradiography. Specificity of the recruited protein kinases was examined by competition with excess nonbiotinylated NFAT binding element. (**b**) The 90 kDa NFAT-associated kinase belongs to the p90 RSK family. Extracts prepared from untreated (–) and treated (+, Ion + PMA) 3T3/L1 cells were preincubated with antibody against RSK (α-RSK). RSK precleared (+) or not (–) extracts were subjected to coupled DNA binding:in-gel kinase assays. The amount of phosphorylation was quantitated by PhosphorImager analysis. (**c**–**d**) Endogenous RSK is recruited to the NFAT:DNA complex upon activation. Biotinylated PPARγ2 proximal NFAT DNA binding element was incubated with untreated (–) or treated (+, Ion + PMA) cell extracts prepared from mouse embryonic fibroblasts (**c**). Effect of MEK1 kinase inhibitor U0126 was also examined. Recruitment of RSK is also assessed using extracts prepared from treated (+, Ion + PMA) or not (–) HEK293 cells transfected with epitope-tagged RSK2 (**d**). Activated (p-RSK) and total RSK (RSK) in the cell extracts, and bound to the immobilized NFAT DNA element were detected by immunoblot assays. Binding of RSK to the PPARγ2 distal NFAT DNA element was also examined (reproduced from ref. 6 with permission of American Society of Microbiology).

4. Phosphate-buffered saline: 137 mM NaCl, 2.7 mM KCl, 10 mM sodium phosphate dibasic, 2 mM potassium phosphate monobasic, pH 7.4 stored at 4°C.
5. Triton lysis buffer (20 mM Tris [pH 7.4], 137 mM NaCl, 2 mM EDTA, 1% Triton X-100, 25 mM γ-glycerophosphate, 1 mM sodium vanadate, 2 mM sodium pyrophosphate,

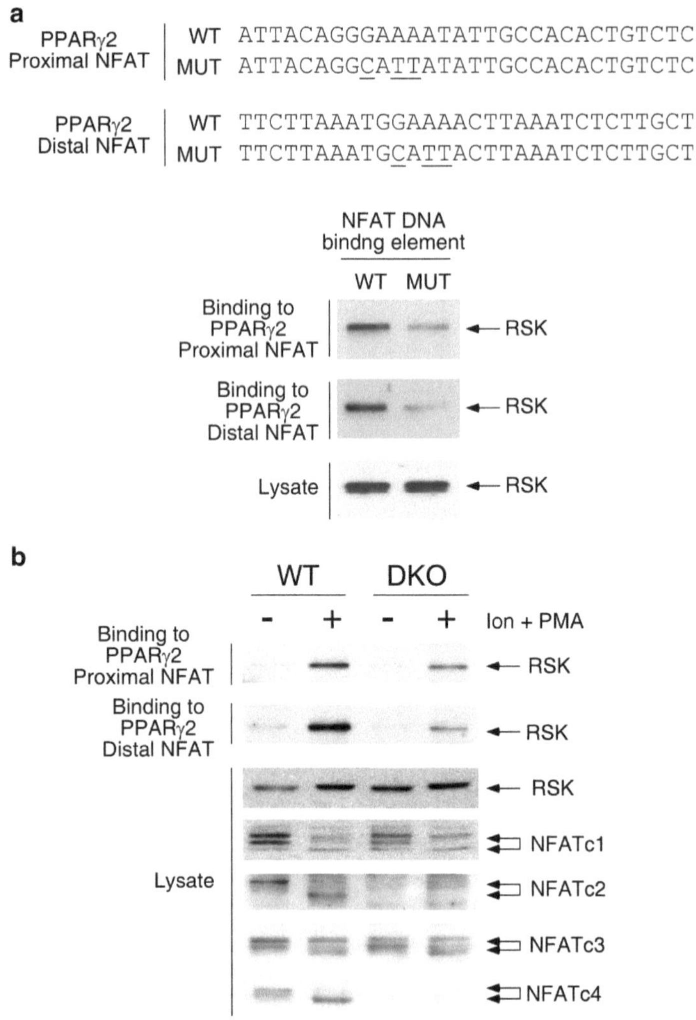

Fig. 2. Requirement of NFAT in the recruitment of RSK. (**a**) Biotinylated PPARγ2 proximal or distal NFAT DNA binding elements were incubated with ionomycin and phorbol ester treated cell extracts prepared from mouse embryonic fibroblasts. Endogenous RSK in the cell extracts, and bound to the immobilized wild-type (WT) NFAT DNA elements were detected by immunoblot assays. Binding of RSK to the mutated (MUT) PPARγ2 proximal or distal NFAT DNA element was also examined. (**b**) Biotinylated PPARγ2 proximal NFAT DNA binding elements were incubated with untreated (–) or ionomycin (Ion, 2 μM) and phorbol ester (PMA, 100 nM) treated (+) cell extracts prepared from wild-type (WT) and *Nfatc2–/– + Nfatc4–/–* (DKO) mouse embryonic fibroblasts. Endogenous RSK in the cell extracts, and bound to the immobilized NFAT DNA element were detected by immunoblot assays. Expression of NFAT proteins was also examined (reproduced from ref. 6 with permission of American Society of Microbiology).

10% glycerol, 1 mM phenylmethylsulfonyl fluoride, 10 μg of leupeptin/ml).

6. Streptavidin agarose resin, Pierce ThermoScientific.

2.2. Immuno-Depletion of p90RSK in Cell Lysate

1. Anti-p90 RSK antibody (R23820), BD Transduction Laboratories.
2. Protein G-Sepharose, Amersham Pharmacia (GE).

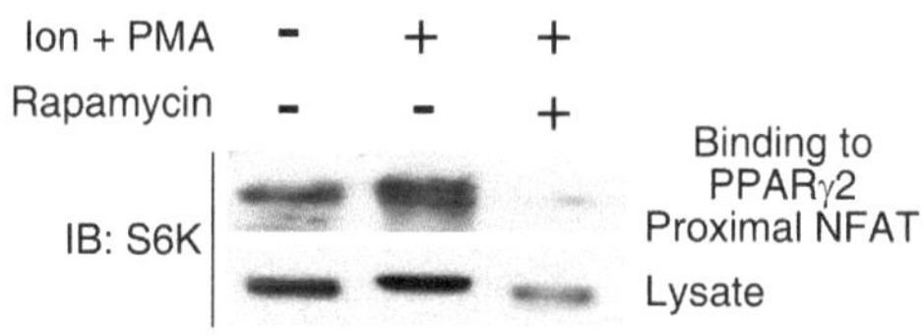

Fig. 3. Ribosomal p70 S6 kinase associates with activated NFAT:DNA complex. The 70 kDa NFAT-associated kinase (Fig. 1) belongs to the p70 S6K family and its binding to the NFAT:DNA complex is sensitive to rapamycin. Biotinylated PPARγ2 proximal NFAT DNA binding element was incubated with untreated (–) or treated (+, Ion + PMA) cell extracts prepared from mouse embryonic fibroblasts. Effect of mTOR kinase inhibitor rapamycin was also examined. S6K in the cell extracts (lysate) and bound to the immobilized NFAT DNA element were detected by immunoblot assays.

2.3. Preparation of Oligonucleotides

Oligonucleotides are dissolved in water at 100 pmol/μl.

1. Biotinylated PPARγ2 proximal NFAT site oligonucleotides:

   Sense: 5′-Biotin-ATTACAGGGAAAATATTGCCACACTGTCTC-3′, Anti-sense: 5′-GAGACAGTGTGGCAATATTTTCCCTCTAAT-3′.
2. Biotinylated PPARγ2 proximal NFAT mutant oligonucleotides: Sense:5′-Biotin-ATTACAGGCATTATATTGCCACACTGTCTC-3′, Anti-sense: 5′-GAGACAGTGTGGCAATATAATGCCTGTAAT-3′.
3. Nonbiotinylated PPARγ2 proximal NFAT site oligonucleotides: Sense:5′-ATTACAGGGAAAATATTGCCACACTGTCTC-3′, Anti-sense: 5′-GAGACAGTGTGGCAATATTTTCCCTCTAAT-3′.

2.4. Affinity DNA-Binding Assay

1. Poly(dI-C), Sigma.
2. Hoefer electrophoresis system for sodium dodecyl sulfate-polyacrylamide gel electrophoresis (SDS-PAGE) or other electrophoresis system, such as Bio-Rad or Invitrogen.
3. Anti-NFATc4 antibody (Santa Cruz catalog no. sc-13036).
4. Anti-phospho-RSK antibody (Santa Cruz catalog no. sc-17033).
5. Anti-p90 RSK antibody (BD Transduction Laboratories catalog no. R23820).
6. Anti-p70 S6 kinase antibody (Santa Cruz catalog no. sc-230).

2.5. In-Gel Kinase Assay

1. Purified recombinant proteins (NFATc4 DNA binding domain and myeline basic protein, Sigma).
2. 20% iso-propanol and 50 mM Tris–HCl buffer, pH 8.0.
3. Denaturing buffer: 6 M guanidine–HCl in 50 mM Tris–HCl, pH 8.0, and 5 mM 2-mercaptoethanol.
4. Renaturing buffer: 0.04% Tween 40 in 50 mM Tris–HCl, pH 8.0, and 5 mM 2-mercaptoethanol.

5. Kinase reaction buffer: 40 mM HEPES, pH 8.0, 2 mM dithiothreitol, 0.1 mM EGTA, and 5 mM $MgCl_2$.
6. [γ-^{32}P]ATP, Amersham Pharmacia GE.
7. Wash buffer: 5% trichloroacetic acid and 1% sodium pyrophosphate.
8. Gel Drier from Bio-Rad.
9. KODAK™ BioMax® MS Intensifying Screen, Perkin Elmer.
10. Kodak BioMax Light autoradiography film.
11. Storm series PhosphorImager system from Molecular Dynamics.

3. Methods

3.1. Preparation of Cell Lysates

Cells were stimulated or not with ionomycin (2 μM) and PMA (100 nM) for 30 min to induce nuclear localization and to activate NFAT. Cells were quickly chilled on ice and washed with ice-colded PBS twice and 1 ml of triton lysis buffer was added. Cells were scraped and incubated for 15 min on ice and centrifuged at 16,000 × *g* in a microcentrifuge for 15 min at 4°C. Supernatant was collected and stored at −80°C until used.

In order to increase the specificity of biotinylated DNA pull down, cell lysates were precleared with streptavidin agarose resin to remove endogenous biotinylated proteins. Proteins that are nonspecifically bound to streptavidin agarose resin will also be removed.

1. Streptavidin agarose resin (20 μl/reaction) was washed with 1 ml Triton lysis buffer for three times to remove organic solvents and preservatives.
2. 500 μg lysate was added to an Eppendorf tube, and total volume up was made up to 500 μl with Triton lysis buffer.
3. The mixture was incubated with rocking at 4°C for 2 h to preclear cell lysates.
4. After 2 h incubation, the mixture was centrifuged at 1,500 × *g* in a microcentrifuge for 5 min at 4°C, and the supernatant was transferred to a new Eppendorf tube.

3.2. Immuno-Depletion of p90RSK in Cell Lysate

Cell lysates were incubated for 5 h at 4°C with 10 μl anti-p90RSK antibody prebound to 20 μl of protein G-Sepharose. Protein G-Sepharose was pelleted by centrifugation at 1,500 × *g* in a microcentrifuge for 5 min at 4°C. Precleared cell lysates were transferred to new Eppendorf tubes and kept at −80°C.

3.3. Preparation of Oligonucleotides

Oligonucleotide encoding the sense strand of PPARγ2 proximal NFAT site and its antisense, complementary oligonucleotides were to prepare double-stranded oligonucleotides for NFAT binding. Fifty microliter of both sense and antisense oligonucleotides were mixed and incubated in a boiling water bath for 3 min and slowly cooled down to room temperature. Double-stranded oligonucleotides were then stored at –20°C (see Notes 2 and 3).

3.4. Affinity DNA-Binding Assay

1. Streptavidin agarose resin precleared cell lysate was incubated with 0.5 μl double-stranded biotinylated PPARγ2 proximal NFAT DNA binding element (10 pmol) at 4°C overnight in the presence of 1 μg of poly(dI-C).
2. To ascertain the specificity of the DNA protein interaction, competition was performed by using tenfold excess (5 μl) of nonbiotinylated NFAT binding element. Nonbiotinylated NFAT binding element with mutation to abolish NFAT or its partner binding can also be used to ascertain specificity (Fig. 4).
3. NFAT:DNA complex was precipitated with 20 μl of streptavidin-agarose at 4°C for an additional 2 h.
4. The complex was washed three times with 1 ml Triton-lysis buffer; the bound proteins were separated by SDS-PAGE and subjected to in-gel kinase assay or immunoblotting analysis with RSK antibodies to confirm the identity (Figs. 1 and 2). Similar immunoblotting analysis was used to identify p70 S6K kinase on the NFAT:DNA complexes (Fig. 3).

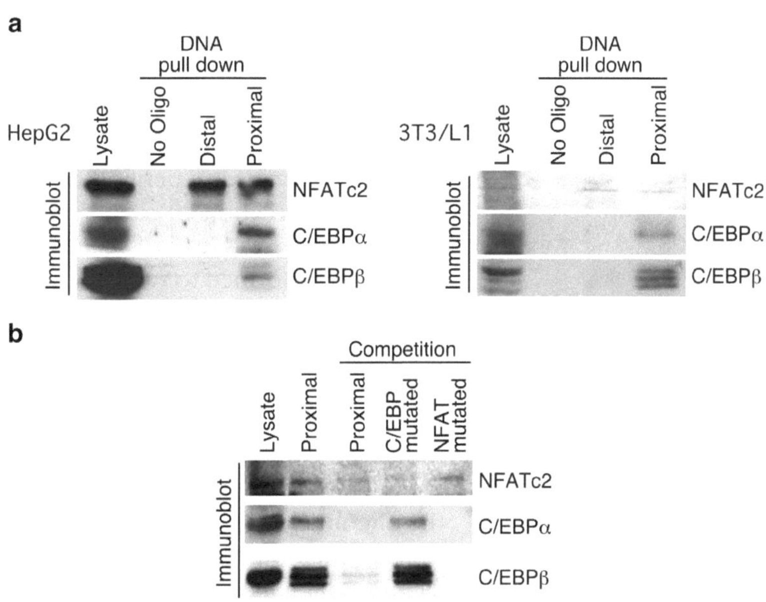

Fig. 4. NFAT and C/EBP interacts with the PPARγ2 promoter. (**a**) Extracts prepared from HepG2 or 3T3/L1 cells were incubated with biotinylated PPARγ2 proximal or distal NFAT oligonucleotides. Protein:DNA complexes were precipitated by streptavidin Sepharose, resolved in SDS-PAGE, and immunoblotting analyses were performed to detect bindings of endogenous NFAT and C/EBP. (**b**) Excess amount of nonbiotinylated PPARγ2-proximal elements (wild type, C/EBP mutated, or NFAT mutated) were used in competition analysis to indicate specificity of NFAT:C/EBP interaction (reproduced from ref. 5 with permission of American Society for Biochemistry and Molecular Biology).

3.5. In-Gel Kinase Assay

1. To prepare the SDS-PAGE for in-gel kinase assay, purified recombinant proteins (NFATc4 DNA binding domain or myelin basic protein, 0.5 mg/ml) were co-polymerized with acrylamide gel matrix (7%), and the prepared gel were used to resolve protein kinases present in the NFAT:DNA precipitates.
2. Resolved gels were soaked in 250 ml 20% iso-propanol and 50 mM Tris–HCl buffer, pH 8.0, for 1 h at room temperature to remove SDS.
3. The gel-bound protein kinases were denaturated in 100 ml of 6 M guanidine–HCl in 50 mM Tris–HCl, pH 8.0, and 5 mM 2-mercaptoethanol for 1 h at room temperature. Gel-bound protein kinases were renaturated in 1 L of 0.04% Tween 40 in 50 mM Tris–HCl, pH 8.0, and 5 mM 2-mercaptoethanol at 4°C overnight with multiple changes of buffer.
4. The gel was first equilibrated in 100 ml kinase reaction buffer for 30 min before the kinase assay. Kinase reaction was carried out in 10 ml kinase reaction buffer containing 50 μM [γ-^{32}P] ATP for 1 h at room temperature.
5. After multiple washes in 250 ml 5% trichloroacetic acid and 1% sodium pyrophosphate at room temperature, gel was dried and phosphorylation was examined by autoradiography and phosphorimaging (see Note 4).

4. Notes

1. Similar to other in vitro purification to identify nuclear co-factors, in vivo confirmation of such regulation and function of the binding partners is critical. Chromatin immunoprecipitations should be performed to validate the presence of such co-factors in the gene locus of interest (6). Association with the co-factors should also be tested upon activation of the transcription factor, such as increase in intracellular calcium to activate NFAT.
2. Only one strand of the double-stranded oligonucleotides should be conjugated with biotin. Labeling both ends of the oligonucleotides with biotin may cause excessive bending of the oligonucleotides upon binding to the streptavidin agarose beads, which will then affect DNA binding of the transcription factors. The length of the oligonucleotides will be dependent on the coverage of the transcription factor on the DNA sequence, which can be assessed by electromobility shift analysis or DNA footprinting.
3. In the case of NFAT-mediated transcription, the formation of composite enhancer element depends largely on NFAT and its partners (5). Binding of NFAT partner is sequence dependent.

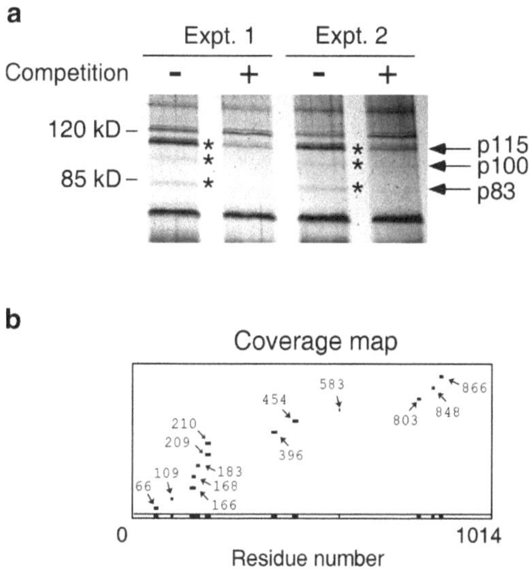

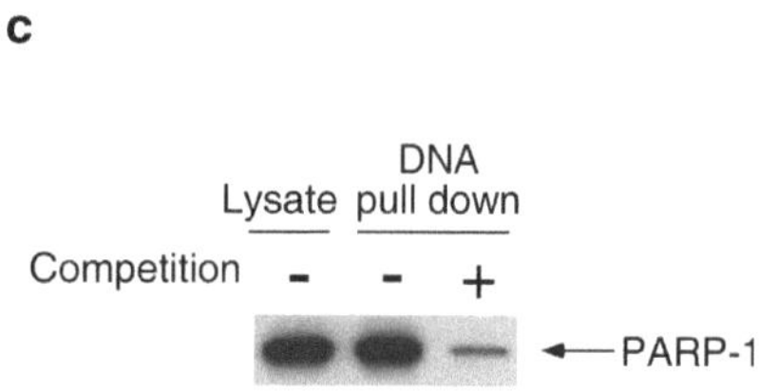

Position	Peptide fragment
66	QPDVEVDGFSELR
109	TLGDFLAEYAK
166	KRDELGFRPEYSASQLK
168	DELGFRPEYSASQLK
183	GFSLLSAEDKEALK
209	KGDEVDGTDEVAKK
210	GDEVDGTDEVAKKK
396	ILTLGKLSQNKDEAK
454	VVCEDFLQDVSASTK
583	YWIFR
803	VVDRDSEEAEVIR
848	YKPFR
866	TTNFAGILSQGLR

Fig. 5. Identification of PARP-1 as NFAT-interacting protein. (**a**) DNA affinity isolation of NFAT interacting proteins using biotinylated NFAT DNA binding elements. Bound proteins (p115, p100, and p83, see *asterisks*) were ascertained by competition with nonbiotinylated NFAT oligonucleotides and revealed by protein staining. Two representatives of binding assays were shown (Expt.1 and Expt.2). (**b**) The 115 kDa NFAT-interacting protein was excised and subjected to tryptic digestion and MALDI-TOF mass spectrometry analysis. Obtained peptide mass was search against the NCBI protein database using the ProFound software. With 99th percentile confidence, computation analysis unambiguously predicted that the 115 kDa NFAT-interacting protein is poly (ADP-ribose) polymerase-1 (PARP-1). (**c**) DNA affinity isolation was performed using NFAT DNA binding elements. The presence of PARP-1 in NFAT:DNA precipitates was detected by immunoblotting analysis. Competition with nonbiotinylated NFAT oligonucleotides was used to ascertain specificity (reproduced from ref. 4 with permission of American Society of Microbiology).

Thus, gene specific enhancer elements can be devised based on their respective sequence to identify distinctive protein:DNA complexes. For example, we found transcription factor C/EBP as critical NFAT partner in regulating transcription of PPARγ2 gene in adipocytes (Fig. 4) while AP1 complex is required for NFAT-mediated IL-2 gene transcription in immune cells. In Schwann cells, NFAT interacts with Sox proteins to mediate myelin gene expression (e.g. Krox20) (7).

4. In addition to in-gel kinase assays, in vitro kinase assays can be used in conjunction with specific inhibitors to elucidate the nature of protein kinase. Other enzymatic assays can also be coupled subsequent to oligonucleotide DNA affinity purification. Many of these enzymatic assays are established and available as detection kits (e.g., Malachite green reaction to detect protein phosphatase and the use of acetylated lysine side chain in HDAC fluorometric activity assay kit). Mass spectrometry can also be carried out to systemically identify components present in the protein:DNA complex (Fig. 5). For example, we have utilized such proteomic approach and found that poly (ADP-ribose)

polymerase-1 (PARP-1) is present in the NFAT:DNA complex (4). We demonstrated that PARP-1 binds to and ribosylates NFAT at the DNA binding domain. Indeed, target disruption of PARP-1 impairs expression of IL-2, which is critically regulated by NFAT.

Acknowledgments

The authors would like to acknowledge the past and current members of the Chow laboratory for their input, discussion, and trouble shooting of the protocol.

References

1. Durand DB, Shaw JP, Bush MR, Replogle RE, Belagaje R and Crabtree GR (1988) Characterization of antigen receptor response elements within the interleukin-2 enhancer. Mol Cell Biol 8:1715–24.
2. Crabtree GR and EN Olson (2002) NFAT signaling: choreographing the social lives of cells. Cell 109 Suppl:S67–79.
3. Hogan PG, Chen L, Nardone J and Rao A (2003) Transcriptional regulation by calcium, calcineurin, and NFAT. Genes Dev 17:2205–32.
4. Olabisi OA, Soto-Nieves N, Nieves E, Yang TT, Yang X, Yu RY, Suk HY, Macian F and Chow CW (2008) Regulation of transcription factor NFAT by ADP-ribosylation. Mol Cell Biol 28:2860–71.
5. Yang TT and Chow CW (2003) Transcription cooperation by NFAT.C/EBP composite enhancer complex. J Biol Chem 278: 15874–85.
6. Yang TT, Xiong Q, Graef IA, Crabtree GR and Chow CW (2005) Recruitment of the extracellular signal-regulated kinase/ribosomal S6 kinase signaling pathway to the NFATc4 transcription activation complex. Mol Cell Biol 25:907–20.
7. Kao SC, Wu H, Xie J, Chang CP, Ranish JA, Graef IA and Crabtree GR (2009) Calcineurin/NFAT Signaling Is Required for Neuregulin-Regulated Schwann Cell Differentiation. Science 323:651–654.

Chapter 6

Chromatin Immunoprecipitation Assay as a Tool for Analyzing Transcription Factor Activity

Padmaja Gade and Dhan V. Kalvakolanu

Abstract

Differential gene expression is facilitated by transcriptional regulatory mechanisms and chromatin modifications through DNA–protein interactions. One of the widely used assays to study this is chromatin immunoprecipitation (ChIP) assay, which enables analysis of association of regulatory molecules to specific promoters and histone modifications in vivo. This is of immense value as ChIP assays can provide glimpse of the regulatory mechanisms involved in gene expression in vivo. This article outlines the general strategies and protocols to study ChIP assays in differential recruitment of transcriptional factors (TFs) and also global analysis of transcription factor recruitment is discussed. Further, the applications of ChIP assays for discovering novel genes that are dependent on specific transcription factors were addressed.

Key words: ChIP, dapk1, IFN-γ, C/EBP-β, Transcription factor activity, Regulation of gene expression, DNA–protein interactions, Transcription factor recruitment

1. Introduction

A myriad of biological processes like gene transcription, DNA replication and recombination, DNA repair, chromosome segregation, chromosomal stability, cell cycle progression, epigenetic silencing, and regulation of gene expression are mediated through protein–DNA interactions. Thus, it is of principal importance to understand the significance of these interactions in driving a biological response.

Techniques commonly used to characterize protein–DNA interactions include electrophoretic mobility shift assay (EMSA) (1), DNAse footprinting (2) and chromatin immunoprecipitation (ChIP) assays. Though, EMSA is a rapid and sensitive method to detect protein–DNA interactions, despite its advantages it has

Ales Vancura (ed.), *Transcriptional Regulation: Methods and Protocols,* Methods in Molecular Biology, vol. 809, DOI 10.1007/978-1-61779-376-9_6,

limitations like the samples are not in chemical equilibrium during the electrophoresis step and also many complexes are significantly more stable in the gel than at free solution. ChIP assays minimize these limitations and provide an efficient tool to determine these protein–DNA interactions occurring in vivo, by immunoprecipitating chromatin with specific antibodies. These assays provide unbiased observations into the chromatin changes occurring in response to extracellular signals or during differentiation and development (3, 4). Although transcription factor recruitment to the promoter is very dynamic, ChIP assays provide an efficient means to study these events.

The ChIP assay has been adopted as a powerful method for the analysis of proteins interacting within a native chromatin environment and is versatile enough for adaptation for a variety of purposes (5, 6). This assay has been utilized in yeast (7), drosophila (8), tetrahymena (9), *Caenorhabditis elegans* (10), various mammalian cell lines, and even on whole mouse embryos (11, 12) for the analysis of low abundance transcription factor binding. In addition to focused study of a single or group of genes, some groups used it for systematic promoter cloning (13, 14) and identification of gene targets using promoter microarrays (7, 15). Also, ChIP has been used to determine the allele-specific transcription factor binding patterns (16) and to measure long-range enhancer binding of specific proteins (17). Additionally, in combination with other techniques, ChIP assays are used to uncover an extraordinarily rich and dynamic chromatin environment (13, 18, 19). Finally, ChIP assay systems can provide useful endpoint measurement for delineation of components within the signal transduction pathways (20).

ChIP assay is a multistep process (Fig. 1) and each step needs to be standardized for obtaining optimum results. In this technique, intact cells are treated with formaldehyde to covalently link protein to DNA (X-ChIP), the nucleoprotein complexes are then sheared either mechanically or by enzymatic digestion. The resultant soluble cross-linked DNA–protein complexes enriched by immunoprecipitation (21). The retrieved complexes are then analyzed by PCR amplification with gene-specific primers to detect and quantify specific DNA targets. The main drawback of ChIP is its inherent variability. This may arise due to variabilities in the cross-linking, immunoprecipitation, and protein–DNA washing efficiencies. Stringent control over experimental conditions may reduce these variabilities. Other limitations include the dependence on the availability of a highly specific antibody and the need for relatively high level as well as broad expression DNA-binding protein of interest. Endogenous ChIP, i.e., without overexpressing these factors, poses additional challenges, including the difficulty of handling large population if the protein of interest is expressed at a low level.

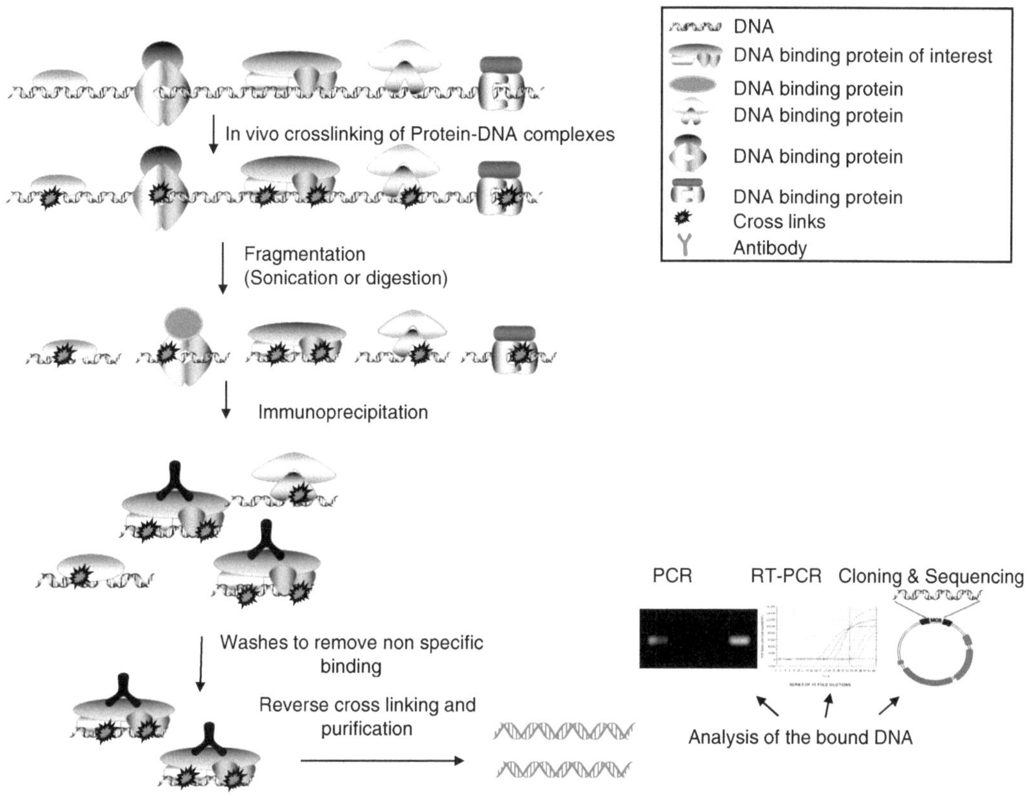

Fig. 1. Schematic representation of the steps involved in chromatin immunoprecipitation (ChIP) assay.

Our group exploited ChIP methodology for fine mapping of transcription factor-binding regions to regulatory elements of a cell death regulatory gene, *DAPK1,* under control of IFN-γ. The region of DNA containing the full potential for specific transcription factor can be mutated to identify the elements present. Loss or enhancement of transcription factor association with active elements within the DNA can be quantified. Mutations can be generated in specific regulatory elements within the DNA regions which prevent transcription factor binding and the mechanism of gene regulation can be defined. Gain or loss of function experiments for these transcription factors can also be quantified using a ChIP assay. For example, transcription factors can be overexpressed or/its expression can be knocked down by RNAi or/knocked out to determine the role of individual factors in the regulation of specific element (22–24). Also, ChIP assay can be used to isolate specific transcription factor gene targets thus identifying both consensus and nonconsensus-binding sites (14). Here, we outline in detail general strategies and an optimized protocols for ChIP (for adherent and suspension cells as the chromatin source) along with troubleshooting guides.

2. Materials

2.1. Cell Culture

1. Mouse embryonic fibroblasts (MEFs) were grown in Dulbecco's modified Eagle's medium (GIBCO) supplemented with 10% fetal bovine serum and 1% antibiotic-antimycotic agent.
2. Mouse IFN-γ (PBL) eliquoted into small volumes and stored at –20°C till use.

2.2. Stock Solutions and Reagents

1. Phosphate-buffered saline (PBS): 10× PBS stock is prepared by adding 1.37 M NaCl, 27 mM KCl, 100 mM Na_2HPO_4, 18 mM KH_2PO_4 (pH should be adjusted to 7.4 with HCl if necessary) and is autoclaved before storing at room temperature. Working solution is prepared by diluting one part with nine parts water.
2. Phenylmethylsulphonylfluoride (PMSF): 1 mM PMSF is prepared by dissolving in isopropanol.
3. Protease inhibitor cocktail: Diluted in water to make 100× stock and frozen in aliquots. Caution: Toxic!!! Should be handled using gloves.
4. RIPA lysis buffer: 20 mM Tris pH 7.4, 1% NP-40, 75 mM NaCl, 1 mM EDTA, 25 mM β glycerophosphate, 10% glycerol. Lysis buffer is supplemented with 1 mM PMSF, 1 mM Na_3VO_4, and 1× protease inhibitor cocktail just before use.
5. ChIP dilution buffer: 0.01% SDS, 1.1%Triton X-100, 1.2 mM EDTA, 16.7 mM Tris–HCl, pH 8.1, 167 mM NaCl. ChIP dilution buffer is supplemented with 1 mM PMSF, 1 mM Na_3VO_4, and 1× protease inhibitor cocktail just before using.
6. C/EBP β antibody (Santa Cruz Biotech) or other specific antibody.
7. Rabbit IgG (Sigma) serves as a negative control. In case the antibody used for ChIP is a monoclonal, it is recommended to use an isotype control antibody.
8. Protein G magnetic beads (Active Motif).
9. Proteinase K diluted to 20 mg/ml in water and frozen (–20oC) in aliquots of 20 ml.
10. Low salt immune complex wash buffer: 0.1% SDS, 1%Triton X-100, 2 mM EDTA, 20 mM Tris–HCl, pH 8.1, 150 mM NaCl. This buffer is stored at 4°C till use and vortexed briefly before use.
11. High salt immune complex wash buffer: 0.1% SDS, 1%Triton X-100, 2 mM EDTA, 20 mM Tris–HCl, pH 8.1, 500 mM NaCl. This buffer is stored at 4oC till use and vortexed briefly before use.

12. LiCl Immune complex wash buffer: 0.25 M LiCl, 1% IGEPAL-CA630, 1% deoxycholic acid, 1 mM EDTA, 10 mM Tris, pH 8.1. This buffer is stored at 4oC till use.
13. TE Buffer: 10 mM Tris–HCl, 1 mM EDTA, pH 8.0. This buffer is stored at 4oC till use.
14. Elution buffer: Prepared just before eluting by combining 1% SDS, 0.1 M $NaHCO_3$ at room temperature.
15. PCR purification kit (QIAGEN) or other similar kits.
16. Oligonucleotide primers: User specific. The lyophilized primers are dissolved in 10 mM Tris–Cl (pH 8.0) to prepare a 100 μM stock.
17. IDTaq™ DNA polymerase Kit (ID Labs) or other similar reagents.
18. SYBR Green Jump start Taq Ready Mix (Sigma) or other similar reagents.

2.3. Equipment

1. Humidified 37°C incubator.
2. Table top centrifuge.
3. Rotating wheel/platform maintained at 4°C and 20°C.
4. Shaking incubator.
5. Branson's sonicator fitted with microtip (or equivalent).
6. Teflon cell scrapers (Fisher).
7. Timer.
8. Variable volume (5–1,000 μl) pipettes and tips.
9. Magnetic stand (Ambion).

3. Methods

In order to establish the ChIP protocol using an antibody against a protein of interest, three critical steps have to be optimized: (1) fixation time, (2) amount of input material, and (3) amount of antibody used. For the optimization step, it is helpful if there is a known target gene for the DNA-binding protein of interest. This serves as a positive control.

1. *Optimization of chemical cross-linking time (fixation time)*. One of the critical steps in ChIP is to capture the DNA–protein interactions within a cellular context, which is achieved using a chemical cross-linking agent, such as formaldehyde (5, 25). The extent of cross-linking is the most important parameter. An excess of cross-linking can result in the reduced antigen availability in chromatin for the antibody. The relative sensitivity

of the antigen epitopes to formaldehyde also should be taken into consideration. This step may be problematic, as the extent of cross-linking may vary from sample to sample. The cross-linking time needs to be empirically determined when setting up ChIP for the first time. Because the handling of materials varies significantly among different users, we suggest each user to optimize the cross-linking time. Since the extent of cross-linking affects IP by a given antibody, the length of cross-linking must be empirically optimized for each antibody initially.

2. *Optimization of the amount of input material.* As DNA-binding proteins are expressed over a wide range in various tissues, the amount of protein of interest available for IP varies widely. For example, for highly expressed proteins, the amount of total protein required for immunoprecipitation will be significantly lower than that of a low abundance protein. Therefore, ChIP must be optimized for the amount of input material (total protein) required for IP using each antibody against the protein of interest. In addition, before beginning to do any ChIP experiment, it is important to determine the expression level or pattern of the protein of interest using western blotting or in situ (Fig. 2). Also in some cases less amount of protein may interact with DNA, although the protein may be widely expressing. In such cases, more ChIP-ed DNA may be required for the detection.
3. *Optimization of amount of the antibody used for ChIP.* The cross-linked DNA–protein complex is immunoprecipitated using a specific antibody against the protein of interest. The quality and specificity of the antibody contributes immensely to the outcome of the ChIP procedure (26). As antibody preparations differ considerably in specificity and titer, the amount of antibody used to immunoprecipitate the cross-linked protein–DNA complex needs to be empirically determined. Different antibodies should be used to optimize the ChIP assay because certain antibodies may efficiently immunoprecipitate the native protein but may not be effective in immunoprecipitating proteins cross-linked to chromatin. Also the specificity of antibody can be checked by Western blot to ensure that it detects the correct size product. An antibody that detects multiple nonspecific bands is not recommended for ChIP. We optimized an antibody against the C/EBP β (Santa Cruz Biotechnology Catalog # sc-150) a rabbit polyclonal antibody for our assays. Before performing ChIP, we ensured its specificity by a Western blot analysis.

Experimental controls and repeats. As with any experiment, proper controls are absolutely necessary for correct interpretation of data obtained from ChIP experiments. Positive controls are difficult to obtain, as the number of experimentally defined direct protein–DNA

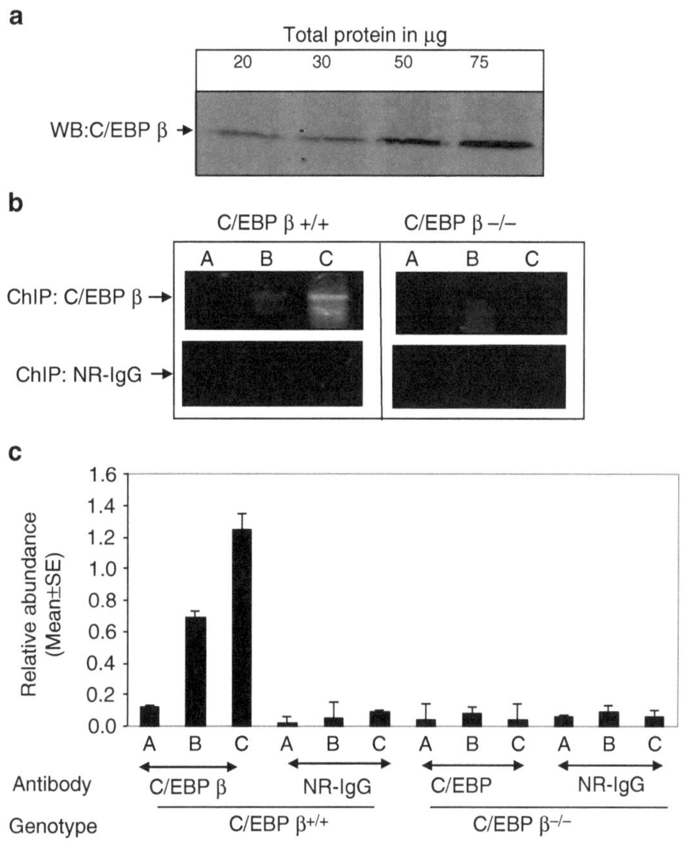

Fig. 2. (**a**) Western blot analysis to determine the expression levels of the C/EBP β protein in wildtype MEFs. The indicated amounts of the total protein was resolved on 10% SDS-PAGE and transferred to PVDF membrane. The membrane was blocked with 5% nonfat dry milk powder and probed with 1:1,000 dilution of the Ab (in 5% milk) for 3 h. Following three washes (10 min each) with 1× TBST, the membrane was incubated in 1:5,000 diluted anti-rabbit secondary Ab (GE) for 1 h. The membrane was washed three times with 1× TBST (10 min each) and Super Signal West Pico detection system (Pierce Biotechnology) was used to visualize the signal (1 min exposure time). The bands highlight the expression level of the protein in MEFS cells. (**b**) Optimization of the input amount for chromatin immunoprecipitation (ChIP). An aliquot of (a) 0.5×10^7, (b) 1×10^7, and (c) 2×10^7 million MEFS cells were cross-linked and sonicated. Cell lysate was immunoprecipitated with 5 μg of C/EBP β or NR-IgG antibody. After reverse cross-linking DNA is purified and 5 μl of it is subjected to PCR with specific primers for 30 cycles. The binding of C/EBP β or NR-IgG to the dapk1 promoter was used to evaluate the efficiency of ChIP. (**c**) Real-time PCR analysis of *dapk1* promoter fragments recovered in ChIP assays performed with the indicated antibodies using chromatin prepared from wild type and C/EBP $\beta^{-/-}$ MEFs. Each bar represents the mean abundance of *dapk1* promoter fragments ± SE of six separate reactions from two independent experiments.

interactions currently available is limited. At least one gene that is known to interact with the transcription factor of interest must be used as a positive control, when available. However, good negative controls lead to proper interpretation of the results. The knock out cell lines of the protein of interest provides specific advantage

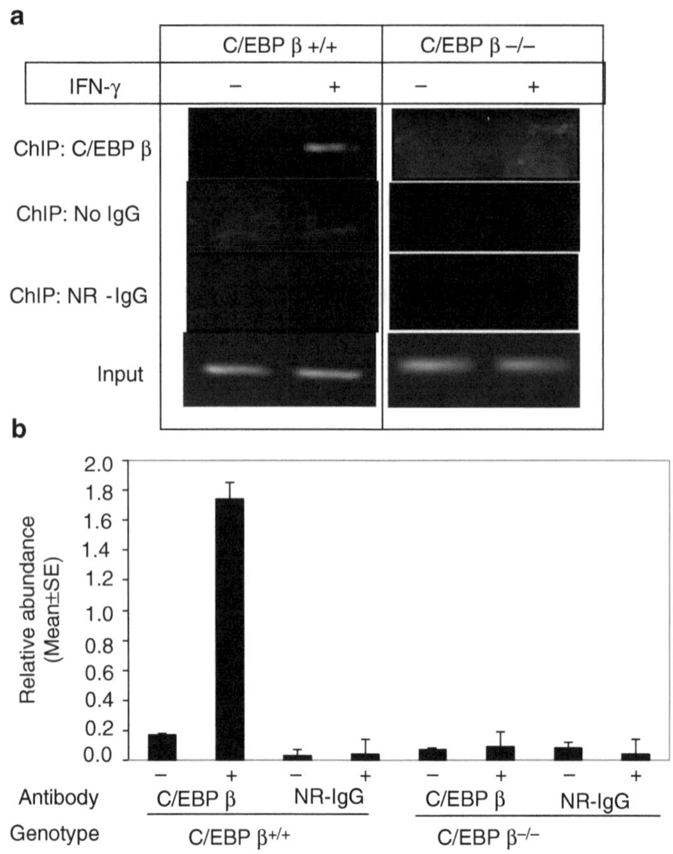

Fig. 3. (**a**) PCR analysis to compare the target *dapk1* promoter fragment enrichment for C/EBP β in C/EBP $\beta^{+/+}$ and C/EBP $\beta^{-/-}$ MEFs with or without IFN-γ treatment (16 h, 500 U/ml). (**b**) Real-time PCR analysis of *dapk1* promoter fragments recovered in ChIP assays performed with the indicated antibodies with or without IFN-γ treatment (16 h, 500 U/ml) using chromatin prepared from wild type and C/EBP $\beta^{-/-}$ MEFs. Each bar represents the mean abundance of *dapk1* promoter fragments ± SE of six separate reactions from two independent experiments.

in ChIP, thus enabling comparison of the degree of target DNA enrichment in biological samples expressing or lacking the protein of interest (Fig. 3). However, such cell lines should be verified for complete lack of the protein of interest, as a partially functional protein (such as a truncated product) may still bind DNA and lead to misinterpretation of the results. Alternatively, RNA*i* may be used to knock down the gene, although care should be taken to judge the efficacy of the RNA*i*. ChIP is a long protocol involving multiple steps. As such, it is important to have multiple technical as well as biological repeats. Technical repeats (triplicates within each experiment to assess the variability due to sample handling) can be generated by performing three separate immunoprecipitation reactions and detections after the cells have been sonicated. The specificity of ChIP reactions should further be ascertained by pre-incubating with the antibody with recombinant antigen or peptide. The following protocol is based on our studies on transcription factor C/EBP-β and its transcriptional co-activator Med1 on gene regulation.

3.1. In Situ Cross-Linking and Harvesting

1. MEFs (2×10^7) were grown in confluence and stimulated or treated with mouse IFNγ or appropriate inducing agent for conditions, where transcriptional activation of the gene of interest has been demonstrated (see Note 1).
2. Proteins were cross-linked to DNA by adding 1% formaldehyde (270 μl of 37% (W/V) formaldehyde) directly to cells growing in 10 ml of growth medium on plate and incubating for 15 min at 37°C (see Note 2) (27). For cross-linking suspended cells, see Subheading 3.7, step 1.
3. The cross-linking reaction is quenched by adding glycine to a final concentration of 0.125 M (141 ml of 1 M glycine per 1 ml of medium) and incubating for 5 min at 37°C (see Note 3).
4. The medium in the plates is aspirated as completely as possible.
5. The cells were scraped and collected by centrifugation ($2{,}000 \times g$ for 5 min at 4°C), then washed twice with cold 1× PBS containing protease inhibitors (1 mM PMSF, and 1× protease inhibitor cocktail).
6. At this stage, cell pellets can be lysed or stored at −80°C up to 1 year.

3.2. Lysis (Must Be Performed on Ice or at 4°C)

1. The cell pellet is lysed in 200 μl of RIPA buffer containing protease inhibitors (and phosphatase inhibitors if needed). The pellet should be suspended by pipetting up and down several times in a microcentrifuge tube (there will be a lot of insoluble material) (see Note 4).
2. The lysate (resuspended cell pellet) is incubated on ice or at 4°C for 15 min and followed with sonication.

3.3. Chromatin Fragmentation

1. The lysate is sonicated to shear the chromatin to 300 and 1,000 base pairs in length keeping the samples ice cold.
2. Our experience with most of the cell lines indicate that chromatin is sheared to the appropriate length with six cycles of 20 s pulses at 30% of maximum power with intermittent cooling on ice for 30 s using Branson's sonicator equipped with a 2 mm tip (see Note 5). However, sonication settings should be determined empirically for each cell type and antibody. (Note: Sonication generates heat. Excessive heat denatures the antigen, which significantly diminishes the detection of chromatin fragments by some antibodies).
3. Digestion with micrococcal nuclease can be done as an alternative, to cleave the chromatin into nucleosomes. This method

does not involve cross-linking step and is referred to as "native chromatin immunoprecipitation." Native ChIP cannot be adopted for these studies as nucleosomes are dynamic and without cross-linking there is chance of their rearrangement during the enzymatic digestion (see Note 6).

4. The sonicated lysates were cleared by centrifuging at 5,700 × *g* in microcentrifuge for 10 min at 4°C. The supernatants were transferred to new tubes and the pellet was discarded.
5. An aliquot of sheared chromatin (equivalent to 0.2 million cells) is transferred to a new microcentrifuge tube and DNA is recovered to determine shearing efficiency.
6. If conditions for sonication are optimized, the lysates can be directly aliquoted for immunoprecipitation or can be stored at −80°C for months.

3.4. Immunoprecipitation

1. Following immunoprecipitation reactions were set up with the equal quantities of the lysates:
   (a) No antibody.
   (b) Nonreactive antibody – rabbit IgG (fraction from the same species in which the specific antibodies were produced).
   (c) Specific antibody – C/EBP β antibody.

   A portion of the supernatant (20% of the sample) is labeled as input and stored to further quantitate the amount of DNA present. If multiple antibodies are to be used with the same DNA preparation, a single mock IP is sufficient as a control for all the antibodies used (see Note 7).
2. The above samples were diluted ten times with ChIP dilution buffer and 5 μg of specific/mock antibody per reaction was added. The mixture was rotated overnight at 4°C by securing the caps tightly.
3. Preclearing ChIP lysates is optional when protein G magnetic beads were being used (for preclearing protocol see Subheading 3.7, step 2) (see Note 8). We highly recommend the use of protein-G-magnetic beads to avoid false positive ChIPs. The chances of sample loss was minimized when Protein-G magnetic beads was used for capturing the ChIP products.
4. The samples were directly incubated with 30 μl of protein G magnetic beads (50% slurry) for 1 h at room temperature with rotation to collect the antibody/protein–DNA complex.
5. The samples with magnetic beads were collected by incubation on "Magnetic stand" for 1 min and the supernatant that contains unbound, nonspecific DNA was carefully aspirated.
6. The samples with beads were washed once with each of the low salt immune complex wash buffer, high salt immune complex

wash buffer, LiCl immune complex wash buffer, and thrice with TE buffer, respectively. A wash consists of resuspending the beads with the specified buffer, rotating for 5 min on a rotating platform at room temperature with 1 ml of the specified buffers, pelleting the beads by incubating for 1 min on magnetic stand, and aspirating the supernatant (see Notes 9 and 10).

3.5. Reverse Cross-linking and Recovery of DNA

1. The protein–DNA complex from the antibody was eluted by adding 250 μl of freshly prepared elution buffer to the above beads. The beads with elution buffer were briefly vortexed and incubated at room temperature for 15 min with rotation. The magnetic beads were pelleted as described above and the supernatant fraction is transferred to new tube and one more elution is repeated. The two supernatants (eluates) were pooled (total volume 500 μl). If the recovered DNA has to be proceeded with cloning see Subheading 3.7, step 3.
2. The pooled eluates and the previously stored input were now reverse cross-linked by incubating at 65°C for 4 h/overnight after adding 20 μl of 5 M NaCl (see Note 11).
3. To the reverse-cross-linked samples 10 μl of 0.5 M EDTA, 20 μl of 1 M Tris–HCl, pH 6.5, and 2 μl of 10 mg/ml proteinase K were added and incubated for 1 h at 45°C. The samples should be handled carefully as some DNA may be lost in subsequent purification steps.
4. DNA in the reverse-cross-linked samples was recovered by column chromatography using Qiagen Qiaquick PCR purification kit. The input and the recovered DNA were recovered with 250 and 50 μl of 10 mM Tris–HCl, pH 8.5, respectively. At this stage, samples are ready for PCR analyses.
5. If heteromeric interaction between different transcription factors binding to a native promoter has to be analyzed, see Note 12.
6. At this step, the samples can be directly amplified to detect the bound DNA or stored for 1 month at -20°C.

3.6. Analysis of the Bound DNA

General strategies of some commonly used techniques for analyzing transcription factor activity were discussed here, including ChIP applications for identifying specific transcription factors (Fig. 4). The detailed protocols for these analyses were out of the realm of this chapter.

1. PCR: The recovered DNA is amplified by PCR using specific primers to the sequence of interest to detect the bound DNA. At least three nonoverlapping primer pairs of about 150 to 200 bp apart for each region of interest should be designed (see Note 13). This also helps for an identification of "real positive" reactions. For PCR, the input (starting material), no antibody, nonreactive antibody, and specific antibody ChIPed

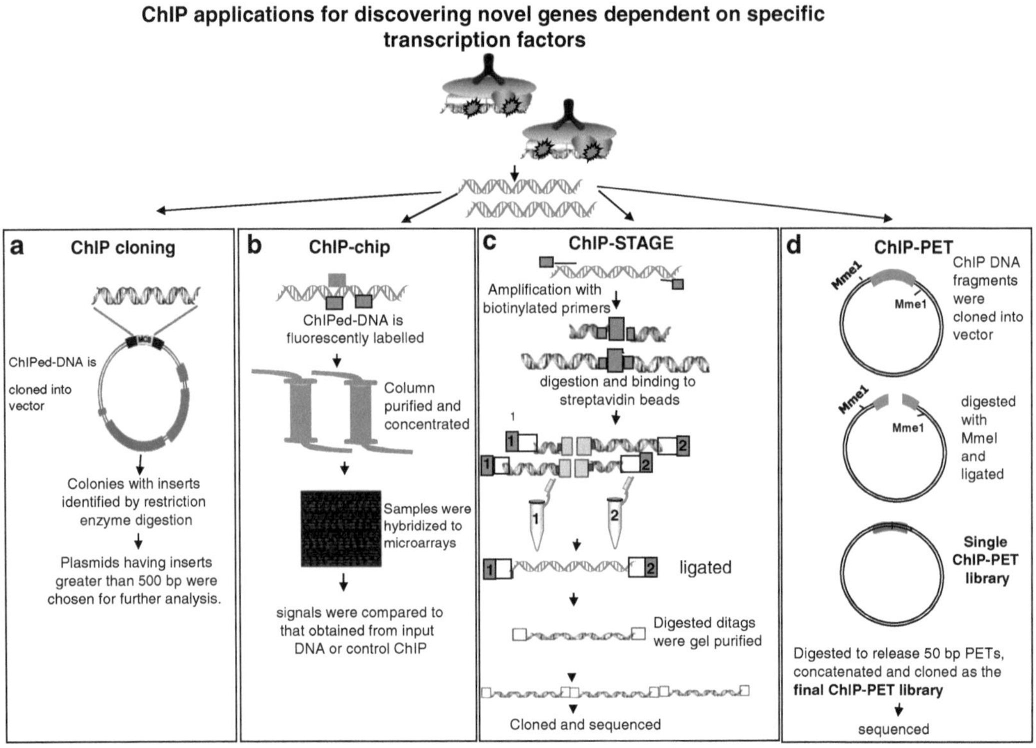

Fig. 4. ChIP applications for discovering novel genes dependent on specific transcription factors (**a**) ChIP cloning: ChIPed DNA was ligated and cloned into vector. Colonies having inserts were identified by restriction enzyme digestion using enzymes in the polylinker. Plasmids with inserts greater than 500 bp were studied further. (**b**) ChIP-chip: The enriched DNA sample is fluorescently labeled, column purified and concentrated. The concentrated samples were hybridized to microarray and signal was compared to that obtained from input DNA or control ChIP. (**c**) ChIP-STAGE: High-throughput sequencing of defined length concatamerized ditags were involved in the STAGE technique. The ChIP DNA was amplified randomly by using biotinylated primers. Amplified DNA fragments were digested and isolated by binding to streptavidin beads. They were ligated separately to one of the two linkers. DNA with linkers were digested and ligated to create 21 bp ditags. Ditags were gel-purified concatamerized, cloned, and sequenced. (**d**) ChIP-PET: ChIP DNA was cloned into vector, which contains two MmeI recognition sites. The ligations were transformed and digested with MmeI and ligated with T4 DNA polymerase. The resulting vector containing a signature tag from each terminal of the ChIP DNA insert was self-ligated and then transformed to form the "single-PET" library. The plasmids from this library were digested to release 50 bp PETs, which were concatenated into long fragments (1–2 kb) and cloned as the final ChIP-PET library for sequencing.

DNA is included. The amount of sample used per reaction must be determined empirically. In general, 5 μl of ChIPed DNA and 1 μl of the input DNA would be sufficient for PCR. PCR conditions vary according to the primers used; generally, 30–35 cycles of amplification will produce in visible bands on agarose gels. The PCR products were resolved on an agarose gel and visualized with ethidium bromide. For analysis, see Note 12.

2. Real-time-PCR (RT-PCR): For RT-PCR, we used SYBR Green Master Mix with 15 μl reactions in 96-well plates on STRATAGENE (default three-step method, 40 cycles) machine. The relative occupancy of the immunoprecipitated factor at a locus is estimated using the following equation 2^(Ct MOCK–Ct

SPECIFIC), where Ct MOCK and Ct SPECIFIC are mean threshold cycles of PCR done in triplicate on DNA samples from mock and specific immunoprecipitations (28) (see Note 14).

3. ChIP cloning: For analysis of the bound DNA by cloning, immunoprecipitated DNA was ligated and cloned into the vector. Colonies having inserts were identified by restriction enzyme digestion using enzymes in the polylinker or by PCR amplification with specific primers. Plasmids having inserts were studied further (14, 29).The putative sites identified can be checked with ChIP assay by synthesizing primers. In spite of some drawbacks with this technique (see Note 15), the main advantage of this protocol lies in that it allows the isolation of known and/or unknown individual target genes.
4. ChIP-chip: Combination of microarray analysis with the ChIP technique led to the improvements in the identification of transcription factor–DNA interactions. After immunoprecipitation, the enriched DNA sample is fluorescently labeled. The control and enriched samples are hybridized to various microarrays and signal was compared to that obtained from input DNA or control ChIP. The two most commonly used microarray formats for the analysis of ChIP-chip samples are PCR product arrays and oligonucleotide arrays. Ideally, to achieve a comprehensive and unbiased glimpse of protein DNA interactions, the DNA microarrays used in ChIP-chip assays should contain elements that represent the entire genome. The in vivo nature of ChIP-chip allows observations into changes in protein–DNA interaction during cellular processes and in response to stimuli (30–33).
5. ChIP-STAGE (for sequence tag analysis of genomic enrichment): Advances in DNA sequencing led to the application of high-throughput DNA sequencing methods to the analysis of protein–DNA interactions. By increasing the number of fragments sequenced, the amount of DNA sampled was increased further enhancing the identification of the enriched DNA fragments. This approach was used to study the regulatory networks of the transcription factors (34, 35).
6. ChIP-PET (For paired-end ditag): In the ChIP-PET method, chromatin bound by DNA-binding proteins is enriched as in ChIP-chip. The enriched DNA sequences are cloned randomly into a plasmid vector, where the insert is flanked by type II restriction enzyme MmeI sites. These plasmids are digested with the restriction enzyme, which cleaves 18 bp into the cloned genomic DNA fragment. The single-stranded plasmid ends are blunt-ended and ligated to form short ditag signature sequences within the plasmid, combining 18 bp from the 5′ end and 18 bp from the 3′ end of the ChIP DNA fragment. Together, the two halves of the ditag signature sequence

represent the full length of the ChIP fragment originally present. The ditags derived from all enriched ChIP DNA fragments are then isolated, concatenated, and cloned into a second plasmid, creating the ChIP-PET library. This library is then extensively sequenced, and the sequences resulting from the enriched ChIP DNA fragments were mapped back to the genome and thus the interacting sites were determined (36, 37).

3.7. Support Protocols for ChIP Assay

1. In vivo cross-linking and harvesting for suspension cells:
   (a) In the fume hood, formaldehyde (37% stock) was added directly to tissue culture media to a final concentration of 1%.
   (b) After adding formaldehyde, the cells were incubated in fume hood on a stir plate for 10′ at RT.
   (c) The cross-linking reaction was stopped by adding powdered glycine directly to flask to a final concentration of 0.125 M continuing to rock/spin for 5′ at RT.
   (d) The cells were collected by centrifugation (2,000 × *g* for 5 min at 4°C), and washing twice with cold 1× PBS containing protease inhibitors (1 mM PMSF, 1× protease inhibitor cocktail).
   (e) The ChIP assay is continued as described in Subheading 3.1 from step 6 onwards.
2. Preclearing ChIP lysates:
   (a) Two micrograms of irrelevant antibody of the same species and isotype as the specific antibody, or normal serum (rabbit is preferred by some researchers) (38) is added to the ChIP lysates.
   (b) The lysates were incubated for 1 h on ice.
   (c) Twenty microliters of protein G magnetic bead slurry was added to the lysates.
   (d) The samples were incubated for 30 min at 4°C with gentle agitation.
   (e) The samples with magnetic beads were incubated for 1 min on magnetic stand and the precleared supernatant is carefully collected.
   (f) The ChIP assay is continued as described in Subheading 3.4 from step 14 onwards.
3. ChIP DNA recovery for cloning:
   (a) The protein–DNA complex from the antibody was eluted by adding 30 μl of freshly prepared elution buffer to the above beads.

(b) The beads with elution buffer were briefly vortexed and incubated at room temperature for 15 min with rotation.

(c) The magnetic beads were pelleted as discussed earlier and the supernatant fraction is transferred to new tube.

(d) The eluate was then diluted with ChIP dilution buffer to 300 μl and proceeded once more with immunoprecipitation as described in Subheading 3.4.

4. Notes

1. The number of cells to be started with has to be determined exponentially and has to be optimized as discussed above. Cell cultures should be healthy and not density arrested prior to cross-linking. In general, we use 2×10^7 cells per antibody per ChIP (fewer cells, as low as 2×10^6 cells, can be used but may result in lower signal to noise ratio).
2. The cross-linking time and formaldehyde concentration can affect both the efficiency of chromatin shearing and the efficiency of precipitating a specific antigen. Shorter cross-linking times (<10 min), lower formaldehyde concentrations or both may improve shearing efficiency; however, for some proteins, especially those that do not directly bind DNA, this might reduce the efficiency of cross-linking and thus the yield of precipitated chromatin. Another point of concern is the loss of antigen epitopes due to denaturation. Formaldehyde is a moderately denaturating agent for proteins and is known to interfere with the secondary and, in particular, tertiary structures, resulting in the unfolding of the protein of interest. The general sensitivity of proteins to formaldehyde in vivo has to be determined empirically.
3. Also, cross-linking for longer periods than 30 min may cause cells to form aggregates that cannot be sonicated efficiently. Quenching with glycine is more critical while working with suspended cells when compared to adherent cells.
4. We employ RIPA buffer to resuspend the cell pellet as it produces less foam during sonication over conventional SDS lysis buffer. The cell material will be very viscous at this stage owing to the release of DNA, and thus will be hard to pipette. The viscosity of the material will decrease after sonication.
5. Sonication conditions must be determined empirically for each cell or tissue type, and sonicator model; optimal average DNA

fragment sizes are 0.3–1 kb. Heating of the sample during sonication, is supposed to denature the proteins so the sample should be held on ice. Foaming should be avoided as it decreases sonication efficiency. Dipping the probe all the way to the bottom avoids foaming. Also 1.5 ml-microcentrifuge tube with conical bottom should be used for sonication as it increases sonication efficiency when compared to 2 ml round bottom ones. Larger volumes for sonication can decrease sonication efficiency. So it is desirable to maintain lower volumes. In our hands, 300–500 μl worked well.

6. The protocol for cell lysis and nuclei preparation in native ChIP assays is described elsewhere (39–41).
7. The amount of antibody added should be in excess of the factor being precipitated, and thus it should be determined empirically for each factor and/or antibody. The incubation time vary from 1 to 12 h and should be determined empirically with each antibody.
8. Preclearing ChIP lysates is a good method to reduce nonspecific backgrounds and eliminate false positives during the assay. The end result will be a lowering of background and an improved signal-to-noise ratio. While it may not be necessary to preclear in all cases, it should be carried out if nonspecific background bands are observed in PCR.
9. Efficient washing is critical to reduce background. The beads should be washed each time with at least 1 ml of the buffer. Transferring the beads to a new tube after the first wash helps to reduce nonspecific bands. The supernatant should be aspirated at each time by placing the tubes on magnetic stand. Care should be taken not to lose beads during the wash steps. If more noise is encountered in negative controls in PCR, then the stringency and/or the number of washings has to be increased.
10. For eliminating false positives during the assay in addition to washings, if two separate (with distinct specificities) antibodies are available for a specific transcription factor, one could ChIP the chromatin with the first antibody. The first round products are re-ChIPed with a second antibody.
11. Reverse cross-linking step is critical as there is chance of sample leakage. The tube caps should be punched and protected by wrapping with parafilm. At this step, the sample can be stored at −20°C and the protocol can be continued next day.
12. To detect the heteromeric interactions between two different transcription factors binding to an element at the native promoter, those two ChIP reactions are first precipitated

with an antibody against factor 1. The immunoprecipitated products from this step are then subjected to ChIP with an antibody against factor 2. Immunoprecipitation reactions with individual factor-specific antibodies and negative control antibodies should be performed to demonstrate the specificity of these reactions. If ChIP reactions with individual antibodies and 2-step ChIP yield similar level of PCR products (as measured by real time PCR), it suggests that two different transcription factors bind to the same promoter elements as heteromer.

13. ChIP PCR primers should be initially tested on genomic DNA for amplification. To achieve high resolution, PCR product size should not be too big usually the amplicon size should be in between 150 and 250 bp. The relative occupancy of a factor at a locus is estimated as the ratio of the intensity of the specific IP band to that of the mock IP band in gel electrophoresis.
14. For real-time-PCR, it is critical that each primer pair yields only one amplicon, which can be determined by analyzing the amplicon dissociation plot, which should yield a single peak. The PCR may then be loaded on a 1% agarose gel to visualize whether a single amplicon is being amplified. Also, Ct values >35 are disregarded and more concentrated DNA should be used in a reanalysis, if necessary. The input sample should ideally be in the 20–30 Ct range and should be adjusted (i.e., diluted) accordingly.
15. In ChIP cloning to decrease the amount of nonspecific DNA, the ChIPed DNA sample may be re-immunoprecipitated, which usually results in a lower amount of precipitated DNA. Also, the size of DNA fragments is critical for ChIP cloning. To reduce the number of nonspecific clones, the chromatin should be sonicated to 1–2 kb fragments and only clones containing DNA fragments of 500 bp or larger should be analyzed further.
16. One should be extremely careful and meticulous throughout this assay. The PCR amplification is highly sensitive. All possible precautions should be taken to prevent contamination from other DNA sources. Amplification in a dedicated space, use of fresh gloves, sterile filter tips, properly calibrated pipettes and clean tubes are highly recommended for this assay.
17. The success of ChIP can be determined from several perspectives and is an art rather than a technique. Sample handling should be appropriate and uniform. For trouble-shooting, see Table 1.

Table 1
Troubleshooting guide for frequent problems encountered with ChIP assays

Problem	Possible reason	Solution
Lower specific IP/ mock IP signal ratio	Antibody is not binding to chromatin due to: (a) Incubation time is too short. (b) Masking of the epitope due to too long cross-linking time. (c) Insufficient amount of chromatin used. (d) Less number of washes leaving high levels of nonspecific binding on beads. (e) Nonspecific binding of the DNA to the agarose beads.	(a) Incubation time of the antibody with the chromatin has to be increased. (b) Cross-linking time has to be decreased. (c) The starting material has to be increased. (d) The specificity of the antibody has to be checked by IP and western blotting. (e) Increase the number/ stringency of the washes and also perform preclearing of the lysates.
Lower PCR signal	(a) Inefficient immunoprecipitation. (b) The DNA may be degraded. (c) The TF is expressed at low abundance. (d) The DNA did not cross-link. (e) The target chosen is not the true target for the DNA binding protein.	(a) The specificity of the antibody has to be checked by IP and western blotting. (b) Proper molecular biology practices should be followed and the samples have to be stored properly. (c) Amount of starting material has to be standardized. (d) Optimal cross-linking time has to be standardized. (e) Known target of the DNA-binding protein may be used to optimize the procedure.
No PCR signal	(a) PCR reaction did not work. (b) Insufficient number of PCR cycles (if not using RT-PCR).	(a) A positive control PCR for master mix and the primers have to be set up with genomic DNA of the same cells. (b) PCR with increasing number of cycles has to be set up until a band is observed.

Acknowledgments

This work is supported by NIH grants CA78282 and CA105005 to D.V.K.

References

1. Fried, M.G. (1989) Measurement of protein-DNA interaction parameters by electrophoresis mobility shift assay. Electrophoresis **10**: 366–76.
2. Christy, R.J., et al. (1991) CCAAT/enhancer binding protein gene promoter: binding of nuclear factors during differentiation of 3T3-L1 preadipocytes. Proc Natl Acad Sci USA **88**: 2593–7.
3. Pillai, S., P. Dasgupta, and S.P. Chellappan (2009) Chromatin immunoprecipitation assays: analyzing transcription factor binding and histone modifications in vivo. Methods Mol Biol **523**: 323–39.
4. Yan, Y., H. Chen, and M. Costa (2004) Chromatin immunoprecipitation assays. Methods Mol Biol **287**: 9–19.
5. Nowak, D.E., B. Tian, and A.R. Brasier (2005) Two-step cross-linking method for identification of NF-kappaB gene network by chromatin immunoprecipitation. Biotechniques **39**: 715–25.
6. Dasgupta, P. and S.P. Chellappan (2007) Chro matin immunoprecipitation assays: molecular analysis of chromatin modification and gene regulation. Methods Mol Biol **383**: 135–52.
7. Kurdistani, S.K. and M. Grunstein (2003) In vivo protein-protein and protein-DNA crosslinking for genomewide binding microarray. Methods **31**: 90–5.
8. Breiling, A., et al. (2001) General transcription factors bind promoters repressed by Polycomb group proteins. Nature **412**: 651–5.
9. Dedon, P.C., et al. (1991) A simplified formaldehyde fixation and immunoprecipitation technique for studying protein-DNA interactions. Anal Biochem **197**: 83–90.
10. Mukhopadhyay, A., et al. (2008) Chromatin immunoprecipitation (ChIP) coupled to detection by quantitative real-time PCR to study transcription factor binding to DNA in Caenorhabditis elegans. Nat Prot **3**: 698–709.
11. Hassan, M.Q., et al. (2004) Dlx3 transcriptional regulation of osteoblast differentiation: temporal recruitment of Msx2, Dlx3, and Dlx5 homeodomain proteins to chromatin of the osteocalcin gene. Mol Cell Biol **24**: 9248–61.
12. Botquin, V., et al. (1998) New POU dimer configuration mediates antagonistic control of an osteopontin preimplantation enhancer by Oct-4 and Sox-2. Genes Dev **12**: 2073–90.
13. Weinmann, A.S. and P.J. Farnham (2002) Identification of unknown target genes of human transcription factors using chromatin immunoprecipitation. Methods **26**: 37–47.
14. Weinmann, A.S., et al. (2001) Use of chromatin immunoprecipitation to clone novel E2F target promoters. Mol Cell Biol **21**: 6820–32.
15. Martone, R., et al. (2003) Distribution of NF-kappaB-binding sites across human chromosome 22. Proc Natl Acad Sci USA **100**: 12247–52.
16. Heckman, C.A. and L.M. Boxer (2002) Allele-specific analysis of transcription factors binding to promoter regions. Methods **26**: 19–26.
17. Johnson, K.D. and E.H. Bresnick (2002) Dissecting long-range transcriptional mechanisms by chromatin immunoprecipitation. Methods **26**: 27–36.
18. Cosma, M.P., T. Tanaka, and K. Nasmyth (1999) Ordered recruitment of transcription and chromatin remodeling factors to a cell cycle- and developmentally regulated promoter. Cell **97**: 299–311.
19. Metivier, R., et al. (2004) Transcriptional complexes engaged by apo-estrogen receptor-alpha isoforms have divergent outcomes. EMBO J **23**: 3653–66.
20. Flanagin, S., et al. (2008) Microplate-based chromatin immunoprecipitation method, Matrix ChIP: a platform to study signaling of complex genomic events. Nucleic Acids Res **36**: e17.
21. Dahl, J.A. and P. Collas (2007) Q2ChIP, a quick and quantitative chromatin immunoprecipitation assay, unravels epigenetic dynamics of developmentally regulated genes in human carcinoma cells. Stem Cells **25**: 1037–46.
22. Gade, P., et al. (2008) Critical role for transcription factor C/EBP-beta in regulating the expression of death-associated protein kinase 1. Mol Cell Biol **28**: 2528–48.
23. Li, H., et al. (2008) The Med1 subunit of transcriptional mediator plays a central role in regulating CCAAT/enhancer-binding protein-beta-driven transcription in response to interferon-gamma. J Biol Chem, **283**: 13077–86.
24. Gade, P., et al. (2009) Down-regulation of the transcriptional mediator subunit Med1 contributes to the loss of expression of metastasis-associated dapk1 in human cancers and cancer cells. Int J Cancer **125**: 1566–74.
25. Orlando, V. (2000) Mapping chromosomal proteins in vivo by formaldehyde-crosslinked-chromatin immunoprecipitation. Trends Biochem Sci **25**: 99–104.
26. Nguyen, J., et al. (2008) Assessment of sera for chromatin-immunoprecipitation. Biotechniques **44**: 66–68.

27. Wells, J. and P.J. Farnham (2002) Characterizing transcription factor binding sites using formaldehyde crosslinking and immunoprecipitation. Methods **26**: 48–56.
28. Nelson, J.D., et al. (2006) Fast chromatin immunoprecipitation assay. Nucleic Acids Res **34**: p. e2.
29. Hug, B.A., et al. (2004) A chromatin immunoprecipitation screen reveals protein kinase Cbeta as a direct RUNX1 target gene. J Biol Chem, **279**: 825–30.
30. Hartman, S.E., et al. (2005) Global changes in STAT target selection and transcription regulation upon interferon treatments. Genes Dev **19**: 2953–68.
31. Odom, D.T., et al. (2004) Control of pancreas and liver gene expression by HNF transcription factors. Science **303**: 1378–81.
32. Euskirchen, G., et al. (2004) CREB binds to multiple loci on human chromosome 22. Mol Cell Biol **24**: 3804–14.
33. Cawley, S., et al. (2004) Unbiased mapping of transcription factor binding sites along human chromosomes 21 and 22 points to widespread regulation of noncoding RNAs. Cell **116**: 499–509.
34. Kim, J., et al. (2005) Mapping DNA-protein interactions in large genomes by sequence tag analysis of genomic enrichment. Nat Methods, **2**: 47–53.
35. Impey, S., et al. (2004) Defining the CREB regulon: a genome-wide analysis of transcription factor regulatory regions. Cell **119**: 1041–54.
36. Wei, C.L., et al. (2006) A global map of p53 transcription-factor binding sites in the human genome. Cell, **124**: 207–19.
37. Loh, Y.H., et al. (2006) The Oct4 and Nanog transcription network regulates pluripotency in mouse embryonic stem cells. Nat Genet **38**: 431–40.
38. Bonifacino, J.S. and E.C. Dell'Angelica (2001) Immunoprecipitation. Curr Protoc Cell Biol, **Chapter** 7: p. Unit 7 2.
39. O'Neill, L.P. and B.M. Turner (2003) Immunoprecipitation of native chromatin: NChIP. Methods **31**: 76–82.
40. Thorne, A.W., F.A. Myers, and T.R. Hebbes (2004) Native chromatin immunoprecipitation. Methods Mol Biol **287**: 21–44.
41. Hebbes, T.R., et al. (1994) Core histone hyperacetylation co-maps with generalized DNase I sensitivity in the chicken beta-globin chromosomal domain. EMBO J, **13**: 1823–30.

Chapter 7

Two-Step Cross-linking for Analysis of Protein–Chromatin Interactions

Bing Tian, Jun Yang, and Allan R. Brasier

Abstract

Eukaryotic gene regulation is controlled, in part, by inducible transcription factor-binding regulatory sequences in a tissue-specific and hormone-responsive manner. The development of methods for the analysis of transcription factor interaction within native chromatin has been a significant advance for the systematic analyses of the timing of gene regulation and studies on the effects of chromatin modifying enzymes on promoter accessibility. Chromatin immunoprecipitation (ChIP) is a specific method involving formaldehyde mediated protein–chromatin fixation to preserve the interaction for subsequent target identification. However, the conventional single-step cross-linking technique does not preserve all protein–DNA interactions, especially for transcription factors in hyper-dynamic equilibrium with chromatin or for coactivator interactions. Here, we describe a versatile, efficient "two-step" XChIP method that involves sequential protein–protein fixation followed by protein–DNA fixation. This method has been used successfully for analysis of chromatin binding for transcription factors (NF-κB, STAT3), polymerases (RNA Pol II), coactivators (CBP/p300, CDK9), and chromatin structural proteins (modified histones). Modifications of DNA extraction and sonication suitable for downstream target identification by quantitative genomic PCR and next generation sequencing are described.

Key words: Chromatin immunoprecipitation, Nuclear factor-κB, Polymerase chain reaction, Next generation sequencing

1. Introduction

Eukaryotic DNA is ordered in a dynamic nucleoprotein complex, "chromatin," that is recognized to play an important mechanism of gene expression control in developmental and physiological stimulation. The first level of chromatin organization is at the nucleosome, a protein structure that regulates access of transcriptional regulators to gene regulatory sequences (1). Chromatin immunoprecipitation (ChIP) is a widely applied technique for the

Ales Vancura (ed.), *Transcriptional Regulation: Methods and Protocols*, Methods in Molecular Biology, vol. 809, DOI 10.1007/978-1-61779-376-9_7,

analysis of native- or cross-linked protein–chromatin interactions. In ChIP, DNA-binding proteins are reversibly fixed to their target DNA. Subsequently, the DNA is sheared, and associated chromatin enriched by immunoprecipitation using specific antibodies directed to protein of interest. The cross-links are then reversed and enrichment of specific DNA is then detected by PCR, array hybridization, or next generation sequencing (NGS).

Although a number of successful studies have employed a ChIP method using a single-step DNA–protein cross-linking reagent (formaldehyde), some highly inducible transcription factors are in hyperdynamic exchange with target DNA and do not cross-link effectively to DNA (2, 3). Similarly, transcriptional coactivators interact with chromatin through protein–protein interactions and therefore do not cross-link well to DNA. To this end, we have optimized a "two-step" XChIP method, where protein–protein interactions are stabilized by cross-linking, followed by formaldehyde-mediated DNA–protein cross-linking.

This method describes our optimization of the two-step XChIP method, a method that has been applied for the identification of NF-κB (4), STAT3 (5), CDK9, CBP/p300, RNA Pol II interactions both in cellulo and in vivo (6). Special modifications of the DNA extraction procedure are needed depending on the downstream method chosen for detection of target gene enrichment (Fig. 1). We focus on the popular quantitative genomic PCR and next generation sequencing, known as "ChIP-Seq" (7).

2. Materials

2.1. Cell Culture and Stimulation

1. Seed the cells 24 h prior to experiment. For epithelial cells, a density of $4–6 \times 10^6$ cells per 100-mm dish works well. This results in ~75% confluence on the day of experiment for adherent cells.
2. Stimulate the cells. For factors requiring stimulation in the absence of serum, cells are changed to growth medium supplemented with 0.5% (w/v) BSA.

2.2. Preparation of Cross-Linking Reagents

1. 0.25 M disuccinimidyl glutarate (DSG) stock (Pierce). Dissolve 50 mg DSG in 100% dimethyl sulfoxide (DMSO) to a concentration of 0.25 M (Table 1). DSG solution is freshly used and excess is discarded. DSG is stored as a dry powder in a dessicated bottle at 4°C.
2. Formaldehyde (FA)-PBS. FA is diluted to 1% final concentration in phosphate-buffered saline (PBS, 137 mM NaCl, 2.7 mM KCl, 4.3 mM Na_2HPO_4, 1.47 mM KH_2PO_4, pH 7.4, Table 2).

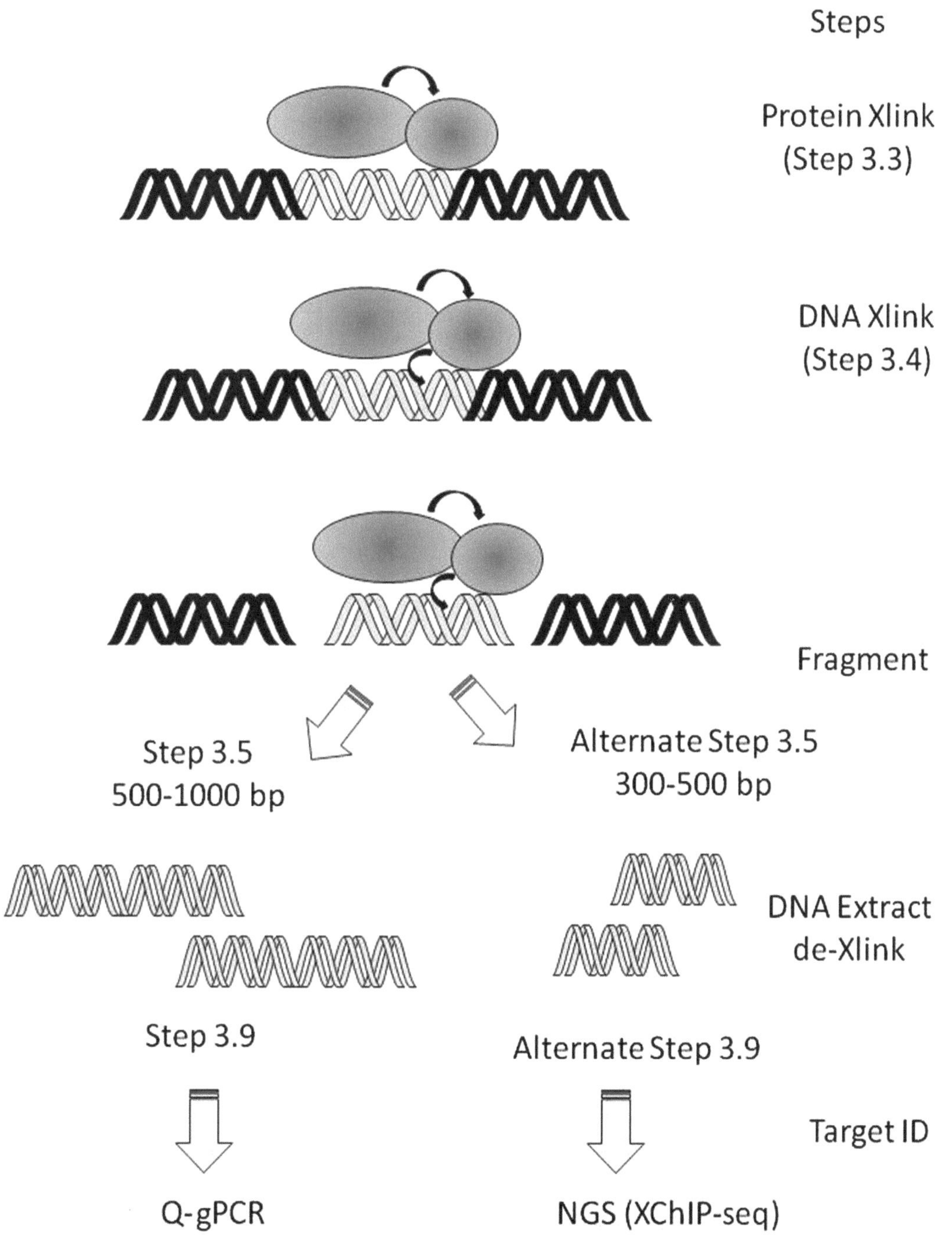

Fig. 1. Overview of the two-step XChIP assay. Shown is a schematic view of the major steps involved in the two-step XChIP assay. Abbreviations: *DMA* dimethyl adipimidate, *DMP* dimethyl pimelimidate, *2HCl DSG* disuccinimidyl glutarate, *DSP* dith iobis(succinimidylpropionate) (Lomant's Reagent), *EGS* ethylene glycol bis(succinimidylsuccinate), *TSAT* tris-(succinimidyl aminotriacetate) (trifunctional).

Table 1
Recipe for DSG solution

Number of plates	DSG (mg)	Vol. DMSO (μL)
2	13	160
4	25	320
6	38	480
12	72	960

Table 2
Recipe for FA-PBS solution

Number of plates	Volume FA	Vol. PBS
2	0.54	20
4	1.08	40
6	1.62	60
12	3.24	120

2.3. Immunoprecipitation Reagents

1. Protease inhibitor cocktail (PIC): Sigma Aldrich Product No. P8340. Stock is stored at −20°C; PIC is added to lysis buffer immediately before use.
2. DynaBeads: Protein A conjugated. Store at 4°C.

2.4. Buffers

1. TE: 10 mM Tris–Cl, pH 7.5, 1 mM EDTA. Store at RT.
2. PBS/$MgCl_2$: 137 mM NaCl, 2.7 mM KCl, 4.3 mM Na_2HPO_4, 1.47 mM KH_2PO_4, pH 7.4, 1 mM $MgCl_2$. Store at RT.
3. Sodium dodecyl sulfate (SDS)-lysis buffer: 1% SDS, 10 mM EDTA, 50 mM Tris–HCl, pH 8.0.
4. Low ionic strength chIP dilution buffer: 50 mM NaCl, 10 mM HEPES, pH 7.4, 1% IGEPAL-CA-630, 10% glycerol. Store at RT. Add PIC immediately before use.
5. LiCl wash buffer: 0.25 M LiCl, 1% IGEPAL CA-630, 1% sodium deoxycholate, 1 mM EDTA, 10 mM Tris–HCl, pH 8.0. Store at RT.
6. High salt wash buffer: 500 mM NaCl, 0.1% SDS, 1% IGEPAL CA-630, 2 mM EDTA, 20 mM Tris–HCl, pH 8.0. Store at RT.
7. Elution buffer: 0.09 M $NaHCO_3$, 1% SDS. Elution buffer is prepared from adding 10% SDS to 0.1 M $NaHCO_3$ immediately before use. Both are stored at RT.

8. Decross-linking mixture. 0.2 M NaCl, 0.1 M EDTA, 0.4 M Tris–HCl, pH 6.8, 0.4 mg/mL proteinase K (diluted from 20 mg/mL stock in H_2O). Decross-linking mixture is prepared immediately before use.

3. Methods

3.1. Experimental Considerations

An overview of the two-step ChIP protocol is shown in Fig. 1. The selection of the cross-linker should be undertaken with some understanding of the chemistry and effective cross-linking length. We have had success with DSG, an irreversible cross-linking agent that cross-links NHS esters with an effective radius of approximately 7 Å. This agent has been useful for highly efficient cross-linking of NF-κB, STAT3, CBP/p300, RNA Pol II, and CDK9 where stimulus – inducible chromatin interactions can be seen (4, 5). A summary of the types of useful cross-linkers, their chemistries, spacing arms, and methods for reversal is shown in Table 3. Some demonstrations of their application have been previously reported (2, 8).

Selection of the appropriate negative control for the immunoprecipitation is an important consideration. We typically include a tube of chromatin immunoprecipitated using pre-immune IgG. This control is important in some approaches for quantification using quantitative real-time genomic PCR (Q-gPCR).

Another critical parameter in the design of the experimental protocol is to decide which type of target identification assay will be employed (Schematically diagrammed in Fig. 1). For downstream

Table 3
Properties of cross-linking reagents. Shown is chemistry, spacer length reversibility and working concentration for popular cross-linking reagents. Abbreviations: DMA dimethyl adipimidate, DMP dimethyl pimelimidate, 2HCl DSG disuccinimidyl glutarate, DSP dith iobis(succinimidylpropionate) (Lomant's Reagent), EGS ethylene glycol bis(succinimidylsuccinate), TSAT tris-(succinimidyl aminotriacetate) (trifunctional)

Cross-linker	Chemistry	Spacer arm (Å)	Reversible?	Working conc. (mM)
Protein–protein				
DMA	Imidoester	8.6	N	10
DMP	Imidoester	9.2	N	10
DSG	NHS-ester	7.7	N	2
DSP	NHS-ester	12	Thiols	2
EGS	NHS-ester	16.1	Hydroxylamine	2
TSAT	NHS-ester	4.2	N	2
Protein–DNA				
Formaldehyde	Methylene bridge	2	65 C + 0.2 M NaCl	1%

analysis using qualitative- or quantitative real-time genomic PCR (Q-gPCR), fragmentation of the chromatin into 500–1,000 bp fragments is optimal. However, for downstream analysis using next generation sequencing, fragmentation into smaller 200–500 bp fragments is preferred. These fragmentation methods are described as alternate protocols (Subheading 3.6). Similarly, the method of decross-linking is different for ChIP-Seq applications, and requires selection of one of the alternate protocols (Subheading 3.11).

3.2. Protein Cross-Linking

1. Wash cells with PBS at room temperature (RT) three times. Tip plate to the side and remove residual solution gently with vacuum aspirator.
2. Add 10 mL of PBS/$MgCl_2$ to each plate. Direct pipette spray to the side of tissue culture dish to avoid disrupting adherent cells (see Note 1).
3. Add 80 μL of freshly prepared 0.25 M DSG solution to each plate with a p200 pipetteman. Rapidly swirl PBS to get the DSG into solution (see Note 2). The final working DSG cross-linking concentration is 2 mM.
4. Incubate at room temperature for 45 min. Under microscopic examination, the cells may become vacuolated, but still should be adherent to plate.

3.3. DNA Cross-Linking

1. Wash with PBS 3×. Tip plate to the side and remove residual solution gently with vacuum aspirator (see Note 3).
2. Add 10 mL of FA-PBS solution to each plate.
3. Incubate cells for 15 min at RT. Formaldehyde reactivity is temperature-dependent; this step is done at 22°C (RT); do not use 37°C.
4. Wash cells three times with PBS (see Note 4).
5. Harvest cells by addition of 1 mL PBS/10-cm^2 plate and scraping with a spatula. Transfer to labeled 1.7-mL Eppendorf tubes.
6. Spin at 900 × *g*, 1 min at RT in a microcentrifuge. Remove PBS carefully not to disturb the cellular pellet.

3.4. Cell Lysis

1. Resuspend pellets in 500 μL of SDS lysis buffer/100 uL PIC mix at room temperature.

3.5. DNA Fragmentation for Quantitative Real-Time Genomic PCR

1. Sonicate the cross-linked cellular lysate four times, 15 s at setting 4 with 10 s break on ice in-between each pulse (Branson sonifier 150, obtained from Branson Ultrasonics, Danbury, CT). Immerse the tip into the middle of the lysate; do not sonicate the surface, as this will make the sample foam. If this occurs, spin sample in microfuge and repeat sonication.

2. Spin at 10.200 × *g*, 10 min at 4°C. Transfer supernatant to a clean, 1.7-mL Eppendorf tube. Be careful not to disturb the pellet.

3.6. (Alternate Protocol) DNA Fragmentation for Chip-Seq

1. The duration and intensity of sonication must be empirically optimized in order to shearing cross-linked chromatin to a size range of 200–500 bp which is ideal for ChIP-seq. It is important to note that the use of two cross-linking agents can make chromatin more resistant to sonication. The concentration of SDS in sonication buffer also has dramatic influence on DNA shearing efficiency. As an example in Fig. 2, we compared DNA fragments obtained from A549 cells after sonication under different SDS concentration (0.1% vs. 1% SDS, Fig. 2a) or different sonication rounds (Fig. 2b). Likewise, over-sonication of chromatin can increase the yield of DNA fragments at right size, but might cause denaturation of proteins and may not produce good signal for ChIP.

3.7. Sizing of Sonicated DNA Fragments

1. Remove 20 μL of sonicated lysate, and add to 200 μL of (200 mM NaCl, 0.5% SDS). Add 1 μL of proteinase K (stock 20 mg/mL in H_2O). Heat at 65°C for 1 h in a heating block.
2. Phenol–chloroform extract the DNA. Add 100 μL of PIC, vortex, spin. Add 1 mL of 100% ethanol, vortex, spin. Dry, run on 1% agarose gel with ethidium bromide.

3.8. Quantification of DNA Recovery

1. Dilute supernatant 1:50 into 100 μL of TE. Read absorbance at 260 nM (A_{260}) in a UV spectrophotometer relative to blank solution (2 μL of SDS-lysis buffer diluted into 100 μL of TE; see Note 5).

3.9. Immunoprecipitation

1. Use equal amounts of lysate (adding equal ODs of DNA) from each treatment for the immunoprecipitation. The maximum amount of lysate added should be no more than 100 μL.
2. Bring total volume up to 900 μL with low ionic strength chip dilution buffer. Add 4 μg of affinity isolated antibody and rotate tubes overnight at 4°C.
3. Dilute herring sperm DNA to a concentration of 100 μg/mL in low ionic strength chip dilution buffer. (Herring Sperm DNA is made at 10 mg/mL in H_2O). Sonicate 3 times, 15 s each pulse. Add 0.5 mL to 40 μL of 50% slurry of prewashed Dynal magnetic beads. Rotate overnight at 4°C.
4. In the morning, wash beads twice with low ionic strength chip buffer, capturing beads on a magnetic stand after each addition of washing buffer. Resuspend to original 50% slurry, and add 40 μL of immunoprecipitation buffer. Rotate 1–4 h, 4°C.

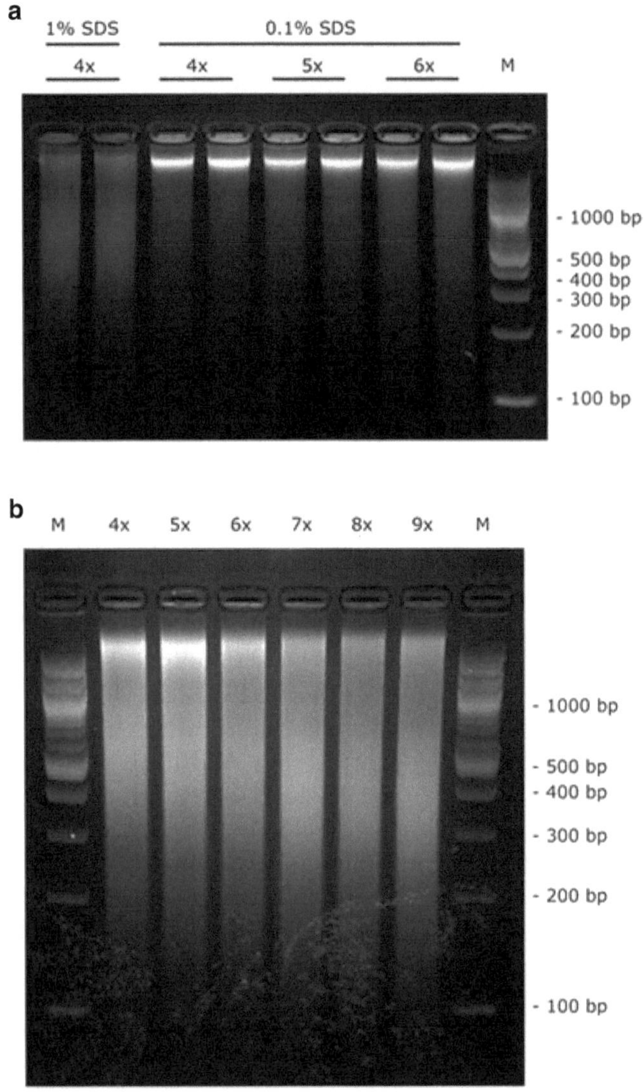

Fig. 2. Sonication efficiency under different conditions. (**a**) Human alveolar epithelial A549 cells were lysed and sonicated in 0.1% or 1% SDS containing buffer using 40% duty cycle and 4-6 rounds for 15 s each. DNA cross-linking was reversed and separated on a 2% agarose gel. 1 kb ladder has been used as a marker of DNA size. Average size of DNA fragments revealed that performing the sonication in a 0.1% SDS containing buffer compromises shearing efficiency. (**b**) A549 cells were lysed in 1% SDS buffer and sonicated for 4-9 rounds. Here, seven rounds of sonication is better for fragmentation into a size range of 200–500 bp.

5. Spin the immunoprecipitates at 2,400 × *g* in microfuge for 1 min at RT. Save 50 μL for input control.
6. Wash beads two times with 500 μL of low ionic strength ChIP dilution buffer (invert tubes before spin.). After first wash, transfer to fresh prelabeled 1.7-mL Eppendorf tube.
7. Wash beads once with 500 μL of high salt buffer.
8. Wash beads once with 500 μL of LiCl buffer.

9. Wash beads 2× with 500 μL of TE buffer.
10. Add 250 μL of elution buffer, vortex, and incubate 15 min at RT.
11. Transfer supernatant to clean Eppendorf tube. Repeat elution with another 50 μL of elution buffer, 15 min at RT. Combine eluates.

3.10. Decross-Link Sample for Q-gPCR

1. Add 51 μL of decross-linking mixture to eluates and incubate at 65°C for 1–4 h.

3.11. (Alternate Protocol) Decross-Link Sample for ChIP-Seq

A high background signal of ChIP-Seq may originate from nondigested RNA since the Klenow fragment used for generating ChIP-seq sequencing library can also react with RNA. Therefore, it is better to include RNase A treatment before proteinase K digestion in order to potentially reduce the RNA caused background (9).

To do this:

1. Add RNase A (Sigma) to the eluate at a final concentration of 20–50 μg/mL. Mix and incubate at 37°C for 30 min.
2. Add proteinase K to a final concentration of 0.4 mg/mL and incubate at 65°C for 1–4 h.

Generally, ChIP DNA as obtained by standard ChIP protocol is of adequate purity for quantitative PCR analysis but often contains contaminants that interfere with subsequent steps required for ChIP-Seq analysis. DNA purification through commercially available clean-up kits, such as Qiagen QiaQuick spin columns, typically eliminates these contaminants and thus ensures a success.

3.12. Decross-Linking Input

1. To process input decross-link in 200 mM NaCl 65°C for 1 h with proteinase K.

3.13. DNA Extraction

1. Add 100 μL of phenol/chloroform to each sample (see Note 6).
2. Vortex/spin at 16,200 × *g* for 3–5 min. (If solution becomes white, add 100 μL of TE, vortex and spin again).
3. Transfer top layer to clean Eppendorf tube.
4. Add 100 μL of chloroform to each sample/vortex/spin and transfer to fresh tube.
5. Add 1 mL of 100% ethanol and 100 μL of 7.5 M ammonium acetate and 3 μL of glycogen (20 mg/mL in H_2O; see Note 7).
6. Incubate at −80°C for 30 min.
7. Spin at 13,000 rpm in a benchtop microfuge for 10 min, 4°C.
8. Wash pellets with 1 mL of 100% ethanol.
9. Spin at 13,000 rpm for 10 min, 4°C.
10. Resuspend pellets in 20–50 μL of TE or H_2O.

3.14. Target Analysis by Quantitative Real-Time Genomic PCR

Q-gPCR is the standardly accepted assay used to accurately measure the amount of enrichment of a particular DNA sequence by ChIP in this protocol. The principle of Q-gPCR is based on detection of a fluorescent signal produced proportionally during amplification of a PCR product. The use of Q-gPCR has several advantages: short turnaround time for data acquisition and analysis; reliable results compared to traditional PCR methods; and it is quantitative. However, as the procedure is very sensitive, there is an increased chance of variability of the results. There are several aspects of Q-gPCR that need to be considered for reliable and reproducible results. They are: (1) primer design using dedicated softwares, some of which are freely available on the Internet; (2) deciding on positive and negative controls; (3) technical and biological repeats, and (4) analyses of the data.

3.14.1. Designing Q-gPCR Primers for Genomic DNA Regions of Interest

1. Primers are designed encompassing the predicted transcription factor-binding site if the latter information is available. As a control, we use primers designed to amplify the 3′-untranslated region (3′-UTR) of the target gene.
2. We regularly use NCBI/Primer-BLAST in NCBI Web site for designing real-time PCR primer pairs that are typically 20–25 mers with a Tm range of 58–63°C (maximum Tm difference 2°C) and that amplify a product of 50–150 bp. The melting temperature is calculated based on the nearest-neighbor thermodynamic parameters. The GC content is set between 40 and 55%.
3. Order primers with standard desalting purification.
4. The Q-gPCR primers are dissolved in TE buffer at 8 μM of stock concentrations. Each pair of primers was examined by making standard curve of CT using a series of diluted human genomic cDNA (Fig. 3). Only the primers that its CT value showed excellent correlation with the concentrations of human genomic cDNA are used for Q-gPCR in this protocol.

3.14.2. Equipment and ChIP-Q-gPCR Reagents

1. Bio-Rad MyiQ Real-Time PCR Detection System is used in this protocol.
2. Bio-Rad iQ SYBR Green Supermix – a prepackaged PCR mix containing dNTPs, 50 U/mL iTaq DNA polymerase, 6 mM $MgCl_2$, SYBR Green I, and 20 nM fluorescein.
3. Mix each primer pair (forward and reverse of experimental or control) in the same tube and prepare a stock concentration of 8 μM each with 10 mM Tris–HCl, pH 8.5.
4. We use purified human genomic cDNA as positive control and negative control is chromatin immunoprecipitated with IgG.
5. Bio-Rad 96-well clear multiplate PCR plate are used; the plate is sealed by Bio-Rad Microseal B film before putting into PCR machine.

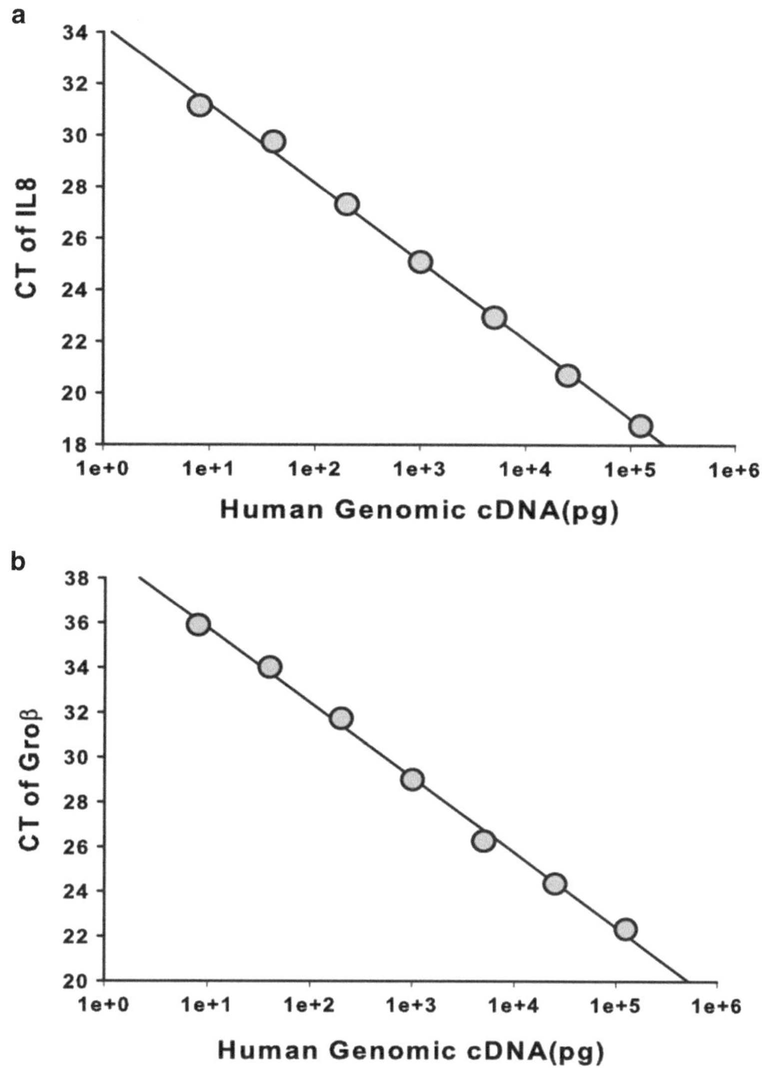

Fig. 3. Dynamic range of Q-RT-gPCR assay. To check the Q-RT-gPCR dynamic range of primers designed in this protocol, each primer pair can be used to amplify a series of diluted human genomic DNA. Plotted is the mean Ct value of duplicate samples for each log-transformed concentration. Note the wide range of linear response. (**a**) IL8 primers; (**b**) Groβ primers.

3.14.3. Real-Time PCR Set Up

1. Set up real-time PCRs in triplicate with the diluted input DNA, as shown in Table 4. The total reaction volume in each tube is 20 μL (see Notes 8–10).
2. Select the two step-melting PCR protocol from iQ5 software window to run Q-gPCR.

Table 4
Recipe for Q-gPCR reaction mixture

Component	Amount/reaction (μL)	Final
Primer mix (stock 8 μM each)	1	400 nM
DNA sample (ChIP or diluted "Input" DNA)	3	
Nuclease-free water	6	
SYBR Green Supermix	10	
Total volume	20	

3.14.4. Data Analysis

1. After the cycling is complete, first visualize the dissociation plot. The dissociation plot for a primer should produce a single sharp peak. More than one peak represents additionally amplified products and the amplification data should not be used.
2. The next step is to adjust the baseline. The default baseline is set from cycles 3–15. To find out whether the baseline needs to be adjusted, it is important to find out which reaction emerges before the default baseline. For this, change the plot to linear view. For reactions that emerge after 15 cycles, no changes have to be made. For highly abundant targets, the end point of the baseline (15 is the default) needs to be adjusted. Usually, the baseline is adjusted to one or two cycles before the earliest amplification.
3. The threshold cycle (Ct; the fractional cycle number at which the fluorescence passes the threshold) value is determined in the experimental report.
4. Export the results into Microsoft Excel for analysis. The Ct values for the technical repeats should show minimal variation (ideally be within half a Ct value of each other).
5. Calculate the ΔCt value (normalized to the input samples) for each sample: ΔCt (Ct (sample) − Ct (input)]. Next, calculate the ΔΔCt (ΔCt (experimental sample) − ΔCt (negative control)) (see Note 11).
6. Calculate the fold difference between experimental sample and negative control using the equation $2^{-\Delta\Delta Ct}$. As indicated earlier, the negative control is DNA immunoprecipitated with control IgG.

3.15. Target Analysis by Next Generation Sequencing

ChIP coupled with NGS, known as ChIP-Seq, enables the unbiased identification of protein–chromatin interactions on a global basis. When compared with microarray analysis of ChIP (ChIP-chip), ChIP-Seq is less expensive, has higher signal/noise ratios and requires lower amounts of ChIP DNA for genome-wide analysis (7). This process is initiated by adding sequencing adapters to blunt-ended ChIP DNA. Subsequently, PCR is performed for limited number of cycles to generate the sequencing library which is ready to be sequenced, e.g., using the procedure developed for the Illumina Genome Analyzer (10). In general, ~10–20 million sequence reads (for 36 cycles of sequencing) are obtained for each sample per run. The sequence of ChIP DNA fragments is then mapped back to the reference genome and further visualized using a genome browser to identify the binding sites.

Although commercial services are currently available for ChIP-seq analysis starting from purified ChIP DNA, many considerations on ChIP DNA preparation have to be taken into account in order to achieve successful ChIP-Seq results.

3.15.1. Materials

1. ChIP DNA sample from Subheading 3.11–3.13 (approximately 10 ng in 42.5 μL of water), the quality should always be validated (e.g., by real-time PCR) before performing the library preparation steps for high-throughput sequencing.
2. NEB Next™ DNA Sample Prep Master Mix Set 1 (New England Biolabs).
3. T4 DNA ligase (Enzymatics).
4. QIAquick PCR purification Kit (Qiagen).
5. Adapter oligonucleotides, Illumina PE PCR primers 1.0 and 2.0.
6. E-Gel® SizeSelect™ 2% Starter kit (Invitrogen).
7. Agilent Technologies 2100 Bioanalyzer.
8. Illumina Genome Analyzer II system and associated equipment, reagents, kits, and software.

3.15.2. End Repair

1. Prepare the following reaction mix:
   - ChIP enriched, Q-gPCR verified DNA (42.5 μL).
   - NEB Next End Repair Reaction Buffer (5 μL).
   - NEB Next End Repair Enzyme Mix (2.5 μL).

   The total volume should be 50 μL.
2. Incubate in a thermal cycler for 30–45 min at 20°C.
3. Purify DNA on QIAquick column and elute in 42 μL of elution buffer.

3.15.3. dA-Tailing

1. Prepare the following reaction mix:
   - End repaired, blunt DNA (42 μL).
   - NEB Next dA-Tailing Reaction Buffer (5 μL).
   - Klenow exo (3 μL).

   The total volume should be 50 μL.
2. Incubate for 30 min at 37°C.
3. Purify the reaction product on QIAquick column and elute in 34 μL of elution buffer.

3.15.4. Adapter Ligation

1. Prepare the following reaction mix:
   - dA-tailed DNA (34 μL).
   - Enzymatics 2× rapid ligation buffer (40 μL).
   - 2.5 μM adapter oligo mix (1 μL).
   - Enzymatics T4 DNA ligase (5 μL).

   The total volume should be 80 μL.
2. Incubate for 15 min at room temperature.
3. Purify DNA on QIAquick column and elute in 50 μL elution buffer.

3.15.5. Gel-Purification of Ligation Products

1. On E-Gel® SizeSelect™ 2% Agarose (Invitrogen), load 100–250 ng of DNA ladder diluted in 5–10 μL into the middle well (1:50, 10 μL).
2. Load 25 μL of samples into every other lane of the gel.
3. Load 25 μL of deionized water into all empty wells.
4. Run SizeSelect 2% (program 8) and set time according to the estimation table.
5. Extract DNA fragment at approximately 400 bp and/or 600 bp in size.

3.15.6. PCR Enrichment

1. Prepare the following reaction mix on ice:
   - Gel extracted DNA (24 μL).
   - 25 μM PCR Primers 1.0 and 2.0 Mix (1 μL).
   - 2× Phusion™ High-Fidelity PCR Master Mix (25 μL).

   The total volume should be 50 μL.
2. Amplify using the following PCR protocol:
   - 30 s at 98°C.
   - 6–18 cycles of 10 s, 98°C; 30 s, 65°C; 30 s, 72°C.
   - 5 min at 72°C.
3. Purify DNA on QIAquick column and elute in 50 μL of elution buffer.

3.15.7. Library Validation

1. Load 1 μL of the resuspended construct and 1 μL of the negative control on an Agilent Technologies 2100 Bioanalyzer using the high-sensitivity DNA1000 kit.
2. Check the size, purity, and concentration of the sample.

3.15.8. Sequencing and Data Analysis

1. Sequence adapter ligated ChIP DNA using the Illumina Genome Analyzer II.
2. The sequence reads obtained after base-calling can be aligned to a reference genome and further visualized using, for example the UCSC or Ensembl genome browser.

3. Several algorithms can be utilized to analyze ChIP-Seq data and identify the enriched regions across the whole genome, thus harbor the protein–chromatin interaction in vivo.

4. Notes

1. Inclusion of divalent cations is important to keep cells adherent during fixation. $CaCl_2$ can be substituted for $MgCl_2$ if necessary.
2. Addition of DSG solution will become milky immediately on contact with PBS, but should become clear with rapid mixing. There should be no precipitate under the microscope.
3. DSG must be efficiently washed prior to addition of formaldehyde; DSG will react with formaldehyde and reduce its effective concentration.
4. We do not specifically neutralize FA; SDS-lysis buffer contains Tris; the free amines in Tris neutralize the cross-linking reaction.
5. A_{260} readings of 0.1–0.8 are generally within the linear range; if the chromatin is outside of this range, adjust the dilutions accordingly and repeat the A_{260} measurement.
6. During phenol–chloroform extraction, Phase Lock Gel Light tubes (Flowgen Bioscience, FPR5101) can be used easily to eliminate interphase-protein contamination and increase the yields of up to 30%.
7. GlycoBlue (Ambion, AM9516), a glycogen covalently linked to a blue dye, can be used instead of glycogen as carrier for the ethanol precipitation. It substantially facilitates nucleic acid precipitation while increasing the size and visibility of the pellet.
8. Eliminate bubble before running PCR because it will affect the final Ct results.
9. Make mixtures for PCR setup to reduce the differences among reactions.
10. If the initial set of 400 nM primer does not work, optimize the primer concentration.
11. There are two common normalization methods for ChIP analysis using Q-gPCR – the percent input method and the fold enrichment method. With the percent input method, signals obtained from the ChIP are divided by signals obtained from an input sample. This input sample represents the amount of chromatin used in the ChIP. Typically, 1% of starting chromatin is used as input. In this case, 6.644 cycles (i.e., $\log_2$ of 100) must be subtracted from the Ct value of diluted input

before calculating delta Ct. The fold enrichment method is also called "signal over background" method. With this method, the ChIP signals are divided by the mock antibody signals, representing the ChIP signal as the fold increase in signal relative to the background signal. The assumption of this method is that the level of background signal is reproducible between different primer sets, samples, and replicate experiments. Combined these two methods together, another method aims to end up with a signal over noise ratio normalized to total input DNA signal.

Acknowledgments

This work was supported, in part, by NIH grants AI062885 (A.R.B.), NHLBI contract BAA-HL-02-04 (A.R.B.), and ES06676 (to K. Elferink, UTMB).

References

1. Kornberg, R.D., and Lorch, Y. (1999) Twenty five years of the nucleosome, fundamental particle of the eukaryote chromosome. *Cell* **98**:285–294.
2. Nowak, D.E., Tian, B., and Brasier, A.R. (2005) Two-step cross-linking method for identification of NF-κB gene network by chromatin immunoprecipitation. *Biotechniques* **39**:715–725.
3. Bosisio, D., Marazzi, I., Agresti, A., Shimizu, N., Bianchi, M.E., and Natoli, G. (2006) A hyper-dynamic equilibrium between promoter-bound and nucleoplasmic dimers controls NFκB-dependent gene activity. *EMBO J* **25**:798–810.
4. Nowak, D.E., Tian, B., Jamaluddin, M., Boldogh, I., Vergara,L. A., Choudhary, S., and Brasier, A.R. (2008) RelA Ser276 phosphorylation is required for activation of a subset of NFκB-dependent genes by recruiting cyclin-dependent kinase 9/cyclin T1 complexes. *Mol Cell Biol* **28**:3623–3638.
5. Hou, T., Ray, S., and Brasier, A.R. (2007) The functional role of an IL-6 inducible CDK9-STAT3 complex in human fibrinogen gene expression. *J. Biol. Chem.* **282**:37091–37102.
6. Ramadoss, P., Chiappini, F., Bilban, M., and Hollenberg,A.N. (2010) Regulation of hepatic six transmembrane epithelial antigen of prostate 4 (STEAP4) expression by STAT3 and CCAAT/enhancer-binding protein+¦. *J. Biol. Chem.* **285**:16453–16466.
7. Johnson, D.S., Mortazavi, A., Myers, R.M., and Wold,B. (2007) Genome-wide mapping of *in vivo* protein-DNA interactions. *Science* **316**:1497–1502.
8. Zeng, P.Y., Vakoc, C.R., Chen, Z.-C., Blobel, G.A., and Berger, S.L. (2006) *In vivo* dual cross-linking for identification of indirect DNA-associate proteins by chromatin Immunoprecipitation. *Biotechniques* **41**:694698.
9. Waldminhaus, T., and Skarstad, K. (2010) ChIP on Chip: surprising results are often artifacts. *BMC Genomics* **11**:414.
10. Schmidt, D., Wilson, M.D., Spyrou, C., Brown, G.D., Hadfield, J., and Odom, D.T. (2009) ChIP-seq: using high-throughput sequencing to discover protein-DNA interactions. *Methods* **48**:240–248.

Chapter 8

Chromatin Immunoprecipitation Analysis of NFκB Transcriptional Regulation by Nuclear IκBα in Human Macrophages

Sitharam Ramaswami, Subrata Manna, Ashish Juvekar, Steven Kennedy, Ales Vancura, and Ivana Vancurova

Abstract

Transcription factor NFκB comprises a family of proteins that serve as crucial regulators of genes involved in host immune and inflammatory responses, cell survival, proliferation, and differentiation. Since transcription of NFκB-dependent genes is increased in numerous inflammatory disorders as well as in many types of cancer and leukemia, inhibition of NFκB-dependent transcription thus represents an important therapeutic target. We have previously shown that in human leukocytes, transcription of NFκB-dependent genes is inhibited by the nuclear translocation and accumulation of IκBα, which can be induced by an inhibitor of CRM1-dependent nuclear export, leptomycin B (LMB). In this chapter, we describe a protocol that uses chromatin immunoprecipitation (ChIP) to analyze the regulation of NFκB recruitment to NFκB-dependent promoters by nuclear IκBα induced by LMB. We show that in lipopolysaccharide (LPS)-stimulated human U-937 macrophages, recruitment of NFκB p65 and p50 proteins to NFκB-dependent promoters of IκBα and cIAP2 genes is suppressed by the LMB-induced nuclear IκBα. Even though in this study we use U-937 macrophages, this protocol should be readily adaptable to analyze the regulation of NFκB recruitment by nuclear IκBα also in other cell types.

Key words: Chromatin immunoprecipitation, IκBα, Leptomycin B, Macrophages, NFκB, Nuclear accumulation, Nuclear transport, Transcriptional regulation

1. Introduction

NFκB is a key transcriptional regulator of genes involved in immune and inflammatory responses, as well as genes regulating cell proliferation and survival (1–5). Sustained activation of NFκB is associated with the pathogenesis of numerous inflammatory disorders, such as sepsis, arthritis, or chronic lung disease, as well as with many types of cancer and leukemia (6–10). Since inhibition of

Ales Vancura (ed.), *Transcriptional Regulation: Methods and Protocols*, Methods in Molecular Biology, vol. 809,
DOI 10.1007/978-1-61779-376-9_8, © Springer Science+Business Media, LLC 2012

NFκB activity reduces pro-inflammatory and anti-apoptotic gene expression, NFκB appears to be one of the most promising targets in the treatment of many inflammatory disorders as well as different types of leukemia and cancer.

The NFκB protein family is composed of five distinct DNA-binding subunits, called NFκB p50, NFκB p52, p65 (RelA), c-Rel, and relB (1–5). These NFκB proteins are characterized by a highly conserved N-terminal region, designated as Rel homology domain, which mediates DNA binding, nuclear localization, subunit dimerization, and interaction with the inhibitory proteins IκB. NFκB subunits can form homo- or heterodimers, and dimerization is a prerequisite for binding to DNA (1–3). In the classical model of NFκB activation, NFκB exists in the cytoplasm of unstimulated cells in an inactive form associated with the inhibitory protein IκBα (1–3). Following cell stimulation by extracellular stimuli, including inflammatory cytokines and bacterial products, IκBα is phosphorylated, ubiquitinated, and selectively degraded by the 26S proteasome (1–5). This results in unmasking of the nuclear localization sequence of NFκB dimers, their translocation to the nucleus and induction of transcription of NFκB-dependent genes. One of the first genes induced following NFκB activation is IκBα, since the IκBα promoter also contains the NFκB binding region (1). This newly synthesized IκBα can then enter the nucleus, remove NFκB from gene promoters, and transport NFκB proteins back to the cytoplasm, representing a crucial regulatory mechanism also called "post-induction repression," which terminates NFκB activation during persistent stimulation and limits the NFκB response (11–14).

In vitro studies of NFκB DNA binding activity using electrophoretic mobility shift assay (EMSA) showed that individual dimers bind various promoter sites with differential affinity (15–17). However, in vivo experiments demonstrated that the ability of specific genes to selectively recruit certain NFκB dimers cannot be predicted on the basis of the in vitro EMSA data or overexpression experiments (18–20). Studies from our laboratory have shown that NFκB activity is inhibited by the nuclear translocation and accumulation of IκBα, which can be induced by two independent mechanisms: the previously described inhibition of nuclear IκBα export by leptomycin B (LMB) (21–24), and by a previously unrecognized novel mechanism consisting of the inhibition of the 26S proteasome (25).

In this chapter, we describe a protocol that uses chromatin immunoprecipitation (ChIP) to analyze the regulation of NFκB recruitment to responsive promoters by the LMB-induced nuclear IκBα. We show that in human U-937 macrophages, recruitment of NFκB p65 and p50 proteins to promoters of IκBα and the anti-apoptotic gene cIAP2 is suppressed by the LMB-induced nuclear IκBα. The main points of this protocol are: (1) cell culture and incubation of differentiated U-937 macrophages with LMB, which induces the nuclear accumulation of IκBα; (2) ChIP analysis of NFκB and IκBα recruitment to NFκB-dependent promoters using p65 and p50 NFκB and IκBα antibodies; and (3) quantitative real-time polymerase chain reaction (PCR) analysis using primers for IκBα and cIAP2 gene promoters.

2. Materials

2.1. Cell Culture

1. U-937 cells obtained from the American Type Culture Collection (Manassas, VA).
2. RPMI medium (Gibco, Bethesda, MD) supplemented with 10% fetal bovine serum (FBS), 2 mM glutamine, 1 mM sodium pyruvate, and penicillin–streptomycin solution.
3. Phorbol 12-myristate 13-acetate (PMA, Calbiochem) is dissolved at 1 mg/ml in dimethyl sulfoxide (DMSO) and stored at −80°C.
4. Lipopolysaccharide (LPS, Sigma, L-4391) is dissolved at 1 mg/ml in sterile water and stored at −80°C.
5. LMB (Sigma, L-2913) (see Note 1).
6. Phosphate buffered-saline (PBS, pH 7.4).
7. 75 cm^2 culture flasks.
8. Standard 6-well plates with clear flat bottom.
9. Teflon cell scrapers.
10. 2 ml microcentrifuge tubes.

2.2. Chromatin Immunoprecipitation

2.2.1. Protein-DNA Cross-Linking

1. 37% formaldehyde solution (Sigma): Aliquot in sterile 15 ml tubes in the laminar flow cabinet and store at a room temperature.
2. 2.5 M glycine solution: Dissolve 3.75 g of glycine in 20 ml of deionized water and store at a room temperature.

2.2.2. Cell Lysis and DNA Shearing

1. PBS, pH 7.4.
2. 100 mM phenylmethylsulfonyl fluoride (PMSF): Dissolve 17 mg of PMSF in 1 ml of absolute ethanol. Store at −20°C (it will not freeze) (see Note 2).
   - Protease inhibitor cocktail for mammalian cell extracts containing pepstatin A, bestatin, leupeptin, aprotinin, 4-(2-aminoethyl)-benzenesulfonyl fluoride (AEBSF) and trans-epoxysuccinyl-L-leucylamido (4-guanidino) butane (E-64), obtained from Sigma (P-8340) (see Note 3).

3. SDS lysis buffer: 1% (w/v) SDS, 10 mM EDTA and 50 mM Tris–HCl, pH 8.1.
4. ChIP dilution buffer: 0.01% (w/v) SDS, 1.1% (v/v) Triton X-100, 1.2 mM EDTA, 16.7 mM Tris–HCl, pH 8.1, and 167 mM NaCl.
5. 15 ml centrifuge tubes.
6. 1.5 ml microcentrifuge tubes.
7. Vibra cell sonicator.

2.2.3. Preclearing and Immunoprecipitation

1. Protein A/G PLUS-Agarose (Santa Cruz Biotechnology).
2. Immunoprecipitating antibodies: NFκB p65 (Santa Cruz Biotechnology, sc-372), NFκB p50 (Santa Cruz Biotechnology, sc-7178) and IκBα (Santa Cruz Biotechnology, sc-371).

2.2.4. Washing

1. Triton® X-100 (Sigma).
2. IGEPAL CA-630 (Sigma).
3. Low salt immune complex wash buffer: 0.1% (w/v) SDS, 1% (v/v) Triton X-100, 2 mM EDTA, 20 mM Tris–HCl, pH 8.1, and 150 mM NaCl. Adjust pH to 8.1. Store at 4°C. Add protease inhibitors (2 mM PMSF and 2% (v/v) protease inhibitor cocktail from Sigma (P-8340)) just before use.
4. High salt immune complex wash buffer: 0.1% (w/v) SDS, 1% (v/v) Triton X-100, 2 mM EDTA, 20 mM Tris–HCl, pH 8.1, and 500 mM NaCl. Adjust pH to 8.1. Store at 4°C. Add protease inhibitors (2 mM PMSF and 2% protease inhibitor cocktail from Sigma (P-8340)) just before use.
5. LiCl immune complex wash buffer: 0.25 M LiCl, 1% (v/v) IGEPAL CA-630, 1% (w/v) deoxycholic acid (Na salt), 0.1% (w/v) SDS, 1 mM EDTA and 10 mM Tris–HCl, pH 8.1. Adjust pH to 8.1. Store at 4°C. Add protease inhibitors (2 mM PMSF and 2% protease inhibitor cocktail from Sigma (P-8340)) just before use.
6. 1× Tris–EDTA (TE) buffer: 1 mM EDTA and 10 mM Tris–HCl, pH 8.1. Adjust pH to 8.1. Store at 4°C.

2.2.5. Elution and Reverse Cross-Linking

1. Elution buffer: 1% SDS and 0.1 M $NaHCO_3$. Add 100 µl of 10% SDS and 100 µl of 1 M $NaHCO_3$ to 0.8 ml of deionized water to make 1 ml of elution buffer. Prepare fresh before the reaction; each sample requires 500 µl of elution buffer (see Note 4).
2. 5 M NaCl: Dissolve 29.2 g of NaCl in 100 ml of deionized water. Store at 4°C.
3. 0.5 M EDTA: Dissolve 18.6 g of ethylenediaminetetraaceticacid (EDTA) disodium salt dihydrate (Sigma) in 100 ml of deionized water. Store at 4°C.

4. 1 M Tris–HCl, pH 6.5: Dissolve 12.1 g of Tris base in 80 ml of deionized water. Adjust pH to 6.5 with 6 N HCl. Bring total volume to 100 ml, filter, and store at 4°C.
5. Proteinase K solution, 20 mg/ml (Invitrogen).
6. 1.5 ml microcentrifuge tubes.

2.2.6. DNA Extraction

1. Phenol: chloroform: isoamyl alcohol 25:24:1, pH 8.0 (Sigma).
2. Chloroform: isoamyl alcohol 24:1 (Sigma).
3. Ethanol, absolute.
4. 70% ethanol: Add 70 ml of absolute ethanol to deionized water to make up a volume of 100 ml. Store at 4°C.
5. GenElute™-LPA (Sigma, 56575).
6. 3 M sodium acetate (NaOAc): Dissolve 6.15 g of anhydrous NaOAc (Sigma) in 25 ml of deionized water. Adjust pH to 5.2 using glacial acetic acid. This stock solution can be stored at room temperature.
7. Nuclease-free water (Ambion).
8. 1.5 ml microcentrifuge tubes.

2.3. Real-Time PCR

1. iQ™ SYBR® Green Supermix (Bio-Rad).
2. iCycler iQ™ PCR Plates, 96 well (Bio-Rad).
3. iCycler iQ™ Optical tape (Bio-Rad).
4. 1.5 ml microcentrifuge tubes.

3. Methods

In this section, we describe the protocol for ChIP analysis of NFκB transcriptional regulation by nuclear IκBα in human U-937 macrophages. However, this protocol can easily be modified and used for other cells as well. In cells that do not exhibit constitutive NFκB activity, such as unstimulated U-937 cells, NFκB activity is first induced by LPS or other pro-inflammatory agents. The nuclear translocation of IκBα is then induced by LMB. Recruitment of NFκB subunits and IκBα to NFκB-responsive promoters is measured by ChIP and analyzed by quantitative real-time PCR. On average, this protocol can be accomplished within 4–6 days.

Figure 1 illustrates recruitment of p65 NFκB, p50 NFκB, and IκBα to the promoter of IκBα in U-937 macrophages stimulated with LPS in the absence and presence of LMB. Both p65 and p50 NFκB are recruited to IκBα promoter within 1 h after LPS stimulation. However, while p65 NFκB recruitment is greatly inhibited by the LMB-induced nuclear IκBα, p50 recruitment is suppressed

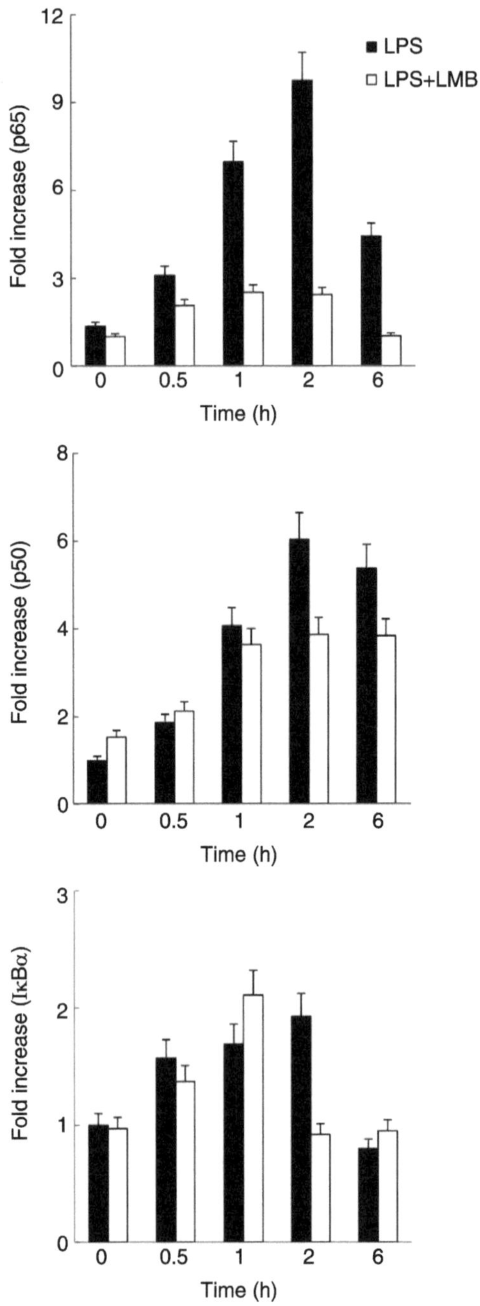

Fig. 1. Recruitment of NFκB p65, NFκB p50 and IκBα proteins to IκBα promoter in LPS-stimulated U-937 macrophages. U-937 cells (5×10^6) were differentiated with 10 ng/ml PMA for 24 h, and stimulated with LPS (1 μg/ml) in the absence (full columns) or presence (empty columns) of 10 nM LMB for 0, 0.5, 1, 2 and 6 h. The in vivo NFκB p65, NFκB p50, and IκBα recruitment to IκBα promoter was measured by ChIP analysis and quantified by real-time PCR. The data are presented as the change in occupancy over the human IGX1A (SA Biosciences) sequence control.

to a lesser extent. As shown in the bottom panel of Fig. 1, IκBα is recruited to the IκBα promoter as well; the maximum recruitment is achieved 1 h after LPS stimulation. This is consistent with the previously observed nuclear accumulation of IκBα in LPS-stimulated U-937 macrophages (23, 24), and with the removal of NFκB from IκBα promoter. These data indicate that in LPS-stimulated U-937 macrophages, both p65 and p50 NFκB are recruited to IκBα promoter – presumably as p65/50 heterodimers, and this recruitment is regulated by nuclear IκBα.

Figure 2 illustrates NFκB and IκBα recruitment to cIAP2 promoter. Similarly as in the case of IκBα promoter, both p65 and p50 NFκB are recruited to cIAP2 promoter and this recruitment is suppressed by the LMB-induced nuclear IκBα. However, in contrast to IκBα, the maximum NFκB recruitment to cIAP2 promoter occurs later, 2 h after LPS stimulation. Together, these data indicate that both NFκB-dependent genes, IκBα and cIAP2, are regulated in LPS-stimulated macrophages by p65/p50 NFκB heterodimers, and this regulation is controlled by the nuclear IκBα.

3.1. Cell Culture

1. Grow U-937 monocytic cells in RPMI medium supplemented with 10% FBS, 2 mM glutamine, 1 mM sodium pyruvate, and penicillin–streptomycin solution to a concentration 1×10^6 cells/ml.
2. Induce differentiation of U-937 cells into macrophages by adding PMA at a final concentration 10 ng/ml. Mix well but gently. Transfer the cells to a 6-well plate; each well will get 5 ml of the above cell suspension.
3. Incubate the plate 1 day in tissue culture incubator (37°C, 5% CO_2 humidified atmosphere), to allow cell differentiation (induced by 24-h treatment with PMA).
4. After 24 h, add LMB at a final concentration of 10 nM and / or LPS at a final concentration of 1 μg/ml. Incubate the cells in tissue culture incubator for the desired times.

3.2. Chromatin Immunoprecipitation

3.2.1. Protein–DNA Cross-Linking

1. At desired times, cross-link proteins to DNA by adding 135 μl of 37% formaldehyde stock solution to 5 ml of cell culture so that the final concentration of formaldehyde is 1%. Perform this step in the laminar flow cabinet.
2. Incubate the cells with formaldehyde for 20 min in tissue culture incubator (37°C, 5% CO_2 humidified atmosphere). Be consistent with the fixation condition for all time points (see Note 5).
3. Stop the formaldehyde induced cross-linking by adding 250 μl of 2.5 M glycine solution into each well so that the final concentration of glycine is 0.125 M.

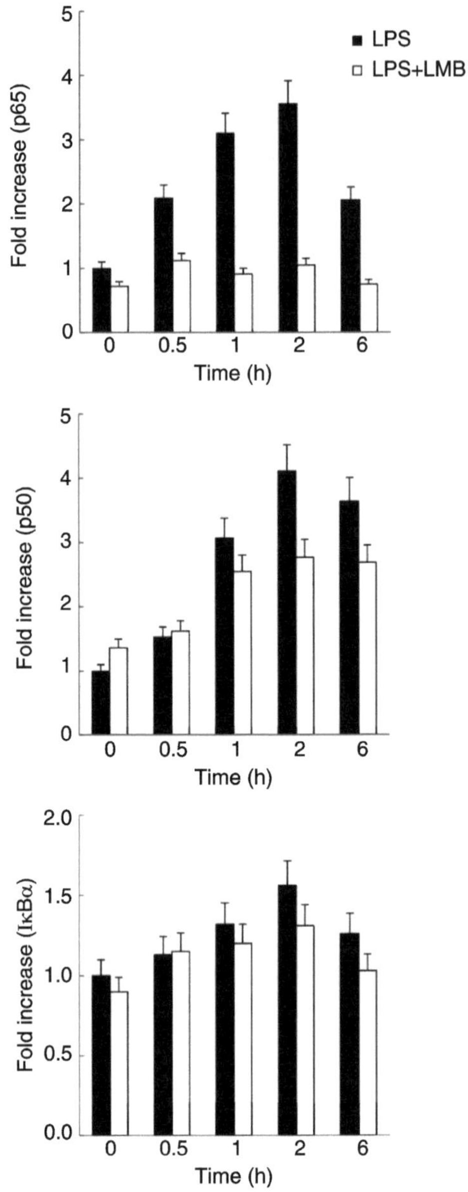

Fig. 2. Recruitment of NFκB p65, NFκB p50, and IκBα proteins to cIAP2 promoter in LPS-stimulated U-937 macrophages. U-937 cells (5×10^6) were differentiated with 10 ng/ml PMA for 24 h, and stimulated with LPS (1 μg/ml) in the absence (full columns) or presence (empty columns) of 10 nM LMB for 0, 0.5, 1, 2, and 6 h. The in vivo NFκB p65, NFκB p50, and IκBα recruitment to cIAP2 promoter was measured by ChIP analysis and quantified by real-time PCR. The data are presented as the change in occupancy over the human IGX1A (SA Biosciences) sequence control.

3.2.2. Cell Lysis and DNA Shearing

1. Scrape the cells from the wells and collect them in 15 ml centrifuge tubes. Centrifuge at $1,700 \times g$ for 4 min in a refrigerated centrifuge (see Note 6).
2. Carefully aspirate the supernatant and remove as much medium as possible.

3. Add 2 ml of ice-cold PBS containing protease inhibitors (2 mM PMSF and 2% protease inhibitor cocktail) into the tubes and centrifuge at 1,700 × *g* for 4 min in a refrigerated centrifuge. Repeat this wash step one more time.
4. Resuspend the pellets in 250 μl of SDS lysis buffer and transfer to a prechilled 1.5 ml microcentrifuge tube. Incubate on ice for 10 min to aid the cell lysis.
5. Sonicate the lysates on ice to shear DNA into 400 to 500 bp fragments (see Note 7).
6. Centrifuge the sonicated samples for 10 min at 12,000 × *g* at 4°C (see Note 6).
7. Transfer the supernatants to new prechilled 2 ml microcentrifuge tubes and discard the pellets.
8. Dilute the sonicated cell supernatants to a final volume of 2 ml by adding 1.75 ml of ChIP dilution buffer containing the protease inhibitors (2 mM PMSF and 2% protease inhibitor cocktail).
9. Set aside a portion of the diluted cell supernatant (100 μl) to quantify the amount of DNA present in different samples. This sample is considered to be your input/starting material and needs to have the protein–DNA cross-links reversed by heating at 65°C for 4 h (see step 6 of the Subheading 3.2.5).

3.2.3. Preclearing and Immunoprecipitation

1. To reduce the nonspecific background, add 80 μl of protein A/G PLUS-agarose slurry to 1.9 ml of the diluted cell supernatant (from step 8 of Subheading 3.2.2) and incubate for 2 h at 4°C while rotating the tubes on a shaker.
2. Pellet the agarose beads by brief centrifugation at 150 × *g* at 4°C and carefully collect the supernatant in a prechilled 2 ml microcentrifuge tube.
3. Add 5 μg of immunoprecipitating antibody per reaction and incubate overnight at 4°C with constant rotation in a shaker (see Note 8).
4. The next day, add 50 μl of protein A/G PLUS-agarose slurry to each sample and incubate for 2 h at 4°C with rotation to collect the specific antibody-protein complexes.
5. Pellet the agarose beads with bound protein complexes by gentle centrifugation (150 × *g* at 4°C for 1 min). Carefully discard the supernatants containing unbound, nonspecific protein–DNA. The A/G agarose pellets now contain the specific protein-antibody complexes.

3.2.4. Washing

1. Add 1 ml of ice-cold low salt immune complex wash buffer containing protease inhibitors (2 mM PMSF and 2% protease inhibitor cocktail) to the tubes containing the protein A/G

agarose antibody-protein complexes. Incubate the complexes at 4°C for 5 min with constant rotation.

2. Pellet the agarose beads by centrifugation (150×*g*, 1 min, 4°C). Carefully discard the supernatants and keep the pellets on ice.
3. Add 1 ml of ice-cold high salt immune complex wash buffer containing the protease inhibitors to protein A/G agarose antibody-protein complexes and incubate 5 min at 4°C while rotating on a shaker.
4. Centrifuge the agarose beads at 150×*g* for 1 min at 4°C. Carefully discard the supernatants and add 1 ml of ice-cold LiCl immune complex wash buffer containing the protease inhibitors to the pellets. Incubate the beads at 4°C for 5 min with constant rotation.
5. Centrifuge the agarose beads at 150×*g* for 1 min at 4°C. Carefully discard the supernatants and add 1 ml of ice-cold 1× TE buffer. Incubate the beads at 4°C for 5 min with constant rotation. Centrifuge as described above and carefully remove the supernatant as much as possible.

3.2.5. Elution and Reverse Cross-Linking

1. Add 250 μl of freshly prepared elution buffer to the pelleted A/G agarose antibody-protein complexes. Vortex briefly (5–10 s) and incubate at room temperature for 12 min with constant rotation.
2. Pellet down the agarose beads by centrifuging at 150×*g* for 1 min at room temperature. Carefully transfer the supernatant (eluate) to a new, labeled 1.5 ml microcentrifuge tube.
3. Add 250 μl of freshly prepared elution buffer to the pellets. Repeat steps 1 and 2 described above to elute the remaining proteins from the agarose beads. Centrifuge at 150×*g* for 1 min at room temperature and transfer the supernatant to 1.5 ml microcentrifuge tube already containing 250 μl of supernatant from the step 2. You should now have 500 μl of the eluate per reaction.
4. Add 20 μl of 5 M NaCl to the combined eluates (500 μl) and 4 μl of 5 M NaCl to the input/starting material (100 μl) from step 9 of Subheading 3.2.2 to reverse the protein–DNA cross-links. Incubate the complexes 4 h in a 65°C water bath.
5. After 4 h, add 10 μl of 0.5 M EDTA, 20 μl of 1 M Tris–HCl, pH 6.5, and 1 μl of 20 mg/ml proteinase K to the combined eluates and incubate for 1 h at 45°C.
6. For the input sample, add 2 μl of 0.5 M EDTA, 4 μl of 1 M Tris–HCl, pH 6.5, and 0.2 μl of 20 mg/ml proteinase K and incubate for 1 h at 45°C.

3.2.6. DNA Extraction

1. Add equal volume of phenol:chloroform:isoamyl alcohol (25:24:1, pH 8.0) to the microcentrifuge tube containing the eluate with protein–DNA cross-links reversed. Vortex vigorously for 5 s. Centrifuge at 12,000 × *g* for 3 min at room temperature. Collect the top (aqueous) layer.
2. Add equal volume of chloroform:isoamyl alcohol (24:1) to the aqueous layer from step 1 (above) in a microcentrifuge tube. Vortex vigorously for 5 s. Centrifuge at 12,000 × *g* for 3 min at room temperature. Collect again the top (aqueous) layer.
3. Add twice the volume of 100% (absolute) ethanol, 1/10th volume of 3 M NaOAc, pH 5.2, and 2 μl of GenElute-LPA to the aqueous layer from step 2 (above) in a microcentrifuge tube. Mix by inverting the tube. Keep the microcentrifuge tube at −20°C for overnight to allow DNA precipitation.
4. The next day, centrifuge the microcentrifuge tube at 12,000 × *g* for 30 min at 4°C. Collect the DNA containing pellet.
5. Wash the pellet with 100 μl of ice-cold 70% ethanol. Centrifuge at 12,000 × *g* for 10 min at 4°C. Collect the pellet. Try to remove as much ethanol as possible.
6. Allow the pellet to air dry.
7. Dissolve the DNA pellet in 50 μl of nuclease-free water. Perform real-time PCR as described below.

3.3. Real-Time PCR

3.3.1. Design and Dilution of the Primers

1. Design all PCR primers with the optimum annealing temperature at 55°C (see Note 9).
2. Prepare a 100 μM stock of each primer, both forward and reverse. From the 100 μM primer stocks, prepare 10 μM working stock solutions for PCR reaction. Each PCR reaction has a volume of 25 μl and will use 1 μl of each 10 μM primer so that the final concentration of each primer in PCR reaction is 0.4 μM.
3. The input DNA should be diluted five times in nuclease-free water before performing PCR reaction. Each PCR reaction (25 μl volume) for the input sample will use 5 μl of five times diluted input DNA.

3.3.2. Setting Up the Real-Time PCR

1. Set up the PCR reaction on ice.
2. Each reaction using immunoprecipitated DNA has a total volume of 25 μl, and is set up as follows (see Note 10):

   SYBR® Green Supermix: 12.5 μl.

   Nuclease-free water: 7.5 μl.

   Forward primer (10 μM): 1.0 μl.

   Reverse primer (10 μM): 1.0 μl.

   DNA sample: 3.0 μl.

3. Each reaction using input DNA has a total volume of 25 μl, and is set up as follows (see Note 11):

   SYBR® Green Supermix: 12.5 μl.

   Nuclease-free water: 5.5 μl.

   Forward primer (10 μM): 1.0 μl.

   Reverse primer (10 μM): 1.0 μl.

   Input DNA sample (five times diluted): 5.0 μl.

4. After loading the master-mix and the DNA into the wells of the PCR plate, seal the plate carefully using an optical tape and continue with the real-time PCR reaction in the thermal cycler.
5. Each reaction for both the immunoprecipitated and input samples should include a positive and negative control (see Note 12).

4. Notes

1. LMB (MW 540.7) can be purchased from Sigma, Calbiochem or Biomol. LMB from Sigma is supplied as a methanol solution at a concentration of 5 μg/ml. LMB inhibits the nuclear export of NES-containing proteins at concentrations 10–20 nM (26, 27). For the final working concentration of 10 nM, the Sigma LMB stock should be diluted 900-fold. The LMB stock solution should be stored at –20°C, and when opened, kept on ice to minimize evaporation.
2. PMSF is unstable in aqueous environment. It is essential that it is dissolved in absolute alcohol (ethanol, methanol, or isopropanol), and is added to buffers in the final working concentration 2 mM just before use.
3. This protease inhibitor cocktail contains protease inhibitors with broad specificity for the inhibition of serine, cysteine, aspartic proteases, and aminopeptidases. It should be stored at –20°C, and added to buffers just before use, in the final concentration 15 μl/10^6 cells.
4. Prepare stock solutions of 10% SDS and 1 M $NaHCO_3$ by dissolving 5 g of SDS in 50 ml of deionized water and 4.2 g of $NaHCO_3$ in 50 ml deionized water, respectively. These stock solutions can be stored at room temperature.
5. For Vibra cell sonicator, use 10-s pulses at setting 40. However, if you use a different sonicator, you may have to calibrate it so that the length of sheared DNA fragments is 400 to 500 bp.
6. It is helpful always to centrifuge the tubes in one position (e.g., when using 1.5 ml microcentrifuge tubes, position them with cap snaps facing toward the center of the rotor). This way you can always expect the pellets to be at the same place.

7. Make sure to keep the samples on ice at all times in between the shearing. In addition, it is also helpful to place the tip of the sonicator in a beaker filled with ice for 30 s in between successive 10-s pulses to ensure that the tip is not overheated.
8. In this protocol, we used 25 μl each of p65 NFκB antibody (sc-372; 200 μg/ml), p50 NFκB antibody (sc-7178; 200 μg/ml), and IκBα antibody (sc-371; 200 μg/ml) to obtain 5 μg of antibody per each immunoprecipitation reaction.
9. The primers for amplifying the NFκB-binding sites at IκBα and cIAP2 promoters were designed to anneal optimally at 55°C. Prepare a 10 μM working stock solution of each primer and store at −20°C.
10. It is convenient to prepare a master-mix by mixing all components (for desired amount of reactions) except for the DNA. Prepare the master-mix in a 1.5 ml microcentrifuge tube. Aliquot 22 μl of the master-mix into each well of the PCR plate. Add 3 μl of DNA sample into each well.
11. Prepare a master-mix containing all of the above except the input DNA in a 1.5 ml microcentrifuge tube. Aliquot 20 μl of the master-mix into each well of the PCR plate. Add 5 μl of input DNA sample (five times diluted) into each well.
12. The positive control primers provide a control for successful ChIP and gene transcription. In this protocol, we used the ChIP-qPCR Human GAPDH Proximal Promoter Positive Control primer (GPH10001C(+)01A, SA Biosciences). The negative primer provides a reference of the amount of nonspecific genomic DNA that co-immunoprecipitates during the procedure. In this protocol, we used ChIP-qPCR Human 1GX1A Negative Control primer (GPH 00001C(−)01A, SA Biosciences).

Acknowledgments

This work was supported by NIH grants GM079581 and AI085497 to I. Vancurova.

References

1. Baeuerle, P. A. and Baltimore, D. (1996) NFκB: ten years after. *Cell* **87**, 13–20.
2. Baldwin, A. S. (2001) The transcription factor NFκB and human disease. *J. Clin. Invest.* **107**, 3–6.
3. Yamamoto, Y. and Gaynor, R. B. (2001) Therapeutic potential of inhibition of the NFκB pathway in the treatment of inflammation and cancer. *J. Clin. Invest.* **107**, 135–142.
4. Ghosh, S. and Karin, M. (2002) Missing pieces in the NFκB puzzle. *Cell* **109**, S81–96.
5. Wan, F. and Lenardo, M. J. (2010) The nuclear signaling of NFκB: current knowledge, new insights, and future perspectives. *Cell Res.* **20**, 24–33.
6. Prasad, S., Ravindran, J. and Aggarwal, B. B. (2010) NFκB and cancer: how intimate is the relationship. *Mol. Cell. Biochem.* **336**, 25–37.

7. Li, Q., Withoff, S. and Verma, I. M. (2005) Inflammation-associated cancer: NFκB is the lynchpin. *Trends Immunol.* 26, 318–325.
8. Grivennikov, S. I., Greten, F. R. and Karin, M. (2010) Immunity, inflammation, and cancer. *Cell* **140**, 883–899.
9. Ghosh, S. and Hayden, M. S. (2008) New regulators of NFκB in inflammation. *Nature Rev. Immunol.* **8**, 837–848.
10. Fan, Y., Dutta, J., Gupta, N., Fan, G. and Gélinas, C. (2008) Regulation of programmed cell death by NFκB and its role in tumorigenesis and therapy. *Adv. Exp. Med. Biol.* **615**, 223–250.
11. Arenzana-Seisdedos, F., Thompson, J., Rodriguez, M. S., Bachelerie, F., Thomas, D. and Hay, R. T. (1995) Inducible nuclear expression of newly synthesized IκBα negatively regulates DNA-binding and transcriptional activities of NF-κB. *Mol. Cell Biol.* **15**, 2689–2696.
12. Arenzana-Seisdedos, F., Turpin, P., Rodriguez, M., Thomas, D., Hay, R. T., Virelizier, J. L. and Dargemont, C. (1997) Nuclear localization of IκBα promotes active transport of NF-κB from the nucleus to the cytoplasm. *J. Cell Sci.* **110**, 369–378.
13. Turpin, P., Hay, R. T. and Dargemont, C. (1999) Characterization of IκBα nuclear import pathway. *J. Biol. Chem.* **274**, 6804–6812.
14. Rodriguez, M., Thompson, J., Hay, R. T. and Dargemont, C. (1999) Nuclear retention of IκBα protects it from signal-induced degradation and inhibits NF-κB transcriptional activation. *J. Biol. Chem.* **274**, 9108–9115.
15. Kunsch, C., Ruben, S. M. and Rosen, C. A. (1992) Selection of optimal kappa B/Rel DNA-binding motifs: interaction of both subunits of NFκB with DNA is required for transcriptional activation. *Mol. Cell Biol.* **12**, 4412–4421.
16. Lin, R., Gewert, D. and Hiscott, J. (1995) Differential transcriptional activation *in vitro* by NFκB/Rel proteins. *J. Biol. Chem.* **270**, 3123–3131.
17. Udalova, I. A., Mott, R., Field, D. and Kwiatkowski, D. (2002) Quantitative prediction of NFκB DNA-protein interactions. *Proc. Natl. Acad. Sci.USA* **99**, 8167–8172.
18. Saccani, S., Pantano, S. and Natoli, G. (2001) Two waves of NFκB recruitment to target promoters. *J. Exp. Med.* **193**, 1351–1359.
19. Saccani, S., Pantano, S. and Natoli, G. (2003) Modulation of NFκB activity by exchange of dimers. *Mol. Cell* **11**, 1563–1574.
20. Saccani, S., Marazzi, I., Beg, A. A. and Natoli, G. (2004) Degradation of promoter-bound p65/RelA is essential for the prompt termination of the NFκB response. *J. Exp. Med.* **200**, 107–113.
21. Vancurova, I., Miskolci, V. and Davidson, D. (2001) NF-κB activation in TNF-stimulated neutrophils is mediated by protein kinase C-δ: Correlation to nuclear IκBα. *J. Biol. Chem.* **276**, 19746–19752.
22. Castro, S., Miskolci, V., Kalasapudi, B., Davidson, D. and Vancurova, I. (2002) NFκB regulation in human neutrophils by nuclear IκBα: Correlation to apoptosis. *J. Immunol.* **169**, 3947–3953.
23. Ghosh, C. C., Vu, H. Y., Mujo, T. and Vancurova I. (2008) Analysis of nucleocytoplasmic shuttling of NFκB proteins in human leukocytes. *Methods Mol. Biol.* **457**, 279–292.
24. Ghosh, C. C., Ramaswami, S., Juvekar, A., Vu, H. Y., Galdieri, L., Davidson, D. and Vancurova, I. (2010) Gene specific repression of proinflammatory cytokines in stimulated human macrophages by nuclear IκBα. *J. Immunol.* **185**, 3685–3693.
25. Vu, H.Y., Juvekar, A., Ghosh, C., Ramaswami S., Le, D. H. and Vancurova, I. (2008) Proteasome inhibitors induce apoptosis of prostate cancer cells by inducing nuclear translocation of IκBα. *Arch. Biochem. Biophys.* **475**, 156–163.
26. Yoshida, M., Nishikawa, M., Nishi, K., Abe, K., Horinouchi, S. and Beppu, T. (1990) Effects of leptomycin B on the cell cycle of fibroblasts and fission yeast cells. *Exp. Cell. Res.* **187**, 150–156.
27. Kudo, N., Wolff, B., Sekimoto, T., Schreiner, E. P., Yoneda, Y., Yanagida, M., Horinouchi, S. and Yoshida, M. (1998) Leptomycin B inhibition of signal-mediated nuclear export by direct binding to CRM1. *Exp. Cell. Res.* **242**, 540–547.

Chapter 9

In Vivo ChIP for the Analysis of Microdissected Tissue Samples

Chris Murgatroyd, Anke Hoffmann, and Dietmar Spengler

Abstract

The development of chromatin immunoprecipitation assays (ChIP) as a tool to examine the interactions between nuclear proteins and DNA has enhanced essentially our understanding of the dynamic association of transcription factors and chromatin modifiers with target DNA sequences. Still in vivo ChIP experiments of the central nervous system continue to represent a challenge given the considerable cellular and functional diversity, which makes the dissection of discrete circumscribed structures highly desirable. Tiny amounts of tissue can result, however, in insufficient quantities of starting material incompatible with many ChIP applications and lead to variable results. Here, we discuss the suitability of currently available ChIP protocols for in vivo ChIP experiments and present a new streamlined protocol that allows the processing of multiple samples with less time on hands.

Key words: ChIP, Chromatin, Immunoprecipitation, Brain, Microdissection

1. Introduction

The interaction of proteins with DNA underlies a host of cellular functions, including the regulation of gene expression. The access to the DNA is critically controlled through the concerted interaction of epigenetic mechanisms, such as DNA methylation and histone modifications, which precedes the dynamic association of transcription factors and chromatin modifiers with their target DNA sequences.

Chromatin immunoprecipitation (ChIP) is the method of choice to investigate DNA–protein interactions inside the cell. It has been used for mapping the localization of histone modifications, transcription factors and other chromatin-associated proteins on a locus-specific and genome-wide scale. The technique is based

Ales Vancura (ed.), *Transcriptional Regulation: Methods and Protocols*, Methods in Molecular Biology, vol. 809, DOI 10.1007/978-1-61779-376-9_9, © Springer Science+Business Media, LLC 2012

on the principle that living cells are treated with formaldehyde to cross-link proteins and DNA within ~2 Å of each other. As a result, proteins are covalently bound to their target sequences on the DNA and provide a snapshot of transcription factor occupancy. Next, the cross-linked DNA is fragmented into pieces of 300–600 bp and protein–DNA complexes are immunoprecipitated by antibodies against the protein of interest (Fig. 1). This allows the enrichment for specific DNA sequences which can then be identified either through amplification by PCR (for review, see ref. 1), hybridization to microarrays (ChIP-on chip) (for review, see ref. 2),

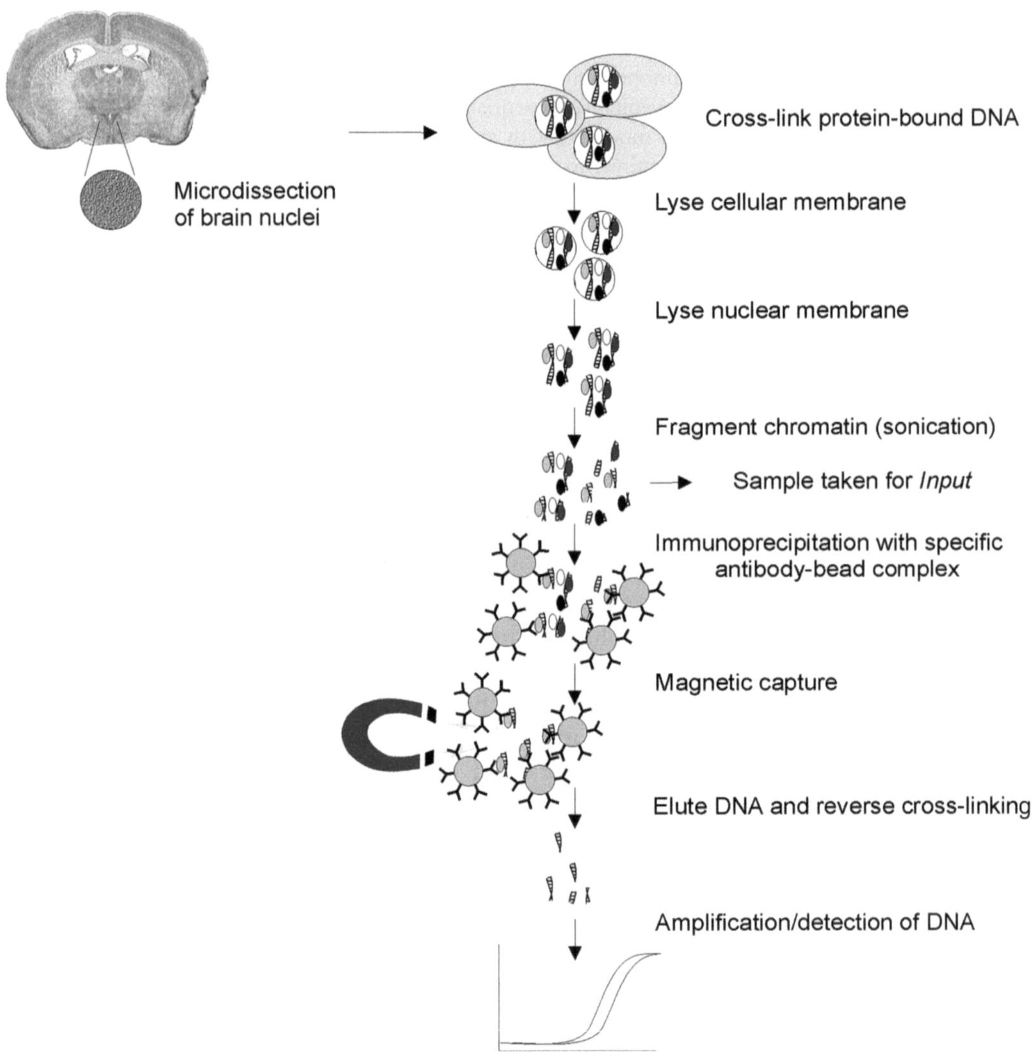

Fig. 1. Outline for the in vivo ChIP procedure. Cells are cross-linked with formaldehyde and DNA is sheared by sonication. Protein–DNA complexes are immunoprecipitated with specific antibodies and enriched DNA is purified for quantitative analysis.

cloning (for review, see ref. 3), or direct high-throughput sequencing (ChIP-seq) (for review, see ref. 4).

The extraordinary complexity of the central nervous system is thought to rely on refined transcriptional networks and hierarchies. The availability of in vivo ChIP technology offers for the first time the opportunity to elucidate the underlying protein and target gene interactions. Neurons of related functions typically cluster in nuclei, which are not only small in size and limited to hundreds to thousands of cells, but still rather heterogeneous due to varying synaptic inputs and spatial organization. This heterogeneity necessitates that brain tissues have to be treated differently to relatively more homogeneous tissues and cell lines by adaption of the ChIP technique to the relatively small cell numbers obtained from tissue punches. For example, one cortex tissue punch (250 μm thick by 1 mm diameter, 0.2 mm^3) equals to approximately 20,000 total cells and 8,000 neurons based on rat cortex cell density (5).

The conventional ChIP assay, pioneered by Alexander Varshavsky and coworkers in the 1980s (6), has undergone various refinements to meet specific demands (for review, see ref. 7). Here, we focus on those that are of particular interest for the application of in vivo ChIP to neuronal tissues (see Table 1). An obvious limitation arises from the limited numbers of cells available for ChIP experiments and the fact that the recovery rate of cross-linked DNA varies from 1 to 10% of total DNA in the starting material. One major step forward was the reduction of the timescale in the "Fast ChIP" protocol (8) by employing an ultrasonic water bath to reduce the antibody incubation step and Chelex resin to elute the immunoprecipitated DNA fragments from the protein A-beads which can then be directly used as PCR template without further purification. However, while this ChIP method can be completed within 1 day, these shortenings reduce overall DNA yield and quality and demand relatively large cell samples ($0.5–1 \times 10^6$) (9). Another method omitted the DNA clean-up step by performing PCR directly on the immunoprecipitated cross-linked DNA (10). However, in our hands PCR detection was unreliable and frequently failed in case of small tissue samples. Moreover, it has been reported that this ChIP technique is not compatible with qPCR analysis dependent on the nature of the carrier beads used since some of them, such as magnetic beads, appear to interfere with quantification of the SYBR Green signal during real-time PCR (11). Combining high-throughput with shortened immunoprecipitation and wash steps the EpiQuick ChIP kit (Epigentek™) is based on a microplate format whereby all steps are performed without sample transfers, opening some potential for automation (12, 13). However, the kit still relies on an input of no less than 500,000 cells per reaction.

To process smaller cell numbers, the carrier ChIP (CChIP) (14) uses carrier chromatin from Drosophila (or any other species sufficiently distant from the species investigated) to reduce reaction

Table 1
Comparison of ChIP methods

Method	Starting material (no. cells)	Duration of procedure	Antibody incubation time/procedure	Wash steps	DNA elution/reversal of cross-linking	Major improvements
Conventional ChiP method (25)	10^6–10^7	3–4 days	Immunoprecipitation overnight, then add protein G agarose (2 h)	1× low salt buffer, 1× high salt buffer, × LiCl buffer, 1× TE buffer	65°C 4 h (1% SDS) followed by proteinase K, digestion then phenol/chloroform and ethanol precipitation	
Fast ChIP (8)	10^7	4–6 h	Immunoprecipitation in ultrasonic water bath 15 min, then add protein A agarose beads	5–6× IP buffer	Chelex 100 resin, 100°C 10 min then prot.K 30 min, then 100°C 10 min. No DNA clean-up	Chelex 100-based DNA isolation procedure. Ultrasonic bath
Rapid ChIP assay (10)	10^8	>1 day	4°C for 3 h followed by magnetic protein G (Dynabeads™)	3× lysis buffer, 3× washing buffer, 3×, TE buffer	Directly use immuopre-cipitates as PCR templates	Elimination of reversal of cross-linking step
EpiQuick™ ChIP	10^6	5 h	Immunoprecipitation on strip well for 1 h	Wash directly on strip well	Prot.K 65°C 15 min, then 65°C 90 min. Elute DNA from strip then purify on spin columns	Strip microplate format for high-throughput
CChIP (14)	100 cells with 5×10^7 SL2 cells as carrier	2–3 days	Immunoprecipitation under native conditions overnight	1× 100 mM NaCl	Room temperature 15 min with 1% SDS followed by phenol/chloroform and ethanol precipitation. DNA detected by radioactive PCR	Very low number of cells can be analyzed

Fast CChIP (16)	Brain punch (1 mm) with 10^7 yeast cells as carrier	4 h	Ultrasonic water bath for 1 h followed by protein G agarose beads	5× IP buffer.	Chelex 100 resin, 100°C 10 min. No DNA clean-up	Combination of Fast ChIP and CChIP methods
Q2ChIP (17)	10^5 cells which can be diluted 1,000-fold	>1 day	4°C for 2 h with magnetic protein A (Dynabeads™)	3× RIPA buffer, 1× TE buffer	68°C for 2 h (SDS elution, cross-link reversal and proteinase K digestion combined in single step)	Combination of cross-link reversal, proteinase K digestion and DNA elution into a single step
MicroChIP (18)	1,000–50,000 cells or tissue punches	>1 day	4°C for 2 h with magnetic protein A (Dynabeads™)	3× RIPA buffer, 1× TE buffer	EITHER 68°C for 2 h (SDS elution, cross-link reversal and proteinase K digestion combined in single step). OR Chelex 100	Suitable for ChIP from small (~1 mm^3) tissue punches
In vivo ChIP	10,000–50,000 cells or tissue punches	2 days	4°C overnight, then add magnetic Dynabeads™ protein G	1× low salt buffer, 1× high salt buffer, 1× LiCl buffer, 1× TE buffer	68°C for 2 h (SDS elution, cross-link reversal and proteinase K digestion combined in a single step) then 95°C for 10 min	

volume and so increase recovery and minimize background signal. However, this demands improved primer design for detection of the target from the foreign background DNA and the protocol is relatively time consuming. Furthermore, in its published form "CChIP" applies to the native ChIP (NChIP) method (15) in which cross-linking is omitted and does not allow for detection of transcription factor interactions. To overcome some of these limitations, a "Fast CChIP" procedure was developed combining the "Fast ChIP" and "CChIP" methods using *Saccharomyces cerevisiae* as a source of carrier chromatin. While this approach allowed to study in vivo transcription factor–DNA interactions in as little as 0.2 mm^3 brain tissue (16) and demonstrated firstly the applicability of ChIP for the study of very small tissue sizes – the procedure still does not overcome the disadvantages associated with the use of carrier chromatin.

To advance further the issue of limited starting material a quick and quantitative ChIP (Q2ChIP) assay was presented that involved chromatin preparation from relatively large cell numbers which could then be diluted and aliquoted for use in many independent chromatin immunoprecipitations. This can be performed from a starting material of only 100,000 cells and appeared suitable for up to 1,000 histone ChIPs or 100 transcription factor ChIPs, allowing many samples to be prepared in parallel and stored, while the whole procedure can be undertaken in a day (17). To scale down further the number of cells needed for the starting material, the "MicroChIP" (or μChIP) assay was developed in which chromatin can be prepared from as little as 1,000 cells without the need for carriers, such as nonspecific chromatin, large polymers, or nonspecific nucleic acid. At the lower range, the assay can also use about 100 cells per ChIP sample and be applied to the study of small (1 mm^3) tissue microdissection samples. In addition, it can be completed within 1 day, depending on the length of the immunoprecipitation time required (18).

Taken together, the original ChIP method has been progressively modified and streamlined over the last 20 years to meet the needs of specific experiments and the amounts of available starting material (see Table 1). We have further customized a ChIP method, combining various steps taken from the previously reported protocols, for use on brain microdissection samples (19). Furthermore, this modified protocol facilitates the handling of multiple samples at the same time and can also be adapted to sequential-ChIP.

Firstly, we perform separate cell and nuclear membrane lyses to increase the relative concentration of chromatin in the sample and use a sonicating water bath to achieve an even distribution of the energy (the energy of sonication in traditional sonicators depends on the depth of the sonication probe in the liquid), increased reproducibility and the possibility to process many samples in parallel. Immunoprecipitations are performed overnight in a ChIP dilution

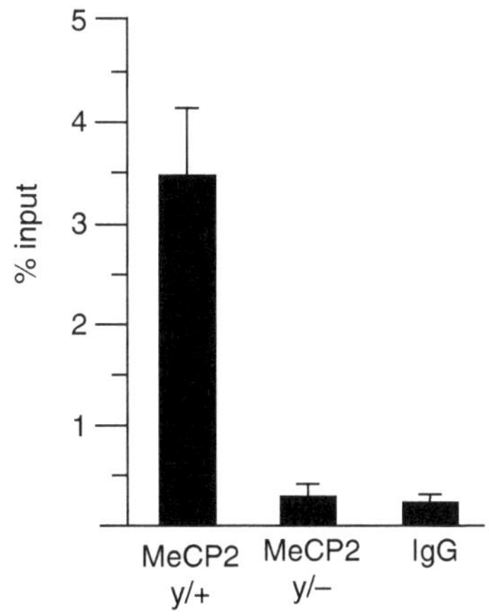

Fig. 2. ChIP analysis of MeCP2 occupancy at the AVP enhancer in the hypothalamic nucleus paraventricularis (PVN). Chromatin samples from PVN tissues of Mecp2$^{(y/+)}$ wild-type or Mecp2$^{(y/-)}$ null male mice (26) were immunoprecipitated with an antibody against the C-terminal of Mecp2 (19). DNA recovered from the ChIP experiments were analyzed by qPCR for the presence of the AVP enhancer region (19). The results support that Mecp2 occupies the AVP enhancer in the PVN of Mecp2$^{(y/+)}$ wild-type mice while no enrichment was measured in Mecp2$^{(y/-)}$ null male mice that served as negative control.

buffer to increase maximum binding followed by incubation with magnetic Dynabeads™ protein G or A (10) and washing with increasingly stringent buffers (1). The elution of DNA from the magnetic beads, reversal of cross-linking and proteinase K digestion are performed in one step (17) followed by protein denaturation at 95°C. The DNA is then further purified through spin columns which we find far more efficient than ethanol precipitation. A series of samples can be simultaneously processed in a highly reproducible manner by this cost-effective method in a reasonably short time-frame. Exemplary, we demonstrate binding of the methyl-binding protein 2 (Mecp2) to the arginine vasopressin (AVP) enhancer in a single brain nucleus (Fig. 2).

2. Materials

2.1. Tissue Disruption and Cross-Linking of Protein–DNA Interactions

1. Cryostat (Leica™ CM3050 S).
2. Brain areas are punched using a sample corer, 1.0 mm diameter (Fine Science Tools™).
3. Dry ice.
4. 37% formaldehyde.
5. 1.25 M glycine.
6. 1× Phosphate-Buffered Saline (PBS).

2.2. Cell Lysis and Sonification of DNA

1. Cell lysis buffer: 10 mM Hepes, pH 7.9, 1.5 mM $MgCl_2$, 10 mM KCl, 0.5% Igepal.
2. Nuclear lysis buffer: 50 mM Tris–HCl, pH 8.1, 1% SDS, 10 mM EDTA.
3. Protease inhibitor cocktail (Sigma™) 100× concentrated mix.
4. Sonicator using indirect sonication principles (for example, Bioruptor, Diagnode™ utilizing a waterbath containing high power ultrasound generating elements located below the tank).

2.3. Immunoprecipitation of Cross-Linked Protein–DNA Interactions

1. ChIP dilution buffer: 16.7 mM Tris–HCl, pH 8.1, 167 mM NaCl, 1.1% Triton X-100, 0.01% SDS, 1.2 mM EDTA, protease inhibitor.
2. Dynabeads Protein G or A (Invitrogen™) which are uniform, superparamagnetic beads 2.8 μm in diameter with recombinant protein G or A covalently coupled to their surface.
3. Magnetic holder for 1.5-ml tubes. We use a stand with place for 24 tubes (Chemagen™) to allow the processing of numerous ChIP samples simultaneously.
4. Primary antibody of choice.
5. Rotating wheel to fit 1.5-ml tubes for continuous tilting and rotation of tubes.
6. Low-salt wash buffer: 20 mM Tris–HCl, pH 8.1, 150 mM NaCl, 0.1% SDS, 1% Triton X-100, 2 mM EDTA.
7. High-salt wash buffer: 20 mM Tris–HCl, pH 8.1, 500 mM NaCl, 0.1% SDS, 1% Triton X-100, 2 mM EDTA.
8. LiCl wash buffer: 10 mM Tris–HCl, pH 8.1, 250 mM LiCl, 1% Igepal, 1% sodium deoxycholate, 1 mM EDTA.
9. TE buffer: 10 mM Tris–HCl, pH 8.0, 1 mM EDTA.

2.4. Elution, Reversal of Cross-Linking, and Purification of DNA

1. Elution buffer: 50 mM Tris–HCl, pH 8, 1 mM EDTA, 1% SDS, 50 mM $NaHCO_3$.
2. 5 M NaCl.
3. RNAse A: 20 mg/ml.
4. 10% SDS.
5. Proteinase K.
6. DNA/PCR purification kit (Ultraclean PCR Clean-Up Kit).

2.5. Amplification of Immunoprecipitated DNA

1. LightCycler FastStart DNA Master Plus Set SYBR Green I (Roche™).
2. Primers.
3. LightCycler (Roche™).

3. Methods

The most commonly used fixative to cross-link protein–DNA complexes within cells is formaldehyde (see Note 1). Upon cell harvest, protein–DNA complexes are fragmented by sonification to obtain DNA fragments in a size range from 200 to 1,000 bp. Chromatin or DNA-binding proteins are isolated by immunoprecipitation with specific antibodies against the proteins of interest. Appropriate negative and positive control antibodies are included to verify the specificity of the precipitation reactions (see Note 2). Complexes are recovered using Dynabeads™ coupled to protein G or protein A (see Note 3). This allows highly efficient recovery of DNA–protein complexes without the need for carrier.

Depending on the size and nature of brain punches, we might pool some samples combining between three and five tissue punches for each ChIP analysis.

3.1. Cross-Linking of Protein–DNA Interactions in Brain Punch

1. Use a tissue punch (20) to isolate brain regions of interest [we used a 1 mm tissue punch to collect hypothalamic nucleus paraventricularis (PVN) from the mouse brain (Fig. 2)]. The punches are collected in 1.5-ml tubes and frozen on dry ice. At this point, the tissue punches may be stored at −80°C for up to 1 year.
2. Disrupt and homogenize the tissue in 200 μl of PBS using a pipette [we pooled three PVN from adult mice for each ChIP (Fig. 2)].
3. To cross-link the DNA and protein, add 5.4 μl of formaldehyde to a final concentration of 1% and incubate at room temperature for 10 min.
4. Terminate the cross-linking reaction by adding 10× glycine (1.25 M) to a final concentration of 0.125 M. Swirl and incubate at room temperature for 5 min.
5. To wash, pellet the cells (800 × *g* for 4 min at 4°C) and remove supernatant. Add 1× PBS (ice cold) (included protease inhibitor, Sigma™ 10 μl to 1 ml), mix, and centrifuge. Repeat this two times.

3.2. Lysis of Cells and Sonication

1. To lyse the cells and release the nuclei, the cells are resuspended in 100 μl of cell lysis buffer with protease inhibitor.
2. Incubate on ice for 15 min, vortex briefly every 5 min.
3. Pellet cells (800 × *g* for 4 min at 4°C) and remove supernatant.
4. To lyse the nuclei, the cells are resuspended in 100 μl of nuclear lysis buffer including protease inhibitor.

5. Sonicate the lysate to shear the chromatin to an average length of 400–500 bp. Be sure to keep the samples cold by performing the procedure on ice during this step. The sonication time and number of pulses should be carefully optimized (see Note 4) and will vary depending on sonicator and cell/tissue type. However, we have found using a Bioruptor (Diagenode™) a setting of 3 × 3 min with 30 s pulses between 30 s pauses with the machine setting on "high power," was optimal for 1 mm tissue punches.
6. Pellet the released DNA and proteins (10,000 × *g* for 10 min at 4°C).
7. Transfer supernatant to fresh tube and measure DNA concentration. The sonicated samples can also be stored (−80°C) at this stage for future use.

3.3. Immunoprecipitation of Cross-Linked Protein–DNA Interactions

1. Transfer 50 μl of the supernatant (from each 100 μl of sonicated sample, 50 μl are used for each ChIP) to a new tube and dilute fivefold with dilution buffer. Save 1% of total supernatant as an input control. This will be the input fraction and is kept at 4°C until the first step of Subheading 3.3.
2. Add the primary antibody (see Note 3) and incubate over night with constant rotation in a cold room at 4°C.
3. Add 20 μl of Dynabeads™ and incubate for a further 1 h at 4°C. Dynabeads™ must be firstly completely resuspended and equilibrated by washing twice with *dilution buffer* before use.
4. Capture the magnetic Dynabead™/protein–DNA complexes by placing the tubes in a magnetic rack for 1 min to separate the beads from the solution and remove and discard the supernatant. We perform this in a cold room.
5. DNA bound magnetic beads are washed, to remove DNA and proteins nonspecifically associated with the Dynabeads™, for 3–5 min with rotation in a cold room. Firstly, the beads are washed with 1 ml of low-salt wash buffer, followed by high-salt wash buffer, LiCl wash buffer and finally TE buffer.

3.4. Elution, Reversal of Cross-Linking, and Purification of DNA

1. Elute the immune complex by adding 100 μl of elution buffer containing 50 μg/ml proteinase K to the magnetic bead/protein–DNA complexes. Mix at 62°C for 2 h to release the protein-bound DNA and reverse the cross-linking. Then, incubate at 95°C for 10 min to denture the proteins and inactivate the proteinase K.
2. The DNA is recovered from the sample through spin column purification using the Ultra Clean PCR Clean-Up Kit in a volume of 50 μl (see Note 5).

3.5. PCR Amplification of DNA

Analyze the results of the immunoprecipitation by real-time PCR; alternatively, it is possible to proceed to cloning for the identification of target genes (for review, see ref. 3) or perform either ChIP-sequence (for review, see ref. 4) or ChIP-on-chip (for review, see ref. 2). In addition, one can easily apply this technique to sequential-ChIP studies (see Note 6).

We perform qPCR on a LightCycler (Roche™) using LightCycler FastStart DNA Master Plus SYBR Green (Roche™). For primer design see Note 7. Determine the amount of precipitated DNA relative to *input* as [(amount of *ChIP DNA*)/(amount of *input DNA*)] × 100. We then express the data as percent (±s.d.) precipitated DNA relative to *input DNA*.

4. Notes

1. *Cross-linking*. Formaldehyde is the most common cross-linking agent used for ChIP. Since efficiency of cross-linking depends on the nature of the protein to cross-link, it is important to consider the nature of cross-linker and duration of cross-linking. To capture weak or transient protein–DNA interactions, a longer incubation time with formaldehyde may be required. However, increasing the cross-linking time can make fragmentation of the chromatin more difficult. Formaldehyde has a relatively short cross-linking spacer arm and cross-links nuclear components located within 2 Å of each other, thus it can be ineffective when analyzing proteins indirectly bound to DNA. To capture indirect associations between DNA and transcriptional cofactors, one could consider longer range cross-linkers, such as dimethyl apidimidate (DMA) (21) to cross-link proteins not in direct contact with DNA, or contrary one might consider cisplatin that cross-links proteins to DNA but not proteins to proteins (1) (for review, see ref. 22).

   It is important to test cross-linking efficiency for each protein of interest. In case cross-linking information is not available, we recommend starting with 1% formaldehyde for 10 min using, as a positive control, a protein known to be cross-linked under these conditions (a histone or RNA polymerase II). Cross-linking conditions are tested by ChIP and PCR analysis of the precipitated DNA. Failure to cross-link will result in no PCR product after ChIP; in this instance, cross-linking time may be increased and/or additional cross-linkers should be tested alone or in combination with formaldehyde.
2. *Controls*. In addition to performing ChIP with the antibody against the protein of interest, one should also perform controls

utilizing an isotype-matched antibody recognizing a nonnuclear and non-DNA-binding protein in addition to omission of antibody. These ChIP results will allow determining the extent of any nonspecific binding. The positive control should refer to a protein known to colocalize with an identified locus in the cell type examined. This may include an antibody to histones or RNA polymerase II.

3. *Antibodies.* The amounts of antibody required can be tested. However, we find that 5 μg of antibody per ChIP mostly suffices for this in vivo ChIP protocol.

   It is important that the epitope recognized by the antibody remains available after cross-linking to succeed in ChIP and is not buried in the protein complex. Antibodies also have to be of very high avidity so that the interaction with the protein will resist to the washing steps. Likewise, the protein to be immunoprecipitated has to cross-link efficiently to the chromatin. In general, the success of a ChIP experiment strongly depends on the availability of ChIP-grade antibodies. Although for some proteins, ChIP-grade antibodies are available from commercial suppliers; this might not be the case for homemade antibodies.

   Nonspecific binding to the Dynabeads™ can occur with increasing incubation times. Therefore, we carry out immunoprecipitation overnight and add the Dynabeads™ only during the last hour of the incubation. To match the Dynabead™ type used (i.e., covalently coupled to protein G or protein A) to the isotype of antibody under study refer to the Invitrogen™ homepage. For instance, most mouse monoclonal antibodies bind strongly to protein G beads while rabbit polyclonal antibodies bind strongly to both protein A and G beads and protein A is more efficient for guinea pig.

4. *Sonication.* Sonication is a critical step because long chromatin fragments are not as efficiently immunoprecipitated while fragments that are too short might be incompatible with PCR. To obtain a good resolution for ChIP, DNA fragments should range between 400 and 500 bp. If the average sonicated DNA fragment exceeds 1,000 bp, then it becomes difficult to distinguish between DNA–protein interactions in a specific genomic location and DNA segments that are inadvertently immunoprecipitated by interaction of transcription factor to regions several hundred base pairs away from the area of interest. Sonication times will vary depending on cell and tissue type while long sonication steps (i.e., more than 30 s) can cause sample overheating and denaturation. Starting with small tissue sizes, it is not possible to visualize chromatin directly by agarose

gel electrophoresis to determine chromatin size following sonication. Therefore, two approaches can be taken: either testing conditions with higher cell numbers or pooling sonications of several tissues, or alternatively using a PCR-based system to test amplification of a ~300 bp product between the tissue samples with reference samples (see ref. 18 for further details). Ensure that all sonication is performed on ice to reduce the chance of overheating.

5. *Purification of DNA*. We find that the spin columns are far more sensitive than ethanol precipitation to obtain higher DNA quality with a broader linear range. To determine the efficiency of the ChIP, one should utilize PCR to amplify a genomic region known to be bound by the protein of interest (positive control) as well as a random genomic area that should not be bound by the protein (negative control).
6. *Sequential-ChIP*. Immunoprecipitates from the in vivo ChIP protocol can be further analyzed in a sequential ChIP (also known as re-ChIP) to determine the in vivo colocalization of proteins interacting or in close contact in a chromatinized template. For this purpose, two independent rounds of immunoprecipitations are performed with different specific antibodies (see ref. 23). A method modified for the in vivo ChIP reported here involves elution of the immunoprecipitated complexes and magnetic bead mixture from the last stage of Subheading 3.3 in 25 μl 10 mM DTT by incubation for 30 min at 37°C with frequent vortexing (24). The supernatant is then diluted in 20 times dilution buffer. Keep 1% of mixture as a second input sample and add 10 μg of a second antibody of interest. Incubate overnight as before and add magnetic beads and follow the subsequent ChIP procedure.
7. *Amplification of DNA*. Design of primers for validating ChIP: (1) 20–24 mers; (2) Tm should be 58–60°C; (3) G/C content should be 40–70%; (4) Avoid runs of an identical nucleotide (e.g., CCCC) especially Gs; (5) Keep amplified fragments between 150 and 250 bp; (6) Test primers on genomic DNA prior to ChIP samples to standardize for efficiency.

Acknowledgment

This work was funded by the European Union (CRESCENDO – European Union Contract number LSHM-CT-2005-018652) and the Deutsche Forschungsgemeinschaft (SP 386/4-2 to D.S.)

References

1. Spencer VA, Sun JM, Li L, Davie JR. (2003). Chromatin immunoprecipitation: a tool for studying histone acetylation and transcription factor binding. *Methods* **31**, 67–75.
2. Yoder SJ, Enkemann SA. (2009). ChIP-on-Chip Analysis methods for Affymetrix Tiling Arrays. *Methods Mol Biol* **523,** 367–381.
3. Ortt K, Sinha S. (2010). Chromatin immunoprecipitation for identifying transcription factor targets in keratinocytes. *Methods Mol Biol* **585**, 159–170.
4. Park PJ. (2009). ChIP-seq: advantages and challenges of a maturing technology. *Nat Rev Genet* **10**, 669–680.
5. Herculano-Houzel S, Lent R. (2005). Isotropic fractionator: a simple, rapid method for the quantification of total cell and neuron numbers in the brain. *J Neurosci* **25**, 2518–2521.
6. Solomon MJ, Larsen PL, Varshavsky A. (1988). Mapping protein-DNA interactions *in vivo* with formaldehyde: evidence that histone H4 is retained on a highly transcribed gene. *Cell* **53**, 937–947.
7. Collas P. (2009). The state-of-the-art of chromatin immunoprecipitation. *Methods Mol Biol* **567**, 1–25.
8. Nelson JD, Denisenko O, Bomsztyk K. (2006). Protocol for the fast chromatin immunoprecipitation (ChIP) method. *Nat Protoc* **1**, 179–185.
9. Nelson J, Denisenko O, Bomsztyk K. (2009). The fast chromatin immunoprecipitation method. *Methods Mol Biol* **567**, 45–57.
10. Kohzaki H, Murakami Y. (2007). Faster and easier chromatin immunoprecipitation assay with high sensitivity. *Proteomics* 7, 10–14.
11. Collas P. (2009). The state-of-the-art of chromatin immunoprecipitation. *Methods Mol Biol* **567**, 1–25.
12. Venteclef N, Haroniti A, Tousaint JJ, Talianidis I, Delerive P. (2008). Regulation of Anti-atherogenic Apolipoprotein M Gene Expression by the Orphan Nuclear Receptor LRH-1. *J Bio Chem* **283**, 3694–3701.
13. Flanagin S, Nelson JD, Castner DG, Denisenko O, Bomsztyk K. (2008). Microplatebased chromatin immunoprecipitation method, Matrix ChIP: a platform to study signaling of complex genomic events. *Nucleic Acids Res* **36**, 17.
14. O'Neill LP, VerMilyea MD, Turner BM. (2006). Epigenitic characterization of the early embryo with a chromatin immunoprecipitation protocol applicable to small cell populations. *Nat Genet* **38**, 835–841.
15. O'Neill LP, Turner BM. (1996). Immunopre cipitation of chromatin. *Methods Enzymol* **274**, 189–197.
16. Hao H, Liu H, Gonye G, Schwaber JS. (2008). A fast carrier chromatin immunoprecipitation method appliciable to microdisected tissue samples. *J Neurosci Methods* **172**, 38–42.
17. Dahl JA, Collas P. (2007). Q2ChIP, a quick and quantitative chromatin immunoprecipitation assay unravels epigenetic dynamics of developmentally regulated genes in human carcinoma cells. *Stem Cells* **25**, 1037–1046.
18. Dahl JA, Collas P. (2008). A rapid micro chromatin immunoprecipitation assay (microChIP). *Nat protocol* **3**, 1032–1045.
19. Murgatroyd C, Patchev AV, Wu Y, Micale V, Bockmühl Y, Fischer D, Holsboer F, Wotjak CT, Almeida OF, Spengler D. (2009). Dynamic DNA methylation programs persistent adverse effects of early-life stress. *Nat Neurosci* **12**, 1559–1566.
20. Palkovits M. (1972). Isolated removal of hypothalamic or other brain nuclei of the rat. *Brain Research* **59**, 449–450.
21. Fujita N, Wade PA. (2004). Use of bifunctional cross-linking reagents in mapping genomic distribution of chromatin remodeling complexes. *Methods* **33**, 81–85.
22. Zeng PY, Vakoc CR, Chen ZC, Blobel GA, Berger SL. (2006) *In vivo* dual cross-linking for identification of indirect DNA-associated proteins by chromatin immunoprecipitation. *Biotechniques*; **41**, 694–698.
23. Furlan-Magaril M, Rincón-Arano H, Recillas-Targa F. (2009). Sequential chromatin immunoprecipitation protocol: ChIP-reChIP. *Methods Mol Biol* **543**, 253–266.
24. Métivier R, Penot G, Hübner MR, Reid G, Brand H, Kos M, Gannon F. (2003). Estrogen receptor-alpha directs ordered, cyclical, and combinatorial recruitment of cofactors on a natural target promoter. *Cell* **115**, 751–763.
25. M. Braunstein, A.B. Rose, S.G. Holmes, C.D. Allis and J.R. Broach. (1993). Transcriptional silencing in yeast is associated with reduced nucleosome acetylation. *Genes Dev* 7, 592–604.
26. Guy J, Hendrich B, Holmes M, Martin JE, Bird A. (2001). A mouse Mecp2-null mutation causes neurological symptoms that mimic Rett syndrome. *Nat Genet* **27**, 322–326.

Chapter 10

Quantification of Protein–DNA Interactions by In Vivo Chromatin Immunoprecipitation in Yeast

Amparo Pascual-Ahuir and Markus Proft

Abstract

Chromatin immunoprecipitation (ChIP) is the most widely used method to measure the interaction of proteins with their target DNA sequences in the living cell. The use of ChIP can address many of the fundamental processes underlying transcription, such as the positioning and modification of nucleosomes, the binding of specific transcription factors to regulatory sequences, the secondary recruitment of chromatin-modifying complexes, and other signalling molecules to chromosomal DNA, and the occupancy of RNA polymerase complexes. ChIP is especially useful to define the dynamic nature of these processes. The basis for ChIP in most applications is the determination of the immunoprecipitation (IP) efficiency of individual genomic regions by comparing the amounts of DNA in the IP sample and the input sample before immunoprecipitation. DNA quantification relies on sensitive methods, such as quantitative PCR (qPCR), in real time. Here, we describe the methodology to perform ChIP in the yeast *Saccharomyces cerevisiae* in combination with qPCR in real time to determine protein–DNA association in vivo.

Key words: ChIP, Chromatin, Transcription, Quantitative PCR in real time, Yeast, Formaldehyde cross-link, EvaGreen

1. Introduction

Transcriptional regulation relies on the dynamic binding of proteins to their targets in chromosomal DNA. The expression level of a specific gene in eukaryotic cells is modulated by the association of histone proteins and their diverse modifications, specific transcription factors, chromatin-modifying complexes, regulatory molecules, such as protein kinases, and the RNA polymerase complexes with their general transcription factors. Chromatin immunoprecipitation (ChIP) is the standard method to determine the association of these proteins with DNA in the living cell (1, 2). The method

Ales Vancura (ed.), *Transcriptional Regulation: Methods and Protocols*, Methods in Molecular Biology, vol. 809, DOI 10.1007/978-1-61779-376-9_10,

consists of an initial cross link of the protein to DNA, typically done by formaldehyde treatment of live cells. The cross-linked chromatin is then purified and fragmented by sonication, which yields the "input" sample. This material is subsequently immunoprecipitated with an antibody that specifically recognizes the protein of interest in order to enrich the DNA fragments originally bound by the protein (immunoprecipitation, "IP", sample). The application of sensitive detection methods, such as quantitative PCR (qPCR), allows to calculate the "IP efficiency" at specific genomic regions, which is defined as the amount of DNA in the IP sample divided by the amount of DNA in the input sample. In this standard ChIP, the IP efficiency for a sequence associated with the protein is higher than for control genomic regions, thus defining the "fold enrichment" of the protein of interest over the control. Apart from the targeted ChIP experiment, large-scale applications of ChIP have been developed for protein location surveys at whole genomes. In this case, the samples are amplified by PCR and then hybridized to microarrays (3, 4) or alternatively analyzed by massive DNA sequencing (5–7).

2. Materials

2.1. Preparation of Cross-Linked Chromatin Fragments

1. Formaldehyde solution 37% (see Note 1).
2. 3 M glycine (see Note 2).
3. TBS buffer: 20 mM Tris–HCl, pH 7.5, 150 mM NaCl (see Note 3).
4. FA lysis buffer: 50 mM Hepes–KOH, pH 7.5, 150 mM NaCl, 1 mM EDTA, 1% Triton X-100, 0.1% sodium deoxycholate, 0.1% SDS, 1 mM PMSF (see Note 4).
5. Glass beads, 0.5-mm diameter.
6. Efficient cell disruptor for small samples, for example Mini-BeadBeater (Biospec Products Inc.).
7. Sonicator with Microtip for small sample volumes (2 ml).

2.2. Chromatin Immunoprecipitation, Elution, and Decross-link

1. FA lysis buffer: 50 mM Hepes–KOH, pH 7.5, 150 mM NaCl, 1 mM EDTA, 1% Triton X-100, 0.1% sodium deoxycholate, 0.1% SDS, 1 mM PMSF (see Note 4).
2. FA lysis buffer containing 0.5 M NaCl (see Note 5).
3. Buffer B: 10 mM Tris–HCl, pH 8.0, 1 mM EDTA, 0.25 M LiCl, 0.5% sodium deoxycholate, 0.5% Nonidet NP40 (=Igepal CA-630).
4. TE buffer: 10 mM Tris–HCl, pH 8.0, 1 mM EDTA.
5. Buffer C: 50 mM Tris–HCl, pH 7.5, 10 mM EDTA, 1% SDS.

6. Pronase 20 mg/ml (Roche cat. No. 10165921001).
7. 4 M LiCl.
8. TE-saturated phenol (e.g. Sigma P4557, see Note 6).
9. Chloroform.
10. GlycoBlue 15 mg/ml (Ambion cat. No. AM9515).
11. Ethanol 95%.
12. Ethanol 70%.
13. Dynabeads Protein A (Invitrogen cat. No. 100.02D) or Protein G (Invitrogen cat. No. 100.04D).
14. Magnetic Particle Concentrator (Invitrogen DynaMag-2, cat. No. 123.21D).
15. PBS+BSA buffer: 140 mM NaCl, 2.7 mM KCl, 10 mM Na_2HPO_4, 1.8 mM KH_2PO_4, pH 7.3, 5 mg/ml BSA (see Note 7).

2.3. Analysis of the Immunoprecipitation Efficiency by Quantitative PCR in Real Time

1. 2× EvaGreen qPCR Basic HS Mix (Biotium cat. No. BTIU31004).
2. Hot start Taq DNA polymerase (e.g. TrueStart Taq DNA polymerase 5 U/μl, Fermentas cat. No. EP0612).
3. ROX (5-(and-6)-carboxyrhodamine 6G succinimidyl ester; Molecular Probes cat. No. C6157, see Note 8).
4. Real-time PCR machine with compatible 96-well reaction plates (e.g. Applied Biosystems 7000 or 7500 real time PCR systems).

3. Methods

3.1. Preparation of Cross-Linked Chromatin Fragments

1. Add formaldehyde to a final concentration of 1% to 40 ml of yeast culture (see Notes 1 and 9).
2. Incubate cell samples for 20 min at room temperature, and move occasionally.
3. Add 6 ml 3 M glycine and incubate for another 5 min at room temperature.
4. Spin down cells by centrifugation (2,500×*g* for 3 min), discard the supernatant, and resuspend pellet in 50 ml ice-cold TBS buffer.
5. Repeat centrifugation and wash cells again in 50 ml, ice-cold TBS buffer.
6. Resuspend cells in 5 ml ice-cold FA lysis buffer, spin down by centrifugation (2,500×*g* for 3 min), and discard the supernatant. At this point, the cell pellets can be stored at –20°C for future use.

7. Resuspend the cells in 1 ml ice-cold FA lysis buffer and transfer samples to 2-ml plastic tubes with screw cap (e.g. Sarstedt cat. No. 72.693.005). Add 0.5-ml glass beads and carefully close the tubes.
8. Disrupt the cells in a Mini-BeadBeater at maximal speed at 4°C. Five rounds of 2.5 min are enough to efficiently lyse the cells. Put your samples on ice for 5 min in between homogenizations.
9. Transfer the lysates to fresh 1.5-ml Eppendorf tubes. Wash the remaining beads with 0.5 ml ice-cold FA lysis and pool the lysates.
10. Centrifuge in a microcentrifuge at 13,000 × *g* for 1 min.
11. Discard the supernatant and resuspend the pellet in 1 ml ice-cold FA lysis buffer (see Note 10).
12. Centrifuge again at 13,000 × *g* for 1 min and resuspend the pellet in 1 ml ice-cold FA lysis buffer.
13. Sonicate the samples to fragment the cross-linked chromatin to an average of 300 bp (see Note 11).
14. Centrifuge in a microcentrifuge at 13,000 × *g* for 10 min at 4°C.
15. Transfer supernatant (contains the soluble chromatin fragments) to a fresh 1.5-ml Eppendorf tube. The samples (total cross-linked chromatin) can be stored at −70°C for future use.

3.2. Chromatin Immunoprecipitation, Elution, and Decross-link

1. The day before the IP, the magnetic beads have to be prepared according to the following steps. You need 20 μl of Dynabeads for each IP.
2. Resuspend the amount of Dynabeads you need for all immunoprecipitations in 1 ml PBS + BSA buffer and incubate for 5 min in a roller. Place the samples into the magnetic rack and discard the supernatant by aspiration.
3. Repeat the wash 2× with the same conditions.
4. Resuspend Dynabeads in 1 ml cold PBS + BSA buffer and add the amount of antibody you need for all IPs (see Note 12). Incubate on a roller overnight at 4°C.
5. The next day, wash the Dynabeads 3× as described before, finally resuspend in the original volume in PBS + BSA buffer, and use for IP.
6. Prepare the IP reactions which contain 0.5 ml total chromatin, 0.5 ml FA lysis buffer, and 20 μl antibody loaded and washed Dynabeads. Incubate on a roller for 60 min at room temperature.
7. Place the samples in the magnetic rack and discard the supernatant by aspiration. Perform the following wash steps in 1-ml

volume: 2× FA lysis buffer; 2× FA lysis buffer containing 0.5 M NaCl; 1× buffer B; 1× buffer TE. For each wash, incubate the samples for 5 min on a roller at room temperature.

8. After the last wash step, resuspend the beads in 250 μl buffer C and elute by incubating the samples at 65°C for 10 min with continuous agitation. Place the samples immediately into the magnetic rack and transfer supernatant to a fresh 1.5-ml Eppendorf tube.
9. At this point, the samples for the total chromatin (without IP) have to be prepared. Mix 50 μl of total cross-linked chromatin with 200 μl TE buffer.
10. Add to all samples (IP and total chromatin) 250 μl TE + 20 μl Pronase (see Note 13).
11. Decross-link the samples by incubating at 42°C for 1 h and at 65°C for 5 h (see Note 14).
12. Add 50 μl 4 M LiCl and 0.5 ml phenol:chloroform (1:1), agitate vigorously, and centrifuge for 10 min at 8,000 × *g*.
13. Transfer supernatant to a fresh 1.5-ml Eppendorf tube and repeat the extraction by adding 0.5 ml chloroform.
14. Centrifuge for 10 min at 8,000 × *g* and transfer supernatant to a fresh tube containing 1 ml 96% ethanol + 25 μl GlycoBlue.
15. Incubate overnight at −20°C.
16. Centrifuge at 13,000 × *g* for 10 min at 4°C and discard supernatant. At this point, a blue-coloured pellet should be visible.
17. Wash pellet with 0.5 ml 70% ethanol, repeat centrifugation, discard supernatant, and dry in a vacuum centrifuge for 5 min.
18. Resuspend IP and total chromatin samples in 100 μl TE buffer. Store samples at −20°C for future use.

3.3. Analysis of the Immunoprecipitation Efficiency by Quantitative PCR in Real Time

1. For quantitative PCR, use 2 μl of undiluted IP sample or 2 μl of 1:100 diluted (TE buffer) total chromatin sample. Prepare 2× EvaGreen master mix by mixing the following components for 100 reactions: 0.5 ml EvaGreen qPCR Basic HS Mix, 20 μl hot-start Taq DNA polymerase (5 U/μl), 6 μl ROX (0.1 mM in DMSO). Up- or downscale for the exact number of reactions you want to run.
2. Each PCR reaction contains 2 μl template DNA (IP or total), 3 μl primer mix (3.3 μM, see Note 15), and 5 μl EvaGreen master mix. Prepare all PCR reactions in triplicate. Transfer the samples to a multiwell plate compatible with your real-time PCR system. Seal the reaction plate and centrifuge briefly (500 × *g*, 30 s).

3. Run the following program for PCR amplification: 1 cycle (2 min, 95°C) followed by 40 cycles (5 s, 95°C; 5 s, 53°C; 30 s, 72°C). Choose the option ROX for the passive dye control.
4. Check the reproducibility of your triplicate amplification curves and eventually eliminate outliers. Determine the mean threshold cycle (Ct) for your triplicate reactions at a constant threshold for all amplifications.
5. For each DNA region tested, calculate the IP efficiency (IE) with the formula: $IE = 1.9^{Ct(total) - Ct(IP)}$ (see Note 16).

4. Notes

1. Formaldehyde is toxic and should not be stored at temperatures <15°C. Wear gloves and work in the hood.
2. Slightly heat (microwave) to allow glycine to dissolve completely. The mixture should be prepared freshly as it precipitates when stored for a longer time.
3. TBS buffer can be prepared as a 10× stock solution.
4. Prepare FA lysis buffer with all components, except PMSF, and store at 4°C. Add the required amount of PMSF from a 100 mM stock solution in isopropanol just before use.
5. Supplement FA lysis buffer with 350 mM NaCl from a 5 M stock to reach a final concentration of 0.5 M.
6. Phenol is highly corrosive and can cause skin burns. Wear gloves and protective clothing while handling phenol in the hood.
7. Prepare PBS buffer as a 10× stock solution without BSA, autoclave, and store at room temperature. Prepare 1× PBS freshly and supplement with solid BSA to a final concentration of 5 mg/ml.
8. Dissolve in dimethylsulphoxide (DMSO) to a final concentration of 10 mM. This stock solution should be stored at −20°C with light protection.
9. The growth conditions and the choice of yeast strains depend on the specific experiment. We routinely use yeast strains which express multiply HA- or myc-epitope-tagged fusion proteins from their chromosomal loci. The protocol described here is scaled in such a way that the initial steps of chromatin preparation can be done in 50-ml Falcon tubes. The formaldehyde treatment causes very rapid protein–DNA and protein–protein cross links and is suitable to resolve dynamic association/dissociation events on DNA in the range of minutes.

10. At this step, the large chromatin complexes are not soluble and accumulate in the pellet.
11. Efficient fragmentation of the chromatin samples is important to obtain good IP signals over the total chromatin control. Using a Dr Hielscher UP200S sonicator with a microtip, we usually apply 6 sonication rounds of 30 s to each sample. The sonicator settings are: cycle 0.75, amplitude 0.75. The samples should be kept on ice for 3 min between the treatments. It is important to completely introduce the sonicator tip into the sample to avoid excessive foam formation. You can check the size of the DNA fragments produced by your particular sonication treatment by resolving an aliquot of the total chromatin after decross link, phenol extraction, and RNase treatment on a 1.5% agarose gel.
12. The amount of antibody required to efficiently immunoprecipitate your protein of interest from cross-linked chromatin in the conditions described here has to be determined experimentally for each individual antibody. We routinely use HA- and myc-epitope-tagged fusion proteins for ChIP applying a total of 25 μg of anti-HA (12CA5, Roche Cat no. 11666606001) or 25 μg of anti-myc (9E10, Roche Cat no. 11667203001) for each IP. The use of protein A is recommended for the IP of rabbit polyclonal and IgG2b-type mouse monoclonal antibodies while protein G is recommended for the IP of all types of mouse monoclonal antibodies.
13. A stock solution of Pronase can be prepared at 20 mg/ml in TE buffer and stored at −20°C.
14. The incubation at 65°C can be performed overnight.
15. The primer design is crucial for a successful analysis of ChIP by qPCR. We strongly recommend the use of specific software for the primer design. We routinely use Primer3 (8) and set the "3′-self compatibility" to 0 to avoid primer-dimer artefacts in the PCR amplification. Primers should be 20–24 nucleotides long, have a T_m of approximately 60°C, and span a genomic region of 75–150 base pairs. Primer pairs can be prepared at 3.3 μM in water and stored at −20°C.
16. If possible, include in the PCR analysis one or several DNA regions to which your protein of interest does not bind. This allows to calculate the fold occupancy for your factor, which is the ratio: IE at the bound region/IE at the unbound region or $1.9^{(Ct(total)-Ct(IP))bound-(Ct(total)-Ct(IP))unbound}$. Another important control is a "no-template" sample which contains the specific primer mix, EvaGreen master mix, but no DNA template to control for unspecific amplification mainly caused by primer-dimer formation.

References

1. Aparicio, O., Geisberg, J. V., Sekinger, E., Yang, A., Moqtaderi, Z., and Struhl, K. (2005) Chromatin immunoprecipitation for determining the association of proteins with specific genomic sequences in vivo. *Curr Protoc Mol Biol Chapter 21*, Unit 21 23.
2. Collas, P. (2010) The current state of chromatin immunoprecipitation. *Mol Biotechnol* **45**, 87–100.
3. Buck, M. J., and Lieb, J. D. (2004) ChIP-chip: considerations for the design, analysis, and application of genome-wide chromatin immunoprecipitation experiments. *Genomics* **83**, 349–360.
4. Ren, B., and Dynlacht, B. D. (2004) Use of chromatin immunoprecipitation assays in genome-wide location analysis of mammalian transcription factors. *Methods Enzymol* **376**, 304–315.
5. Johnson, D. S., Mortazavi, A., Myers, R. M., and Wold, B. (2007) Genome-wide mapping of in vivo protein-DNA interactions. *Science* **316**, 1497–1502.
6. Park, P. J. (2009) ChIP-seq: advantages and challenges of a maturing technology. *Nat Rev Genet* **10**, 669–680.
7. Robertson, G., Hirst, M., Bainbridge, M., Bilenky, M., Zhao, Y., Zeng, T., Euskirchen, G., Bernier, B., Varhol, R., Delaney, A., Thiessen, N., Griffith, O. L., He, A., Marra, M., Snyder, M., and Jones, S. (2007) Genome-wide profiles of STAT1 DNA association using chromatin immunoprecipitation and massively parallel sequencing. *Nat Methods* **4**, 651–657.
8. Rozen, S., and Skaletsky, H. (2000) Primer3 on the WWW for general users and for biologist programmers. *Methods Mol Biol* **132**, 365–386.

Chapter 11

Mapping Protein–DNA Interactions Using ChIP-Sequencing

Charles E. Massie and Ian G. Mills

Abstract

Chromatin immunoprecipitation (ChIP) allows enrichment of genomic regions which are associated with specific transcription factors, histone modifications, and indeed any other epitopes which are present on chromatin. The original ChIP methods used site-specific PCR and Southern blotting to confirm which regions of the genome were enriched, on a candidate basis. The combination of ChIP with genomic tiling arrays (ChIP-chip) allowed a more unbiased approach to map ChIP-enriched sites. However, limitations of microarray probe design and probe number have a detrimental impact on the coverage, resolution, sensitivity, and cost of whole-genome tiling microarray sets for higher eukaryotes with large genomes. The combination of ChIP with high-throughput sequencing technology has allowed more comprehensive surveys of genome occupancy, greater resolution, and lower cost for whole genome coverage. Herein, we provide a comparison of high-throughput sequencing platforms and a survey of ChIP-seq analysis tools, discuss experimental design, and describe a detailed ChIP-seq method.

Key words: Chromatin immunoprecipitation, High-throughput sequencing, "Next-generation" sequencing, Illumina (Solexa) sequencing, Transcription, Cancer, Nuclear hormone receptor, Androgen receptor

1. Introduction

Chromatin immunoprecipitation (ChIP) has been used to map specific histone modifications (1) and sites of protein occupancy (2–4), providing insights into the dynamic changes associated with gene transcription and identifying the target genes of transcription factors and other regulatory proteins. The ChIP method utilizes specific antibodies to allow enrichment of the genomic regions associated with epitopes corresponding to chromatin marks or chromatin-bound proteins. Since many of these interactions are transient, protein–DNA interactions are cross-linked using formaldehyde, allowing their isolation by ChIP. Chromatin is then

Ales Vancura (ed.), *Transcriptional Regulation: Methods and Protocols*, Methods in Molecular Biology, vol. 809,
DOI 10.1007/978-1-61779-376-9_11, © Springer Science+Business Media, LLC 2012

fragmented to allow separation of genomic fragments bound by the transcription factor of interest away from those which are not bound. Following antibody enrichment (immunoprecipitation), formaldehyde cross links are reversed and enriched DNA fragments are then purified. ChIP-enriched genomic DNA fragments can be mapped to the genome and quantified using genomic microarrays (ChIP-chip), direct sequencing (ChIP-seq), or quantitative PCR. Each of these approaches offers specific advantages and disadvantages. For example, the application of quantitative PCR to assess ChIP-enriched DNA fragments requires a priory knowledge of ideally both the target genes and binding sequences which are associated with a given transcription factor. Genomic tiling microarrays have been widely used to map transcription factor-binding sites (ChIP-chip), with genomic coverage targeted to specific regions of interest (5), gene promoter regions (3), or whole non-repetitive genomes (4), and have allowed assessment of histone marks in as few as 10^3 cells (6) and transcription factor binding from $>10^4$cells (7). However, ChIP-chip approaches require many rounds of amplification to yield sufficient DNA quantities for microarray hybridization and are limited by the restraints of microarray probe design, resulting in uneven genomic coverage and low resolution of DNA-binding sites. Given the large size of higher eukaryote genomes (e.g. human genome $\sim 3 \times 10^9$ bp), genome tiling microarrays covering the whole non-repetitive genome are printed on 7–38 individual microarrays (depending on the platform and probe spacing); thus, the amount of ChIP DNA required and the cost of each replicate are high. The combination of ChIP with second-generation sequencing technologies (ChIP-seq) offered an attractive alternative to ChIP-chip, requiring fewer amplification steps, providing more complete genome coverage, increasing the resolution of DNA-binding sites and reducing the cost of whole genome coverage (Table 1) (8). Three second-generation sequencing platforms have been developed (Table 1), although platforms which yield many short sequence reads and give greater special resolution are more appropriate for ChIP-seq analysis (Table 1). Most ChIP-seq studies to date have used Illumina (Solexa) sequencing to identify ChIP-enriched DNA fragments, and these studies have shown the benefits of greater genomic coverage and special resolution in the study of transcriptional regulation and chromatin organization.

However, this relatively new technology is not without its problems; specifically, ChIP-seq approaches require access to expensive specialist equipment, have poorly defined standards for data analysis, and recommend an order of magnitude more starting material than ChIP-chip approaches. A recent study reported a new ChIP-seq method using the Heliscope single-molecule sequencing platform (Helicos) allowing successful genome-wide coverage from low cell numbers, without the need for amplification steps (9).

Table 1
Comparison of high-throughput sequencing platforms

Platform	Read length	Coverage	ChIP-seq advantages	Approximate cost/ genome coverage
Illumina G2 Illumina HiSeq	35–250 bp 35–100 bp SE/PE 26-	10–95 Gb/run 26–200 Gb/run	Established protocols, widely used	Not listed ~30× coverage of human genome in a single run for under $10,000 (USD)
Roche/454 Genome Sequencer FLX Titanium	400 bp average (50–500 bp)	400–600 Mb/run	Least suitable for ChIP-seq	Not listed (high)
ABI/Solid	35–75 bp	~300 Gb/run	Deep coverage	$3,000 cost/ genome
Helicos Heliscope	25–55 bp reads	25–35 Gb/run	No amplification, less GC bias	Not listed (low, limited sample preparation required)

However, this platform has not yet been extensively tested in comparison with other platforms and in our hands; ChIP-seq using Illumina (Solexa) sequencing allows identification of transcription factor-binding sites from small amounts of starting material (IM and CM unpublished data). The problem of data analysis is related to the relatively recent development of these second-generation sequencing technologies and many analysis tools have been developed (Table 2). However, the performance of these analysis packages varies greatly, as highlighted by a recent community challenge which reported a detailed comparison of 11 currently available ChIP-seq analysis packages (30). We have attempted to provide a comprehensive list of all available analysis tools and the studies which have tested these different methods (Table 2); however, there is no clear consensus to recommend which analysis tools may be most appropriate for a given data set. Therefore, we would recommend comparing the results of two or more analysis tools and cross-checking the analysis results with the aligned sequence tag pile-ups to ensure that false-positive and false-negative rates are low.

However, a more fundamental question remains with regard to genome-wide coverage using any genome-wide ChIP methods (ChIP-chip or ChIP-seq), since most studies resort to taking arbitrary windows around genes of 10–100 kb in an attempt to link

Table 2
Compendium of ChIP-seq analysis tools currently available

Package	Download site	Description	Citation
Cisgenome	http://www.biostat.jhsph.edu/~hji/cisgenome/	Conditional binomial model	(10)
ERANGE/CSPF	http://woldlab.caltech.edu/rnaseq/	Poisson distribution	(11)
Fseq	http://www.genome.duke.edu/labs/furey/software/fseq	Continuous seq-tag density estimation	(12)
Genomatix	http://www.genomatix.de/online_help/help_regionminer/ChIPSeqWorkflow.htm	Commercial uses: NGS Analyzer or MACS	commercial
MACS	http://liulab.dfci.harvard.edu/MACS/	Empirically models fragment lengths and uses a dynamic Poisson distribution	(13)
Kharchenko et al. (CSPP)	http://compbio.med.harvard.edu/Supplements/ChIP-seq	Poisson distribution	(14)
Partek	http://www.partek.com/software	Commercial	commercial
PICS	http://www.bioconductor.org/packages/2.6/bioc/html/PICS.html	Probabilistic inference of ChIP-Seq using an empirical Bayes mixture model approach	(15)
SWEMBL	http://www.ebi.ac.uk/~swilder/SWEMBL/	Markov chain in both strand directions, using penalties proportional to the number of reads	(16)
USeq	http://useq.sourceforge.net/	Suite of applications for analyzing Illumina, SOLiD, and 454 data	(17)
BayesPeak	http://www.compbio.group.cam.ac.uk/Resources/BayesPeak/csbayespeak.html	Negative binomial distribution, Bayesian posterior probabilities	(18)
HPeak	http://www.sph.umich.edu/csg/qin/HPeak/	Hidden Markov model-based approach	(19)

XSET	Not listed	Simple height criteria following eXtension of single-end tags (XSET)	(20)
QuEST	http://mendel.stanford.edu/sidowlab/downloads/quest/	Kernel density estimation	(21)
FindPeaks	http://vancouvershortr.sourceforge.net/	Simple height criteria	(22)
SISSR	http://www.rajajothi.com/sissrs/	Compares reads on different strands	(23)
PeakSeq	http://www.gersteinlab.org/proj/PeakSeq/	Sample normalization binomial distribution	(24)
Mikkelsen et al.	Not listed	Supervised hidden Markov Model *p*-values from permutations	(25)
ChIPDiff	http://cmb.gis.a-star.edu.sg/ChIPSeq/tools.htm	Hidden Markov model	(26)
ChIP-Seq	http://ccg.vital-it.ch/chipseq/	Modular online analysis tool	(1)
ChromaSig	http:// bioinformatics-renlab.ucsd.edu/rentrac/wiki/ChromaSig	Unsupervised learning method (probabilistic method)	(27)
PoissonMixtureModel	Not listed	Mixture model-based analysis method for RNAP II analysis	(28)
Seqminer	http:// bips.u-strasbg.fr/seqminer/tiki-index.php	Allows comparisons between reference genome and multiple ChIP-seq datasets	
Sole-Search	http:// havoc.genomecenter.ucdavis.edu/	One sample *t*-test	(29)

transcription factor-binding sites to their functional targets. Therefore, most studies could provide the same level of utilized data using genomic tiling microarrays focussed on protein-coding genes and miRNA loci. This situation is likely to continue until information about the three-dimensional structure of chromatin in the nucleus is mapped for a given cell type, opening the door for future studies and more comprehensive analyses of the existing data sets.

2. Materials

The materials and methods described below are adapted from previously published protocols (2, 3, 31) and provide a focussed description of positive- and negative-control experiments to measure androgen-stimulated AR binding using ChIP in combination with direct Solexa sequencing (Illumina). However, these methods are also more generally applicable to the study of AR in other contexts and also the study of other chromatin-bound factors.

2.1. Cell Culture and Cross Linking

1. RPMI supplemented with 10% fetal bovine serum (FBS).
2. Phenol red-free RPMI supplemented with 10% charcoal dextran-stripped FBS.
3. AR ligands (e.g. DHT or the synthetic androgen R1881).
4. 11% formaldehyde in 50 mM HEPES–KOH, pH 7.5, 100 mM NaCl, 1 mM EDTA, 0.5 mM EGTA (see Note 1).
5. Formaldehyde-quenching solution of 2.5 M glycine.

2.2. Harvesting Cells and Sonication

1. Cell scrapers.
2. Phosphate-buffered saline (PBS) supplemented with Complete protease inhibitors (Roche).
3. Rotating tube mixer at 4°C.
4. Sonicating water bath (e.g. Diagenode Bioruptor) or probe sonicator (see Note 2).
5. ChIP cell lysis buffer: 50 mM HEPES–KOH, pH 7.5, 140 mM NaCl, 1 mM EDTA, 10% glycerol, 0.5% Igepal (NP-40), 0.25% Triton X-100, 1× SIGMAFAST protease inhibitors (Sigma).
6. ChIP nuclei wash buffer: 10 mM Tris–HCl, pH 7.5, 200 mM NaCl, 1 mM EDTA, 0.5 mM EGTA.
7. ChIP nuclear lysis buffer: 10 mM Tris–HCl, pH 7.5, 100 mM NaCl, 1 mM EDTA, 0.5 mM EGTA, 0.1% sodium deoxycholate, 0.5% SDS (see Note 3).
8. 10% Triton X-100.
9. Agarose.

2.3. Immuno-precipitation

1. Protein A/G Dynal magnetic beads (Invitrogen) and magnetic tube rack (see Note 4).
2. Antibodies to target proteins (e.g. rabbit anti-AR N20, Santa Cruz) (see Note 5 for resources listing validated ChIP-grade antibodies).
3. PBS supplemented with 0.5% BSA.
4. RIPA ChIP wash buffer: 50 mM HEPES, pH 7.6, 1 mM EDTA, 0.5 M LiCl, 1% Igepal (NP-40), 0.7% sodium deoxycholate.
5. TE with 50 mM NaCl: 10 mM Tris–HCl, pH 8.0, 1 mM EDTA, 50 mM NaCl.
6. Elution buffer: 1% SDS, 0.1 M $NaHCO_3$.

2.4. DNA Isolation

1. TE: 10 mM Tris–HCl, pH 8.0, 1 mM EDTA.
2. RNase A, 1 mg/ml (DNase-free).
3. Proteinase K, 20 mg/ml.
4. 5 M NaCl.
5. Glycogen (Roche) or suitable carrier for precipitation.
6. Phenol:chloroform:isoamyl alcohol (25:24:1).
7. Isopropanol.
8. 75% ethanol.
9. 10 mM Tris–HCl, pH 8.0.

2.5. Analysis of Enrichment

2.5.1. Real-Time PCR

1. Oligonucleotide primers to genomic regions of interest.
2. Sybr Green PCR master mix (Applied Biosystems).
3. Optical PCR plates and adhesive covers compatible with the real-time PCR instrument.

2.5.2. ChIP-Seq

1. T4 DNA polymerase.
2. Klenow DNA polymerase.
3. T4 polynucleotide kinase.
4. dNTP mix (10 mM).
5. DNA Clean and Concentrator-5 Kit (Zymo Research).
6. Klenow fragment (3′→5′ exo-minus, 5 U/μl).
7. dATP (1 mM).
8. Illumina oligonucleotide adapters.
9. T4 DNA ligase.
10. Phusion DNA polymerase.
11. Illumina oligonucleotide primers 1.1 and 2.1.
12. Dedicated "clean" electrophoresis equipment.

13. High purity agarose (e.g. Low Range Ultra Agarose, Bio-Rad).
14. Sybr safe DNA stain.
15. Dark reader transilluminator (Clare Chemical).
16. Qiagen MinElute Gel Extraction Kit.
17. Agilent Bioanalyser (see Note 6).

3. Methods

The ChIP-seq method described below has been successfully used to map AR-binding sites, binding sites for other transcriptional regulators, and RNAP II occupancy using Illumina (Solexa) sequencing. However, the basic ChIP method presented could equally be applied to the study of other factors and the same experimental design, quality control tests, and analysis tools can be used in combination with other sequencing platforms (using the appropriate library preparation methods).

The AR is activated by androgen stimulation, allowing the same antibody to be used in both positive- and negative-control conditions, thus providing an ideal control for antibody specificity; however, other controls are widely used for non-inducible systems or steady-state measurements (see Note 7).

3.1. Cell Culture and Cross Linking

1. Maintain LNCaP cells in RPMI supplemented with 10% FBS in cell culture incubators (5% CO_2 at 37°C) and passage at a dilution of 1:3 when approaching confluence with trypsin/EDTA. For ChIP-seq assays, we use 10^7 cells for each ChIP reaction.
2. When cells are ~70% confluent, aspirate media from culture flasks, wash cells with PBS, and replace media with phenol red-free RPMI supplemented with 10% charcoal dextran-stripped FBS.
3. After 72 h, replace cell culture media with media supplemented with androgens (e.g. 1 mM R1881) or an equal volume of ethanol (vehicle) and return cells to the incubator for 4 h.
4. For every 10 ml of culture media, add 1/10th volume of 11% formaldehyde solution to cell culture media. Incubate flasks at room temperature for 10 min. (see Note 1).
5. Quench the formaldehyde cross-linking reaction by adding 1/20th volume of 2.5 M glycine solution directly to culture media (to give a final concentration of 125 mM) and incubate for 5 min at room temperature.

3.2. Harvesting Cells and Sonication

1. Transfer flasks to ice, remove media, and wash cells twice with ice-cold PBS supplemented with protease inhibitors.
2. Aspirate PBS from cells and harvest cells using a cell scraper. Transfer cells to a 15-ml tube using a wide-bore pipette tip and centrifuge cells at 1,000 × *g* for 3 min at 4°C.
3. Aspirate residual PBS and add 5 ml of ChIP cell lysis buffer per 5×10^6 cells. Incubate on a rotary tube mixer at low speed for 10 min at 4°C. Centrifuge at 1,200 × *g* for 5 min in a bench-top centrifuge and discard the supernatant.
4. Re-suspend pellet in 5 ml of ChIP nuclei wash buffer, and incubate on a rotary tube mixer at low speed for 5 min at 4°C. Centrifuge at 1,200 × *g* for 5 min in a bench-top centrifuge and discard the supernatant.
5. Re-suspend pellet in 1 ml of ChIP nuclear lysis buffer. Split nuclear lysate into 4× 250-μl aliquots and sonicate for 15 min at maximum power in a Bioruptor sonicator (Diagenode) to fragment chromatin to an average length of 500 bp (see Note 2). Re-pool sonicated lysates, add 100 μl of 10% Triton X-100, and centrifuge in a bench-top microfuge at 14,000 × *g* for 10 min at 4°C. Transfer the supernatant to a 15-ml tube and add 2 ml of ChIP nuclear lysis buffer and 200 μl of 10% Triton X-100.
6. Take 50 μl from each sample for the total genomic input control. Assess the extent of sonication by electrophoresis on a 1% agarose gel, after reversing formaldehyde cross links (*see* below). A smear should be visible, with the majority of fragments between 250 bp and 1 kb.

3.3. Immunoprecipitation

1. Aliquot 100 μl of protein-A Dynal beads per ChIP reaction and wash three times with 1 ml of PBS–BSA (0.5%), collecting beads on a magnetic rack in between washes. Re-suspend beads in 250 μl of PBS–BSA, add 7.5 μg of AR N20 antibody (Santa Cruz), and incubate overnight on a rotary mixer at 4°C. Wash antibody–bead complexes three times with 1 ml of PBS–BSA, using a magnetic rack, and re-suspend in 100 μl of PBS–BSA.
2. Add 100 μl of antibody–bead complexes to the 3 ml of sonicated lysates in 15-ml tubes and incubate overnight at 4°C on a rotary mixer at low speed.
3. Working in a cold room at 4°C, transfer chromatin–antibody–bead complexes to 1.5-ml centrifuge tubes by sequentially adding 1 ml of the mixture to tubes on a magnetic rack and discarding the supernatant.
4. Wash beads six times with 1 ml of ice-cold RIPA ChIP wash buffer, taking care to fully re-suspend beads during washes and using a magnetic rack to immobilize beads between washes.

5. Wash beads once with 1 ml of ice-cold TE supplemented with 50 mM NaCl, immobilize beads using a magnetic rack, and discard the supernatant. Centrifuge at 3,000 × *g* in a bench-top microfuge for 2 min at 4°C and discard the residual supernatant.
6. Elute ChIP material by adding 200 μl of ChIP elution buffer and incubating at room temperature for 1 min with vigorous mixing. Centrifuge samples at 300 × *g* for 3 min to pellet beads and transfer the supernatant to a fresh tube. Repeat the elution and combine eluates.

3.4. DNA Isolation

1. Reverse protein–DNA cross links of total genomic input control and ChIP samples by adding NaCl to a final concentration of 200 mM and incubate at 65°C overnight.
2. Digest contaminating RNA by adding RNase A (20 μg/ml final concentration). Incubate at 37°C for 30 min.
3. Digest proteins by adding EDTA (10 mM final concentration), Tris–HCl, pH 6.7 (20 mM final concentration), and proteinase K (80 μg/ml final concentration). Incubate at 55°C for 1 h.
4. Recover DNA by adding an equal volume (~500 μl) of phenol:chloroform:isoamyl alcohol, vortex vigorously, and centrifuge at 13,000 × *g* for 15 min. Carefully transfer the aqueous phase to a fresh tube, add 20 μg of glycogen and an equal volume of isopropanol. Vortex vigorously and centrifuge at 13,000 × *g* for 15 min. Discard supernatant and add 1 volume of 75 % ethanol per volume of isopropanol used. Centrifuge at 7,000 × *g* for 8 min, discard supernatant, and air dry pellet. Re-suspend pellet in 60 μl of 10 mM Tris–HCl, pH 8.0.

3.5. Analysis of Enrichment

Two methods for quantifying ChIP enrichment are described below: ChIP-quantitative PCR and ChIP-seq. When designing and analysing ChIP experiments, it is important to bear in mind that ChIP enriches rather than isolates genomic targets. Therefore, while the enriched protein-bound DNA fragments are highly enriched, the majority of DNA isolated from a ChIP reaction is likely to comprise fragments not bound by the protein of interest purely because these DNA fragments constitute such a large proportion of the genome.

3.5.1. Real-Time PCR

ChIP, together with quantitative real-time PCR, is currently the "gold standard" for assessing or confirming ChIP enrichment of specific genomic targets and as such is a useful tool for assessing ChIP efficiency prior to genome-wide analysis and for confirming the results of genome-wide analyses. When analysing ChIP enrichment using real-time PCR, it is necessary to compare the test ChIP with a control ChIP (see Note 3) to assess specific enrichment over background. It is also necessary to compare the candidate genomic region (i.e. the region believed to be bound by the protein of interest) with a control genomic region which is not bound by the protein. The control genomic region allows an assessment of the non-specific DNA from each ChIP and can be used to normalize between the test and control ChIPs. Finally, in order to avoid bias caused by differences in PCR efficiency between test and control PCR reactions, it is advisable to use a serial dilution of input material as a standard curve for each PCR reaction (e.g. 1× to 1/128).

1. Into an optical PCR plate, aliquot 1 μl of ChIP DNA and standard curve samples in triplicate for each genomic region to be analyzed by real-time PCR (usually, a minimum of two PCR reactions, for the candidate region and the control region).
2. Mix 10 pmol of each primer with water and SybrGreen Master mix to a 1× final concentration in a final volume of 10 μl.
3. Aliquot PCR mix onto ChIP DNA, seal plates with adhesive covers, and centrifuge briefly.
4. Use the PCR conditions suggested for use with the SybrGreen mix used (e.g. hot start: 50°C 2 min, 95°C 10 min (95°C 15 s, 60°C 1 min), repeat 40 times). The addition of a dissociation curve at the end of the PCR allows an assessment of the specificity of the PCR.
5. Specific enrichment by ChIP can be assessed using the equation:

$$\text{Relative enrichment} = \text{ControlEff}^{(\text{control sample Ct}-\text{ test sample Ct})} / \text{TestEff}^{(\text{control sample Ct}-\text{test sample Ct})}$$

Here, "ControlEff" is the efficiency of the control PCR and "TestEff" is the efficiency of the test PCR, both calculated using the formula: $10^{(1/-\text{slope of standard curve})}$. "Control sample Ct" and "test sample Ct" are the cycle thresholds (Ct) at which the PCR reactions for control or test samples become exponential.

3.5.2. Direct Sequencing

The combination of sequencing-based approaches with ChIP circumvents many of the problems associated with ChIP-chip (e.g. probe design, probe specificity, genome coverage, and bias introduced by amplification). Comparison of ChIP-seq and ChIP-chip for the STAT1 transcription factor revealed a 64–71% overlap between the binding sites identified by both techniques, although ChIP-seq found 3.8-fold more binding sites in total suggesting that it is the more sensitive method (32). The emergence of third-generation, single-molecule sequencing platforms, such as the Heliscope system (Helicos), and the promise of nanopore technology (33) suggest that in the near future the amount of starting material required will be greatly reduced and post-ChIP amplification steps may no longer be required. However, for most applications, where starting material is not severely limited (e.g. larger tissue samples or cell lines), and given the ever-increasing capacity and falling cost of second-generation sequencing technologies, it seems likely that this will remain the platform of choice for ChIP-seq studies. Accordingly, the method outlined below describes the ChIP-seq library preparation for the second-generation Solexa (Illumina) sequencing platform.

1. Blunt DNA fragments from 50 μl of the ChIP material and 50 ng of total genomic input control using 15U T4 DNA polymerase, 5U Klenow DNA polymerase, and 50U T4 polynucleotide kinase in 1× T4 DNA ligase buffer. Incubate for 30 min at 20°C. Clean up DNA using the Zymogen DNA Clean and Concentrate-5 kit, eluting in 32 μl of preheated EB.
2. Add A-overhangs to 32 μl of blunted ChIP and input DNA samples using 15U Klenow 3′→5′ exo-minus, with 200 μM dATP in Klenow buffer. Incubate at 37°C for 30 min. Clean up DNA using the Zymogen DNA Clean and Concentrate-5 kit, eluting in 8 μl of preheated EB.
3. Ligate Illumina adapters to 8 μl of ChIP and input DNA using 2 μl of adapters and 2.5 μl of Quick T4 DNA ligase (NEB) in 1× Quick DNA ligase buffer. Incubate at room temperature for 15 min. Clean up DNA using the Zymogen DNA Clean and Concentrate-5 kit, eluting in 23 μl of preheated EB.
4. Amplify and enrich adapter-ligated DNA fragments by PCR using 23 μl of ChIP and input DNA, 1 μl of Illumina PCR primer 1.1, 1 μl of Illumina PCR primer 2.1, 0.5 μl of Phusion DNA polymerase, 1× high-fidelity PCR buffer, and 200 μM dNTP. Amplify using the following conditions: 98°C 30 s, (98°C 10 s, 65°C 30 s, 72°C 30 s), repeat 17 times, 72°C 5 min. Clean up amplified material using a Qiagen PCR purification kit, eluting DNA in 30 μl of preheated EB.
5. Size select ChIP and total genomic input material using agarose gel electrophoresis. Pour a 2% TAE agarose gel with 1× SybrSafe DNA stain. Load DNA ladder, ChIP, and input samples using only glycerol (12% final glycerol concentration). Use specific electrophoresis equipment for library preparation and run only one library per gel. Run gel at 120 for 45 min, visualize on a Dark Reader transilluminator, and excise the 200–300 bp part of the DNA smear. Purify the DNA using a Qiagen MiniElute Gel Extraction Kit, eluting in 15 μl of preheated EB (see Note 8).
6. Measure DNA concentration on an Agilent Bioanalyser chip and proceed with sequencing on an Illumina Genome Analyzer.
7. Base calling from raw image files, quality control of sequence reads, alignment of short sequence reads to the reference genome, removal of exact duplicate reads, and peak calling

require the implementation of an Illumina bioinformatics pipeline in combination with other bioinformatics programs (e.g. MAQ and MACS), requiring significant bioinformatics support (see Note 9). With regard to the analysis of filtered and aligned data, there are many ChIP-seq peak calling packages (Table 2) and 11 have recently been directly compared (12). Careful implementation of multiple analysis packages (e.g. the best performing from direct comparisons: MACS 1.3.5, USeq, Partek, SWEMBL, or BPC) should provide the most reliable peak calls. These analysis tools require specialist bioinformatics support and so simplified Web-based analysis tools offer a useful alternative to wet lab scientists (e.g. Cistrome or Sole Search) (34, 35), although it is likely that these tools may be most useful for first-pass analysis.

4. Notes

1. Cross-linking of protein–DNA interactions is commonly used when studying transcription factor binding. Formaldehyde is most widely used for this purpose and produces covalent cross links between amino or imino groups, which are within 2 Å from each other. It is also possible to use other cross-linking agents, such as imidoesters or NHD-esters (e.g. dimethyl pimelimidate (DMP) or disuccinimidyl glutarate (DSG)) in combination with formaldehyde, to increase the efficiency of cross linking, which may be most applicable to low-abundance DNA-binding proteins (36). The use of imidoesters or NHS-esters as cross-linkers also provides an opportunity to alter the resolution from 2 to 20 Å, depending on the spacer length of the ester used (36). It is possible to perform ChIP without cross linking (i.e. native ChIP); however, this is only suitable for proteins which bind stably to DNA and is mainly used in ChIP assays for histones (37).
2. It is essential to optimize sonication conditions for each cell type and sonicator, since this step defines the resolution of ChIP. It is advisable to test a range of conditions, including length of pulse, number of pulses, and amplitude of sonication. The efficiency of sonication should be assessed by resolving a sample of total chromatin after sonication and decross-linking using agarose gel electrophoresis.
3. Sodium dodecyl sulphate (SDS) is liable to precipitate and can cause foaming when used with probe-based sonicators. Therefore, *N*-lauroyl sarcosine can be substituted with SDS to avoid these problems. However, in our experience, when using a water bath sonicator, SDS or *N*-lauroyl sarcosine works equally well.

4. Dynal magnetic protein-A/G beads have lower non-specific DNA binding, reducing background, and increasing specific enrichment compared to either agarose or sepharose beads. In general, protein-A or protein-G beads may be used for ChIP using rabbit antibodies and protein-G may be used for ChIP using mouse, sheep, and goat IgG1 antibodies. However, we use an equal mixture of protein-A and protein-G beads to allow comparison of ChIP using antibodies from different species.
5. A number of research groups and companies provide searchable databases and compendia of validated ChIP-grade antibodies (38–44).
6. Quantitation of adapter-ligated ChIP DNA in sequencing libraries is an essential step to determine the loading on sequencing flow cells. It is common to use the Agilent Bioanalyser system to quantify ChIP-seq libraries following limited PCR enrichment of adapter-ligated DNA fragments. However, quantitative PCR-based methods have also been described, which utilize the adapter sequences to allow accurate quantitation of only DNA fragments which are attached to sequencing adapters (45).
7. There are many control ChIP experiments which can be used as a reference to assess specific enrichment in the test ChIP. The choice of which control to use depends on the system under investigation and the question to be addressed. Many studies use an IgG ChIP control to assess non-specific enrichment caused by protein–DNA complexes binding to beads or IgG. However, this control does not account for any "off-target" binding of the specific antibody used for ChIP. In order to assess the specific enrichment by a ChIP antibody, it is necessary to compare isogenic cells in which the protein of interest is not bound to DNA or alternatively lacks the target protein completely. In the case of NHRs, it is possible to compare hormone-deprived cells to cells stimulated with the specific NHR ligand resulting in nuclear translocation and DNA binding (e.g. androgen treatment to activate the AR). Where possible, the best controls for ChIP may be isogenic cells which are null for the target protein (e.g. have targeted deletions of the gene encoding the target protein) or alternatively RNAi "knock-down" of the target protein.
8. A recent methods paper has comprehensively described improvements to the preparation of Illumina sequencing libraries (45). This included a suggestion that heating dsDNA during library preparation may have an impact on GC bias in the resultant library. Therefore, the increased yields afforded by preheating elution buffers in DNA purification columns and gel extraction kits may be counter balanced by the cost of introducing experimental noise into this sensitive system, with implications for the fidelity of ChIP-seq libraries.

9. A recent review discussed the characteristics of ChIP-seq analysis software packages (46) and many of the currently available analysis tools were compared side-by-side in a ChIP-seq community challenge (30). The top scoring analysis tools were MACS 1.3.5, USeq, Partek, SWEMBL, and BPC, which used alone or in combination should provide the most reliable peak calls. However, the published comparisons may not accurately model all types of ChIP-seq data sets, not all analysis tools were included in these comparisons (Table 2), and it remains possible that some analysis tools may perform better than others for certain ChIP-seq profiles.

Acknowledgements

C. E. Massie is a postdoctoral researcher at the Department of Haematology, University of Cambridge. I. G. Mills is a research group leader within the Centre for Molecular Medicine (Norway), a Visiting Scientist at Cancer Research UK and an Honorary Senior Visiting Research Fellow within the Department of Oncology at the University of Cambridge.

References

1. Barski, A., Cuddapah, S., Cui, K., Roh, T. Y., Schones, D. E., Wang, Z., Wei, G., Chepelev, I., and Zhao, K. (2007) High-resolution profiling of histone methylations in the human genome, *Cell 129*, 823–837.
2. Carroll, J. S., Meyer, C. A., Song, J., Li, W., Geistlinger, T. R., Eeckhoute, J., Brodsky, A. S., Keeton, E. K., Fertuck, K. C., Hall, G. F., Wang, Q., Bekiranov, S., Sementchenko, V., Fox, E. A., Silver, P. A., Gingeras, T. R., Liu, X. S., and Brown, M. (2006) Genome-wide analysis of estrogen receptor binding sites, *Nature Genetics 38*, 1289–1297.
3. Massie, C. E., Adryan, B., Barbosa-Morais, N. L., Lynch, A. G., Tran, M. G., Neal, D. E., and Mills, I. G. (2007) New androgen receptor genomic targets show an interaction with the ETS1 transcription factor, *EMBO Reports 8*, 871–878.
4. Wang, Q., Li, W., Zhang, Y., Yuan, X., Xu, K., Yu, J., Chen, Z., Beroukhim, R., Wang, H., Lupien, M., Wu, T., Regan, M. M., Meyer, C. A., Carroll, J. S., Manrai, A. K., Janne, O. A., Balk, S. P., Mehra, R., Han, B., Chinnaiyan, A. M., Rubin, M. A., True, L., Fiorentino, M., Fiore, C., Loda, M., Kantoff, P. W., Liu, X. S., and Brown, M. (2009) Androgen receptor regulates a distinct transcription program in androgen-independent prostate cancer, *Cell 138*, 245–256.
5. Bolton, E. C., So, A. Y., Chaivorapol, C., Haqq, C. M., Li, H., and Yamamoto, K. R. (2007) Cell- and gene-specific regulation of primary target genes by the androgen receptor, *Genes Dev 21*, 2005–2017.
6. Dahl, J. A., Reiner, A. H., and Collas, P. (2009) Fast genomic muChIP-chip from 1,000 cells, *Genome Biol 10*, R13.
7. Acevedo, L. G., Iniguez, A. L., Holster, H. L., Zhang, X., Green, R., and Farnham, P. J. (2007) Genome-scale ChIP-chip analysis using 10,000 human cells, *BioTechniques 43*, 791–797.
8. Johnson, D. S., Mortazavi, A., Myers, R. M., and Wold, B. (2007) Genome-wide mapping of in vivo protein-DNA interactions, *Science 316*, 1497–1502.
9. Goren, A., Ozsolak, F., Shoresh, N., Ku, M., Adli, M., Hart, C., Gymrek, M., Zuk, O., Regev, A., Milos, P. M., and Bernstein, B. E. (2010) Chromatin profiling by directly sequencing small quantities of immunoprecipitated DNA, *Nature Methods 7*, 47–49.
10. Ji, H., Jiang, H., Ma, W., Johnson, D. S., Myers, R. M., and Wong, W. H. (2008) An integrated software system for analyzing

ChIP-chip and ChIP-seq data, *Nat Biotechnol 26*, 1293–1300.
11. Mortazavi, A., Williams, B. A., McCue, K., Schaeffer, L., and Wold, B. (2008) Mapping and quantifying mammalian transcriptomes by RNA-Seq, *Nature methods 5*, 621–628.
12. Boyle, A. P., Guinney, J., Crawford, G. E., and Furey, T. S. (2008) F-Seq: a feature density estimator for high-throughput sequence tags, *Bioinformatics 24*, 2537–2538.
13. Zhang, Y., Liu, T., Meyer, C. A., Eeckhoute, J., Johnson, D. S., Bernstein, B. E., Nussbaum, C., Myers, R. M., Brown, M., Li, W., and Liu, X. S. (2008) Model-based analysis of ChIP-Seq (MACS), *Genome Biol 9*, R137.
14. Kharchenko, P. V., Tolstorukov, M. Y., and Park, P. J. (2008) Design and analysis of ChIP-seq experiments for DNA-binding proteins, *Nat Biotechnol 26*, 1351–1359.
15. Zhang, X., Robertson, G., Krzywinski, M., Ning, K., Droit, A., Jones, S., and Gottardo, R. PICS: Probabilistic Inference for ChIP-seq, *Biometrics.*
16. Schmidt, D., Schwalie, P. C., Ross-Innes, C. S., Hurtado, A., Brown, G. D., Carroll, J. S., Flicek, P., and Odom, D. T. A CTCF-independent role for cohesin in tissue-specific transcription, *Genome Res 20*, 578–588.
17. Nix, D. A., Courdy, S. J., and Boucher, K. M. (2008) Empirical methods for controlling false positives and estimating confidence in ChIP-Seq peaks, *BMC Bioinformatics 9*, 523.
18. Spyrou, C., Stark, R., Lynch, A. G., and Tavare, S. (2009) BayesPeak: Bayesian analysis of ChIP-seq data, *BMC Bioinformatics 10*, 299.
19. Qin, Z. S., Yu, J., Shen, J., Maher, C. A., Hu, M., Kalyana-Sundaram, S., and Chinnaiyan, A. M. HPeak: an HMM-based algorithm for defining read-enriched regions in ChIP-Seq data, *BMC Bioinformatics 11*, 369.
20. Robertson, G., Hirst, M., Bainbridge, M., Bilenky, M., Zhao, Y., Zeng, T., Euskirchen, G., Bernier, B., Varhol, R., Delaney, A., Thiessen, N., Griffith, O. L., He, A., Marra, M., Snyder, M., and Jones, S. (2007) Genome-wide profiles of STAT1 DNA association using chromatin immunoprecipitation and massively parallel sequencing, *Nature methods 4*, 651–657.
21. Valouev, A., Johnson, D. S., Sundquist, A., Medina, C., Anton, E., Batzoglou, S., Myers, R. M., and Sidow, A. (2008) Genome-wide analysis of transcription factor binding sites based on ChIP-Seq data, *Nature methods 5*, 829–834.
22. Fejes, A. P., Robertson, G., Bilenky, M., Varhol, R., Bainbridge, M., and Jones, S. J. (2008) FindPeaks 3.1: a tool for identifying areas of enrichment from massively parallel short-read sequencing technology, *Bioinformatics 24*, 1729–1730.
23. Jothi, R., Cuddapah, S., Barski, A., Cui, K., and Zhao, K. (2008) Genome-wide identification of in vivo protein-DNA binding sites from ChIP-Seq data, *Nucleic Acids Res 36*, 5221–5231.
24. Rozowsky, J., Euskirchen, G., Auerbach, R. K., Zhang, Z. D., Gibson, T., Bjornson, R., Carriero, N., Snyder, M., and Gerstein, M. B. (2009) PeakSeq enables systematic scoring of ChIP-seq experiments relative to controls, *Nat Biotechnol 27*, 66–75.
25. Ku, M., Jaffe, D. B., Issac, B., Lieberman, E., Giannoukos, G., Alvarez, P., Brockman, W., Kim, T. K., Koche, R. P., Lee, W., Mendenhall, E., O'Donovan, A., Presser, A., Russ, C., Xie, X., Meissner, A., Wernig, M., Jaenisch, R., Nusbaum, C., Lander, E. S., and Bernstein, B. E. (2007) Genome-wide maps of chromatin state in pluripotent and lineage-committed cells, *Nature 448*, 553–560.
26. Xu, H., Wei, C. L., Lin, F., and Sung, W. K. (2008) An HMM approach to genome-wide identification of differential histone modification sites from ChIP-seq data, *Bioinformatics 24*, 2344–2349.
27. Hon, G., Ren, B., and Wang, W. (2008) ChromaSig: a probabilistic approach to finding common chromatin signatures in the human genome, *PLoS Comput Biol 4*, e1000201.
28. Feng, W., Liu, Y., Wu, J., Nephew, K. P., Huang, T. H., and Li, L. (2008) A Poisson mixture model to identify changes in RNA polymerase II binding quantity using high-throughput sequencing technology, *BMC Genomics 9 Suppl 2*, S23.
29. Blahnik, K. R., Dou, L., O'Geen, H., McPhillips, T., Xu, X., Cao, A. R., Iyengar, S., Nicolet, C. M., Ludascher, B., Korf, I., and Farnham, P. J. Sole-Search: an integrated analysis program for peak detection and functional annotation using ChIP-seq data, *Nucleic Acids Res 38*, e13.
30. http://sourceforge.net/projects/useq/files/CommunityChIPSeqChallenge/. ChIP-seq community chal.
31. Schmidt, D., Wilson, M. D., Spyrou, C., Brown, G. D., Hadfield, J., and Odom, D. T. (2009) ChIP-seq: using high-throughput sequencing to discover protein-DNA interactions, *Methods (San Diego, Calif) 48*, 240–248.

32. Robertson, G., Hirst, M., Bainbridge, M., Bilenky, M., Zhao, Y., Zeng, T., Euskirchen, G., Bernier, B., Varhol, R., Delaney, A., Thiessen, N., Griffith, O. L., He, A., Marra, M., Snyder, M., and Jones, S. (2007) Genome-wide profiles of STAT1 DNA association using chromatin immunoprecipitation and massively parallel sequencing, *Nature Methods 4*, 651–657.
33. http://www.nanoporetech.com/. Nanopore.
34. http://cistrome.dfci.harvard.edu/ap/. Cistrome.
35. http://chipseq.genomecenter.ucdavis.edu/cgi-bin/chipseq.cgi. Sole Search.
36. Nowak, D. E., Tian, B., and Brasier, A. R. (2005) Two-step cross-linking method for identification of NF-kappaB gene network by chromatin immunoprecipitation, *BioTechniques 39*, 715–725.
37. West, A. G., Huang, S., Gaszner, M., Litt, M. D., and Felsenfeld, G. (2004) Recruitment of histone modifications by USF proteins at a vertebrate barrier element, *Molecular Cell 16*, 453–463.
38. http://www.chiponchip.org/Antibody/chip.html. Compendium of ChIP grade antibodies.
39. http://www.abcam.com/index.html?c=917. Abcam ChIP grade antibodies.
40. http://www.diagenode.com/en/topics/antibodies/antibodies.php. Diagenode ChIP antibodies.
41. http://www.cellsignal.com/technologies/chip.html. Cell Signalling ChIP antibodies.
42. http://www.activemotif.com/catalog/18/chip-validated-antibodies.html. Active motif ChIP antibodies.
43. http://www.millipore.com/microsites/search.do?q=&filterProductTypes=taxonomy%3a%5e73UUAR%2f73UUB8&module=antibody&tabValue=ANTIBODY&filter=61964%3a%5e%22Epigenetics+%26+Nuclear+Function%22%24&filter=61965%3a%5e%22Chromatin+Biology%22%24&filter=60679%3a%5e%22Chromatin+Immunoprecipitation+(ChIP)%22%24&show=10#0:0. Millipore ChIP antibodies.
44. http://www.invitrogen.com/site/us/en/home/Products-and-Services/Applications/RNAi-Epigenetics-and-Gene-Regulation/Chromatin-Remodeling/Chromatin-Immunoprecipitation-ChIP/antibodies-for-chip.html. Invitrogen ChIP antibodies.
45. Quail, M. A., Kozarewa, I., Smith, F., Scally, A., Stephens, P. J., Durbin, R., Swerdlow, H., and Turner, D. J. (2008) A large genome center's improvements to the Illumina sequencing system, *Nature Methods 5*, 1005–1010.
46. Pepke, S., Wold, B., and Mortazavi, A. (2009) Computation for ChIP-seq and RNA-seq studies, *Nature Methods 6*, S22–32.

Chapter 12

ChIP and Re-ChIP Assays: Investigating Interactions Between Regulatory Proteins, Histone Modifications, and the DNA Sequences to Which They Bind

Agnieszka D. Truax and Susanna F. Greer

Abstract

Chromatin immunoprecipitation (ChIP) assays were developed in order to comprehensively describe physiological interactions between DNA sequences, transcriptional regulators, and the modification status of associated chromatin. In ChIP assays, living cells are treated with chemical cross-linkers to covalently bind proteins to each other and to their DNA targets. Once cross-linked to associated proteins, chromatin is extracted and fragmented by sonication and protein–DNA complexes are isolated using specific antibodies against a target protein. The cross-links that bind proteins to DNA are then reversed, and purified DNA fragments are analyzed by qPCR to determine if a specific sequence is present. As DNA regulatory elements frequently rely on the interaction of multiple transcription factors and cofactors to regulate gene expression, Re-ChIP methods were developed to allow for the identification of multiple (concurrently binding) proteins on a single DNA sequence. Re-ChIP assays have enabled the analysis of multiple, simultaneous, posttranslational modifications to histones in order to determine the combinatorial pattern of modifications associated with transcriptional status of a gene. Together, ChIP and Re-ChIP have contributed to the elucidation of the epigenetic code-regulating gene expression and have enhanced our understanding of physiological binding of proteins to DNA targets. The protocols that follow describe general strategies used to perform ChIP and Re-ChIP assays for the study of specific protein–DNA interactions.

Key words: Chromatin, Gene regulation, Transcription, Chromatin immunoprecipitation

1. Introduction

Protein and DNA interactions in the nucleus regulate DNA compaction and the accessibility and binding of regulatory DNA sequences with transcriptional regulators and cofactors. Elucidation of the modification status of chromatin and the in vivo association of transcription factors with regulatory regions of DNA are essential to understanding the distinct stages of transcription, cell division,

Ales Vancura (ed.), *Transcriptional Regulation: Methods and Protocols*, Methods in Molecular Biology, vol. 809,
DOI 10.1007/978-1-61779-376-9_12, © Springer Science+Business Media, LLC 2012

and DNA replication and repair. Analyses of the transcription factors bound to DNA were pioneered with electrophorectic mobility shift assays (1, 2). While these assays provided the ability to study protein–DNA interactions in vitro, they did not provide information about the binding status of proteins to DNA in intact cells. A more useful for studying protein and DNA interactions is the chromatin immunoprecipitation or ChIP assay. Development of ChIP assays was pioneered by Alexander Varshavsky and coworkers in the 1980s (3, 4) and has since proven to be a powerful tool for characterizing DNA–protein and protein–protein interactions as ChIP allows determination of specific protein associations to particular genomic regions. To date, ChIP assays have been utilized in analysis of histone modifications and have contributed significantly to the discovery of the histone code and the epigenetic regulation of gene expression.

In traditional ChIP assays, live cells are treated to covalently cross-link protein and DNA (5, 6). Following cross-linking, cells are lysed and chromatin is sheared by sonication (5) or is treated with micrococcal nucleases (7, 8). An antibody against the protein of interest is then used to immunoprecipitate fragments of DNA bound to protein. Associated DNA is sheared and incubated with low pH buffer to reverse the covalent cross-links, protease is added to digest isolated protein, and the associated DNA is isolated and analyzed for the presence of specific DNA sequences as shown in Fig. 1.

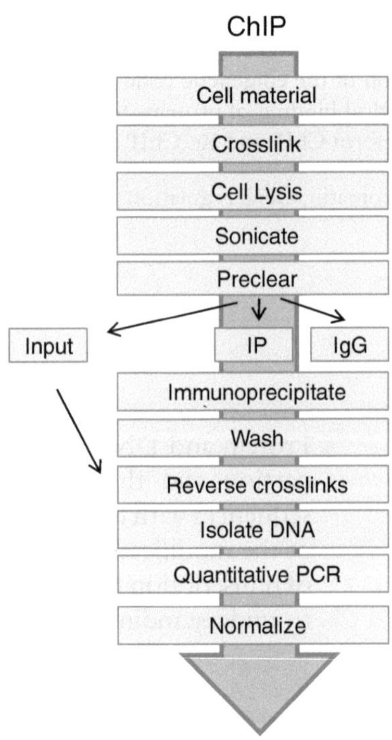

Fig. 1. ChIP protocol as described in the text.

There are two traditional types of ChIP assays based on technical differences in preparation of cross-linked chromatin. In cross-linked ChIP (XChIP), which is frequently employed for mapping DNA targets of transcription factors or other chromatin-associated proteins, cross-linking is accomplished with formaldehyde or UV light (9, 10). Cross-linked chromatin is then sheared by sonication to 500–1,000 base pairs (bp) fragments. Alternatively, Native ChIP (NChIP) is often utilized to target histone modifiers, where micrococcal nuclease treatment is employed to cleave chromatin into oligonucleosomes (7, 11). Of the two methods, the advantage of formaldehyde cross-linking is that cross-links are rapidly formed and are fully reversible (5, 12). In addition to formaldehyde, when analyzing proteins that do not directly bind to DNA or that are part of a multiprotein complex, use of a secondary cross-linker like disuccinimidyl glutarate (DSG) may be advantageous. In this way, double cross-linking significantly augments the ChIP signal compared to formaldehyde-only cross-linking by increasing intermolecular cross-linking (13). Following primary cross-linking, the protein–protein conjugation is cross-linked to DNA using formaldehyde as the second cross-linker. A purification step subsequently removes proteins from the DNA fraction through extensive digestion with proteinase K and mild heat treatments.

On the basis of ChIP, Re-ChIP assays were developed to allow for sequential immunoprecipitations determining higher order protein–protein interactions in the context of chromatin. Re-ChIP assays employ sequential immunoprecipitation reactions and are based on the premise that chromatin–protein complexes associated with an immunoprecipitated protein of interest can be re-immunoprecipitated in a subsequent immunoprecipitation reaction with an antibody against a second protein of interest present in the complex. DNA recovered in the second immunoprecipitation reaction is recovered, purified, and analyzed as shown in Fig. 2. When analyzing ChIP and Re-ChIP precipitates, PCR or quantitative PCR (QPCR) methods are used. If PCR is employed, the intensity of a DNA band visible on an agarose gel indicates the amount of DNA bound to the protein/s of interest. However, if quantification of the ChIP signal is desired, QPCR should be employed. QPCR calculates the amount of DNA associated with protein/proteins of interest based on kinetics of the QPCR, where the accumulation of PCR product is analyzed at each reaction cycle. As such, QPCR data are normalized against input samples. In the protocols that follow, we present optimized ChIP and Re-ChIP protocols with additional strategies to optimize experimental design.

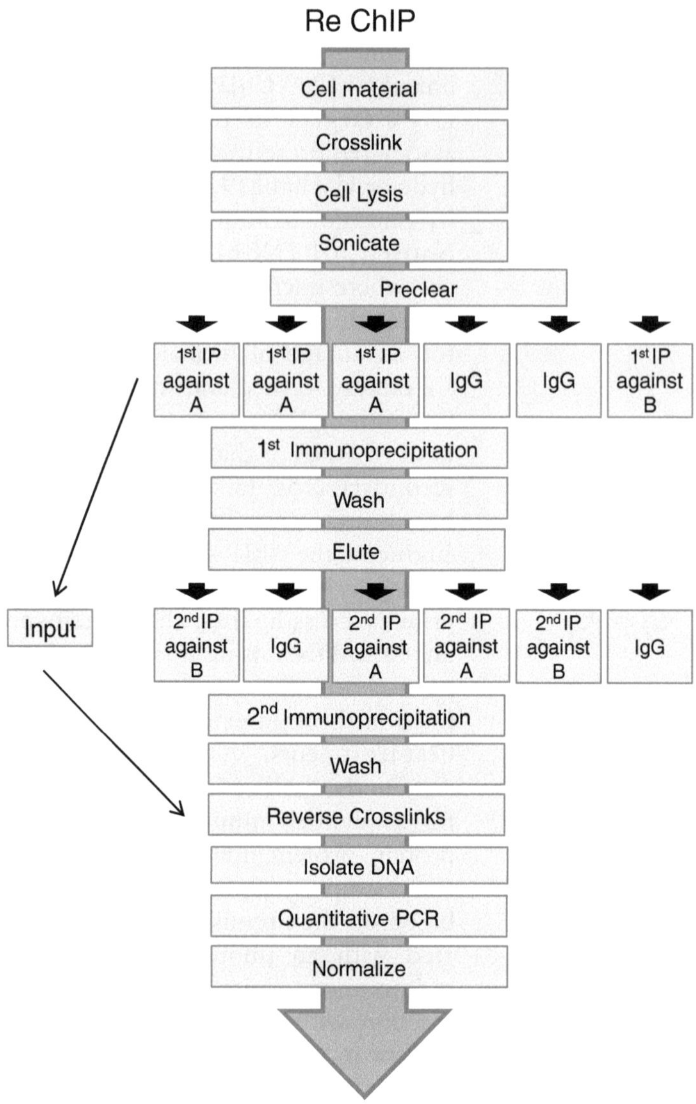

Fig. 2. Re-ChIP protocol as described in the text (various positive and negative controls for Re-ChIP assays are explained in details in Notes 3 and 7).

2. Materials

2.1. Cell Culture

1. HeLa (human epithelial) cells from ATCC (Manassas, VA): Maintain in high-glucose Dulbecco's modified Eagle medium (DMEM); culture in 150-mm dishes.

2.2. ChIP and Re-ChIP

1. 1× phosphate-buffered saline (PBS).
2. 37% Formaldehyde, Molecular Biology Reagent Grade.

3. 1 M glycine: 3.75 g in 50 ml of dH_2O, sterile filtered, store at room temperature.
4. 5 M NaCl.
5. Protease Inhibitor Stock: Prepare 50× stock by dissolving one protease inhibitor tablet (Roche cat# 11873580001) in 1 ml of nuclease-free water.
6. 70% ethanol.
7. Salmon sperm (Millipore), recombinant Protein A covalently bound to agarose by alkylamine linkage, store at 4°C. In addition to Protein A, other immunoglobulin-binding proteins, such as Protein G and Protein A/G, can be used to detect immunoglobulins.
8. Proteinase K (Roche), store at 4°C.
9. Phenol–chloroform–isopropanol mix (25:24:1).
10. Glycogen.
11. Independently designed and HPLC-purified primers and probes.
12. ChIP-validated antibodies.
13. Isotype control antibody.

2.3. Buffers

1. SDS lysis buffer: 1% SDS, 10 mM EDTA, 50 mM Tris–HCl, pH 8.0, dH_2O supplemented with protease inhibitor (Roche) immediately prior to use. Store at room temperature.
2. ChIP dilution buffer: 0.01% SDS, 1.1% Triton X-100, 1.2 mM EDTA, 16.7 mM Tris–HCl, pH 8.0, 167 mM NaCl, dH_2O supplemented with protease inhibitor (Roche) immediately prior to use. Store at 4°C.
3. Low-salt wash buffer: 0.1% SDS, 1% Triton X-100, 2 mM EDTA, 20 mM Tris–HCl, pH 8.0, 150 mM NaCl, dH_2O. Store at 4°C.
4. High-salt wash buffer: 0.1% SDS, 1% Triton X-100, 2 mM EDTA, 20 mM Tris–HCl, pH 8.0, 500 mM NaCl, dH_2O. Store at 4°C.
5. Lithium chloride (LiCl) wash buffer: 0.25 M LiCl, 1% NP40, 1% DOC, 1 mM EDTA, 10 mM Tris–HCl, pH 8.0, dH_2O. Store at 4°C.
6. TE buffer: 10 mM Tris–HCl, pH 8.0, 2 mM EDTA. Store at 4°C.
7. SDS elution buffer: 1% SDS, 0.1 M $NaHCO_3$, dH_2O. Prepare fresh, do not store.
8. Re-ChIP wash buffer: 2 mM EDTA, 500 mM NaCl, 0.1% SDS, 1% NP40, dH_2O. Store at 4°C.

9. Re-ChIP elution buffer: 1× TE, 2% SDS, 15 mM DTT, supplemented with protease inhibitor (Roche) immediately prior to use. Prepare fresh, do not store.

2.4. Instruments and Software

1. 7900HT Sequence Detection System by Applied Biosystem.
2. Tabletop swing bucket centrifuge.
3. Heat block or 45–65°C incubator.
4. Rocking platform.

2.5. Dual Cross-Linking ChIP

1. DSG is readily hydrolyzed, should be protected from moisture, and prepared immediately before use. A stock of 0.5 M DSG is dissolved in DMSO (0.0163 g DSG/100 μl of DMSO) and used at final concentration 2 mM.

3. Methods

3.1. Preparation of Cells

1. ChIP assays require copious amounts of starting material. A minimum of 1×10^6 cells is recommended for each immunoprecipitation; however, the number of cells necessary for performing ChIP assays can extend to 200 million cells and must be empirically determined. In our protocols, we typically plate $2–3 \times 10^6$ HeLa cells (or other adherent cell line) on 15-cm plates in 20 ml of media (see Note 1).
2. Stimulate or treat cells to obtain desired transcriptional status of the gene/s of interest.

3.2. Cell Lysis, Sonication, and Immunoprecipitation (as Graphed in Fig. 1)

1. Following stimulation, cross-link by adding formaldehyde to a final concentration of 1% (541 μl of 37% formaldehyde/20 ml media) directly into the tissue culture plate. Incubate for 10 min at room temperature in order to cross-link DNA to protein (see Note 2).
2. To terminate cross-linking, add glycine to a final concentration of 0.125 M (3.15 ml/20 ml media). Incubate for 5 min at room temperature.
3. Scrape the cells, transfer the cells in media to 50-ml conical tubes, and place on ice. If the cells you are using are nonadherent, collect them along with the media in a 50-ml conical tube.
4. Using a tabletop centrifuge with a swing bucket rotor, centrifuge cells at 820 × *g* for 5 min at 4°C. Wash the cell pellet once with 5 ml of cold 1× PBS and centrifuge at 820 × *g* for 5 min at 4°C. Discard the supernatant and proceed to step 5 or, alternatively, snap freeze the pellet in liquid nitrogen and store at −80°C for up to 2 months.

5. Using a Pipetman, completely resuspend the pellet in 400 μl of SDS lysis buffer (to which protease inhibitor has been added) and transfer the homogenized samples to 1.5-ml microcentrifuge tubes. Lyse samples on ice for 20 min.
6. Following initial lysis, sonicate the samples to complete cellular lysis and to shear DNA into 500–1,000 bp fragments. The time and number of pulses necessary to shear the DNA depend on the cell type used, the extent of cross-linking, and the instrument used for sonication. We have observed that for most adherent cells 2–3 cycles of sonication at a power of 3 for 30 s on a Biologics Ultrasonic Homogenizer (Model 150 V/T) is sufficient to shear chromatin. However, the number of pulses needed to sufficiently shear chromatin depends on cell density and type and must be empirically determined as cell type and density are altered (see Note 3). Following each round of sonication, samples should be placed on ice.
7. Sonication efficiency is easily analyzed by agarose gel electrophoresis to confirm appropriate shearing of chromatin. Typically, 5 μl of the sonicated samples can be visualized on a 1% agarose gel. Evenly sonicated samples run as a single band visible at the 500–1,000 bp range. If samples run as a higher molecular weight smear or if DNA is visible in the wells of the gel, the sonication is incomplete and samples should be resonicated as above (see Note 4).
8. To pellet cellular debris, centrifuge sonicated samples in a microcentrifuge at 17,000 × *g* for 10 min at 4°C and transfer the supernatant to 1.5-ml microcentrifuge tubes. Proceed directly to preclearing or the samples can be snap frozen and stored at −80°C for up to 2 months.
9. To reduce nonspecific binding and background noise, preclear samples with 60 μl of salmon sperm/protein A slurry for 30–60 min at 4°C with rotation. Pellet beads by centrifugation in a microcentrifuge at 600 × *g* for 1 min and collect supernatant fractions in 1.5-ml microcentrifuge tubes.
10. Prior to adding immunoprecipitating antibody, transfer 10 μl of the supernatant to 1.5-ml microcentrifuge tubes and freeze at −80°C. These samples serve as input controls and, as they are indicative of the total amount of chromatin present in individual ChIP reactions, are used to normalize data derived from immunoprecipitated samples (see Note 5).
11. The remaining supernatant is divided into a minimum of two immunoprecipitation reactions: an experimental sample to which 5–15 μg of antibody is added and a control sample to which the same amount of nonspecific antibody control is added as shown in Fig. 1. The amount of immunoprecipitating antibody added depends on quality, affinity, and specificity of

the antibody (see Note 6). Following addition of antibody, immunoprecipitation reactions are diluted to 1 ml with ChIP dilution buffer and are incubated with antibody overnight at 4°C with rotation (see Note 5).

3.3. Elution of Associated Protein–DNA Complexes and Reversal of Protein–DNA Cross-Linking

1. Recover antibody-bound protein–DNA complexes by adding an appropriate volume of salmon sperm DNA/protein A, G, or A/G agarose slurry for 1–2 h at 4°C with rotation. The choice and volume of agarose slurry are dictated by the isotype of the immunoprecipitating antibody; however, 60 μl of salmon sperm DNA/protein A agarose slurry is used for most protocols in our laboratory.
2. Pellet beads by a brief centrifugation at 600 × *g* in a microcentrifuge for 1 min, aspirate and discard the supernatant fraction.
3. Wash the bead pellets for 3 min on a rotating platform at 4°C with 1 ml of the following solutions (four washes total). Between washes, pellet the beads by centrifugation at 600 × *g* in a microcentrifuge for 1 min and discard the supernatant fractions. Perform aspirations with care to avoid dislodging/aspirating the bead pellets:
   (a) Low-salt wash buffer
   (b) High-salt wash buffer
   (c) LiCl wash buffer
   (d) TE buffer
4. Following the wash steps, elute protein–DNA from the agarose beads by adding 250 μl of freshly prepared SDS elution buffer to the pelleted agarose beads. Briefly vortex the beads in elution buffer and incubate the samples at room temperature with rotation for 15–30 min. Pellet the agarose beads by centrifugation at 600 × *g* in a microcentrifuge for 1 min and carefully transfer the supernatant fraction (eluate) to a 1.5-ml microcentrifuge tube. Repeat the elution step with an addition of 250 μl of SDS elution buffer. Following the second elution, combine the eluates. The total volume following elution should total 500 μl. These are your immunoprecipitation and control samples.
5. Thaw the frozen input samples on ice and add 490 μl of ChIP dilution buffer to each sample.
6. To reverse protein–DNA cross-links, add 20 μl of 5 M NaCl to the immunoprecipitation, control, and input samples and incubate rotating overnight at 65°C.

3.4. Extraction of Recovered DNA

1. To digest isolated protein, add 10 μl of 500 mM EDTA, 20 μl of Tris–HCl, pH 7.2, and 1 μl of proteinase K to each sample and incubate rotating for 1 h at 45°C.

2. DNA can be extracted by using either phenol–chloroform or by using commercially available columns. Although column isolation is easier, we have found that in ChIP assays, where nonhistone proteins are isolated, commercial column recovery provides insufficient material for analysis. To recover isolated DNA by phenol–chloroform extraction and ethanol precipitation, add 550 μl of phenol–chloroform–isoamyl to each sample, mix ten times by inversion (do not vortex), and centrifuge at room temperature in a microcentrifuge for 5 min at 17,000 × *g*. Transfer the aqueous phase to a 1.5-ml microcentrifuge tube; add 550 μl of chloroform to each sample, mix by inversion, and centrifuge at room temperature at 17,000 × *g* in a microcentrifuge for 5 min.
3. Transfer the aqueous phase to 1.5-ml centrifuge tubes; add 1 μl of glycogen (to enhance visualization of the DNA pellet) and 1 ml of 100% ethanol, mix well and store at −80°C overnight.

3.5. Isolation and Analysis of Immunoprecipitated DNA

1. To isolate DNA, samples are thaw frozen on ice and centrifuged at 17,000 × *g* in a microcentrifuge at room temperature for 20 min.
2. To wash the DNA pellets, carefully aspirate the supernatant and add 1 ml of 70% ethanol.
3. Centrifuge the washed DNA pellets at 17,000 × *g* in a microcentrifuge for 20 min at room temperature.
4. Carefully remove the ethanol by aspiration and air dry pellets for 5–10 min.
5. Resuspend isolated DNA in 30 μl of TE and incubate at 37°C for 1 h to completely dissolve. Resuspended DNA is stored at −20°C.
6. To quantitate isolated DNA, run quantitative real-time PCRs with primers and probes targeted for the gene of interest. QPCR data are normalized to the total input samples and graphed as percentage of input or fold enrichment.

3.6. Re-ChIP (as Graphed in Fig. 2 see Notes 7 and 8)

1. Re-ChIP assays follow the ChIP protocol above to step 11 of Subheading 3.2. Following initial overnight immunoprecipitation, wash protein–DNA–bead complexes three times with ChIP washing buffer, followed by double wash with 1× TE buffer.
2. Elute the washed immunoprecipitated protein–DNA complexes by incubation for 30 min at 37°C in 75 μl of Re-ChIP elution buffer.
3. Centrifuge for 3 min at 600 × *g* in a microcentrifuge to collect the beads at the bottom of the Eppendorf tube.

4. Following centrifugation, isolate the supernatant and dilute sample 20 times (to a final volume of 1.5 ml) with ChIP dilution buffer supplemented with 50 μg of BSA and protease inhibitor. Proceed with a second immunoprecipitation reaction with antibodies against the protein of interest and with an irrelevant antibody control. Following sequential immunoprecipitations, continue with Subheading 3.3 in the ChIP protocol.

3.7. Dual Cross-Linking Addendum to ChIP and Re-ChIP (see Note 9)

Follow ChIP/Re-ChIP protocol as in Subheading 3.1. At completion of Subheading 3.1, follow the steps below to enhance protein/protein/DNA cross-linking.

1. Scrape the cells on tissue culture plates; transfer the cells in media to 50-ml conical tubes and place on ice.
2. Centrifuge cells at 1,000 × *g* for 5 min at 4°C, wash with 5 ml of 1× PBS, and recentrifuge.
3. Resuspend cell pellets in 5 ml of 1× PBS, pH 8.0, with 1 mM $MgCl_2$.
4. Following resuspension, cross-link samples by adding 20 μl of 0.5 M DSG directly into the 50-ml conical tube. Incubate for 30 min at room temperature on a 3D rotator to cross-link DNA to protein.
5. To quench the cross-linking reaction, add 100 μl of 1 M Tris–HCl, pH 7.4, to the 50-ml conical tubes and rock for an additional 10 min.
6. Pellet the cross-linked cells by centrifugation at 1,000 × *g* for 5 min at 4°C, carefully remove the supernatant by aspiration (the pellet easily becomes dislodged), and resuspend cells in 10 ml of 1× PBS.
7. For double cross-linking, add 270 μl of formaldehyde (37%, final concentration 1%), and rock samples at room temperature for 10 min.
8. To terminate cross-linking, add glycine to a final concentration of 0.125 M (3.15 ml/20 ml media). Incubate for 10 min at room temperature.
9. Using a tabletop centrifuge with a swing bucket rotor, centrifuge cells at 1,000 × *g* for 5 min at 4°C. Wash the cell pellet once with 5 ml of cold 1× PBS. Carefully discard the supernatant (pellets easily become dislodged) and either proceed to Subheading 3.2, step 5, or snap freeze the pellet in liquid nitrogen and store at −80°C for up to 2 months.
10. Proceed with ChIP or Re-ChIP protocol, Subheading 3.2, step 5.

4. Notes

1. The number of cells plated on the first day of the experiment varies according to cell type and protein to be immunoprecipitated, i.e., isolation of endogenous proteins may require more cells while isolation of overexpressed proteins may require fewer cells. The number of cells necessary is also dependent on the quality of the immunoprecipitating antibody and the binding frequency of the immunoprecipitated protein to DNA. Once an appropriate cell number is determined, keep the cell number and sonication conditions constant in each experiment to ensure that experimental results are comparable from experiment to experiment.
2. Protein/DNA cross-linking times range from a low of 10 min to a high of several hours. The extent of cross-linking is one of the most important parameters of the assay and appropriate cross-linking conditions must be determined and optimized for every new experiment. Too little cross-linking results in an inability to isolate protein–DNA complexes while excessive cross-linking can lead to a loss of a material, reduced antigen availability in chromatin, and resistance to sonication (5).
3. In conventional ChIP protocols, cells or tissues are cross-linked by treatment with formaldehyde. To increase the degree of protein–protein cross-linking, cross-linking can be performed first with DSG followed by secondary formaldehyde cross-linking (13). When determining the extent of cross-linking necessary in an experiment, it is important to remember that cross-linking proteins may decrease their biological activity upon conjugation as a result of conformational changes in proteins.
4. Chromatin sheared to a size of 500–1,000 bp is usually used for ChIP and Re-ChIP experiments. In general, sonication efficiency can be improved by use of small sonication volumes and V-bottom, rather than round-bottom, microcentrifuge tube. It should also be noted that shearing is inefficient if the sample emulsifies. Emulsification can be avoided by using a lower shearing power. If the sample emulsifies, discontinue the sonication and centrifuge the sample at 4°C for 5 min at 8,000 rpm in a microcentrifuge to remove trapped air. To prevent overheating and denaturation of chromatin, samples should be kept on ice and sonications should be performed discontinuously (example: sonicate for 30 s, place the sonicated sample on ice for 30 s; repeat as necessary). Finally, to ensure equal sonication of all samples, immerse the sonicator tip to the same level in all samples and do not touch the tip of the sonicator to the walls or to the bottom of the tube.

5. Inclusion of proper controls is critical in reducing background signals in ChIP and Re-ChIP assays. One classical control is performed by including a control immunoprecipitation using an isotype control antibody. For an isotype control, the IgG used for the control should be generated in the same species in which the ChIP or Re-ChIP antibody was generated. Another widely used negative control is a bead-only control. Bead-only controls allow identification of nonspecific binding of protein/DNA complexes to the beads used to precipitate the complex of interest. In addition to negative controls, investigators new to ChIP and Re-ChIP should also consider using positive-control antibodies against well-characterized transcription factors and/or histone modifications. Including positive controls can be very important when setting up new studies as it allows the investigator to determine if the ChIP or Re-ChIP was successfully carried out. Another crucial control is the input sample; this sample indicates the total amount of chromatin used in the ChIP reaction. The input aliquot is taken just prior to preclearing, and is simultaneously treated with immunoprecipitation reactions to reverse cross-links, isolate and analyze. Input samples are then used to normalize immunoprecipitation reactions (see Figs. 1 and 2).
6. Antibodies used for immunoprecipitation are important contributors to the success of ChIP and Re-ChIP assays; not all commercially available antibodies can effectively immnunoprecipitate protein–DNA complexes. Antibodies that perform well in Westerns, supershifts, and other applications may not work well in ChIP and Re-ChIP. ChIP and Re-ChIP assays require highly specific antibodies that recognize epitopes in free solution and also under fixed conditions. Antibodies can be available as polyclonal and monoclonal preparations. Monoclonal antibodies have a higher specificity compared to polyclonals, but polyclonal sera may recognize multiple epitopes, thus increasing the signal of detection and the efficiency of immunoprecipitation. If an antibody has been commercially tested or published to successfully immunoprecipitate proteins cross-linked to chromatin, the antibody is termed "chip grade." Five micrograms of an antibody of this type is the optimum amount recommended for ChIP and Re-ChIP assays by most companies. However, depending on the antibody and manufacturer's recommendations, 2–3 μg could be sufficient, or as much as 15 μg required, in order to obtain sufficient signal. To avoid the need to perform the large number of controls required for an untested antibody, we recommend using antibodies that have already been successfully used in ChIP and Re-ChIP assays. Validated ChIP-grade antibodies are available from multiple companies, including Abcam,

Santa Cruz Biotechnology Inc, Active Motif, and Upstate (Millipore Corporation).

7. Re-ChIP assays are designed to determine whether two given factors localize at the same genomic region. It is, therefore, crucial to include an independent negative control, where a control antibody is used for the first immunoprecipitation and then a second round of immunoprecipitation follows with an antibody against proteins A and B of interest in two separate samples. Additionally, in Re-ChIPs, if the first immunoprecipitation is not successful, the second immunoprecipitation will not work. Therefore, a Re-ChIP experiment should also include positive and negative controls for the first immunoprecipitation step. In this manner, if the second immunoprecipitation fails, the investigator will be able to determine if the first immunoprecipitation was successful via the positive control. As the first IP is performed on noneluted chromatin, the positive-control antibody can be directed against any protein expected to be present in prepared chromatin. It is also strongly advised to perform a "no-antibody" control during the second IP to validate that successful second immunoprecipitation is only the result of the second antibody and not by any carryover from the first antibody. Thus, it is recommended to perform each first immunoprecipitation in duplicate in order that the eluted chromatin samples can be used in the second immunoprecipitation with both an experimental antibody and with a negative control. Finally, the first antibody can be used as a positive control in the second immunoprecipitation. Since the chromatin used in the second immunoprecipitation was eluted by the first antibody, the chromatin can be re-immunoprecipitated by the same antibody as shown in Fig. 2.
8. Outcomes for Re-ChIPs are one of the three possibilities: (a) complete co-occupancy, (b) no co-occupancy, (c) partial co-occupancy of the two factors being tested. Complete co-occupancy (a) occurs only when protein A and B always associate with the same DNA fragment and neither of the proteins is found on the DNA in the absence of the other. There are two types of partial co-occupancy (b). The first case occurs when protein B always binds to DNA in the presence of protein A while protein A can be found on DNA in the absence of protein B. The second case occurs when proteins A and B can bind independently of one another; they just happen to co-occupy the same genomic region. This possibility occurs when protein A and B do not interact with each other and recognize different target sequences in an enhancer or other transcriptional regulatory region of a promoter. No co-occupancy occurs when protein

A and B associate with DNA, but their associations are mutually exclusive.

9. Cells are at room temperature for extended period of time, so be very quick and careful during aspirations as warm cells do not adhere well to plastic tubes. Double cross-linking procedures require multiple steps, each of which can involve a loss of material; we, therefore, use 15-ml conical tubes to decrease the surface area for adhesion and always use extra caution during the aspiration steps.

Acknowledgments

The authors thank Greer lab members, Julie Morgan and Jodi Osborn, for editing the manuscript. Studies in the author's laboratory are supported by grants from the American Cancer Society, the Georgia Cancer Coalition (to S.F. Greer), and the Georgia State University Molecular Basis of Disease Program (to A. D. Truax).

References

1. Garner, M. M., and Revzin, A. (1981) A gel electrophoresis method for quantifying the binding of proteins to specific DNA regions: application to components of the Escherichia coli lactose operon regulatory system, *Nucleic Acids Res 9*, 3047–3060.
2. Fried, M., and Crothers, D. M. (1981) Equilibria and kinetics of lac repressor-operator interactions by polyacrylamide gel electrophoresis, *Nucleic Acids Res 9*, 6505–6525.
3. Solomon, M. J., and Varshavsky, A. (1985) Formaldehyde-mediated DNA-protein cross-linking: a probe for in vivo chromatin structures, *Proc Natl Acad Sci USA 82*, 6470–6474.
4. Solomon, M. J., Larsen, P. L., and Varshavsky, A. (1988) Mapping protein-DNA interactions in vivo with formaldehyde: evidence that histone H4 is retained on a highly transcribed gene, *Cell 53*, 937–947.
5. Orlando, V. (2000) Mapping chromosomal proteins in vivo by formaldehyde-crosslinked-chromatin immunoprecipitation, *Trends Biochem Sci 25*, 99–104.
6. Wells, J., and Farnham, P. J. (2002) Characterizing transcription factor binding sites using formaldehyde crosslinking and immunoprecipitation, *Methods 26*, 48–56.
7. Umlauf, D., Goto, Y., and Feil, R. (2004) Site-specific analysis of histone methylation and acetylation, *Methods Mol Biol 287*, 99–120.
8. Litt, M. D., Simpson, M., Recillas-Targa, F., Prioleau, M. N., and Felsenfeld, G. (2001) Transitions in histone acetylation reveal boundaries of three separately regulated neighboring loci, *EMBO J 20*, 2224–2235.
9. Gilmour, D. S., and Lis, J. T. (1984) Detecting protein-DNA interactions in vivo: distribution of RNA polymerase on specific bacterial genes, *Proc Natl Acad Sci USA 81*, 4275–4279.
10. Kuo, M. H., and Allis, C. D. (1999) In vivo cross-linking and immunoprecipitation for studying dynamic Protein:DNA associations in a chromatin environment, *Methods 19*, 425–433.
11. Pillai, S., and Chellappan, S. P. (2009) ChIP on chip assays: genome-wide analysis of transcription factor binding and histone modifications, *Methods Mol Biol 523*, 341–366.
12. Sutherland, B. W., Toews, J., and Kast, J. (2008) Utility of formaldehyde cross-linking and mass spectrometry in the study of protein-protein interactions, *J Mass Spectrom 43*, 699–715.
13. Carlsson, J., Drevin, H., and Axen, R. (1978) Protein thiolation and reversible protein-protein conjugation. N-Succinimidyl 3-(2-pyridyldithio)propionate, a new heterobifunctional reagent, *Biochem J 173*, 723–737.

Chapter 13

Transcriptional Regulation of Genes via Hypoxia-Inducible Factor

Olga Roche and Michael Ohh

Abstract

Hypoxia-inducible factor (HIF) is the principal transcription factor that regulates adaptive physiologic responses to compromised oxygen tension. von Hippel–Lindau (VHL) tumor-suppressor protein binds and ubiquitylates the catalytic α subunit of HIF in an oxygen-dependent manner for rapid destruction via the 26S proteasome, thereby establishing VHL as a critical negative regulator of HIF. Mutations in *VHL* cause VHL disease, which is frequently characterized by the overexpression of HIFα and the development of tumors in multiple organ systems, including the central nervous system and the kidney. Here, we describe classical experimental approaches to demonstrate and validate HIF-responsive transcriptional regulation of genes.

Key words: Hypoxia, HIF, HRE, EMSA, Real-time PCR

1. Introduction

Cellular adaptation to hypoxia is regulated by the hypoxia-inducible factor (HIF) family of heterodimeric transcription factors (1). In mammals, there are three HIFα subunits that are encoded by three different genes: HIF1α, HIF2α, also known as endothelial PAS (EPAS) domain, and the least characterized HIF3α. In the presence of oxygen, HIFα is hydroxylated on conserved prolines within the oxygen-dependent degradation (ODD) domain by a family of prolyl hydroxylases (PHDs) (2). This modified form of HIFα is recognized and bound by von Hippel–Lindau (VHL) tumor-suppressor protein, which serves as the recognition component of a Cullin-based E3 ubiquitin ligase complex called ECV (Elongins/Cul2/VHL) that polyubiquitylates HIFα for subsequent 26S proteasome-mediated destruction (3–5). Under reduced oxygen tension, HIFα remains unmodified and escapes ECV recognition

Ales Vancura (ed.), *Transcriptional Regulation: Methods and Protocols*, Methods in Molecular Biology, vol. 809,
DOI 10.1007/978-1-61779-376-9_13, © Springer Science+Business Media, LLC 2012

(3–5). Thus, under hypoxia, HIFα is stabilized, recruits ARNT and the transcriptional coactivator p300, and binds to the hypoxia-responsive elements (HREs; 5′-RCGTG-3′) within enhancers and promoters of numerous hypoxia-inducible genes, such as *vascular endothelial growth factor* (*VEGF*), *erythropoietin* (*EPO*), and *glucose transporter-1* (*GLUT1*), to trigger angiogenesis, erythropoiesis, and anaerobic metabolism, respectively (1, 2).

HIF not only plays an important role in normal physiologic response to reduced oxygen tension, but also has critical roles in oncogenesis as inferred from the positive correlations observed between the extent of HIFα expression and disease aggressiveness or resistance to radiation and chemotherapy (1). Perhaps, the most direct association between tumor-causing mutations and HIF is through mutations in *VHL*, which cause the autosomal dominant VHL cancer syndrome characterized by the development of highly vascular tumors, such as retinal hemangioblastoma and clear-cell renal cell carcinoma (CCRCC) (2). The hypervascular nature of tumors is due, in large part, to the HIF-mediated induction of VEGF production, which plays a key role in normal physiologic blood vessel formation, as well as pathologic tumor angiogenesis.

New presumed hypoxia-regulated genes are continually being added to the growing list of putative hypoxia-inducible genes. Electrophoretic mobility shift assay (EMSA) and quantitative real-time PCR allow determination of whether the oxygen-dependent transcriptional regulation of a given gene is mediated by the engagement of HIF to the HRE within the enhancer/promoter of a suspected target gene. Here, we describe these methods using *VEGF* as a model HIF target gene.

2. Materials

2.1. Cell Culture and Extraction of RNA

1. Dulbecco's modified Eagle's medium (DMEM) (Gibco) supplemented with 10% fetal bovine serum (FBS), 1:100 penicillin–streptomycin solution (Sigma), and 0.25 mg/ml G418 (Sigma).
2. RNeasy kit (Qiagen). Kit is stored at room temperature.

2.2. DNA Removal Step and RNA Quantification

1. DNA-free Kit (Ambion). Includes DNase enzyme, DNase buffer, and DNase inactivation reagent.
2. 10 mM Tris–HCl, pH 7.5.
3. UVettes (Eppendorf).
4. RNase/DNase free 1.5-ml microtubes (Progene).
5. Spectrophotometer.

2.3. Reverse Transcription of RNA

1. Random Hexamer primer (Fermentas).
2. SuperScript II reverse transcriptase kit (Invitrogen). It contains DTT, reverse transcriptase enzyme, and 5× first-strand buffer. Store at −20°C.
3. 100 mM dNTP set (Invitrogen). An equimolar mix of the four dNTPs is diluted to a concentration of 10 mM in ultrapure DNase/RNase-free distilled water (Invitrogen). Aliquot and store at −20°C.
4. 0.2-ml thin wall tube (Axygen).
5. Thermal cycler PTC-200 (MJ Research).

2.4. Real-Time PCR

1. Human genomic DNA (100 μg/500 μl; Roche).
2. *VEGF* primers (5′-CTCTCTCCCTGATCGGTGACA-3′ and 5′-GGAGGGCAGAGCTGAGTGTTA-3′). Reference gene, *U1snRNP70*, primers (5-CAACGACAGCCGAGACATGTA-3′ and 5′-AGCCTCCATCAAATACCCATT-3′) (see Notes 1–3). Dilute primers in DNase/RNase-free distilled water to a concentration of 50 μM. Prepare a working stock mixing 50 μl of each forward and reverse primer.
3. Platinum Taq DNA Polymerase kit (Invitrogen). Includes the Platinum Taq enzyme, 10× PCR buffer, and 50 mM $MgCl_2$. Store at −20°C.
4. 100 mM dNTP set (Invitrogen). Prepare as described below.
5. SYBR green (Invitrogen) is a reporter dye that binds to double-stranded DNA. SYBR green is diluted in DNase/RNase-free distilled water 1:1,000. Aliquot and store at −20°C protected from the light.
6. ROX Reference Dye (Invitrogen). ROX provides an internal fluorescence reference to which the reporter dye signal is normalized during data analysis.
7. 384-Well clear optical reaction plates (Applied Biosystems).
8. ABI Prism 7900HT sequence detection system (Applied Biosystems).
9. SDS 2.3 software (Applied Biosystems).

2.5. Preparation of the EMSA Probes

1. HRE forward and reverse oligos are 24 nucleotides long and contain the HRE from *VEGF* (5′-CACAGTGCATACGTGGGC TCCAAC-3′). Scrambled oligos are random oligos that contain the same number of nucleotides as HRE oligos, but do not contain any HRE sequences (5′-CGATTGATACCCTAGAT TCAGGGAG-3′). Oligos are dissolved in DNase/RNase-free distilled water at a concentration of 300 ng/μl. Store at −20°C.
2. Dilute Spermidine (Sigma) in water to 10 mM.

3. Gamma ^{32}P adenosine 5′-triphosphate (Perkin Elmer). Gamma ^{32}P is a β emitter, so necessary precautions must be taken. Shielding is required and handling time should be kept to a minimum (experiments should, therefore, be planned carefully in advance). Lab coat and two pairs of gloves should be worn when handling gamma ^{32}P. Geiger counter should be used continually to monitor possible contaminations. Moreover, a dosimeter must be worn. Follow the rules of your institute for disposing radioactive waste.
4. T4 polynucleotide kinase (PNK) and 10× PNK buffer (NEB).
5. Illustra Probequant G-50 Micro column (GE Healthcare).
6. Liquid scintillation cocktail for aqueous samples (Beckman Coulter).
7. 20-ml Disposable scintillation vials (Fisherbrand).

2.6. Polyacrylamide Gel Electrophoresis

1. Tris–Borate–EDTA (TBE) buffer (5×): 0.5 M Tris base, 0.45 M boric acid, and 0.01 M EDTA. Store at room temperature. Running buffer (0.5× TBE) is prepared by diluting 200 ml of 5× TBE with 1,800 ml of water in a measuring cylinder. Store at 4°C.
2. Ammonium persulfate (APS) is dissolved in water to a 10% concentration. Store at 4°C.
3. Thirty percent acrylamide/bisacrylamide (Bio-Rad) solution (37.5:1 with 2.6% C) (this is a neurotoxin when unpolymerized, and care must be taken to minimize the chance of exposure) is stored at 4°C. *N,N,N,N′*-tetramethylethylenediamine (TEMED; Sigma) is stored at room temperature.

2.7. Binding Reaction

1. TNT T7 quick-coupled (Promega).
2. pcDNA3, HA-HIF1α, HA-HIF2α, and ARNT plasmids (5).
3. Dilute salmon testes DNA (Sigma) in water to a final concentration of 0.1 μg/μl. Aliquot and store at –20°C.
4. 10× binding buffer: 40% glycerol, 50 mM Tris, pH 7.5,100 mM KCl, 10 mM DTT, 2 mg/ml BSA, and 0.2% Triton X-100. Vortex, aliquot, and store at –20°C.
5. HA antibody (Santa Cruz Biotechnology). Store at 4°C.

2.8. Gel Loading and EMSA Developing

1. 5× DNA loading dye (Thermo Scientific). Aliquot and store at –20°C.
2. Grade 703 blotting paper (VWR).
3. BioMax MR film (Kodak).

3. Methods

3.1. Extraction of RNA

1. The adherent RCC4 cell line derived from renal carcinoma cells devoid of VHL ($VH^{-/-}$; $HIF1\alpha$+/+; $HIF2\alpha$+/+) and VHL-reconstituted RCC4 cell line (RCC4-VHL) are passaged in 100-mm plates when approaching confluence using trypsin/EDTA. One 100-mm plate is required for each experimental condition. A 1:4 split of the cells provides experimental cultures that are approaching confluence after 4 days. At this point, the cultures are rinsed once with 1× PBS.
2. Extract RNA with the RNeasy kit following manufacturer's instructions (see Note 4). In the final step, elute RNA with 30 μl of RNase-free water provided in the kit.

3.2. Removal of DNA and Quantification of RNA Concentration

For real-time PCR, it is necessary to remove any contaminating DNA in the sample to avoid amplification of DNA instead of RNA.

1. Add to the RNA sample 3 μl of DNase I buffer.
2. Add 0.5 μl of DNase I enzyme to digest DNA, mix, and incubate for 15 min at 37°C. Centrifuge the sample for 15 s at 845×*g* in a microcentrifuge, add another 0.5 μl of DNase I, and incubate for 15 min at 37°C.
3. To inactivate the DNase enzyme, add 5 μl of DNase-inactivating reagent. Mix and incubate for 1 min, mix again, and incubate for another minute. Centrifuge for 1 min at 16,000×*g* in a microcentrifuge. The DNase-inactivating reagent is on the bottom of the tube. Transfer the supernatant to a new 1.5-ml tube.
4. Prepare a dilution of 2 μl of the RNA samples (RCC4-Mock, RCC4-VHL) in 98 μl of 10 mM Tris. Prepare a blank control containing only Tris for calibration of the machine. Transfer the samples to UVettes. Measure the concentration and quality of the RNA in the spectrophotometer using the program for RNA (see Note 5).

3.3. Reverse Transcription of RNA

1. The reverse transcription reaction starts with a hot start. In a 0.2-ml tube, add 2 μg of RNA, 1 μl of random hexamer, and DNase/RNase-free distilled water to a final volume of 12 μl. Program a hot start in the thermal cycler: 10 min at 70°C followed by 30 min at 4°C and load both tubes. When the PCR block has been at 4°C for 15 min, take the tubes from the block and centrifuge for 15 s at 845g in a microcentrifuge to collect the entire samples in the bottom of the tube.
2. Add to both 0.2-ml tubes 4 μl of 5× first-strand buffer, 2 μl of 0.1 M DTT, 1 μl of the 10 mM dNTPs mix, and 1 μl of the superscript enzyme. A master mix should be prepared to eliminate variations in the composition of the reaction between tubes.

The cDNA synthesis is performed for 1.5 h at 42°C, followed by 15 min at 70°C, and 4°C forever.

3. Dilute the cDNA samples 1:10 in DNase/RNase-free distilled water. These cDNA samples can be stored at 4°C for a short period and –20°C for a long term.

3.4. Real-Time PCR

1. An absolute standard curve method is used. The absolute quantification determines the input copy number of the transcript of interest by relating the PCR signal to a standard curve. The standard curve is prepared with dilutions of human genomic DNA (6) (see Note 6).
2. A nontemplate control (water) is run for each primer. Each pair of cDNA or genomic DNA dilution primer is run in triplicate in the plate.
3. Prepare a 1:10 dilution of the previously 1:10 diluted cDNAs. 5 μl of 1:100 diluted cDNA is necessary for each well.
4. Genomic DNA dilutions are prepared in DNase/RNase-free distilled water. An intermediate dilution 1:31 of the genomic DNA is made. A second dilution 1:4 is prepared. This dilution contains 2,700 copies of each gene [the mass of the haploid human genome is 3.5 pg, and therefore 1 ng of genomic DNA contains 286 copies of a single-copy gene (http://www.genomesize.com)]. Four serial dilutions 1:3 are prepared from the second dilution (these dilutions contains 900, 300, 100, and 33.3 copies of each gene) (see Note 7).
5. Prepare a master mix for both genes *VEGF* and *U1snRNP70* taking into account that each well contains 1 μl of 10× PCR buffer, 0.6 μl of 50 mM $MgCl_2$, 0.2 μl of 10 mM dNTPs, 2.42 μl of DNase/RNase-free distilled water, 0.3 μl of 1:1,000 SYBR Green, 0.25 μl of ROX reference dye, and 0.05 μl of Platinum Taq polymerase. Divide in 2 aliquots and add the corresponding primer mix, taking into account that for each well 0.175 μl of 50 μM primer (forward plus reverse) is added.
6. The plate is loaded and 5 μl of the cDNA or genomic DNA sample are loaded in the corresponding wells, followed by 5 μl of the master mix containing the primer.
7. Program in the ABI Prism 7900HT: 95°C for 3 min, 40 cycles of 95°C for 10 s, 65°C for 15 s, and 72°C for 20 s, and 1 cycle of 95°C for 15 s, 60°C for 15 s, and 95°C for 15 s. The last cycle allows calculating the melting temperature of the products obtained in the PCR (see Note 8).
8. An SDS file (Applied Biosystems) is created to store the data and later analyzed. Load the plate in the ABI Prism 7900HT and start the reaction.

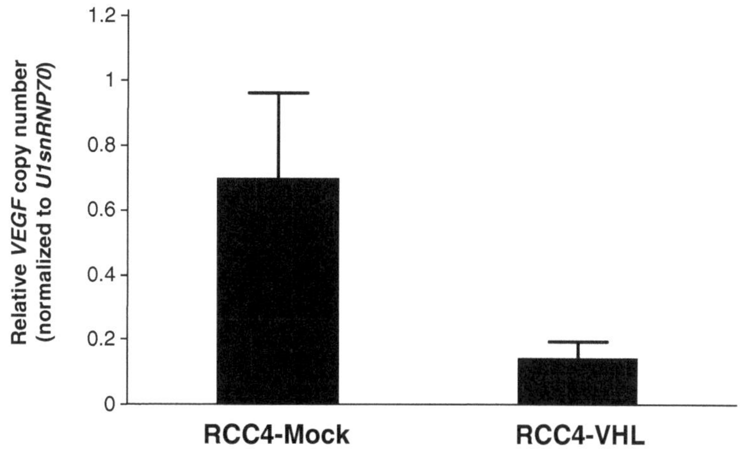

Fig. 1. Expression of *VEGF* was measured by quantitative real-time PCR in RCC4-Mock and RCC4-VHL cells and normalized to *U1snRNP70* mRNA level.

9. Results are analyzed using the SDS 2.3 software (Applied Biosystems). An average copy number of the triplicates for RCC4-Mock and RCC4-VHL is obtained (see Note 9). Expression of *VEGF* is calculated dividing the number of copies of *VEGF* in RCC4-Mock or RCC4-VHL by the corresponding copy number of *U1snRNP70* in the respective sample. An example of the results obtained is shown in Fig. 1.

3.5. Preparation of HRE Probe for EMSA

1. Forward and reverse oligos are labeled in separated reactions. For each labeling reaction, add in a 1.5-ml tube: 12 μl of DNase/RNase-free distilled water, 2 μl of oligo, 2.5 μl of PNK buffer, 2.5 μl of spermidine, and 1 μl of T4 PNK. In a designated radioactive area, add to each tube 5 μl of gamma ^{32}P-ATP (see Note 10). Incubate at 37°C for 1 h in a heat block. After incubation, centrifuge samples in a microcentrifuge for several seconds to eliminate the risk of radioactive contamination when the tubes are opened.
2. To eliminate the unincorporated gamma ^{32}P-ATP, a purification step using the G-50 microcolumns is performed. First, centrifuge the new columns for 1 min at 845g in a microcentrifuge. Load the forward and the reverse oligos labeling reactions to these columns. Centrifuge again the columns for 2 min at 845g. Transfer the labeled oligos to a new 1.5-ml tube. Throw out the columns containing the unincorporated gamma ^{32}P-ATP to the radioactive waste.
3. In a scintillation vial, add 7 ml of scintillation liquid and 2 μl of the labeled mixture of oligos and measure the counts per million in the sample using a scintillation counter. Dilute labeled oligos to a concentration of 100,000 cpm/μl.

4. For annealing the forward- and reverse-labeled oligos, heat the mixture of both at 95°C for 5 min in a heat block. After 5 min, the heat block is turned down to 37°C and the reactions leaved overnight. Annealed oligos (probe) are stored at −20°C. In a nonradioactive area, 20 μl of forward and 20 μl of reverse oligos are mixed and annealed the same way that is described for radioactive oligos. Annealed oligos are stored at −20°C.

3.6. In Vitro Translation of Proteins and Binding Reaction

1. pcDNA3, HA-HIF1α, HA-HIF2α, and ARNT plasmids are in vitro translated in different tubes. Prepare fresh in vitro translated proteins by adding to a 1.5-ml tube 1.5 μg of the corresponding plasmid, 15 μl of TNT T7, and 1.5 μl of cold methionine. Incubate for 90 min at 30°C in a heat block. Centrifuge samples for several seconds to collect all of the samples at the bottom of the tube.
2. In addition to the binding reaction of HA-HIF1α or 2α/ARNT complex to the HRE, two control competition experiments are performed by adding 250× molar excess of unlabeled HRE or scrambled oligos. Moreover, a control for the binding specificity of HIF to the HRE probe is done by adding a specific antibody against the protein tag (example, HA) or to HIF1/2α itself, which should generate a "supershift" profile. The negative control is performed with in vitro-translated empty plasmid (pcDNA3).
3. In a nonradioactive area, add to a 1.5-ml tube for each reaction: water to adjust the final volume of the reaction to 25 μl, 2.5 μl of 10× binding buffer, 2 μl of salmon testes DNA, if it is required the unlabeled HRE or the unlabeled scrambled oligos, and 3 μl of the corresponding in vitro-translated proteins (see Note 11). Take the 1.5-ml tube to the radioactive area, transfer the 1.5-ml tube to a Plexiglass rack, and add 1 μl of the labeled HRE probe (100,000 cpm) to each tube (see Note 12). Incubate for 15 min at room temperature, then add to the corresponding tubes 4 μl of the anti-HA antibody, and incubate for another 15 min. After incubation, add 5 μl of loading dye to each binding reaction.

3.7. Polyacrylamide Gel Electrophoresis

1. These instructions assume the use of 18×20-cm gels. It is important to clean the glass using mild detergent, and then dry and clean with 75% ethanol before use to eliminate traces of SDS.
2. Prepare a 2-mm-thick, 5% gel by mixing 5 ml of 5× TBE, 36 ml of water, 8.33 ml of acrylamide/bisacrylamide solution, 500 μl of APS, and 100 μl of TEMED. Pour the gel and insert the 20-well comb (see Notes 13 and 14).
3. When the gel is polymerized, pre-run the gel with cold 0.5× TBE at 150 V for 90 min in a cold room.

4. Take the samples to the cold room inside a Plexiglass rack. Load the samples in the gel and run at 200 V for 4 h behind a Plexiglass shield.
5. Disconnect the gel unit from the power supply. Transfer the running buffer to a specific container for gamma ^{32}P liquids with a 25-ml pipette.
6. Disassemble the gel, lay it on top of a sheet of blotting paper, cover with transparent plastic film, and place in an X-ray film cassette.
7. Take the gel in the X-ray film cassette to the gel dryer. A clean sheet of blotting paper is put over the drying machine and on top of it the gel (see Note 15). Gel is dried for 2 h at 72°C.
8. Put the gel on an X-ray film cassette with the MR film. The cassette is stored at –80°C. Film is developed 24 h later (see Note 16). An example of the EMSA result is shown in Fig. 2.

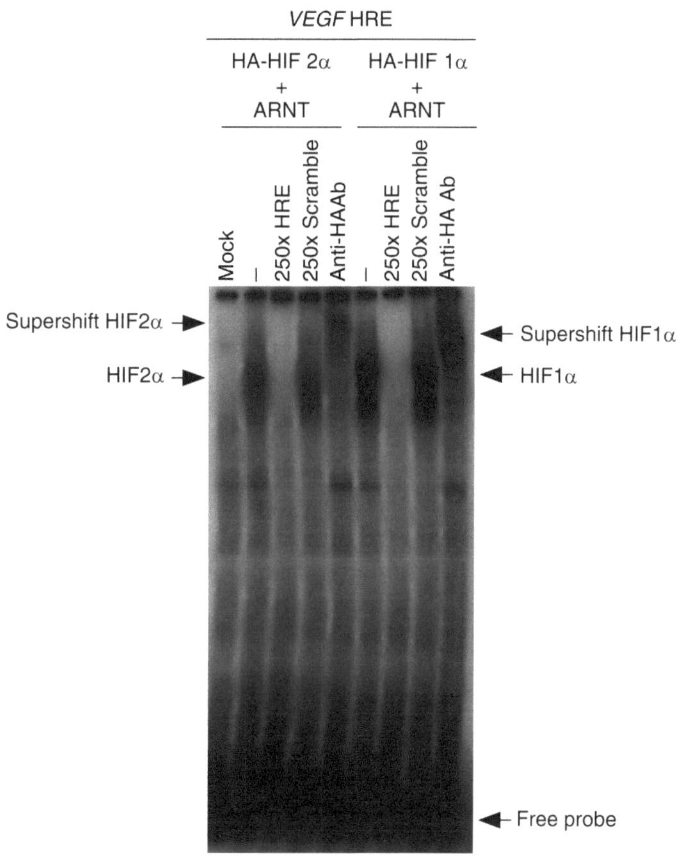

Fig. 2. EMSA of in vitro-translated HIF1 (HA-HIF1α/ARNT) and HIF2 (HA-HIF2α/ARNT) complexes. HIF1 and HIF2 complexes bind to ^{32}P-labeled *VEGF* HRE probe (*lanes 2* and *6*). Competitions with 250× molar excess of unlabeled wild-type *VEGF* HRE (*lanes 3* and *7*) and competition with 250× unlabeled scrambled probe (*lanes 4* and *8*) are shown. HIF complex bound to *VEGF* HRE supershifted with an anti-HA antibody (*lanes 5* and *9*).

4. Notes

1. Other reference genes can also be used, such as β-*actin* or *glyceraldehyde 3 phosphate dehydrogenase*. These genes are higher expressed than *U1snRNP70* and as a consequence the copy number is outside the standard curve and lower dilutions of genomic DNA should be used.
2. The reference gene should have similar copy number in all the cell types used in an experiment. Thus, if different cell lines are involved in a study, then several reference genes should be analyzed to determine which of the genes has the most appropriate and consistent expression for a given cell type.
3. To design primers to amplify other HIF target genes, it has to be taken into account that as genomic DNA is used to prepare the standard curve forward and reverse primers for each gene should align in the same exon to ensure that the PCR amplicons from the cDNA sample and the genomic DNA are the same.
4. RNases are very stable. Certain precautions must be taken into account to minimize the risk of RNases in your samples that degrade test RNA. Before starting RNA extraction, rinse the pipettes, microcentrifuge, and tube racks with 0.1 M NaOH and 1 mM EDTA solution. Gloves should be changed after touching exposed skin or common surfaces. RNase-free tips should be used.
5. The 260/280 absorbance ratio for RNA should be between 1.9 and 2.1.
6. A cDNA that has been previously characterized can also be used for making the standard curve.
7. Dilutions of genomic DNA can be prepared fresh for each and every experiment, but also bigger volumes can be prepared and stored at 4°C so that the same dilutions can be used in different experiments reducing variations in results between experiments.
8. A trick that allows monitoring in which well samples were added, and therefore making easier to load the plate, is to dilute the cDNA samples and the dilutions of genomic DNA in yellow water while the mix with the rest of the components of the reaction is prepared in blue water. Yellow water and blue water are prepared by adding 1 drop of food dye in 25 ml of DNase/RNase-free distilled water. The colored water is aliquoted and stored at −20°C.
9. SDS software allows viewing the raw data. If the SYBR green signal is close to ROX signal, a new aliquot of SYBR should be used.

10. The slope of the standard curve measures the efficiency of the PCR (slope = $10^{(-1/\text{slope})} - 1$). Hence, an efficiency close to 100% means a slope between –3.1 and –3.6 (http://www3.appliedbiosystems.com/cms/groups/mcb_marketing/documents/generaldocuments/cms_040377.pdf).
11. Half-life of gamma ^{32}P-ATP is 14 days. As a consequence, gamma ^{32}P-ATP older than 2 weeks should not be used since signal generated would be too low to detect the binding of HA-HIFα/ARNT complex to the HRE probe.
12. Instead of salmon testes DNA, DNA poly dI-dC can be used (Thermo scientific).
13. A titration experiment with different quantities of in vitro-translated proteins should be done prior to the experiment to determine the exact quantity necessary to have a strong signal in the EMSA.
14. Polyacrylamide gels can be prepared in advance and stored at 4°C in a resealable plastic bag with wet paper towels on top of the wells.
15. Reusing blotting paper on top of the gel dryer can result in black spots later on the X-ray film.
16. If the signal is not strong enough, the time of film exposure can be longer, days, weeks, or even a month.

Acknowledgments

This work was supported by grants from the Canadian Institutes of Health Research (CIHR; MOP77718) and the Canadian Cancer Society (16056 and 18460). O. Roche is a recipient of a CIHR postdoctoral fellowship. M. Ohh is a Canada Research Chair.

References

1. Semnza GL (2003). Targeting HIF-1 for cancer therapy. Nat Rev Cancer 3:721–32.
2. Kaelin WG Jr (2002). Molecular basis of the VHL hereditary cancer syndrome. Nat Rev Cancer 2: 673–682.
3. Ivan M, Kondo K, Yang H, Kim W, Valiando J, Ohh M, Salic A, Asara JM, Lane WS, Kaelin WG Jr (2001). Hifalpha targeted for VHL-mediated destruction by praline hydroxylation: implication for O_2 sensing. Science 292: 449–51.
4. Jaakkola P, Mole DR, Tian YM, Wilson MI, Gielbert J, Gaskell SJ, von Kriegsheim A, Hebestreit HF, Mukherji M, Schofield CJ, Maxwell PH, Pugh CW, Ratcliffe PJ (2001). Targeting of HIF-alpha to the von Hippel-Lindau ubiquitylation complex by O_2-regulated prolyl hydroxylation. Science 292: 468–472.
5. Ohh M, Park CW, Ivan M, Hoffman MA, Kim TY, Haung LE, Pavletich N, Chau V, Kaelin WG (2000). Ubiquitination of hypoxia-inducible factor requires direct binding to the beta-domain of the von Hippel-Lindau protein. Nat Cell Biol 2: 121–3.
6. Yun JJ, Heisler LE, Hwang IL, Wilkins O, Lau SK, Hyrcza M, Jayabalasingham B, Jin J, McLaurin J, Tsao M, Der SD (2006). Genomic DNA functions as universal external standard in quantitative real-time PCR. Nucleic Acids Res 34(12): e85.

Chapter 14

Exchange Protein Directly Activated by Cyclic AMP-1-Regulated Recruitment of CCAAT/Enhancer-Binding Proteins to the Suppressor of Cytokine Signaling-3 Promoter

William A. Sands, Hayley D. Woolson, Stephen J. Yarwood, and Timothy M. Palmer

Abstract

The ability of prototypical second messenger cyclic AMP (cAMP) to positively control transcription of the somatostatin gene was pivotal to the original identification of the transcription factor cAMP response element-binding protein. However, it is now clear that alternative intracellular cAMP sensors, of which the exchange protein directly activated by cAMP (Epac) proteins have been studied most intensively, also initiate transcription of key genes in response to cAMP elevation. For example, we have demonstrated in vascular endothelial cells that activation of Epac1 is necessary for cAMP-mobilizing agents to trigger the induction of the gene-encoding suppressor of cytokine signaling-3 (SOCS-3), a potent inhibitor of interleukin (IL)-6 signaling. This is achieved through the recruitment of CCAAT/enhancer-binding protein (C/EBP) transcription factors to the SOCS-3 promoter.

Here, we describe in detail how to identify and measure cAMP-mediated recruitment of a specific C/EBP isoform to a candidate regulator region of the SOCS-3 promoter in vascular endothelial cells in vitro. We also describe the RNA interference strategies with which we identified a role for Epac1 and SOCS-3 in being responsible for mediating the inhibitory effect of cAMP elevation on IL-6 signaling.

Key words: Cyclic AMP, Epac, Chromatin immunoprecipitation, Endothelial cells, RNA interference

1. Introduction

The vascular endothelium has been strongly implicated in the pathology of many disease states that involve aberrant platelet activation or excessive leukocyte infiltration into underlying tissue, including atherosclerosis and sepsis (1, 2). Conversion of the

Ales Vancura (ed.), *Transcriptional Regulation: Methods and Protocols*, Methods in Molecular Biology, vol. 809,
DOI 10.1007/978-1-61779-376-9_14, © Springer Science+Business Media, LLC 2012

endothelium from a predominantly anti-inflammatory/anticoagulant state to a proinflammatory/fibrotic phenotype is a complex process triggered by multiple chemical stimuli (e.g., cytokines, pathogen-derived molecules) as well as mechanical stress and hypoxia. Key changes include the mobilization of gene transcription programs that control the induction of chemokines, such as monocyte chemoattractant protein-1 (MCP-1)/CCL2 and IL-8/CXCL8, proinflammatory cytokines, such as tumor necrosis factor (TNF)-α, interleukin (IL)-1β, and IL-6, and adhesion molecules, like E-selectin and VCAM-1. This results in the recruitment and activation of circulating leukocytes from the circulation prior to their arrest on the endothelium and subsequent migration, or diapedesis, into underlying tissue (3, 4).

Even though the anti-inflammatory effects of the prototypical, intracellular, second messenger cyclic AMP (cAMP) are well-established, the molecular mechanisms underlying them remain largely unclear, particularly for cytokines that utilize the JAK/STAT pathway. We have, therefore, undertaken a systematic examination of cAMP's effects on IL-6 signaling, particularly in the context of vascular EC function since these cells are an important target for IL-6 in both normal and disease states. These results have demonstrated that activation of both the JAK-STAT and Raf-MEK-ERK1,2 pathways in response to this cytokine is severely compromised by exposure to cAMP-elevating drugs. This is achieved by the ability of cAMP to trigger the induction of the gene for suppressor of cytokine signaling-3 (SOCS-3), a potent inhibitory regulator of IL-6 signaling which targets gp130, the signal-transducing component of the IL-6 receptor signaling complex. However, rather than proceeding through cAMP-dependent protein kinase A (PKA) and subsequent phosphorylation of transcription factor cAMP-response element-binding (CREB) protein, SOCS-3 induction requires a previously unappreciated pathway involving activation of the alternative cAMP sensor exchange directly activated by cAMP-1 (Epac1) and subsequent mobilization of CCAAT/enhancer-binding proteins (C/EBPs) to the SOCS-3 promoter (5–8).

In this article, we provide protocols for delivery of small interfering RNAs (siRNAs) into a model human umbilical vein EC (HUVEC) system, an approach which has proved invaluable in allowing dissection of this new pathway. In addition, we describe in detail a protocol that we used to detect recruitment of the C/EBPβ isoform to a specific region of the human SOCS-3 promoter in response to either global elevation of intracellular cAMP levels or selective activation of Epac1.

2. Materials

2.1. Cell Culture and Activators of cAMP Signaling

1. HUVECs (pooled cryopreserved cells from Lonza; cat. no. CC-2519).
2. Cultures maintained in endothelial growth medium-2 (EGM-2) supplemented with EGM-2 supplements and growth factors (2% (v/v) fetal bovine serum, hydrocortisone, human fibroblast growth factor, vascular endothelial growth factor, R^3-insulin-like growth factor, ascorbic acid, human epidermal growth factor, gentamicin, amphotericin-B, and heparin) (Lonza; cat. no. CC-33162). Hereafter, this is referred to as complete EGM-2.
3. Solutions of endothelial grade trypsin–EDTA and HEPES-buffered saline for subculturing.
4. The diterpene adenylyl cyclase activator forskolin is reconstituted to a stock concentration of 10 mM in dimethyl sulfoxide (DMSO) and stored in single-use aliquots at −20°C.
5. The phosphodiesterase 4-selective inhibitor rolipram, which blocks cAMP hydrolysis to 5′-AMP, is reconstituted to a stock concentration of 10 mM in DMSO and stored in single-use aliquots at −20°C.
6. The Epac-selective activator 8-(4-chlorophenylthio)-2′-*O*-methyl-cAMP (termed "007") (9) (BioLog) is reconstituted to a stock concentration of 10 mM in DMSO and stored in single-use aliquots at −20°C.

2.2. Chromatin Immuno-precipitation

1. Cell lysis buffer: 5 mM PIPES, pH 8.0, 85 mM potassium chloride, 0.5% (v/v) NP-40 supplemented with protease inhibitors (0.1 mM phenymethylsulfonyl fluoride (PMSF), 1 μg/ml soybean trypsin inhibitor, and 1 μg/ml benzamidine). The buffer minus protease inhibitors can be stored at 4°C for several months. On the day of the experiment, the volume of buffer required can be taken from the stock and protease inhibitors added fresh from concentrated stocks (see Note 1).
2. 37% (w/v) formaldehyde solution stored at room temperature.
3. Nuclear lysis buffer: 50 mM Tris–HCl, pH 8.1, 10 mM EDTA, 1% (w/v) SDS supplemented with protease inhibitors (see item 1). The buffer minus protease inhibitors can be stored at 4°C for several months. On the day of the experiment, the volume of buffer required can be taken from the stock and protease inhibitors added fresh from concentrated stocks (see Note 1).

4. Immunoprecipitation (IP) antibody: Rabbit polyclonal anti-C/EBPβ antibody (Santa Cruz Biotechnology, sc-150X). The antibody can be stored at 4°C for several months.
5. Protein A-Sepharose 4B Fast Flow beads in suspension with 20% (v/v) ethanol (Sigma). This can be stored for several months at 4°C, but care should be taken to ensure that the buffer does not evaporate over time (see Note 2).
6. Immunoprecipitation buffer: 20 mM Tris–HCl, pH 8.1, 2 mM EDTA, 150 mM sodium chloride, 1% (v/v) Triton X-100, 0.01% (w/v) SDS supplemented with protease inhibitors (see Note 1). The buffer minus protease inhibitors can be stored at 4°C for several months. On the day of the experiment, the volume of buffer required can be taken from the stock and protease inhibitors added fresh from concentrated stocks (see Note 1).
7. Wash buffer 1: 20 mM Tris–HCl, pH 8.1, 2 mM EDTA, 150 mM sodium chloride, 1% (v/v) Triton X-100, 0.1% (w/v) SDS supplemented with protease inhibitors (see item 1). The buffer minus protease inhibitors can be stored at 4°C for several months. On the day of the experiment, the volume of buffer required can be taken from the stock and protease inhibitors added fresh from concentrated stocks (see Note 1).
8. Wash buffer 2: 20 mM Tris–HCl, pH 8.1, 2 mM EDTA, 500 mM sodium chloride, 1% (v/v) Triton X-100, 0.1% (w/v) SDS supplemented with protease inhibitors (see item 1). The buffer minus protease inhibitors can be stored at 4°C for several months. On the day of the experiment, the volume of buffer required can be taken from the stock and protease inhibitors added fresh from concentrated stocks (see Note 1).
9. Wash buffer 3: 10 mM Tris–HCl, pH 8.1, 1 mM EDTA, 250 mM lithium chloride, 1% (v/v) NP-40, 1% (w/v) sodium deoxycholate supplemented with protease inhibitors (see item 1). The buffer minus protease inhibitors can be stored at 4°C for several months. On the day of the experiment, the volume of buffer required can be taken from the stock and protease inhibitors added fresh from concentrated stocks (see Note 1).
10. Elution buffer: 1% (w/v) SDS, 100 mM sodium bicarbonate. The buffer should be made up freshly as required.
11. Proteinase K (Promega): Lyophilized powder is reconstituted to a concentration of 10 mg/ml in 50 mM Tris–HCl, pH 8, 10 mM calcium chloride for storage below −20°C in single-use aliquots, which are stable for at least 6 months.
12. TE, pH 8.0: 10 mM Tris–HCl, pH 8.0, 1 mM EDTA. Stable at room temperature for at least 6 months.

2.3. siRNA-Mediated Target Gene Knockdown in HUVECs

1. Opti-MEM reduced serum medium (Invitrogen).
2. HiPerFect transfection reagent (Qiagen).
3. siRNAs: Epac1/RAPGEF3 siGENOME SMARTpool siRNA (Dharmacon RNA Technologies, cat. no. M-007676-00), SOCS-3 siRNA pool (Santa Cruz Biotechnology, cat. no. sc-41000), and control non-targeting siRNA-A (Santa Cruz Biotechnology, cat. no. sc-37007) are each reconstituted to a stock concentration of 10 μM in RNase-free water and stored in single-use aliquots at −20°C.

3. Methods

3.1. ChIP Assay

Standard transcription factor assays, such as electrophoretic mobility shift assays (EMSAs) or reporter gene assays, provide only limited information about the general state of transcription factor activation in cells. Specifically, because they are global indicators of transcription factor activity, they provide only limited information about what is happening at specific promoter/enhancer elements in intact cells. To gain more detailed information on what is happening at the level of intact promoter/enhancer elements, *ch*romatin *i*mmuno*p*recipitation (ChIP) assays can be employed. These allow investigation of the dynamics of transcription factor binding and epigenetic modifications at specific promoters in intact cells following various treatments. To achieve this, primers are designed that span the sequence of interest, for example a transcription factor site on a particular promoter. Occupancy of the site of interest is then assessed by immunoprecipitating the desired antigen and amplifying the DNA sequence within the promoter of interest by PCR. The basic principle is straightforward: cells are fixed and nuclei are prepared. Nuclear chromatin is then sheared into small fragments by probe sonication: this is a crucial step as the resolution of the assay depends on efficient shearing. It is also important to note that optimal shearing conditions for individual cell types must be determined empirically. Sheared chromatin is then immunoprecipitated with relevant antibodies and the cross-links are broken, followed by digestion of associated proteins. Promoter element enrichment under each experimental condition is then determined by quantitative PCR. Here, we describe a protocol we have developed to investigate cAMP-regulated binding of C/EBPβ to the human SOCS-3 promoter in HUVECs. Further expansion of this approach to ChIP-Seq technology ultimately allows unbiased genome-wide identification of Epac-regulated C/EBP-binding sites (to within 50 bp) throughout the human genome (10).

1. On day 1, passage HUVECs at approximately 70–80% confluence from T-75 flasks into 6-well dishes for experimental work and into a separate T-75 tissue culture flask to maintain and propagate the cell line (see Note 3). The spent complete EGM-2 medium from the flask is first decanted into a sterile 50-ml plastic centrifuge tube. Monolayers are washed with 2-ml/flask of HEPES-buffered saline and aspirated to remove spent medium and nonadherent cells. Endothelial-grade trypsin is then added (1 ml/flask) and the flask placed back into the cell culture incubator to allow cells to detach. Once all the cells have been detached by sharply tapping the flask (this can be determined by examining the cells in the flask under a microscope), the resuspended cells are added to the 50-ml tube containing the spent complete EGM-2 to neutralize the trypsin. HUVECs are then pelleted by centrifugation at $200 \times g$ for 5 min at room temperature and then resuspended in 1 ml of fresh complete EGM-2 and cell density determined using a hemocytometer. If necessary, additional complete EGM-2 is added to the cell suspension to ensure that the resuspended HUVECs achieve a density of 0.5×10^6 cells/ml and 1 ml of this suspension is then added per well of a 6-well tissue culture plate and left overnight in a cell culture incubator to adhere (standardization is important for consistency; see Note 4). For maintenance of the cell line, 1.2×10^6 cells from the remaining suspension can be seeded into a fresh T-75 flask and complete EGM-2 added to give a final volume of medium of 10 ml.
2. On day 2, replace medium in each well with 1 ml of fresh complete EGM-2 and stimulate the now-confluent cell monolayers with either forskolin (typical effective final concentration range 10–100 μM) with or without rolipram (typical effective final concentration 100 μM) or Epac-selective activator 007 (typical effective final concentration range 50–100 μM) and the appropriate vehicle controls (see Note 5).
3. Terminate the reaction by directly adding 27 μl/well of 37% (w/v) formaldehyde cross-linking reagent (final concentration in well = 1% (w/v)) and swirl when adding to ensure even distribution. Incubate at room temperature for 10 min with shaking: longer incubation times are possible, but should be tested empirically as they may affect subsequent shearing of chromatin and/or antigen recognition.
4. To stop the cross-linking reaction, add 0.1 ml/well of 1.25 M glycine and incubate for a further 10 min at room temperature with shaking.
5. Carefully wash the cell monolayers twice with 5 ml/well of phosphate-buffered saline (PBS) (minus calcium and magnesium) and detach the cells with 1 ml/well of endothelial-grade trypsin using the procedure described in step 1. Once detached,

transfer the cell suspension to a 14-ml sterile centrifuge tube containing 5 ml of complete EGM-2 to neutralize the action of the trypsin and generate a cell pellet by centrifugation at 200 × *g* for 5 min at room temperature.

6. Discard the supernatant and resuspend the cells in 1 ml of ice-cold PBS supplemented with protease inhibitors (0.1 mM PMSF, 10 μg/ml soybean trypsin inhibitor, and 10 μg/ml benzamidine – see Note 1). Transfer the whole sample to a chilled 1.5-ml microcentrifuge tube and centrifuge at 200 × *g* for 5 min in a refrigerated microcentrifuge set at 4°C.
7. Discard the supernatant and thoroughly suspend the cell pellet in 0.5 ml of lysis buffer by gentle pipetting to avoid frothing. It is important that this step is sufficiently thorough to disperse the sample but if clumps persist, the sample may be passed through a 25-guage needle or gently disrupted up using a B Dounce homogenizer.
8. Centrifuge the sample at 200 × *g* for 5 min in a refrigerated microcentrifuge set at 4°C and discard the supernatant. To the remaining isolated nuclear pellet, add 0.2 ml of nuclear lysis buffer and leave on ice for 10 min.
9. The sample must now be sonicated. For our experiments in HUVECs, isolated nuclei were sonicated (10 × 10 second pulses) using a microtip probe fitted to a Branson Sonifer 250 probe sonicator set at maximum (see Note 6). Samples are left to cool on ice for at least 1 min between each pulse. This protocol is sufficient to shear genomic DNA into fragment sizes ranging from 400 to 600 bp (see Note 7). In order to standardize/normalize your results, 10% of each sample could be taken and reserved at this stage. All samples can be stored at –70°C at this point if necessary.
10. Determine the protein concentration in each sample. We use a bicinchoninic acid (BCA) protein assay (11) for which we take duplicate 2-μl aliquots from each experimental sample. Samples are then equalized for protein content (we have found that 0.1–0.2 mg/sample is best for our purposes, but it is recommended that you optimize protein content for your own application).
11. Equalize sample volume to 0.15 ml with nuclear lysis buffer and then add 1.35 ml of IP buffer (see Note 8).
12. To reduce background signal, the sample is precleared by the addition of 20 μl of 50% (v/v) slurry of protein A–Sepharose beads (see Note 9) and incubation at 4°C on a rotating wheel for 30 min. Samples are then microcentrifuged at 4°C and the supernatant transferred to a separate prechilled microcentrifuge tube.

13. Add antibody to sample. For immunoprecipitation of C/EBPβ, we add 1 μg/sample of anti-C/EBPβ antibody (see Note 10). Samples are then incubated on a rotating wheel at 4°C for 2 h. It is absolutely essential to control for nonspecific immunoprecipitation by performing identical immunoprecipitations in parallel using equivalent amounts of matched control antibodies.
14. Add 30 μl of 50% (v/v) slurry of pretreated protein A–Sepharose beads per sample and incubate on a rotating wheel at 4°C for a further 1 h (see Note 11).
15. Capture immune complexes by taking the samples off the wheel and allowing the beads to settle by gravity by leaving on ice for up to 5 min, then centrifuging at 100 × *g* for 5 min in a refrigerated microcentrifuge set at 4°C before leaving to settle for a further 30 s on ice, and very carefully removing as much supernatant as possible using a P1000 followed by a 1-ml syringe to which a 25-guage needle is attached.
16. Wash the samples sequentially using 1 ml/sample of wash buffer 1, wash buffer 2, and then wash buffer 3 followed by two further washes with TE, pH 8.0 (see Note 12).
17. After carefully removing all of the remaining supernatant using a 25-guage needle, elute the precipitated chromatin from the beads by adding 0.25 ml/sample of elution buffer and incubate samples at room temperature for 15–30 min on a rotating wheel. Then, centrifuge the samples at 100 × *g* for 2 min at room temperature and transfer all of the supernatant to a screw-capped 2-ml microcentrifuge tube. Repeat this step and combine the eluates so that each sample has a final volume of 0.5 ml.
18. Reverse the formaldehyde cross-links by adding 20 μl of 5 M sodium chloride to each sample and incubate at 65°C for 4 h using a heating block.
19. Digest proteins in each sample by adding 20 μl of 1 M Tris–HCl, pH 6.5, and 2 μl proteinase K (stock concentration 10 mg/ml, thus final concentration is 37 μg/ml). Incubate samples at 45°C for 1 h.
20. Add 1.25 ml of prechilled absolute ethanol to each sample and precipitate the DNA for 1 h at −70°C or overnight at −20°C. Centrifuge the samples at 14,000 × *g* for 30 min and allow the pellets to air dry. Resuspend each pellet in 175 μl of TE, pH 8.0, and extract DNA by adding 0.3 ml of phenol:chloroform:isoamyl alcohol (25:24:1) saturated with 10 mM Tris–HCl, pH 8.0, and 1 mM EDTA. Once the upper layer has been carefully removed to another microfuge tube, a further 175 μl of TE, pH 8.0, can be added to the phenol:chloroform:isoamyl alcohol to completely extract the DNA in the sample.

Both fractions are pooled and 30 μl of 5 M sodium chloride supplemented with 5 μg of glycogen added before DNA in each sample is reprecipitated by adding 2.5 volumes of prechilled absolute ethanol as described earlier. After centrifugation, the pellets are air dried and resuspended in no more than 50 μl of 175 μl TE, pH 8.0, or sterile water. We then use between 2 and 4 μl per sample for subsequent quantitative real-time PCR amplification reactions.

21. To determine the relative amount of target DNA precipitated, and thus the level of the antigen of interest bound to the promoter under investigation, we use SYBR Green detection of PCR products. The reactions are performed in quadruplicate in a 96-well plate in a DNA Engine Opticon 2 Real-Time Cycler that incorporates a two-color detection system (Bio-Rad Laboratories) (see Note 7 for more detailed PCR detection-related information).

The amount of DNA present in the sample is determined by taking the cycle threshold (Ct) values that are generated and calculating the amount of DNA from a standard curve made from parallel PCRs using known amounts of target DNA. Results can also be normalized to the amount of chromatin you started with. To do this, the 10% sample ("Input") taken in step 7 is amplified in parallel with the rest of the samples from step 13 onward. Any nonspecific signal obtained for the IgG control immunoprecipitation should also be subtracted from the test ChIP samples. For both the input and immunoprecipitated samples, care must be taken to ensure that the amount of sample used for each reaction lies within the linear range of the standard curve.

3.2. Transfection-Mediated siRNA Delivery into HUVECs

Ultimately, it is important to determine whether the induction of the SOCS-3 gene by a cAMP–Epac1–C/EBP pathway has any functional significance in ECs. The most straightforward strategy to test this is to determine the effects of blocking SOCS-3 induction on its well-characterized ability to inhibit downstream signaling by a soluble IL-6 receptor-α/IL-6 trans-signaling complex. To this end, we have optimized protocols for introducing small interfering RNAs (siRNAs) into HUVECs to not only assess the effects of blocking SOCS-3 induction, but also test the importance of other proteins involved in this cAMP-activated pathway. Again, the principle is straightforward: exogenous synthetic siRNA duplexes are introduced into HUVECs by transient transfection prior to assessment of target protein knockdown and its functional consequences.

1. On day 1, passage HUVECs at approximately 70–80% confluence from T-75 flasks into 6-well dishes for experimental work and into a separate T-75 tissue culture flask to maintain and propagate the cell line (see Note 3). The spent complete EGM-2 medium from the flask is first decanted into a sterile 50-ml

plastic centrifuge tube. Monolayers are washed with 2 ml/flask of HEPES-buffered saline and aspirated to remove spent medium and nonadherent cells. Endothelial-grade trypsin is then added (1 ml/flask) and the flask placed back into the cell culture incubator to allow cells to detach. Once all the cells have been detached by sharply tapping the flask, the resuspended cells are added to the 50-ml tube containing the spent complete EGM-2 to neutralize the trypsin. HUVECs are then pelleted by centrifugation at $200 \times g$ for 5 min at room temperature and resuspended in 1 ml of fresh complete EGM-2, and cell density is determined using a hemocytometer. If necessary, additional complete EGM-2 is added to the cell suspension to ensure that the resuspended HUVECs achieve a density of 0.2×10^6 cells/ml. Add 1 ml of suspension to each well of a 6-well tissue culture plate and leave overnight in the cell culture incubator to adhere. For maintenance of the cell line, 1.2×10^6 cells from the remaining suspension can be seeded into a fresh T-75 flask and complete EGM-2 added to give a final volume of medium of 10 ml.

2. On the morning of day 2, the HUVECs should have reached 70–80% confluence. Aspirate the culture medium and replace with fresh complete EGM-2 (0.7 ml/well).
3. Thaw sufficient aliquots of nontargeting control and test siRNAs for the experiments and dilute to the desired concentrations in 0.1 ml of Opti-MEM in sterile microcentrifuge tubes (see Note 13). Mix thoroughly by briefly vortexing.
4. Add 12 μl of HiPerFect transfection reagent to each tube of siRNA/Opti-MEM and mix gently by pipetting up and down five times. Incubate the sample at room temperature for 5–10 min (but no longer) to allow the formation of transfection reagent–siRNA complexes.
5. Add the siRNA–HiPerFect–Opti-MEM mix dropwise to the appropriate well of HUVECs, and gently tilt the plate back and forward and from side to side five or six times to evenly distribute the mix over the cells. Return the plate to the cell culture incubator.
6. After 3 h, add 1.6 ml/well of complete EGM-2 and return to the cell culture incubator.
7. On day 3, check HUVEC cultures under the light microscope to ensure that siRNA transfection has had no obvious deleterious effects on cell viability (see Note 14). If the siRNA-transfected cells look in similar good condition to nontransfected control HUVECs, return to the incubator.
8. On day 4, perform experimental treatment, cell lysis, and subsequent assessment of target gene knockdown and effects on function by SDS-PAGE and immunoblotting (see Fig. 1 for an example).

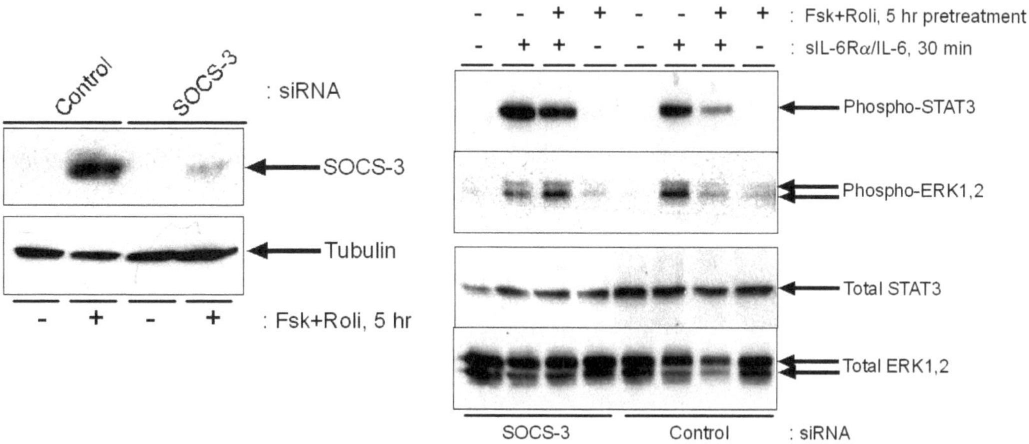

Fig. 1. Effect of SOCS-3 knockdown on cAMP-mediated inhibition of IL-6 *trans*-signaling in HUVECs in vitro. *Left panel*: HUVECs were transfected with nontargeting control and SOCS-3-specific siRNAs prior to treatment with forskolin + rolipram (Fsk + Roli) for 5 h as indicated. Soluble cell extracts equalized for protein content were then fractionated by SDS-PAGE for immunoblotting with anti-SOCS-3 and tubulin antibodies. *Right panel*: HUVECs from the same transfection experiment were pretreated with or without Fsk + Roli for 5 h followed by exposure to a soluble IL-6 receptor α (sIL-6Rα)/IL-6 trans-signaling complex for 30 min as indicated. Soluble cell extracts equalized for protein content were then fractionated by SDS-PAGE for immunoblotting with the indicated antibodies against total and functionally active/phosphorylated versions of signal transducer and activator of transcription 3 (STAT3) and extracellular signal-regulated kinase 1,2 (ERK1,2). RNAi-mediated suppression of SOCS-3 induction is observed to abolish the inhibitory effect of Fsk + Roli pretreatment on sIL-6Rα/IL-6 stimulation of STAT3 phosphorylation on Tyr705 and ERK1,2 phosphorylation on Thr202/Tyr204 (ERK1 numbering: Thr183/Tyr185 on ERK2). Reproduced from ref. 8 with permission from Elsevier.

4. Notes

1. Stocks of soybean trypsin inhibitor and benzamidine (each 10 mg/ml) can be made up in deionized water and stored in single-use aliquots at –20°C for at least 6 months. PMSF must be made up fresh from powder to a concentration of 10 mM in methanol on the day of the experiment before being added to the appropriate buffer to produce a final concentration of 0.1 mM.
2. To minimize evaporation once opened, stock bottles of protein A–Sepharose beads should be capped tightly and the seal sheathed with tightly wrapped Nescofilm after each use. However, should significant evaporation occur over time, make up 20% (v/v) ethanol in PBS and add enough of this solution to the bottle so that the ratio of bead to buffer volume is maintained at approximately 1:1.
3. Culture and propagate the HUVECs in T-75 cell culture flasks in 10 ml of complete EGM-2 medium. The cells should be maintained in a humidified 37°C cell culture incubator with 5% (v/v) CO_2.

4. It is very important that the specific conditions must be determined empirically for different cell types. In the case of primary cells like HUVECs, it is also important to note whether responses differ between different passages, as primary cells can dedifferentiate in culture. For the experiments described here, we have used HUVECs from passages 2 until 5.
5. In our hands, a treatment with forskolin produces a robust increase in C/EBPβ recruitment to the SOCS-3 promoter that is detectable at 30 min. Selective activation of Epac using 007 produces a more transient response that peaks between 5 and 15 min after stimulation (6).
6. We strongly recommend determining empirically the best conditions for your particular application. For the sake of reproducibility, the sonicator microprobe should not be pitted: some tips can be reground to ensure smoothness. We recommend that you initially collect chromatin samples after various cycles and check their size range by agarose gel electrophoresis after reversal of cross-links and proteinase K digestion. It is also recommended that you take a fraction of your experimental sample for analysis by agarose gel electrophoresis to check for efficient shearing.
7. Primers should be designed to lie on either side of the site of interest. The amplicons should also be in the order of 200 bp and can be designed using standard software, such as Primer3 (http://primer3.sourceforge.net/). PCRs should also be optimized to improve sensitivity and/or specificity before commencing ChIP experiments. We used the following primers spanning an area on the human SOCS-3 promoter predicted to comprise multiple consensus C/EBP-binding sites by the TESS transcription factor prediction program (http://www.cbil.upenn.edu/cgi-bin/tess/tess): forward 5′-CTGTCGCC AGGTTGGAGT-3′ and reverse 5′-GGTCAGGAGATCGAG ACCAT-3′. This generated a single predicted 197-bp product, which was verified by agarose gel electrophoresis. PCRs (25 μl, final volume and a final Mg^{2+} concentration of 1.5 mM) are performed in quadruplicate using a DNA Engine Opticon II real-time, two-color PCR detection system (Bio-Rad Laboratories) and employed 36 cycles of denaturation at 94°C (30 s), annealing at 50°C (20 s) and extension at 72°C (10 s). Melting curves are determined using the following parameters: 95°C cooling to 70°C, and ramping to 90°C at 0.2°C/s. For other cell types/applications, you may have to vary the number of cells used and change the size of the culture vessel, as the sensitivity of the assay is determined by the efficiency of your precipitating antibody and your PCR.
8. Some applications may benefit from the addition of protein phosphatase inhibitors at the following final concentrations:

sodium orthovanadate (100 μM), sodium phosphate (10 mM), and sodium fluoride (10 mM).

9. We advise cutting the bottom end from tips when pipetting protein A–Sepharose beads and frequently resuspending the stock bottle in between additions to ensure that each tube receives the same amount of beads.
10. It is important to make sure that you use sufficient amount of antibody to precipitate as much of your antigen from the chromatin solution as possible to maximize detection of differences between experimental samples. You can check this by performing pilot immunoprecipitations under a range of conditions beforehand and checking efficiency by SDS-PAGE and immunoblotting. Make sure that you load a fraction of your input sample as a control.
11. A 50% (v/v) slurry of Protein A–Sepharose beads are prepared by washing three times in 1 ml of wash IP buffer and incubated for 30 min at 4°C with rotation with 1 ml of IP buffer supplemented with 1 mg/ml immunoglobulin-free bovine serum albumin and 1 mg/ml salmon sperm DNA. The beads are then washed three more times with 1 ml of wash IP buffer and resuspended in the same buffer to their original volume.
12. You may need to optimize salt concentrations for your own particular application. If high background signal is a problem, this may be overcome by transferring the volume of the final wash step to a fresh microcentrifuge tube. Care must be taken not to lose any of your beads to the sides of your pipette tip: we find that this can be reduced by repeatedly pipetting a solution of 0.1% (w/v) bovine serum albumin in TE, pH 8.0. The tips are then washed by repeatedly pipetting in 1 ml of TE, pH 8.0; this procedure significantly reduces bead loss.
13. We strongly recommend that you optimize conditions for optimal target gene knockdown for your particular application. When testing new siRNAs, we initially test a range of concentrations (1–200 nM) to determine the lowest concentrations capable of producing the most effective knockdown. For SOCS-3 and Epac1, we have been able to routinely obtain 80–90% knockdown of the target protein versus levels in nontargeted siRNA-transfected HUVECs (5–8).

    When performing such experiments, it is essential (as with all siRNA experiments) to include control transfections using equivalent concentrations of nontargeting siRNAs and also mock-transfected cells, i.e., cells treated with transfection reagent alone without any siRNAs. Such controls are critical to ensure that any functional consequences observed as a result of target gene knockdown are specific and also not simply due to nonspecific siRNA-mediated activation of innate immune signaling proteins, such as PKR (12).

14. HUVECs are relatively flat cells that normally display what is termed a "cobblestone morphology" when cultured in vitro. It is important to check any deleterious effect of siRNA delivery on gross indicators of cell viability, such as cell rounding and detachment, under the microscope before proceeding with functional assays.

Acknowledgments

This work was supported by project grants from the British Heart Foundation, the UK Biotechnology and Biological Sciences Research Council, and Chest Heart and Stroke Scotland to TMP and SJY and a PhD studentship from the British Heart Foundation to HDW.

References

1. Hansson, G.K., and Libby, P. (2006) The immune response in atherosclerosis: a double-edged sword. *Nat. Rev. Immunol.* **6**, 508–519.
2. Schouten, M., Wiersinga, W.J., Levi, M, and van der Poll, T. (2008) Inflammation, endothelium, and coagulation in sepsis. *J. Leukoc. Biol.* **83,** 536–545.
3. Hansson, G.K., Robertson, A.K., and Söderberg-Nauclér, C. (2006) Inflammation and atherosclerosis. *Annu. Rev. Pathol.* **1**, 297–329.
4. Choi EY, Santoso S, and Chavakis T. (2009) Mechanisms of neutrophil transendothelial migration. *Front. Biosci.* **14**, 1596–1605.
5. Sands, W.A., Woolson, H.D., Milne, G.R., Rutherford, C., and Palmer T.M. (2006) Exchange protein activated by cyclic AMP (Epac)-mediated induction of suppressor of cytokine signaling 3 (SOCS-3) in vascular endothelial cells. *Mol. Cell Biol.* **26**, 6333–6346.
6. Yarwood, S.J., Borland, G., Sands, W.A., and Palmer, T.M. (2008) Identification of CCAAT/enhancer-binding proteins as exchange protein activated by cAMP-activated transcription factors that mediate the induction of the SOCS-3 gene. *J. Biol. Chem.* **283**, 6843–6853.
7. Borland, G., Bird, R.J., Palmer, T.M., and Yarwood, S.J. (2009) Activation of protein kinase Calpha by EPAC1 is required for the ERK- and CCAAT/enhancer-binding protein beta-dependent induction of the SOCS-3 gene by cyclic AMP in COS1 cells. *J Biol Chem.* **284**, 17391–17403.
8. Woolson, H.D., Thomson, V.S., Rutherford, C., Yarwood, S.J., and Palmer, T.M. (2009) Selective inhibition of cytokine-activated extracellular signal-regulated kinase by cyclic AMP via Epac1-dependent induction of suppressor of cytokine signalling-3. *Cell. Signal.* **21**, 1706–1715.
9. Enserink, J.M., Christensen, A.E., de Rooij, J., van Triest, M., Schwede, F., Genieser, H.G., Døskeland, S.O., Blank, J.L., and Bos, J.L. (2002) A novel Epac-specific cAMP analogue demonstrates independent regulation of Rap1 and ERK. *Nat. Cell Biol.* **4**, 901–906.
10. Johnson, D.S., Mortazavi, A., Myers, R M., and Wold, B. (2007) Genome-wide mapping of in vivo protein-DNA interactions. *Science* **316**, 1497–502.
11. Smith, P.K., Krohn, R.I., Hermanson, G.T., Mallia, A.K., Gartner, F.H., Provenzano, M.D., Fujimoto, E.K., Goeke, N.M., Olson, B.J., and Klenk, D.C. (1985) Measurement of protein using bicinchoninic acid. *Anal Biochem.* **150**, 76–85.
12. Sledz, C.A., and Williams, B.R. (2004) RNA interference and double-stranded-RNA-activated pathways. *Biochem. Soc. Trans.* **32**, 952–956.

Part II

Chromatin Structure

Chapter 15

Computational Analysis of Promoter Elements and Chromatin Features in Yeast

John J. Wyrick

Abstract

Regulatory elements in promoter sequences typically function as binding sites for transcription factor proteins and thus are critical determinants of gene transcription. There is growing evidence that chromatin features, such as histone modifications or nucleosome positions, also have important roles in transcriptional regulation. Recent functional genomics and computational studies have yielded extensive datasets cataloging transcription factor binding sites (TFBS) and chromatin features, such as nucleosome positions, throughout the yeast genome. However, much of this data can be difficult to navigate or analyze efficiently. This chapter describes practical methods for the visualization, data mining, and statistical analysis of yeast promoter elements and chromatin features using two Web-accessible bioinformatics databases: ChromatinDB and Ceres.

Key words: Transcription factor binding sites, TFBS, Nucleosome positions, Histone modifications, Data visualization, Data mining, Statistical analysis, *Saccharomyces cerevisiae*

1. Introduction

Gene transcription is regulated by transcription factor proteins, which bind to short DNA sequences typically found in proximal promoter regions. Our knowledge of transcription factor binding sites (TFBS), and their function, is perhaps greatest for the model yeast *Saccharomyces cerevisiae*. Pioneering studies have used chromatin immunoprecipitation microarray (ChIP-chip) experiments to map the genomic binding sites of more than 200 different transcription factor proteins in yeast (1, 2). In parallel, comparative genomics approaches have been used to analyze the promoter sequences of related yeast species to identify conserved TFBS in *S. cerevisiae* (1, 3–5). A number of bioinformatics tools and databases have

Ales Vancura (ed.), *Transcriptional Regulation: Methods and Protocols*, Methods in Molecular Biology, vol. 809,
DOI 10.1007/978-1-61779-376-9_15,

been developed to analyze these TFBS datasets, including the *S. cerevisiae* Promoter Database (SCPD) (6), MYBS (7), YEASTRACT (8, 9), and Ceres (see below).

However, chromatin structure and histone modifications also play important roles in transcription. For example, histone lysine acetylation and methylation have been strongly associated with gene transcription (10, 11), and these modifications function to regulate the expression or silencing of many yeast genes (12, 13). Similarly, the positioning of nucleosomes along genomic DNA can have an important role in regulating yeast gene expression (14). For example, many yeast genes contain a nucleosome-free or depleted region in their proximal promoter sequence (15–17). Intriguingly, nucleosome-free regions are enriched for predicted, conserved, and bound TFBS (15, 18–20). Indeed, we have found that TFBS located in nucleosome-free regions are more likely to have functional roles in transcriptional regulation (20). These and related studies (e.g., refs. 21, 22) have highlighted the importance of chromatin in transcriptional regulation, and emphasize the potential significance of integrating DNA sequence information, such as promoter elements, with chromatin data.

Groundbreaking studies by multiple groups have mapped global patterns of histone modifications (23–30) and nucleosome positions (18, 31–36) throughout the yeast genome. These data provide an important resource for investigating the role of chromatin in yeast gene expression. The ChromatinDB database (37) provides a user-friendly and Web-accessible interface to many of these global histone modification datasets. Moreover, we have recently developed the Ceres database (20) to enable the integrated analysis of yeast TFBS in the context of nucleosome positioning data. In this chapter, we describe practical uses of these databases and associated software tools for the analysis of yeast promoter elements and chromatin features.

2. Materials

The Ceres and ChromatinDB databases can be readily accessed using an Internet-connected computer and Web browser. For the optimum user experience, we recommend using the Mozilla Firefox Web browser. A free version of Firefox can be downloaded from the following Web site: http://www.mozilla.com/firefox/. However, most popular Web browsers should be readily compatible with the Ceres and ChromatinDB databases. Javascript should be enabled in the Web browser to ensure full functionality with these databases.

3. Methods

In this chapter, we first describe how one can use the freely available Ceres database to visualize promoter elements and chromatin features for user-selected genes. Second, we describe data mining methods to search for genes containing user-specified promoter elements, and we analyze the histone modification patterns in these promoters using the ChromatinDB database. Third, we discuss methods to search for promoter elements located in specific chromatin contexts (e.g., a nucleosome-free region). Finally, we describe procedures for the statistical analysis of enriched promoter elements or chromatin features among user-selected gene sets.

3.1. Visualization of Promoter Elements and Chromatin Features

Overview: The Ceres database provides flexible tools for visualizing promoter elements such as predicted, conserved, or bound TFBS, and chromatin features such as nucleosome positions. As an example, the first part of this section describes how to visualize the promoter features of the *GAL1* gene, which encodes a key enzyme in galactose metabolism and is regulated by the Gal4 transcription factor (38, 39). It is important to note, however, that the Ceres visualization tool can be used to visualize the promoters of up to 100 user-selected genes simultaneously. The second part of this section describes how to visualize changes in nucleosome positioning in the promoter region of the *UBC4* gene.

1. Use your Web browser to navigate to the Web site for the Ceres database (20), which can be found at the following Web address: http://wyrickserver.smb.wsu.edu/Ceres/.
2. Near the top of the Ceres home page, click the link labeled "Visualization." Alternatively, you can directly navigate to the visualization page by typing in the URL: http://wyrickserver.smb.wsu.edu/cgi-bin/Ceres/cgi/visualize_select.pl.
3. The Gene Selection page allows the user to input a list of up to 100 gene names in the text box labeled "Accessions" (see Note 1). The users can also select various options that will affect the type of data displayed or the parameters used to display the data (see Note 2). Ensure that the "Compact" Visualization type is selected.
4. Type in the gene name *GAL1* into the Accessions text box, and click the "Display" button near the bottom of the page.
5. A portion of the resulting visualization output is shown in Fig. 1. The image shown near the top of Fig. 1 depicts TFBS (see Note 3) and nucleosome positions in the promoter of the YBR020W gene, which is the systematic name of *GAL1*. The TFBS are depicted as colored vertical lines; the nucleosome positions are depicted as shaded ovals. The coding region of

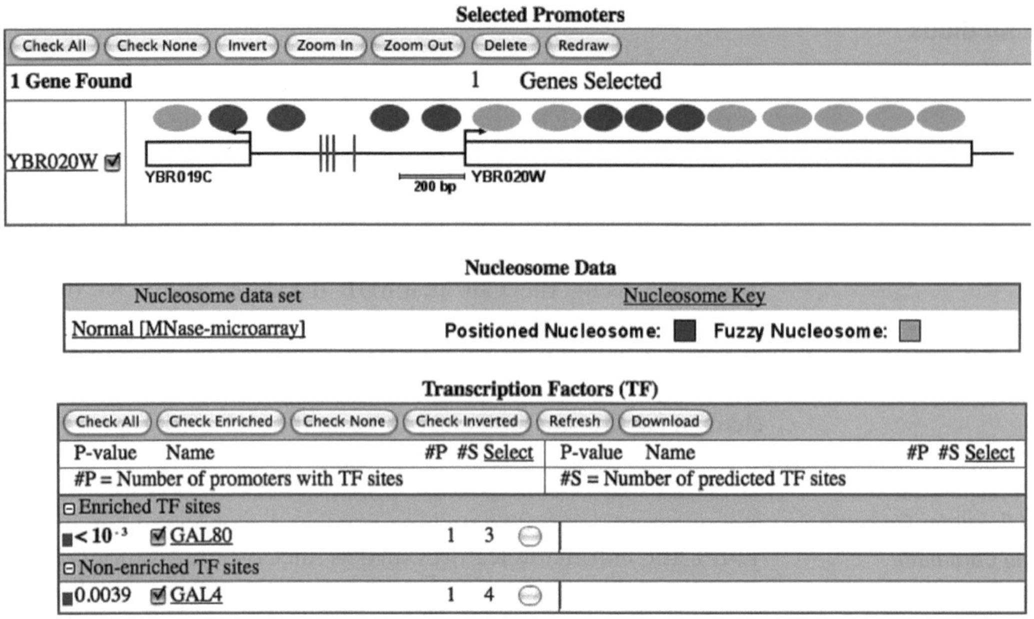

Fig. 1. Visualization of bound TFBS and nucleosome positions in the *GAL1*/YBR020W promoter. The start of the *GAL1*/YBR020W gene is indicated with an *arrow*, and a *black outline rectangle* encompasses the coding region. The *horizontal line* to the *left* of the YBR020W gene represents the promoter sequence of *GAL1*. The *positions* of TFBS in the *GAL1* promoter sequence are indicated as *vertical colored lines*. The *color* of the *lines* indicates the identity of the TFBS, which can be determined by inspection of the *color key* in the Transcription Factors table. The TFBS depicted correspond to Gal4 and Gal80 binding sites (note that three of the *GAL4* TFBS are hidden *behind GAL80* TFBS in the promoter image). The location of the upstream gene (i.e., *GAL10*/YBR019C) is also depicted as it falls within the selected upstream range of the *GAL1*/YBR020W promoter sequence. The *gray-colored ovals above* the YBR020W gene/promoter correspond to the locations of positioned nucleosomes in this genomic region. Note that the *original colors* of this and subsequent figures have been modified for better display in *grayscale*.

the *GAL1*/YBR020W gene is also depicted (see Note 4). In addition, tables are displayed to provide a key to the nucleosome dataset and TFBS displayed (Fig. 1). The visualization image can be saved by right clicking (or control-clicking with a Macintosh computer) on the image, and selecting the save option.

6. Click the "Check None" button in the "Transcription Factor" table and then click the button next to the *GAL4* transcription factor. Only the *GAL4* transcription factor should be selected/checked.
7. Click the "Redraw" button above the promoter image. Now only TFBS for the selected transcription factor (i.e., *GAL4*) should be displayed in the promoter image (see Note 5).
8. Click the "Zoom In" button near the top of the display. This will increase the size and magnification of the *GAL1*/YBR020W promoter image.
9. Hover the pointer over one of the Gal4 binding sites in the promoter image. A pop-up window should appear indicating

the relative position of that particular *GAL4* binding site in the *GAL1*/YBR020W promoter (e.g., GAL4 at −352 to −336 bp upstream of the gene start). A complete list of the positions and sequences of the TFBS in this promoter can be obtained by clicking the "Motifs" button near the top of the page (not shown in Fig. 1).

10. Click on one of the *GAL4* TFBS in the promoter image. A popup window should appear with information about the *GAL4* TFBS, including the log odds matrix of the binding consensus, and a PubMed link to the relevant publication source.
11. Hover the pointer over one of the light gray nucleosomes above the *GAL1*/YBR020W gene display. A pop-up window should appear listing the relative position of the nucleosome relative to the gene start (e.g., Fuzzy Nuc at +27 to +179; see Note 6).
12. Ceres can also be used to visualize two nucleosome datasets simultaneously. Go to the Chromatin Viewer page by clicking on the "Chromatin Viewer" link near the top of the page, or by typing in the following URL into your Web browser: http://wyrickserver.smb.wsu.edu/cgi-bin/Ceres/cgi/chromatin_select.pl
13. In the "Visualization Type" menu, select the "Promoter Display (max 20 genes)" option.
14. In the "Nucleosome Data Set #1" menu, select "Normal [MNase-seq]". In the "Nucleosome Data Set #2" menu, select "Heat Shock [MNase-seq]" (see Note 7).
15. In the "Upstream Range" menu, select "1500" as the Maximum bp upstream.
16. Type in the gene name *UBC4* in the Gene Accessions text box. Click the "Display" button.
17. The resulting Web page shows the promoter image for the *UBC4*/YBR082C gene. In this case, two nucleosome datasets are displayed as colored ovals (see Note 8). The top set of nucleosome positions (i.e., ovals) is colored red and corresponds to nucleosome positions mapped in heat-shocked cells (e.g., "Heat Shock [MNase-seq]" dataset). The bottom set of nucleosome positions is colored blue, and corresponds to nucleosome positions in cells grown under normal growth conditions. Clicking on the "Normal [MNase-seq]" or "Heat Shock [MNase-seq]" links in the "Nucleosome Data" table will open a new page with more detailed information about these chromatin datasets.
18. Click the "Check None" button in the "Transcription Factor" table and then click the button next to the *HSF1* transcription factor. Only the *HSF1* transcription factor should be selected/checked.

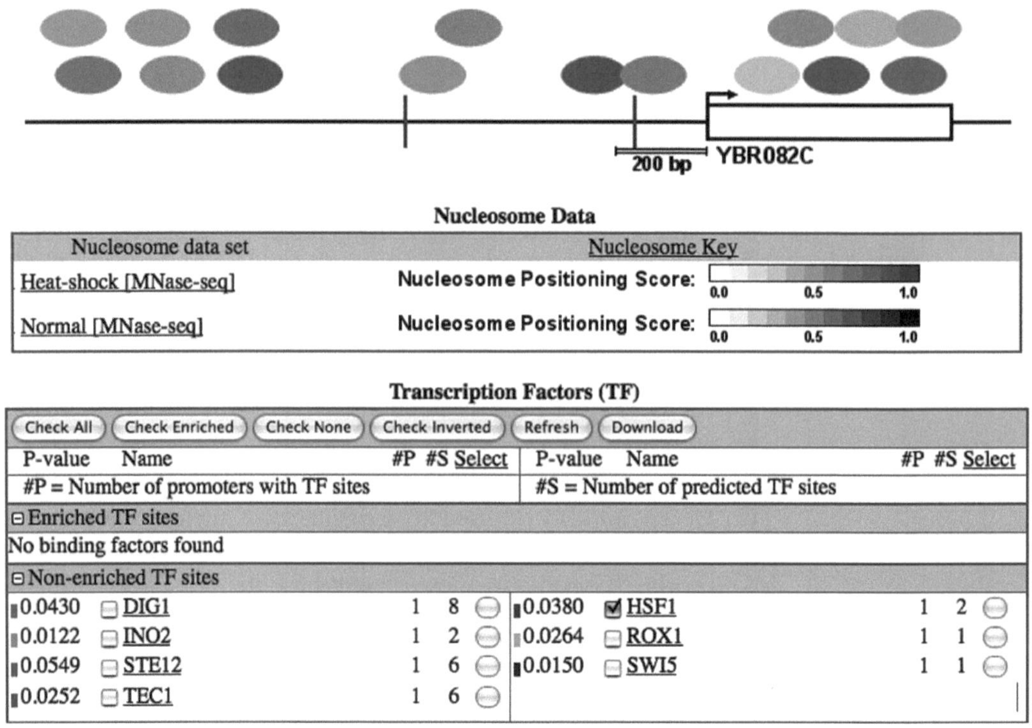

Fig. 2. Visualization of multiple nucleosome datasets in the *UBC4*/YBR082C gene. The *uppermost set* of nucleosome positions (i.e., *ovals*) correspond to Heat Shock [MNase-seq] dataset, while the *lower set* of nucleosome positions correspond to the Normal [MNase-seq] dataset (31). The *intensity of color* in each *oval* corresponds to the positioning *score* for each nucleosome (see "Nucleosome Key" in the figure). Note that only the *HSF1* binding sites are displayed in this image.

19. Click the "Redraw" button above the promoter image. Now only TFBS for the selected transcription factor (i.e., *HSF1*) should be displayed in the promoter image. A portion of the resulting visualization output is shown in Fig. 2. It is interesting to observe that significant changes in nucleosome positioning upon heat shock occur adjacent to two *HSF1* TFBS in the *UBC4*/YBR082C promoter (see Fig. 2).

3.2. Using the Data Mining Tool to Search for Promoters with Selected TFBS

Overview: The Ceres database provides data mining tools to search for genes containing selected TFBS in their promoter sequence. In this section, we describe how to identify target genes that contain a bound *UME6* TFBS in their promoter sequences. *UME6* is a key regulator of genes expressed in yeast meiosis/sporulation (40, 41). A similar approach can be used to identify the target genes of any combination of yeast transcription factors. We also describe how to search for genes that are annotated in a particular functional category (e.g., sporulation), and use the ChromatinDB database to analyze for enriched or depleted histone modifications in the promoter regions of these genes.

1. Click on the link near the top of the page that is labeled "Data Mining". Alternatively, go to the URL: http://wyrickserver.smb.wsu.edu/cgi-bin/Ceres/cgi/datamine_select.pl.
2. In the menu labeled "Transcription Factors" scroll down and select *UME6*. With *UME6* highlighted, click the "Add TFs" button. *UME6* should now be listed in the "Selected TFs" text box.
3. Under "Mining Options," in the "Upstream Range" menu, select "500" as the maximum bp upstream in which to search for *UME6* TFBS. Click the "Submit" button.
4. The subsequent results Web page should list 162 genes that were identified in the search query for *UME6* TFBS. Near the bottom of the page (scroll down) is a table of enriched gene ontology terms ($P < 10^{-4}$; see Note 9). This table is shown in Fig. 3. Inspection of the enriched gene ontology categories in Fig. 3 indicates that many of the genes containing a bound *UME6* TFBS function in yeast meiosis/sporulation, as expected.
5. Scroll up to the "Selected Promoters" table, and click on the underlined name "YHR153C". This should open a new window with the information page for *SPO16*/YHR153C in the Saccharomyces Genome Database (42, 43). Note that the *SPO16*/YHR153C gene is a known functional target of the *UME6* transcription factor (44).
6. Return to the Data Mining gene selection page: http://wyrickserver.smb.wsu.edu/cgi-bin/Ceres/cgi/datamine_select.pl.
7. In the menu labeled "Transcription Factors" scroll down and select *SUM1*, and click the "Add TFs" button. Repeat this process for *UME6*. *SUM1* and *UME6* should now be listed in the "Selected TFs" text box. With multiple transcription factors selected, the data mining tool uses AND logic to find promoters that contain both *SUM1* and *UME6* binding sites.
8. Under the "TF Binding Site Options", select "Medium" Site Conservation and "Medium" ChIP binding. Using this criteria, the data mining query will only identify TFBS that are conserved in one or more other yeast species (i.e., "Medium" site conservation),

Enriched Gene Ontology Terms

P-value	GO Term (Download)	Number of Genes	GO ID	Select
$< 10^{-5}$	condensed nuclear chromosome	6	794	
$< 10^{-5}$	peroxisomal matrix	5	5782	
$< 10^{-10}$	meiosis	23	7126	
$< 10^{-4}$	synapsis	4	7129	
$< 10^{-8}$	reciprocal meiotic recombination	11	7131	
$< 10^{-4}$	regulation of reciprocal meiotic recombination	3	10520	
$< 10^{-7}$	sporulation resulting in formation of a cellular spore	14	30435	

Fig. 3. Enriched gene ontology terms identified in the set of genes containing a bound *UME6* TFBS.

and whose promoter is bound by the respective transcription factor with "Medium" confidence in a ChIP-chip experiment ($P<0.005$).

9. Click the "Submit" button. In the resulting Web page, a single gene (i.e., YHR124W) should be identified by this query.
10. Click on the name/link for YH124W gene. This should bring up a new window containing the SGD information page for the *NDT80*/YHR124W gene. *NDT80* is a key transcriptional regulator of meiosis-specific gene expression and is itself regulated by the Sum1 and Ume6 transcription factors (45).
11. Return to the Data Mining gene selection page: http://wyrickserver.smb.wsu.edu/cgi-bin/Ceres/cgi/datamine_select.pl.
12. Click the button labeled "Gene Ontology". This activates the Gene Ontology search function.
13. Enter the search term "Sporulation" next to the "Search" button, and click the "Search" button. Select the Gene Ontology (GO) term "sporulation resulting in formation of a cellular spore" and click the "Add Terms" button. The GO term "sporulation resulting in formation of a cellular spore" should now be listed in the "Selected Terms" text box. Click the "Submit" button.
14. The subsequent results Web page should list 101 genes that were identified in the search query for the GO term. Click the "Download" button adjacent to the "Selected Promoters" table. This should open a new window containing the list of 101 genes that were identified in this data mining query. Copy this list of genes.
15. Open a new Web browser window, and navigate to the Web site for the ChromatinDB database: http://www.bioinformatics2.wsu.edu/ChromatinDB/.
16. Paste the list of gene names from step 14 into the text box labeled "Gene Accessions." Keep all of the default options (see Note 10), and click the "Display" button.
17. A portion of the resulting Web page is shown in Fig. 4. The enrichment or depletion of histone acetylation or methylation marks is displayed in the graph. The table lists which histone modifications are enriched or depleted in the selected gene set by giving the average rank percentile and calculated *P*-value of enrichment (see Note 11). Note that many histone posttranslational modifications, particularly histone acetylation, are depleted from the promoter regions of genes involved in sporulation under normal (i.e., nonsporulation) growth conditions.

3.3. Using the Data Mining Tool to Search for TFBS in a Defined Chromatin Context

Overview: The Ceres data mining tool can also be used to identify TFBS located in specific chromatin contexts. For example, the binding sites of many yeast transcriptional activators are enriched in nucleosome-free regions of promoters. By contrast, we have

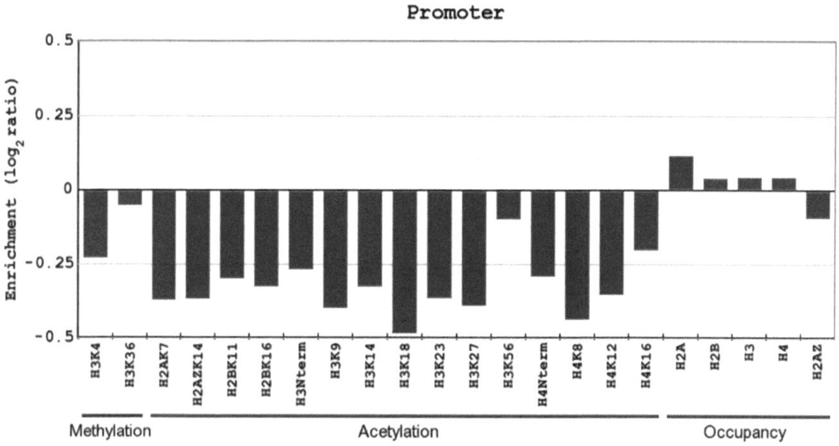

Chromatin Enrichment/Depletion

Promoter				
Histone Modification	Rank Percentile	Trend	P-value	Ranked Gene List
H3K4 Di-Methylation	34.2 %	**Depleted**	**$<10^{-7}$**	
H3K36 Di-Methylation	49.4 %	none	0.8488	
H2AK7 Acetylation	27.8 %	**Depleted**	**$<10^{-4}$**	
H2AZK14 Acetylation	37.2 %	none	$<10^{-3}$	
H2BK11 Acetylation	26.5 %	**Depleted**	**$<10^{-5}$**	
H2BK16 Acetylation	25.2 %	**Depleted**	**$<10^{-6}$**	
H3Nterm Acetylation	33.0 %	**Depleted**	**$<10^{-8}$**	
H3K9 Acetylation	27.4 %	**Depleted**	**$<10^{-5}$**	
H3K14 Acetylation	27.5 %	**Depleted**	**$<10^{-5}$**	
H3K18 Acetylation	26.8 %	**Depleted**	**$<10^{-5}$**	
H3K23 Acetylation	27.6 %	**Depleted**	**$<10^{-5}$**	
H3K27 Acetylation	26.7 %	**Depleted**	**$<10^{-5}$**	
H3K56 Acetylation	43.2 %	none	0.0403	
H4Nterm Acetylation	31.1 %	**Depleted**	**$<10^{-10}$**	
H4K8 Acetylation	21.7 %	**Depleted**	**$<10^{-7}$**	
H4K12 Acetylation	24.9 %	**Depleted**	**$<10^{-6}$**	
H4K16 Acetylation	31.9 %	none	$<10^{-3}$	
H2A Occupancy	57.1 %	none	0.0361	
H2B Occupancy	51.3 %	none	0.6493	
H3 Occupancy	55.9 %	none	0.0525	
H4 Occupancy	54.1 %	none	0.1735	
H2AZ Occupancy	43.9 %	none	0.0719	

Fig. 4. Visualization and statistical analysis of histone posttranslational modifications (and histone occupancy) in the promoter regions of genes in the sporulation gene ontology category. The *top panel* graphically displays the nucleosome-normalized enrichment ($\log_2$ ratio) derived from published ChIP-chip experiments. The *bottom table* summarizes the statistical analysis of enriched or depleted histone posttranslational modifications or histone occupancy in the selected genes.

found that the binding sites of many transcriptional repressors are enriched in linker DNA regions (see Note 12 for further explanation of these chromatin environment categories). Hence, the use of these chromatin environment categories can significantly enhance searches for functional TFBS. This section describes how to use the chromatin environment-specific search tool to find functional binding sites of the *BAS1* transcription factor, which regulates genes that function in histidine and nucleotide biosynthesis in yeast (46–48).

1. Go to the to the Data Mining page by clicking the link labeled "Data Mining" near the top of the page, or by typing in the following URL into your Web browser: http://wyrickserver.smb.wsu.edu/cgi-bin/Ceres/cgi/datamine_select.pl.
2. In the menu labeled "Transcription Factors" select *BAS1*, and click the "Add TFs" button. *BAS1* should now be listed in the "Selected TFs" text box.
3. Under the "Nucleosome Data Set" menu, make sure that the "Normal [MNase-microarray]" dataset is selected.
4. Under the "Chromatin Environment" menu, select the "Nucleosome Free Region" option. This option limits the data mining query to *BAS1* binding sites that are located in a nucleosome-free region.
5. Click the "Submit" button. The subsequent results Web page should list 31 genes that were identified in the search query for *BAS1* TFBS in a nucleosome-free region. Many of these genes (e.g., *ADE2*, *ADE3*, *ADE4*, *ADE5/7*, *ADE6*, *ADE8*, *ADE12*, *ADE13*, *ADE17*, *GCV2*, *HIS4*, *MTD1*, *SHM2*) are known *BAS1* targets.
6. Click the "Check None" button in the "Transcription Factor" table. Then click the button next to the *BAS1* transcription factor. Only the *BAS1* transcription factor should be selected/checked.
7. Click the "Display" button in the "Selected Promoters" table. The resulting Web page shows the promoter images of the 31 target genes, highlighting the location of the *BAS1* TFBS in each promoter. Importantly, at least one *BAS1* binding site in each promoter is located in a nucleosome-free region.
8. Return to the Data Mining gene selection page: http://wyrickserver.smb.wsu.edu/cgi-bin/Ceres/cgi/datamine_select.pl.
9. Again, select *BAS1* by clicking the "Add TFs" button. In this case, select the "Nucleosome Interior" Chromatin Environment. Click the "Submit" button.
10. The resulting Web page should list 12 genes that were identified in the search query for Bas1 TFBS in a nucleosome interior chromatin environment. Unlike the previous query, this list contains few known *BAS1* target genes (only *ADE8*/YDR408C and *ADE17*/YMR120C).
11. Click the "Check None" button in the "Transcription Factor" table, then click the button next to the *BAS1* transcription factor. Only the *BAS1* transcription factor should be selected/checked.

12. In the Selected Promoters Table, click the "Check None" button, and then click the buttons next to the YDR408C and YMR120C genes. Only YDR408C and YMR120C genes should be selected.
13. Click the "Display" button. The resulting page should show the promoter images of the *ADE8*/YDR408C and *ADE17*/YMR120C genes. It is intriguing that in each target gene promoter, there is not only a Bas1 binding site in a nucleosome interior region but also one or more additional Bas1 binding sites in a nucleosome-free region.

3.4. Statistical Analysis of Promoter Elements and Chromatin Features

Overview: The Ceres database can be used to identify enriched or depleted promoter elements or chromatin features in user-selected gene sets. In this section, we will analyze the promoter sequences of yeast proteasome genes for enriched TFBS. We will also analyze the nucleosome positioning of these genes. Similar methods can be used to investigate other gene sets, such as coexpressed genes identified in expression profiling experiments, for example.

1. Go to the to the Gene Set Analysis page by clicking the link labeled "Gene Set Analysis" near the top of the page, or by typing in the following URL into your Web browser: http://wyrickserver.smb.wsu.edu/cgi-bin/Ceres/cgi/geneset_select.pl.
2. Click on the link labeled "Sample Gene List." This should open a page containing a sample list of 33 genes that are classified in the "proteasome complex", "proteasome core complex", or "proteasome regulatory particle" GO categories.
3. Copy this list of 33 genes and paste into the "Gene Accessions" text box in the Gene Set Analysis page. Under the "Upstream Range" menu, select the option "500" maximum bp upstream. Click the "Submit" button.
4. In the resulting analysis page, click the button labeled "Check Enriched" in the Transcription Factors table. This selects transcription factors whose binding sites are significantly enriched in this gene set (see Note 13). In this case, the *REB1* and *RPN4* TFBS are significantly enriched in the promoter regions of this gene set (Fig. 5a), suggesting these transcription factors may play an important role in the regulation of proteasome genes. Indeed, *RPN4* is known to regulate the expression of many proteasome genes (49, 50). The significance of enrichment (i.e., *P*-value) is shown to the left of the name of each transcription factor (Fig. 5a). The number of promoters (#P) containing a TFBS, and the total number of TFBS in this gene set (#S) are also listed (Fig. 5a).
5. Click the button labeled "Site Environments" located in the "Chromatin Analysis Tools" table.
6. The resulting output shows the enrichment of TFBS located in differing chromatin environments among the selected gene set

a

Transcription Factors (TF)

Check All	Check Enriched	Check None	Check Inverted	Refresh	Download

P-value	Name	#P	#S	Select	P-value	Name	#P	#S	Select
#P = Number of promoters with TF sites					#S = Number of predicted TF sites				
⊟ Enriched TF sites									
$< 10^{-4}$	☑ REB1	9	12	◯	$< 10^{-10}$	☑ RPN4	17	18	◯

b

Binding Factor	TF Chromatin Environments in Normal [MNase-microarray]						Details
	Internal		Linker		Nucleosome Free		
	promoters (%)	p-value	promoters (%)	p-value	promoters (%)	p-value	
REB1	1 (3.0%)	0.163	2 (6.0%)	0.037	**7 (21.2%)**	**$< 10^{-3}$**	details
RPN4	0 (0%)	1	2 (6.0%)	0.002	**15 (45.4%)**	**$< 10^{-9}$**	details

Fig. 5. (**a**) Summary of enriched TFBS in the promoter regions of proteasome genes. (**b**) Analysis of enriched TFBS located in specific chromatin environment categories. Note that only bound TFBS (Medium binding criteria) were analyzed.

(Fig. 5b). In this case, the *RPN4* and *REB1* binding sites are preferentially enriched in nucleosome-free chromatin environments in this gene set.

7. In the "Transcription Factors" table, click the button to the right of the *RPN4* transcription factor (under the "*Select*" column). This button selects (i.e., checks) the 17 promoters that contain a bound *RPN4* TFBS.

8. In the "Chromatin Analysis Tools" table, select or check the box labeled "Calculate Correlation (SLOW)". This option calculates the correlation coefficient of the nucleosome profiles of the selected genes. Unselect (or uncheck) the button labeled use "Use nucleosome positioning score". In this example, this option should not be selected.

9. Click the button labeled "Graph" in the "Chromatin Analysis Tools" table. The resulting graph (Fig. 6a) shows the average nucleosome profile of the proteasome genes containing a bound *RPN4* TFBS (see Note 14). The chromatin profiles of these genes are somewhat correlated ($r=0.436$; Fig. 6a) and contain a significant nucleosome depleted region in the –200 to –80 bp region of their promoter.

10. In the "Selected Promoters" table, click the button labeled "Invert". This selects the 16 genes in this gene set that do not contain a bound *RPN4* TFBS in their promoter region.

11. Click the button labeled "Graph" in the "Chromatin Analysis Tools" table. The resulting graph (Fig. 6b) shows the average nucleosome profile of the proteasome genes that do not contain a bound *RPN4* TFBS. Note that the chromatin profiles of these genes are less well correlated ($r=0.290$; Fig. 6b) and appear to differ from the chromatin profile of the genes containing a bound *RPN4* TFBS (Fig. 6a).

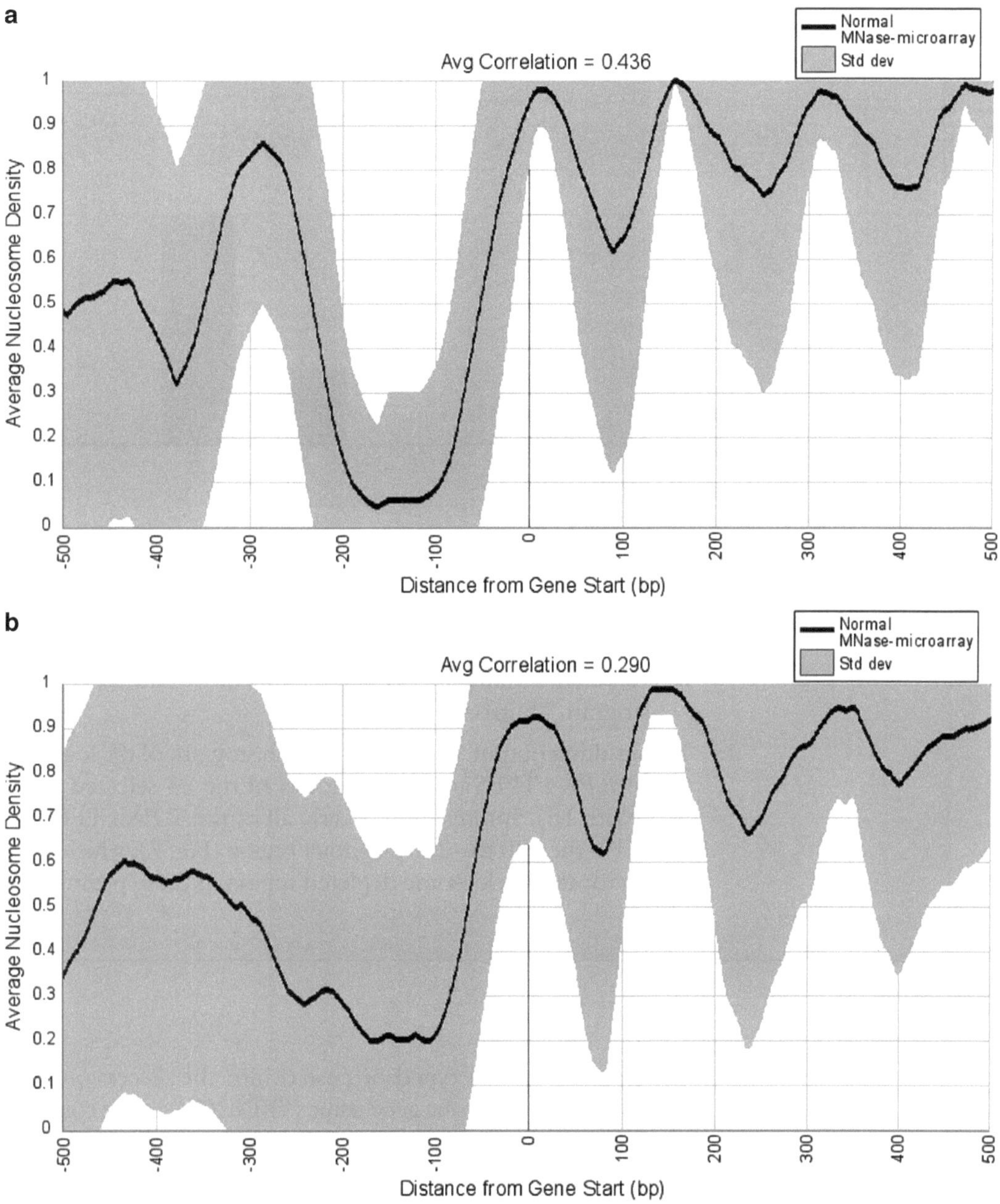

Fig. 6. (**a**) Plot of average nucleosome density among proteasome genes containing a bound *RPN4* TFBS. The *black line* indicates the average nucleosome density among the selected genes. The *gray shading* denotes the standard deviation of the nucleosome density among the selected genes. (**b**) Plot of average nucleosome density among proteasome genes that do not contain a bound *RPN4* TFBS.

12. In the "Selected Promoters" table, click the button labeled "Check All". This selects all 33 promoters.
13. Click the "Check None" button in the "Transcription Factor" table. Then click the button next to the *RPN4* transcription factor. Only the *RPN4* transcription factor should be selected/checked.

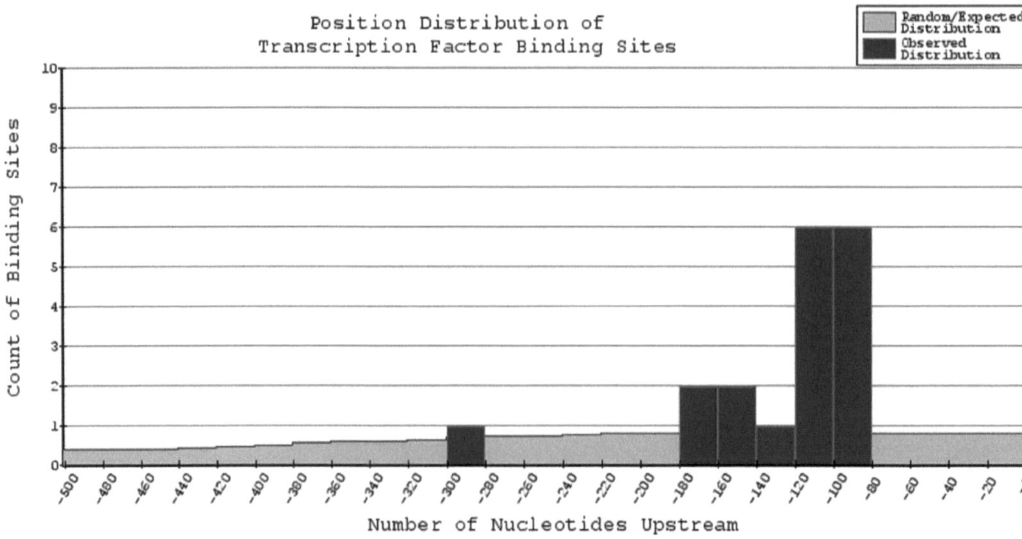

Fig. 7. Histogram plot of the positions of *RPN4* TFBS among the promoters of proteasome genes.

14. Under the Histograms table, enter the value 10 in the text box labeled "Manually set max value of y axis". Click the "Histogram" button.
15. The resulting output (Fig. 7) displays a histogram of the locations of the *RPN4* TFBS in the promoters of the 33 selected genes (see Note 15). Interestingly, nearly all of the *RPN4* TFBS are located in the –80 to –200 promoter region (Fig. 7), which coincides with the nucleosome depleted region in these promoters.

4. Notes

1. Gene names can be typed or pasted into the Accessions text box. Either the systematic gene name (YBR020W) or the common gene name (*GAL1*) can be entered in the Accessions text box. Each gene name should occupy a single line (i.e., hit the return key after each gene name).
2. Key options include the categories of TFBS to display. These are listed under the TF Binding Site Options menu items. For example, the "Medium" Site conservation option only displays TFBS that are conserved in one or more other yeast species. The "Medium" ChIP Binding option only displays TFBS in which the promoter is bound by the respective transcription factor protein with a P-value < 0.005 (based on ChIP-chip data from ref. 1). In the example shown in Fig. 1, the "Medium" ChIP Binding option was selected (this is the default option in

Ceres), so only "bound" TFBS (e.g., *GAL4* and *GAL80*) are shown. Other key options include the length of promoter sequence to display (default is 1,000 bp), and which nucleosome dataset is selected (in this case "Normal [MNase-microarray]," which is based on nucleosome positioning data published in ref. 18).

3. Note that in the "Compact" display (default option), overlapping TFBS can be hidden or difficult to distinguish. Additionally, in the Compact display the width of the line is not proportional to the size of the binding site (i.e., all binding sites are displayed with vertical lines/hashes of the same width). The "Cartoon" display option works better to display overlapping TFBS and provides proportionally sized binding sites.
4. The location of the coding region is based on gene coordinates downloaded from the Saccharomyces Genome Database (42, 43). The coordinates typically comprise the open reading frame of the gene. Introns are not distinguished graphically and are included as part of the gene rectangle. The arrow at the beginning of the coding region corresponds to the gene start (typically the start of the open reading frame), not necessarily the start of transcription.
5. Three of the *GAL4* TFBS were hidden behind *GAL80* TFBS. Gal80 associates with DNA by binding to Gal4. Hence, *GAL80* TFBS overlap with *GAL4* binding sites. The "Cartoon" display option is a good choice to visualize overlapping TFBS.
6. The term "Fuzzy Nucleosome" indicates that the nucleosome is relatively weakly positioned; by contrast, the term "Positioned Nucleosome" indicates that the nucleosome is relatively strongly positioned (18). Fuzzy and positioned nucleosomes are shaded in light gray and dark gray, respectively (see the Nucleosome Key in Fig. 1). The MNase-seq nucleosome datasets in Ceres use a quantitative nucleosome positioning score (31) instead of the qualitative terms fuzzy and positioned.
7. The "Heat Shock [MNase-seq]" dataset maps the nucleosome positions in yeast cells following incubation at 39°C for 15 min (31).
8. In this display, the color of the nucleosome ovals corresponds to how well positioned (or fuzzy) each nucleosome is (31), which we call the nucleosome positioning score. We plan to include in future versions of the Ceres database the option to display a nucleosome score related to the probability of finding a nucleosome at that location (31).
9. The significance of enrichment for each Gene Ontology term is calculated using a hypergeometric probability distribution. For this calculation, the total number of genes is based on the total number of gene entries in the Ceres database (including dubious genes, tRNA genes, etc.). For more details, see the supplemental methods of ref. 20.

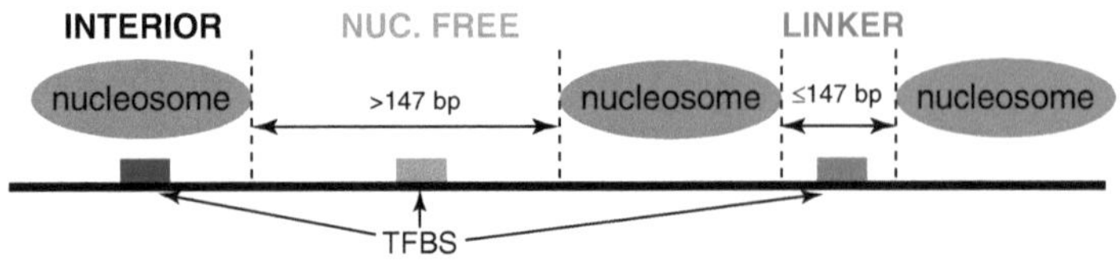

Fig. 8. Cartoon depicting the chromatin environment classification of TFBS based on their location relative to nucleosomes.

10. The ChromatinDB Visualization options enable the user to examine specific categories of histone modifications ("Histone Modification Type"), to normalize the histone modification data with nucleosome occupancy values ("Data Type"), to display data for promoters or coding regions ("Gene Region"), and to select the *P*-value threshold and correction for statistical analysis.
11. ChromatinDB uses a nonparametric Wilcoxon rank sum test to identify enriched or depleted chromatin features (37). In the current version of ChromatinDB, if two genes have a shared promoter region, the histone modification data for that promoter region is separately assigned to both genes. Hence, if two genes with a shared promoter are both included in the input gene list, the average histone modification data for that promoter is essentially given twice the weight in the average $\log_2$ ratio calculation and statistical analysis.
12. A TFBS whose midpoint is located within a nucleosome is classified as having a "nucleosome interior" chromatin environment (Fig. 8). A TFBS located entirely within a nucleosome-free region of more than 147 bp is classified as having a "nucleosome-free" chromatin environment (Fig. 8). All other TFBS, which are generally located between closely spaced nucleosomes (or partially overlapping with a nucleosome), are classified as having a "linker" chromatin environment. For more details, see ref. 20.
13. The significance of enrichment for each TFBS is calculated using a hypergeometric probability distribution. For simplicity, the total number of genes used in this calculation is based on the total number of gene entries in the Ceres database (including dubious genes, tRNA genes, etc.). For more details, see the supplemental methods of ref. 20. The default significance threshold is $P < 10^{-4}$. This threshold can be changed using the *P*-value thresholds options in the Gene Set Analysis page. Note that this analysis tool does not correct for multiple hypothesis testing. This correction should be accounted for when selecting the appropriate *P*-value threshold.

14. If the "Use nucleosome positioning score" option is selected, then the graph displays a plot of the average nucleosome positioning score. For the "Normal [MNase-microarray]" dataset (18), the positioned nucleosomes were assigned a score of 0.7 and the fuzzy nucleosomes were assigned a score of 0.33 (for more details, see ref. 20). The nucleosome positioning scores for the MNase-seq datasets are derived from ref. 31. If the "Use nucleosome positioning score" option is not selected/checked, as in this example, then the graph displays the average nucleosome density and ignores the nucleosome positioning scores.
15. If multiple TFBS are selected/checked, then the histogram tool shows the cumulative distribution of all selected TFBS.

References

1. Harbison CT, Gordon DB, Lee TI, Rinaldi NJ, Macisaac KD, Danford TW et al. (2004) Transcriptional regulatory code of a eukaryotic genome. *Nature* **431**:99–104
2. Lee TI, Rinaldi NJ, Robert F, Odom DT, Bar-Joseph Z, Gerber GK et al. (2002) Transcriptional regulatory networks in Saccharomyces cerevisiae. *Science* **298**:799–804
3. Cliften P, Sudarsanam P, Desikan A, Fulton L, Fulton B, Majors J et al. (2003) Finding functional features in Saccharomyces genomes by phylogenetic footprinting. *Science* **301**:71–76
4. Kellis M, Patterson N, Endrizzi M, Birren B, Lander ES (2003) Sequencing and comparison of yeast species to identify genes and regulatory elements. *Nature* **423**:241–254
5. MacIsaac KD, Wang T, Gordon DB, Gifford DK, Stormo GD, Fraenkel E (2006) An improved map of conserved regulatory sites for Saccharomyces cerevisiae. *BMC Bioinformatics* 7:113
6. Zhu J, Zhang MQ (1999) SCPD: a promoter database of the yeast Saccharomyces cerevisiae. *Bioinformatics* **15**:607–611
7. Tsai HK, Chou MY, Shih CH, Huang GT, Chang TH, Li WH (2007) MYBS: a comprehensive web server for mining transcription factor binding sites in yeast. *Nucleic Acids Res* **35**:W221–226
8. Monteiro PT, Mendes ND, Teixeira MC, d'Orey S, Tenreiro S, Mira NP et al. (2008) YEASTRACT-DISCOVERER: new tools to improve the analysis of transcriptional regulatory associations in Saccharomyces cerevisiae. *Nucleic Acids Res* **36**:D132–136
9. Teixeira MC, Monteiro P, Jain P, Tenreiro S, Fernandes AR, Mira NP et al. (2006) The YEASTRACT database: a tool for the analysis of transcription regulatory associations in Saccharomyces cerevisiae. *Nucleic Acids Res* **34**:D446–451
10. Li B, Carey M, Workman JL (2007) The role of chromatin during transcription. *Cell* **128**:707–719
11. Kouzarides T (2007) Chromatin modifications and their function. *Cell* **128**:693–705
12. Kurdistani SK, Grunstein M (2003) Histone acetylation and deacetylation in yeast. *Nat Rev Mol Cell Biol* **4**:276–284
13. Shilatifard A (2006) Chromatin modifications by methylation and ubiquitination: implications in the regulation of gene expression. *Annu Rev Biochem* **75**:243–269
14. Rando OJ, Ahmad K (2007) Rules and regulation in the primary structure of chromatin. *Curr Opin Cell Biol* **19**:250–256
15. Yuan GC, Liu YJ, Dion MF, Slack MD, Wu LF, Altschuler SJ et al. (2005) Genome-scale identification of nucleosome positions in S. cerevisiae. *Science* **309**:626–630
16. Rando OJ (2006) Chromatin structure in the genomics era. *Trends Genet*
17. Rando OJ, Chang HY (2009) Genome-wide views of chromatin structure. *Annu Rev Biochem* **78**:245–271
18. Lee W, Tillo D, Bray N, Morse RH, Davis RW, Hughes TR et al. (2007) A high-resolution atlas of nucleosome occupancy in yeast. *Nat Genet* **39**:1235–1244
19. Narlikar L, Gordan R, Hartemink AJ (2007) A nucleosome-guided map of transcription factor binding sites in yeast. *PLoS Comput Biol* **3**:e215
20. Morris RT, O'Connor TR, Wyrick JJ (2010) Ceres: software for the integrated analysis of transcription factor binding sites and nucleosome positions in Saccharomyces cerevisiae. *Bioinformatics* **26**:168–174

21. Kim HD, O'Shea EK (2008) A quantitative model of transcription factor-activated gene expression. *Nat Struct Mol Biol* **15**:1192–1198
22. Lam FH, Steger DJ, O'Shea EK (2008) Chromatin decouples promoter threshold from dynamic range. *Nature* **453**:246–250
23. Robyr D, Suka Y, Xenarios I, Kurdistani SK, Wang A, Suka N et al. (2002) Microarray deacetylation maps determine genome-wide functions for yeast histone deacetylases. *Cell* **109**:437–446
24. Bernstein BE, Humphrey EL, Erlich RL, Schneider R, Bouman P, Liu JS et al. (2002) Methylation of histone H3 Lys 4 in coding regions of active genes. *Proc Natl Acad Sci USA* **99**:8695–8700
25. Kurdistani SK, Tavazoie S, Grunstein M (2004) Mapping global histone acetylation patterns to gene expression. *Cell* **117**:721–733
26. Rao B, Shibata Y, Strahl BD, Lieb JD (2005) Dimethylation of histone H3 at lysine 36 demarcates regulatory and nonregulatory chromatin genome-wide. *Mol Cell Biol* **25**:9447–9459
27. Xu F, Zhang K, Grunstein M (2005) Acetylation in histone H3 globular domain regulates gene expression in yeast. *Cell* **121**:375–385
28. Pokholok DK, Harbison CT, Levine S, Cole M, Hannett NM, Lee TI et al. (2005) Genome-wide map of nucleosome acetylation and methylation in yeast. *Cell* **122**:517–527
29. Millar CB, Grunstein M (2006) Genome-wide patterns of histone modifications in yeast. *Nat Rev Mol Cell Biol* **7**:657–666
30. Millar CB, Xu F, Zhang K, Grunstein M (2006) Acetylation of H2AZ Lys 14 is associated with genome-wide gene activity in yeast. *Genes Dev* **20**:711–722
31. Shivaswamy S, Bhinge A, Zhao Y, Jones S, Hirst M, Iyer VR (2008) Dynamic remodeling of individual nucleosomes across a eukaryotic genome in response to transcriptional perturbation. *PLoS Biol* **6**:e65
32. Mavrich TN, Ioshikhes IP, Venters BJ, Jiang C, Tomsho LP, Qi J et al. (2008) A barrier nucleosome model for statistical positioning of nucleosomes throughout the yeast genome. *Genome Res* **18**:1073–1083
33. Kaplan N, Moore IK, Fondufe-Mittendorf Y, Gossett AJ, Tillo D, Field Y et al. (2009) The DNA-encoded nucleosome organization of a eukaryotic genome. *Nature* **458**:362–366
34. Zawadzki KA, Morozov AV, Broach JR (2009) Chromatin-dependent transcription factor accessibility rather than nucleosome remodeling predominates during global transcriptional restructuring in Saccharomyces cerevisiae. *Mol Biol Cell* **20**:3503–3513
35. Weiner A, Hughes A, Yassour M, Rando OJ, Friedman N (2010) High-resolution nucleosome mapping reveals transcription-dependent promoter packaging. *Genome Res* **20**:90–100
36. Whitehouse I, Rando OJ, Delrow J, Tsukiyama T (2007) Chromatin remodelling at promoters suppresses antisense transcription. *Nature* **450**:1031–1035
37. O'Connor TR, Wyrick JJ (2007) ChromatinDB: a database of genome-wide histone modification patterns for Saccharomyces cerevisiae. *Bioinformatics* **23**:1828–1830
38. Bram RJ, Kornberg RD (1985) Specific protein binding to far upstream activating sequences in polymerase II promoters. *Proc Natl Acad Sci USA* **82**:43–47
39. Giniger E, Varnum SM, Ptashne M (1985) Specific DNA binding of GAL4, a positive regulatory protein of yeast. *Cell* **40**:767–774
40. Bowdish KS, Mitchell AP (1993) Bipartite structure of an early meiotic upstream activation sequence from Saccharomyces cerevisiae. *Mol Cell Biol* **13**:2172–2181
41. Strich R, Surosky RT, Steber C, Dubois E, Messenguy F, Esposito RE (1994) UME6 is a key regulator of nitrogen repression and meiotic development. *Genes Dev* **8**:796–810
42. Cherry JM, Adler C, Ball C, Chervitz SA, Dwight SS, Hester ET et al. (1998) SGD: Saccharomyces Genome Database. *Nucleic Acids Res* **26**:73–79
43. Engel SR, Balakrishnan R, Binkley G, Christie KR, Costanzo MC, Dwight SS et al. (2010) Saccharomyces Genome Database provides mutant phenotype data. *Nucleic Acids Res* **38**:D433–436
44. Williams RM, Primig M, Washburn BK, Winzeler EA, Bellis M, Sarrauste de Menthiere C et al. (2002) The Ume6 regulon coordinates metabolic and meiotic gene expression in yeast. *Proc Natl Acad Sci USA* **99**:13431–13436
45. Pak J, Segall J (2002) Regulation of the premiddle and middle phases of expression of the NDT80 gene during sporulation of Saccharomyces cerevisiae. *Mol Cell Biol* **22**:6417–6429
46. Tice-Baldwin K, Fink GR, Arndt KT (1989) BAS1 has a Myb motif and activates HIS4 transcription only in combination with BAS2. *Science* **246**:931–935
47. Daignan-Fornier B, Fink GR (1992) Coregulation of purine and histidine biosynthesis by the transcriptional activators BAS1 and BAS2. *Proc Natl Acad Sci USA* **89**:6746–6750

48. Denis V, Boucherie H, Monribot C, Daignan-Fornier B (1998) Role of the myb-like protein bas1p in Saccharomyces cerevisiae: a proteome analysis. *Mol Microbiol* **30**:557–566

49. Mannhaupt G, Schnall R, Karpov V, Vetter I, Feldmann H (1999) Rpn4p acts as a transcription factor by binding to PACE, a nonamer box found upstream of 26 S proteasomal and other genes in yeast. *FEBS Lett* **450**:27–34

50. Xie Y, Varshavsky A (2001) RPN4 is a ligand, substrate, and transcriptional regulator of the 26S proteasome: a negative feedback circuit. *Proc Natl Acad Sci USA* **98**: 3056–3061

Chapter 16

Chromatin Affinity Purification

Ryoko Harada and Alain Nepveu

Abstract

Chromatin immunoprecipitation (ChIP) has become an essential assay in the field of transcriptional regulation. It is currently the most popular method to monitor the in vivo interaction between a protein and specific genomic sites. The method can also serve to identify novel transcriptional targets when the immunoprecipitated chromatin, sometimes called chipped DNA, is used either as a probe in hybridization experiments with microarrays of genomic DNA (ChIP-chip) or as template in DNA sequencing (ChIP-Seq). ChIP assays rely on the availability of good antibodies that can specifically and efficiently immunoprecipitate the protein under study even after cross-linking. However, good antibodies are not always available. To circumvent this problem, we have developed and validated the method of chromatin affinity purification (ChAP). The subsequent microarray analysis is then referred to as ChAP-chip. In brief, the protein under study is expressed together with two tags in order to allow the purification of chromatin by tandem affinity purification. To ensure that only true targets are identified, it is important to express the recombinant tagged-protein at physiological level. This requirement is not trivial as most expression vectors are designed to express proteins at high levels. We found most convenient to use an inducible retroviral vector in the absence of inducer and transactivator protein. We describe the procedure to generate cells stably expressing recombinant tagged-proteins at physiological level and then to purify the associated chromatin by affinity purification. Targets identified in this manner were validated in independent ChAP assays as well as in ChIP assays using antibodies against the endogenous protein.

Key words: Chromatin immunoprecipitation, Affinity chromatography, Tandem affinity purification, Transcription factors, Promoter, Gene regulation, Genome, Genomic microarray

1. Introduction

Chromatin immunoprecipitation (ChIP) necessitates antibodies that work well in immunoprecipitation, can still recognize the protein following cross-linking and are specific for the protein under study. Unfortunately, these conditions are not always met. One common problem is that the epitope is modified or becomes inaccessible following the treatment of cells with a cross-linking

Ales Vancura (ed.), *Transcriptional Regulation: Methods and Protocols*, Methods in Molecular Biology, vol. 809,
DOI 10.1007/978-1-61779-376-9_16,

agent. For this reason, it is important in preliminary experiments to compare the efficiency of immunoprecipitation before and after cross-linking. Another problem is when available antibodies can recognize several isoforms of a protein. In this case, it is not possible to identify genomic binding sites that are specific to a given isoform. We encountered this issue with the CUX1 transcription factor (1, 2). An alternative approach to ChIP is to perform retroviral infections to generate populations of cells that express physiological levels of a recombinant protein with two epitope tags and then to purify the chromatin by tandem affinity purification (TAP) (3, 4). The purified chromatin can then be amplified and labeled with a fluorophore to serve as a probe in microarray hybridizations (5–7). The purified chromatin can also serve as a template in PCR assays to measure the recruitment of the recombinant protein to specific genomic sites (8).

2. Materials

2.1. Cell Culture

1. Hs578T human breast cells stably carrying either a retroviral vector expressing p110-Tag^2 or the same, empty vector (see Note 1).
2. Dulbecco's modified Eagle's medium (DMEM) supplemented with 5% fetal bovine serum (FBS), glutamine, and penicillin–streptomycin.
3. 150-mm Culture plates.

2.2. Cross-Linking Cells

1. Formaldehyde. Formaldehyde is toxic and a potential carcinogen. It should be handled in a fume hood.
2. 1.25 M glycine.
3. Phosphate-buffered saline (PBS).
4. Teflon cell scrapers.
5. Centrifuge tubes.

2.3. Cell Lysis and Sonication

Protease inhibitors should be added immediately prior to use: Complete EDTA-free protease inhibitor tablets (Roche Applied Science) and 1 mM PMSF (final concentration).

1. Lysis buffer 1: 50 mM HEPES–KOH, pH 7.5, 140 mM NaCl, 1 mM EDTA, 10% glycerol, 0.5% NP40, 0.25% Triton X-100, one tablet/10 ml Complete EDTA-free protease inhibitor tablets and 1 mM PMSF.
2. Lysis buffer 2: 10 mM Tris–HCl, pH 8.0, 200 mM NaCl, 1 mM EDTA, 0.5 mM EGTA, one tablet/10 ml Complete EDTA-free protease inhibitor tablets, and 1 mM PMSF.

3. RIPA-M buffer: 20 mM Tris–HCl, pH 7.4, 2 mM EDTA, 2 mM EGTA, 150 mM NaCl, 1% NP40, 0.5% deoxycholate, 0.1% SDS, one tablet/10 ml Complete EDTA-free protease inhibitor tablets, and 1 mM PMSF.
4. Sonicator (VIRTIS Company model VIRSONIC 60).

2.4. IgG Beads Binding

1. IgG beads (IgG Sepharose 6 fast flow, GE Healthcare).
2. TST buffer: 50 mM Tris–HCl buffer, pH 7.6, 150 mM NaCl, 0.05% Tween 20.
3. 10 mg/ml tRNA.

2.5. TEV Cleavage

1. TEV enzyme (AcTEV protease, Invitrogen).
2. IPP buffer: 10 mM Tris–HCl, pH 8.0, 100 mM NaCl, 0.1% NP40, 10% glycerol.
3. A micro Bio-Spin Chromatography Column (Bio-Rad).
4. TEV buffer: 10 mM Tris–HCl, pH 8.0, 100 mM NaCl, 0.5 mM EDTA, 1 mM DTT. Add DTT immediately prior to use.

2.6. Calmodulin Column

1. Calmodulin binding buffer (CBB): 10 mM Tris–HCl, pH 8.0, 100 mM NaCl, 1 mM imidazole, 1 mM Mg acetate, 2 mM $CaCl_2$, 0.1% NP40, two tablets/10 ml Complete EDTA-free protease inhibitor tablets, 2 mM PMSF, 20 mM β-mercaptoethanol. Add protease inhibitor tablets, PMSF, and β-mercaptoethanol immediately prior to use.
2. Calmodulin elution buffer (CEB): 10 mM Tris–HCl, pH 8.0, 100 mM NaCl, 1 mM imidazole, 1 mM Mg acetate, 2 mM EGTA, 10 mM β-mercaptoethanol.
3. Glass beads 212–300 μm (Sigma-Aldrich).

2.7. SDS-Polyacrylamide Gel Electrophoresis (SDS-PAGE)

1. 1.5 M Tris–HCl, pH 8.8.
2. 0.5 M Tris–HCl, pH 6.8.
3. 10% SDS.
4. 30% (w/v) Acrylamide/0.8% (w/v) bis-acrylamide.
5. 10% (w/v) Ammonium persulfate (APS).
6. TEMED.
7. Water-saturated isopropanol.
8. Running buffer: 25 mM Tris-base, 200 mM glycine, 0.1% (w/v) SDS.
9. Prestained molecular weight markers: Kaleidoscope markers (Bio-Rad).

2.8. Western Blotting

1. Bio-Rad transblot apparatus.
2. PVDF membrane (Amersham Hybond-P PVDF membrane).
3. Filter paper (Whatman).

4. Transfer buffer: 25 mM Tris-base, 200 mM glycine, 20% methanol.
5. TBS-T: Tris-buffered saline with 0.1% Tween 20.
6. Blocking buffer: 15% skim milk, 3% BSA in TBS-T.
7. Primary antibody (rabbit antibody against the calmodulin binding peptide (CBP) tag, see Note 2).
8. Secondary antibody (peroxidase conjugated anti-rabbit antibody).
9. Enhanced chemiluminescent (ECL) reagents (GE Healthcare).

2.9. RNase A/Proteinase K Treatment

1. TE: 10 mM Tris–HCl, pH 8.0, 1 mM EDTA.
2. 10 mg/ml DNase-free RNase A.
3. 10 mg/ml Proteinase K.
4. Chloropane: 49% phenol, 49% chloroform, 2% isoamyl alcohol.
5. 20 mg/ml Glycogen.
6. 3 M Sodium acetate, pH 5.5.
7. Ethanol.

2.10. ChAP-qPCR

1. 10× PCR buffer (Gibco/BRL).
2. 10 mM dNTP.
3. Taq polymerase (Gibco/BRL).
4. 10 μM primers.
5. 50 mM $MgCl_2$.

2.11. ChAP-chip Sample Preparation

1. 10× T4 DNA Pol buffer (NEB).
2. 10 mM dNTP.
3. T4 DNA polymerase (NEB).
4. 10 mg/ml Bovine serum albumin (BSA).
5. 10× T4 DNA ligase buffer (NEB).
6. T4 DNA ligase (NEB).
7. JW102: 5′-GCGGTGACCCGGGAGATCTGAATTC-3′.
8. JW103: 5′-GAATTCAGATC-3′.
9. 10× Thermopol buffer.
10. Taq polymerase (Invitrogen).

3. Methods

CUX1 is translated into a protein of 1,505 amino acids, p200 CUX1, which is then proteolytically processed into a shorter protein of approximately 800 amino acids, p110 CUX1 (Fig. 1b; reviewed in ref. 2). Consequently, the entire primary sequence of p110 is

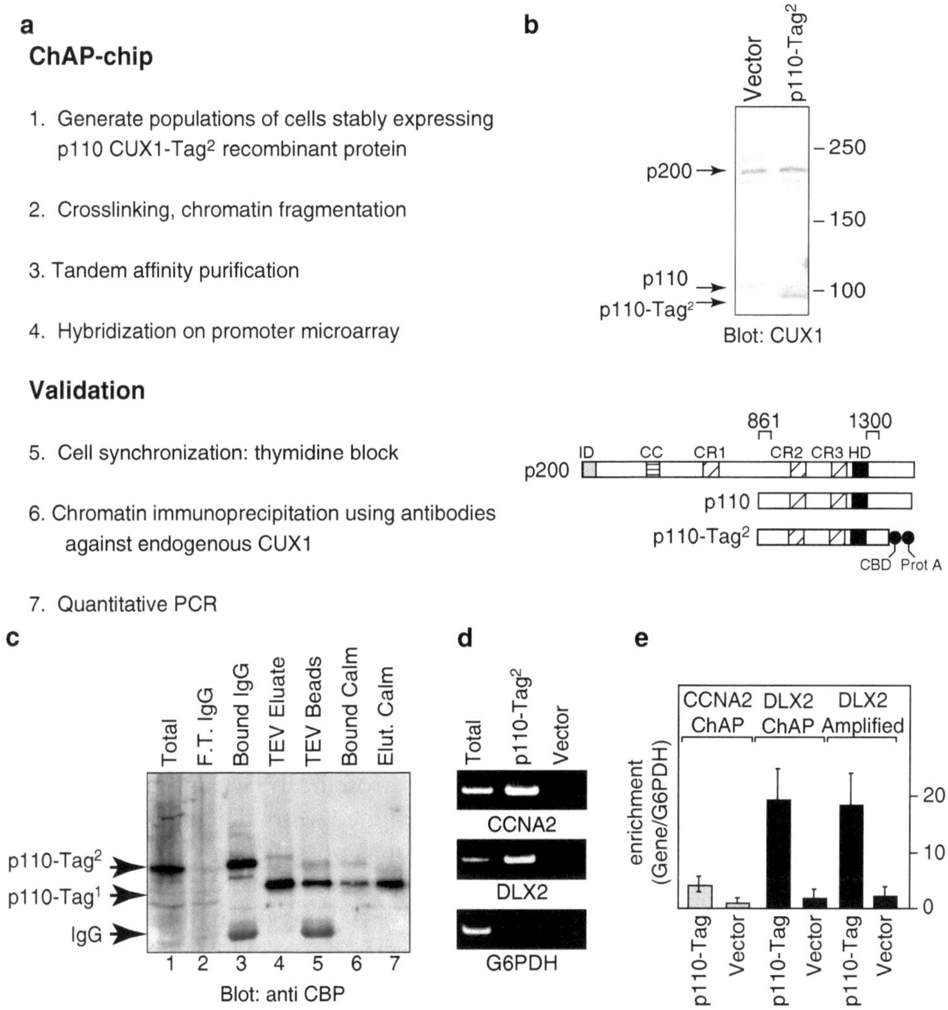

Fig. 1. Strategy for the identification of transcriptional targets of p110 CUX1. (**a**) The methodology used to identify gene targets of p110 CUX1 is summarized in a flowchart and is described in the text. (**b**) HeLa cells were infected with a retroviral vector expressing a recombinant p110 CUX1 protein with two tags at its C terminus. Nuclear extracts were prepared from each population of cells and analyzed by western blot using the 861 and 1,300 CUX1 antibodies. Below is a schematic representation of CUX1 proteins with some of the functional domains: *ID* inhibitory domain; *CC* coiled-coil; *CR1*, *CR2*, and *CR3* Cut repeat 1, 2 and 3; *HD* homeodomain; *CBD* calmodulin-binding domain; *Prot A* protein A. The regions recognized by the 861 and 1,300 antibodies are shown. (**c**) Protein samples from each step of the Taptag purification were analyzed by western blot using the anti-calmodulin-binding protein epitope (CBP) Tag antibody. Nuclear extract (*lane 1*); IgG beads flowthrough (F.T., *lane 2*); or bound (*lane 3*); after TEV digestion, cleaved and eluted from IgG beads (*lane 4*) or still bound to IgG beads (*lane 5*); bound to calmodulin beads (*lane 6*) and eluted with EGTA (*lane 7*). Note that digestion with TEV removes one tag and reduces the size of the recombinant protein. (**d**) Chromatin from Hs578T/p110-Tag[2] and Hs578T/vector cells was submitted to tandem affinity purification (TAP) and analyzed by PCR using primers specific for the CCNA2, DLX2, and G6PDH gene promoters. Representative data from three independent ChAP experiments are presented. (**e**) The purified chromatin from Hs578T/p110-Tag[2] cells was amplified by ligation-mediated PCR prior to the hybridization. The enrichment level of the CCNA2 and DLX2 gene promoters was measured by quantitative real-time PCR (qPCR) before and after LM-PCR. The results represent the mean ± SD from three independent ChAP experiments and their amplification. Reprinted from ref. 1 by permission from Oxford University Press.

contained within p200 and there is currently no CUX1 antibody that can recognize p110 at the exclusion of p200. To investigate the genomic binding sites of p110 CUX1, we therefore developed the method of ChAP (1). The protein was expressed with two epitope tags at its C terminus, p110-Tag^2, and the chromatin was purified using the well-established method of TAP (3, 4). We refer to this procedure as ChAP and the subsequent microarray analysis as ChAP-chip (Fig. 1).

3.1. Preparation of Cells Stably Expressing Physiological Levels of the Recombinant Protein

Particular attention must be given to the level of expression of the recombinant, tagged protein. Indeed, if the recombinant protein is expressed at higher levels than the endogenous protein, it will occupy all the genomic sites typically bound by the endogenous protein and it may also bind to other, lower affinity binding sites that are not functionally relevant. To achieve the right level of expression in each cell of the population, careful consideration must be given to the choice of expression vector and the method to introduce the vector into cells. Note that DNA transfections are not ideal since stably transfected cells typically contain concatemers of plasmids at several integration sites. In contrast, retroviral infections will produce cells that contain one or only a few copies of the retrovirus.

We quickly realized that is was not easy to express moderate or low levels of a recombinant protein. In fact, most if not all expression vectors have been designed with the idea of expressing as much of the desired protein as possible. After a few unsuccessful attempts at crippling some expression vectors, we resorted to use the pRevTRE retroviral vector (Clontech) in the absence of a transactivator. This vector harbors the minimal CMV promoter together with a tetracycline responsive element that can mediate activation by a fusion protein containing the tetracycline repressor and an activation domain. The DNA binding activity of this transactivator, whether in the tet-on or the tet-off version, is modulated by the presence of tetracycline or a derivative thereof. The vector is designed to be silent when the transactivator cannot bind DNA, but activated when the transactivator binds DNA. One can find figures in the literature showing that the basal level of expression is virtually undetectable (9–11). Several clones must be screened, however, in order to identify clones with very low basal and high induced expression. In fact, most clones exhibit a very low, yet detectable basal level of expression which is appropriate for the analysis of binding site occupancy. We found it most convenient to produce high-titer retroviruses with which we could quickly generate populations of stably infected cells. Detailed protocols for the production of retroviruses can be found in the company's manuals (http://www.cellbiolabs.com/lentiviral-packaging-systems).

Once a population of cells stably expressing the desired protein has been obtained, it should be tested to ensure that protein expression is at the right level in the entire population and does not vary among individual cells or clones. Western blot analysis confirmed that p110-Tag^2 was expressed at relatively low level in our population of HeLa/p110-Tag^2 cells (Fig. 1b; see Note 3). In addition, we isolated six independent HeLa/ p110-Tag^2 clones and verified that expression did not vary significantly among individual clones (data not shown). Protein expression in individual cells can also be monitored in indirect immunofluorescence using antibodies against the protein of interest or one of the tags (see Note 4).

3.2. Cross-Linking Treatment

ChAP should be carried out in parallel with two populations of cells: one population that stably express the protein under study and another population stably carrying the empty vector. The chromatin purified from the latter cells will serve as a control to ensure that the enrichment of specific genomic sequences is not due to an artifact.

1. Cultivate exponentially growing Hs578T cells that carry either a retroviral vector expressing p110-Tag^2 or the same, empty vector. We use 20–40 plates for each purification of p110-Tag^2 proteins (see Note 5). However, the number of plates required should be determined empirically, since purification efficiencies vary from one protein to another. It is easier to handle no more than eight plates for cross-linking at a time.
2. Add formaldehyde to the medium to reach a 1% final concentration. Incubate for 10 min at room temperature.
3. Stop the reaction by adding 1.25 M glycine solution to a final concentration of 125 mM and incubate for a minimum of 5 min.
4. Aspirate medium and wash the cells twice with cold PBS and harvest them into a 50-ml centrifuge tube by scraping.
5. Resuspend the cells in an appropriate volume of cold PBS and spin at 500 × *g* for 5 min at 4°C.
6. Remove the supernatant and repeat step 5.
7. Remove the supernatant, resuspend cells in a smaller volume of PBS to transfer into a 15-ml centrifuge tube and spin at 4°C, 500 × *g* for 5 min.
8. Remove the supernatant and continue to Subheading 3.3 or store the cell pellet at –80°C.

3.3. Cell Lysis and Sonication

1. Volumes of lysis buffers 1 and 2 need to be adjusted according to the number of cells and the size of cell pellets to be used. We typically lyse cells from ten plates in 10 ml of the lysis buffer 1. This volume and the volumes below are for this number of cells. Volumes should be adjusted if a smaller or higher number of cells are used.

2. Incubate for 10 min at 4°C and spin at 500×*g* for 5 min. Remove the supernatant, resuspend the cell pellet in 10 ml of the lysis buffer 2. Incubate 10 min at 4°C and spin at 500×*g* for 5 min.
3. Remove the supernatant, resuspend the pellet in 5 ml of RIPA-M buffer. Let it stand for 10 min on ice.
4. Aliquot 1.5-ml samples into 2-ml microfuge tubes.
5. Sonicate on ice-cold water ten times, for 30 s, at level 5 (VIRTIS Company model VIRSONIC 60). Wait at least 1 min between each sonication to avoid over-heating. Then centrifuge for 10 min at 15,800×*g* at 4°C.
6. Combine the supernatant and transfer to a 15-ml centrifuge tube. It is ready for binding to IgG beads.
7. Keep the following aliquots:
   - 2% Total input (100 μl). Store at −80°C.
   - 0.3% Total input (15 μl) for Western blot analysis (Sample #1 total input). Store at −80°C.
   - 5 μl to verify DNA shearing efficiency as described below.

   DNA shearing efficiency verification
   1. Add 5 μl of H_2O to a 5-μl aliquot.
   2. Add RNase A to a final concentration of 50 μg/ml, incubate at 37°C for 1 h.
   3. Add Proteinase K to a final concentration of 50 μg/ml and SDS to 0.5%, incubate at 55°C for 3 h.
   4. Decross-linking: Incubate at 65°C for 6 h.
   5. Add 10 μl of chloropane, vortex, spin, take the upper phase.
   6. Run on a 0.7% agarose gel with molecular weight markers from 0.1 to 2 kbp.

3.4. IgG Beads Binding

1. Take 150 μl of IgG bead slurry (IgG Sepharose 6 fast flow, Amersham) in a 1.5-ml microfuge tube. Wash with 1 ml of TST buffer in order to remove any traces of ethanol. Spin the beads at 1,900×*g* (4,500 rpm) in microcentrifuge for 2 min.
2. Wash the IgG beads twice with 1 ml of RIPA-M buffer.
3. Add the beads to the cell lysate in a 15-ml centrifuge tube. Add tRNA to a final concentration of 100 μg/ml to reduce the background and rotate overnight at 4°C.

3.5. TEV Cleavage

1. Spin the tube at 500×*g* for 5 min. Take a 15-μl aliquot of the supernatant for Western blot analysis and store it at −80°C (Sample #2: 0.3% unbound after IgG beads).

2. Aspirate the remaining supernatant, add 1 ml of IPP buffer (do not add protease inhibitor) and transfer the beads to a 1.5-ml microfuge tube. Rotate 5 min and spin 2 min at 1,900 × *g*.
3. Wash the beads with 1 ml of IPP buffer three more times.
4. Add 1 ml of TEV buffer (do not add protease inhibitor) to the tube. Mix and take a 6-μl aliquot of the bead slurry for Western blot analysis and store at −80°C (Sample #3: 0.6% of bound IgG beads).
5. Spin the beads at 1,900 × *g* for 2 min, aspirate the buffer, add 300 μl of TEV buffer and 3 μl of TEV protease (10 U/μl, total 30 U). The cleavage reaction is done by incubating the tube at 16°C for 2 h. It is important to mix the beads once in a while (every 20 min).

3.6. Calmodulin Beads Binding

1. After TEV cleavage, mix the reaction by pipetting up and down, and load onto a micro Bio-Spin Chromatography Column (Bio-Rad). Collect the eluate (300 μl), which contains the cleaved protein. Add the following chemicals to reach the final concentrations of the CBB buffer: 500 mM imidazole solution (0.6 μl) to a final concentration of 1 mM, 500 mM Mg acetate (0.6 μl) to 1 mM final concentration, and 500 mM $CaCl_2$ (1.2 μl) to 2 mM final concentration.
2. Wash the column with 300 μl of CBB buffer and collect the eluate, which contains the cleaved protein. Combine the eluates (300 + 300 μl). Take a 4-μl aliquot of the eluate for Western blot analysis and store at −80°C (Sample #4: 0.66% of TEV cleaved).
3. Cap the end of the column and add 1,000 μl of TEV buffer. Take a 6-μl aliquot of the beads plus the supernatant for Western blot analysis and store at −80°C (Sample #5: 0.6% of bound IgG beads but not cleaved).
4. Preparation of calmodulin beads: Pipette 100 μl of calmodulin bead slurry into a 1.5-ml microfuge tube. Add 1.5 ml of CBB buffer, rotate for 5 min, and spin the beads at 1,900 × *g* for 2 min. Repeat the wash one more time.
5. Add the 600 μl of protein eluate to the calmodulin beads. Add BSA and tRNA to a final concentration of 100 μg/ml each. Rotate the tube at 4°C for 3 h.

3.7. Elution from the Calmodulin Column

1. Spin down the calmodulin beads at 1,900 × *g* for 2 min. Wash the calmodulin beads three times with 1 ml of CBB buffer, each time rotating the tubes for 5 min. Before spinning down the beads after the last wash, take a 20-μl aliquot of the bead slurry and store at −80°C for protein analysis (#6: 2% of calmodulin bound). While doing the washes, prepare the column as indicated in steps 2 and 3.

2. Set up a microcolumn using a siliconized 200 µl tip. Slightly twist the end of the tip using forceps and add 10 µl of 212–300 µm glass beads. Wash the beads once with 200 µl of H_2O and once with 200 µl of CBB buffer. At this stage, it may be necessary to accelerate the flow rate by cutting the end of the tip. An ideal flow rate is approximately one drop per 10 s, such that the entire 200 µl goes through in 1 or 2 min. If the flow rate is too fast, one can try to pinch the tip further or even start a new column. Make sure that the beads are wet at all time.
3. Add 200 µl of CBB buffer to the beads and transfer them into the microcolumn. Wash once the microfuge tube with 100 µl of CBB buffer and transfer to the microcolumn.
4. Add 25 µl of CEB buffer to the microcolumn and discard the flowthrough.
5. Add 150 µl of CEB buffer to the microcolumn. Collect the eluate, take a 5-µl aliquot and store at −80°C (#7: 3.3% of eluate).

3.8. SDS-PAGE and Western Blot Analysis

To monitor the efficiency of the affinity purification, aliquots taken throughout the purification procedure must be verified by SDS-PAGE and immunoblotting analysis using antibodies against the protein of interest or the CBP tag (see Note 4), as shown in Fig. 1c.

3.9. Decross-Linking of the Affinity-Purified Chromatin

1. Thaw the 2% total (200 µl) and add 3 volume of CEB buffer (600 µl).
2. Incubate the total chromatin input and the purified chromatin at 65°C overnight. This can be done in a water bath. Alternatively, this incubation may be carried out in a PCR machine, in which case one would divide the samples in multiple smaller tubes that fit into the machine.

3.10. RNase A/Proteinase K Treatment and Ethanol Precipitation

1. Add equal volume of TE to the input chromatin and the purified chromatin.
2. Add RNase A at a final concentration of 0.2 mg/ml. Incubate at 37°C for 1–2 h.
3. Add Proteinase K to each sample at a final concentration of 0.2 mg/ml. Incubate at 55°C for 2 h.
4. Add equal volume of chloropane to both the input chromatin and the purified chromatin, vortex and spin at 16,000 × *g* for 2 min.
5. Transfer the aqueous phase to new microfuge tubes. Repeat the chloropane extraction once more.
6. Precipitation of the DNA: add 1 µl of glycogen and 1/10th volume of acetic acid to each tube, vortex, and add 2.5 volumes of ethanol. Keep on ice for 10 min. Spin and resuspend the pellet in 200 µl of DNase-free water. Store at −20°C.

3.11. Enrichment Quantification by qPCR (ChAP-qPCR)

At this stage, it is important to be able to evaluate the chromatin that has been purified. To do this, we perform conventional PCR or quantitative real-time PCR assays to measure the enrichment of genomic regions that are known to be bound by the protein under investigation. Ideally, more than one genomic binding site should be analyzed as well as genomic regions that are not bound by the protein. Moreover, the assays should be carried out in parallel with chromatin isolated from cells that harbor the empty vector. Representative conventional and quantitative real-time PCR assays are shown in Fig. 1d, e.

We typically perform qPCRs as follows:

1. Prepare two PCR tubes for each genomic site to be tested: one for the input DNA and one for the purified DNA (see Note 6).
2. Mix the following components in a PCR tube:

10× PCR buffer	2.0 μl
10 mM dNTP	0.4 μl
Taq polymerase	0.2 μl
10 μM Primers	1.2 μl
50 mM $MgCl_2$	0.5 μl
H_2O	14.7 μl
Total	19.0 μl

3. Dilute 2% total input DNA to 0.02%. Add 1 μl of the diluted total input DNA to one tube, and 1 μl of purified DNA from cells expressing the tagged transcription factor to a second tube.
4. Place the tubes in PCR machine and run at the appropriate temperature (typically 60°C) and number of cycles (typically 30 cycles).
5. Take 7 μl of the PCR, mix with dye, and run on 1% agarose gel. Representative conventional and quantitative real-time PCR assays are shown in Fig. 1d, e.

Another type of control to verify the specificity of the purified chromatin is to do scanning ChAP analysis whereby PCR assays are performed using primers for different regions of a single locus, as shown in Fig. 2. One should observe enrichment of one or a few, but not all, regions of a locus.

3.12. Filling-in Reaction Using T4 DNA Polymerase

Once we are satisfied that the purified chromatin displays specific enrichment for known targets of the transcription factor, we can then proceed to prepare the DNA for hybridization onto a genomic microarray. Typically, we (1) do a fill-in reaction with DNA polymerase T4 in order to obtain blunt DNA fragments, (2) do a ligation with defined linkers, (3) perform a PCR amplification using primer pairs that are complimentary to the linkers.

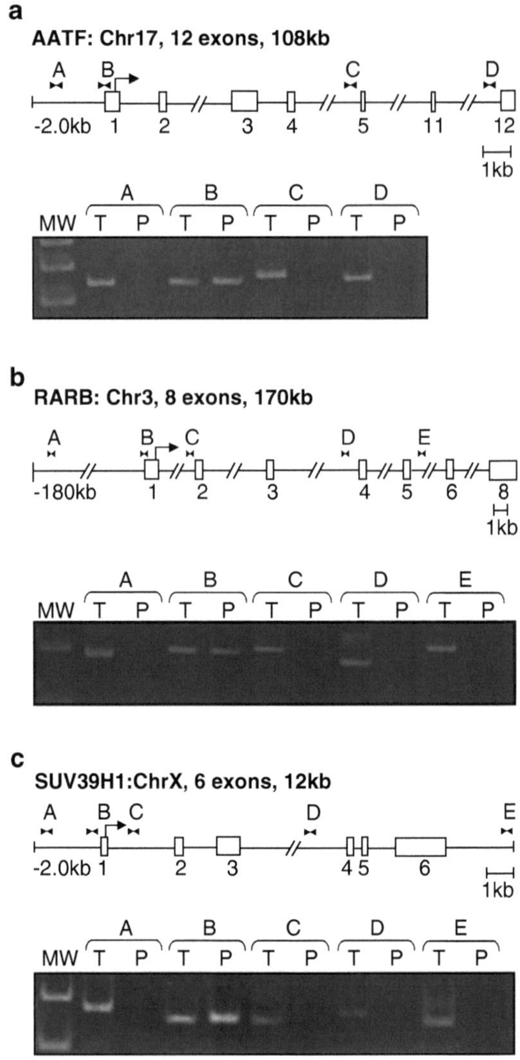

Fig. 2. Scanning ChAP analysis of the AAFF, RARB, and SUV39H1 genes. Chromatin from Hs578T/p110-Tag[2] cells was submitted to TAP and analyzed by PCR using primers specific for different regions of the following loci: AAFF (**a**), RARB (**b**), and SUV39H1 (**c**). Templates for the PCRs were 0.1% total input DNA (T) or ChAP-purified DNA (AP). Exons are numbered and the transcription start sites are indicated by *arrows.*

1. Dilute the 2% total input sample to 0.005%. Take 55-μl aliquots of the 0.005% total input and of the purified DNA.
2. Add the following ingredients and incubate at 12°C for 20 min:

10× T4 Pol buffer	11.0 μl
10 mg/ml BSA	0.5 μl
10 mM dNTP	1.0 μl
T4 DNA polymerase	0.2 μl

H_2O	42.3 µl
DNA	55.0 µl
Total	100.0 µl

3. Purify with chloropane once and then precipitate with ethanol. Resuspend in 10 µl of H_2O.

3.13. Blunt End Ligation

1. Add the following ingredients into the total input or the affinity purified DNA sample. The final volume will be 20 µl (10 + 10 µl):

10× T4 DNA Ligase buffer	2.0 µl
15 µM Annealed Linkers (JW102, JW103)	2.6 µl
T4 DNA ligase	0.5 µl
H_2O	4.9 µl
Total	10.0 µl

2. Incubate at 16°C overnight.

3.14. LM-PCR (Ligation-Mediated PCR)

1. Prepare the following solutions:

Sol A	
H_2O	14.2 µl
10× Thermopol buffer	4.0 µl
dNTP (10 mM)	1.3 µl
oligo JW102 (100 µM)	0.5 µl
Sol B (enzyme mix)	
H_2O	8.0 µl
10× Thermopol buffer	1.0 µl
Taq polymerase	1.0 µl

2. Mix the Sol A with the 20 µl of total input or affinity purified DNA from Subheading 3.13 in a PCR tube.
3. Run the PCR under the following conditions:

   Lid temp 106°C:

PCR

Step 1: 55°C for 4 min
Step 2: 72°C for 3 min
Step 3: 95°C for 2 min
Step 4: 95°C for 30 s
Step 5: 60°C for 30 s

PCR

Step 6: 72°C for 2 min go to **step 4**, 24 times
Step 7: 72°C for 4 min
Step 8: 4°C hold

4. Midway to step 1, add Sol B.
5. After amplification, run 2 μl of the reaction on 1% agarose gel to see the amplification products. You should get DNA around 250–350 bp. Purify with chloropane and precipitate with ethanol. Resuspend in 25 μl of DNase-free water.
6. In order to ensure that the enrichment level is not changed following LM-PCR amplification of the purified DNA, qPCR is performed. A representative result of real-time qPCR is shown in Fig. 1e.

3.15. Labeling and Microarray Hybridization (ChAP-chip)

Amplified DNA samples are processed for labeling and then microarray hybridization. We originally used homemade promoter microarrays and reagents. However, commercially available kits (Agilent and Affymetrix) are easy to use and are highly consistent in quality. We recommend using these kits for genomic chip hybridization. Amplified DNA from empty vector control cells should also be labeled and used for microarray hybridization in order to reduce the background. Such experiments led to the identification, and removal from our list of putative targets, of 19 genes with a *p*-value under 0.005 (see supplemental Table 5 in ref. 1).

3.16. Validation of Transcriptional Targets

Genomic sequences identified in ChAP-chip or ChAP-seq must be validated in independent experiments. Essentially, the ChAP is repeated and the purified chromatin is used in quantitative PCR assays to measure the degree of enrichment of specific sequences in the "chapped chromatin." As an additional control, we have also performed ChIP using antibodies that recognize endogenous proteins. CUX1 antibodies recognize several isoforms of the protein and therefore cannot demonstrate that a specific isoform binds to a given genomic site. Yet, ChIP with these antibodies were able to confirm that one or more isoforms or the protein were present at specific genomic sites. For example, from the microarray analysis we randomly selected 28 genes with different *p*-values and we verified the recruitment of CUX1 proteins to these genes using independent "chipped" DNA. As seen in Fig. 3, genes with lower *p*-values were also highly enriched in "chipped" DNA samples. These results confirmed that the method of ChAP can serve to identify in vivo binding sites for a transcription factor.

Gene	p-values	enrichment
EEF2	9E-14	++
RARB	2E-13	++
MTBP	2E-13	++
DNAJC1	7E-12	++
TBP	4E-09	++
DCP1B	8E-08	++
CARF	2E-07	+
TCEAL1	2E-07	++
PAPA-1	3E-07	++
SDCCAG1	8E-07	++
ORC1L	3E-05	++
CXCL1	0.0005	++
DEDD	0.0005	++
CD2AP	0.0021	+/-
WNT4	0.0032	+
HIS1H2AE/HIST1H2BG	0.0033	++
CDC25C	0.0046	+
PSMD4	0.0047	+/-
AP1M1	0.0048	+
QP-C	0.0052	+
HIST1H2A/HIS1H2BL	0.0052	–
FAPP2	0.0105	+/-
GADD45G	0.0107	–
AARSL	0.0107	++
RP1	0.0108	–
SFPQ	0.0108	+
p25	0.037	–
UCHL5/SSA2	0.0164	+/-

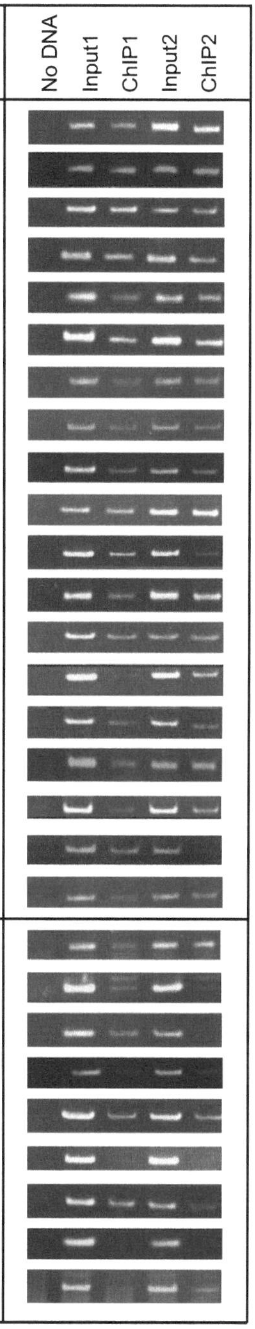

Fig. 3. PCR analysis of target promoters with various *p*-values. Chromatin from Hs578T/p110-Tag^2 cells was submitted to TAP and used for microarray hybridization (ChAP-chip). Note that genes were considered "bound" when the binding *p*-value in the error model was <0.005. From the obtained microarray results, randomly chosen 28 genes with different *p*-values were analyzed for their enrichment with two independent antibody ChIPped DNAs (ChIP1 and ChIP2) from Hs578T cells. Each PCR was performed using primers specific for each promoter and input DNA (0.02%) was used as control. The region to be amplified was chosen to be approximately in the middle of the sequence spotted on the location array.

3.17. Functional Importance of Protein–DNA Interactions

Notwithstanding the strength of the data demonstrating that a protein binds to a genomic site, the demonstration that this interaction is functional, i.e., that it leads to transcriptional regulation, requires additional work. A solid case should be built on evidence obtained from multiple experimental approaches. Commonly used assays include, but at not limited to, measuring expression of the putative target following siRNA-mediated inhibition, or alternatively overexpression, of the transcription factor. We direct the reader to an excellent discussion of how the relevance of a protein–DNA interaction can be assessed (12).

4. Notes

1. This protocol can be adapted to many other cell lines, including suspension cells which can be more easily attained in large amounts. We have used the following cell lines: NIH3T3, HeLa, HEK293, K562, HEL, Ramos, U266, and RPMI8266. For suspension cells, rinse/collect cells using a tabletop centrifuge ($500 \times g$ for 5 min).
2. If available, antibodies against the protein of interest can be used instead.
3. We recommend freezing several tubes of the original cell population, since expression levels of the tagged protein can change in long-term cultures.
4. If one of the epitope tags is protein A, any rabbit antibody will recognize the tagged protein. Following cleavage with the TEV protease, however, the protein A tag is removed.
5. Ten plates of Hs578T cells approximately correspond to 1×10^8 cells.
6. As negative controls, PCRs with affinity-purified DNA from empty vector control cells or without adding DNA will be performed. Alternatively, PCRs will be done using primer pairs to amplify the genomic regions that are not bound by the transcription factor of interest.

Acknowledgments

We acknowledge the expert technical assistance of Ms. Ginette Bérubé and Mr. Lam Leduy in the preparation of retroviral vectors. This research was supported by grant #019389 from the Canadian Cancer Society to A.Nepveu.

References

1. Harada, R., Vadnais, C., Sansregret, L., Leduy, L., Berube, G., Robert, F., and Nepveu, A. (2008) Genome-wide location analysis and expression studies reveal a role for p110 CUX1 in the activation of DNA replication genes, *Nucleic Acids Res 36*, 189–202.
2. Sansregret, L., and Nepveu, A. (2008) The multiple roles of CUX1: Insights from mouse models and cell-based assays, *Gene 412*, 84–94.
3. Rigaut, G., Shevchenko, A., Rutz, B., Wilm, M., Mann, M., and Seraphin, B. (1999) A generic protein purification method for protein complex characterization and proteome exploration, *Nat Biotechnol 17*, 1030–1032.
4. Puig, O., Caspary, F., Rigaut, G., Rutz, B., Bouveret, E., Bragado-Nilsson, E., Wilm, M., and Seraphin, B. (2001) The tandem affinity purification (TAP) method: a general procedure of protein complex purification, *Methods 24*, 218–229.
5. Buck, M. J., and Lieb, J. D. (2004) ChIP-chip: considerations for the design, analysis, and application of genome-wide chromatin immunoprecipitation experiments, *Genomics 83*, 349–360.
6. Oberley, M. J., Tsao, J., Yau, P., and Farnham, P. J. (2004) High-throughput screening of chromatin immunoprecipitates using CpG-island microarrays, *Methods Enzymol. 376*, 315–334.
7. Rodriguez, B. A., and Huang, T. H. (2005) Tilling the chromatin landscape: emerging methods for the discovery and profiling of protein-DNA interactions, *Biochem Cell Biol 83*, 525–534.
8. Taneyhill, L. A., and Adams, M. S. (2008) Investigating regulatory factors and their DNA binding affinities through real time quantitative PCR (RT-QPCR) and chromatin immunoprecipitation (ChIP) assays, *Methods Cell Biol 87*, 367–389.
9. Geck, P., Maffini, M. V., Szelei, J., Sonnenschein, C., and Soto, A. M. (2000) Androgen-induced proliferative quiescence in prostate cancer cells: the role of AS3 as its mediator, *Proc Natl Acad Sci USA 97*, 10185–10190.
10. Saqr, H. E., Omran, O., Dasgupta, S., Yu, R. K., Oblinger, J. L., and Yates, A. J. (2006) Endogenous GD3 ganglioside induces apoptosis in U-1242 MG glioma cells, *J Neurochem 96*, 1301–1314.
11. Sipione, S., Rigamonti, D., Valenza, M., Zuccato, C., Conti, L., Pritchard, J., Kooperberg, C., Olson, J. M., and Cattaneo, E. (2002) Early transcriptional profiles in huntingtin-inducible striatal cells by microarray analyses, *Hum Mol Genet 11*, 1953–1965.
12. Carey, M. F., Peterson, C. L., and Smale, S. T. (2009) Chapter 8: Confirming the Functional Importance of a Protein–DNA Interaction, in *Transcriptional Regulation in Eukaryotes: Concepts, Strategies, and Techniques, Second Edition*, Cold Spring Harbor Laboratory Press, Cold Spring Harbor.

Chapter 17

Determination of Histone Acetylation Status by Chromatin Immunoprecipitation

Luciano Galdieri, John Moon, and Ales Vancura

Abstract

Histone acetylation is the most studied posttranslation modification of nucleosomes. Understanding the mechanisms involved in global and promoter-specific histone acetylation will shed light on the control of transcriptional regulation. Chromatin immunoprecipitation is a powerful technique to study protein–DNA interactions in vivo. Proteins and DNA are cross-linked with formaldehyde, cells are lysed, and DNA is sheared by sonication. Protein–DNA complexes are immunoprecipitated with antibodies specific for total and acetylated histones and the relative occupancy of acetylated and total histones at selected loci is assessed by real-time PCR of the purified DNA.

Key words: Histones, Cross-linking, Immunoprecipitation, *Saccharomyces cerevisiae*, Real-time PCR

1. Introduction

The fundamental unit of chromatin is the nucleosome, consisting of DNA wrapped around a protein octamer containing two copies of each histone proteins H3, H4, H2A, and H2B (1). The histones have two distinct domains: the core globular domain and the N-terminal tail that protrudes from the core. The histone proteins are altered by a variety of posttranslation modifications including acetylation, methylation, ubiquitination, sumoylation, and phosphorylation (2). Because the structure of the nucleosome limits the access to the DNA, cells rely on chromatin modifying complexes to make the DNA more accessible to replication and transcription machineries.

Histone acetylation and deacetylation of lysine residues is one of the most studied posttranslation modification (3, 4) and has

Ales Vancura (ed.), *Transcriptional Regulation: Methods and Protocols*, Methods in Molecular Biology, vol. 809,
DOI 10.1007/978-1-61779-376-9_17, © Springer Science+Business Media, LLC 2012

been shown to regulate transcription by altering the chromatin structure (5) and creating binding sites for regulatory proteins (6, 7). Histone acetylation status is regulated by interplay between histone acetyltranferases (HATs) and histone deacetylases (HDACs) (8). Pertubation of the balanced action of HATs and HDACs alters the expression pattern of genes involved in cellular growth, resulting in tumorigenesis (9, 10).

Histone acetylation is conserved from yeast to mammalian cells (11, 12). Due to the manageability of yeast genome, *Saccharomyces cerevisiae* is a good model to study the effects of histone acetylation on transcription. Commercial availability of antibodies that recognize total histones as well as histones that are singly or multiply acetylated makes chromatin immunoprecipitation (ChIP) a method of choice for the determination of total acetylation levels or acetylation levels per nucleosome at virtually any loci in the genome. ChIP can depict protein–DNA interaction in vivo under different conditions since formaldehyde cross-links protein and DNA shortly after its addition and inactivates majority of enzymes that could affect composition of these complexes, such as proteases, nucleases, HATs, and HDACs.

2. Materials

2.1. Yeast Strains, Growth Media, DNA–Protein Cross-Link, and Spheroplasting

1. *S. cerevisiae* wild-type strain W303 (MATa, ade2-1, his3-1,15, leu2-3,112, trp1-1, ura3-1) or strain derived by genetic manipulation.
2. Yeast extract/peptone/dextrose (YPD) media: 1% (w/v) yeast extract, 2% (w/v) peptone, 2% (w/v) dextrose.
3. Formaldehyde: 37% (v/v).
4. Glycine: 2.5 M.
5. Spheroplast buffer: 40 mM potassium phosphate, pH 7.5, 1.4 M sorbitol.
6. Yeast lytic enzyme from *Arthobacter luteus* (Sigma, St. Louis, MO): 700 U/mL.
7. TBS buffer: 25 mM Tris–HCl, pH 7.4, 137 mM NaCl.

2.2. Cell Lysis and Immunoprecipitation

1. Lysis Buffer (FA): 50 mM HEPES–KOH, pH 7.5, 150 mM NaCl, 1 mM EDTA, 1% (v/v) Triton X-100.
2. Protease inhibitors: Protease inhibitor cocktail (Sigma, Cat. No. P8215) as a 50× stock, Complete protease inhibitor cocktail tablets (Roche, Cat. No. 11 836 145 001), dissolving one tablet in 2 mL of water yields a 25× stock.
3. Phenylmethylsulfonyl fluoride (PMSF): 100 mM in isopropanol.
4. Sodium deoxycholate: 1% (w/v).

5. 0.5 mm glass beads (BioSpec Products, Bartlesville, OK), acid washed (see Note 1).
6. Protein A/G PLUS agarose (Santa Cruz, Santa Cruz, CA).
7. Bovine serum albumin (BSA) fraction V: 5 mg/mL in TBS.
8. Antibodies: anti-histone H3 (Abcam, Cat. No. ab1791), anti-acetyl-histone H3 (Lys14) (Millipore, Cat. No. 07-353), anti-hyperacetylated histone H4 (Penta) (Millipore, Cat. No. 06-946), and anti-histone H4 (Cell signaling, Cat. No. 2592).

2.3. Cross-Link Reversal and DNA Recovery

1. FA buffer supplemented with NaCl: 50 mM HEPES–KOH, pH 7.5, 500 mM NaCl, 1 mM EDTA, 1% (v/v) Triton X-100.
2. Tris–HCl: 10 mM, pH 6.5 and 10 mM, pH 8.0.
3. LiCl detergent solution: 250 mM LiCl, 0.5% (v/v) IGEPAL CA-630, 0.5% (w/v) sodium deoxycholate, 1 mM EDTA, 10 mM Tris–HCl, pH 8.0.
4. ChIP extraction buffer: 1% (w/v) SDS, 0.1 M $NaHCO_3$.
5. Sodium dodecylsulfate (SDS): 10% (w/v).
6. Proteinase K (Sigma): 20 mg/mL.
7. Equilibrated Phenol for DNA extraction (see Note 2).
8. Chloroform/isoamyl alcohol (1:24) solution.
9. Sodium acetate: 3 M, pH 5.2.
10. Ethanol: 100% (v/v) and 70% (v/v).
11. Glycogen (Roche): 20 mg/mL.
12. Tris–EDTA solution: 10 mM Tris–HCl, pH 8.0, 1 mM EDTA.

2.4. Real-Time PCR

1. Nuclease-free water.
2. Primer pairs.
3. IQ SYBR Green Supermix (Bio-Rad).
4. 96-Well PCR plate (Bio-Rad).
5. Microseal "B" Film (Bio-Rad).

3. Methods

To obtain reproducible results, it is important to cross-link the cells rapidly and keep the samples cold for most of the protocol. The antibodies that recognize the total and acetylated lysines allow good detection, however, in this protocol we increased the amount of cells in order to generate a concentrated lysate to increase sensitivity and to perform up to four immunoprecipitations with the same sample.

3.1. Cell Growth, Protein Cross-Linking, and Spheroplasting

1. Grow yeast culture in 20 mL YPD overnight at 30°C at 250 rpm. Determine A_{600nm} of this culture and inoculate three 1-L flasks containing 200 mL YPD to get an A_{600nm} of 0.1. Grow the culture until A_{600nm} is 0.8–1.0 (see Note 3).
2. Add formaldehyde to a final concentration of 1% in each flask (5.4 mL of 37% formaldehyde to each 200 mL culture) and incubate for 15 min at 30°C at 250 rpm. This will allow the proteins to cross-link with the DNA (see Note 4).
3. Add glycine to a final concentration of 0.3 M (24 mL of the 2.5 M solution for each 200 mL culture) and incubate as previously for 5 min. This step is necessary to stop the cross-linking reaction.
4. Harvest the cells by successive centrifugation in four conical tubes (50 mL) for 5 min at 3,000 × *g* at 4°C.
5. Combine the four pellets into one tube with 40 mL spheroplast buffer containing 1% β-mercaptoethanol.
6. Centrifuge the cells again for 5 min at 3,000 × *g* at 4°C.
7. Resuspend the pellet in 10 mL of the spheroplast buffer containing 0.4% β-mercaptoethanol.
8. Measure A_{600nm} after diluting the sample 100×.
9. Digest the cell wall by adding 1 mL of yeast lytic enzyme in a shaking water bath at 37°C at 100 rpm until 80–90% digestion (see Note 5).
10. Centrifuge the spheroplast for 5 min at 1,000 × *g* at 4°C and wash them two times with 40 mL of ice-cold spheroplast buffer and two times with 40 mL of ice-cold TBS buffer. Centrifuge the spheroplasts for 5 min at 1,000 × *g* at 4°C between washes (see Note 6).
11. Resuspend the pellet with 6 mL of TBS and aliquot in eight 1.5 mL microcentrifuge tubes.
12. Centrifuge the tubes for 3 min at 1,000 × *g* at 4°C. Remove the supernatant and store pellets at −70°C for immunoprecipitation.

3.2. Spheroplast Lysis, DNA Shearing, and Immunoprecipitation

1. Start by setting up the protein A/G PLUS agarose antibodies adsorption (see Note 7). Calculate the volume of agarose necessary for each sample: use 50 μL for preclearing and 100 μL for immunoprecipitation of each sample.
2. Add 1 mL of BSA to the protein A/G PLUS agarose and allow it to rotate for 15 min in a rotator at 4°C. Centrifuge for 1 min at 1,000 × *g* at 4°C. Discard the supernatant and repeat the procedure. Make up to the starting volume of protein A/G PLUS agarose with the BSA solution.

3. Preadsorb each 100 μL of protein A/G PLUS agarose with 6 μg of antibody. Incubate for 5–8 h in the rotator at 4°C. Start lysis of the spheroplasts.
4. Thaw spheroplasts on ice and wash them with 1 mL of ice-cold FA buffer supplemented with 0.1% sodium deoxycholate, 1 mM PMSF, 1× protease inhibitor (Sigma), and 1× protease inhibitor (Roche) (see Note 8). Centrifuge for 3 min at 1,000 × *g* at 4°C.
5. Remove the supernatant and resuspend the spheroplasts in 200 μL of ice-cold 1× FA buffer with supplements (0.1% sodium deoxycholate, 1 mM PMSF, 1× protease inhibitor (Sigma), and 1× protease inhibitor (Roche)). Add about 0.3 mL of glass beads. Vortex the tubes 20 times in 15 s burst. Incubate the tubes for 1 min on ice in between bursts (see Note 9).
6. In order to remove the cell lysate from the tube without the glass beads, flame a needle (BD 22G1 PrecisionGlide™ Needle preferred) and puncture the microcentrifuge tube at the bottom and a second time near the top of the tube. Place the microcentrifuge tube into a 15-mL conical tube, screw cap on and centrifuge for 1 min at 2,000 × *g* at 4°C. Combine the eight samples in one 15 mL conical tube.
7. In order to shear the DNA to about 500 bp fragments, use a sonicator (Branson Model 450 is preferred) to apply ultrasound five times with 15% output and five times with 20% output to each sample (see Note 10). Incubate samples for 1 min on ice between bursts. High temperature due to sonication might denature protein and even reverse the proteins–DNA cross-links.
8. Transfer lysate to microcentrifuge tubes and centrifuge for 30 min at 13,000 × *g* at 4°C to remove cell debris. Combine supernatants in a clean 15 mL conical tube. At this step about 2 mL of lysate should be obtained, enough for four immunoprecipitations. Save 100 μL of the lysate at −70°C as your input or whole cell extract (WCE).
9. Prepare four aliquots (one for each immunoprecipitation) of lysate by diluting 500 μL of lysate with 500 μL of 1× FA buffer with supplements (0.1% sodium deoxycholate, 1 mM PMSF, 1× protease inhibitor (Sigma), and 1× protease inhibitor (Roche)). Preclear diluted lysates with 50 μL of protein A/G PLUS agarose in the rotator for 1 h at 4°C (see Note 11).
10. Centrifuge the samples for 30 min at 13,000 × *g* at 4°C to remove precipitates and protein A/G agarose beads. Transfer each supernatant to a new microcentrifuge tube and add 100 μL of protein A/G PLUS agarose beads preadsorbed with the antibodies. Each microcentrifuge tube should contain a different antibody. Incubate in a rotator overnight at 4°C.

3.3. Cross-Link Reversal and DNA Recovery

1. Harvest protein A/G PLUS agarose by centrifugation for 1 min at 100×*g* at 4°C. Supernatants are discarded and beads are incubated for 5 min in a rotator at 4°C in each of the following buffers: FA buffer with supplements (once), FA buffer with supplements and NaCl (once), LiCl detergent solution (once), Tris–EDTA solution (two times). Collect protein A/G PLUS agarose beads after each incubation by centrifugation for 1 min at 100×*g* at 4°C.
2. Extract DNA–protein complexes from the protein A/G PLUS agarose using the ChIP extraction buffer. Add 100 μL of the buffer to each tube. Vortex the samples for 10 s at the highest speed and incubated at room temperature for 10 min (at this point proteolysis is inconsequential). Quickly spin down the beads and transfer the supernatant to a fresh labeled microcentrifuge tube. Repeat the process two more times to have 300 μL final extract volume.
3. In order to reverse the formaldehyde cross-links, add 15 μL of 5 M NaCl to the combined 300 μL extract and incubate for 6 h at 65°C. Thaw the WCE, supplement it with 1 μL of 5 M NaCl and incubate for 6 h at 65°C.
4. In order to digest the proteins, add the following solutions to the combined extract: 6 μL of 0.5 M EDTA, pH 8.0, 12 μL of 1 M Tris–HCl, pH 6.5, and 1 μL proteinase K. To the WCE, the following solutions are added: 10 μL 10% SDS, 2 μL of 0.5 M EDTA pH 8.0, 10 μL of 1 M Tris–HCl pH 8.0, 2 μL of proteinase K. Incubate samples at 42°C for 1 h.
5. Perform phenol/chloroform extraction to purify the DNA. Add 300 μL of phenol to the immunoprecipitated samples. Vortex vigorously for 5 s and centrifuge for 5 min at 13,500×*g* at room temperature. Transfer the upper aqueous phase to a new microcentrifuge tube and discard the bottom phase. Add 150 μL of phenol and 150 μL of chloroform/isoamyl alcohol. Vortex vigorously and centrifuge as before. Transfer the upper aqueous phase to a new microcentrifuge tube and discard the bottom phase. Add 300 μL of chloroform/isoamyl alcohol. Vortex and centrifuge as before. At this step, high efficiency precipitation of the DNA is performed. Transfer the upper aqueous phase to a new microcentrifuge tube and discard the bottom phase. Add 40 μL of 3 M sodium acetate pH 5.2, 800 μL of 100% ethanol, and 1 μL of glycogen. Precipitate the DNA at −20°C overnight.
6. Purify the DNA in WCE by phenol/chloroform extraction. Add 100 μL of phenol to 100 μL of WCE. Vortex vigorously for 5 s and centrifuge for 5 min at 13,500×*g* at room temperature. Transfer the upper aqueous phase to a new microcentrifuge tube and discard the bottom phase. Add 50 μL of phenol and

50 μL of chloroform/isoamyl alcohol. Vortex vigorously and centrifuge as before. Transfer the upper aqueous phase to a new microcentrifuge tube and discard the bottom phase. Add 100 μL of chloroform/isoamyl alcohol. Vortex and centrifuge as before. For a higher efficiency DNA precipitation transfer the upper aqueous phase to a new microcentrifuge tube and discard the bottom phase. Add 13.3 μL of 3 M sodium acetate pH 5.2, 270 μL of 100% ethanol and 1 μL of glycogen. Precipitate the DNA at −20°C overnight.

7. Centrifuge the immunoprecipitation and WCE samples for 30 min at 13,500 × *g* at 4°C. Discard the supernatant and wash the pellets with 1 mL of 70% ethanol. Centrifuge for 15 min at 13,500 × *g* at 4°C. Discard the supernatant and air dry the pellet for 20 min. Resuspend the pellet in 50 μL of Tris–EDTA, pH 8.0. Store at −70°C. The DNA is ready to be analyzed by PCR.

3.4. Real-Time PCR Analysis with SYBR Green

1. This protocol was optimized for a 25-μL PCR reaction to increase savings on reagent costs. Prepare a master mix containing 12.5 μL of SYBR Green supermix and 5 pmol of each primer. Make the volume up to 22 μL with nuclease-free water for each sample (see Note 12).
2. Dilute sample DNA 1:5 and WCE sample 1:100.
3. Place 3 μL of each diluted sample DNA and diluted WCE (see Note 13) in to the appropriate well in the 96-well plate. Each sample should be run in triplicate to allow removal of outliers.
4. Add 22 μL of the master mix in each well of the 96-well PCR plate. Pipetting the master mix only once in the well increases reproducibility. Ensure that all the liquid is at the bottom of the well.
5. Seal the plate with the microseal film.
6. Program the real-time PCR machine as follows: 1 cycle: 10 min 95°C, 40 cycles: 30 s 95°C, 30 s 55–60°C, 30 s 72°C. At the end, determine the melting curve of all the samples.

3.5. Data Analysis

1. The PCR baseline is determined as the cycle that precedes the point where the PCR amplification curve starts increasing linearly; this value is subtracted from each sample. The threshold is determined by calculating the second derivative of each trace from the graph and looking for the point of maximum curvature (see Note 14). The software will display for each well on the plate at which cycle the fluorescence reached the threshold. This value is referred to as *Ct value*.
2. Using a spreadsheet program, calculate the average *Ct value* for each sample for each primer. Subtract the average *Ct value* of the WCE from the average *Ct value* of the immunoprecipitated histone ($Ct^{WCE\text{-}IP}$).

Table 1
Relative occupancy of acetylated histone H3, acetylated histone H4, histone H3, and histone H4 at the *SUC2* promoter

	AcH3	AcH4	H3	H4	WCE
Average *Ct value*	20.73	21.18	20.73	28.14	21.49
SUC2 primers					
Average *Ct value*	20.74	21.53	19.32	26.96	20.79
POL1 primers					
$Ct^{WCE\text{-}IP}$	0.76	0.31	0.76	-6.65	
SUC2 primers					
$Ct^{WCE\text{-}IP}$	0.05	-0.74	1.47	-6.17	
POL1 primers					
$^{EXP\text{-}REF}Ct^{WCE\text{-}IP}$	0.71	1.05	-0.71	-0.48	
$2^{\wedge}\ {}^{EXP\text{-}REF}Ct^{WCE\text{-}IP}$	1.64	2.07	0.61	0.72	

Note: Primer pair for *POL1* was used as a reference

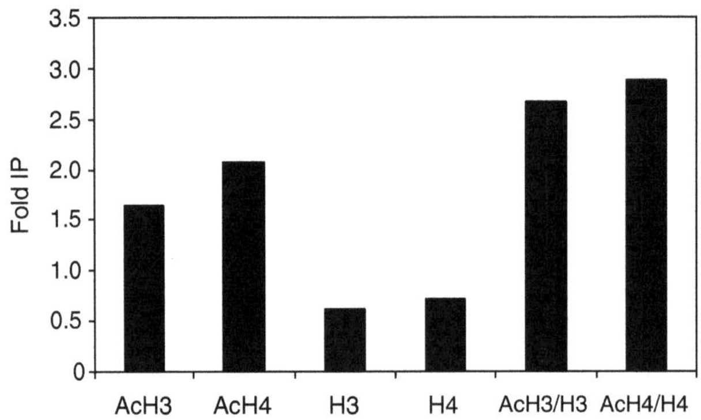

Fig. 1. Levels of total and acetylated histones H3 and H4 at the *SUC2* promoter. Wild-type cells were grown to $A_{600nm} = 0.8$ and subjected to ChIP protocol. The resulting DNA was amplified by real-time PCR with SYBR green detection using primers for the *SUC2* promoter. Results show the occupancy of the acetylated form as well as total levels of histones H3 and H4 relative to the WCE, using *POL1* promoter as a reference. Levels of acetylated histones H3 and H4 per nucleosome at the *SUC2* promoter were calculated by dividing the occupancy of the acetylated histone H3 or H4 by the occupancy of the total histone H3 or H4.

3. Now, subtract $Ct^{WCE\text{-}IP}$ for the experimental primer from the $Ct^{WCE\text{-}IP}$ for the reference primer ($^{EXP\text{-}REF}Ct^{WCE\text{-}IP}$).
4. Finally, calculate 2 power to $^{EXP\text{-}REF}Ct^{WCE\text{-}IP}$ (see Note 15). Repeat the procedure for each acetylated and total histone H3 or H4 immunoprecipitations (an example of the calculation is shown in Table 1).

5. The result represents the occupancy of the immunoprecipitated acetylated or total histone H3 or H4 relative to a control DNA region (Fig. 1).
6. The acetylation of histones depends on the degree of occupancy of nucleosomes in the DNA. In order to correct the acetylation levels for the nucleosome content, divide the result for acetylated histone by the result for total histone (Fig. 1). Perform this calculation for both H3 and H4 histones. The results represent the acetylation per nucleosome.

4. Notes

1. Glass beads are soaked in concentrated nitric acid for 1 h then rinsed thoroughly until pH 6.0. Dry the beads for 30 min at 120°C.
2. Phenol is melted at 68°C. Add equal volume of 1 M Tris–HCl, pH 8.0. Stir for 30 min. Let it sit at room temperature until two phases are formed. Remove the upper aqueous phase and measure the pH. Repeat the process until pH > 7.8. Remove the final aqueous phase and add 10% the final volume of 0.1 M Tris–HCl pH 8.0 supplemented with 0.2% β-mercaptoethanol (13).
3. Selective preculture medium should be used to grow cells containing plasmids and limit the growth on YPD main culture to two generations to avoid excessive loss of plasmid.
4. Amino and imino groups of side chains of amino acid residues of proteins and nucleic acids react with formaldehyde forming a Schiff base. This base will react with another amino group, cross-linking proteins, and nucleic acids (14).
5. Check A_{600nm} every 15 min. If digestion does not proceed well, add more yeast lytic enzyme. Complete digestion within 1 h. Calculate the % digestion as follows:

$$\% \text{ digestion} = \frac{A_{600\,\text{nm}}\text{before digestion} - A_{600\,\text{nm}}\text{after digestion}}{A_{600\,\text{nm}}\text{before digestion}} \times 100.$$

6. Spheroplasts are fragile, therefore resuspend it slowly and avoid using vortex. Keep it on ice all the time.
7. Setting up adsorption of antibodies to the beads before starting the cell lysis is important since antibodies and beads should incubate together for 5–8 h. During incubation time, proceed with cell lysis and DNA shearing.
8. Protease inhibitors and sodium deoxycholate are not stable in solution, therefore add them fresh to the 1× FA buffer. Prepare the amount of 1× FA buffer with supplements necessary for the day.

9. It is important to incubate the tubes for 1 min in between bursts to keep the sample cold during the entire process to avoid proteolysis.
10. If a different sonicator apparatus is used, it is important to check the size of chromatin fragments before continuing with the protocol. After sonication, remove a 100-μL aliquot and incubate 1 h at 42°C with 10 μL of proteinase K with subsequent incubation at 65°C for 6 h. Purify the DNA with phenol–chloroform and analyze it by electrophoresis in a 1% agarose gel. DNA fragments should be of an average size of 500 bp.
11. This protocol was optimized for increased sensitivity and reproducibility of acetylated and total histones H3 and H4 detection, therefore a concentrated lysate is generated. Preclearing the lysate reduces the background due to unspecific binding of proteins to the protein A/G PLUS agarose.
12. For each sample, always run a reaction with the primer pair for the locus of interest as well as a reference primer for a DNA region that it is not affected by the different conditions tested. The *Ct value* for the reference primer can be considered background.
13. Different primers have different efficiencies, therefore each primer pair should be analyzed for the immunoprecipitated sample and the WCE for accurate quantification of relative binding.
14. Software like MyIQ™ (Bio-Rad) calculate the baseline cycle and threshold value by default.
15. In a perfect PCR reaction, the number of amplified molecules doubles after each cycle, therefore with a slope of 2. However, for more accurate results, the slope should be calculated for each primer pair.

References

1. Luger K, Mader AW, Richmond RK, Sargent DF, Richmond TJ (1997) Crystal structure of the nucleosome core particle at 2.8Å resolution. Nature 389:251–260
2. Berger SL (2002) Histone modificatins in transcriptional regulation. Curr Opin Genet Dev 12:142–148
3. Kurdistani SK, Tavazoie S, Grunstein M (2004) Mapping global histone acetylating patterns to gene expression. Cell 117:721–733
4. Pokholok DK, Harbison CT, Levine S et al (2005) Genome-wide map of nucleosome acetylation and methylation in yeast. Cell 122:517–527
5. Tse C, Sera T, Wolffe AP, Hansen JC (1998) Disruption of higher-order folding core histone acetylation dramatically enhances transcription of nucleosomal arrays by RNA polymerase III. Mol Cell Biol 18:4629–4638
6. Dhalluin C, Carlson JE, Zeng L et al. (1999) Structure and ligand of a histone acetyltransferase bromodomain. Nature 399:491–496
7. Owen DJ, Ornaghi P, Yang JC et al (2000) The structural basis for the recognition of acetylated histone H4 by the bromodomain of histone acteyltransferase Gcn5p. EMBO J 19:6141–6149
8. Kurdistani SK, Grunstein M (2003) Histone acetylation and deacetylation in yeast. Nat Rev Mol Cell Biol 4:276–284
9. Shukla V, Vaissiere T, Herceg Z (2008) Histone acetylation and chromatin signature in stem cell identity and cancer. Mutation Res 637:1–15

10. Lafon A, Chang CS, Scott EM, Jacobson SJ, Pillus L (2007) MYST opportunities for growth control: yeast genes illuminate human cancer functions. Oncogene 26:5373–5394
11. Wang L, Mizzen C, Ying C et al (1997) Histone acetyltransferase activity is conserved between yeast and human GCN5 and is required for complementation of growth and transcriptional activation. Mol Cell Biol 17: 519–527
12. Doyon Y, Selleck W, Lane W et al (2004) Structural and functional conservation of the NuA4 histone acetyltransferase complex from yeast to humans. Mol Cell Biol 24:1884–1896
13. Sambrook J (1989) Molecular cloning: a laboratory manual. Cold Spring Harbor Laboratory Press, New York
14. Orlando V, Strutt H, Paro R (1997) Analysis of chromatin structure by *in vivo* formaldehyde cross-linking. Methods 11:205–214

Chapter 18

Immunostaining of *Drosophila* Polytene Chromosomes to Investigate Recruitment of Chromatin-Binding Proteins

Magdalena Murawska and Alexander Brehm

Abstract

Gene transcription is a complex process that involves a large number of proteins. These proteins can be brought to their target genes by a variety of different mechanisms: many transcription factors interact with specific DNA sequences in promoters or enhancers, several epigenetic regulators bind histones bearing specific modifications, elongation factors and some RNA processing factors bind to the transcribing RNA polymerase, and other factors interact directly with nascent transcripts or noncoding RNA. Immunostaining of *Drosophila* polytene chromosomes allows the genome-wide localization of factors involved at different stages of transcriptional regulation. In this chapter, we present protocols that adapt the general technique to probe different recruitment mechanisms employed by these factors, including specific interactions with phosphorylated RNA polymerase II and RNA-mediated chromatin associations.

Key words: Immunostaining, Polytene chromosomes, Chromatin, Heat-shock genes, RNA, Flavopiridol

1. Introduction

During development of *Drosophila melanogaster* third instar larvae, the genome of salivary gland cells undergoes repeated rounds of endoreplication. As a result, polytene chromosomes that consist of approximately 1,000 synapsed sister chromatids are produced. The centromeric regions, however, remain underreplicated, bundle together, and form the so-called chromocenter. These giant chromosomes can be easily studied under a light microscope and display a characteristic banding pattern (1). Staining of polytene chromosomes with DNA-binding dyes allows the visualization of heterochromatin and euchromatin (Figs. 1–3). Heterochromatin is concentrated at the chromocenter, the telomeres, and encompasses

Ales Vancura (ed.), *Transcriptional Regulation: Methods and Protocols*, Methods in Molecular Biology, vol. 809, DOI 10.1007/978-1-61779-376-9_18, © Springer Science+Business Media, LLC 2012

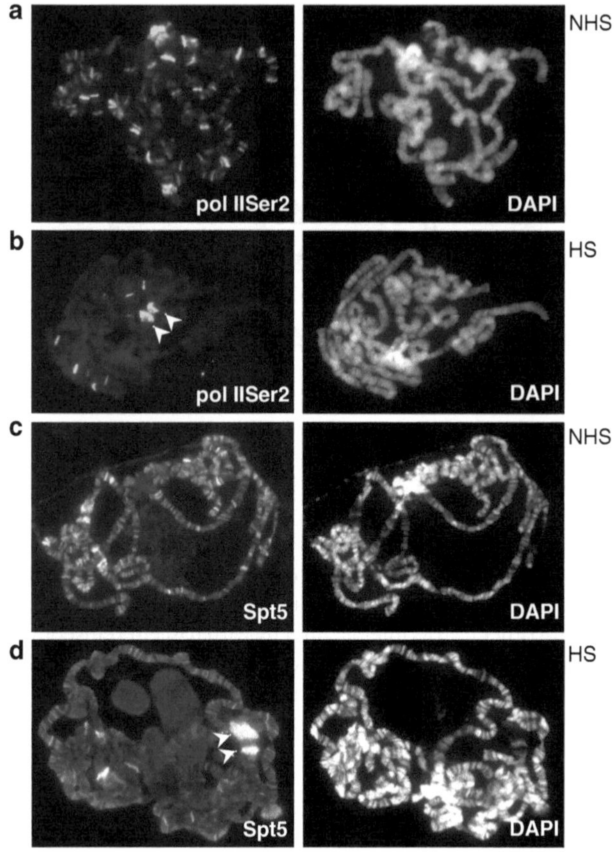

Fig. 1. Redistribution of RNA polymerase II and the elongation factor Spt5 to heat-shock loci on polytene chromosomes. Polytene chromosomes from wild-type flies (OrR) were processed directly without heat shock (NHS, panels **a**, **c**) or treated with heat shock at 37°C for 30 min (HS, panels **b**, **d**). Polytenes were stained with antibodies against RNA pol II Ser 2P (panels **a**, **b**) or against the elongation factor – Spt5 (panels **c**, **d**). Right column shows DNA stained with DAPI. Arrows indicate *hsp70* loci. Polytene chromosomes are shown at the same magnification.

most of the fourth chromosome. In addition, the second, third, and the sex chromosomes show densely stained (heterochromatic) bands along the chromosome arms. These are divided by weakly stained (euchromatic) interbands, which contain most actively transcribed genes.

Transcriptionally active "puffs" represent chromatin regions with a highly relaxed chromatin structure that exhibit exceptionally high transcriptional activity. Examples include ecdysone-dependent genes that become active during larval development and actively transcribed heat-shock genes. The heat-shock genes have become a popular model to study fundamental issues of gene transcription. They offer three main advantages: (a) their activation is under full experimental control and is achieved by a simple temperature shift;

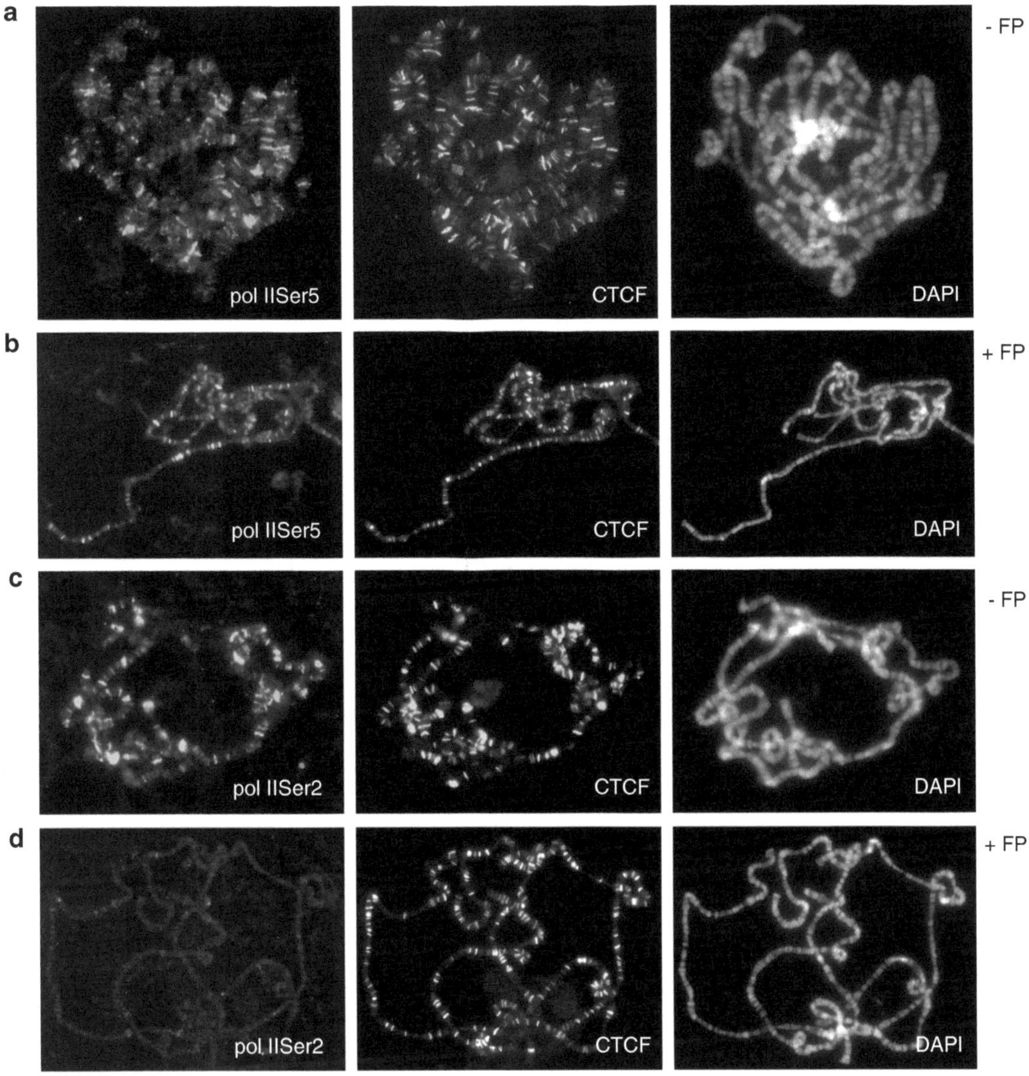

Fig. 2. Flavopiridol inhibits phosphorylation of serine 2 at RNA polymerase II without affecting serine 5 phosphorylation. Dissected salivary glands were treated with 500 nM flavopiridol (+FP, **b**, **d**) or incubated with 50% Schneider medium (–FP, **a**,**c**) for 25 min and processed for immunofluorescence. Ser 2P, but not Ser 5P, forms of RNA Pol II decreases dramatically upon flavopiridol treatment (compare **b** and **d**). Unrelated protein – CTCF – was used for staining as a negative control.

(b) heat-shock gene activation is a rapid process; and (c) heat-shock genes are activated to extremely high transcription levels facilitating the detection of heat-shock gene transcripts. The heat-shock genes, in particular the paradigmatic *hsp70* genes, have been used, among other things, to demonstrate the first instance of a "paused" polymerase, to measure elongation rates, to determine the role of RNA polymerase II phosphorylation, and to define chromatin changes upon gene activation (2, 3). As a result, they are one of the best-understood model genes that we have at our disposal.

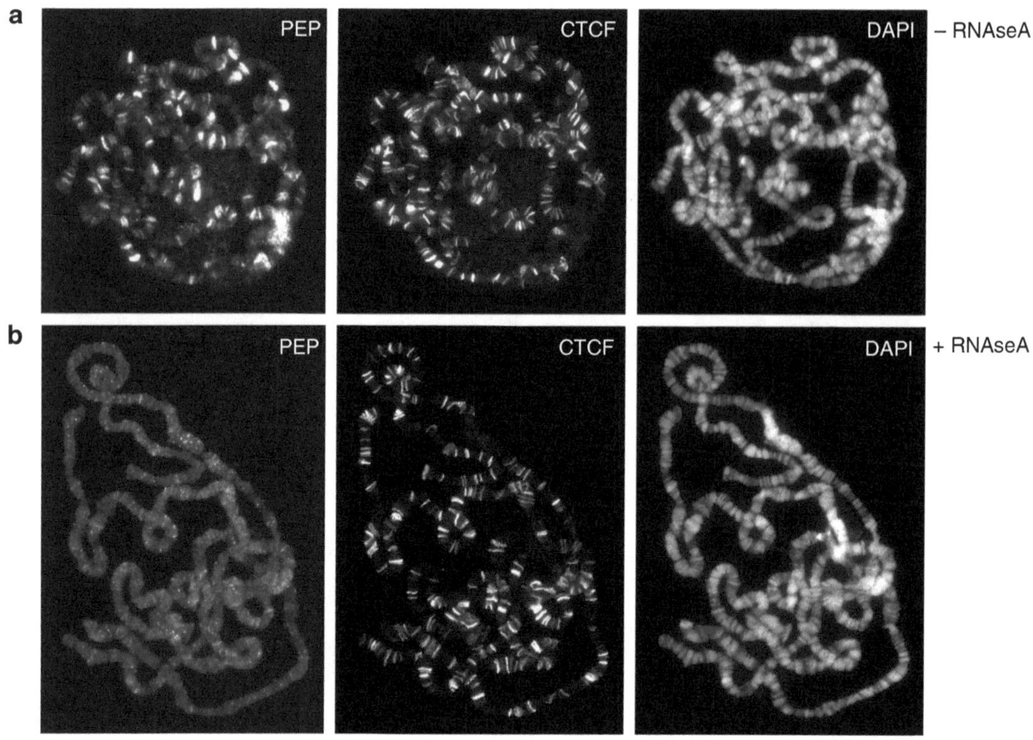

Fig. 3. RNase A treatment affects PEP binding to polytene chromosomes. Salivary glands were permeabilized and incubated without (**a**) or with RNase A (**b**) and processed for immunofluorescence. Upon RNase A treatment, signals from RNA-binding protein – PEP – disappear, whereas DNA-binding protein – CTCF – stays unaffected. DAPI staining indicates that RNase treatment of chromosomes does not disturb their structure.

Indirect immunofluorescence on polytene chromosomes ("polytene immunostaining" for short) allows the localization of factor-binding sites on *Drosophila* chromosomes (4, 5). The advantages of this method are its low cost, relative speed and simplicity, and the ability to investigate an entire genome at once, albeit at a lower resolution than more elaborate techniques, such as ChIP-chip and ChIP-seq.

Polytene immunostaining is predominantly used to study chromosome-associated factors regulating gene expression. It allows to map the distribution of factors across the genome. In addition, it is possible to characterize DNA-binding sites occupied by a given factor in vivo, for example by introducing transgenes containing binding elements or mutated versions thereof. Finally, it is employed again by use of mutant or transgenic flies to map protein domains or activities required for chromosome association.

In addition to these classical approaches, polytene immunostaining can be combined with other protocols to address more specific questions. In this chapter, we present three such variations of the classical polytene staining technique that can be used to analyze proteins directly involved in the transcription process.

We first describe a protocol to rapidly and strongly induce the heat-shock genes in polytene chromosomes. We explain how to visualize the rapid redistribution of RNA polymerase II to activated heat-shock genes that occurs in vivo. This protocol can be used to determine if a factor of interest accumulates at active heat-shock genes, like RNA polymerase II does, or if its chromatin association remains unaltered. This simple assay allows researchers to determine if the heat-shock genes are a good model system to investigate their favorite factor.

Then, we discuss a treatment that modulates the phosphorylation level of RNA polymerase II. This makes it possible to analyze if the association of a given factor with actively transcribed genes depends on RNA polymerase phosphorylation.

Finally, we present a protocol to remove RNA from polytene chromosomes by gentle RNAse treatment prior to polytene immunostaining. The role of coding and noncoding RNA in mediating chromatin association of proteins is increasingly appreciated, and the number of factors that have been shown to require an RNA component for chromatin binding is increasing rapidly (6–9). The protocol we present allows the experimenter to determine if RNA is required for a given factor to bind to chromosomes or not.

2. Materials

2.1. Dissection of Salivary Glands

1. General fly maintenance (fly food, vials, binocular; for details, see ref. 4).
2. Forceps (135 mm, Dumont).
3. Phosphate-buffered saline (PBS): 137 mM NaCl, 2.7 mM KCl, 10 mM Na_2HPO_4, 2 mM KH_2PO_4, pH 7.4.
4. PBS with 0.1% Triton-X 100 for permeabilization of glands (RNase protocol).
5. RNase A (DNAse free, QIAGEN; RNase protocol).
6. Binocular.
7. Schneider's Drosophila medium (Gibco).
8. Flavopiridol (Sigma): Dissolved in deionized water to a concentration of 1 mg/ml, kept at 4°C in the fridge. For salivary gland treatment, this stock solution is freshly diluted to 500 nM in 50% Schneider medium. This solution must not be kept on ice and can only be used once.

2.2. Polytene Chromosome Squashes

1. Pencil with rubber at the end.
2. Glass slides with frosted end (76×26 mm, Thermo Scientific).
3. Coverslips (22×22 mm and 22×40 mm, Thermo Scientific).

4. Dimethyldichlorosilane (PlusOne Repel-Silane ES (GE Healthcare)) to treat coverslips.
5. Glass dishes to collect and wash larvae.
6. Whatman paper: Thick Whatman paper is needed during the squashing procedure, thin Whatman paper is used to remove excess liquid.
7. Dewar to hold liquid nitrogen, safety glasses.
8. Scalpel to remove coverslips.
9. Coplin jars (57×90 mm, Hellendahl) to collect glass slides with polytene squashes.
10. Fixation solution: Prepare freshly by mixing 500 μl of deionized water, 450 μl of acetic acid, 50 μl of 10% formaldehyde (methanol free); keep on ice. Solution is stable for up to 2 h (see Note 1).
11. Light microscope.

2.3. Immunostaining

1. Humidity chamber: A plastic box with moistened tissue.
2. Aluminum foil.
3. Fluoromount – G (SouthernBiotech, No: 0100-01).
4. Nail polish to seal coverslips.
5. Box to collect processed glass slides.
6. 5% dry milk in PBS (blocking solution and washing solution used after primary antibody incubation).
7. 5% dry milk and 2% normal goat serum (Sigma) in PBS (antibody solution).
8. Wash A/B: PBS, 300/400 mM NaCl, 0.2% NP-40, 0.2% Tween 20 (washing solutions used after secondary antibody incubation).
9. Secondary antibodies coupled to fluorophores (for example, Alexa-fluor 488 or 546; Invitrogen).
10. DAPI solution (0.2 μg/ml; Invitrogen) for counterstaining DNA. DAPI solution is stored in the dark at 4°C. It can be used for up to 2 months.
11. RNA polymerase II antibodies (Covance): H5 monoclonal mouse (Ser 2P specific), H14 mouse monoclonal (Ser 5P specific).

3. Methods

3.1. Main Immunostaining Protocol

Dissection of salivary glands

1. Drosophila larvae are cultured under standard conditions at 26°C (*Drosophila* Protocols). Third instar wandering larvae are

selected for dissection. Larvae should not be grown under overcrowded conditions to allow the full development of large glands and polytene chromosomes. This can be ensured by transferring the flies to fresh vials every 1–2 days.

2. Two or more third instar larvae are washed in PBS in a glass dish. They are then transferred into a drop of PBS on a coverslip. Dissection takes place under a binocular. The tip of the mouth hook is grasped with one pair of forceps. A second pair of forceps is used to keep the larva in place. The larva is then pulled apart to dissect out the salivary glands (for a detailed description, see *Drosophila* Protocols). Using the forceps, the mouth hook, brain, and imaginal discs are separated from the glands. In addition, as much fat tissue as possible should be removed as it can interfere with antibody staining.
3. A drop of fixation solution is placed onto a coverslip. The salivary glands are transferred into this solution. Glands are fixed for 5 min at room temperature (see Note 2).

Polytene squashing

4. A microscope slide is put on top of the coverslip with the fixed glands in fixation solution. The coverslip is tapped with the eraser side of a pencil to squash the chromosomes. While tapping, the coverslip is gently moved back and forth. Excess fixation solution is removed with a thin Whatman paper. The squashed chromosomes are quickly checked under the light microscope (phase contrast) at 40× magnification.
5. Slides with sufficiently spread out chromosomes are placed with the coverslip side down on a thick Whatman paper. A second piece of Whatman paper is placed on top. Chromosomes are flattened by applying strong pressure with a thumb onto coverslip. Horizontal movements of the coverslip must be avoided as these break the chromosomes.
6. Each glass slide is immediately transferred into liquid nitrogen for 30 s. The glass slide is taken out of liquid nitrogen, the coverslip is removed with a scalpel, and the glass slide is placed into a Coplin jar containing PBS.
7. Slides can be processed immediately or kept in PBS at 4°C overnight.

Immunostaining

8. Glass slides are washed with PBS in the Coplin jar for 10 min while rotating. PBS is replaced by blocking solution (5% milk in PBS) and gentle rotation is continued for 30 min. In the meantime, primary antibody dilutions are prepared (see Note 3). We routinely use appropriate dilutions of antibodies in 5% milk in PBS supplemented with 2% normal goat serum (NGS) to reduce nonspecific protein binding.

9. After blocking, slides are rinsed with PBS and excess liquid is carefully removed with a tissue. Slides are put in a humidity chamber. 40 μl of primary antibody solution is transferred onto the glass slide to cover the chromosomes completely. Chromosomes are then covered with a coverslip (22 × 40 mm; trapping of air bubbles must be avoided). The humidity chamber is covered with aluminum foil and can be kept up to 2 days at 4°C.
10. Coverslips are removed and glass slides are rinsed in PBS, and then washed for 5 min three times in PBS containing 5% milk in a Coplin jar while rotating.
11. Glass slides are rinsed in PBS and excess liquid is carefully removed with a tissue. Slides are put in a humidity chamber. 40 μl of appropriate secondary antibody coupled to a fluorophor (we use Alexa 488 or Alexa 546) is diluted 1:200 in PBS containing 5% milk and 2% NGS. The secondary antibody solution is then transferred onto the glass slide to cover the chromosomes completely. Chromosomes are then covered with a coverslip (22 × 40 mm; trapping of air bubbles must be avoided). The humidity chamber is covered with aluminum foil (to avoid fluorophore bleaching) and incubated for 1 h in the dark.
12. Coverslips are removed and glass slides are rinsed in PBS, and then washed with buffers A and B for 10 min in the dark.
13. Glass slides are rinsed in PBS and excess liquid is carefully removed with a tissue. Slides are put in a humidity chamber. 40 μl of DAPI solution is transferred onto the glass slide to cover the chromosomes completely. Chromosomes are then covered with a coverslip (22 × 40 mm; trapping of air bubbles must be avoided). The humidity chamber is covered with aluminum foil (to avoid fluorophore bleaching) and incubated for 4–5 min at room temperature. Slides are rinsed once in PBS and then washed in PBS for 10 min.

Mounting and microscopic analysis

14. Glass slides are taken out of PBS and carefully dried by placing the edge of the slide on a tissue. Fluoromount is applied to the coverslip (22 × 40 mm) and the coverslip is carefully placed onto the glass slide avoiding trapping of air bubbles. Slides are sealed on all four edges with a nail polish. Slides are then left in the dark for a few minutes until the nail polish has dried. Slides can then be analyzed with a fluorescence microscope (see Fig. 1a, c) or can be stored at 4°C in the dark for weeks.

3.2. Analysis of Factor Association with Active Heat-Shock Genes

Factors involved in *Drosophila* heat-shock gene activation can be readily investigated by immunostaining polytene chromosomes from larvae that have been subjected to a temperature shift (10).

1. Third instar larvae are washed in PBS and transferred into microcentrifuge tubes (5–6 larvae). The lids of the tubes are pierced with a needle to ensure oxygen supply during heat shock.
2. The tubes are then placed for 30 min in a water bath at 37°C. Note: Heat shock can be induced in a shorter time, such as 2.5 min (11).
3. Larvae are taken out and immediately dissected in a drop of PBS and processed further as described in the main protocol (steps 2–14 of Subheading 3.1). Figure 1 shows an example of the redistribution of RNA polymerase II and the elongation factor Spt5 to active heat-shock genes.

3.3. Analysis of the Role of RNA Polymerase II Phosphorylation for Factor Binding to Active Genes

The C-terminal domain (CTD) of RNA polymerase II becomes phosphorylated as it progresses from initiation to elongation (12). The CTD is composed of a heptamer sequence, which contains serine residues at the second and fifth position (Ser 2 and Ser 5, respectively). Initiating RNA polymerase II becomes phosphorylated at Ser 5. As RNA polymerase II changes into elongation mode, Ser 2 becomes phosphorylated as well. The phosphorylated serines are required for recruitment of chromatin-modifying and RNA-processing complexes during transcription. Ser 5 phosphorylation has been implicated in binding the Set1 histone methyltransferase, capping enzyme and splicing factors. Ser 2 phosphorylation has been linked to recruitment of Set2 histone methyltransferase and polyadenylation factors. By treating salivary glands with flavopiridol, a specific inhibitor of the P-TEFb kinase, which is responsible for Ser 2 phosphorylation, Ser 2 phosphorylation levels of RNA polymerase II can be greatly decreased, whereas Ser 5 phosphorylation remains unaffected (13). Using this treatment in conjunction with polytene immunostaining makes it possible to assess how important Ser 2 phosphorylation is for the association of factors with actively transcribed genes.

1. Third instar larvae are washed in PBS and salivary glands are dissected in a drop of 50% Schneider medium (diluted with deionized water without serum and antibiotics) as described in the main protocol (step 2 of Subheading 3.1).
2. Glands are incubated in a drop of 500 nM flavopiridol solution for 25 min. Flavopiridol is diluted in 50% medium and prepared always fresh. Do not keep on ice as it precipitates.
3. Transfer to fixation solution, fix, squash, and immunostain as described above (steps 3–14 of Subheading 3.1). As a control, use antibodies against RNA Pol II Ser 2 and Ser 5 (Covance).

   Figure 2 shows polytene chromosomes that have been treated with flavopiridol. Chromosomes were immunostained with phosphorylation-specific RNA polymerase II antibodies and a control antibody recognizing the DNA-binding protein CTCF.

3.4. Analysis of the Role of RNA in Factor Chromatin Association

In this protocol, salivary glands are permeabilized and incubated with RNase A prior to immunostaining (14). Several versions of this general strategy have been reported (6–9). For salivary gland dissection, follow the steps 1–2 of Subheading 3.1.

1. Place 25 ml of permeabilization solution on a coverslip. Transfer two pairs of salivary glands into it. Incubate for 2 min at room temperature.
2. Transfer glands into RNase A solution (PBS with 0.5 mg/ml RNAse A; the optimal concentration of RNase A should be determined in pilot experiments). Incubate glands for 8–10 min (see Notes 4 and 5). In parallel control experiments, incubate glands only in PBS.
3. Continue with the step 3 of the Protocol 3.1.

   Figure 3 shows RNase-treated and control polytene chromosomes immunostained with antibodies against the RNA-binding protein PEP and the DNA-binding protein CTCF.

4. Notes

1. The final concentration of formaldehyde is 0.5%, which is less than that used in most protocols (between 1.85 up and 5%). However, for most of the antibodies tested in our lab, these conditions work well.
2. Fixation conditions should be optimized for each protein of interest, for example, by varying formaldehyde concentration and incubation time.
3. When using an antibody for the first time, different concentrations have to be tested. If it is known at which concentration an antibody works in Western blots, then the concentration used for immunostaining should be 100-fold higher.
4. Prolonged incubation in RNAse A solution can result in overdigestion of chromosomes, which makes them resistant to squashing.
5. As a negative control, it is important to immunostain a protein, which is known to bind chromatin in an RNA-independent manner. We have used the zinc finger DNA-binding protein CTCF or histone H3 for this purpose. However, any DNA-binding protein can be used as a control. If RNase treatment affects the immunostaining pattern observed with antibodies directed against these proteins, this indicates general, unspecific effects on chromatin structure due to overdigestion of chromosomes with RNase.

Acknowledgments

We are grateful to R. Renkawitz and H. Saumweber and Fred Winston for the gift of antibodies and to R. Renkawitz-Pohl for access to microscopes. M.M. was supported by a fellowship from the International Research Training Group 1384. Work in A.B.'s lab is supported by the Deutsche Forschungsgemeinschaft.

References

1. Bridges, C. B. (1935) Salivary chromosome maps with a key to the banding of the chromosomes of *Drosophila melanogaster. J Heredity* **26**, 60–64.
2. Petesch, S. J., Lis J. T. (2008) Rapid, transcription-independent loss of nucleosomes over a large chromatin domain at Hsp70 loci. *Cell* **134**, 74–84.
3. Core, L. J, Lis, J. T. (2008) Transcription regulation through promoter- proximal pausing of RNA polymerase II. *Science* **319**, 1791–1792.
4. Paro, R. (2000) Mapping Protein Distributions on Polytene Chromosomes by Immunostaining. In: Sullivan, W., Ashburner, M., Hawley, R.S. (eds), Drosophila Protocols, Cold Spring Harbor Laboratory Press, 131–140.
5. Schwartz, B. E., Werner, J. K., Lis, J. T. (2004) Indirect immuno fluorescent labeling of *Drosophila* polytene chromosomes: visualizing protein interactions with chromatin *in vivo. Methods Enzymol.* **376**, 393–404.
6. Piacentini, L., Fanti, L., Berloco, M., Perrini, B., Pimpinelli S. (2003) Heterochromatin protein 1 (HP1) is associated with induced gene expression in *Drosophila* euchromatin. *J Cell Biol* **161**, 707–714.
7. Kiesler, E., Hase, M. E., Brodin, D., Visa N. (2005) Hrp59, an hnRNP protein in *Chironomus* and *Drosophila,* binds to exonic splicing enhancers and is required for expression of a subset of mRNAs. *J Cell Biol* **168**, 1013–1025.
8. Tyagi, A., Ryme, J., Brodin, D., Ostlund, Farrants, A. K., Visa, N. (2009) SWI/SNF Associates with Nascent Pre-mRNPs and Regulates Alternative Pre-mRNA. *PLoS Genet* **5**, e1000470.
9. Akhtar, A., Zink, D., Becker, P.B. (2000) Chromodomains are protein- RNA interaction modules. *Nature,* **407**, 405–409.
10. Johansen, K. M., Cai, W., Deng, H., Bao, X., Zhang, W., Girton, J., Johansen J. (2009) Polytene chromosome squash methods for studying transcription and epigenetic chromatin modification in Drosophila using antibodies. *Methods* **48**, 387–397.
11. Boehm, A. K., Saunders, A., Werner, J., Lis, J. T. (2003) Transcription factor and polymerase recruitment, modification, and movement on dhsp70 in vivo in the minutes following heat shock. *Mol Cell Biol* **23**, 7628–7637.
12. Buratowski, S. (2009) Progression through the RNA polymerase II CTD cycle. *Mol Cell* **36**, 541–546.
13. Ni, Z., Schwartz, B. E, Werner, J., Suarez, J. R., Lis, J. T. (2004) Coordination of transcription, RNA processing, and surveillance by P-TEFb kinase on heat shock genes. *Mol Cell* **13**, 55–65.
14. Richter, L., Bone, J. R., Kuroda, M. I. (1996) RNA-dependent association of the *Drosophila* maleless protein with the male X chromosome. *Genes Cells* **1**, 325–336.

Chapter 19

Detection of Transcriptional Activators, Co-activators, and Chromatin Remodeling by Chromatin Immunoprecipitation Coupled with Real-Time PCR

Tamara Y. Erkina and Alexandre M. Erkine

Abstract

Investigation of DNA–protein interactions is a key approach in understanding mechanisms of gene regulation. The method described allows detection of dynamic DNA–protein interactions occurring at gene promoters in living cells during the time scale of seconds and minutes. The combination of chromatin immunoprecipitation with real-time PCR allows for detection of changes in activator and co-activator content of any promoter during transcriptional activation. The described method is most applicable to investigation of processes resulting in nucleosome loss at gene promoters during the induction of transcription. The approach is also applicable to any dynamic process involving DNA–protein interactions.

Key words: ChIP, Protein–DNA interactions, Transcription, Chromatin, Histones, HSF, Heat shock proteins, RNA polymerase II, Q-PCR

1. Introduction

Genomic processes are often characterized by dynamic changes in DNA–protein contacts. Investigation of these interactions in vivo is key in understanding the regulation of genomic processes. A number of existing methods targeting investigation of DNA–protein interaction such as different variants of footprinting, indirect end-labeling in nucleosome mapping, electrophoretic mobility shift assay (EMSA), and others are labor intensive, require implementation of radioactive isotopes, and seldom address these interactions in vivo. We improved the currently existing methods of chromatin immunoprecipitation (ChIP) coupled with real-time quantitative PCR by (1) eliminating the need for radioactive isotope utilization,

Ales Vancura (ed.), *Transcriptional Regulation: Methods and Protocols*, Methods in Molecular Biology, vol. 809,
DOI 10.1007/978-1-61779-376-9_19, © Springer Science+Business Media, LLC 2012

(2) lowering the signal-to-noise ratio by utilization of magnetic beads conjugated with protein A, protein G, or protein A/G, and (3) shortening the experimental procedure so that final graph representation of data can be obtained in 2–3 days.

The protocol can be separated into the following major stages: (1) preparation of the yeast culture and formaldehyde cross-linking in vivo, (2) preparation of cell lysates, (3) immunoprecipitation of the DNA–protein complexes, (4) purification of the immunoprecipitated DNA, (5) quantitative real-time PCR, and (6) data processing and histogram building. We have successfully implemented this protocol for analysis of activator and co-activator binding dynamics, as well as determination of the kinetics of nucleosome loss at promoters of *S. cerevisiae* heat shock genes (1–3). The protocol was also successfully used for mammalian cell cultures.

2. Materials

2.1. Preparation of the Yeast Culture and Formaldehyde Cross-linking In Vivo

1. Yeast extract/peptone/dextrose (YPD) medium for growth of yeast cells: 1% bacto-yeast extract, 2% bacto-peptone, and 2% dextrose. For adenine mutants, add 40 mg/l of adenine for faster growth. Prepare 50 ml aliquots in 250-ml flasks for individual time points, or 300 ml aliquots in 1,000-ml flasks for time course experiments (see Fig. 1).

   Autoclave and store at room temperature.
2. Formaldehyde: 37% solution.
3. NaN_3: 2 M.
4. Glycine: 2.5 M.
5. TBS buffer: 20 mM Tris–HCl, pH 7.5, 200 mM NaCl.
6. Lysis buffer I: 50 mM HEPES–KOH, pH 7.5, 150 mM NaCl, 1 mM EDTA, 1% Triton X-100, 0.1% sodium deoxycholate. Use during cell harvesting and during preparation of the cell lysate with the protein inhibitor cocktail at a dilution of 1/100; and during washes of magnetic beads at a 1/1,000 dilution. Prepare 1 l; store at 4°C.
7. A protease inhibitor set (protease inhibitor cocktail set IV, Calbiochem) designed for yeast cell lysates.

2.2. Preparation of Cell Lysate

1. Zirconium Grinding Beads (0.5 mm Zirconia/Silica Beads, BioSpec Products, Inc.)
2. Ceramic Spheres, 1/4 in. (Q-Biogene, Cat.#6540-422).
3. Elution buffer: 25 mM Tris–HCl, pH 7.5, 5 mM EDTA, 0.5% SDS.

	1	2	3	4	5	6	7	8	9	10	11	12
a												
b		input HSP12	nhs (0') HSP12	0.5' HSP12	1' HSP12	2' HSP12	4' HSP12	8' HSP12	16' HSP12	32' HSP12	64' HSP12	No DNA HSP12
c		input HSP12	nhs (0') HSP12	0.5' HSP12	1' HSP12	2' HSP12	4' HSP12	8' HSP12	16' HSP12	32' HSP12	64' HSP12	No DNA HSP12
d		input HSP12	nhs (0') HSP12	0.5' HSP12	1' HSP12	2' HSP12	4' HSP12	8' HSP12	16' HSP12	32' HSP12	64' HSP12	No DNA HSP12
e		input PHO5	nhs (0') PHO5	0.5' PHO5	1' PHO5	2' PHO5	4' PHO5	8' PHO5	16' PHO5	32' PHO5	64' PHO5	No DNA PHO5
f		input PHO5	nhs (0') PHO5	0.5' PHO5	1' PHO5	2' PHO5	4' PHO5	8' PHO5	16' PHO5	32' PHO5	64' PHO5	No DNA PHO5
g		input PHO5	nhs (0') PHO5	0.5' PHO5	1' PHO5	2' PHO5	4' PHO5	8' PHO5	16' PHO5	32' PHO5	64' PHO5	No DNA PHO5
h												

Fig. 1. A setup example for a 96-well plate addressing measurement of an antigen abundance at the *HSP12* locus relative to the *PHO5* locus during the heat shock time course experiment. Rows B, C, D, and E, F, G are triplicate repeats. Column 2 is an input control. Column 12 is a no-DNA (primer–dimer formation) control. Columns 3 is a nonheat-shock pretreatment sample, and columns 4–11 are samples taken after specific time periods of heat shock. Column 12 should show no amplification values. Differences in values between wells 2B, 2C, 2D and wells 2E, 2F, 2G will be the difference in potential for primers of amplifying *HSP12* and *PHO5*. This difference should be used for normalization of the values of columns B, C, D against columns E, F, G. Differences between the values of wells 3 and 11 for *PHO5* should not be drastic, and will show fluctuations in the amount of DNA between samples. These should be used for normalization of signals from the *HSP12* locus (similar to the loading control in northern or western blotting techniques). Normalized differences between wells 3 and 11 for *HSP12* will show the change in abundance of an antigen at the *HSP12* locus during the time course of heat shock (see Fig. 2).

4. Pronase: 20 μg/μl; make several 1 ml aliquots.
5. DNA purification kit (MinElute purification kit, Qiagen).

2.3. Immunoprecipitation of the DNA–Protein Complexes

1. Protein A, protein G, and protein A/G magnetic beads (Dynabeads protein A, Invitrogen).
2. A primary antibody against the protein of interest.
3. Lysis buffer II: 50 mM HEPES–KOH, pH 7.5, 500 mM NaCl, 1 mM EDTA, 1% Triton X-100, 0.1% sodium deoxycholate. Use during washes of magnetic beads with protein inhibitor cocktail at a 1/1,000 dilution. Prepare 500 ml; store at 4°C.
4. Lysis buffer III: 10 mM Tris–HCl, pH 7.5, 250 mM LiCl, 1 mM EDTA, 0.5% Triton X-100, 0.5% sodium deoxycholate.
5. TE buffer: 10 mM Tris–HCl, pH 7.5; 0.5 mM EDTA.
6. Elution buffer same as in section 2.2.

2.4. Quantitative Real-Time PCR

1. Master mix for quantitative real-time PCR (Absolute SYBR Green Rox Mix, Thermo Fisher Scientific).

3. Methods

3.1. Preparation of the Yeast Culture and Formaldehyde Cross-linking In Vivo

The protocol below is adopted for heat shock kinetic experiments studying dynamic DNA–protein interactions at yeast heat shock gene promoters (see Fig. 2). It can be easily modified for other experimental protocols.

1. Plate the strain of interest on the corresponding medium (YPD, or drop-out selective medium). If a knockout strain is used, it is a good idea to use the parental strain in parallel as a control.
2. Inoculate 5 ml of appropriate medium with a single colony and grow to saturation.
3. Inoculate the appropriate volume (50 ml for single time point, or 300 ml for a time course experiment). Grow the yeast cells in rich medium (YPD) to <0.7 OD_{600}; in synthetic media to <0.5 OD_{600} (see Note 1).
4. Prepare 50-ml tubes for nonheat shock samples, and the appropriate amount of tubes corresponding to the number of time points of heat shock. In each tube dispense 1.08 ml of 37% formaldehyde (final concentration 1%) and 0.4 ml of 2 M NaN_3 (final concentration 20 mM) for a final volume of 40 ml of cell culture per tube. Place the tubes for heat shock samples into a 39°C incubator, and the nonheat shock sample into a 30°C incubator.
5. Dispense 20 ml of YPD or corresponding drop-out medium and 20 ml of the cell culture into a tube containing formaldehyde and NaN_3. Start a timer to record the cross-linking time of the control nonheat shock sample. To the remaining cell culture, add an equal volume of hot YPD medium (prewarmed to 50°C; the temperature of the cell culture has to reach 39°C instantaneously) and start the timer. At appropriate heat shock time points, dispense 40 ml of heat-shocked cell culture into the corresponding, previously prepared 50-ml tubes with formaldehyde and NaN_3.
6. After 10 min of cross-linking at the appropriate temperature, add 4.4 ml of 2.5 M glycine (to a final concentration of 250 mM). Continue incubation for another 5 min, then place on ice. Glycine stops the cross-linking reaction by providing an excess of RNH_2 groups.
7. After all samples are collected, pellet the cells by centrifugation at 5,000 × *g* at 4°C for 5 min. Discard the supernatant fractions.
8. Wash the cell pellets twice with 50 ml of TBS (resuspend the cells in TBS, pellet by centrifugation at 5,000 × *g* at 4°C for 5 min, discard the supernatant fraction, repeat the wash).

9. Resuspend the cells in 1 ml of lysis buffer I containing protease inhibitors at a dilution of 1/100. You can stop the procedure here by −80°C freeze down or proceed to the preparation of cell lysates.

3.2. Preparation of Cell Lysates

1. Prepare 2-ml orange-cap tubes (FastPrep tubes, MP Biomedicals) containing ~700 μl of Zirconium grinding Beads (0.5 mm Zirconia/Silica Beads, BioSpec Products, Inc.) and one 1/4 in. Ceramic Sphere (Q-Biogene Cat#6540-422). Utilization of the ceramic sphere significantly increases cell disruption efficiency. The total weight of each tube with the cap, beads, and the sphere should be 3.5 g. Into each tube add a cell suspension in lysis buffer I containing protein inhibitors (at a 1/100 dilution). Fill the tubes with lysis buffer I so that no air is left in the tubes (foam can lead to incomplete disruption of cells and/or protein denaturation), and break the cells by vortexing in a FastPrep shaker at 4°C (speed 6 for 40 s, three times, with 10 min rest at 4°C between cycles). Examine the suspension under microscope to be sure that no more than 10% unbroken cells remain. The other 90% of the sample cells should look like empty shells clamped to each other (see Note 2).
2. Punch the top and bottom of the tubes containing the disrupted cells with a hot needle, drop the 2-ml tubes bottom up into 15-ml tubes, and pulse spin. Remove the 2-ml tubes with the zirconium beads remaining inside of the large tubes using forceps and discard them. If desired, the zirconium beads and ceramic spheres can be washed and reused. If the 2-ml tubes still contain cell lysate, repeat the procedure in other 15-ml tubes and combine the lysates.
3. Spin the 15-ml tubes containing cell lysate at 10,000 × *g* for 20 min at 4°C. Discard the supernatant fraction containing cytoplasmic proteins. These proteins are not cross-linked to the DNA and, if not removed, will later decrease the efficiency of immunoprecipitation.
4. Add 1 ml of lysis buffer I containing protein inhibitors (protease inhibitor cocktail set IV, Calbiochem) at a dilution of 1/100.
5. Sonicate the suspension (Branson Sonifier 450; titan microtip probe, output 40%, constant duty cycle-100%) for 20 s, repeating six times, with at least 5 min intervals on ice between each 20-s pulse (see Note 3).
6. To remove cell debris, centrifuge the suspension at 10,000 × *g* for 20 min at 4°C. The cross-linked DNA–protein complexes are now in the supernatant fraction. Transfer the supernatant fraction to fresh 2-ml microcentrifuge tubes on ice. The supernatant solutions can be frozen at −80°C or processed immediately for immunoprecipitation.

7. Determine the size of DNA fragments. The sonication step should shear chromatin DNA to an average size of 0.5 kb. The size of the DNA fragments should be checked by agarose gel electrophoresis. To prepare DNA samples, take 100 μl of cell lysate, add 80 μl of elution buffer, then add 20 μl of pronase (20 mg/mL). Incubate at 42°C for 2 h, and then at 65°C for at least 6 h or overnight to reverse the protein–DNA cross-linking. Clean the DNA using a Qiagen kit (MinElute cleanup kit). (Follow the protocol included with the kit.) The final sample volume is 30 μl; use 5 μl for the agarose gel electrophoresis analysis. The DNA should look like a smear, with the average fragment size of 0.5 kb. If the fragment size is out of range (larger fragments), sonication of samples should be repeated to achieve the desired fragment size (two cycles, output 40%, constant duty cycle-100%) for 20 s, see step 5 above.

3.3. Immunoprecipitation of the DNA–Protein Complexes

1. To the 100 μl of lysate (step 6 above), which is equivalent to 2.5×10^7 cells of original cell culture, add the primary antibody against the protein of interest (usually 1–5 μl). Incubate on a rotating wheel for at least 2 h at RT or overnight at 4°C (see Note 4).
2. Select the appropriate magnetic beads. Use Protein A-magnetic beads (Dynabeads protein A, Invitrogen), or alternatively use Protein G or protein A/G magnetic beads. The choice will be determined by the species of primary antibodies being used for immunoprecipitation, as they can have different affinities to protein A or G. Prepare the magnetic beads by washing according to the manufacturer's manual. Use 20 μl of a 50% bead suspension for each 100 μl of cell lysate (see Note 5).
3. Add 20 μl of protein A beads (50% suspension in lysis buffer I) to the mixture and incubate on a rotating wheel for 2–3 h at RT.
4. The DNA–protein complexes should now be on the beads. Pellet the beads in the magnetic rack (Dynal MPC-S, Thermo Fisher Scientific) and discard the supernatant fraction (see Note 6).
5. Add 1 ml of lysis buffer I containing protein inhibitor cocktail at a 1/1,000 dilution to the beads, and wash beads on a rotating wheel for 10 min at RT.
6. Pellet the beads on the magnet rack and discard the supernatant fraction again. Add lysis buffer I and repeat steps 5 and 6.
7. Wash the beads twice with 1 ml of lysis buffer II containing protein inhibitor cocktail at a 1/1,000 dilution as in steps 5 and 6. Pellet the beads between each wash and carefully remove the supernatant fractions.

8. Wash the beads twice with 1 ml of ice-cold buffer III containing protein inhibitor cocktail at a 1/1,000 dilution as in steps 5 and 6. Pellet the beads between each wash and carefully remove the supernatant fractions.
9. Wash the beads with 1 ml of TE buffer twice. Pellet the beads between each wash, and carefully remove the supernatant fractions.

3.4. Purification of the Immunoprecipitated DNA

1. Elute the DNA–protein complexes from the magnetic beads by adding 80 μl of elution buffer (0.5% SDS, 25 mM Tris–HCl, pH 7.5, 5 mM EDTA) to the samples and incubating for 10 min.
2. Pellet the beads on the magnetic rack. The DNA–protein complexes are now eluted from beads. Transfer the supernatant fraction to new tubes (the beads can be discarded). The resulting solutions are the IP samples.
3. Prepare input samples. Take 20 μl of the cell lysate (Subheading 3.2, step 6) and add 60 μl of elution buffer. These are now the INPUT samples. Follow the steps below in the same manner as was done for the IP samples from step 2.
4. Since we are now interested only in DNA, proteins should be destroyed. Add 20 μl of pronase (20 mg/mL) to the samples, including the input samples. Incubate all the samples at 42°C for 2 h, then incubate the samples at 65°C for at least 6 h or overnight to reverse the protein–DNA cross-linking.
5. Clean the DNA using a Qiagen kit (MinElute cleanup kit). Follow the protocol included with the kit. The final sample volume should be 30 μl. For a better yield of DNA, we recommend combining three consecutive 10 μl elutions from the Qiagen cleanup columns into the same collection tube.
6. Dilute the IP samples to 300 μl and the input samples to 900 μl with highest quality water. Store the samples at −20°C.

3.5. Quantitative Real-Time PCR

Preliminary work has to be performed, to design primers that avoid any internal hairpins or homo- and hetero-duplex formation, using DNASTAR LaserGene 8 software. For each region of amplification, at least four different primers (two forward and two reverse) should be designed and tested. The amplicon size should be 100–150 bp. At least four different primer pair combinations (from four designed primers) are usually tested in real-time PCR reactions as described below using an input DNA sample, and separately without a DNA template. The final primer pair should not exhibit an extra peak in the melting curve protocol of real-time PCR setup, and should not show any quantitative amplification after 35 PCR cycles without the presence of a template.

1. Quantitative real-time PCR reactions are performed using Master mix (Absolute SYBR Green Rox Mix AB-1163/A, Thermo Fisher Scientific) according to the manufacturer's recommendations. If the volume of the wells in the plate is 500 μl, then the reaction volume should be 20 μl; otherwise, if the volume of wells is 200 μl or less, a 10-μl reaction volume should be used. Prepare the master mix considering the number of reactions to be run, plus an extra 10%. Add the primers into the master mix to a concentration of 140 nM so that the final concentration of primers in the reaction will be 70 nM. see Fig. 1 for an example of the plate setup.
2. For each reaction, 10 μl of DNA sample from the previous step must been utilized. Input samples are first diluted tenfold, and then used similar to the IP samples. Primer concentration for each reaction is 70 nM. Each reaction volume should be 20 μl; and each sample reaction should be done in triplicate. Additionally, each sample in triplicate is similarly amplified with normalization primers (*PHO5* or intergenic chromosome V region). These additional amplifications for each sample will serve as internal DNA loading controls and for sample quality assessment.
3. Seal the plate with the optical plate cover. To bring down all the reaction mix components and to eliminate bubbles, centrifuge the plates at 1,000 × *g* at room temperature, checking to be sure that all bubbles are eliminated. Place the plate into the real-time PCR apparatus (e.g., the ABI7300).
4. The PCR program that should be used is: the preliminary step at 94°C for 15 min; then 35 cycles of 94°C for 30 s, 60°C for 30 s, and 72°C for 30 s; and finally the melting curve reading.

3.6. Data Processing and Histogram Building

Ideally, PCR reactions should reach a threshold by 20–25 cycles. If the threshold is not reached by around 30 cycles or more, one of several potential explanations is possible: (1) the antibodies used are not efficient, (2) DNA was lost during the process, or (3) the reaction mix was prepared incorrectly. The best indicator of a healthy situation is when the triplicate reaction curves coincide.

Signals from individual DNA regions must be normalized against signals from either the *PHO5* promoter region or the intergenic region of chromosome V, and then against the INPUT samples (see Fig. 1). Normalization against the *PHO5* promoter region or the intergenic region of chromosome V counterbalances any fluctuations between samples in the amount of DNA (similar to internal loading normalization used during blotting procedures). Normalization against input signal (which is already normalized internally against *PHO5* or the intergenic region of chromosome V) allows actual abundance measurement. The choice of *PHO5* for normalization is stipulated by the notion that this locus is known

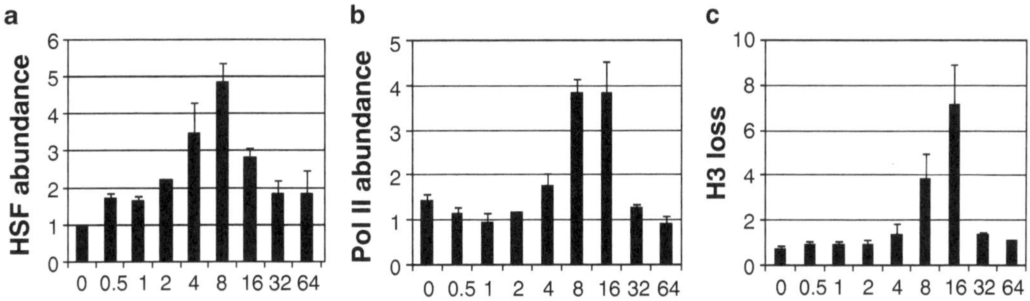

Fig. 2. Histograms representing change in abundance of HSF and RNA polymerase II, as well as loss of histone H3, at the *HSP12* promoter during the time course of heat shock. (**a**) ChIPs using an antibody against HSF performed in the WT strain. *Y*-axis: abundance of HSF. *X*-axis: time after heat shock (0–64 min). All real-time PCR values were normalized relative to the input and to the background values of the intergenic region of chromosome V. Values represent mean ± SD. ($n \geq 3$). (**b**) The same as panel **a**, except the antibody used was against Pol II. (**c**) The same as panel **a** except the antibody used was against the C terminus of histone H3. The results of histone ChIP experiments are traditionally reported as a drop (1–0 or 100–0%) in histone content during gene activation. This form of presentation restricts analysis of the data, since the closer the values are to 0, the more difficult it is to see fine differences. Alternatively, the results can be presented as an inverse value, which represents the degree of histone displacement and changes from 1 to ∞. This presentation format better conveys fine differential chromatin remodeling, and has been utilized by us previously (1). While having the above-mentioned advantages, this form of presentation has a tendency to deemphasize the significant histone losses (50–75% of initial level) seen in most of the mutants bearing inactivations of individual chromatin remodeling activities (1, 2, 5).

to contain stably positioned nucleosomes and they do not change position during heat shock (4). This normalization is especially useful when a chromatin dynamic is under investigation (Fig. 2c).

One of the questions often raised concerns normalization against the signal from the region that does not contain the protein (antigen) of interest by default (e.g., Pol II in the intergenic region of chromosome V). Our observation is that in this case, the background signal is proportional to the amount of total DNA in the sample, and normalization compensates for the fluctuation of sample DNA contents.

The following are examples of primer sequences: (coordinates are relative to the first ATG codon of the open reading frame): *PHO5* (−214 to −192; −20 to −48), *HSP12* (−304 to −279; −82 to −107), *HSP82* (−193 to −167; −37 to −69), *SSA4* (−307 to −279; −70 to −98), and chromosome V intergenic region (5′-GCAATCAACATCTG AAGAAAA GAAAGTAGT-3′, and 5′-CATAATCTGCTGAAAAA TGGCGTAAAT-3′).

4. Notes

1. If the culture is overgrown to more than 1.5 OD_{600}, the procedure should be started over. Less overgrown cultures can be reinoculated in fresh warm media so that at least two cell divisions (3–4 h) occur before reaching the target OD_{600}. If a small number of

cultures is going to be grown (2–4), then several flasks could be inoculated for overnight growth with different amounts of culture (e.g., 10, 30, 100 μl for fast growing WT or 100, 300, 1,000 μl for slow growing mutants).

2. It is highly recommended NOT to lengthen the cell disruption steps and NOT to overheat the samples because at this point in the procedure the proteins are prone to degradation.
3. Foaming of the solution during sonication results in unequal shearing of DNA samples. To prevent foaming, keep the end of a sonicator tip near the bottom of the sample tubes. Do not push the bottom of the tube with the sonicator tip, which could easily lead to puncturing a hole in the tube and loss of the sample. Placing the tip end near the surface induces foaming. If foaming occurs – do not continue. Instead, spin down sample, remove the foam, and then continue the sonication.
4. Appropriate antibody concentrations should be predetermined by performing preliminary experiments to maximize the level of specific co-immunoprecipitated DNA, and to minimize the precipitation of nonspecifically bound DNA. As a rough approximation, the dilution of the primary antibody at this step is 10–100-fold lower than the dilution of this antibody when used for Western blotting detection of the antigen. It is a common mistake to overload cell lysate samples with antibody, which often leads to a loss of immunoprecipitation specificity. Best results are usually obtained with a polyclonal antibody as they are typically less prone to a loss of specificity due to epitope masking caused by formaldehyde cross-linking.
5. For multiple samples, begin with the total necessary amount of beads, wash them, then dilute them with lysis buffer I. Dispense them into the individual IP tubes, constantly suspending beads, so that equal amount of beads is aliquoted in each tube.
6. Following the washing steps as outlined is critical for the entire experiment; follow them carefully and thoroughly. Each subsequent step (Subheading 3.3, steps 5–9) removes nonspecifically bound DNA–protein complexes that, without washing, will give an overwhelming background signal during the downstream quantitative PCR steps. These rigorous washing steps diminish the overall signal at the end of the process, but simultaneously make it more specific.

Acknowledgment

This work was supported by an NSF grant (MCB-1029254).

References

1. Erkina, T. Y., and Erkine, A. M. (2006) Displacement of histones at promoters of Saccharomyces cerevisiae heat shock genes is differentially associated with histone H3 acetylation. *Mol Cell Biol* **26**, 7587–600.
2. Erkina, T. Y., Tschetter, P. A., and Erkine, A. M. (2008) Different requirements of the SWI/SNF complex for robust nucleosome displacement at promoters of heat shock factor and Msn2- and Msn4-regulated heat shock genes. *Mol Cell Biol* **28**, 1207–17.
3. Erkina, T. Y., Zou, Y., Freeling, S., Vorobyev, V. I., and Erkine, A. M. (2010) Functional interplay between chromatin remodeling complexes RSC, SWI/SNF and ISWI in regulation of yeast heat shock genes. *Nucleic Acids Res* **38**, 1441–9.
4. Erkine, A. M., and Gross, D. S. (2003) Dynamic chromatin alterations triggered by natural and synthetic activation domains. *J Biol Chem* **278**, 7755–64.
5. Barbaric, S., Luckenbach, T., Schmid, A., Blaschke, D., Horz, W., and Korber, P. (2007) Redundancy of chromatin remodeling pathways for the induction of the yeast PHO5 promoter in vivo. *J Biol Chem* **282**, 27610–21.

Chapter 20

Chromatin Endogenous Cleavage and Psoralen Crosslinking Assays to Analyze rRNA Gene Chromatin In Vivo

Joachim Griesenbeck, Manuel Wittner, Romain Charton, and Antonio Conconi

Abstract

In eukaryotes, multiple copies of ribosomal RNA (rRNA) genes co-exist in two different chromatin states: actively transcribed (nucleosome depleted) chromatin, and nontranscribed (nucleosomal) chromatin. The presence of two rRNA gene populations compromises the interpretation of analyses obtained by the standard biochemical methods that are used to study chromatin structure (e.g., nuclease digestion and chromatin immunoprecipitation). Here, we provide a protocol to investigate the specific association of proteins with the two rRNA gene chromatin populations in vivo, using *Saccharomyces cerevisiae* as a model eukaryote.

Key words: Chromatin, Chromatin endogenous cleavage, Psoralen crosslinking, rDNA, Ribosomal DNA, Ribosomal RNA genes, RNA polymerase I

1. Introduction

Ribosomal genes (rRNA genes, or rDNA) located in the nucleolus are transcribed at very high rate by RNA polymerase I contributing to up to 60% of total cellular transcription in exponentially growing cells (1). The rRNA genes are present in multiple copies, often more than 100, clustered in arrays of tandemly arranged transcription units. Within these clusters, rRNA genes exist in two different chromatin states throughout the cell cycle (2). However, only a portion of the genes is actively transcribed and depleted of nucleosomes, the rest is nontranscribed and assembled in nucleosomes (2–4). Thus, the interpretation of results obtained from analyses of rRNA gene chromatin by digestion with nucleases

Ales Vancura (ed.), *Transcriptional Regulation: Methods and Protocols*, Methods in Molecular Biology, vol. 809, DOI 10.1007/978-1-61779-376-9_20, © Springer Science+Business Media, LLC 2012

followed by standard Southern blot assay, or by chromatin immunoprecipitation (ChIP), is difficult because these techniques cannot distinguish between the two different populations of rRNA genes (5). As a consequence the outcome represents only an average of a heterogeneous mixture of chromatin states. To overcome this limitation, we developed a method that selectively determines the protein composition of chromatin at actively transcribed or nontranscribed rRNA genes, in the model organism *Saccharomyces cerevisiae* hereafter called yeast. To this end chromatin endogenous cleavage (ChEC), established by the group of Ulrich Laemmli (6), is used in combination with psoralen photocrosslinking developed in the group of José Sogo (7).

The ChEC assay is based on a genetic modification of the chromosomal locus of interest, whereby the corresponding gene product is expressed as a fusion protein with a C-terminal Micrococcal Nuclease (MNase). MNase is an endonuclease that requires calcium as co-factor. Importantly, the MNase fused to the target protein is inactive in yeast cells, because the concentration of calcium is low in both cytoplasm and nucleoplasm. Thus, to determine the composition of rDNA chromatin, cells are first treated with formaldehyde, which crosslinks the MNase-tagged protein to its chromosomal binding sites. Then nuclei are isolated and incubated in a buffer containing calcium, which activates the MNase. The active MNase introduces nicks and DNA double strand cuts in the proximity of its binding site. The cleavage sites can be mapped to any locus and at high resolution by agarose gel electrophoresis and Southern blot (see ref. 6 and Fig. 1), or by primer extension (8).

Psoralen photocrosslinking was used to demonstrate the co-existence of two distinct rDNA chromatin populations (2). Psoralen (Trioxsalen, 4,5′,8-trimethylpsoralen) and its derivatives photoreact preferentially with double-stranded nucleic acids to form interstrand crosslinks. In chromatin, the extent of crosslinking is inhibited by the presence of nucleosomes. It was observed that psoralen photocrosslinking is more efficient in chromatin of transcribed rRNA genes than in chromatin of nontranscribed rRNA genes. Consequently, the two different populations can be separated by native agarose gel electrophoresis because purified psoralen-crosslinked rDNA fragments derived from transcribed rRNA genes migrate slower than fragments derived from nontranscribed genes. These and other experiments led to the conclusion that psoralen accessible rRNA genes are transcribed and devoid of nucleosomes, whereas psoralen inaccessible rRNA genes are transcriptionally inactive and assembled into nucleosomes (5).

The combination of ChEC with psoralen photocrosslinking (ChEC/psoralen assay) allows to selectively analyze the protein composition of transcribed and nontranscribed rRNA genes (3, 9). In these experiments, psoralen photocrosslinking is performed with nuclei after ChEC. As a result, either the psoralen accessible

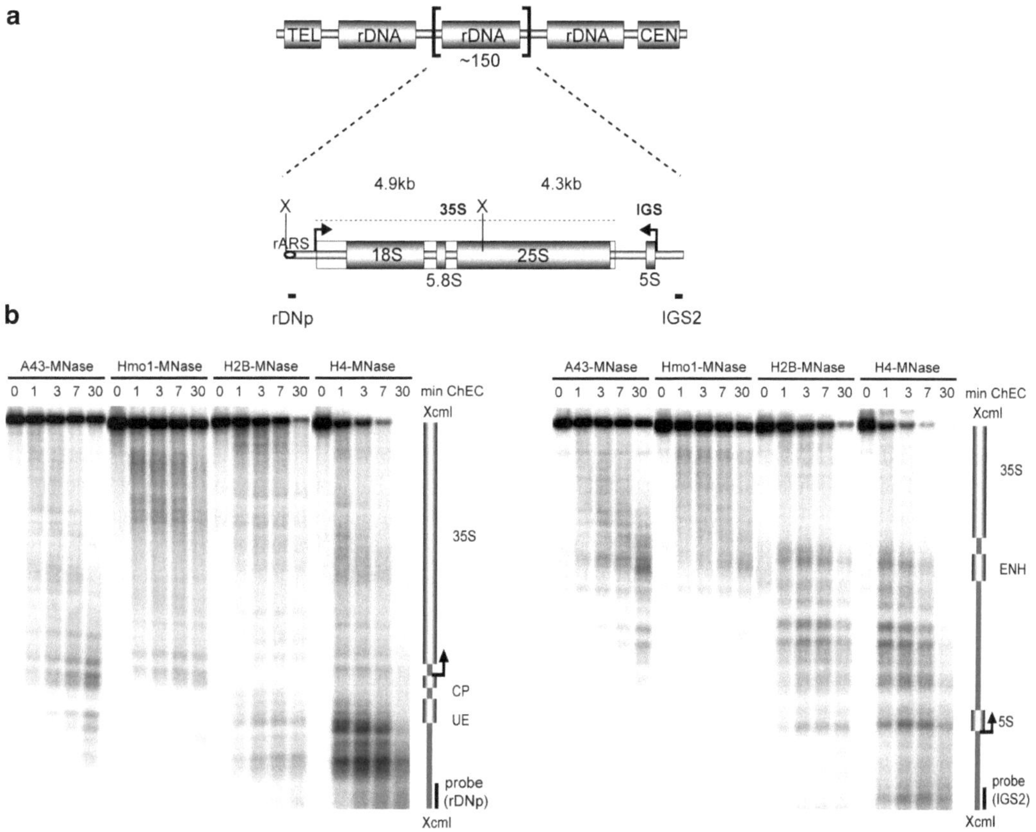

Fig. 1. ChEC analysis of nuclei derived from strains expressing A43-, Hmo1-, H2B-, and H4-MNase fusion proteins. (**a**) Schematic representation of the yeast rDNA locus. *CEN* centromere, *TEL* telomere, *IGS* intergenic spacer region, *rARS* ribosomal autonomous replication sequence; 35S, 25S, 18S, 5.8S, rRNA coding sequences transcribed by RNA polymerase I; 5S, rRNA coding sequence transcribed by RNA polymerase III; X, *XcmI* restriction sites (the size of fragments obtained is indicated on the *top*); rDNp, IGS2, probes used in the Southern blot; *arrows* mark transcription start sites. (**b**) Yeast strains expressing the indicated MNase fusion proteins were subjected to ChEC for the times indicated above each lane. DNA was isolated, digested with *XcmI* and analyzed by Southern blot (indirect end labeling) using probes detecting a 4.9-kb large fragment comprising the promoter and parts of the 35S rRNA coding sequence and a 4.3-kb large fragment visualizing a fragment of the IGS and the 35S rRNA gene coding sequence. A cartoon of the rDNA region visualized on the autoradiograph is indicated on the *right*. Features are as described above. *CP* core promoter, *UE* upstream element, *ENH* enhancer.

or the psoralen inaccessible rDNA is degraded if the MNase fusion protein is a specific component of the transcribed or of the nontranscribed rRNA genes, respectively (Fig. 2). To date, ChEC/psoralen assays have provided evidence for the presence of RNA polymerase I and the HMG-box protein Hmo1 at transcribed rRNA genes. In addition, the same experiments have shown that histone proteins are present at nontranscribed rRNA genes but largely depleted at transcribed rRNA genes (4). In conclusion, the assay presented here can be applied to determine changes in general protein/DNA interactions during transcription, replication, and repair. Virtually any protein can be tagged and its association determined in situ (3, 4).

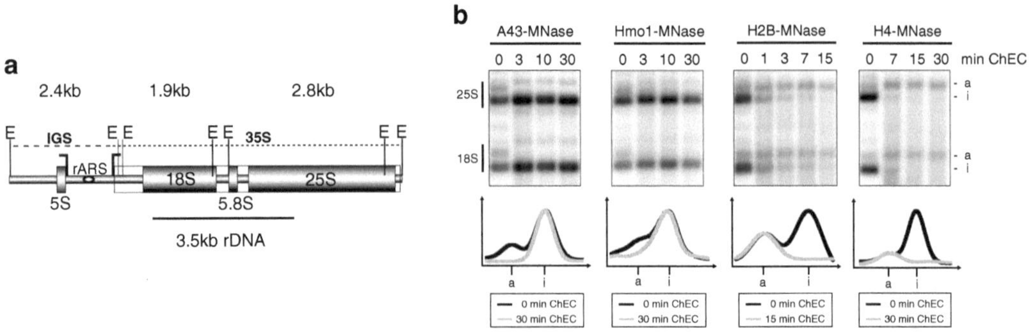

Fig. 2. ChEC/psoralen analysis of nuclei derived from strains expressing A43-, Hmo1-, H2B-, and H4-MNase fusion proteins. (**a**) Schematic representation of the yeast rDNA locus. Features are as described in the legend to Fig. 1a. E, *Eco*RI restriction sites (the size of fragments obtained is indicated on the *top*); 3.5 kb rDNA, probe used in the Southern blot. (**b**) Yeast strains expressing the indicated MNase fusion proteins were subjected to ChEC/psoralen analysis. ChEC was performed for the times indicated above each lane. DNA was isolated, digested with *Eco*RI, and analyzed by Southern blot using probes detecting fragments of the 18S (1.9 kb) and 25S rRNA coding region (2,8 kb). Graphs below depict radioactivity profiles in lanes before or after activation of the MNase for the period indicated below each plot (for details *see* ref. 9). Positions of rDNA fragments derived from psoralen-accessible (a) or psoralen-inaccessible (i) rRNA genes are indicated.

2. Materials

2.1. Yeast Cultures, Formaldehyde Fixation, and Isolation of Nuclei

1. Yeast Peptone Dextrose (YPD), or Yeast Peptone Adenine Dextrose (YPAD) medium (see Notes 1 and 2).
2. 37% Formaldehyde.
3. 2.5 M glycine, stored at room temperature.
4. 1 M stock spermine solution, stored in aliquots at −20°C.
5. 0.5 M stock spermidine tetrahydrochloride solution, stored in aliquots at −20°C.
6. Buffer A: 15 mM Tris–HCl, pH 7.4, 80 mM KCl, 2 mM EDTA, pH8, 0.5 mM spermidine, 0.2 mM spermine. Spermidine and spermine should be supplemented fresh prior to use (buffer without addition of polyamines can be stored at room temperature but should be cooled to 4°C prior to use).
7. Protease inhibitors (100×): Benzamidine (33 mg/ml), PMSF (17 mg/ml), dissolved in ethanol and stored in aliquots at −20°C.
8. Glass beads (diameter 0.75–1.0 mm), stored at −20°C.
9. Vibrax-VXR (IKA), rotary shaker.

2.2. Chromatin Endogenous Cleavage and Psoralen Photocrosslinking

1. Buffer Ag: 15 mM Tris–HCl, pH 7.4, 80 mM KCl, 0.1 mM EGTA, pH 8.0, 0.5 mM spermidine, 0.2 mM spermine. For addition of spermidine and spermine, see subheading 2.1, item 6.
2. Thermomixer® Dry Block Heating Shaker (Eppendorf).

3. 0.1 M $CaCl_2$, stored at room temperature.
4. IRN buffer: 50 mM Tris–HCl, pH 8.0, 20 mM EDTA, pH 8.0, 0.5 M NaCl, stored at room temperature.
5. 4,5′,8-Trimethylpsoralen (psoralen, trioxsalen), dissolved in ethanol (0.2 mg/ml). Solutions are stable for up to 2 months at 4°C in the dark, (store in dark glass vials). Stir solution for approximately 30 min prior to use.
6. 15 W Blacklight blue lamps (315–400 nm) (Sankyo-Denki).
7. RNase A (20 mg/ml), dissolved in 50 mM Tris–HCl, pH 8.0, 10 mM EDTA, stored at room temperature.
8. 10% sodium dodecyl sulfate (SDS), stored at room temperature.
9. Proteinase K (20 mg/ml), dissolved in 50 mM Tris–HCl, pH 8.0, 1 mM $CaCl_2$ and stored in aliquots at −20°C.
10. Roti®-Phenol/Chloroform/Isoamyl alcohol, (Roth) stored at 4°C in the dark.

2.3. Agarose gel Electrophoresis and Southern Blotting

1. Tris–Borate–EDTA (TBE): 89 mM Tris, 89 mM boric acid, 2 mM EDTA.
2. UltraPure™ Agarose (Invitrogen).
3. 10× DNA loading buffer: 0.25% bromophenol blue (w/v), 0.25% xylene cyanol (w/v), 40% glycerol (w/v), stored at room temperature.
4. SYBR® Safe DNA stain (Invitrogen).
5. UV-Transilluminator 20×40 cm (INTAS).
6. Denaturing solution: 0.5 M NaOH, 1.5 M NaCl, stored at room temperature.
7. Blotting buffer: 1 M ammonium acetate, stored at room temperature.
8. Blotting paper MN 827 B (Macherey-Nagel).
9. Membrane Positive™ (3×0.3 m) (MP Biomedicals).

2.4. Probe Preparation, Membrane Hybridization, Washing, and Membrane Stripping

1. RadPrime DNA Labeling System (Invitrogen).
2. ProbeQuant™ G-50 Micro Columns (GE Healthcare).
3. UltraPure™ Salmon Sperm DNA (10 mg/ml) (Invitrogen).
4. Hybridization buffer: 0.5 M sodium phosphate, pH 7.2, 7% SDS, stored at room temperature (preheat to 65°C before use).
5. 20× Saline–Sodium Citrate (SSC): 17.5% sodium chloride, 8.8% sodium citrate dehydrate, stored at room temperature.
6. Rinse buffer: 3×SSC, 0.1% SDS, stored at room temperature.
7. Wash buffer 1: 0.3×SSC, 0.1% SDS, stored at room temperature.
8. Wash buffer 2: 0.1×SSC, 0.1% SDS, stored at room temperature.

9. Wash buffer 3: 0.1×SSC, 1.5% SDS, stored at room temperature.
10. 20× Saline–Sodium Phosphate–EDTA (SSPE): 3 M sodium chloride, 0.2 M sodium phosphate, pH 7.4, 0.02 M EDTA.

3. Methods

Although the method is described for yeast, similar assays can be envisioned to study protein/DNA interactions in higher eukaryotes. However, since the expression level of MNase fusion proteins influences considerably the endogenous cleavage, it is not advised to use ectopic expression systems. Also, it is unclear if variation in the intracellular calcium concentration in higher eukaryotes could lead to a premature activation of the MNase activity. Therefore, Laemmli and co-workers developed a technique, called chromatin immunocleavage (ChIC), which is not based on the co-expression of MNase fusion proteins in the cell (6). ChIC relies on the availability of specific antibodies against the proteins of interest. In ChIC experiments, nuclei from formaldehyde fixed cells are incubated with the specific antibody. Thereafter, the recombinant MNase protein fused to Protein A from *Staphylococcus aureus* is added, which tethers the MNase to the heavy chain of the specifically bound antibodies. To date ChIC has only been applied to yeast, but we believe that it can be applied in combination with psoralen photocrosslinking to higher eukaryotes as well.

3.1. Formaldehyde Fixation and Isolation of Nuclei

1. Yeast cells expressing the MNase fusion protein of interest are grown to 0.5 OD_{600}. Formaldehyde is added to a final concentration of 1% and cells are incubated for 15 min at growth temperature with shaking (see Note 3).
2. Quenching of formaldehyde: glycine is added to 125 mM final concentration for at least 5 min at room temperature.
3. Cells are harvested in 50-ml tubes (4,200 ×*g*, 5 min at 4°C in a microcentrifuge), suspended in 1 ml of water, and transferred into 1.5-ml microtubes. Yeast cells are collected by centrifugation (16,000 ×*g*, 1 min at 4°C in a microcentrifuge; discard the supernatant), and the pellets are frozen in liquid nitrogen and stored at −20°C. Alternatively, cells can be used immediately without freezing.
4. All the following steps are performed on ice or in the cold room at 4°C.
5. Cells are washed in 0.6 ml of buffer A and 1× Protease inhibitors and then collected by centrifugation (16,000 ×*g*, 2 min at 4°C in a microcentrifuge; discard supernatant). The washing step is repeated three times.

6. Finally, cells are suspended in 350 μl of buffer A, 1× Protease inhibitors and ~500 μl of glass beads are added. Note: There must be enough buffer solution to cover the beads by a thin layer. Cell disruption is done for 10 min at maximum speed on the Vibrax shaker at 4°C.
7. To collect the cell lysates, the bottom and cap of microtubes are pierced with a hot needle and placed in a 15-ml tube. After centrifugation (130 ×*g*, 1 min at 4°C in a microcentrifuge) the glass beads remain in the microtubes and are discarded. The crude cell lysates, which are collected in the 15-ml tubes, are transferred into new 1.5-ml microtubes and centrifuged at 16,000 ×*g* for 2 min at 4°C (microcentrifuge).
8. The supernatants are discarded and the pellets (containing crude nuclei) are suspended in 0.6 ml of buffer A and 1× Protease inhibitors. After another centrifugation step (16,000 ×*g* for 2 min at 4°C) supernatants are removed (see Note 4). The nuclei pellets can be frozen in liquid nitrogen and stored at −80°C or used immediately in a ChEC assay.

3.2. Chromatin Endogenous Cleavage

1. Nuclei isolated from formaldehyde-crosslinked cells are suspended in buffer Ag and 1× Protease inhibitors in a total volume of 550–600 μl (see Note 5), and pre-incubated at 30°C, thermomixer set at 750 rpm, for 3 min.
2. After vigorous vortexing, two 100 μl aliquots of the nuclei suspensions are transferred to new tubes and used as controls (0 min ChEC): no calcium is added.
3. For the rest of the nuclei suspensions, MNase-fusion proteins are activated by the addition of $CaCl_2$ (2 μl of 0.1 M stock solution per 100 μl reaction volume; the final concentration is 2 mM) and incubated at 30°C with constant shaking (thermomixer set at 750 rpm).
4. At different time intervals, 100 μl aliquots are taken (see Notes 6 and 7), transferred to 1.5-ml microtubes containing 100 μl of IRN buffer and mixed to stop the MNase activity. Samples mixed with IRN can be kept at room temperature. At the end of the time course, 100 μl of IRN are added to the "0 min ChEC" aliquots (see item 2).
5. For ChEC/psoralen analyses, continue to Subheading 3.3. For ChEC analyses, follow the steps described in Subheading 3.4.

3.3. Chromatin Endogenous Cleavage: Psoralen Photocrosslinking Assay

1. Both the ChEC/psoralen assay and ChEC alone can be performed in parallel. Samples are split into 1/3 (67 μl) for the ChEC/psoralen and 2/3 (133 μl) for the ChEC assay.
2. For the ChEC/psoralen assay, 67 μl of the nuclei suspension are mixed with 133 μl of IRN buffer and transferred to a 24-well plate (see Note 8).

3. 10 μl of 4,5′,8-trimethylpsoralen (psoralen) (0.2 mg/ml) are added and samples are incubated for 5 min on ice and in the dark. Subsequently, the suspensions are irradiated with UVA light (315–400 nm) at a distance of 2–3 cm for 5 min on ice (see Note 9). As a mock control, ethanol is added instead of psoralen.
4. The psoralen crosslinking step is repeated three times (total addition of psoralen: 40 μl; total UV irradiation time: 20 min).
5. The samples are transferred into 1.5-ml microtubes.

3.4. DNA Extraction and Restriction Enzyme Digestion

1. Nuclei are treated with 2 μl of RNase A (20 mg/ml), mixed, and incubated at 37°C for at least 1 h.
2. 10 μl of 10% SDS and 2 μl of Proteinase K (20 mg/ml) are added, mixed, and samples are incubated for 1 h at 56°C.
3. Formaldehyde crosslink is reversed by incubation at 65°C overnight.
4. DNA is extracted with phenol–chloroform–isoamyl alcohol.
5. 1× volume of IRN and 2.5× volume of ethanol are added and DNA is precipitated at −20°C for at least 20 min.
6. DNA is collected by centrifugation at 16,000 × *g* for 20 min at 4°C.
7. DNA pellets are washed with 150 μl of 70% ethanol and centrifugation is repeated as in step 6.
8. DNA is dried for 5–10 min at room temperature and resuspended in 25 μl of TE buffer. 12 μl of each sample are digested with the appropriate restriction enzyme in a final volume of 20 μl as recommended by the manufacturer overnight (see Note 10).

3.5. Agarose gel Electrophoresis and Southern Blotting

1. 10× DNA loading buffer is added to the samples prior to loading on a 1% agarose gel (250 ml, 15 × 20 cm) (see Note 11).
2. DNA prepared from the ChEC assay is separated at 6 V/cm during 6 h in TBE running buffer. DNA prepared from the ChEC/psoralen assay is separated at 6 V/cm for 8 h in the same running buffer.
3. After electrophoresis, the agarose gels used to separate psoralen crosslinked DNA are irradiated 2 min with 254 nm UV light (UV-Transilluminator, INTAS) on both sides (see Note 12).
4. Gels are equilibrated in ~500 ml Denaturing solution for 15 min on a rotary shaker at low rpm. The Denaturing solution is changed and the equilibration step is repeated once.

5. Thereafter, gels are equilibrated in ~500 ml of Blotting buffer, two times 15 min each on a rotary shaker at low rpm with buffer change.
6. DNA is transferred to a positively charged nylon membrane overnight, with the gel upside down (10).
7. The membrane is air-dried and DNA is covalently linked to the membrane surface by UV irradiation (254 nm; 30 J/cm^2).

3.6. Probe Preparation, Membrane Hybridization, Washing, and Membrane Stripping

1. The membrane is prehybridized with 50 ml of Hybridization buffer for 1 h at 65°C.
2. DNA-labeling is performed according to the instructions of the RadPrime DNA Labeling System (Invitrogen) with [α-^{32}P]dATP.
3. The radiolabeled probe is purified from free [α-^{32}P]dATP using ProbeQuant™ G-50 Micro Columns (GE Healthcare) according to the protocol provided by the manufacturer.
4. Salmon sperm DNA is added to the radiolabeled probe in a screw cap microtube. The amount of salmon sperm DNA is adjusted to result in a final concentration of 0.1 μg/ml in the hybridization solution.
5. The radiolabeled probe/salmon sperm DNA is incubated at 95°C in a thermomixer for 10 min and then quickly chilled on ice.
6. The prehybridization solution is discarded and replaced by 15 ml of fresh Hybridization buffer (preheated to 65°C). The probe is added and hybridization is performed overnight at 65°C.
7. The hybridization solution is discarded, and the membrane is rinsed once with 50 ml of Rinse buffer.
8. Washings are done twice each with 50 ml of Wash buffer 1, followed by Wash buffer 2 and 3 for 15 min at 65°C.
9. The membrane is dried, wrapped in plastic, and exposed to a phosphor imaging screen.
10. For membrane stripping before re-probing, 50–100 ml of Strip buffer are brought to the boil in a microwave oven and poured into a hybridization tube containing the membrane. The tube is rolled for 15 min in a hybridization oven at 80°C. This step is repeated until no more radioactivity is detected on the membrane (usually two to three times).
11. The membrane is rinsed with 2×SSPE and can be used for hybridization with another probe. Alternatively, the membrane is stored at room temperature.

4. Notes

1. Yeast Peptone Dextrose (YPD) medium is used as growth medium and supplemented with adenine if the strain is auxotrophic for adenine.
2. Unless stated otherwise, all solutions should be prepared in water that has a resistivity of 18.2 MΩ-cm and total organic content of less than five parts per billion. This standard is referred to as "water" in this text.
3. The temperature plays a very important role during formaldehyde crosslinking and, thus, on the outcome of ChEC assays. At high temperature, crosslinking of proteins to proteins and proteins to DNA is more efficient than at lower temperatures. Therefore, variation in the kinetics of DNA degradation by MNase tagged proteins may be a consequence of the temperature at which formaldehyde crosslinking was performed. The most likely explanations for this phenomenon are as follows. (1) At low temperatures the crosslink is weak, increasing the occurrence of nonspecific degradation. (2) MNase activity could be compromised by excessive crosslinking at higher temperatures.
4. Buffer A contains 2 mM EDTA, which inhibits premature MNase activation if Ca^{2+} is released from the endoplasmic reticulum. The EDTA concentration can be increased up to 4 mM without affecting the quality of the experiment.
5. The volume to suspend crude nuclei is calculated such that it exceeds the summed volume of all samples of the ChEC time course by at least 50 μl.
6. The incubation time in $CaCl_2$ containing buffer is carefully chosen and depends on the protein that is tagged with MNase. Namely, for very abundant proteins it is advisable to add naked, linear plasmid DNA to the nuclei to have an exogenous control for unspecific digestion. It is conceivable that MNase fusion proteins cut DNA surrounding their binding site and, thus, releasing themselves in the solvent and cutting more DNA nonspecifically, like free MNases. After Southern blotting the membrane can be hybridized with the plasmid-specific probe to monitor the amount of nonspecific degradation.
7. It is important to mix the suspension before taking aliquots because nuclei sediment.
8. It is advisable to use 24-well noncoated tissue plates to avoid loss of nuclei.
9. Remove the lid of the 24-well plate prior to UV irradiation since plastic filters UV light.

10. For the mock ethanol control (no psoralen) it is sufficient to digest 1 μl of DNA solution.
11. For ChEC, it is useful to prepare gels containing dyes for nucleic acids like ethidium bromide or SYBR® Safe. However, dyes cannot be used for gels that are employed to analyze DNA from ChEC/psoralen assay, since these DNA intercalating molecules mitigate the shift in migration promoted by psoralen crosslinking.
12. This is a crucial step. Irradiation with short wave UV (254 nm) reverses the covalent bonds between psoralen and pyrimidines on the opposite DNA strands. This allows DNA denaturation prior to Southern blotting and, thus, the hybridization with the radioactively labeled single-stranded probe.

Acknowledgments

This work was supported by a mobility program accompanied by the Bavarian Research Alliance GmbH as an assignment from the Bavarian State Chancellery to J.G. and by the Ministère des Relations Internationales du Québec to A.C. M.W. is a recipient of a fellowship from the Elitenetzwerk Bayern and R.C. is supported by the Natural Sciences and Engineering Research Council of Canada (NSERC No. 326873-2009) to A.C.

References

1. Warner, J. R. (1999) The economics of ribosome biosynthesis in yeast. *Trends Biochem. Sci 24*, 437–440.
2. Conconi, A., Widmer, R.M., Koller, T., and Sogo, J. M. (1989) Two different chromatin structures coexist in ribosomal RNA genes throughout the cell cycle. *Cell 57*, 753–61.
3. Merz, K., Hondele, M., Goetze, H., Gmelch, K., Stoeckl, U., and Griesenbeck, J. (2008) Actively transcribed rRNA genes in S. cerevisiae are organized in a specialized chromatin associated with the high-mobility group protein Hmo1 and are largely devoid of histone molecules. *Genes Dev 22*, 1190–204.
4. Dammann, R., Lucchini, R., Koller, T., and Sogo, J. M. (1993) Chromatin structures and transcription of rDNA in yeast Saccharomyces cerevisiae. *Nucleic Acids Res 21*, 2331–8.
5. Toussaint, M., Levasseur, G., Tremblay, M., Paquette, M., and Conconi, A. (2005) Psoralen photocrosslinking, a tool to study the chromatin structure of RNA polymerase I--transcribed ribosomal genes. *Biochem Cell Biol 83*, 449–59.
6. Schmid, M., Durussel, T., and Laemmli, U. K. (2004) ChIC and ChEC; genomic mapping of chromatin proteins. *Mol Cell 16*, 147–57.
7. Sogo, J. M., Ness, P. J., Widmer, R. M., Parish, R. W., and Koller, T. (1984) Psoralen-crosslinking of DNA as a probe for the structure of active nucleolar chromatin. *J. Mol. Biol 178*, 897–919.
8. Schmid, M., Arib, G., Laemmli, C., Nishikawa, J., Durussel, T., and Laemmli, U. K. (2006) Nup-PI: the nucleopore-promoter interaction of genes in yeast. *Mol. Cell 21*, 379–391.
9. Goetze, H., Wittner, M., Hamperl, S., Hondele, M., Merz, K., Stoeckl, U., and Griesenbeck, J. (2010) Alternative Chromatin Structures of the 35S rRNA Genes in Saccharomyces cerevisiae Provide a Molecular Basis for the Selective Recruitment of RNA Polymerases I and II. *Mol. Cell. Biol 30*, 2028–2045.
10. Sambrook, J., and Russell, D. W. (2006) Southern Blotting: Capillary Transfer of DNA to Membranes. *Cold Spring Harb Protoc 2006*, pdb.prot4040.

Chapter 21

UV-Induced DNA Damage and DNA Repair in Ribosomal Genes Chromatin

Julie Pelloux, Maxime Tremblay, Raymund J. Wellinger, and Antonio Conconi

Abstract

Cyclobutane pyrimidine dimers (CPDs) and (6,4) pyrimidine–pyrimidone dimers are the major DNA lesions (or photoproducts) induced by ultraviolet light and are removed by the nucleotide excision repair (NER) pathway. If not repaired, DNA damage can lead to genome instability. The genome is organized into nuclear domains with distinct functions and chromatin structures. Although studies on NER in all chromosomal contexts are important to understand the mechanisms of genome maintenance, we focused on NER in the nucleolus. The attractive feature of the rDNA locus is its chromatin structure; not all rRNA genes are transcribed and both active (no nucleosomes) and inactive (nucleosomes) rRNA genes coexist in the nucleolus. These characteristics allow for direct comparison of NER in two very different chromatin structures. Yeast is used as a model system and the methods employed are as follows: nuclei isolation, restriction enzyme digestion of chromatin to release active rRNA genes, psoralen cross-linking, T4-endonuclease-V enzyme to detect CPDs and CPDs repair over relatively large stretches of DNA, and primer extension to follow DNA damage and repair at nucleotide level. Using this approach, we have shown that NER is faster in nonnucleosomes vs. nucleosomes rDNA, that the formation of CPDs promotes changes in the active rDNA chromatin, and that NER is coupled to rRNA genes transcription.

Key words: Chromatin, Cyclobutane pyrimidine dimer (CPDs), Nucleolus, Nucleotide excision repair, Photoproducts [CPDs and (6,4)s], Psoralen cross-linking, Pyrimidine–pyrimidone dimers (6,4), rDNA, Ribosomal genes, RNA polymerase I, T4 endonuclease V, Ultraviolet light radiation

1. Introduction

Environmental agents such as ultraviolet light, several genotoxic chemicals, and pollutants cause DNA damage. If not repaired, DNA damage can lead to mutations and increased risk for cancer (1). UV light induces the formation of cyclobutane pyrimidine dimers (CPD) and (6,4) pyrimidine–pyrimidone dimers, which represents 70–80% and ~20% of the total UV photoproducts, respectively.

Ales Vancura (ed.), *Transcriptional Regulation: Methods and Protocols*, Methods in Molecular Biology, vol. 809, DOI 10.1007/978-1-61779-376-9_21, © Springer Science+Business Media, LLC 2012

Nucleotide excision repair (NER) is a large multienzymatic complex that removes CPDs from the DNA in five steps: recognition of the lesion, incision of the damaged DNA strand at both sides of the lesion, excision of a ~30 nt DNA fragment containing the lesion, filling the gap by DNA synthesis, and DNA ligation of the synthesized patch (2). NER is divided in two subpathways: global genome repair (GG-NER) that repairs nontranscribed DNA and the nontranscribed strand of active genes, and transcription-coupled repair (TC-NER) that repairs the transcribed strand of active genes (3). NER is conserved from yeast to human, and studies in yeast have made major contributions in elucidating the mechanisms of NER in human (4). We use the yeast *Saccharomyces cerevisiae* as a biological model to study NER in chromatin, but the same techniques can be used to study DNA repair in other organisms.

The genome is organized into nuclear domains that favor distinct chromatin structures (e.g., highly repetitive sequences, centromeres, telomeres, noncoding and coding sequences, and the nucleolus). Hence, the NER enzymes must recognize DNA lesions, remove them, and synthesize a DNA patch within all these domains (5). In the nucleolus, ribosomal genes (rRNA genes, or rDNA) are present in multiple copies clustered in tandem repeats, and in fast growing cells, rRNA genes are transcribed at a very high rate by RNA polymerase I (6, 7). Within the clusters, rRNA genes exist in two different chromatin structures: active (no nucleosome rDNA) and inactive (nucleosome rDNA) (8, 9). Thus, the existence of the two rDNA-chromatin states allows for direct comparison of DNA repair in nucleosome vs. nonnucleosome DNA, and to search for putative chromatin remodeling factors that could promote NER in nucleosomes (10).

Yeast (or any eukaryotic cell) nuclei are readily isolated and digested with restriction enzymes to release active rRNA genes, which separate them from inactive rRNA genes. Commercially available T4-endonuclease-V enzyme nicks DNA specifically at CPD sites, which is then denatured and separated by alkaline agarose gel electrophoresis. Southern blotting allows detection of DNA nicks at CPD sites, and their removal during repair, in DNA stretches of ~1 kb or longer. Primer extension by Taq polymerase, which stops at CPDs and other DNA lesions, is used to measure DNA damage and DNA repair at the nucleotide level (11, 12).

2. Materials

2.1. Yeast Culture and UV Irradiation

1. Yeast Extract Peptone Dextrose (YEPD: 1% yeast extract, 2% peptone, and 2% glucose) (see Note 1).
2. Phosphate Buffered Saline solution (PBS: 137 mM NaCl, 2.7 mM KCl, 10 mM Na_2HPO_4, 2 mM KH_2PO_4, pH 7.4).

3. Gold fluorescent light bulb.
4. PCR hood with germicidal UVC light.
5. UVX digital radiometer with sensor model UVX-25 (UVP Inc. Upland, CA).
6. Plastic trays used for irradiation of yeast cells in liquid suspension.
7. Nuclei Isolation Buffer [NIB: 50 mM morpholinepropanesulfonic acid (MOPS), pH 8.0, 150 mM potassium acetate, 2 mM $MgCl_2$, 17% glycerol, 0.5 mM spermine, and 0.15 mM spermidine] (see Note 2).
8. 0.1 M spermine and 0.1 M spermidine (see Note 2).
9. Glass beads, acid-washed (425–600 μm, Sigma-Aldrich) (see Note 3).
10. 14-ml Falcon polypropylene Round-Bottom Tube.

2.2. Isolation of Nuclei and Release of Active rDNA Chromatin

1. 10% Triton X-100.
2. Restriction enzymes (e.g., *EcoRI*) and buffer (New England BioLabs).
3. Thermomixer (Eppendorf, Hamburg, Germany).
4. 0.5 M EDTA, pH 8.0.
5. TE Buffer: 10 mM Tris–HCl, pH 8.0, 1 mM EDTA.

2.3. DNA Isolation and Release of Inactive rDNA

1. 3 M sodium acetate, pH 5.2 (NaOAc).
2. 10% sodium dodecyl sulfate (SDS).
3. Phenol-chloroform-isoamyl alcohol (25:24:1) solution equilibrated to pH 9.8 with 1 M Tris–HCl, pH 9.8 (see Note 4).
4. 1 M Tris–HCl, pH 9.8.
5. Chloroform.
6. 70% ethanol and 100% isopropanol, stored at −20°C.
7. RNAse A (10 μg/ml) (see Note 5).
8. Restriction enzymes (e.g., *Hind*III or *Nhe*I) and buffer (New England BioLabs).

2.4. T4-Endonuclease-V Digestion

1. 100% ethanol, stored at −20°C.
2. T4-Endonuclease-V buffer (50 mM Tris–HCl, pH 7.5, 5 mM EDTA, pH 8.0).
3. T4-Endonuclease-V enzyme 20 U/μl.
4. 6× Alkaline loading buffer (300 mM NaOH, 6 mM EDTA, 18% (w/v) ficoll (type 400), 0.15% (w/v) bromocresol green, 0.25% Xylene cyanol). Aliquots are stored at −20°C.

2.5. Psoralen Photo-cross-linking

1. 4,5′,8-Trimethylpsoralen (Psoralen; Sigma-Aldrich) is dissolved in 100% ethanol (0.2 mg/ml). The solution is stable for up to 2 months at 4°C in the dark. Stir solution for 30 min before use (see Note 6).
2. 15 W Blacklight blue lamps (315–400 nm) (Sankyo-Denki).
3. Proteinase K (20 mg/ml) dissolved in 1 mM $CaCl_2$ and 50 mM Tris–HCl, pH 8.0.
4. 0.2 M $CaCl_2$.
5. Saturated NaCl solution.
6. Loading dye for native gel electrophoresis: 10% glycerol, 0.05% (w/v) bromophenol blue, 0.05% (w/v) xylene cyanol; stored at room temperature.

2.6. Agarose Gel Electrophoresis and Southern Blot

1. Horizontal gel electrophoresis apparatus (20×25 cm) with a 16-slot comb.
2. Agarose, analytical grade.
3. Solution for alkaline agarose gels: 0.05 M NaCl and 1 mM EDTA; prepared before use.
4. Solutions of 10 M NaOH and of 4 M NaCl.
5. Alkaline running buffer (50 mM NaOH, 1 mM EDTA); prepared before use.
6. Tris acetate–EDTA buffer (TAE: 48 mM Tris acetate, 1.2 mM EDTA, pH 8.0).
7. Hybond-XL nylon membrane (GE Healthcare).

2.7. DNA Probes, Filter Membrane Hybridization and Washing

1. Oligonucleotides diluted to 100 ng/μl in H_2O.
2. 10× T4-polynucleotide kinase buffer (100 mM $MgCl_2$, 700 mM Tris–HCl, pH 7.4).
3. 0.1 M dithiothreitol (DTT).
4. γ-ATP^{32} (3,000 Ci/mmol) (PerkinElmer).
5. T4 polynucleotide kinase.
6. ProbeQuant™ G-50 Micro Columns (GE Healthcare).
7. 0.5 M Na_2HPO_4, pH 7.5.
8. Hybridization solution: 0.5 M Na_2HPO_4, 1 mM EDTA, 1 ng/ml tRNA, 7% SDS, 10 mg/ml bovine serum albumin, stored in 20-ml aliquots at −20°C.
9. Wash solutions: "A" (0.5 M Na_2HPO_4, 1% SDS), "B" (0.25 M Na_2HPO_4, 0.5% SDS), "C" (0.1 M Na_2HPO_4, 0.5% SDS), and "D" (0.05 M Na_2HPO_4, 0.5% SDS).

2.8. Labeling of DNA Oligos for Primer Extension

1. Oligonucleotides diluted to 10 pmol/μl in H_2O.
2. 10× T4-polynucleotide kinase (PNK) buffer (100 mM $MgCl_2$, 700 mM Tris–HCl, pH 7.4).

3. γ-ATP^{32} (3 000 Ci/mmol; 10 mCi/ml) (PerkinElmer).
4. T4 polynucleotide kinase.
5. TE buffer: 10 mM Tris–HCl, pH 8.0, 1 mM EDTA.
6. ProbeQuant™ G-25 & G-50 Micro Columns (GE Healthcare).
7. Biodegradable Counting Scintillant (GE Healthcare).
8. Wallac 1409 DSA Liquid Scintillation Counter (PerkinElmer).

2.9. Primer Extension

1. DNA template (50 ng/μl) (see Note 7).
2. Taq polymerase (5 U/μl).
3. 10× Taq polymerase buffer (500 mM KCl, 15 mM $MgCl_2$, 100 mM Tris–HCl, pH 9.0).
4. 10 mM dNTP solution (Wisent).
5. Salmon sperm (2 mg/ml).
6. 3 M sodium acetate solution, pH 5.2 (NaOAc).
7. 70 and 100% ethanol (EtOH), kept at −20°C.
8. Loading buffer (98% deionized formamide, 10 mM EDTA, pH 8.0, 0.025% xylene cyanol FF, 0.025% bromophenol blue).

2.10. Denaturing Polyacrylamide Gel Electrophoresis

1. Acrylamide–bis-acrylamide 19:1 40% solution.
2. 10% ammonium persulfate (APS) (see Note 8).
3. TEMED.
4. Urea.
5. 10× Tris Borate EDTA buffer (890 mM Tris base, 890 mM boric acid, 20 mM EDTA, pH 8.0).
6. Sigmacote (Sigma Aldrich).
7. Whatman DE81 paper.

2.11. Quantification of DNA Damage and Plotting of DNA Repair Curves

1. PhosphorImager (GE-Healthcare).
2. ImageQuant (GE-Healthcare) and Peakfit (Systat Software Inc.).

3. Methods

The experimental design to separate active from inactive rRNA genes and to measure repair of CPDs in the two rDNA chromatin fractions is shown in Fig. 1. The irradiation of yeast cells, extraction of nuclei and part of the DNA isolation must be done in a dark room under gold fluorescent light, and all samples must be kept in the dark to avoid the activation of photolyase (13). Yeast are harvested before UV irradiation (−UV), soon after (time 0 repair) and at defined repair times (e.g., 0.5, 1, 2, and 4 h) post UV irradiation.

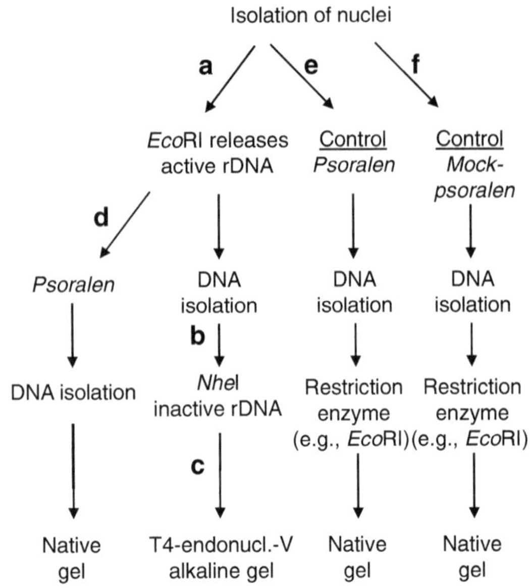

Fig. 1. Experimental design.

Nuclei are isolated and digested with a restriction enzyme (e.g., *Eco*RI; see Fig. 1a) to release active rDNA chromatin (nucleosomes are absent from the active rRNA genes, making them accessible to the restriction endonucleases). The inactive rDNA chromatin is analyzed by a second restriction enzyme digestion (e.g., *Hind*III or *Nhe*I; Fig. 1b) of DNA that has been isolated from half of each sample of *Eco*RI-digested nuclei. The *Eco*RI sites are located within two *Nhe*I (or *Hind*III) sites (Fig. 2a) and, therefore, the complete *Nhe*I band (~4.4 kb) can originate only from rDNA that was not cleaved by *Eco*RI. That is, the complete *Nhe*I fragment derives from inactive rRNA genes where *Eco*RI accessibility is inhibited by the presence of nucleosomes. To measure NER in DNA fragments that originate from the active and inactive rRNA genes, the DNA is then treated with T4-endonuclease-V (Fig. 1c), which nicks DNA at CPD sites, and analyzed on denaturing alkaline agarose gels. The double restriction enzyme digestion allows for direct analysis of T4-endonuclease-V cuts in the active (*Eco*RI-band) and inactive (*Nhe*I-band) rDNA (Fig. 2b) (14–16). To monitor that indeed *Eco*RI releases only active rRNA genes, the second half of each sample of *Eco*RI-digested nuclei is photo-cross-linked with psoralen (Fig. 1d). The DNA is purified and analyzed on native agarose gels. Psoralen cross-linking is used to monitor the ratio of active and inactive rDNA chromatin. With this technique, active rRNA genes bind more psoralen than inactive rRNA genes. Consequently, psoralen-cross-linked rDNA fragments from active genes have a slower migration on gels than fragments from inactive genes (Fig. 2d) (9).

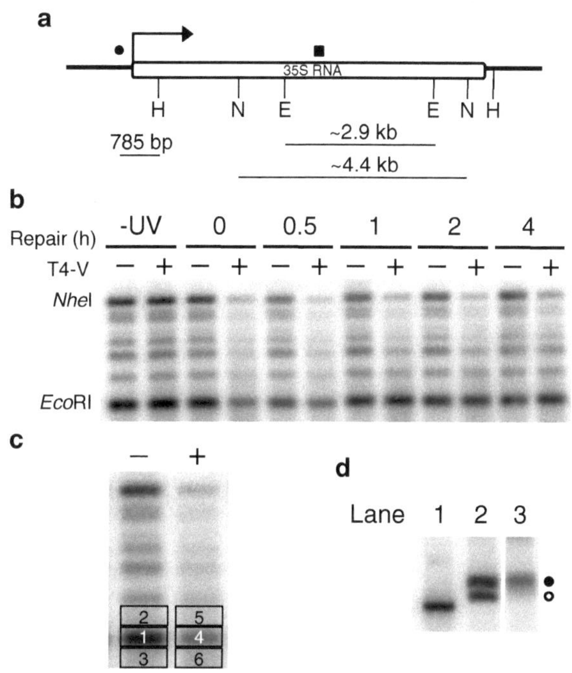

Fig. 2. (**a**) Map of the yeast 35 S rRNA gene. The *arrow* shows the start and direction of transcription. The *filled square* and *circle* represent the labeled oligonucleotides used in Fig. 2**b–d** and in Fig. 3, respectively. H indicates the position of the two *Hind*III restriction sites, N indicates the position of the two *Nhe*I restriction sites, and E the positions of two (out of seven) *Eco*RI sites in the rDNA locus. (**b**) Southern blot of alkaline denaturing ~1% agarose gel electrophoresis, which shows the T4-endonuclease-V-based DNA repair analyses of active (*Eco*RI) and inactive (*Nhe*I) rDNA. DNA samples were mock treated (–) or treated with T4-endonuclease-V (+); –UV are data from nonirradiated cells and, 0–4, from UV irradiated cells at 180 J/m^2 that were harvested after the indicated repair times (in hours). (**c**) Example of measurements of CPDs in the *Eco*RI DNA fragment at time 0 h repair; (–) and (+) are as described in (**b**). (**d**) Southern blot of native ~1% agarose gel electrophoresis, which shows the DNA band retardation after psoralen cross-linking. DNA was isolated from: non-cross-linked nuclei and digested with *Eco*RI (*lane 1*); cross-linked nuclei and digested with *Eco*RI (*lane 2*); *Eco*RI digested first and then cross-linked nuclei (*lane 3*). Labels denote active rDNA (*filled circle*) and inactive rDNA (*open circle*).

The *Taq* polymerase extension arrest at dimers and this property is exploited to map the photoproducts at nucleotide resolution (Fig. 3) (11). As presented here, the technique is used to map photodimers in *total* (active + inactive) rRNA genes. To map each photoproduct independently in the active and inactive rRNA genes, the two rDNA-populations are fractionated on preparative agarose gel electrophoresis prior to the primer extension assay, as described in (12).

3.1. Yeast Culture and UV Irradiation

1. Yeast is grown exponentially in 1 l of YEPD media overnight at 30°C to $OD_{660} \approx 0.4$.
2. Cells are collected by centrifugation at 1,800 × *g* for 2 min and resuspended in 300 ml of ice-cold PBS (see Note 9).

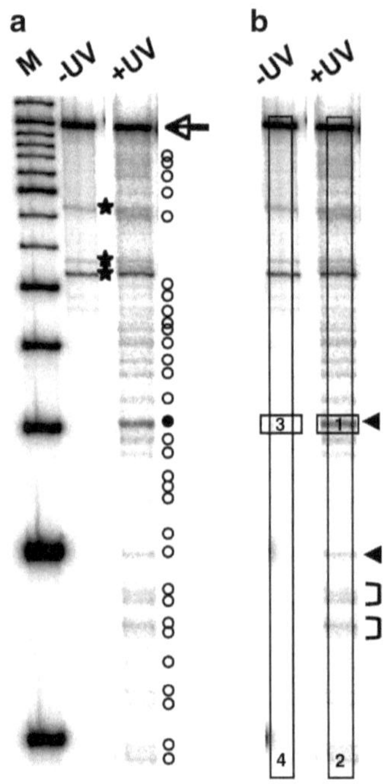

Fig. 3. (**a**) DNA was isolated from nonirradiated (–UV) and irradiated (+UV) cells at 180 J/m^2, and digested with *Hind*III. The end-labeled oligonucleotide was annealed to the bottom strand 270 bp upstream of the transcription initiation site and was extended by the Taq polymerase to the *Hind*III restriction site (see Fig. 2a; 785 bp). M: DNA ladder (1,350, 916, 766, 700, 650, 600, 550, 500, 450, 400, 350, 300, 250, 200 bases). Labels denote: natural stops (*stars*), stops at photoproducts (*circles*), a hotspot for photoproduct formation (*filled circle*) and undamaged full length (785 bases) DNA fragment (*arrow*). (**b**) Example of measurement for the frequency of a single photoproduct.

3. An aliquot of 45 ml is placed in a 50 ml conical tube labeled "–UV" and kept on ice.
4. The intensity of the UV lamp is monitored with the radiometer and the time required to irradiate at the specific UV dose (e.g., 180 J/m^2) is calculated as following:

   [Dose (J/m^2)/Intensity (μW/cm^2)] × 100 = Time (seconds) (see Note 10).

   Yeast suspensions are irradiated in plastic trays (see Note 11).
5. After irradiation, the cells are collected in an Erlenmeyer. An aliquot of 45 ml is placed in a 50-ml conical tube labeled "0 repair" and kept on ice.
6. The remaining cells (~210 ml) are collected by centrifugation at 1,800 × *g* for 2 min and resuspended in 250 ml of prewarmed (30°C) YEPD. The cells are incubated at 30°C under continuous shaking for increasing repair times (e.g., 0.5, 1, 2, and 4 h).

7. The "–UV" and "0 repair" samples are collected by centrifugation at 1,800 × *g* for 2 min. The pellets are resuspended with 1.5 ml of NIB and transferred in 14 ml round-bottom tubes containing each 1.5 ml of glass beads (acid-washed) that have been rinsed twice with NIB. The samples are stored at −80°C (in the dark).
8. For each repair time point, collect 50 ml of cells and proceed as described in steps 7–8.

3.2. Isolation of Nuclei and Release of Active rDNA Chromatin

1. The samples are thawed slowly on ice and in the dark; it takes about 15 min.
2. 45 μl of 10% Triton X-100 is added to each sample. The cells are broken-opened with a vortex set at maximum speed for 30 s, with 30 s gaps on ice. The vortex procedure is repeated 16 times.
3. Transfer the cell lysates (each sample is ~2 ml) in a new 14-ml tube prepared on ice. The glass beads are rinsed twice with 1 ml of NIB and collected with the rest of the cell lysates. At the end, there should be ~4 ml of nuclei lysate for each sample.
4. After centrifugation at 1,350 × *g* for 2.5 min, the supernatants (nuclei) are carefully transferred into new 14-ml tubes. Then, 2 ml of NIB are added to each pellet (cell debris), resuspended by vortexing, centrifuged again at 1,350 × *g* for 2.5 min and the supernatants are collected. At the end, the NIB volume for each nuclei preparation is ~6 ml.
5. The nuclei are collected by centrifugation at 10,000 × *g* and 4°C for 5 min.
6. The translucent nuclei pellets are resuspended in 800 μl of NIB with a 200 μl pipette set at ~170 μl (see Note 12). Nuclei suspensions are separated in two aliquots of 400 μl and stored at −80°C (in the dark).
7. The nuclei (400 μl aliquots) are thawed on ice and collected by centrifugation at 10,000 × *g* for 5 min at 4°C in the dark. The supernatants are carefully removed and the nuclei pellets are resuspended in 500 μl of 1× *Eco*RI buffer with a pipette (see Note 13).
8. 5 μl of *EcoRI* (5 U/μl) are added to each tube, followed by a quick gentle vortex and the incubation is done at 37°C for 15 min in the thermomixer set at 550 RPM.
9. The reaction is stopped by the addition of 10 μl of 0.5 M EDTA and chilled on ice.
10. 100 μl aliquots of each nuclei suspension are used for the psoralen-cross-linking assay (The psoralen photo-cross-linking assay is done to ascertain that active rDNA is efficiently released by restriction enzyme digestion, Fig. 2), 700 μl of NIB is added for a final volume of 800 μl and the samples are stored at −80°C.

3.3. DNA Isolation and Release of Inactive rDNA

1. To each of the 400 μl of nuclei suspensions in *Eco*RI buffer (with EDTA) the following solutions are added in the following order: 100 μl of TE, 225 μl of 3 M NaOAc, and 35 μl of 10% SDS. The samples are mixed after the addition of each solution.
2. One volume of equilibrated phenol–chloroform–isoamyl alcohol is added, and the phenol extraction is done by shaking the samples for 2–3 min (no vortex), followed by centrifugation at 10,000 × *g* for 3 min. The aqueous phase is collected avoiding taking too much of the interphase. This is followed by a second phenol extraction and by one extraction with chloroform (NB: after the first phenol the photolyase is inactivated and the rest of the extraction can be done under normal light conditions).
3. The DNA is precipitated by the addition of 1 volume (~760 μl) of isopropanol (100%), followed by incubation for at least 1 h at −20°C, and collected by centrifugation at 10,000 × *g* for 20 min at 4°C.
4. Without drying, the DNA pellets are resuspended in 500 μl of TE, 10 μl of RNase A (10 mg/ml) and incubated at 37°C for 1 h.
5. To each sample, 225 μl of 3 M NaOAc, pH 5.2, and 35 μl of 10% SDS are added in the order described. The samples are mixed after the addition of each solution. The organic extraction is done as in step 2 and the DNA is precipitated as in step 3.
6. After centrifugation, the supernatant is removed and 1 ml of 70% ethanol is added to wash the DNA pellets, followed by centrifugation at 10,000 × *g* for 5 min.
7. The DNA pellets are air-dried and resuspended in 100 μl of H_2O. Then, 2 μl of each sample are used to determine the quality and quantity of DNA by mini-agarose gel electrophoresis.
8. A second restriction enzyme digestion (e.g., with *Nhe*I or *Hind*III) is performed to release the inactive rDNA, following the manufacturer's protocol.
9. The DNA is extracted once with phenol–chloroform–isoamyl alcohol and once with chloroform, followed by precipitation in ethanol.

3.4. T4-Endonuclease-V Digestion

1. DNA samples are collected by centrifugation at 10,000 × *g* and at 4°C for 20 min, washed in 70% ethanol and air-dried.
2. Each DNA sample is resuspended in 65 μl of T4-endonuclease-V buffer by tapping the tubes repeatedly, followed by incubation at 37°C under continuous mixing for 1 h (thermomixer set at 900 RPM).
3. 1 μl of T4-endonuclease-V enzyme is diluted with 34 μl of T4-endonuclease-V buffer and kept on ice. Two 0.6 ml tubes

for each sample, labeled "+T4" and "–T4", are prepared on ice; 2 μl of diluted T4-endonuclease-V enzyme is added to the + T4 labeled tubes and 2 μl of T4-endonuclease-V buffer is added to the –T4 labeled tubes. To each tube, 15 μl of DNA is added, mixed (no vortex), and quick spun. The T4-endonuclease-V incubation is done at 37°C for 1.5 h.

4. The T4-endonuclease-V digestion is stopped with the addition of 3 μl of 6× alkaline loading buffer and chilled at 4°C until the alkaline agarose gel is ready for loading (see Subheading 3.6).

3.5. Psoralen Photo-cross-linking

1. Nuclei prepared for psoralen cross-linking are stored in 800 μl of NIB at –80°C and in the dark (Subheading 3.2, step 10). The samples are thawed on ice, together with 400 μl of nuclei suspension that will be used as psoralen control (Figs. 1e and 2d; lane 2) and as psoralen mock treated control (Figs. 1f and 2d; lane1). These control nuclei are diluted with 1.2 ml of NIB and separated in two samples of 800 μl each. The nuclei suspension are added to the wells of a 24-well plate (noncoated tissue plate) and kept on ice.
2. 10 μl of psoralen (0.2 mg/ml) is added to the nuclei of each well, except for the mock treated control. After mixing the nuclei with a pipette, the nuclei are incubated on ice and in the dark for 5 min. Photo-cross-linking is done on ice with the UVA lamp (315–400 nm) placed at a distance of 2–3 cm for 5 min.
3. Step 2 is repeated three more times. Thereafter, the nuclei are collected in 1.5-ml tubes.
4. To isolate the DNA, nuclei are collected by centrifugation at 10,000 × *g* and 4°C for 10 min. The supernatant is removed and the nuclei pellets are resuspend in 400 μl of TE (the nuclei are clumpy and it takes some time to resuspend them). Then, 9 μl of 10% SDS are added, followed by 5 μl of 0.2 M $CaCl_2$ and 10 μl of proteinase K (20 mg/ml). The samples are mixed after the addition of each solution and incubated for 1 h at 55°C. After the incubation, 450 μl of saturated NaCl is added and mixed with a vortex for 10 s, followed by centrifugation for 10 min at 10,000 × *g* and 4°C. The supernatants are transferred in 2-ml tubes, 1 volume of 100% isopropanol is added and the samples are incubated at –20°C for at least 30 min. The nucleic acids are precipitated by centrifugation for 20 min at 10,000 × *g* and 4°C.
5. The supernatants are carefully removed and the wet pellets are resuspended in 500 μl of TE and 10 μl of RNase A (10 mg/ml). After incubation at 37°C for 1 h, 225 μl of 3 M NaOAc and 35 μl of 10% SDS are added to each sample, taking care to mix the samples after addition of each solution. The DNA is extracted once with phenol–chloroform–isoamyl alcohol and once with chloroform, followed by precipitation in ethanol as described previously.

6. After removing the supernatants, the DNA pellets are air-dried and resuspended in 100 μl of H_2O. To evaluate the quality and quantity of DNA, 2 μl of each sample is analyzed on an agarose minigel.
7. The controls DNA from psoralen cross-linked nuclei and from mock treated nuclei (Fig. 1e and f) are digested with *Eco*RI as recommended by the manufacturer. After organic extraction, once with phenol–chloroform–isoamyl (pH ~9.8) and once with chloroform, and ethanol precipitation, the DNA pellets are resuspended in 20 μl of loading buffer for native agarose gel electropheresis.

3.6. Agarose Gel Electrophoresis and Southern Blot

3.6.1. Native Agarose Gel Electrophoresis

1. For a 20×25 cm gel cast, prepare 280 ml of 1% agarose in 1× TAE buffer (do not use ethidium bromide, or any other DNA intercalating dyes, as it inhibits the shift in migration of DNA induced by psoralen cross-linking). The agarose gel is left to solidify for at least 1.5 h. About 2 l of 1× TAE buffer are needed and the DNA samples are loaded on the gel as following: (1) DNA isolated from mock treated nuclei (no psoralen cross-linking) and then *Eco*RI digested; (2) DNA isolated from psoralen cross-linked nuclei and then *Eco*RI digested and (3) DNA isolated from *Eco*RI digested nuclei prior to psoralen cross-linking. Gel migration is run for 22 h at 65 V with constant recirculation of the running buffer (Fig. 2d).
2. At the end of the electrophoresis, the DNA is decross-linked by exposing the gel to for 7 min at 1,600 μW/cm^2 (~6,700 J/m^2) to 254 nm UV (decross-linking of psoralen cross-linked DNA is necessary to facilitate DNA hybridization with radiolabeled probes).
3. The gel is equilibrated in 0.4 M NaOH for 15 min and the DNA is transferred to the nylon membrane overnight by the alkaline procedure described in the manufacturer protocol. The membrane is then air-dried and the DNA is covalently bound to the membrane by UVC irradiation as described by the manufacturer.

3.6.2. Denaturing Alkaline Agarose Gel Electrophoresis

1. Place 2.8 g of agarose and a magnetic stirrer bar in an Erlenmeyer flask, together with 280 ml of 50 mM NaCl and 1 mM EDTA, and heat to dissolve the agarose.
2. The agarose is cooled to ~55°C with continuous slow stirring to avoid the formation of bubbles, and poured into the gel tray. The gel is left at room temperature for at least 2 h before it is equilibrated with 2 l of alkaline running buffer. Equilibration is done overnight (see Note 14).
3. The equilibration buffer is replaced with fresh running buffer and the samples are loaded. Load 20 μl of each sample in the order of –UV, 0, 1/2, 1, 2, and 4 h repair, alternating-T4-endonuclease-V and +T4 (Fig. 2b).

4. Electrophoresis is done at 35 V for 43 h with continuous recirculation of the buffer. The pump is started when the samples have migrated into the gel (~ 1 h running time at 35 V), and buffer recirculation is done at slow speed to avoid the formation of bubbles. After 22 h from the start, the buffer is replaced with freshly made running buffer and electrophoresis is run for another 21 h.
5. The agarose gel is prepared for Southern blotting by soaking it in 0.4 M NaOH for 15 min before transferring the DNA to the nylon membrane. The transfer is done overnight by the alkaline procedure described in the manufacturer protocol. At the end of the transfer, the membrane is air-dried and the DNA is covalently bound to the membrane by UVC irradiation as described by the manufacturer.

3.7. DNA Probes, Filter Membrane Hybridization and Washing

1. DNA damage and repair are measured in the transcribed and nontranscribed strands of rRNA genes using single stranded oligonucleotides that are designed to be complementary to one or the other strand. The oligonucleotides are end-labeled by the T4 polynucleotide kinase (T4 PNK) as following: to 12 μl of H_2O, 1 μl of oligonucleotide (100 ng/μl), 2 μl of 10× T4 PNK buffer, 1 μl of 0.1 M DTT, 3 μl of γ-ATP^{32}, and 1 μl of T4 PNK are added and mixed. After a quick spin, the kinase reaction is done at 37°C for 1 h. Thereafter, the T4 PNK enzyme is inactivated at 68°C for 10 min, 80 μl of TE is added, and the free nucleotides are removed with a ProbeQuant™ G-50 Micro Column as specified by the manufacturer.
2. The nylon membrane is first prehybridized at 37°C with 0.5 M Na_2HPO_4 for 15 min and then with the hybridization solution at 37°C for 1 h. 30 μl of radiolabeled oligonucleotide are added to 1 ml of hybridization solution and the mixture is poured to the prehybridizing membrane. Hybridization is done at 37°C overnight.
3. The filter membranes are washed as following: the hybridization solution is removed and washing solution "A" is added, following by incubation at 37°C for 5 min. Solution "A" is removed and the membrane is washed one more time with new solution "A". The procedure is repeated first with solution "B" and then with solutions "C" and "D". After air-drying, the membrane is exposed to a PhosphorImager screen.

3.8. Labeling of DNA Oligos for Primer Extension and Marker

1. The kinase reaction is prepared as following: to 1 μl of oligonucleotide (10 pmol), 2 μl of 10× T4 PNK buffer, 1 μl of T4 PNK, 3 μl of γ-ATP^{32}, and 13 μl of H_2O are added.
2. The reaction is performed for 1 h at 37°C. At the end, the sample is diluted with 30 μl of TE buffer to obtain a final volume of 50 μl.

3. The nonincorporated nucleotides are removed with ProbeQuant™ G-25 Micro Column as specified by the manufacturer.
4. To monitor the labeling reaction, 1 μl of radiolabeled oligonucleotides is added to 10 ml of Biodegradable Counting Scintillant liquid. Normally, readings are between 150,000 and 200,000 cpm (see Note 15).
5. The marker is a 50-bp DNA ladder from New England Biolabs (Fig. 3a; M) and is end-labeled as following: to 1 μl (100 ng) of DNA ladder, 5 μl of γ-ATP^{32}, 2 μl of 10× T4 PNK buffer, 1 μl of T4 PNK, and 11 μl of H_2O are added. The reaction is done for 1 h at 37°C. To stop the reaction, 30 μl of TE buffer is added and the sample is incubated for 10 min at 65°C. Nonincorporated nucleotides are removed with ProbeQuant™ G-50 Micro Column.

3.9. Primer Extension

1. Genomic DNA is isolated from nonirradiated and irradiated yeast cultures, and digested with the appropriate restriction enzyme (e.g., *Hind*III). Add to 2 μl of DNA (50 ng), 4 μl of 10× Taq buffer, 3 μl of radiolabeled oligonucleotides (0.6 pmol), 1 μl of dNTP, and water to bring the final volume to 39 μl.
2. The samples are denatured at 95°C for 10 min and rapidly chilled on ice for 10 min. The primer extension is started with the addition of 1 μl of Taq polymerase (5U).
3. The thermocycling for the primer extension is as following: 95°C for 5 min; 30 cycles of denaturation (95°C for 45 s), annealing (55°C for 5 min) and extension (72°C for 5 min); one final extension (72°C for 10 min) (see Note 16).
4. The products of the primer extension are precipitated with 4 μl (1/10th volume) of 3 M NaOAc, 1 μl of salmon sperm (see Note 17), 112.5 μl (2.5× volume) of ice cold 100% EtOH and incubated at –20°C for 1 h. The DNA sample is collected by centrifugation at 10,000 × *g* for 20 min at 4°C. The supernatant is removed with a pipette (see Note 18) and the pellet rinsed with 150 μl of 70% EtOH. After centrifugation for 5 min, the 70% EtOH is removed and the sample is dried before it is resuspended in 5 μl of loading buffer.

3.10. Denaturing Polyacrylamide Gel Electrophoresis

1. 50 ml of polyacrylamide solution is prepared for a gel cast of 20 × 45 cm. Namely, 21 g of urea is dissolved in 7.5 ml of acrylamide solution, 5 ml of 10× TBE, and water to a final volume of 50 ml. Thereafter, 300 μl of 10% APS and 22 μl of TEMED are added and the acrylamide is poured into the cast. The polymerization reaction is done overnight.
2. Gel is prerun until it reaches the temperature of 50°C. The wells are rinsed and the samples are loaded. Gel-run is done at ~2,000 V, 30 mA, and 60 W for 1 h and 30 min.

3. At the end of the electrophoresis, the gel is dried on Whatman DE81 paper and exposed with a PhosphorImager screen.

3.11. Quantification of DNA Damage and Plotting of DNA Repair Curves

3.11.1. T4-Endonuclease-V Essay

1. An example of the T4-endonuclease-V essay is shown in Fig. 2b. Quantification using ImageQuant software is done as following a box is drawn tightly around, but not touching, the signal in the –T4 lane (Fig. 2c; box 1). Five duplicated boxes are pasted on the + T4 lane (box 4) and above and below of the bands (boxes 2, 3, 5, 6). The signals measured above and below the bands are used for background correction (boxes 2, 3 to correct box 1; boxes 5 and 6 to correct box 4). The same procedure is repeated for the signals corresponding to the active (*Eco*RI) and inactive (*Nhe*I) DNA fragments, from –UV to 4 h repair.
2. The density of each box (in total there will be 36 boxes) is transferred to an Excel spreadsheet. The density (D) of each band (e.g., *Eco*RI and *Nhe*I) is corrected by subtracting the average of the corresponding background-boxes. For example to correct (D) of box 1, the background (B) = (box2 + box3)/2 is calculated. For each band, the absolute signal (S) = (D – B) is calculated for both –T4 (S_{-T4}) (box 1) and + T4 (S_{+T4}) (box 4) samples.
3. The content of CPDs in a defined restriction DNA fragment, and for each repair time point, is calculated using the Poisson expression: CPDs/DNA fragment = –ln (S_{+T4}/S_{-T4}).
4. Percent of CPD removal is plotted over the repair incubation time, whereby the values measured for 0 h repair correspond to 0% repair. The resulting percentage indicates the amount of repair in the fragment as compared to the initial damage. Examples of DNA repair curves are found in (14–16).

3.11.2. Quantification DNA Damage at the Nucleotide Level

1. Quantification using ImageQuant software is done as following: a box is drawn tightly around, but not touching, the signal of a single photoproduct (Fig. 3b, filled triangle; box 1), or a cluster of photoproducts (Fig. 3b, bracket). Another box, designed as large as possible, is drawn around the entire + UV lane (box 2). Duplicates, box 3 and box 4, are pasted on the –UV lane and used for background correction.
2. The density of each box is transferred to an Excel spreadsheet. The frequency (F) of a single photoproduct (or cluster of photoproducts) is analyzed by measuring the intensity of the signal of the photoproduct (or cluster of photoproducts) divided by the signal that is measured for the whole lane. The correction for signal noise (background) is done by subtracting of the ratio of signal densities that are measured in the –UV lane: F = (box 1/box 2) – (box 3/box 4).

3. The percent of repair for each photoproduct is plotted over the incubation time, whereby the values measured for 0 h repair correspond to 100% damage (or 0% repair): Percent of Repair $=100-100\times(F_{(\text{repair time point})}/F_{(\text{time }0)})$. Examples of DNA repair curves are found in (12).

4. Notes

1. Autoclave YEP and 4% glucose solution separately.
2. The Nuclei Isolation Buffer is prepared without spermine and spermidine and stored at −20°C. Spermine and spermidine are stored at −20°C and are added before use.
3. Glass beads are soaked in 0.25 M HCl overnight with gentle stirring. Thereafter, they are rinsed repeatedly in H_2O until the pH of water is neutral. Glass beads are air-dried before autoclaving. Glass beads can be reused several times.
4. The preparation of phenol–chloroform–isoamylalcohol is done in the fume hood. The solution is saturated with water and stored at 4°C. The equilibration to pH 9.8 is done shortly before use to avoid the oxidation of phenol, which damages the DNA.
5. RNase A (powder) is resuspended in Tris–HCl, pH 7.5, to 10 mg/ml, and boiled for 30 min to inhibit DNase activities. Aliquots are stored at −20°C.
6. Be careful! Psoralen is a DNA intercalating agent; wear gloves and work in the fume hood.
7. The minimal amount of DNA depends on the locus of interest. For example, 50 ng of total genomic DNA are sufficient to study the yeast rDNA. For a single copy gene, we recommend using 1 μg.
8. Aliquots of 10% APS are stored at −20°C.
9. YEPD filters UV light. Therefore, the cells are well rinsed in PBS and irradiation is done in PBS.
10. The UV emission increases as the lamps warm up. To avoid fluctuations during irradiation, the lamps are turned on and prewarmed 5 min before irradiation
11. To avoid uneven irradiation, the thickness of the yeast suspension in PBS must be equal or below to 2 mm, and the maximal cell concentration must be 1.8×10^9 cells/ml.
12. Nuclei are hard to resuspend. Pipette up and down slowly and avoid taking nuclei "clumps" inside of the pipette tip.

13. All NIB solution must be removed for optimal *Eco*RI digestion. Clumps of nuclei also hinder restriction enzymes digestions.
14. A poor solidification of the agarose gel results in less sharp DNA bands. Also, DNA transfers better to the filter membrane if the gel is well equilibrated in alkaline buffer.
15. Labeling of oligo DNA primers is a critical step. In order to optimize the amount of radiolabeled nucleotides, we use equimolar amounts of oligonucleotides (10 pmol) and γ-ATP^{32} (3.33 pmol/μl).
16. The thermocycling conditions will take into consideration the melting temperature of oligonucleotides.
17. Salmon sperm DNA is added as carrier to precipitate the DNA if its concentration is less or equal to 1 ng/μl.
18. The precipitation is monitored by measuring and comparing the radioactivity in supernatants vs. pellets.

Acknowledgments

We thank Dr. Bell for critical reading of the manuscript and the Centre de recherche clinique Étienne-le Bel du CHUS. This work was supported by the grant from the Nat. Sci. Eng. Res. Coun. Can. (NSERC).

References

1. Hoeijmakers JH (2001) Genome maintenance mechanisms for preventing cancer. *Nature* **411**: 366–374
2. Friedberg EC, Walker CG et al (2005) Mechanism of nucleotide excision repair in eukaryotes. In: Friedberg EC et al (ed) DNA repair and mutagenesis, 2nd edn. ASM press, Washington
3. Hanawalt PC, Spivak G (2008) Transcription-coupled DNA repair: two decades of progress and surprises. *Nat Rev Mol Cell Biol* **9**: 958–970
4. Prakash S, Prakash L (2000) Nucleotide excision repair in yeast. *Mutat Res* **451**:13–24
5. Smerdon MJ, Conconi A (1999) Modulation of DNA damage and DNA repair in chromatin. In: Moldave K (ed) Progress in nucleic acid research and molecular biology. Academic Press, San Diego
6. McStay B, Grummt I (2008) The epigenetics of rRNA genes: from molecular to chromosome biology. *Annu Rev Cell Dev Biol* **24**:131–157
7. Moss T, Langlois F et al (2007) A housekeeper with power of attorney: the rRNA genes in ribosome biogenesis. *Cell Mol Life Sci* **64**:29–49
8. Sogo JM, Thoma F (2003) The structure of rDNA chromatin. In: Olson MOJ (ed) The nucleolus. Springer, New York
9. Toussaint M, Levasseur G et al (2005) Psoralen photocrosslinking, a tool to study the chromatin structure of RNA polymerase I transcribed ribosomal genes. *Biochem Cell Biol* **83**:449–459
10. Conconi A (2005) The yeast rDNA locus: a model system to study DNA repair in chromatin. *DNA Repair* **4**:897–908
11. Wellinger RE, Thoma F (1996) Taq DNA polymerase blockage at pyrimidine dimers. *Nucleic Acids Res* **24**:1578–1579
12. Meier A, Thoma F (2005) RNA polymerase I transcription factors in active yeast rRNA gene promoters enhance UV damage formation and inhibit repair. *Mol Cell Biol* **25**:1586–1595

13. Sancar A (2008) Structure and function of photolyase and in vivo enzymology: 50th anniversary. *J Biol Chem* **283**: 32153–32157
14. Conconi A, Bespalov VA et al (2002) Transcription-coupled repair in RNA polymerase I-transcribed genes of yeast. *Proc Natl Acad Sci USA* **99**:649–654
15. Conconi A, Paquette M et al (2005) Repair-independent chromatin assembly onto active ribosomal genes in yeast after UV irradiation. *Mol Cell Biol* **25**:9773–9783
16. Tremblay M, Teng Y et al (2008) Complementary roles of yeast Rad4p and Rad34p in nucleotide excision repair of active and inactive rRNA gene chromatin. *Mol Cell Biol* **28**:7504–7513

Chapter 22

Analysis of *SUC2* Promoter Structure by Nucleosome Scanning

Jennifer Chang and Ales Vancura

Abstract

Chromatin remodeling is a key mode of transcriptional regulation, and studying the nucleosome positioning at promoters is an important means to understand how genes are regulated. Nucleosome scanning is a convenient method to study nucleosome positioning. Yeast cells are converted to spheroplasts and nuclei are isolated. The nuclei are then digested by micrococcal nuclease to yield mononucleosome-sized DNA. Using a set of overlapping primers that cover the entire promoter, quantitative real-time PCR is performed using the mononucleosome DNA as the template. The nucleosome enrichment for each primer is calculated to yield a map of nucleosome occupancy across the promoter.

Key words: *Saccharomyces cerevisiae*, Nucleosome scanning, Chromatin, Promoter structure, Transcription, PLC1, Inositol polyphosphates

1. Introduction

The eukaryotic genome is highly organized and compacted into chromatin within the nucleus. Because the packaging of DNA into chromatin retards certain fundamental cellular processes such as DNA replication and transcription, cells rely on chromatin remodeling complexes to make chromatin more accessible to the transcriptional machinery. These remodeling complexes make the chromatin more accessible by altering the structure or content of the nucleosomes, stimulating them into an "active" state, translocating nucleosomes, or replacing canonical histones with histone variants (1, 2). The chromatin remodeling complexes use the energy released from the hydrolysis of ATP to remodel chromatin and contain a highly conserved ATPase subunit (3, 4).

Ales Vancura (ed.), *Transcriptional Regulation: Methods and Protocols*, Methods in Molecular Biology, vol. 809,
DOI 10.1007/978-1-61779-376-9_22, © Springer Science+Business Media, LLC 2012

The SWI/SNF superfamily contains four major subclasses of chromatin remodeling complexes: the SWI2/SNF2 subfamily (of which RSC is also a member), CHD, INO80, and ISWI families of chromatin-remodeling complexes (3, 4).The ATPase subunits are represented by Swi2p/Snf2p of the SWI/SNF complex, by Swr1p of the SWR complex, and by Sth1p of the RSC complex in yeast and by Brahma and by BRG-1 and BRM in the SWI/SNF-related complexes in Drosophila and humans (3).

The *SUC2* gene of *Saccharomyces cerevisiae* encodes the sucrose-hydrolyzing enzyme invertase (5, 6). The *SUC2* promoter has been widely used in studies of glucose regulation and chromatin structure. Initial micrococcal nuclease (MNase) digestion experiments showed the presence of four well-positioned nucleosomes occupying the length of the *SUC2* promoter in its repressed state (7, 8). However, upon induction, the nucleosomes are remodeled. This remodeling is SWI/SNF dependent (9–11). The *SUC2* promoter has such a strong need for chromatin remodeling due to the placement of the four promoter nucleosomes. Nucleosomes occlude both the TATA box, preventing TBP binding, and the UAS, which contains other *cis*-regulatory sites (12). The UAS contains binding sites for the repressor Mig1p, which in turn recruits the Tup1p-Ssn6p corepressor complex (13, 14). Under inducing conditions Mig1p is converted into an activator of gene expression (14).

Traditional techniques of analyzing nucleosome positioning require MNase digestion followed by Southern blotting. The positions of the nucleosomes must be inferred from the protected chromatin sites. Here, we present the method of nucleosome scanning, which does not require inferring the positions of nucleosomes. Rather, nucleosome-dense regions appear as peaks, whereas nucleosome-poor regions appear as valleys.

2. Materials

2.1. Yeast Strains, Growth Media, and SUC2 Induction

1. *S. cerevisiae* wild-type strain W303 (*MAT*a *ade2-1 his3-1,15 leu2-3,112, trp1-1, ura3-1*) and mutant strains that are derivatives of W303: HL1-3 (15) (*MAT*a *plc1::URA3*), DY2348 (16) (*MAT*a *swi2::HIS3*), A0003 (17) (*MAT*α *ipk2::HIS3*), A0004 (18) (*MAT*α *ipk1::kanMX*), and LSY507 (19) (*MAT*α *kcs1::HIS3*).
2. Yeast extract/peptone/dextrose (YPD) medium for normal growth and *SUC2*-repressing conditions: 1% bacto-yeast extract, 2% bacto-peptone, and 2% dextrose.
3. Medium for *SUC2*-inducing conditions: 1% bacto-yeast extract, 2% bacto-peptone, and 0.05% dextrose.

2.2. Cell Lysis, Micrococcal Nuclease (MNase) Digestion of Chromatin, and DNA Extraction

1. Spheroplast buffer: 1.4 M sorbitol, 40 mM potassium phosphate, pH 7.5.
2. Yeast lytic enzyme (lyticase from *Arthrobacter luteus*, Sigma no. L2524, ≥2,000 U/mg protein): 700 U/ml in 10% sucrose, 10 mM Tris–HCl, pH 7.5.
3. Phenylmethylsulfonyl fluoride (PMSF): 100 mM in ethanol. This is a 100× stock.
4. Protease inhibitors: protease inhibitor cocktail for use with fungal and yeast extracts (Sigma no. P8215, this is a 50× stock) and complete, mini protease inhibitor cocktail tablets (Roche no. 11836153001; dissolve 1 tablet in 2 ml water to get a 25× stock).
5. SPC buffer: 1 M sorbitol, 20 mM PIPES, pH 8.3, 0.1 mM $CaCl_2$.
6. Ficoll buffer: 9% Ficoll-400, 20 mM PIPES, 0.1 mM $CaCl_2$, 1 mM PMSF, pH 9.0.
7. Micrococcal nuclease (MNase): ≥6,000 units per mg protein (Worthington Biochemicals no. LS004797). Dissolve entire bottle of lyophilized powder (15 kU) in 300 μl of buffer (5 mM Tris–HCl, pH 7.5, 0.01 mM $CaCl_2$) and make 20 μl aliquots. Store aliquots at −20°C and discard after 3–4 freeze–thaws.
8. MNase stop solution: 10 mM EDTA, pH 8.0, 1% SDS.
9. Mononucleosome fragment purification: QIAquick gel extraction kit (Qiagen).
10. Phenol for DNA extraction: equilibrate with Tris–HCl to pH 8.0.
11. Chloroform for DNA extraction: 24 parts chloroform to 1 part isoamyl alcohol.
12. RNase A: 0.5 mg/ml RNase, DNase-free from bovine pancreas (Roche).

2.3. Quantitative Real-Time PCR

1. *SUC2* promoter nucleosome scanning primers: set of 24 overlapping pairs of primers, each of which generates a 107 ± 15 bp PCR product and whose PCR product overlaps by 30 ± 10 bp. Sequences are available upon request.
2. *PHO5-TATA* control primer pair: primers designed against the nucleosome that occupies the *TATA* box of the *PHO5* promoter. This control primer is used to determine the nucleosome enrichment for any individual nucleosome scanning primer. Sequence of sense primer is 5′-CCATTTGGGATAAGGGTA AACATC-3′. Sequence of antisense primer 5-AGAGATGAAG CCATACTAACCTCG-3.
3. PCR reaction mixture: iQ SYBR Green Supermix for Single-Color Real-Time PCR Detection System (Bio-Rad): The supermix

from Bio-Rad contains all necessary reagents, including DNA polymerase, for a PCR reaction.

3. Methods

This method of nucleosome scanning gives a higher resolution mapping of nucleosome positioning than does the traditional method of MNase digestion followed by Southern blot analysis. As nucleosome scanning relies on quantitative PCR, it is more high-throughput than the traditional method and does not require the use of dangerous radioisotopes. Our focus is on studying the role of *PLC1* and InsPs in chromatin structure but this method can be applied to any yeast strain or model organism from which mono-nucleosome DNA can be easily purified.

In *S. cerevisiae*, phosphoinositide-specific phospholipase C (Plc1p encoded by the *PLC1* gene) is the sole source of inositol polyphosphates (InsPs). Plc1p cleaves phosphatidylinositol 4,5-bisphosphate (PtdIns(4,5)P_2) into two second messengers: inositol 1,4,5-trisphosphate ($InsP_3$) and 1,2-diacylglycerol (DAG) (20). Despite being the sole source of intracellular InsPs, deletion of *PLC1* is not lethal but does result in a number of aberrant phenotypes (20–23).

Pc1p along with InsPs kinases constitute a nuclear signaling pathway (24, 25) and the various InsPs have different signaling roles (17, 18, 26–29). $InsP_4$, $InsP_5$, and $InsP_6$ have been shown to regulate transcription by modulating the activity of chromatin remodeling complexes both in vivo and in vitro (27, 28). In vitro nucleosome mobilization by the SWI/SNF complex is stimulated by $InsP_4$ and $InsP_5$, whereas $InsP_6$ inhibits nucleosome mobilization by chromatin remodeling complexes (28). Induction of the phosphate-responsive gene *PHO5* is impaired in *plc1Δ* and *ipk2Δ/arg82Δ* mutants (27). The SWI/SNF and INO80 chromatin remodeling complexes are not efficiently recruited to the *PHO5* promoter in these mutants, thus suggesting that InsPs can regulate transcription through chromatin remodeling (27). Furthermore, Plc1p localizes to centromeric loci and plays a role in establishing proper centromeric chromatin structure (15, 30, 31). MNase digestion has also revealed that global chromatin structure in *plc1Δ* and other InsPs kinase mutants is altered (see Fig. 1). Thus, we were interested in studying the nucleosome positioning of a SWI2-regulated promoter such as *SUC2*. However, this technique can be easily applied to the study of any promoter or genomic region of interest (32–36).

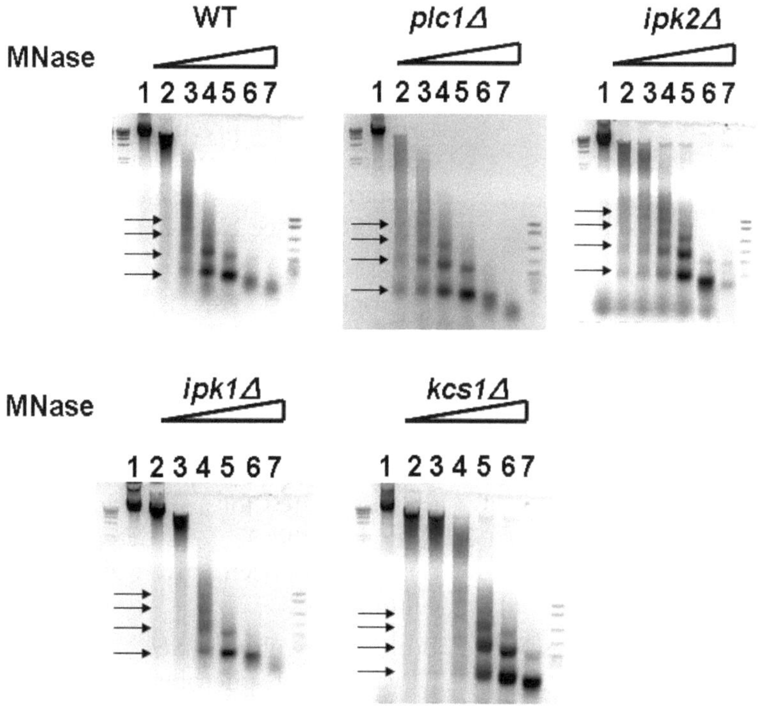

Fig. 1. InsPs affect the global structure of chromatin. Nuclear fraction from wild-type, *plc1Δ*, *ipk2Δ*, *ipk1Δ*, and *kcs1Δ* strains were incubated with increasing concentrations of MNase (lanes 1–7: 0, 0.1, 0.3, 1, 3, 10, and 30 U/ml). DNA was phenol-chloroform extracted and run on a 2% agarose gel. Arrows indicate positions of mono-, di- ,tri- and tetranucleosomes. Deletion of *PLC1* results in chromatin being more accessible to MNase compared to the chromatin from the wild-type strain. Even at very low concentrations of enzyme (0.1 U/ml, lane 2), bands corresponding to the mono-, di-, tri- and tetranucleosomes can be seen in the *plc1Δ* mutant, whereas in wild-type, the total chromatin has barely been digested. There is more digestion of the chromatin from the *ipk2Δ* mutant versus wild-type but the MNase-susceptible phenotype is less pronounced than in the *plc1Δ* mutant. At low concentrations of MNase (0.1 and 0.3 U/ml, lanes 2 and 3), the chromatin in the *ipk2Δ* mutant has been digested more as compared to wild-type as the nucleosome pattern is already clearly distinct even at such low concentrations. However, at the highest concentrations of MNase (10 and 30 U/ml, lanes 5 and 6), the chromatin from the *ipk2Δ* mutant appears to be more resistant to digestion than either the wild-type or *plc1Δ* mutant, as a mononucleosome band can still be seen. This difference could be attributed to the fact that that *ipk2Δ* lacks two different species of inositol polyphosphates, $InsP_4$ and $InsP_5$, but accumulates $InsP_3$. $InsP_4$ and $InsP_5$ could have varying effects on the digestion pattern. Another nonmutually exclusive possibility is that accumulation of $InsP_3$ in the *ipk2Δ* mutant is counteracting the lack of $InsP_4$ and $InsP_5$. Both the *ipk1Δ* and *kcs1Δ* mutants are more resistant to digestion than that of wild-type (with *kcs1Δ* being the more resistant of the two), as the bulk of the chromatin in the mutants is still in the high molecular weight range under the two lowest MNase concentrations (0.1 and 0.3 U/ml, lanes 2 and 3). This suggests a role for the inositol pyrophosphates in chromatin structure as well.

3.1. Yeast Cell Growth and SUC2 Induction

1. All growth is carried out at 30°C with shaking (250 rpm).
2. Yeast culture is first grown in 50 ml of normal YPD media (2% glucose) overnight. A_{600nm} of this preculture is determined and a volume of this preculture is added to 800 ml (200 ml in 4 separate 1 l flasks; see Note 1) of normal YPD media so that the starting $A_{600nm} = 0.375$. The culture is allowed to grow for 2 doublings (see Note 2) so that the $A_{600nm} = 1.5$.
3. Half (400 ml) of the culture is immediately processed for spheroplasting. This is the *SUC2*-repressed culture. The remaining half of the culture is centrifuged 5 min, 3,000 × *g* and the cell pellet is washed with sterile ddH_2O (see Note 3) and centrifuged again. The pellet is resuspended in 400 ml of inducing medium (1% bacto-yeast extract, 2% bacto-peptone, and 0.05% dextrose), the A_{600nm} is determined, and the culture then incubated for an additional hour at 30°C with shaking. After 1 h, the A_{600nm} is determined again to ensure that the cells have not been actively dividing. This is the *SUC2*-induced culture.
4. Both sets of cultures (*SUC2*-repressed and *SUC2*-induced) are harvested and spheroplasted in the same way. The culture is transferred to 50-ml conical tubes and centrifuged for 5 min at 3,000 × *g*. The supernatant is carefully poured off and the cell pellet is resuspended in 40 ml of ddH_2O and transferred to a new 50-ml conical tube. The cells are centrifuged for 5 min, 3,000 × *g*.
5. The pellet is resuspended in 40 ml of ice-cold spheroplast buffer. The suspension is centrifuged 5 min, 3,000 × *g*. The pellet is resuspended in 20 ml of spheroplast buffer supplemented with 200 μl of 2-mercaptoethanol. The A_{600nm} is determined and the cells are converted to spheroplasts with the addition of 1 ml of yeast lytic enzyme. Digestion is allowed to proceed at 37°C with shaking (250 rpm). A_{600nm} is determined every 15 min until it has dropped by 75–90% of the original A_{600nm}. Depending on the yeast strain used, digestion generally occurs between 30 and 60 min (see Note 4).
6. The spheroplasts are centrifuged at 5 min, 800 × *g*, at 4°C and the supernatant is very carefully removed (see Note 5). The spheroplasts are washed two times each with 40 ml of ice-cold spheroplast buffer, followed by two times with 40 ml of ice-cold SPC buffer. After the last wash, the supernatant is poured off and the pellet is resuspended in the remaining SPC buffer. Any spheroplasts that are not used for MNase digestion should be stored at −80°C.

3.2. MNase Digestion

1. The spheroplasts are resuspended in 2 ml of ice-cold SPC buffer. Then 20 ml of ice-cold Ficoll buffer (see Note 6) supplemented with protease inhibitors (Sigma and Roche) and PMSF is added to the resuspended cells. The spheroplasts are slowly and gently inverted for 1 min to mix and then placed on ice for 10 min without disturbance.
2. The lysed spheroplasts are centrifuged for 20 min at 20,000 × *g* at 4°C. The supernatant is poured off and the pellet containing the nuclei is washed with 5 ml of ice-cold SPC buffer supplemented with protease inhibitors and PMSF. The sample is centrifuged again for 10 min at 20,000 × *g* at 4°C.
3. All but 2 ml of the SPC buffer wash is removed. The pellet containing the nuclei is resuspended in this remaining SPC buffer (an additional 1 ml of buffer may be added if the pellet is particularly difficult to resuspend) and 200 μl aliquots are made. Aliquots should be stored at −80°C.
4. For MNase digestion of total chromatin from wild-type, *plc1Δ*, *ipk2Δ*, *ipk1Δ*, and *kcs1Δ* strains, 200 μl aliquots of the nuclear fraction are treated with 0, 0.1, 0.3, 1, 3, 10, and 30 U/ml of MNase for 10 min at 37°C with gentle shaking (invert the microcentrifuge tubes 4–6× every 2 min).
5. For nucleosome scanning of *SUC2*-repressed and *SUC2*-induced wild-type, *plc1Δ*, and *swi2Δ* cells, 200 μl aliquots were treated with 0, 2, 4, 8, 10, and 12 U/ml of MNase for 10 min at 37°C with gentle shaking (invert the microcentrifuge tubes 4–6× every 2 min).
6. The MNase digestion reaction is terminated with 1/10 the reaction volume (20 μl) of MNase stop solution.
7. DNA is extracted with 1× volume of phenol. The microcentrifuge tube is vortexed at the highest setting for 1 min to make sure that the nuclear fraction and phenol mix well. The tubes are centrifuged on a benchtop microcentrifuge at its maximum speed (usually 13,000–14,000 × *g*) for 5 min. The upper aqueous layer is removed, 1× volume of phenol/chloroform–isoamyl alcohol (24:1) is added and the tubes are vortexed and centrifuged. The upper aqueous layer is removed again, 1× volume of chloroform–isoamyl alcohol is added, the tubes are vortexed and centrifuged. The upper aqueous layer is transferred to a new microcentrifuge tube and the DNA is precipitated with 2.5× volume of 100% ice-cold ethanol and 1/10 volume of 3 M sodium acetate pH 5.2 for 1 h at −20°C. Tubes are centrifuged at 14,000 × *g* at 4°C for 30 min . The ethanol is carefully poured off and the DNA pellet is washed with 1 ml of 70% ethanol, and centrifuged again for 15 min. The 70% ethanol is poured off and the DNA is allowed to air-dry at room temperature (~15 min).

8. The DNA is resuspended in 50 μl of TE (10 mM Tris–HCl, 1 mM EDTA, pH 8.0) and 3 μl of RNase is added. The DNA is RNase-treated for 1 h at 37°C and 4 μl is run on a 2% agarose gel. Unused DNA is stored at −20°C.
9. For nucleosome scanning, the MNase concentration that yielded predominantly mononucleosome DNA (~147 bp) was scaled up on multiple aliquots of nuclear fraction and MNase digested as described above.

3.3. Nucleosome Scanning and Quantitative Real-Time PCR

1. The entire scaled-up MNase reaction that yielded predominantly mononucleosomes was run on a 2% gel. The mononucleosome band (~147 bp) was cut out of the gel and purified using the QIAquick Gel Extraction Kit from Qiagen. This DNA was diluted tenfold (see Note 7) and used as the template for quantitative real-time PCR.
2. Nucleosome scanning of the *SUC2* promoter was performed based on the method of Sekinger et al. (32) with a few modifications. For *SUC2*, we used a set of 24 overlapping pairs of primers to cover a 740 bp long region of the promoter (7, 8), each of which generated a 107 ± 15 bp PCR product (see Note 8). The primers were located in 30 ± 10 bp intervals (see Fig. 2).
3. Quantitative PCR was performed by using real-time PCR (25 μl reaction mixture) using the iQ SYBR Green Supermix Single Color Real-Time PCR Detection System from Bio-Rad (see Note 9).

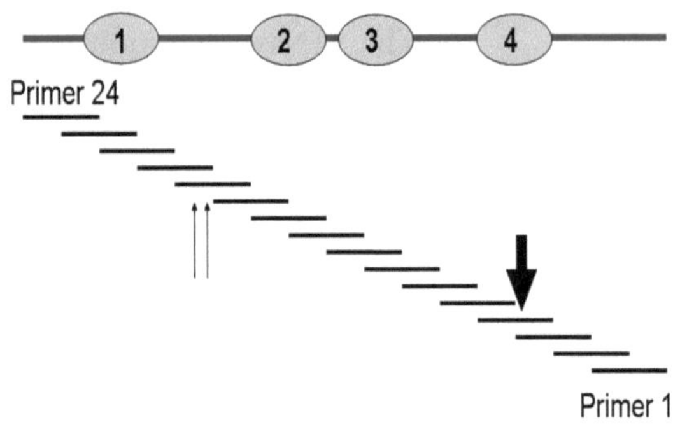

Fig. 2. An overview of nucleosome scanning. Quantitative real-time PCR is performed using a set of overlapping pairs of primers to cover a target region of DNA. Each primer pair generates a 107 ± 15 bp PCR product (see Note 8). Each primer pair overlaps by a 30 ± 10 bp interval. Therefore, each new primer contributes an additional 30 bp of sequence to the target region. The nucleosome enrichment level of a given DNA region is normalized against a control primer, which is arbitrarily set to 1. This control primer amplifies a DNA region known to be occupied by a nucleosome. All other in vivo nucleosome DNA enrichment values are presented relative to this standard. Numbered ovals represent nucleosomes. *Short black horizontal lines* represent overlapping primer pairs. *Thick black arrow* pointing downward indicates a primer pair that would show nucleosome enrichment. *Double thin arrows* pointing upward indicate a primer pair that would show nucleosome depletion.

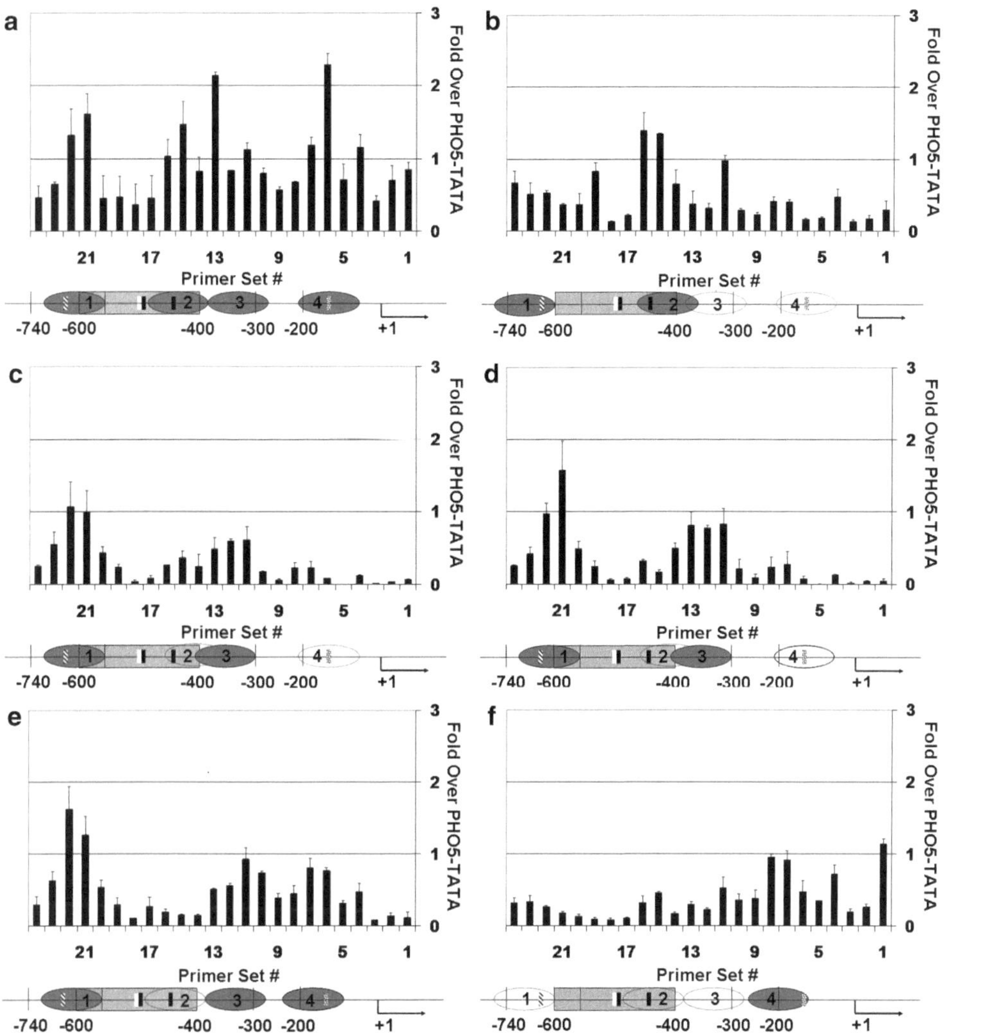

Fig. 3. Nucleosome positioning at the *SUC2* promoter in wild-type, *swi2Δ*, and *plc1Δ* cells under repressing and inducing conditions. The cells were grown in the presence of high glucose (2%). Fold occupancy represent the fold enrichment over the control primer *PHO5-TATA*. Diagram represents positions of the nucleosomes (*circles*) and selected *cis*-regulatory elements (*small bars*). Closed circles represent positions of nucleosomes in repressed wild-type or nucleosomes repositioned upon induction in wild-type or nucleosomes in the mutant strains with different positioning relative to the wild-type repressed state. *Open circles* represent nucleosomes that are displaced. +1 and *arrow* represent transcription start site. Long *light gray rectangle* represents the UAS. Diagonal bar represents the Sko1p binding site. The white bar represents the Gcr1p binding site. The *black bars* represent the two Mig1p binding sites. The *speckled bar* represents the TATA box. (**a**) wild-type, high glucose, repressed state. The position of the 4 nucleosomes can be clearly seen as peaks in the graph and correlate well with previously published results (9, 12, 13, 37). Interestingly, nucleosomes 2 and 3 appear as one large overlapping peak. (**b**) wild-type, low glucose, induced state. Chromatin remodeling of the *SUC2* promoter can clearly be seen as nucleosomes 3 and 4 have disappeared and nucleosomes 1 and 2 have shifted to make the UAS more accessible. This corresponds with previously published results (9, 12, 38) that show that there is increased MNase digestion at that the positions originally occupied by nucleosomes 1, 2, and 3 in the wild-type repressed state. (**c**) *swi2Δ*, high glucose, repressed state. The *swi2Δ* mutant does not resemble wild-type in its repressed state. A previous study has indicated that the promoter in the *swi2Δ* mutant is identical to that of wild-type in its repressed state (12), while other studies have shown that there are differences in MNase cleavage between the two (10, 38). In our results, only two nucleosomes can be detected (nucleosomes 1 and 3). It was surprising to observe that nucleosome 4, which occludes the TATA box, is not present either before or after induction. This is in contrast to previous findings that showed protection from MNase

4. The PCR cycling conditions were as follows: 1 cycle of 2.5 min 94°C; 40 cycles of 20 s 94°C, 30 s 55°C, 30 s 72°C; 1 cycle of 15 min 72°C; hold at 4°C (see Note 10).
5. The nucleosomal DNA enrichment level of a given DNA region was normalized against the control primer *PHO5-TATA* (see Note 11), which was arbitrarily set to 1. All other in vivo nucleosome DNA enrichment values are presented relative to this standard. PCR efficiency for each primer pair was normalized with purified genomic DNA (see Fig. 3).

4. Notes

1. All 800 ml of yeast culture may be grown in a larger volume flask that is at least 5× the volume of the culture grown. However, in our experience, larger volume cultures tend to contaminate more easily especially for slow-growing strains than smaller volumes. To save time and effort, we use multiple 200 ml cultures to obtain the total number of cells that we need. Multiple smaller volume cultures also make it unnecessary to divide the *SUC2*-induced cultures from the repressed cultures later on.
2. Alternatively, the culture can be grown overnight so that it undergoes >2 doublings but still reaches a final $A_{600nm} = 1.5$. This method is recommended for slow-growing strains such as *swi2Δ* and *plc1Δ*.
3. All traces of the original media must be removed as this is a glucose-rich media. In order to induce *SUC2*, the cells must be completely in a low glucose growth environment.

Fig. 3. (continued) cleavage in the TATA region in *swi2Δ* (10, 12). (**d**) *swi2Δ*, low glucose, induced state. Our results indicate that there is no remodeling of the *SUC2* promoter in *swi2Δ*. The promoter is not altered after the cells are switched to the inducing state. The nucleosomes are not repositioned as they should be, supporting the fact that *SUC2* expression requires the chromatin remodeling of its promoter by SWI/SNF. (**e**) *plc1Δ*, high glucose, repressed state. The promoter in the *plc1Δ* mutant in its repressed state differs from both that of wild-type and *swi2Δ*. In total, three nucleosomes are detectable prior to induction, with one nucleosome (nucleosome 4) partially occluding the TATA box (unlike in *swi2Δ*). Both Mig1p binding sites are fully accessible. (**f**) *plc1Δ*, low glucose, repressed state. As *plc1Δ* cells are capable of expressing *SUC2*, this suggests that the *SUC2* promoter is remodeled in the absence of InsPs. When shifted to low glucose conditions, the *SUC2* promoter is indeed remodeled in the *plc1Δ* mutant. The promoter does not resemble the promoter in wild-type cells after induction but it is clearly different from the *plc1Δ* promoter in its repressed state, as it would appear that all but one of the nucleosomes (nucleosome 4) have been evicted. Alternatively, the nucleosomes could be remodeled in such a way that they are more loosely organized and, therefore, more difficult to detect. It is surprising that nucleosome 4 is still present as it partially occludes the TATA box and nucleosome competition for TBP binding is one of the major reasons why the *SUC2* promoter is remodeled. The presence of a nucleosome at this location is usually a transcriptional deterrent. However, what might be detected here is not a nucleosome but rather part of the transcriptional machinery, such as TBP.

4. 10 μl of the digestion reaction is added to 1 ml of water in a plastic spectrophotometer cuvette. The cuvette is covered with parafilm and inverted several times to mix thoroughly. The reading is taken after the A_{600nm} has stabilized. If digestion does not proceed well, add an additional 1 ml of yeast lytic enzyme.
5. Once yeast cells have been converted to spheroplasts, all subsequent centrifugation steps are carried out at 5 min, 800 × *g*, 4°C as the spheroplasts are very fragile. Spheroplasts should never be vortexed. They should be resuspended carefully by pipetting up and down until the mixture is uniform.
6. Ficoll buffer should be prepared beforehand as it is extremely viscous and takes time to fully dissolve. It should be stored at 4°C and taken out of refrigeration the day of use and stirred right before use to make sure the Ficoll concentration is uniform. The spheroplast pellet will be difficult to resuspend in the Ficoll buffer due to its viscosity. The pellet can be stirred first and then pipetted up and down to resuspend.
7. A tenfold dilution of our mononucleosome DNA worked best under our PCR conditions. However, it is recommended to make several different dilutions (up to 100-fold) of the purified mononucleosome DNA and perform several test runs before settling on the final amount of DNA to use as template.
8. In our case, 24 pairs of primers were sufficient to cover the 740-bp long region of the *SUC2* promoter that we were interested in. However, the total number of primers used in the set will vary based on the length of the region to be studied.
9. The iQ SYBR Green Supermix Single Color Real-Time PCR Detection System from Bio-Rad contains all necessary reagents, including DNA polymerase, for a PCR reaction and is supplied as a 2× stock. All that needs to be added to run the PCR reaction are template DNA and gene-specific primers.
10. The melt curve of the PCR products may also be programmed into the PCR program at this time. Calculating the melt curve may be helpful to identify nonspecific or contaminating PCR products. To add the melt curve to the PCR program, after the final extension step at 72°C and before the hold at 4°C, program 80 cycles of 10 s at 55°C + 0.5°C.
11. Any primer that amplifies a well-characterized DNA region known to be occupied by a nucleosome can be used as the control primer. As we studied the *SUC2* promoter, we used *PHO5-TATA* as our control.

References

1. Erkina, T. Y., and Erkine, A. M. (2006) Displacement of histones at promoters of Saccharomyces cerevisiae heat shock genes is differentially associated with histone H3 acetylation. *Mol Cell Biol 26*, 7587–600.
2. Rando, O. J., and Chang, H. Y. (2009) Genome-Wide Views of Chromatin Structure. *Annual Review of Biochemistry 78*, 245–271.
3. Hogan, C., and Varga-Weisz, P. (2007) The regulation of ATP-dependent nucleosome remodelling factors. *Mutat Res 618*, 41–51.
4. Varga-Weisz, P. D., and Becker, P. B. (2006) Regulation of higher-order chromatin structures by nucleosome-remodelling factors. *Current Opinion in Genetics & Development 16*, 151–156.
5. Carlson, M., Taussig, R., Kustu, S., and Botstein, D. (1983) The secreted form of invertase in Saccharomyces cerevisiae is synthesized from mRNA encoding a signal sequence. *Mol Cell Biol 3*, 439–47.
6. Taussig, R., and Carlson, M. (1983) Nucleotide sequence of the yeast SUC2 gene for invertase. *Nucleic Acids Res 11*, 1943–54.
7. Matallana, E., Franco, L., and Perez-Ortin, J. E. (1992) Chromatin structure of the yeast SUC2 promoter in regulatory mutants. *Mol Gen Genet 231*, 395–400.
8. Perez-Ortin, J. E., Estruch, F., Matallana, E., and Franco, L. (1987) Fine analysis of the chromatin structure of the yeast SUC2 gene and of its changes upon derepression. Comparison between the chromosomal and plasmid-inserted genes. *Nucleic Acids Res 15*, 6937–56.
9. Gavin, I. M., and Simpson, R. T. (1997) Interplay of yeast global transcriptional regulators Ssn6p-Tup1p and Swi-Snf and their effect on chromatin structure. *EMBO J 16*, 6263–71.
10. Hirschhorn, J. N., Brown, S. A., Clark, C. D., and Winston, F. (1992) Evidence that SNF2/SWI2 and SNF5 activate transcription in yeast by altering chromatin structure. *Genes Dev 6*, 2288–98.
11. Sudarsanam, P., Cao, Y., Wu, L., Laurent, B. C., and Winston, F. (1999) The nucleosome remodeling complex, Snf/Swi, is required for the maintenance of transcription in vivo and is partially redundant with the histone acetyltransferase, Gcn5. *EMBO J 18*, 3101–6.
12. Wu, L., and Winston, F. (1997) Evidence that Snf-Swi controls chromatin structure over both the TATA and UAS regions of the SUC2 promoter in Saccharomyces cerevisiae. *Nucleic Acids Res 25*, 4230–4.
13. Bu, Y., and Schmidt, M. C. (1998) Identi fication of cis-acting elements in the SUC2 promoter of Saccharomyces cerevisiae required for activation of transcription. *Nucleic Acids Res 26*, 1002–9.
14. Treitel, M. A., and Carlson, M. (1995) Repression by SSN6-TUP1 is directed by MIG1, a repressor/activator protein. *Proc Natl Acad Sci USA 92*, 3132–6.
15. DeLillo, N., Romero, C., Lin, H., and Vancura, A. (2003) Genetic evidence for a role of phospholipase C at the budding yeast kinetochore. *Mol Genet Genomics 269*, 261–70.
16. Yu, Y., Eriksson, P., and Stillman, D. J. (2000) Architectural transcription factors and the SAGA complex function in parallel pathways to activate transcription. *Mol Cell Biol 20*, 2350–7.
17. Odom, A. R., Stahlberg, A., Wente, S. R., and York, J. D. (2000) A role for nuclear inositol 1,4,5-trisphosphate kinase in transcriptional control. *Science 287*, 2026–9.
18. York, J. D., Odom, A. R., Murphy, R., Ives, E. B., and Wente, S. R. (1999) A phospholipase C-dependent inositol polyphosphate kinase pathway required for efficient messenger RNA export. *Science 285*, 96–100.
19. York, S. J., Armbruster, B. N., Greenwell, P., Petes, T. D., and York, J. D. (2005) Inositol diphosphate signaling regulates telomere length. *J Biol Chem 280*, 4264–9.
20. Flick, J. S., and Thorner, J. (1993) Genetic and biochemical characterization of a phosphatidylinositol-specific phospholipase C in Saccharomyces cerevisiae. *Mol Cell Biol 13*, 5861–76.
21. Romero, C., Desai, P., DeLillo, N., and Vancura, A. (2006) Expression of FLR1 transporter requires phospholipase C and is repressed by Mediator. *J Biol Chem 281*, 5677–85.
22. Yoko-o, T., Matsui, Y., Yagisawa, H., Nojima, H., Uno, I., and Toh-e, A. (1993) The putative phosphoinositide-specific phospholipase C gene, PLC1, of the yeast Saccharomyces cerevisiae is important for cell growth. *Proc Natl Acad Sci USA 90*, 1804–8.
23. Payne, W. E., and Fitzgerald-Hayes, M. (1993) A mutation in PLC1, a candidate phosphoinositide-specific phospholipase C gene from Saccharomyces cerevisiae, causes aberrant mitotic chromosome segregation. *Mol Cell Biol 13*, 4351–64.
24. Wera, S., Bergsma, J. C., and Thevelein, J. M. (2001) Phosphoinositides in yeast: genetically tractable signalling. *FEMS Yeast Res 1*, 9–13.
25. Irvine, R. (2007) Cell signaling. The art of the soluble. *Science 316*, 845–6.
26. Mulugu, S., Bai, W., Fridy, P. C., Bastidas, R. J., Otto, J. C., Dollins, D. E., Haystead, T. A., Ribeiro, A. A., and York, J. D. (2007) A conserved

family of enzymes that phosphorylate inositol hexakisphosphate. *Science 316*, 106–9.

27. Steger, D. J., Haswell, E. S., Miller, A. L., Wente, S. R., and O'Shea, E. K. (2003) Regulation of chromatin remodeling by inositol polyphosphates. *Science 299*, 114–6.
28. Shen, X., Xiao, H., Ranallo, R., Wu, W. H., and Wu, C. (2003) Modulation of ATP-dependent chromatin-remodeling complexes by inositol polyphosphates. *Science 299*, 112–4.
29. Lee, Y. S., Mulugu, S., York, J. D., and O'Shea, E. K. (2007) Regulation of a cyclin-CDK-CDK inhibitor complex by inositol pyrophosphates. *Science 316*, 109–12.
30. Lin, H., Choi, J. H., Hasek, J., DeLillo, N., Lou, W., and Vancura, A. (2000) Phospholipase C is involved in kinetochore function in Saccharomyces cerevisiae. *Mol Cell Biol 20*, 3597–607.
31. Desai, P., Guha, N., Galdieri, L., Hadi, S., and Vancura, A. (2009) Plc1p is required for proper chromatin structure and activity of the kinetochore in Saccharomyces cerevisiae by facilitating recruitment of the RSC complex. *Mol Genet Genomics 281*, 511–23.
32. Sekinger, E. A., Moqtaderi, Z., and Struhl, K. (2005) Intrinsic histone-DNA interactions and low nucleosome density are important for preferential accessibility of promoter regions in yeast. *Molecular Cell 18*, 735–748.
33. Bai, L., Charvin, G., Siggia, E. D., and Cross, F. R. (2010) Nucleosome-Depleted Regions in Cell-Cycle-Regulated Promoters Ensure Reliable Gene Expression in Every Cell Cycle. *Developmental cell 18*, 544–555.
34. Kondilis-Mangum, H. D., Cobb, R. M., Osipovich, O., Srivatsan, S., Oltz, E. M., and Krangel, M. S. (2010) Transcription-Dependent Mobilization of Nucleosomes at Accessible TCR Gene Segments In Vivo. *Journal of Immunology 184*, 6970–6977.
35. Preti, M., Ribeyre, C., Pascali, C., Bosio, M. C., Cortelazzi, B., Rougemont, J., Guarnera, E., Naef, F., Shore, D., and Dieci, G. (2010) The Telomere-Binding Protein Tbf1 Demarcates snoRNA Gene Promoters in Saccharomyces cerevisiae. *Molecular Cell 38*, 614–620.
36. Burgos-Rivera, B., Ruzicka, D. R., Deal, R. B., McKinney, E. C., King-Reid, L., and Meagher, R. B. (2008) ACTIN DEPOLYMERIZING FACTOR9 controls development and gene expression in Arabidopsis. *Plant Molecular Biology 68*, 619–632.
37. Boukaba, A., Georgieva, E. I., Myers, F. A., Thorne, A. W., Lopez-Rodas, G., Crane-Robinson, C., and Franco, L. (2004) A short-range gradient of histone H3 acetylation and Tup1p redistribution at the promoter of the Saccharomyces cerevisiae SUC2 gene. *J Biol Chem 279*, 7678–84.
38. Fleming, A. B., and Pennings, S. (2007) Tup1-Ssn6 and Swi-Snf remodelling activities influence long-range chromatin organization upstream of the yeast SUC2 gene. Nucleic Acids Res *35*, 5520–31.

Chapter 23

Chromatin Immunoprecipitation of Mouse Embryos

Anne K. Voss, Mathew P. Dixon, Tamara McLennan, Andrew J. Kueh, and Tim Thomas

Abstract

During prenatal development, a large number of different cell types are formed, the vast majority of which contain identical genetic material. The basis of the great variety in cell phenotype and function is the differential expression of the approximately 25,000 genes in the mammalian genome. Transcriptional activity is regulated at many levels by proteins, including members of the basal transcriptional apparatus, DNA-binding transcription factors, and chromatin-binding proteins. Importantly, chromatin structure dictates the availability of a specific genomic locus for transcriptional activation as well as the efficiency, with which transcription can occur. Chromatin immunoprecipitation (ChIP) is a method to assess if chromatin modifications or proteins are present at a specific locus. ChIP involves the cross linking of DNA and associated proteins and immunoprecipitation using specific antibodies to DNA-associated proteins followed by examination of the co-precipitated DNA sequences or proteins. In the last few years, ChIP has become an essential technique for scientists studying transcriptional regulation and chromatin structure. Using ChIP on mouse embryos, we can document the presence or absence of specific proteins and chromatin modifications at genomic loci in vivo during mammalian development. Here, we describe a ChIP technique adapted for mouse embryos.

Key words: ChIP, Histones, Transcription factors, Activators, Repressors, Chromatin-modifying enzymes

1. Introduction

The genetic material consists of the DNA double helix, which is wrapped 1.6 times around a protein core consisting of a $(H3–H4)_2$ tetramer flanked on two sides by H2A–H2B dimers (1). The DNA and the eight core histones form the smallest chromatin particle, the nucleosome. Nucleosomes can be arranged in an open conformation as beads on a string or packaged in tight proximity to each

Ales Vancura (ed.), *Transcriptional Regulation: Methods and Protocols*, Methods in Molecular Biology, vol. 809,
DOI 10.1007/978-1-61779-376-9_23, © Springer Science+Business Media, LLC 2012

other forming 30-nm fibres. The 30-nm fibres are further organised into fibres of increasing diameter from 60 to 80 nm at interphase to 500–750 nm in metaphase chromosomes (2, 3). The open conformation, euchromatin, is permissive for transcriptional activation, whereas the closed conformation, including the highly condensed heterochromatin, generally silences gene activity, although exceptions to this rule exist (4). Whether chromatin is present in an open or closed conformation depends to a large part on posttranslational modifications to the core histone, in particular histone acetylation (5). To examine the molecular effects of certain interventions, such as mutation of a particular chromatin modifier or DNA-binding transcription factor, on their putative or known target genes in vivo, it is desirable to assess if the intervention causes changes in chromatin structure. In order to understand the changes that take place during embryonic development, it is necessary to perform chromatin immunoprecipitation (ChIP) experiments on embryonic tissues immediately after dissection. The following ChIP protocol was developed based on the detailed instructions accompanying Upstate cat#17-295, but was modified so that it could be used on mouse embryos (6). In principle, a ChIP experiment consists of two parts: (a) immunoprecipitation of DNA sequences and associated proteins and (b) analysis of the chromatin immunoprecipitate, i.e. the analysis of the precipitated DNA or proteins. Analysis of the precipitated DNA is typically performed by quantitative PCR (qPCR) or by one of the mass sequencing techniques. All analysis techniques rely on the quality of the chromatin immunoprecipitate. The protocol described here is concerned with generating high-quality ChIP-DNA, which can be analysed using qPCR or mass sequencing.

2. Materials

2.1. Embryo Recovery

1. Autoclaved MT-PBS (16 mM Na_2HPO_4 $2H_2O$, 4 mM NaH_2PO_4 H_2O, 149 mM NaCl, pH to 7.4).
2. Sterile, 50-ml conical tube (Falcon).
3. Dissection instruments: Two pairs of fine forceps, a pair of fine scissors, sterilised in 95% ethanol for 3 min and air dried.
4. Stereo-dissection microscope.
5. Sterile, 10-cm bacterial Petri dishes.
6. Autoclaved 1.5-ml microcentrifuge tubes.
7. 1,000-μl micropipette.
8. Autoclaved tips for 1,000-μl micropipette.

2.2. Cross Linking of DNA and Associated Proteins

1. Sterile, 50-ml conical tubes.
2. MilliQ water, >18 MΩ/cm^2, autoclaved.
3. Paraformaldehyde.
4. 1 N KOH (corrosive!).
5. 1.375 M glycine in autoclaved milliQ water.
6. Vortexer.
7. Rocking platform or roller mixer at room temperature.
8. Autoclaved MT-PBS.
9. Complete protease inhibitors (Roche, 11 873 580 001, EDTA free).
10. 0.5 M sodium butyrate (Sigma, B5887, histone deacetylase inhibitor).
11. Dry ice or liquid nitrogen.

2.2.1. Preparation of Fresh 18.5% Formaldehyde

Safety warning:
This is a potentially hazardous procedure. Read material safety instructions for all reagents before proceeding. Formaldehyde should be handled in a fume hood. Wear full-face shield, gloves, mask, and lab coat. In case of contact with eyes, flush immediately and continually for 15 min. Follow protocol exactly and with caution. Full-face shield is essential for the heating step.

1. Add 4.8 ml of autoclaved milliQ water to a 50-ml conical tube.
2. Add 0.925 g of paraformaldehyde.
3. Add 35 μl of 1 N KOH.

   Full-face shield is required for heating step. Use fume hood with shield down.
4. Cap tightly and place in a 400–500-ml glass beaker containing 200 ml of water.
5. Microwave until the water in the beaker begins to boil. Do not allow the water to boil vigorously. Overheating may cause the tube to leak. Remove the tube with caution.
6. Vortex paraformaldehyde solution.
7. Repeat heating and vortexing until paraformaldehyde is in solution.
8. Cool and retain solution on ice; use within 2 h.

2.2.2. Preparation of 100× Protease Inhibitors

1. Dissolve 1 tablet of complete protease inhibitors (Roche, 11 873 580 001, EDTA free) in 400 μl of autoclaved water.

2.2.3. Preparation of MT-PBS with Fresh Inhibitors

1. 38.8 ml MT-PBS.
2. 400 μl of 100× protease inhibitors (1× final concentration).

3. 800 μl of 0.5 M sodium butyrate (10 mM final concentration, when aiming to detect histone acetylation status).
4. Add protease inhibitors just before use.

2.3. Fragmentation of the Chromatin

1. Autoclaved milliQ water.
2. 0.5 M EDTA, pH 8.0.
3. 1 M Tris–HCl, pH 8.0.
4. 10% (w/v) SDS in autoclaved milliQ water.
5. Autoclaved 5 M NaCl.
6. 10% (v/v) Triton X-100 in autoclaved milliQ water.
7. SDS lysis buffer: 1% SDS, 10 mM EDTA, 50 mM Tris–HCl, pH 8.1.
8. SDS lysis buffer with fresh inhibitors (see 2.3.1).
9. Cooling tube rack.
10. Branson 250 sonifier with a stepped microtip or equivalent.
11. Autoclaved 1.5-ml microcentrifuge tubes (Eppendorf).
12. 1-ml cuvettes for spectrophotometry at 260 and 280 nm.
13. Autoclaved 2-ml microcentrifuge tubes (Eppendorf).
14. ChIP dilution buffer: 0.01% SDS, 1.1% Triton X-100, 1.2 mM EDTA, 16.7 mM Tris–HCl, pH 8.1, 167 mM NaCl.
15. RNase A, 10 mg/ml in H_2O (Sigma); DNase inactivated at 95°C for 10 min.

2.3.1. Preparation of SDS Lysis Buffer with Fresh Inhibitors

1. 38.8 ml SDS lysis buffer.
2. 400 μl of 100× protease inhibitors (1× final concentration).
3. 800 μl of 0.5 M sodium butyrate (10 mM final concentration, when aiming to detect histone acetylation status).
4. Add protease inhibitors just before use.

2.3.2. Preparation of ChIP Dilution Buffer with Fresh Inhibitors

1. 38.8 ml ChIP dilution buffer.
2. 400 μl of 100× protease inhibitors (1× final concentration).
3. 800 μl of 0.5 M sodium butyrate (10 mM final concentration, when aiming to detect histone acetylation status).
4. Add protease inhibitors just before use.

2.4. Chromatin Immunoprecipitation

1. TE buffer: 10 mM Tris–HCl, pH 8.0, 1 mM EDTA, pH 8.0.
2. 50% slurry of pre-absorbed protein A agarose beads/salmon sperm DNA/BSA (protein A for rabbit polyclonal ChIP antibodies; note relative affinity of protein A versus G to antibodies of different species and isotypes).

3. Low-salt, immune complex wash buffer: 0.1% SDS, 1% Triton X-100, 2 mM EDTA, 20 mM Tris–HCl, pH 8.1, 150 mM NaCl (no inhibitors).
4. High-salt, immune complex wash buffer: 0.1% SDS, 1% Triton X-100, 2 mM EDTA, 20 mM Tris–HCl, pH 8.1, 500 mM NaCl (no inhibitors).
5. 10% Nonidet P-40 in autoclaved milliQ water (or IGEPAL-CA630).
6. 10% sodium deoxycholate in autoclaved milliQ water.
7. LiCl immune complex wash buffer: 0.25 M LiCl, 1% IGEPAL-CA630, 1% sodium deoxycholate, 1 mM EDTA, 10 mM Tris–HCl, pH 8.1 (no inhibitors).
8. Freshly prepared elution buffer: 1% SDS, 0.1 M $NaHCO_3$ (no inhibitors).
9. Rocking platform or roller mixer at 4°C.
10. Rocking platform or roller mixer at room temperature.
11. Autoclaved 2-ml microcentrifuge tubes (Eppendorf).

2.4.1. Pre-absorbed Protein A Agarose Beads

Prepare a day in advance to allow the BSA and DNA to absorb to the resin.

1. Centrifuge 1.5-ml vial containing 500 µl of compact with protein A agarose beads at 2,500 rpm in a Heraeus Biofuge Pico, rotor #3325 [593 × *g*] for 1 min in a microcentrifuge.
2. Remove and discard supernatant.
3. Wash beads with autoclaved MT-PBS pipetting with a wide-bore tip in order not to damage the beads.
4. Centrifuge vial with beads at 2,500 rpm [593 × *g*] for 1 min in a microcentrifuge.
5. Remove and discard supernatant.
6. Wash beads with autoclaved MT-PBS.
7. Centrifuge vial with beads at 2,500 rpm [593 × *g*] for 1 min as above.
8. Remove and discard supernatant.
9. Wash beads with autoclaved TE, pH 8.0.
10. Remove and discard supernatant.
11. Add autoclaved TE, pH 8.0, to beads 1:1 (v/v; i.e. 50% slurry).
12. Add 500 µg BSA (purified BSA 100×, NEB, B90015, stock 10 mg/ml) per 500 µl of *compact* bead volume.
13. Add 200 µg sonicated salmon sperm DNA per 500 µl of *compact* bead volume (stock 10 mg/ml, Stratagene).
14. Incubate overnight on roller at 4°C to allow the BSA and DNA to absorb to the resin.

2.5. Reversal of Cross Linking and DNA Recovery

1. Autoclaved 5 M NaCl.
2. Autoclaved 0.5 M EDTA, pH 8.0.
3. Autoclaved 1 M Tris–HCl, pH 6.5.
4. Proteinase K 10 mg/ml in autoclaved milliQ water.
5. Phenol, equilibrated to pH 6.5.
6. Chloroform.
7. Isoamyl alcohol.
8. Phenol/chloroform/isoamyl alcohol (50:49:1, v/v/v).
9. Chloroform/isoamyl alcohol (49:1, v/v).
10. Autoclaved 3 M sodium acetate, pH 5.2, in milliQ water.
11. 100% ethanol.
12. 70% ethanol in autoclaved milliQ water.
13. Autoclaved 10 mM Tris–HCl, pH 7.4.
14. Shaking heating block, 45 and 65°C.

Safety warning:
Read material safety instructions for all reagents before proceeding. Avoid inhalation of fumes and contact with phenol and chloroform. Wear eye protection, lab coat, and gloves. Handle in a fume hood with care. Discard phenol/chloroform according to local regulations.

3. Methods

Consideration of appropriate controls. Consideration should be given to the appropriate controls for specific experimental settings. ChIP is commonly conducted to determine the presence of a particular protein, a DNA-binding transcription factor, or a chromatin-binding transcriptional regulator at a specific locus or to determine the genome-wide distribution of such proteins. Commonly used controls are (a) an isotype control, (b) anti-RNA polymerase II control, (c) no-antibody control, and (d) input control. The isotype control is a control antibody of the same immunoglobulin isotype as the specific antibody, but not detecting an antigen in the sample material. This is a negative control detecting non-specific binding of the control antibody to chromatin-associated proteins. It does not necessarily produce the same non-specific component as the experimental antibody. The anti-RNA polymerase II antibody serves as a positive control and is expected to precipitate transcriptionally active genes and within these the transcription start site with preference. The no-antibody control yields chromatin that is precipitated non-specifically by the protein A agarose beads. The input control is derived from the same sample, but has not undergone the precipitation step. Comparison of abundance of genomic regions between input and precipitate (provided the same total amount of

DNA is used for quantitation of both) allows assessment of the enrichment of specific genomic regions by precipitation with an antibody directed against a particular chromatin-associated protein.

If specifically modified histone residues are the primary interest, an antibody precipitating all modified and unmodifed versions of the same histone is often used as a control. When starting material from different experimental groups is used, for example mouse embryos that are either wild type or homozygous for a loss-of-function mutation, wild-type material serves as a control for the mutant material.

When analysing the genomic material in the chromatin immunoprecipitate by qPCR, further controls can be introduced. These aim to assess (a) the efficiency of the qPCR reaction and (b) the overall efficiency of the ChIP. The efficiency of the PCR reaction can be assessed by analysing fivefold or tenfold dilution standard curves for each pair of PCR amplification primers and within sample by assessing the accumulation of the PCR product during each PCR cycle (7, 8). The overall efficiency of the ChIP can be assessed if the effects of the experimental variable are expected to result in pronounced differences, e.g. treatment with histone deacetylase inhibitors in the case of detecting specific, acetylated histone residues. Similarly, mutation of loci encoding a histone-modifying enzyme can be expected to result in detectable differences in histone acetylation (6). The presence of unaffected or mildly affected loci in the precipitate can be quantified and can serve as normaliser loci for the affected loci (8). Calibrator samples can serve as controls for experimental groups (8). For example, if wild-type mouse embryos serve as controls for embryos that are mutant for a particular DNA-binding transcription factor X, then a known target gene Y that is downregulated in the embryos mutant for X would be an affected locus and would be expected to exhibit a reduction in transcriptionally active chromatin marks. In contrast, an unrelated housekeeping gene that is expressed at normal levels in both wild-type and mutant embryos would be expected to have normal levels of transcriptionally active chromatin marks and can serve as a normaliser control. The great advantage of such a normaliser control is that it is a within-sample control and allows controlling for variability during the immunoprecipitation procedure, which none of the other types of controls addresses.

The most critical aspects of the method. We found that three aspects of the procedure were most essential for success: (a) consistency of chromatin fragmentation, (b) avoiding loss of material during the procedure, and (c) the specificity of the antibody. Chromatin fragmentation can be achieved by sonication and enzymatic cleavage. In Notes 1–3, sonication using two different sonication apparatuses and one enzymatic method are discussed. The optimal chromatin fragment size depends on the aim of the experiment. If the survey of a large region of DNA by qPCR is planned, less-frequent cleavage of chromatin may be desirable. A fragment peak size of 1,000 bp would allow the assessment of an average of 1 kb per qPCR primer pair, whereas fragment size of 150 bp would require seven primer pairs to cover the same genomic region. Moreover, transcription factor-binding sites are largely depleted of nucleosomes (9). There fore, fragmentation between most nucleosomes may disrupt a particular target sequence and so impair its detection by PCR. If, however, the aim is to examine chromatin modifications or protein association with a specific, short DNA sequence or mass sequencing of the precipitated DNA is planned, then short chromatin fragments may be preferable. The primer pairs are typically designed to amplify approximately 100 bp in both cases. Avoiding loss of material throughout the procedure is critical if a precipitate-internal control is impractical. A growing number of ChIP-grade antibodies are commercially available. However, good-quality antibodies that display high specificity in immunofluorescence and immunoblotting, but have not been tested in ChIP, may also be used successfully.

3.1. Embryo Recovery

The following procedure is generally subject to animal ethics regulations and a permit specifying the procedure must be obtained prior to performing the experiment.

Mouse embryos from E8.5 to E11.5 have been used successfully in this protocol. The total DNA yield in the immunoprecipitate depends on the number of genomic sites occupied by the protein of interest and the developmental stage of the embryo. An E8.5 embryo precipitated with an antibody directed against acetylated histone 3 lysine 14 yields sufficient DNA to perform assessment of six genomic sites in triplicate qPCR repeats.

1. Prepare fresh 18.5% paraformaldehyde for Subheading 3.2.
2. Sacrifice pregnant dam by cervical dislocation.
3. Dissect uterus.
4. Place uterus into 50-ml conical tube containing sterile MT-PBS.
5. Transfer uterus to 10-cm Petri dish with sterile MT-PBS.
6. Dissect embryos.
7. If required, retain yolk sac in additional labelled 1.5-ml tube for genotyping.
8. Transfer E8.5–E10.5 embryos into labelled 1.5-ml tubes.
9. Place embryo containing tubes on ice.
10. Proceed immediately to Subheading 3.2.

3.2. Cross Linking of DNA and Associated Proteins

1. Add 1 ml of MT-PBS to each embryo.
2. Keep embryos on ice.
3. Gently triturate approximately ten times to break up the embryo using a 1,000-µl micropipette.
4. Add 56 µl of fresh 18.5% formaldehyde (final concentration 1%) and invert five times to mix.

5. Incubate on rocking platform for 15 min at room temperature.
6. Add 106 μl of 1.375 M glycine to stop cross-linking reaction and invert five times to mix.
7. Incubate on rocking platform for 5 min.
8. Centrifuge for 4 min at 5,000 rpm [2,375×*g*] in microcentrifuge.
9. Aspirate and discard supernatant. (At each step, remove as much medium as possible without disturbing the cell pellet).
10. Wash cells using ice-cold MT-PBS with fresh inhibitors (complete protease inhibitors and, in addition for detection of histone acetylation, histone deacetylase inhibitor sodium butyrate).
11. Centrifuge for 4 min at 5,000 rpm [2,375×*g*] in microcentrifuge.
12. Aspirate and discard supernatant.
13. Wash cells using ice-cold MT-PBS with inhibitors.
14. Centrifuge for 4 min at 5,000 rpm [2,375×*g*] in microcentrifuge.
15. Aspirate and discard supernatant.
16. Snap freeze on dry ice or in liquid nitrogen.
17. Store at −80°C.

3.3. Fragmentation of the Chromatin

1. Defrost cross-linked cell pellet on ice.
2. Resuspend cell pellet in 200 μl of SDS lysis buffer with inhibitors (add protease inhibitors to PBS just prior to use).
3. Incubate for 10 min on ice (any SDS that precipitates during this time is dissolved in the next step).
4. Sonicate lysate on ice two times at 10% power for 11 pulses on 90% duty cycle with the stepped microprobe (see Notes 1–3 for details of sonication, alternative methods, and optimisation of chromatin fragmentation). *Safety warning: Ear protection must be worn during sonication.*
5. Rest on ice while processing other samples.
6. Repeat steps 4 and 5 two times resulting in a total of 6× 11 pulses of sonication per sample.
7. Centrifuge samples for 10 min at 13,000 rpm [16,060×*g*] at 4°C in microcentrifuge.
8. Transfer the supernatant to a new 1.5-ml microcentrifuge tube and discard pellet.
9. Withdraw 20 μl of each sample and dilute 1:50 with autoclaved milliQ water.
10. Measure the absorbance of a 1/50 dilution at 260 and 280.
11. Use the OD_{260} reading to adjust concentrations (see Note 4; adjusting the chromatin material based on optical density).
12. Transfer 200 μl of sonicated and concentration adjusted material to a fresh 2-ml microcentrifuge tube.

13. Add 1.8 ml of ChIP dilution buffer with fresh inhibitors.
14. Mix.
15. Proceed immediately to Subheading 3.4.
16. Retain 1–5% (20–100 μl, larger volumes from younger embryos) of sonicated, concentration-adjusted material as "input", i.e. starting material before precipitation at −20°C (see Note 5; uses and processing of input material).

3.4. Chromatin Immunoprecipitation

Steps 1–4 are aimed to reduce non-specific background ("pre-clearing").

1. To 2 ml of sonicated, diluted supernatant (from Subheading 3.3, step 14 above), add 75 μl of protein A agarose beads/salmon sperm DNA/BSA (50% slurry).
2. Incubate for 30 min at 4°C with agitation on rocking platform.
3. Centrifuge at 4,000 rpm [1,520 × *g*] for 1 min to pellet agarose beads.
4. Transfer the supernatant fraction to a new 2-ml microcentrifuge tube and discard pelleted beads.
5. Add the immunoprecipitating antibody to the 2-ml supernatant fraction and incubate overnight at 4°C with agitation on rocking platform. The amount of antibody to be added needs to be determined empirically. A good starting amount is between 2 and 10 μg.
6. Add 60 μl of protein A agarose beads/salmon sperm DNA/BSA (50% slurry).
7. Incubate for 1 h at 4°C with agitation on rocking platform to bind the antibody/chromatin complex.
8. Gently centrifuge at 2,500 rpm [593 × *g*] at 4°C for 1 min to pellet the agarose bead-bound antibody/chromatin complex.
9. Carefully remove the supernatant, which contains unbound chromatin. Retain the pellet for the next step. The unbound chromatin fraction can be retained for comparison of precipitated versus non-precipitated chromatin (see Note 6; uses of unbound chromatin fraction). Sequentially wash the protein A agarose beads/antibody/chromatin complex in the following manner.
10. Wash the protein A agarose beads/antibody/chromatin complex for 5 min on a roller mixer at 4°C with 1 ml of low-salt, immune complex wash buffer.
11. Gently centrifuge at 2,500 rpm [593 × *g*] at 4°C for 1 min to pellet the agarose bead-bound antibody/chromatin complex.
12. Carefully remove and discard the supernatant.

13. Wash the protein A agarose beads/antibody/chromatin complex for 5 min on a roller mixer at 4°C with 1 ml of high-salt, immune complex wash buffer.
14. Centrifuge at 2,500 rpm [593 × *g*] at 4°C for 1 min to pellet the agarose bead-bound antibody/chromatin complex.
15. Remove and discard the supernatant.
16. Wash the protein A agarose beads/antibody/chromatin complex for 5 min on a roller mixer at 4°C with 1 ml of LiCl immune complex wash buffer.
17. Centrifuge at 2,500 rpm [593 × *g*] at 4°C for 1 min to pellet the agarose bead-bound antibody/chromatin complex.
18. Remove and discard the supernatant.
19. Wash the protein A agarose bead-bound antibody/chromatin complex for 5 min on a roller mixer at 4°C with 1 ml of TE buffer.
20. Centrifuge at 2,500 rpm [593 × *g*] at 4°C for 1 min to pellet the agarose bead-bound antibody/chromatin complex.
21. Remove and discard the supernatant.
22. Wash the protein A agarose bead-bound antibody/chromatin complex for 5 min on a roller mixer at 4°C with 1 ml of TE buffer.
23. Centrifuge at 2,500 rpm [593 × *g*] at 4°C for 1 min to pellet the agarose bead-bound antibody/chromatin complex.
24. Remove and discard the supernatant.

    This bead-bound fraction could be used directly for the analysis of chromatin-precipitated proteins or as outlined in the following steps to recover the chromatin-precipitated DNA and associated proteins.
25. Add 250 μl of freshly prepared elution buffer to the pelleted protein A agarose beads/antibody/chromatin complex.
26. Invert several times and vortex briefly to mix.
27. Incubate at room temperature for 15 min on the roller mixer with regular inversion.
28. Centrifuge at 4,000 rpm [1,520 × *g*] for 1 min in microcentrifuge.
29. Carefully transfer the supernatant fraction (eluate) to a fresh tube.
30. To the pelleted protein A agarose beads, add 250 μl elution buffer to elute a second time.
31. Invert several times and vortex briefly to mix.
32. Incubate at room temperature for 15 min on the roller mixer with regular inversion.
33. Centrifuge at 4,000 rpm [1,520 × *g*] for 1 min as above.

34. Carefully transfer the supernatant fraction (eluate) to combine eluates (total volume = ~500 μl) and discard beads.
35. Centrifuge combined eluates at 4,000 rpm [1,520 × *g*] for 2 min to pellet any residual beads.
36. Carefully transfer the combined eluates to a new tube and discard beads (see Note 7; alternative precipitation method).

3.5. Reversal of Cross Linking and DNA Extraction

1. Add 20 μl of 5 M NaCl to the combined eluates (500 μl).
2. Mix by vortexing.
3. Heat at 65°C for 4 h to reverse histone-DNA cross linking.
4. Optional: Store sample at −20°C and continue the next day.
5. Add 10 μl of 0.5 M EDTA, 20 μl of 1 M Tris–HCl, pH 6.5, and 2 μl of 10 mg/ml proteinase K to the combined eluates.
6. Mix by vortexing.
7. Incubate for 1 h at 45°C.
8. Add an equal volume of phenol/chloroform/isoamyl alcohol (50:49:1).
9. Mix by vortexing.
10. Centrifuge at 13,000 rpm [16,060 × *g*] for 5 min in microcentrifuge.
11. Recover aqueous top phase into a fresh 2-ml microcentrifuge tube. Do not disturb interface.
12. Add an equal volume of chloroform/isoamyl alcohol (49:1).
13. Centrifuge at 13,000 rpm [16,060 × *g*] for 5 min as above.
14. Recover aqueous top phase into a fresh 2-ml microcentrifuge tube. Do not disturb interface.
15. Add 1/10 volume of 3 M sodium acetate, pH 5.2 (55 μl).
16. Mix by vortexing.
17. Optional: Add 20 μg glycogen (1 μl of 20 mg/ml). Mix by vortexing (see Note 8).
18. Add 2.2 volumes of 100% ethanol (1,320 μl).
19. Mix by vortexing.
20. Incubate at −20°C overnight.
21. Centrifuge at 13,000 rpm [16,060 × *g*] for 15 min.
22. Remove and discard supernatant without touching the pellet.
23. Wash pellet and tube with 1 ml 70% ethanol.
24. Centrifuge at 13,000 rpm [16,060 × *g*] for 10 min.
25. Remove and discard supernatant without touching the pellet.
26. Air dry pellet briefly.
27. Resuspend in 20–100 μl of 10 mM Tris–HCl, pH 7.4, depending on the age of the embryo and the expected abundance of chromatin- or DNA-binding sites of the precipitation target.

28. Mix by vortexing.
29. Inspect visually for dissolution.
30. Store at −20°C.

This DNA material is suitable for quantitation by a variety of methods, including qPCR.

4. Notes

1. *Chromatin fragmentation using a probe sonicator.* Sonication generates a considerable amount of heat, which can lead to disruption of protein–DNA interactions and to degradation, if the samples are not kept cool. Sonication is carried out with the tubes held in a Perspex tube rack containing an ice water bath. This keeps the samples cool while allowing the samples to be viewed from the side. The probe is positioned into the sample tube below the surface of the liquid close to the bottom of the tube, but without touching the tube. Operation of the sonication probe too close to the surface of the liquid should be avoided, as this can lead to foaming of the sample, which can lead to loss of material. Wash sonicator probe between samples by sonicating autoclaved milliQ water in a 50-ml conical tube. These conditions work well using the Branson 250 sonifier with a stepped microtip (Fig. 1). Other cell types or sonicators require optimisation (see Note 3).
2. *Other chromatin fragmentation methods.* We have conducted several tests using a sonication bath (Bioruptor) instead of a sonicator probe inserted into the sample. While the sonication bath yielded excellent results when using small numbers of cells

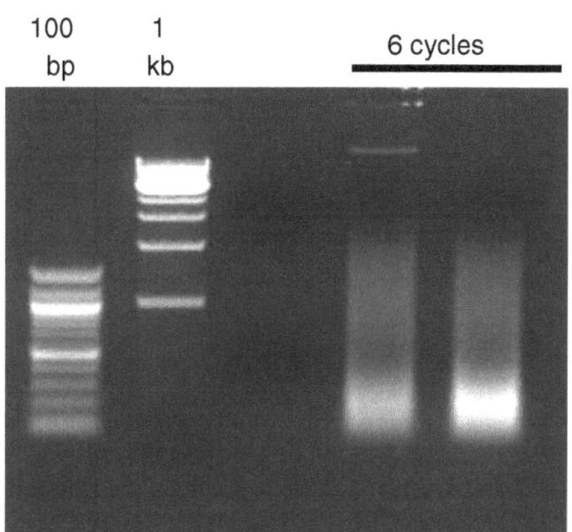

Fig. 1. E10.5 embryos dissociated mechanically, cross-linked, and fragmented using the Branson 250 sonifier with a stepped microtip for the number of 11-pulse cycles indicated above the lanes. Material from one-tenth of an E10.5 embryo loaded per lane. 100 bp and 1 kb indicate 100 bp and 1 kb ladder.

in single-cell suspension, the chromatin of whole, dissociated E10.5 embryos was not fragmented consistently in the sonication bath (Figs. 2–4). A whole E10.5 embryo may yield too much material for efficient and consistent fragmentation in the sonication bath. We have successfully used an enzymatic chromatin fragmentation method (Millipore, 17–375) on E8.5 mouse embryos (Fig. 5). The majority of DNA fragments yielded using either of these alternative methods were approximately 150 bp, suggesting that the chromatin was cleaved between most nucleosomes. Chromatin fragment size was easily controlled varying the number of cycles using the Branson 250 sonifier with a stepped microtip. The optimal fragment size depends on the aim of the project (see discussion in Subheading 3, Methods).

3. *Optimisation of chromatin fragmentation*. Optimisation of chromatin fragmentation is conducted by comparing different sonication conditions, for example vary the number of cycles of sonication, cycle length, power setting, etc.

   (a) Follow procedures from Subheadings 3.1 and 3.2.

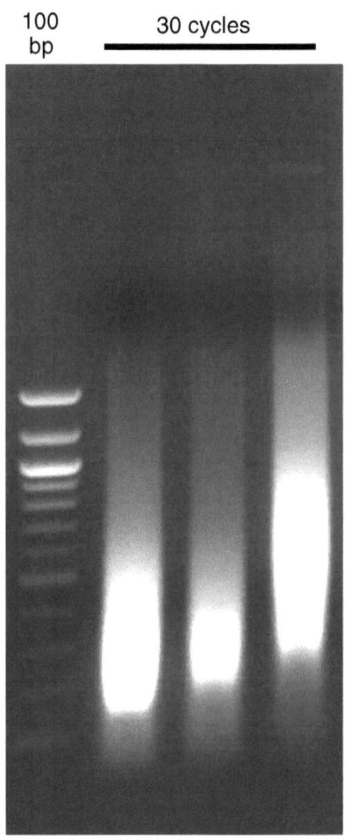

Fig. 2. E10.5 embryos dissociated mechanically (pipetting) and subjected to chromatin fragmentation using a sonication bath set to maximal power for thirty 30 s on/30 s off (1 min) cycles. Material from one E10.5 embryo loaded per lane. Note the inconsistent range of fragment sizes and inconsistent recovery of material. 100 bp indicates 100 bp ladder.

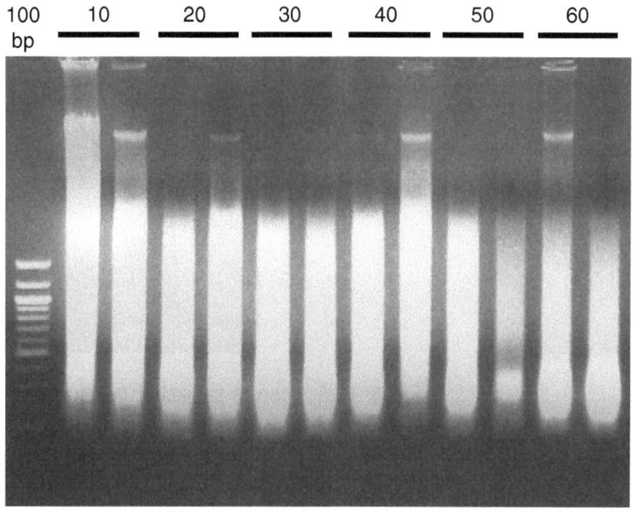

Fig. 3. E10.5 embryos dissociated mechanically, cross-linked, and fragmented using a sonication bath for the number of 30 s on/30 s off (1 min) cycles indicated above. Material from one E10.5 embryo loaded per lane. 100 bp indicates 100 bp ladder.

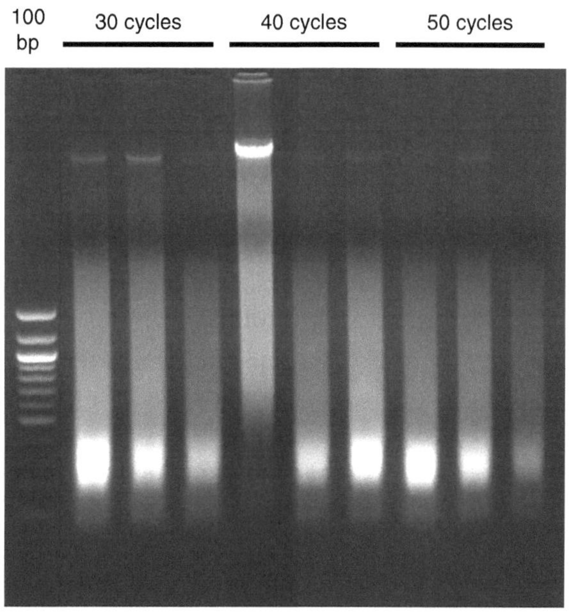

Fig. 4. E10.5 embryos dissociated enzymatically (trypsin) and subjected to chromatin fragmentation using a sonication bath set to maximal power for the number of 30 s on/30 s off (1 min) cycles as indicated above the lanes (one embryo per sample). Note narrow peak of fragment size around 300 bp, the inconsistent fragmentation of the three samples processed for 40 cycles, and the inconsistency in material recovery within each treatment group. Material from one E10.5 embryo loaded per lane. 100 bp indicates 100 bp ladder.

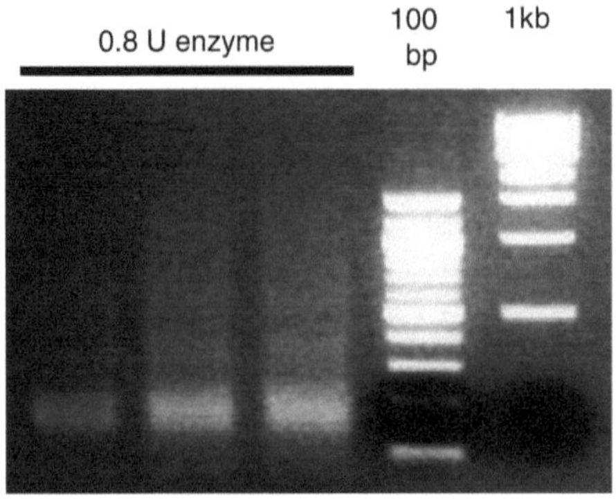

Fig. 5. E8.5 embryos subjected to an enzymatic chromatin preparation kit (Upstate, Millipore, 17–375) according to the manufacturer's instructions. The E8.5 embryos in the three lanes from left to right were increasingly developmentally advanced and, therefore, yielded increasing amounts of material. Material from one-third of an E8.5 embryo loaded per lane. 100 bp and 1 kb indicate 100 bp and 1 kb ladder.

(b) Retain two samples as an unsheared DNA control. Resuspend these in 200 μl of SDS lysis buffer.

(c) With the remaining samples, follow procedure from Subheading 3.3, steps 1–8, but vary sonication conditions. Use two samples for each sonication condition as replicates.

(d) Reverse cross linking as in procedure from Subheading 3.5, steps 1–3.

(e) Add 1 μl of DNase-free RNase A and incubate at 37°C for 30 min.

(f) Extract DNA as in procedure from Subheading 3.5, steps 5–7, 15, 16, and 18–29 (i.e. without phenol/chloroform/isoamyl alcohol extraction and without glycogen addition).

(g) Separate DNA fragments on a 1% agarose/ethidium bromide gel alongside 100 bp to 5 kb DNA fragment size marker.

(h) Photograph ethidium bromide-stained DNA.

(i) Examine fragment sizes.

Fragment sizes ranging from 200 bp to 2 kb with a peak fragment size of 1 kb are suitable for many applications and allow the sampling of the genome in tiles of an average size of 1 kb. For other applications, smaller or larger fragment sizes may be desirable and these can be produced by varying the sonication conditions.

4. *Adjusting the chromatin material based on optical density.* The optical density is an approximate assessment of the relative

concentration of material between samples. As the samples contain cross-linked DNA and protein, more exact measurements are not practical. Use the OD_{260} reading to adjust concentrations such that equal amounts and concentration of material can be used in the immunoprecipitation step. Make up differences in volume with SDS lysis buffer as required. Mix samples by inverting. Correct all samples to the level of the most dilute sample. However, if one sample is significantly lower than all others, it may be preferable to discard this sample rather than diluting the other samples excessively.

5. *Uses and processing of the input material.* One half each of the input material can be used to determine (a) the success of chromatin fragmentation by gel electrophoresis (see Notes 1–3 and Figs. 1–5) and (b) enrichment of genomic loci in the immunoprecipitate over the input material (total genome).

   For assessments of chromatin fragmentation, the cross linking has to be reversed, as in Note 3, (d) to (i), but with reduced volumes, e.g. 20 μl of input made up to 250 μl with elution buffer.

   For determining enrichment of genomic loci in the immunoprecipitate over input, the cross linking has to be reversed, the DNA in the input material has to be purified, as in Subheading 3.5. To facilitate processing of input samples, dilute in elution buffer. Typically, input samples are made to 250 μl with elution buffer and processed as half the volumes stated in Subheading 3.5. If enrichment over input is to be determined, both input and precipitate require spectrophotometric assessment of DNA concentration.

6. *Uses and processing the unbound chromatin fraction.* Like the input sample above, the unbound fraction has to undergo reversal of cross linking and DNA recovery. The unbound chromatin fraction is expected to be depleted of proteins and DNA fragments that are specifically enriched in the precipitate.

7. *Alternative precipitation method.* In young embryos (E8.5), the liquid volumes that cannot be retrieved from between the beads in the protein A/G–agarose bead precipitation method can substantially reduce the yield. We have used magnetic beads (Millipore, 16–661) as a successful alternative, if the starting material was limiting.

8. *Effects of the addition of glycogen during the sodium acetate/ethanol precipitation step on subsequent quantitation of DNA by qPCR.* Addition of glycogen to the material after reversal of the cross linking can improve the yield during DNA precipitation (Fig. 6).

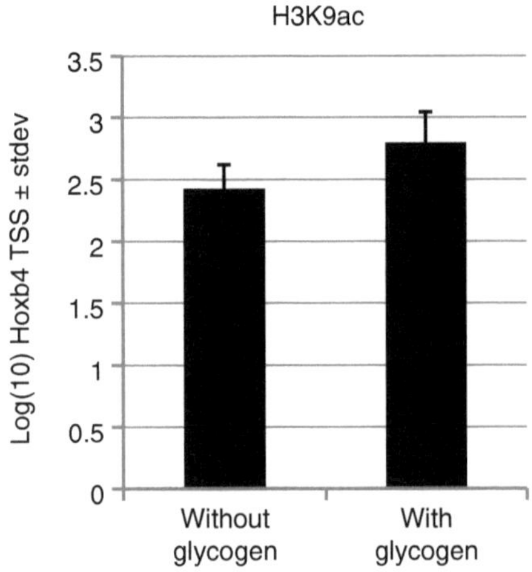

Fig. 6. Q-PCR detection of the transcription start site of the *Hoxb4* gene in anti-acetylated histone H3 lysine 9 (H3K9ac) ChIP material subjected to DNA precipitation in the presence or absence of glycogen. In this case, glycogen neither appear to improve yield dramatically nor to affect the efficiency of the qPCR reaction significantly.

References

1. Luger, K., Mader, A. W., Richmond, R. K., Sargent, D. F., and Richmond, T. J. (1997) Crystal structure of the nucleosome core particle at 2.8 A resolution, *Nature 389*, 251–260.
2. Kireeva, N., Lakonishok, M., Kireev, I., Hirano, T., and Belmont, A. S. (2004) Visualization of early chromosome condensation: a hierarchical folding, axial glue model of chromosome structure, *J Cell Biol 166*, 775–785.
3. Tremethick, D. J. (2007) Higher-order structures of chromatin: the elusive 30 nm fiber, *Cell 128*, 651–654.
4. Schones, D. E., and Zhao, K. (2008) Genome-wide approaches to studying chromatin modifications, *Nat Rev Genet 9*, 179–191.
5. Shogren-Knaak, M., Ishii, H., Sun, J. M., Pazin, M. J., Davie, J. R., and Peterson, C. L. (2006) Histone H4-K16 acetylation controls chromatin structure and protein interactions, *Science 311*, 844–847.
6. Voss, A. K., Collin, C., Dixon, M. P., and Thomas, T. (2009) Moz and retinoic acid coordinately regulate H3K9 acetylation, Hox gene expression, and segment identity, *Dev Cell 17*, 674–686.
7. Pfaffl, M. W. (2004) Quantification strategies in real-time PCR, in *A-Z of quantitative PCR* (Bustin, S. A., Ed.), International University Line (IUL), La Jolla, CA, USA.
8. Taneyhill, L. A., and Adams, M. S. (2008) Investigating regulatory factors and their DNA binding affinities through real time quantitative PCR (RT-QPCR) and chromatin immunoprecipitation (ChIP) assays, *Methods Cell Biol 87*, 367–389.
9. Henikoff, S. (2008) Nucleosome destabilization in the epigenetic regulation of gene expression, *Nat Rev Genet 9*, 15–26.

Chapter 24

Chromatin Immunoprecipitation in Mouse Hippocampal Cells and Tissues

Badi Sri Sailaja, Takumi Takizawa, and Eran Meshorer

Abstract

Chromatin immunoprecipitation (ChIP) has been developed for studying protein–DNA interactions and has been extensively used for mapping the localization of posttranslationally modified histones, histone variants, transcription factors, or chromatin modifying enzymes at a given locus or on a genome-wide scale. ChIP methods have been modified and improved over the years to fit a variety of different cell types and tissues. Here, we present a detailed protocol for hippocampal ChIP, of both minced tissue and enzyme-separated hippocampal cells. This protocol enables to study chromatin–protein interactions in a specified population of hippocampal cells, allowing to study chromatin regulation in the central nervous system in a variety of conditions and disorders. Our assay has been developed for histone modifications but is suited for any chromatin binding protein for which specific ChIP-grade antibodies are available.

Key words: Chromatin immunoprecipitation, ChIP, Hippocampus, Histone modifications, Brain, Transcription, Neurons, Chromatin

1. Introduction

The eukaryotic genome is packaged as a highly compacted DNA–protein complex termed chromatin that forms primarily during DNA replication. The newly synthesized DNA is wrapped around a histone octamer particle, composed of two copies each of the core histones H2A, H2B, H3, and H4, which together with the DNA, form the core subunit of chromatin – the nucleosome. Two adjacent nucleo somes are connected through linker DNA, which is usually bound by the linker histone H1. Dynamic changes in chromatin structure, which are mainly manifested through chromatin modifications, mark an important contribution to the regulation of gene expression in dividing as well as nondividing cells, such as neurons (1–3).

Ales Vancura (ed.), *Transcriptional Regulation: Methods and Protocols*, Methods in Molecular Biology, vol. 809,
DOI 10.1007/978-1-61779-376-9_24, © Springer Science+Business Media, LLC 2012

In particular, acetylation/deacetylation and methylation/demethylation of specific lysine residues on nucleosomal histone proteins, as well as methylation/demethylation of cytosines in the DNA, are ways by which chromatin structure can influence ongoing transcription. Such chromatin-related changes may lead, in turn, to altered synaptic plasticity. Studying the interactions of nuclear proteins with DNA has been immensely improved by the advent of chromatin immunoprecipitation (ChIP) (4–6). ChIP provides a reliable method to monitor the presence of proteins and protein modifications (especially histones) on chromatin. ChIP coupled with real-time PCR (qPCR) has become the gold standard to determine chromatin–protein interactions (7) and has been continuously improved and modified to fit a variety of different cell types and tissues (8–12). In recent years, the combination of ChIP with DNA microarrays (ChIP-chip) or high-throughput sequencing (ChIP-Seq) technologies has enabled the profiling of histone modifications, histone variants, and transcription factor occupancy on a genome-wide scale (13). Here, we provide a refined ChIP protocol for hippocampal tissue and a modified version for hippocampal cells, which can be used as template for real-time PCR, tiling arrays, or high-throughput sequencing.

Hippocampal ChIP involves the surgical dissection and mincing of the hippocampus. ChIP can be performed on the entire hippocampus, on specific hippocampal cell types (i.e., neurons, cholinergic neurons, etc.) after FACS sorting, and also on cultured hippocampal neurons following single cell separation of the tissue (Fig. 1). In all cases, chromatin is purified and cross-linked. The isolated chromatin is then reduced to smaller fragments by sonication, and chromatin fragments in the form of protein–DNA complexes are precipitated using specific antibodies. The precipitated DNA is purified and can be used as a template for real-time PCR (ChIP), microarrays (ChIP-chip) or deep sequencing (ChIP-seq).

2. Materials

2.1. Chemicals

1. 36.5% Formaldehyde: Molecular biology grade formaldehyde solution is used. Dilute with PBS to make it 1% (add 270 μl of formaldehyde to 10 ml PBS). Formaldehyde is hazardous and should be handled inside a chemical hood.
2. Glycine: 2.5 M glycine stock is prepared by dissolving 18.76 g of tissue culture grade glycine in 100 ml of deionized water. Dilute in 10 ml of deionized water by adding 500 μl of 2.5 M glycine to make a final concentration of 0.125 M.
3. Protein A/G Agarose beads: Use protein A Agarose beads or magnetic beads, which can reduce some background and

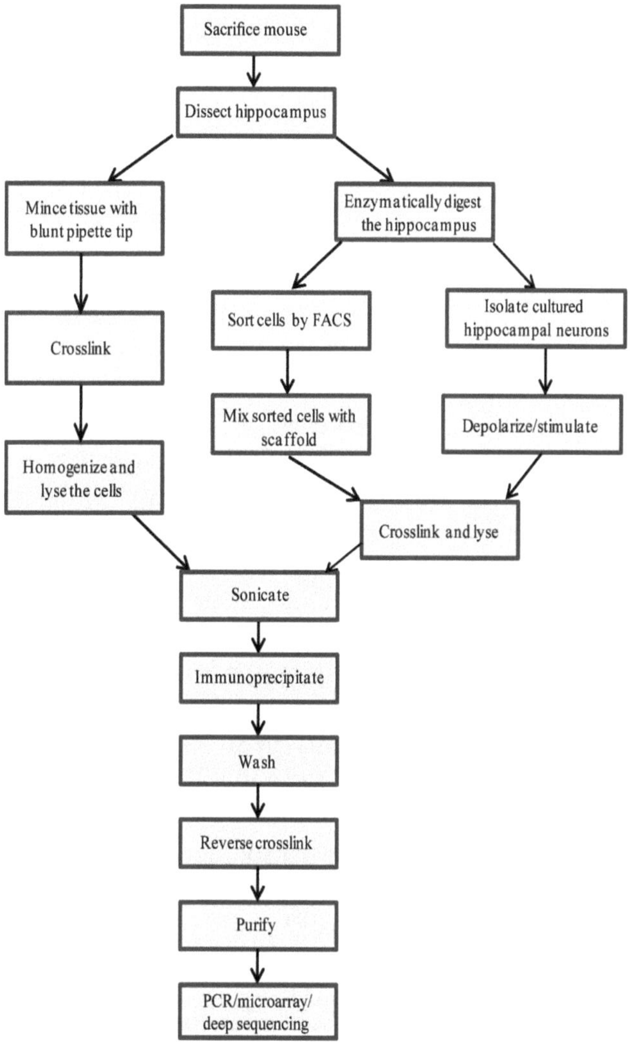

Fig. 1. A flowchart of the three ChIP protocols on hippocampal cells and tissue.

loss of sample. For magnetic beads, use Dynabeads Protein G (Invitrogen) for rabbit polyclonal primary antibodies and Dynabeads M-280 anti-Mouse IgG for mouse monoclonal antibodies.

4. If using magnetic beads, use DynaMag-2 magnet (Invitrogen).
5. Dounce homogenizer (Kontes): Use pestle B for hippocampal tissue.
6. PBS (phosphate buffered saline): Dissolve 137 mM NaCl, 2.7 mM KCl, 10 mM sodium phosphate dibasic, 2 mM potassium phosphate monobasic in 1 l of deionized water to make it 1× PBS and adjust pH to 7.4 with HCl or NaOH.
7. PMSF: 200 mM PMSF stock is prepared by dissolving 0.348 g of PMSF in 10 ml of absolute ethanol or methanol. Aliquot

into 1.5 ml microcentrifuge tubes and store at −20°C. The half-life of thawed PMSF is 30 min, so add before use.

8. 3 M sodium acetate, pH 5.2: Dissolve 20.4 g of sodium acetate in 50 ml of deionized water. Adjust pH to 5.2 with HCl.
9. Aprotinin, leupeptin, and pepstatin: Prepare 1 mg/ml stocks and store as 1 ml aliquots in −20°C.
10. RNase: Ribonuclease (Sigma) 10 mg/ml stock solution is dissolved in 10 mM sodium acetate buffer, pH 5.2. Heat to 100°C for 15 min, allow to cool to room temperature, then adjust to pH 7.4 with 0.1 volumes of 1 M Tris–HCl, pH 7.4. Aliquot RNase into 1.5-ml microcentrifuge tubes and store at −20°C.
11. Proteinase K: Prepare 10 mg/ml stock solution of Proteinase K (Sigma), aliquot and store in −20°C.
12. Phenol:chloroform: Molecular biology grade phenol–chloroform–isoamyl alcohol (25:24:1) saturated with 10 mM Tris, pH 8.0, 1 mM EDTA (Sigma) is used. It is hazardous; therefore, it needs to be handled inside a chemical hood. Store at 4°C.
13. Papain solution: Combine 2 mM DL-cysteine hydrochloride, 46 mM D (+) glucose, 10 mg BSA, and 15,000 U of DNaseI (Sigma) in 30 ml of PBS. Filter-sterilize, prepare 3 ml aliquots, and store in −20°C until use. Add 100 μl of papain (Sigma) just before use.
14. DNase I solution: Dissolve 10,000 U of DNase I (Sigma) and 12 mM $MgSO_4 \cdot 7 H_2O$ in 20 ml of HBSS. Aliquot and store in −20°C.
15. Trypsin solution: Dissove 2 mM DL-cysteine hydrochloride, 46 mM D-glucose, 10 mg BSA, and 15,000 DNase I (Sigma) in 30 ml of PBS. Filter (0.22 μm), aliquot (3 ml), and store in −20°C. Add 0.1% (v/v) trypsin EDTA solution just before use.
16. Plating medium: Use alpha MEM medium (Invitrogen) supplemented with 10% FBS and antibiotic–antimycotic mix (Invitrogen) 1:100.
17. Maintenance medium: Use Neurobasal medium (Invitrogen) supplemented with B-27 serum-free supplement 1:50 (Invitrogen), GlutaMax diluted 1:100 (Invitrogen) and antibiotic–antimycotic mix diluted 1:100.
18. Poly-L-lysine (PLL) coated dishes: Dissolve 1 mg/ml PLL hydrobromide (Sigma) into borate buffer (1.24 g boric acid, 1.9 g borax, 400 ml of deionized water), aliquot, and store at −20°C until use. Add 10 μg/ml PLL hydrobromide (diluted 1:100 in borate buffer) onto culture dishes and keep overnight at room temperature. Wash three times with deionized water and dry completely.

2.2. Buffers

1. Lysis buffer: 10 mM EDTA, 1% (w/v) SDS, 50 mM Tris–HCl, pH 7.5. Add the following protease inhibitors fresh to the lysis buffer just before use: 1 mM PMSF, 10 mg/ml aprotinin, 10 mg/ml leupeptin, and 10 mg/ml pepstatin (see Note 1).
2. ChIP dilution buffer: 0.01% (w/v) SDS, 1.1% (v/v) Triton, 1.2 mM EDTA, 16.7 mM Tris–HCl, pH 8.1, and 167 mM NaCl. Add the protease inhibitors fresh to the ChIP dilution buffer just before use.
3. Low salt wash buffer: 0.1% SDS, 1% Triton, 2 mM EDTA, 20 mM Tris–HCl, pH 8.1, 15 mM NaCl.
4. High salt wash buffer: 0.1% SDS, 1% Triton, 2 mM EDTA, 20 mM Tris–HCl, pH 8.1, 500 mM NaCl.
5. LiCl wash buffer: 0.25 M LiCl, 1% (v/v) NP-40, 1% deoxycholic acid, 1 mM EDTA, 10 mM Tris–HCl, pH 8.1.
6. Elution buffer: 1% SDS, 0.1 M $NaHCO_3$.
7. TE: 10 mM Tris–HCl, pH 7.5, 1 mM EDTA.

3. Methods

3.1. ChIP on Mouse Hippocampus

3.1.1. Cross-Linking, Lysis, and Sonication

1. Dissect the hippocampus from an anesthetized and decapitated mouse. At this point, tissue can be snap-frozen in liquid nitrogen and stored in −80°C.
2. If frozen, thaw tissue on ice. Crush or mince the entire fresh/thawed hippocampus with a blunt 1,000 μl tip. If tissue is crushed using a mortar and pestle and liquid nitrogen, it leads to a significant loss of sample. To make a blunt tip, quickly heat the end of the tip in fire and immediately smoothen it inside a 1.5-ml microcentrifuge tube (see Note 2).
3. Fix the crushed tissue with 1% formaldehyde and incubate with gentle agitation for 15 min at room temperature to cross-link DNA to bound proteins (see Note 3). Centrifuge the sample at 440 × *g* for 5 min at room temperature in a swing-out rotor.
4. Add 0.125 M glycine from 2.5 M stock solution to terminate cross-linking. Incubate the sample in the centrifuge tube for 5 min at room temperature and centrifuge at 440 × *g* for 5 min at room temperature in a swing-out rotor.
5. Wash with cold PBS (2–3 times). Add 1 mM PMSF, 1 μg/ml aprotinin, 1 μg/ml pepstatin, and 1 μg/ml leupeptin just before use and centrifuge the cells at 440 × *g* for 4 min at 4°C in a swing-out rotor.
6. Aspirate the supernatant with the pipette and resuspend the pellet in 1 ml of ice-cold lysis buffer with freshly added protease inhibitors and PMSF for 15 min on ice and homogenize the samples

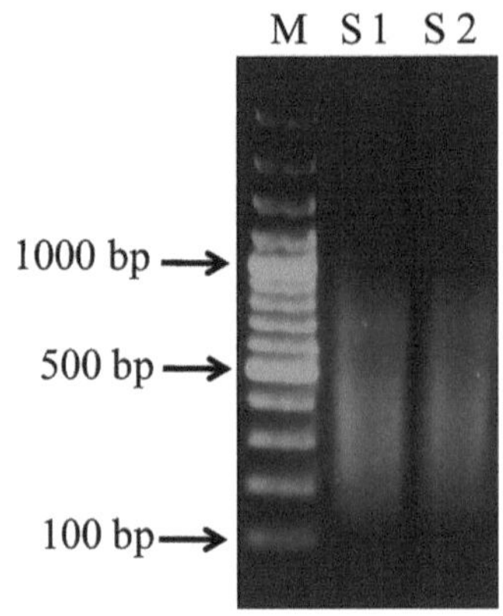

Fig. 2. Chromatin shearing of hippocampus. Chromatin is sonicated to fragments of 200–1,000 bp, which are electrophoresed on 1.2% agarose gels. M denotes a 100 bp ladder; S1 and S2 denote the sonicated samples.

slowly (10 strokes on ice) with dounce homogenizer and tight pestle. After homogenization, transfer the sample to the 1.5-ml microcentrifuge tube carefully to avoid sample loss in this critical step and centrifuge at 2,260 × *g* for 5 min in a swing-out rotor.

7. Aspirate the supernatant and resuspend the pellet in ice-cold lysis buffer (250–600 μl).
8. Sonicate for 20 min using Bioruptor sonicator with 30 s on/off pulses twice for 10 min each while replenishing the ice and water. Alternatively, sonicate 3× for 7 min each (21 min total). Centrifuge at 20,780 × *g* for 30 min at 4°C on fixed-angle rotor (see Note 4).
9. Collect the supernatant carefully and transfer to another 1.5-ml microcentrifuge tube. Chromatin can be snap-frozen in liquid nitrogen and kept in −80°C
10. To evaluate shearing, add 75 μl of deionized water and 4 μl of 5 M NaCl to 25 μl of supernatant and reverse cross-link at 65°C (4 h to overnight). Run the sheared chromatin on 1% agarose gel (Fig. 2).

3.1.2. Immunoprecipitation and DNA Purification

1. The chromatin should be thawed on ice, if previously frozen. Save 20% of pre- immunoprecipitated lysate as input for later normalization (step 7). The rest of the chromatin should be divided into three microcentrifuge tubes (for IgG, a positive control and the antibody of choice) for immunoprecipitation, each containing 10–60 μg of chromatin. Add ChIP dilution buffer to a final volume of 1.5 ml.
2. To reduce non specific background, pre-clear the 1.5 ml diluted cell supernatant with 40 μl of Protein A-Agarose beads for 45 min at 4°C with agitation (add beads with a pre-cut tip) and

pellet beads by centrifugation at 4,420 × *g* for 1 min and collect the supernatant fraction (see Note 5). In case of magnetic beads, use magnet to collect the supernatant instead of centrifuge.

3. Add antibody (1–10 μg) to the 1.5 ml of supernatant and incubate overnight at 4°C with rotation to collect the antibody complex (see Note 6). Pellet beads by gentle centrifugation at 110 × *g* for 1 min at 4°C. Carefully remove the supernatant that contains the unbound DNA.
4. Add 60 μl of salmon sperm DNA/Protein A-Agarose slurry with a precut tip and incubate for 1 h at 4°C with rotation to collect the antibody–histone complex and pellet beads by gentle centrifugation at 110 × *g* for 1 min at 4°C. Carefully remove the supernatant that contains the unbound DNA.
5. The beads should be washed for 2 min on rotating platform with 1 ml of each buffer listed.
   (a) Low salt wash buffer
   (b) High salt buffer
   (c) LiCl buffer
   (d) 1× TE

   Centrifuge at 110 × *g* at 4°C for 1 min and remove the supernatant between each wash (see Note 7).
6. Add 125 μl of freshly prepared elution buffer to the beads, vortex briefly, and incubate at room temperature for 15 min with rotation. Spin down beads at 440 g at room temperature for 1 min and transfer the eluate to another 1.5-ml microcentrifuge tube. Repeat elution to make a final volume of 250 μl.
7. Add 20 μl of 5 M NaCl to the combined eluates and reverse histone–DNA cross-links by heating at 65°C for 2–16 h. Include input material from step 1 here (add 12 μl of 5 M NaCl to 150 μl of input). Add 5 μl of 0.5 M EDTA (3 μl to input), 10 μl of 1 M Tris–HCl, pH 6.5 (6 μl to the input), 1 μl of 10 mg/ml RNase and 2 μl of 10 mg/ml proteinase K to the combined eluates and incubate for 1 h at 45°C.
8. Recover DNA by adding equal volumes of phenol–chloroform (see Note 8). Add 1/10 volume of 3 M sodium acetate, pH 5.2, and 2 volumes of isopropanol and keep in −80°C for 1 h to overnight. Wash pellets with 70% ethanol and air-dry. Resuspend the pellet in deionized water or TE (60–100 μl), and measure the concentration of the purified DNA. Alternatively, use DNA cleanup kit (Qiagen). The purified DNA can now be used as a template for real-time PCR or for high-throughput sequencing/hybridization on tiling arrays.

3.2. ChIP on Mouse Hippocampal Neurons by Enzymatic Hippocampal Cell Dissociation

In this method, the hippocampus is subjected to enzymatic digestion with trypsin or papain. To perform ChIP on a specified population of neurons, cell sorting can be performed using FACS ARIA sorter (or equivalent). Sorted cells should be mixed with scaffold cells of a different organism (with unrelated DNA), such as insect cells.

3.2.1. Preparation of Cells: Papain/Trypsin Digestion

1. Dissect hippocampi and wash with HBSS. Add papain solution (or trypsin solution) and incubate at 37°C for 15 min. Add 5 ml of plating medium to stop the digestion and centrifuge at 110 × *g* for 5 min and discard supernatant.
2. Add DNase I solution with glass pipette (5 ml). Use a 200-μl tip at the tip of a glass pipette and triturate until no visible chunk is seen. Add 5 ml of plating medium and pass through 70 μm filter to a 50 ml falcon tube. Centrifuge at 110 × *g* for 5 min.
3. Collect supernatant very carefully. Suspend the cells in 5 ml of plating medium and centrifuge at 110 × *g* for 5 min. Collect supernatant very carefully.
4. Suspend the cells in 1× PBS and wash with PBS. Indirect immunostaining protocol of FACS can be followed according to Abcam's protocol and sorting of the cells can be done with the specific antibodies using FACS ARIA sorter.

3.2.2. Cross-Linking, Cell Lysis and Sonication

1. After cell sorting, mix with proper scaffold and start with one to five million total cells.
2. Cross-link proteins to DNA by adding 1% formaldehyde (270 μl of 37% formaldehyde into 10 ml of medium). Incubate for 10 min at room temperature with rotation and centrifuge at 440 × *g* for 5 min at room temperature on a swing-out rotor.
3. Add 0.125 M glycine (500 μl of 2.5 M glycine into 10 ml of medium). Incubate for 10 min at room temperature with rotation and centrifuge at 440 × *g* for 5 min at room temperature on a swing-out rotor.
4. Wash cells twice with ice-cold PBS. Add 5 ml of ice-cold PBS containing protease inhibitor cocktail (1 mM PMSF, 1 μg/ml aprotinin, 1 μg/ml leupeptin, 1 μg/ml pepstatin A). Add PMSF just prior to use. Centrifuge for 5 min, 110 × *g* at 4°C.
5. Warm lysis buffer to room temperature and add protease inhibitors just before use. Resuspend the pellet in ice-cold lysis buffer (250–600 μl) (250 μl to two million cells).
6. Proceed to the sonication step 8 of Subheading 3.1.1.

3.2.3. Immunoprecipitaion and DNA Purification

Refer to Subheading 3.1.2.

3.3. ChIP on Cultured Primary Hippocampal Neurons

3.3.1. Culture Isolated Hippocampal Neurons (see Note 9)

1. After triturating hippocampal neurons (follow steps 1 and 2 of Subheading 3.2.1), suspend the cells in plating medium and determine the cell density. The total yield should be around $0.8–1.0 \times 10^6$ cells per embryo. Plate cells onto Poly-L-lysine (PLL) coated dishes. Our standard density is $1.6–1.8 \times 10^6$ cells per 100-mm dish. We usually use three 100-mm dishes for one histone antibody but the number should be optimized for each antibody (Fig. 3).
2. After 3–4 h, change the medium to maintenance medium.
3. On the third day of culture, replace one half of the volume with fresh maintenance medium containing 10 μM of cytosine arabinoside (1-β-D- arabinofuranosylcytosine).
4. Replace one half of the medium every 3 days. After more than 10 days in culture, intertwined network of dendrites and axons is developed. Cells are ready to be used for ChIP. Neurons can be depolarized with KCl or a GABA-A receptor antagonist, bicuculline, or stimulated with a variety of ligands. The culture has the advantage of being easy to manipulate for studying neuronal responses to various stimuli.

3.3.2. Cross-Linking, Cell Lysis, and Sonication

1. Cross-link cells by directly adding 1% formaldehyde. Incubate at room temperature for 10 min.

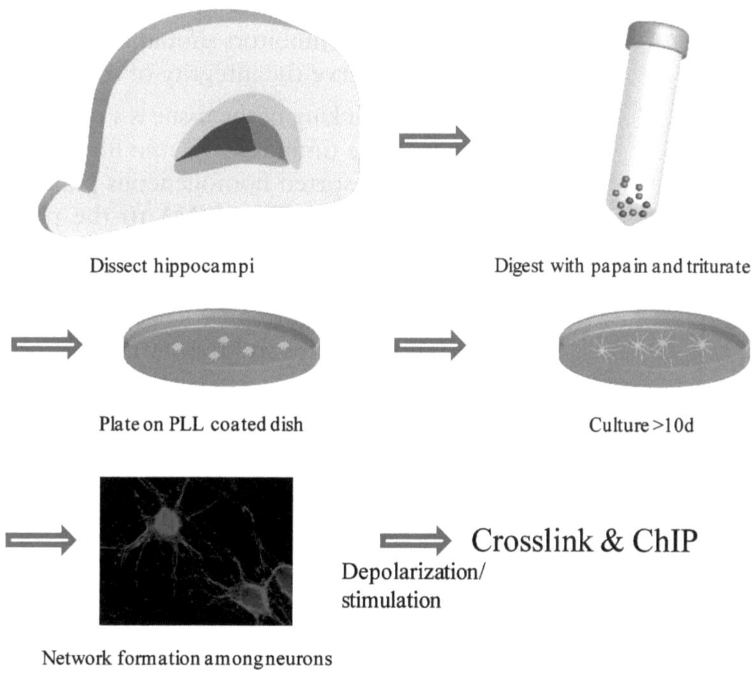

Fig. 3. Culture of hippocampal neurons. Hippocampi are digested and cell suspension is plated on dishes. After culturing for more than 10 days, neurons develop intricated connections and can be depolarized or stimulated.

2. Add 0.125 M glycine and incubate for 10 min at room temperature.
3. Wash the cells with ice-cold PBS twice and then add PBS (2 ml per 100 mm-dish) containing protease inhibitor cocktail. Collect the cells into a 15-ml tube by scraping off with cell lifter.
4. Spin down the cells by centrifugation for 10 min, 110×*g* at 4°C.
5. Proceed to step 5 of Subheading 3.2.2.

3.3.3. Immunoprecipitation and DNA Purification

Refer to Subheading 3.1.2.

4. Notes

1. ChIP assays have been extensively used to study histone acetylation and methylation patterns at a variety of genomic loci. To analyze histone acetylation, add 10 mM sodium butyrate to the lysis buffer, dilution and elution buffers. Sodium butyrate prevents the loss of histone acetylation during sample preparation by inhibition of histone deacetylase activity.
2. Mincing of tissue is done by the pipette blunt tip to mince the tissue before cross-linking. Tissue can also be cut into small pieces with a scalpel. Perform all steps of cell lysis and immunoprecipitation on ice or at 4°C to prevent chromatin degradation. Protease inhibitors should be added fresh wherever mentioned to preserve the integrity of the proteins.
3. Cross-linking of the tissue is a crucial step, and we have standardized the timings as 15 min for hippocampal tissue and 10 min for the sorted homogeneous population of hippocampal cells. Cross-linking the DNA to the protein by formaldehyde for longer time may mask critical protein isotopes. Since the hippocampus is very small, one should be very careful not to lose the tissue during processing, especially when douncing with homogenizer and transferring to the microcentrifuge tube.
4. Sonication is an important step, especially when a specific DNA fragment size is required. Fragment size usually varies from 200 to 1,000 bp. We use Bioruptor sonicator (Diagenode), which is used for sonication of chromatin (or any high molecular weight nucleic acid) into defined size fragments and delivers highly reproducible results. It is particularly well suited for chromatin immunoprecipitation assays. As an alternative to sonication, digestion with micrococcal nuclease (MNase) can also be done to cleave the chromatin into oligonucleosomes or mononucleosomes. This method is normally used for native chromatin immunoprecipitation (14) when chromatin is not fixed. However, this technique is less favorable with

formaldehyde cross-linking, since cross-linked chromatin is inefficiently cleaved by nucleases.

5. We find that the best beads to use are protein A-Agarose beads from Millipore. ChIP assays can be performed with protein A, protein G, or protein A/G beads. The choice of beads depends on the isotype of the antibody to be used for the immunoprecipitation. If unknown, use a mixture of protein A/G beads.
6. Standardization of antibodies needs to be done prior to the experiment. Good quality and proper amount of antibodies should be used. Usually try lower amounts such as 1–2 μg and proceed to higher amounts if necessary (5–10 μg).
7. Washing steps are more crucial after the immunoprecipitation reactions. The washes should be efficient and stringent to reduce nonspecific background. The number of washes can be increased depending on the antibody. After washing, one should be extremely careful while handling the supernatant.
8. Phenol and chloroform are volatile, toxic, and inflammable and should be used with caution inside a chemical hood. During phenol–chloroform (P/C) precipitation, add an equal volume of P/C to sample, mix well but do not vortex. Use aqueous (upper) layer. If the sample appears dirty, repeat P/C step again until the interface is fairly clean.
9. The protocol for culturing hippocampal neurons is adapted from Banker's method (15).

Acknowledgments

EM is a Joseph H. and Belle R. Braun Senior Lecturer in Life Sciences and is supported by the Israel Science Foundation (ISF 215/07 and 943/09), the Israel Ministry of Health (6007), the European Union (IRG-206872) and the National Institute for Psychobiology in Israel. TT is supported by the CDA of the International Human Frontier Science Program Organization. BSS is a Safra fellow.

References

1. Huang, L. T., Holmes, G. L., Lai, M. C., Hung, P. L., Wang, C. L., Wang, T. J., Yang, C. H., Liou, C. W., and Yang, S. N. (2002) Maternal deprivation stress exacerbates cognitive deficits in immature rats with recurrent seizures. *Epilepsia* **43**, 1141–1148.
2. Crosio, C., Heitz, E., Allis, C. D., Borrelli, E., and Sassone-Corsi, P. (2003) Chromatin remodeling and neuronal response: multiple signaling pathways induce specific histone H3 modifications and early gene expression in hippocampal neurons. *J Cell Sci* **116**, 4905–4914.
3. Guan, Z., Giustetto, M., Lomvardas, S., Kim, J. H., Miniaci, M. C., Schwartz, J. H., Thanos, D., and Kandel, E. R. (2002) Integration of long-term-memory-related synaptic plasticity involves bidirectional regulation of gene expression and chromatin structure. *Cell* **111**, 483–493.

4. Dedon, P. C., Soults, J. A., Allis, C. D., and Gorovsky, M. A. (1991) A simplified formaldehyde fixation and immunoprecipitation technique for studying protein-DNA interactions. *Anal Biochem* **197**, 83–90.
5. Oneill, L. P., and Turner, B. M. (1996) Immunoprecipitation of chromatin. *Methods Enzymol* **274**, 189–197.
6. Kuo, M. H., and Allis, C. D. (1999) In vivo cross-linking and immunoprecipitation for studying dynamic protein: DNA associations in chromatin environment. *Methods* **19**, 425–433.
7. Jenuwein, T., and Allis, C. D. (2001) Translating the histone code. *Science* **293**, 1074–1080.
8. Nelson, J. D., Denisenko, O., and Bomsztyk, K. (2006) Protocol for the fast chromatin immunoprecipitation (ChIP) method. *Nat Protoc* **1**, 179–185.
9. Attema, J. L., Papathanasiou, P., Forsberg, E. C., Xu, J., Smale, S. T., and Weissman, I. L. (2007) Epigenetic characterization of hematopoietic stem cell differentiation using miniChIP and bisulfite sequencing analysis. *Proc Natl Acad Sci USA*. **104**, 12371–12376.
10. Collas, P., and Dahl, J. A. (2008) Chop it, ChIP it, check it: the current status of chromatin immunoprecipitation. *Frontiers in Bioscience* **13**, 929–943.
11. Dahl, J. A., and Collas, P. (2008) mu ChIP - a rapid micro chromatin immunoprecipitation assay for small cell samples and biopsies. *Nucleic Acids Res* **36**, e15.
12. Dahl, J. A., and Collas, P. (2007) A quick and quantitative chromatin immunoprecipitation assay for small cell samples. *Front Biosci.* **12**, 4925–4931.
13. Barski, A., and Zhao, K. (2009) Genomic Location Analysis by ChIP-Seq. *J Cell Biochem* **107**, 11–18.
14. Thorne, A. W., Myers, F. A., and Hebbes, T. R. (2004) Native chromatin immunoprecipitation. *Methods Mol Biol* **287**, 21–44.
15. Kaech, S., and Banker, G. (2006) Culturing hippocampal neurons. *Nat Protoc* **1**, 2406–2415.

Part III

Chromatin Modifying Complexes

Chapter 25

Approaches for Studying Nucleosome Movement by ATP-Dependent Chromatin Remodeling Complexes

Swetansu K. Hota and Blaine Bartholomew

Abstract

Packaging DNA into compact chromatin enables eukaryotic cells to organize and regulate their genome. Packaging is achieved by wrapping ~146–147 bp of DNA around a histone octamer to form a nucleosome, the basic unit of chromatin. Chromatin is a barrier of the bound DNA to factors involved in DNA-dependent processes such as transcription, replication, recombination, and repair. Several multisubunit protein complexes can move nucleosome to different positions on DNA utilizing energy derived from ATP hydrolysis and thereby facilitate access to DNA. Several methods are described for measuring nucleosome movement both in vivo and in vitro which provide important insights into the remodeling process.

Key words: Nucleosome position, Indirect end-labeling, Chromatin remodeling, Histone-DNA cross-linking, SWI/SNF, ISW2

1. Introduction

ATP-dependent chromatin remodelers range in size and complexity from a 1 to 17 subunit complex (1, 2). A common characteristic of these remodelers is mobilizing nucleosomes in order to change their position on DNA (3). The mobilization of nucleosomes can be subsequently used in different contexts to disassemble nucleosomes or to exchange out parts of the nucleosome for different histone variants (4, 5). The model system used in our example is *Saccharomyces cerevisiae* which affords us the two main advantages of exceptionally powerful molecular genetics and a great system in which to do detailed biochemical analysis. This chapter focuses on methods that help determine the direction and distance of nucleosome movement by chromatin remodelers.

Ales Vancura (ed.), *Transcriptional Regulation: Methods and Protocols*, Methods in Molecular Biology, vol. 809,
DOI 10.1007/978-1-61779-376-9_25,

We describe a method called indirect end labeling that is used to observe nucleosome movement in vivo. More precise measurements are made in vitro using a technique called site-directed mapping that maps nucleosome movement with single base pair (bp) resolution. Site-directed mapping provides more than just nucleosome position on DNA as changes in histone–DNA contacts can be mapped at multiple positions throughout the nucleosome. The dynamics of the remodeling process is carefully dissected by doing time-lapsed experiments to find how these contacts are moved and changed at different steps in the remodeling process.

In the protocols given we generally focus on its application to the ISW2 and SWI/SNF remodelers. These techniques are not restricted to these particular complexes and are only given as examples that should be able to be applied to many other complexes. In some cases we have tried other remodelers than those depicted, but are unable to show them because of length concerns and because the technique itself was not altered in any significant way from that given.

2. Materials

2.1. Mapping Nucleosome Position and Movement In Vivo by Indirect End Labeling

2.1.1. Cell Growth and Chromatin Preparation

1. YPD medium: 1% Yeast extract, 2% peptone and 2% dextrose, 0.008% adenine sulfate in E-pure water (reagent grade bacteria-free water with 17.9–18.3 MΩ cm resistance), autoclaved at 121°C for 20 min.
2. 1.1 M Sorbitol: Prepare in E-pure water, filter sterilize using 0.22-μm filters and store at 4°C.
3. YLE: 50 mg/ml Zymolyase, 1 M sorbitol, and 5 mM β-mercaptoethanol, make single use 1-ml aliquots and store at –20°C.
4. Spheroplast digestion buffer (SDB): 1 M sorbitol, 50 mM NaCl, 10 mM Tris–HCl, pH 7.5, 5 mM $MgCl_2$, 1 mM $CaCl_2$, 1 mM β-mercaptoethanol, 0.5 mM Spermidine, 0.075% v/v Nonidet P40. The solution is filter sterilized using a 0.22-μm filter and stored at 4°C.
5. Micrococcal nuclease: 15 U/μl in 50% glycerol, aliquot 50 μl each and store at –20°C. Follow manufacturer's recommendation for reaction, storage, and stopping conditions.
6. Stop solution: 5% SDS, 250 mM EDTA. It is made freshly by mixing equal volumes of 10% SDS and 500 mM EDTA. Keep the solution at 37°C to stop the SDS from precipitating.

2.1.2. DNA Purification

1. 1:1 Mixture of phenol (pH 8.0, equilibrated) and chloroform: Add equal volume of Tris–HCl, pH 8.0 equilibrated phenol and chloroform, mix well using a magnetic stirrer and store at 4°C. Use the lower phenol/chloroform layer and avoid the upper Tris layer.

2. RNase I_f (50,000 U/ml, New England Biolabs).
3. 7.5 M Ammonium acetate: Prepare in E-pure water and store at room temperature.
4. TE: 10 mM Tris–HCl, pH 8.0, 1 mM EDTA.
5. Restriction endonucleases: To mark the boundary of chromatin region to be analyzed and to create DNA markers.

2.1.3. Southern Blotting for Detection of Nucleosomal Ladder

1. Optikinase™ (USB).
2. γ^{32}P ATP (6,000 Ci/mmol, PerkinElmer).
3. 10× DNA loading dye: 0.25% Bromophenol blue, 50% glycerol.
4. Tris–borate–EDTA buffer (TBE): 89 mM Tris, 88.9 mM boric acid, and 2 mM EDTA. To make 10× TBE, add 108 g of Tris, 55 g of boric acid, and 100 ml of 0.2 M EDTA. Bring to 1 l with E-pureH_2O.
5. Denaturation solution: 1.5 M NaCl and 0.5 M NaOH, store at room temperature.
6. Neutralization solution: 1.5 M NaCl, 0.5 M Tris–HCl, pH 7.0, and 1 mM EDTA, store at room temperature.
7. Nylon membrane (MagnaGraph Nylon transfer membrane, 0.45 μm).
8. 20× SSC: 3 M NaCl and 0.3 M sodium citrate, adjust pH to 7.0 with 1 M HCl.
9. UV cross-linker.
10. Microcon ultracel YM3 (Amicon Biopharmaceuticals) filter membranes.
11. Prehybridization solution: 1.5× SSC, 5× Denhardt's solution (available as 50× stock from USB), 0.1% SDS. Prepare fresh by diluting 20× SSC, 50× Denhardt's solution and 10% SDS in E-pure water. Filter through 0.45 μm membranes.
12. Wash solution: 2× SSC and 0.1% SDS, prepare this solution from 20× SSC and 10% SDS in E-pure water and store at room temperature.

2.2. Mapping Nucleosome Position and Movement In Vitro by Site-Directed Mapping

2.2.1. DNA Probe Preparation, Nucleosome Reconstitution, and Gel Analysis

1. γ-P^{32} ATP (6,000 Ci/mmol, PerkinElmer).
2. 10× Optikinase buffer and Optikinase™ enzyme (USB).
3. dATP, dTTP, dGTP, and dCTP.
4. pGEM 3Z-601 plasmid DNA with 601 high-affinity nucleosome positioning sequence (6).
5. PCR purification kit.
6. Histone octamer with unique cysteine residues engineered to be close to nucleosomal DNA.
7. Sonicated salmon sperm DNA (10 mg/ml).

8. TE: 10 mM Tris–HCl, pH 8.0, 1 mM EDTA.
9. Nucleosome reconstitution buffer: 25 mM Tris–HCl, pH 7.5.
10. TBE buffer: 89 mM Tris, 88.9 mM boric acid, and 2 mM EDTA.
11. 10% Ammonium persulfate (APS): Prepare by dissolving solid APS in E-pure water and store at −20°C in 1-ml aliquots.
12. 4% native PAGE: 4% acrylamide (35:36, acrylamide:bis-acrylamide) gel in 1× Tris–borate buffer. Add TEMED and 10% (APS) to polymerize.
13. 6× DNA loading dye: 0.25% xylene cyanol, 0.25% bromophenol blue, and 30% glycerol.

2.2.2. APB Modification and Checking Cross-Linking Efficiency

1. 100 mM Azidophenacyl bromide (APB): 24 mg of APB in 1 ml of di-methyl formamide (DMF; 99.8% pure, Aldrich).
2. Transilluminator with 312 nm UV (Spectroline BI-O-Vision, UV/White Light Transilluminator).
3. 17% SDS-PAGE.
4. Buffer H-0.1: 25 mM Na–HEPES, pH 7.8, 100 mM NaCl, 0.5 mM EGTA pH 8.0, 0.1 mM EDTA, pH 8.0, 2 mM $MgCl_2$, 20% glycerol, 0.02% NP-40, and 1 mM PMSF.
5. BSA: Bovine serum albumin (10 mg/ml).
6. 1× SDS sample loading buffer: 2% SDS, 5% BME, 62.5 mM Tris–HCl, pH 6.8, and 0.088% bromophenol blue.

2.2.3. Mobilization and Nucleosome Position Mapping

1. ATP: 100 mM lithium salt solution of ATP, pH 7.0. Store at −20°C.
2. Adenosine-5′-*O*-(3-thiotriphosphate) or γS ATP: 100 mM lithium salt solution of γS ATP, pH 7.0, >90% pure, store at −20°C.
3. 5% Native PAGE: 5% Acrylamide (60:1, acrylamide:bis-acrylamide ratio) gel in 0.2× TBE buffer. Add TEMED and 10% APS to polymerize at room temperature.
4. ST Buffer: 30 mM NaCl and 20 mM Tris–HCl, pH 8.0.
5. Tris–SDS: 1 M Tris–HCl, pH 8.0, and 1% SDS.
6. Phenol (equilibrated, pH 8.0):chloroform (4:1): Add four volumes of Tris–HCl, pH 8.0, equilibrated phenol to one volume of chloroform, mix well using a magnetic stirrer and store at 4°C. Use the lower phenol–chloroform layer.
7. Alkaline cleavage buffer: 2% SDS, 20 mM ammonium acetate, 0.1 mM Na–EDTA in E-pure water, store at room temperature.
8. DNA sample loading buffer: 0.025% Bromophenol blue, 0.025% xylene cyanol, 5 mM EDTA, and 95% deionized formamide.

9. 6.5% Denaturing gel: 6.5% Acrylamide (35.36:1, acrylamide:bis-acrylamide ratio) gel in 1× TBE buffer with 8 M urea. Add TEMED and 10% APS and leave at room temperature to polymerize.
10. Whatman 3-mm chromatography paper.

2.3. Detection of Nucleosome Movement by Gel Shift Assays

1. 20 cm × 20 cm × 0.1 cm native gel with 20- to 25-well comb.
2. 4% Native polyacrylamide (acrylamide:bis acrylamide ratio, 35.36:1 or 79:1) suitable for analysis of protein complexes binding to nucleosome.
3. 5% Native polyacrylamide (acrylamide:bis acrylamide ratio, 60:1) suitable for analysis of different translational positions of nucleosomes.
4. TEMED (N,N,N',N'-tetramethylethylenediamine).
5. 10% APS in E-pure H_2O; aliquot in 1-ml tubes and store at −20°C.

3. Methods

3.1. Mapping Nucleosome Position and Movement In Vivo by Indirect End Labeling

Indirect end labeling is a method to analyze DNA binding sites or positioned nucleosomes between two restriction sites in chromosomal DNA (7). After lysing cells, native chromatin is treated with micrococcal nucleases (MNase) to cleave DNA at nucleosomal free sites or other sites not tightly bound by other factors. The genomic region of interest is visualized by removing all proteins from DNA and digesting with a restriction endonuclease at a site toward the edge of the region of interest. Southern blotting is used to visualize the particular genomic fragment from this region and the partial cleavage products due to the MNase digestion are also observed. This technique can be used to determine nucleosome position at a specific locus with and without a particular chromatin remodeler present to find how the remodeler affects nucleosome position in vivo.

3.1.1. Cell Growth and Chromatin Preparation

1. Grow 100-ml yeast cells to OD_{600} of 1.0 (see Note 1).
2. Harvest 50 ml of cells by spinning at 3,000 × *g* for 5 min, resuspend with 1 ml media at 4°C.
3. Transfer to a microfuge tube in ice, centrifuge at 5,000 × *g*, and remove the supernatant.
4. Gently resuspend the pellet in 950 μl of YLE, mix by inverting for 15–60 s.
5. Spin at 14,000 × *g* for 5 s and remove YLE.
6. Gently add 950 μl of 1 M sorbitol, invert tube twice, and spin for 5 s at 180° to the original position. Remove and discard sorbitol solution.

7. Repeat the wash once more keeping the microfuge tube back in original position.
8. Resuspend the spheroplast in 1.2 ml of spheroplast digestion buffer (SDB).
9. Transfer 200 μl each to microfuge tubes containing appropriate amount of MNase (0, 0.5, 1, 2, or 4 μl of 15 U/μl stock), mix gently, and incubate at 37°C for 4 min (see Note 2).
10. Terminate the reaction by adding 20 μl of stop solution and mix vigorously.
11. Spin briefly and store the lysates at –20°C.

3.1.2. DNA Purification

1. Add 200 μl of phenol:chloroform (1:1) to each tube from Subheading 3.1.1, step 11, containing the MNase-digested chromatin, vortex for 5 s, and spin at 14,000 × *g* (see Note 3).
2. Transfer aqueous phase to a microfuge tube containing 1 μl or 50 U of RNase I_f (NEB), mix well, and incubate at 37°C for 30 min.
3. Repeat phenol/chloroform extraction.
4. Transfer aqueous phage to new tube; add 40 μl of 7.5 M ammonium acetate and 260 μl of 100% isopropanol, mix well, spin at 14,000 × *g* for 10 min, and discard the supernatant.
5. Add 190 μl of 80% ethanol, spin for 5 s, discard the supernatant, spin again and discard the supernatant.
6. Dry the pellet and resuspend in 20 μl of TE.
7. Follow steps 8–12 to prepare naked DNA control MNase digests.
8. Add 300 μl of SDB to 18 μl of the purified DNA sample from chromatin which was not digested with MNase.
9. Add 2 μl of 1:20 dilution of 15 U/μl MNase, mix quickly, and incubate at 22°C for 30–60 s.
10. Add 300 μl of phenol–chloroform, vortex 5 s, spin at 14,000 × *g* for 5 min, and collect the supernatant.
11. Add 60 μl of 7.5 M ammonium acetate, 390 μl of isopropanol, mix well. Spin at 14,000 × *g* for 10 min, wash pellet with 80% ethanol, and air-dry the pellet.
12. Resuspend the pellet in 20 μl of TE.
13. Follow steps 14–16 to prepare a DNA ladder/marker.
14. Digest 5 μl of undigested DNA with 20 U of selected restriction enzyme for 3 h in a reaction volume of 20 μl (see Note 4).
15. Add 80 μl of TE followed by 100 μl of phenol–chloroform, vortex for 5 s and spin at 14,000 × *g* for 5 min.

16. Pool aqueous phases and precipitate DNAs with 0.2 volumes of 7.5 M ammonium acetate and 1.3 volume of isopropanol, wash with 80% ethanol, air-dry, and resuspend the pellet in 18 μl of TE.

3.1.3. Southern Blotting for Detection of Nucleosomal Ladder

1. Prepare a mix containing 2 μl of water, 2.5 μl of 10× restriction enzyme buffer, and 2.5 μl of restriction enzyme. These restriction enzymes are chosen to cut at the boundary region of chromatin to be analyzed for nucleosome positions.
2. Add 7 μl of restriction enzyme mix to 18 μl of chromatin/naked DNA/marker digest, mix well, and incubate for 3 h/overnight at appropriate temperature.
3. Add 2.5 μl of 10× DNA loading dye to each digest to make it 27.5 μl.
4. Load 25 μl out of the 27.5 μl of chromatin/naked DNA restriction digest and 3–4 μl of marker digest to 1.5% agarose gel.
5. Run at 6 mA/cm till bromophenol blue is 1 cm of the bottom.
6. Mark/clip the orientation of the gel, stain in ethidium bromide, and scan the gel.
7. Denature DNA in the gel by soaking the gel in denaturation solution (1.5 M NaCl/0.5 M NaOH) for 15 min and repeat (see Note 5).
8. Rinse with distilled water and neutralize by soaking gel neutralization solution (1.5 M NaCl, 0.5 M Tris–HCl, pH 7.0, and 1 mM EDTA) for 15 min and repeat once.
9. Blot overnight with 20× SSC to a nylon membrane (see Note 6).
10. Rinse with 2× SSC and UV cross-link at 1,200 MJ.
11. End label 5 pmol of primer by using Optikinase and γ^{32}P ATP and purify the primer by washing three times with 500 μl of TE in Microcon ultracel YM3 filter membranes at room temperature.
12. Prehybridize blot in 10–20 ml of prehybridization solution at 64°C for 1 h with rotation (see Note 7).
13. Boil probe in 250–500 μl of 2.5 mg/ml salmon sperm DNA for 5 min and quench on ice for 2 min. Add probe–DNA mix to prehybridization solution and hybridize blot at 64°C for 12–18 h with rotation.
14. Wash probe once with 100 ml of wash solution at 60°C for 20 min and twice in 100 ml of wash solution at 64°C for 15 min.
15. Seal the blot in a thin plastic bag and expose to a phosphorimager screen, and scan the image.

3.2. Mapping Nucleosome Position and Movement In Vitro by Site-Directed Mapping

Site-directed mapping measures the interactions of specific histone residues with DNA at base pair resolution. This approach requires attachment of photoreactive aryl azide to cysteine residues that have been engineered at key points within nucleosomes (8–10). The histone protein is cross-linked to the proximal DNA with ultraviolet light and DNA is cleaved at the cross-linked site. The cleavage site can be determined by gel electrophoresis and therefore map the site to which a histone amino acid is proximal to a particular nucleotide in DNA. There are 3–4 different amino acid positions within each of the four histone proteins that have been used for this purpose. In histone H2A alanine residue 45 when changed to cysteine and modified, cleaves DNA 39 bp from the dyad on the strand that is in the 5′–3′ direction from the dyad axis. Serine 53 of histone H2B when changed to cysteine and modified cleaves the DNA strand 53 bp from the dyad axis in the 3′–5′ direction. Both of these modified sites have a high degree of strand specificity which is not as prevalent for comparable sites in histone H4 and H3. When serine 45 of histone H4 is replaced with cysteine and modified it cleaves DNA 2 and 6 bp from the 3′ and 5′ sides of the dyad axis, respectively. The site 2 bp from the dyad is more efficiently cleaved than is the site 6 bp from the dyad axis. Likewise when methionine 120 of histone H3 is replaced with cysteine and modified it cleaves with the same specificity as does residue 45 of H4. These modified cysteines can be used to monitor the initial nucleosome position and map the changes in nucleosome position that occur during and after ATP-dependent remodeling. A map of the path of DNA around the histone octamer can also be made by comparing the region(s) of DNA proximal to these different sites in the histone octamer. These sites provide a total of six distinct sites together from which to compare the length of DNA placed between these sites. If the length is as expected from the known nucleosome structure then the DNA is wrapped normally around the histone octamer. If the length of DNA between the sites has increased due to remodeling, it suggests the formation of bulges within the nucleosome.

3.2.1. DNA Probes, Nucleosome Reconstitution, and Gel Analysis

1. Prepare a 4% native polyacrylamide gel and prerun at constant 100 V (for 10 cm length × 20 cm width gels) with 0.5× TBE at 4°C in dark for 1 h. The current on the gel at the start of prerun should be 16–18 mA and will reduce to 10–12 mA at the end of prerunning.
2. Phosphorylate 20 pmol of forward or reverse oligonucleotides at 5′ end with 12 μl of (6,000 Ci/mmol) γ-P^{32} radiolabel ATP using 10 U of Optikinase in 1× Optikinase buffer in a 30-μl of reaction volume. Incubate at 37°C for 45 min and inactivate Optikinase enzyme by heating at 65°C for 20 min.

3. Set up a 100-μl PCR with 2 ng of plasmid template having "601" nucleosome positioning sequence, 1 μl of recombinant *Pfu* DNA polymerase or appropriate DNA polymerase, 200 μM dNTPs, 20 pmol of labeled forward or reverse oligonucleotide, 25 pmol of either reverse or forward cold oligonucleotides in 1× *Pfu* or appropriate DNA polymerase reaction buffer. Excess unlabeled oligonucleotide utilizes all radiolabeled oligonucleotide in the PCR.
4. Clean PCR product by using manufacturer's recommendation for PCR purification with premade silica-based columns. Analyze the PCR product/probe DNA on a 4% native polyacrylamide gel and 6.5% denaturing polyacrylamide to ensure single PCR product. It is important to use both gel systems because in native gel secondary structure of the probe DNA can be seen due to difference in their mobility while in denaturing gel spurious PCR products can be detected.
5. Nucleosomes are reconstituted with starting volumes of 10 μl. To a microfuge tube add 10 μg of salmon sperm DNA, 100–200 fmol of radioactive end labeled DNA probe and NaCl (final concentration 1.9 M).
6. Incubate at 37°C for 10 min and add 10–15 μg of octamer (optimal octamer amount determined empirically) and make up the volume to 10 μl. Incubate at 37°C for 25 min.
7. Add 3.2, 6, 9, and 39 μl of reconstitution buffer to dilute final NaCl concentration to 1.3, 1, 0.78, and 0.283 M in a final 67.2 μl reconstitution. Each dilution step follows incubation of the reaction at 37°C for 10 min.
8. Mix 1 μl of nucleosome with 3 μl of 10% glycerol and load 4 μl to 4% native PAGE (already prerun, use fresh 0.5× TBE and flush wells before loading samples), run at constant 200 V at 4°C for 1½ to 2 h. Use DNA loading dye to track and run the gel till bromophenol blue is 5/6th of the distance to the bottom of the gel.

3.2.2. APB Modification and Checking Cross-linking Efficiency

1. Follow all the steps below in room illuminated with only 20 W incandescent lights.
2. Modification premix is made by adding 6 μl of freshly made 100 mM APB, 6 μl of di-methyl formamide, and 50 μl of 80% glycerol.
3. Immediately add 4.1 μl of modification premix to 64 μl of nucleosome from step 7 of Subheading 3.2.1.
4. Incubate at 25°C for 3 h in dark.
5. Pour a 17% SDS-PAGE.
6. Quench the modification reaction by adding BME to a final concentration of 1 mM.

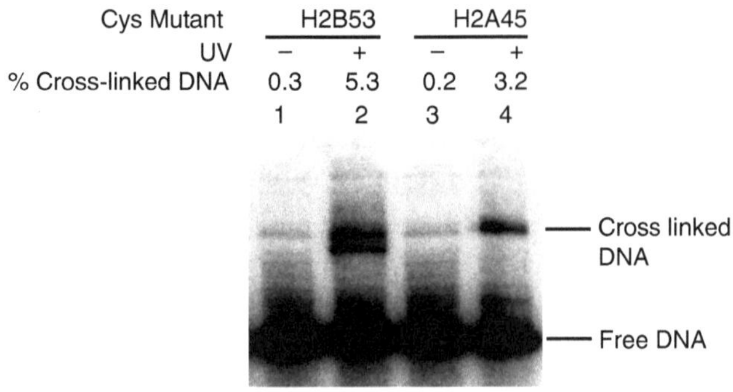

Fig. 1. The efficiency of modified histone cross-linking to DNA determined by SDS-PAGE. Nucleosomes had histone H2B or H2A with cysteine replacing serine 53 or alanine 45, respectively. Cysteine was modified by incubating the nucleosomes with APB and cross-linked for 3 min at 310 nm. The uncross-linked and cross-linked samples were analyzed on a 17% SDS-PAGE, dried, and visualized by phosphorimaging.

7. To check cross-linking efficiency, prepare two sets of reaction each containing 1 μl of nucleosome, 6 μl of H-0.1 buffer, 1 μl of 33 mM $MgCl_2$, and 1.5 μl of 1 mg/ml BSA in a total volume of 15 μl.
8. To cross-link APB modified cysteines to nearest DNA, UV irradiate one set with 312 nm using UV transilluminator for 3 min from a distance of 10 cm.
9. Add 8 μl of 1× SDS sample loading buffer to 2 μl each of uncross-linked and cross-linked samples and incubate at 37°C for 3 min. Spin briefly and load all 10 μl of sample on to a 17% SDS-PAGE. Run till the bromophenol blue dye reaches the bottom of the gel.
10. Take out the gel, mark the orientation and transfer to the surface of blotting paper, cover with Saran Wrap, and dry under vacuum at 80°C for 2 h.
11. Expose to phosphorimager screen and scan image.
12. Quantify the percentage of cross-linking and compare with uncross-linked samples. Typical cross-linking efficiency is 5–7% (Fig. 1).

3.2.3. Mobilization and Nucleosome Position Mapping

1. Follow all the steps below in a darkened room with 20 W incandescent light bulbs.
2. To a microfuge tube on ice, add 30 μl of buffer H-0.1, 5 μl of 33 mM $MgCl_2$, 7.5 μl of 10 mM BME, 7.5 μl of 1 mg/ml BSA, 6.5 μl of 10 mM ATP, 5 μl of APB-modified nucleosome (880 fmol/μl), substoichiometric amount of remodeling enzyme (200 fmol to 1 pmol) and water to make 80 μl (see Note 8).

3. Make another reaction without ATP or remodeling enzyme to determine initial nucleosome position.
4. Incubate the reaction at 30°C for 30 min.
5. Stop the reaction by adding 3 μl of 5 mg/ml salmon sperm DNA and 5 mM γ-SATP.
6. Load 3 μl in 5% native PAGE to check for nucleosome remodeling.
7. Following gel loading UV cross-link the sample(s) at 310 nM for 3 min using a transilluminator.
8. Dilute remaining 80 microliter reaction to 300 microliter by adding 220 μl of ST buffer.
9. Add 3 μl of 10% SDS to give a final concentration of 0.1% SDS and vortex to mix the reaction.
10. Heat the reaction at 70°C for 20 min and cool down the reaction to room temperature for 15 min.
11. Add 300 μl of phenol/chloroform (4:1) and vortex vigorously.
12. Spin at 13,000 × *g* for 10 min. Be careful not to disturb the interface containing the cross-linked histone and nucleosomal DNA.
13. Work above the interface and using a small pipettor remove 270 μl of supernatant. This supernatant will be noncross-linked DNA. Leave 20–30 μl of supernatant so as not to disturb the interface as it contains the desired cross-linked DNA.
14. Wash phenol layer three times with Tris–SDS (1 M Tris–HCl, pH 8.0, containing 1.0% SDS). Each time add 280 μl of Tris–SDS, vortex and spin (13,000 × *g* for 10 min), and pipet out 280 μl leaving 20–30 μl in the interface. Record radioactive signal by monitoring with a survey meter, the signal of the wash should decrease.
15. Add 30 μl of 3 M sodium acetate, 1 μl of 10 mg/ml salmon sperm DNA as carrier, and 750 μl of ice-cold 100% ethanol. Vortex and precipitate in ice at dark overnight.
16. Prepare 2 ml of alkaline cleavage buffer.
17. Centrifuge the samples at 15,000 × *g* for 40 min at 4°C.
18. Wash pellets two times with 380 μl of 75% ice-cold ethanol. Do not disturb the pellets. Check loss of signal in washes by survey meter.
19. Air-dry pellet for 5–10 min and look visually for dryness. Do not dry by heat or vacuum. Take care not to over dry as pellets will stick to microfuge.
20. Fully resuspend in 100 μl of alkaline cleavage buffer. Vortex for 30 min to 1 h and check radioactive signal of the buffer. Maximum signal should be in the cleavage buffer and minimum in the tube.

21. Transfer to a fresh microfuge tube, heat at 90°C for 20 min, and give a brief spin.
22. Add 5 μl of 2.0 M NaOH per 100 μl (final concentration 0.1 M) and heat at 90°C for 45 min.
23. During the incubation, premix 0.5 μl of 10 mg/ml salmon sperm DNA and 2 μl of 1 M $MgCl_2$ per reaction.
24. After incubation, add 100 μl of 20 mM Tris–HCl, pH 8.0, 6 μl of 2 M HCl, 2.5 μl of salmon sperm DNA and $MgCl_2$ mix, and 470 μl of ice-cold 100% ethanol to each reaction. Mix well by vortexing.
25. Precipitate at −20°C overnight.
26. After overnight precipitation, centrifuge the samples at 15,000 × *g* for 45 min at 4°C.
27. Wash once with ice-cold 75% ethanol. Be very careful not to disturb the pellet. Air-dry for 5–10 min.
28. Resuspend in 8 μl of DNA sample loading buffer and vortex for 30 min to resuspend the pellet. The entire radioactive signal should be in sample loading buffer with minimal signal on the tube.
29. Load 4 μl on to a 40 cm × 20 cm × 0.4 mm, 6.5% denaturing urea gel along with a DNA sequencing ladder of the same probe DNA. Run at constant 40 W power till the bromophenol blue reaches bottom of the gel. Optionally, a second set of samples may be loaded along with the sequencing ladder after 40 min of first sample loading to resolve both small and larger fragments.
30. Pry open the gel plates, transfer the gel to Whatman paper and cover with Saran Wrap. Vacuum dry at 80°C for 1 h, expose to phosphorimager screen for 2–14 days and scan the phosphorimager screen.

3.3. Detection of Nucleosome Movement by Gel Shift Assays

Electrophoretic mobility shift assays or gel shifts assays are commonly used to study protein–DNA associations. We regularly use both native agarose and polyacrylamide gel shift assays to study formation of nucleosomes, protein binding to nucleosomes, and movement of nucleosome on DNA. Different ratios of acrylamide to bis-acrylamide are used in preparing native gels suitable for either analysis of different translational position of nucleosomes or binding of proteins to nucleosomes.

1. A native gel with 4% polyacrylamide for analysis of protein binding to nucleosome or 5% polyacrylamide for analysis of translational positions of nucleosome is prepared beforehand.
2. Wash wells with a syringe and prerun for at least 2 h with 0.5× TBE (for 4% native gel) or 0.2× TBE for 5% high-resolution gel. We usually monitor the starting current of 18–20 mA for 20 cm × 20 cm × 0.1 cm 4% native gel with 0.5× TBE that

reduces to 10–12 after 2 h of prerunning. Similarly 10–12 mA for 5% gel reduces to 5–6 mA after 2 h of prerun.

3. Change buffer and wash the wells again before loading samples. Load samples at a lower voltage so as to ensure that they enter gel as soon as loaded and there is not a significant migration difference between the first and last loaded samples. Run the gel at constant 200 V for 4 h at 4°C for 4% native gel. Native 5% gels are run at a constant 200 V and 4°C, but for 5 h 30 min with continuous buffer recirculation by using peristaltic pumps. It is important to recirculate 0.2× TBE buffer between the two chambers of the rig so as to maintain the buffering capacity.
4. Carefully pry open the glass plates and transfer gel to filter paper.
5. Cover the gel with Saran Wrap, remove air bubbles, and dry at 80°C for 1 h.
6. Expose the dried gel to a phosphorimager screen (Fuji films) for required amount of time and scan in a phosphorimager.

4. Notes

1. After the cells reach OD_{600} of 1.0, immediately transfer them to ice. All further cell manipulations are carried in ice. Before harvesting the cells precool centrifuge rotor to 4°C and keep all the reagents in ice.
2. A titration of micrococcal nuclease, temperature and time of incubation can be optimized for different strains to have desired partial digestion of chromatin and not under or over digestion of chromatin.
3. DNA purification steps 1–6 are performed at room temperature.
4. Select restriction sites so as to have bands at least 500 bp separate from each other and serve as a good reference marker. Follow the respective restriction endonuclease manufacture protocol for reaction condition such as addition of BSA and temperature. Steps 15 to 16 of subheading 3.1.2 of marker digest are at room temperature.
5. Put the gel in a glass tray with copious amount of denaturation or neutralization buffer so as to immerse it completely. Put the tray in an orbital shaker and shake it for the required amount of time.
6. Both capillary transfer and vacuum transfer are equally efficient in transferring the DNA fragments to nylon membrane. After transfer, the agarose gel can be restained with ethidium bromide. Loss of all smaller fragments and significant amount of larger fragment are a good sign of DNA transfer from gel to membrane.

7. Please make sure that the entire blot is in contact with the prehybridization and hybridization solution.
8. The translational position of intermediates as well as final remodeled products can be measured by slowing down the reaction. We usually reduce the rate of remodeling reaction by using a lower concentration of ATP and changing the reaction temperature to 18°C. The ability to slow the remodeling reaction can be followed by samples on a 5% high-resolution native-PAGE.

References

1. Clapier, C. R. and Cairns, B. R. (2009) The biology of chromatin remodeling complexes. *Annu Rev Biochem.* **78**, 273–304.
2. Gangaraju, V. K. and Bartholomew, B. (2007) Mechanisms of ATP dependent chromatin remodeling. *Mutat Res.* **618**, 3–17.
3. Kagalwala, M. N., Glaus, B. J., Dang, W., Zofall, M. and Bartholomew, B. (2004) Topography of the ISW2-nucleosome complex: insights into nucleosome spacing and chromatin remodeling. *EMBO J.* **23**, 2092–2104.
4. Dechassa, M. L., Sabri, A., Pondugula, S., Kassabov, S. R., Chatterjee, N., Kladde, M. P. and Bartholomew, B. (2010) SWI/SNF has intrinsic nucleosome disassembly activity that is dependent on adjacent nucleosomes. *Mol Cell.* **38**, 590–602.
5. Mizuguchi, G., Shen, X., Landry, J., Wu, W. H., Sen, S. and Wu, C. (2004) ATP-driven exchange of histone H2AZ variant catalyzed by SWR1 chromatin remodeling complex. *Science.* **303**, 343–348.
6. Lowary, P. T. and Widom, J. (1998) New DNA sequence rules for high affinity binding to histone octamer and sequence-directed nucleosome positioning. *J Mol Biol.* **276**, 19–42.
7. Kent, N. A., Karabetsou, N., Politis, P. K. and Mellor, J. (2001) In vivo chromatin remodeling by yeast ISWI homologs Isw1p and Isw2p. *Genes Dev.* **15**, 619–626.
8. Flaus, A., Luger, K., Tan, S. and Richmond, T. J. (1996) Mapping nucleosome position at single base-pair resolution by using site-directed hydroxyl radicals. *Proc Natl Acad Sci USA.* **93**, 1370–1375.
9. Kassabov, S. R. and Bartholomew, B. (2004) Site-directed histone-DNA contact mapping for analysis of nucleosome dynamics. *Methods Enzymol.* **375**, 193–210.
10. Kassabov, S. R., Henry, N. M., Zofall, M., Tsukiyama, T. and Bartholomew, B. (2002) High-resolution mapping of changes in histone-DNA contacts of nucleosomes remodeled by ISW2. *Mol Cell Biol.* **22**, 7524–7534.

Chapter 26

Mapping Protein–DNA and Protein–Protein Interactions of ATP-Dependent Chromatin Remodelers

Swetansu K. Hota, Mekonnen Lemma Dechassa, Punit Prasad, and Blaine Bartholomew

Abstract

Chromatin plays a key regulatory role in several DNA-dependent processes as it regulates DNA access to different protein factors. Several multisubunit protein complexes interact, modify, or mobilize nucleosomes: the basic unit of chromatin, from its original location in an ATP-dependent manner to facilitate processes, such as transcription, replication, repair, and recombination. Knowledge of the interactions of chromatin remodelers with nucleosomes is a crucial requirement to understand the mechanism of chromatin remodeling. Here, we describe several methods to analyze the interactions of multisubunit chromatin-remodeling enzymes with nucleosomes.

Key words: Nucleosome, Chromatin remodeling, SWI/SNF, ISW2, DNA–protein interactions, DNA and histone cross-linking, DNA footprinting

1. Introduction

ATP-dependent chromatin remodelers range in size and complexity from a 1 to a 17 subunit complex (1, 2). A common characteristic of these remodelers is mobilizing nucleosomes in order to change their position on DNA (3). The mobilization of nucleosomes can be subsequently used in different contexts to disassemble nucleosomes or to exchange parts of the nucleosome for different histone variants (4, 5).

The size of some of these complexes makes it a demanding task to find the domains of these large macromolecular complexes that interact with chromatin. The approaches described in this chapter have the advantage of not requiring large amounts of complex like that required for most structural studies, such as NMR and X-ray

Ales Vancura (ed.), *Transcriptional Regulation: Methods and Protocols*, Methods in Molecular Biology, vol. 809, DOI 10.1007/978-1-61779-376-9_26, © Springer Science+Business Media, LLC 2012

diffraction. And yet these techniques provide detailed information about the complex that can be further tested and refined. Generally, the first step in our approach to study these interactions is to use DNA footprinting to determine the regions of nucleosomal and extranucleosomal DNA that are tightly bound by the remodeler. For this purpose, hydroxyl radical generated by Fe-EDTA is used as the nucleolytic agent to map only those sites that are tightly bound (3, 4). Next, we are able to find both the subunit(s) that are tightly bound at these sites and others more loosely associated with DNA through an approach called site-directed DNA cross-linking which transfers a label to the target protein (4, 6). We design modified DNA such that a photoactivated cross-linker is incorporated at a specific DNA site along with a radiolabel. A radiolabel is transferred to the target protein for ease in identifying the cross-linked protein after cross-linking the bound protein. There are two different methods for modifying DNA that are described in which the cross-linker is incorporated either in the phosphate backbone or to the nucleotide base. Each of these has different spatial constraints and, therefore, scans different parts of the DNA tertiary structure. Since not all interactions are restricted to the DNA component of nucleosomes, we have also designed a similar approach for site-specific attachment of photocross-linkers to histone proteins (4). The fundamental design is essentially the same, which involves attachment of photocross-linker to a cysteine and the transfer of radiolabel to the target cross-linked protein. Like DNA footprinting, these techniques can be used to examine the dynamics of the remodelers in time-lapsed experiments.

Another approach is described that uses a type of protein footprinting technique to map the interactions of remodelers with free DNA or with nucleosomes. In this approach, a derivative of Fe-EDTA is tethered either to the phosphate backbone of DNA or to the histone proteins for what is called site-directed proteolysis (7–10). The hydroxyl radical is generated at a localized site within DNA or histone protein that causes proteolytic cleavage of the adjoining protein. The cleavage site is mapped by having either the N- or C-terminal end of the target protein epitope tagged and finding the sizes of the cleaved epitope-tagged proteolytic fragments by immunoblotting. The main advantage of this approach is that it pinpoints the domain or region of the particular subunit that is associated either with histones or DNA.

In the protocols given, we generally focus on their application to the ISW2 and SWI/SNF remodelers. These techniques are not restricted to these particular complexes and are only given as examples that should be able to be applied to many other complexes. In some cases, we have tried other remodelers than those depicted, but are unable to show them because of length concerns and because the technique itself was not altered in any significant way from that given.

2. Materials

2.1. Mapping DNA–Protein Interactions by DNA Footprinting

2.1.1. DNA Probe Synthesis by PCR and Nucleosome Reconstitution

1. γ-P^{32} ATP (6,000 Ci/mmol).
2. 10× Optikinase buffer and Optikinase™ enzyme (USB).
3. dATP, dTTP, dGTP, and dCTP: 100 mM solution of sodium salt, >99% triphosphates from Fisher Scientific.
4. pGEM 3Z-601 plasmid DNA (11).
5. PCR purification kit.
6. Reconstitution buffer: 1 M Tris–HCl, pH 7.5, store in −20°C.

2.1.2. Immobilization of DNA Template

1. Plasmid p-199-1 having "601" nucleosome-positioning sequence (NPS) (6).
2. Restriction enzymes: NdeI, BamHI, and EcoR*I*.
3. Bio-14-dATP (Invitrogen Inc.).
4. Bio-11-dUTP (Enzo lifesciences).
5. Klenow DNA polymerase (3′ to 5′ exonuclease negative) from NEB. The 3′ to 5′ exonuclease negative (exo-) version of the enzyme is preferred so that the enzyme does not chew back annealed primers in the absence of sufficient dNTPs.
6. T4 DNA ligase: 2,000,000 U/ml supplied in a buffer containing 50 mM KCl, 10 mM Tris–HCl, pH 7.4, 0.1 mM EDTA, 1 mM DTT, and 50% glycerol from NEB. We use a T4 DNA ligase storage buffer containing 1 mM BME instead of DTT to dilute the T4 DNA ligase to 5,000 U/ml as BME very poorly reduces aryl azide groups in the photoreactive azido-benzoyl dUTP (AB-dUTP).
7. T4 DNA ligase storage buffer: 10 mM Tris–HCl, pH 7.4, 50 mM KCl, 5 mM β-mercaptoethanol, 0.1 mM EDTA, and 50% glycerol.
8. 1× D1+T buffer: 30 mM Tris–HCl, pH 8.0, 50 mM KCl, 7 mM $MgCl_2$, 1 mM β-mercaptoethanol, and 0.05% Tween-20.
9. M-280 streptavidin-coated Dyanbeads (Dynal).
10. Magnetic separation stands for DNA bead isolation (Promega).
11. BW buffer: 5 mM Tris–HCl, pH 7.5, 0.5 mM EDTA, pH 8.0, and 1 M sodium chloride.

2.1.3. Hydroxyl Radical DNA Footprinting

1. Ammonium ferrous sulfate ($FeNH_4SO_4$): Prepare fresh 3.33 mM $FeNH_4SO_4$ in E-pure water and discard any remaining solution after use.
2. Sodium ascorbate: Prepare fresh 66.6 mM sodium ascorbate in E-pure water.
3. Thiourea: Prepare 1 M thiourea in E-pure water and store at 4°C. Solutions 1 to 3 can be prepared in 1 M concentration

and stored at –20°C after wrapping in aluminum foil. Discard any remaining working stock solution after hydroxyl radical reaction.

4. Ammonium acetate: Prepare 6 M ammonium acetate in E-pure water. Store at room temperature.
5. Hydrogen peroxide: 30% stock solutions are wrapped with aluminum foil and stored at 4°C.
6. Spin columns (Pierce).
7. S-200 Sephacryl resin (GE Healthcare).
8. Formamide dye: 95% formamide, 0.063% bromophenol blue, and 0.063% xylene cyanol.

2.2. Mapping DNA–Protein Interactions by DNA Cross-Linking

1. 5′ phosphorylated and unphosphorylated oligonucleotides (50 nmol scale); the oligonucleotides were 20–25 nucleotides in length.
2. Synthesis of oligonucleotides with phosphorothioate incorporated between the third and fourth nucleotide from the 5′ end (100-nmol scale).
3. 100 mM tetraethylammonium bicarbonate (TEAB). Triethylamine (99% pure, Sigma–Aldrich) is used and if yellowish in color is redistilled to ensure purity. Dilute with water for a 1.5 M stock solution. The triethylamine does not readily mix with water, but upon adjusting the pH to 8.0 by bubbling carbon dioxide through the solution it does go into solution. Store TEAB at 4°C.
4. TE: 10 mM Tris–HCl, pH 8.0, and 1 mM EDTA.
5. 100 mM para-azidophenacyl bromide (APB) in dimethyl formamide (DMF, 99.8% pure from Aldrich).
6. TE-saturated isobutanol: In a glass bottle, mix an equal volume of TE and isobutanol for 1–2 h, allow to separate, and use the upper layer. Store at room temperature.
7. TE-saturated ethylether: Mix an equal volume of TE and ethyl ether, allow to separate, and use upper layer.
8. T4 DNA polymerase, 5 units/μl.
9. Deionized formamide (99%, Sigma): To deionize formamide, add 5 g of mixed bed resin AG 501-X8 with indicator (BioRad) to 100 ml of formamide, stir for 1–2 h at room temperature, make a 1 ml aliquot, and store at –20°C.
10. AB-dUTP, AB-dCTP, and AB-dTTP.
11. α-P^{32} ATP, α-P^{32} GTP, and α-P^{32} CTP (6,000 Ci/mmol).
12. 6.5% denaturing gel: 6.5% acrylamide (35.5:1 acrylamide to bisacrylamide), 8 M urea in 1× Tris–borate EDTA buffer. Add 100 μl each of TEMED and 10% ammonium persulfate and allow for 30–45 min at room temperature to polymerize.

13. 10% denaturing gel: 10% acrylamide (35.5:1 acrylamide to bisacrylamide), 8 M urea in 1× Tris–borate EDTA buffer. Add 100 μl each of TEMED and 10% ammonium persulfate and allow for 30–45 min at room temperature to polymerize.
14. Spectroline BI-O-VISION UV/white light transilluminator.
15. DNaseI (2 units/μl) from Ambion (see Note 1).
16. S1 nuclease (100 units/μl) from Fermentas (see Note 1).
17. Zinc acetate solution: 0.5 M glacial acetic acid, 12.5 mM ZnCl. Make fresh zinc acetate solution by adding 6 μl of glacial acetic acid with 2.5 μl of 1 M ZnCl.
18. 5× denaturing buffer for SDS-PAGE: 25% β-mercaptoethanol, 10% SDS, 0.3 M Tris–HCl, pH 6.8, and 0.4% bromophenol blue.

2.3. Mapping Protein Interactions with Nucleosomes by Histone Site-Directed Cross-Linking and Label Transfer

2.3.1. DNA for Nucleosome Reconstitution

1. Biotin-tagged forward and a reverse oligonucleotide primer for amplifying a 276 bp biotin-tagged 601 nucleosome-positioning DNA from p159-1-G-27 plasmid (4).
2. Plasmid DNA template: p159-1-G-27 plasmid (4).
3. PCR HotMaster Mix (Eppendorf).
4. Microcon-50 centrifugal filters (Amicon).
5. TE: 10 mM Tris–HCl, pH 8.0, 1 mM EDTA.

2.3.2. Histone Octamers with Site-Specific Cysteines

1. *Xenopus laevis* recombinant H2A, H2B, H3, and H4 histone proteins purified from *Escherichia coli* and the histone protein in water with 1 mM BME were lyophilized and the powder was stored at −20°C (12).
2. Single-cysteine mutant *X. laevis* recombinant H2A, H2B, H3, and H4 histone proteins are generated by side-directed mutagenesis.
3. Histone unfolding buffer: 6 M guanidinium–HCl, 20 mM Tris–HCl, pH 7.5, 5 mM dithiothreitol (DTT).
4. Octamer refolding buffer: 2 M NaCl, 10 mM Tris–HCl, pH 7.5, 1 mM EDTA, 5 mM 2-mercaptoethanol (BME).
5. Amicon-50 Ultra filters (Millipore).

2.3.3. Coupling of ^{125}I-PEAS to Cysteine-Containing Nucleosomes

1. *N*-((2-pyridyldithio) ethyl)-4-azidosalicylamide (PEAS, Invitrogen) is dissolved in dimethylsulfoxide (DMSO) and stored as a 30 mM stock solution in dark at −20°C (see Note 7).
2. IODO-GEN iodination tubes (Pierce) precoated with a mild oxidation reagent 1,3,4,6-tetrachloro-3α, 6α-diphenylglycouril.
3. Na^{125}I (17 Ci/mg, PerkinElmer) (see Notes 9–12).
4. Slide-A-lyzer MINI Dialysis Units (Thermo Scientific).

5. DMSO (99.9% pure, spectrophotometric grade, Sigma–Aldrich).
6. Sephadex G-25 resin (GE Life Sciences).
7. Nucleosome storage buffer: 10 mM Tris–HCl, pH 7.5, 1 mM EDTA, 0.01% NP40, 5% glycerol, 100 mM NaCl.
8. 100 mM sodium phosphate buffer, pH 7.4. Store at room temperature.
9. 80 mM methionine: Dissolve 0.60 g of methionine in 50 ml of water and autoclave.
10. 0.45 mg/ml tyrosine (2.5 mM): Dissolve 23 mg of tyrosine in 50 ml of water and autoclave; the solubility of tyrosine in water is 0.45 mg/ml at 25°C and pH 3.2–7.5; it can be difficult to completely dissolve.
11. 4% native PAGE (35.36:1 acrylamide-to-bis-acrylamide ratio) in 0.5× TBE buffer.
12. 4× SDS-loading buffer without reducing reagent: 8% SDS, 0.24 M Tris–HCl, pH 6.8, 0.32% bromophenol blue, 20% glycerol.
13. Resolving buffer (5×): 1.88 M Tris–HCl, pH 8.8, 0.5% SDS. Store at room temperature.
14. Stacking buffer (8×): 1 M Tris–HCl, pH 6.8, 0.8% SDS. Store at room temperature.
15. 16% SDS-polyacrylamide gel: 16% acrylamide (37.5: 1, acrylamide-to-bis-acrylamide ratio) in 0.37 M Tris–HCl, pH 8.8, and 0.1% SDS. Add 6.7 ml of 30% acrylamide, 2.5 ml of 5× resolving buffer and 3.3 ml of water, and 20 μl each of TEMED and 10% APS to make resolving gel. Cover surface with isobutanol for 30–45 min at room temperature and rinse with water. Make stacking gel by adding 1 ml each of 30% acrylamide and 8× stacking buffer, 6 ml of water, and 30 μl each of TEMED and 10% APS.
16. Gel staining dyes: Coomassie blue-250, SYPRO Ruby (Molecular Probes), ethidium bromide, and SYBR Gold (Invitrogen).

2.3.4. Photoaffinity Cross-Linking and Label Transfer

1. Chromatin-remodeling enzymes (SWI/SNF, ISW1a and b) purified from yeast by immunoaffinity purification using anti-FLAG M2 antibodies (Sigma) (4).
2. Recombinant Gal4-VP16 protein purified from *E. coli* (13).
3. 10× SWI/SNF binding buffer: 200 mM Na-HEPES, pH 7.8, 30 mM $MgCl_2$, 0.8% NP40, 17% glycerol, 2 mM PMSF, 20 mM BME.
4. SWI/SNF storage buffer: 20 mM Na-HEPES, pH 7.8, 15% glycerol, 200 mM NaCl, 0.1% Tween-20.
5. 4–12% Bis–Tris SDS-polyacrylamide gels or 4-20% Tris–glycine SDS-polyacrylamide gels.

(a) 4–12% Bis–Tris SDS-PAGE: Prepare 4% Bis–Tris by adding 1.88 ml of Bis–Tris (2.5 M Bis–Tris, 1.5 M HCl, adjust to pH 6.4), 1.25 ml of acrylamide (38.8% acrylamide to 1.2% bis-acrylamide), and 20 μl each of TEMED and 10% APS and adjust volume to 12.5 ml with water. Similarly, prepare 12% Bis–Tris by using 3.75 ml of acrylamide. Using a gradient mixture, pour both 4 and 12% Bis–Tris to form a gradient resolving PAGE. Run the gel in 1× MES or MOPS SDS running buffer (50 mM MES or MOPS, 50 mM Tris, 0.1% SDS, and 1 mM EDTA).

(b) 4–20% Tris–glycine SDS-PAGE: Prepare 4% resolving gel mix by adding 1.7 ml of 30% acrylamide (37.5:1, acrylamide-to-bis-acrylamide ratio), 2.5 ml of 5× resolving buffer, and 20 μl each of TEMED and 10% APS to a final volume of 12.5 ml. Similarly, prepare 20% resolving gel mix by adding 8.3 ml of 30% acrylamide. Using a gradient mixture, pour both 4 and 20% resolving gel mix to form a gradient resolving PAGE. Run the gel in 1× Tris–glycine buffer (25 mM Tris, 200 mM glycine, and 0.1% SDS).

6. 4× SDS-loading buffer: 8% SDS, 20% BME, 0.24 M Tris–HCl, pH 6.8, 0.32% bromophenol blue, 20% glycerol.
7. M-280 Streptavidin Dynabeads (Dynal).
8. Magnetic separation stand (Promega).
9. 4% native PAGE (79:1, acrylamide-to-bis-acrylamide ratio) for SWI/SNF–nucleosomes binding.
10. Fixing solution: 50% methanol, 7% glacial acetic acid.
11. Washing solution: 10% methanol, 7% glacial acetic acid.

2.4. Mapping SWI/SNF Interactions with Nucleosomes by Site-Directed Proteolysis Using FeBABE

2.4.1. Phosphorothioate-DNA

1. Biotin-tagged forward and reverse primers to amplify 276 bp biotin-tagged phosphorothioate-DNA from plasmid template p159-1-G-27 plasmid (4).
2. Plasmid DNA template: p159-1-G-27 plasmid (4).
3. AmpliTaq Gold DNA polymerase (Applied Biosciences).
4. Deoxycytidine α-thiotriphosphate (dCTTPαS) (Glenn Research).
5. Microcon YM-50 (Amicon).

2.4.2. Coupling FeBABE to DNA and Nucleosomes

1. FeBABE: (Fe(III) (S)-1-(p-Bromoacetamido-benzyl)ethylene diamine tetraacetic acid) (Pierce) is dissolved in conjugation buffer to give 8.5 mM concentration (see Note 15).
2. 2-iminothiolane (2-IT) (Pierce) is dissolved in conjugation buffer to a final concentration of 4.25 mM (see Note 14).
3. Conjugation buffer: 10 mM MOPS, pH 8.0, 1 mM EDTA, 5% glycerol, 0.01% NP40, 0.5 mM PMSF (see Note 13).

2.4.3. Protein Cleavage with Tethered FeBABE

1. Cleavage buffer: 10 mM MOPS, pH 8.0, 1 mM EDTA, 10% glycerol, 0.01% NP40, 0.5 mM PMSF.
2. 0.48% H_2O_2 with 10 mM EDTA at pH 7. Hydrogen peroxide is a 30% stock, stored at 4°C in dark (see Note 15).
3. 50 mM sodium ascorbate with 10 mM EDTA at pH 7(see Note 15).
4. 10× SWI/SNF binding buffer: 200 mM Na-HEPES, pH 7.8, 30 mM $MgCl_2$, 0.8% NP40, 17% glycerol, 2 mM PMSF, 20 mM BME.
5. SWI/SNF storage buffer: 20 mM Na-HEPES, pH 7.8, 15% glycerol, 200 mM NaCl, 0.1% Tween-20.
6. Anti-HA high affinity antibodies from clone 3F10 (Roche).
7. SWI/SNF complex, in which the Snf2 is HA epitope tagged and purified from yeast by immunoaffinity.
8. 2× SDS-PAGE loading buffer: 125 mM Tris–HCl, pH 6.8, 4% SDS, 1.4 M BME, and 2.6 mM bromophenol blue.
9. M-280 Streptavidin Dynabeads (Dynal).
10. Magnetic separation stand (Promega).
11. 1× W&B buffer: 5 mM Tris–HCl, pH 7.5, 0.5 mM EDTA, 1 M NaCl.

2.4.4. In Vitro Translation of Truncated Snf2 with His_6–HA Tags

1. Forward Snf2 primers with upstream T7 promoter sequence (tcgatgcatactaatacgactcactatagggagagccaccATG, the T7 promoter sequence is underlined; translation-start ATG shown in capital letters).
2. Reverse Snf2 primer has a poly (T) sequence at its 5′ end consisting of 30 Ts.
3. PCR template: His_6–HA-tagged Snf2 cloned into pET-21b+.
4. Methionine ^{35}S (1,175 Ci/mmol or 43.48 TBq/mmol, PerkinElmer).
5. T7 coupled in vitro reticulocyte lysate transcription and translation kit (Promega), which contains TNT T7 Quick Master Mix, methionine, and nuclease-free water. The kit is stored at −70°C.

3. Methods

3.1. Mapping DNA–Protein Interactions by DNA Footprinting

DNA–protein interactions can be mapped using DNA footprinting, which gives an overall picture of the organization of a protein interacting with the DNA by protecting DNA from cleavage by certain reagents (3, 4, 14). DNA footprinting by hydroxyl radical is a preferred method as it cleaves DNA nonspecifically, giving single base pair resolution and requires tight binding of the protein to its particular DNA target site. Another advantage of this method is that hydroxyl radicals can be easily generated with a wide range of buffers and salt concentrations using readily available chemical reagents. However, hydroxyl radicals can be easily quenched by the presence of glycerol in the reaction and hence higher glycerol concentrations are not permitted. Hydroxyl radicals are relatively small and highly reactive and can abstract a proton from the sugar backbone causing phosphodiester bond breakage. In this section, we describe a method for DNA probe preparation, hydroxyl radical footprinting, and analysis of DNA footprinting data.

3.1.1. DNA Probe Synthesis by PCR and Nucleosome Reconstitution

1. Prepare a 4% native polyacrylamide gel, pre-run at constant 100 V (for 10×20-cm gels) with 0.5× TBE for 1 h.
2. Phosphorylate 20 pmol of forward or reverse oligonucleotides at 5′ end with 12 μl of 6,000 Ci/mmol, γ-P^{32} radiolabel ATP using 10 units of Optikinase in 1× Optikinase buffer in a 30-μl reaction volume. Incubate at 37°C for 45 min and inactivate Optikinase enzyme by incubating at 65°C for 20 min.
3. Set up a 100 μl PCR reaction with 12 ng of plasmid template having "601" NPS, 200 μM dNTPs, 20 pmol of labeled forward or reverse oligonucleotide, 25 pmol of either reverse or forward cold oligonucleotides in 1× PCR buffer, and appropriate DNA polymerase.
4. Clean PCR product by using manufacturer's recommendation for purification using a PCR purification kit.
5. Analyze the PCR product/probe DNA on a 4% native polyacrylamide gel and 6.5% denaturing polyacrylamide to ensure a uniform PCR product. It is important to use both the gel systems because in native gel secondary structure of the probe DNA can be seen due to difference in their mobility while in denaturing gel spurious PCR products can be determined.
6. To reconstitute nucleosomes, add 10 μg of salmon sperm DNA, 100–200 fmol of radioactive end-labeled DNA probe, and NaCl (final concentration 1.9 M) to a microfuge tube, incubate at 37°C for 10 min, and add 10–15 μg of octamer (determined empirically) to final volume of 10 μl. Incubate at 37°C for 25 min.
7. Add 3.2, 6, 9, and 39 μl of reconstitution buffer to dilute final NaCl concentration to 1.3 M, 1.0 M, 0.78 M, and 0.283 M to a final 67.2 μl reconstitution. Each dilution step follows incubation of the reaction at 37°C for 10 min.

3.1.2. Immobilization of DNA Template for Gapped DNA Probe Synthesis

1. DNA template is prepared by digesting p-199-1 plasmid (500 pmol) with NdeI followed by incorporation of biotinylated nucleotides, Bio-14-dATP (Invitrogen), and Bio-11-dUTP (Enzo) with Klenow exonuclease negative DNA polymerase (NEB) (3).

2. Unincorporated biotinylated nucleotides are removed using Microcon YM-30 filters followed by washing three times with 400 μl of TE, pH 8.0.
3. The samples are eluted three times by adding 100 μl of TE and inverting the filter into a fresh Eppendorf tube.
4. Eluted samples are digested with 2,000 units of BamHI in 400 μl.
5. DNA template (434 bp) is immobilized with magnetic beads coated with streptavidin from Dynal by incubating at 30°C for 45 min with intermittent mixing (30 s mixing at 1,200 rpm and 30 s pause) in Thermomixer (Eppendorf) to avoid settling of magnetic beads at the bottom of Eppendorf tubes.
6. The magnetic beads are washed two times with 200 μl of BW buffer. Washes and the immobilized DNA templates are preserved for agarose gel analysis. Figure 1 shows the biotinylated

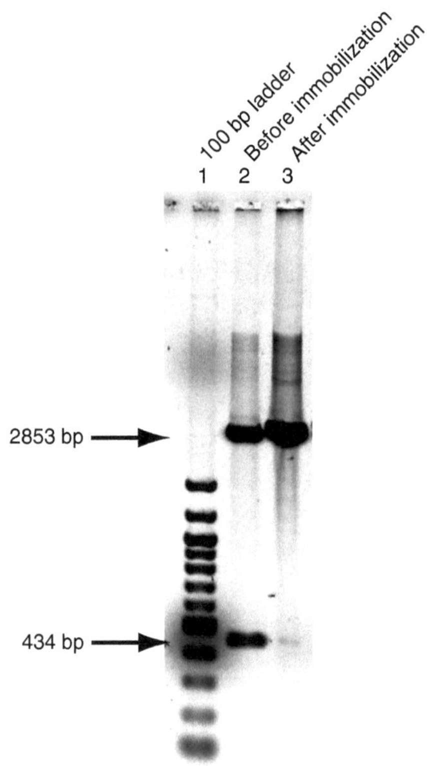

Fig. 1. Efficient immobilization of DNA template to magnetic streptavidin beads. *Lanes 2* and *3* show biotinylated p-199-1 DNA digested with NdeI and BamHI. *Lane 2* shows biotinylated DNA before immobilization, where 434 bp and 2,853 bp DNA fragments are seen. After immobilization in *lane 3*, there is disappearance of the biotinylated 434 bp DNA fragment and the larger DNA fragment remains unbound.

p-199-1 DNA before immobilization, where 434 bp DNA fragment having 601 NPS is seen along with 2,853 bp of plasmid backbone. After immobilization, 434 bp DNA is not seen on the agarose gel, indicating its complete immobilization.

7. The nonbiotinylated DNA strand is stripped by 0.1 M NaOH, and the single-stranded DNA beads are neutralized with 0.1 M HCl followed by bead equilibration in buffer 1× D1+T and stored at 4°C (15).

3.1.3. DNA Probe Synthesis with and Without Gap for DNA Footprinting

1. DNA probes are synthesized in which 40 pmol of 5′ phosphorylated oligonucleotide −31/−9, having two deoxyuridine residues at −24/−25 is annealed to 4 pmol of single-stranded immobilized DNA along with 40 pmol of upstream oligonucleotide −105/−82. The nomenclature of the oligonucleotide is represented by its position on the DNA template with respect to dyad axis present on the 601 NPS.
2. Oligonucleotides are annealed in 1× D1+T buffer by heating at 70°C for 3 min and then lowering the temperature by 10°C every 10 min and finally incubating at 37°C for 30 min with intermittent mixing. Intermittent mixing is done in a Thermomixer (Eppendorf) with 30 s pause and 30 s vortexing at 1,200 rpm.
3. The beads are washed with 1× D1+T buffer twice to remove excess oligonucleotides and then resuspended in 20 μl of 1× D1+T buffer.
4. Extension and ligation of annealed oligonucleotides are done with 6 U of T4 DNA polymerase and 50 U of T4 DNA ligase in a buffer containing 50 μg of BSA, 500 μM of dNTP mix, and 1 mM of ATP in 40 μl with intermittent mixing at 37°C for 60 min.
5. Beads are washed three times with TE buffer, pH 8.0, containing 0.1% SDS, once with 1× D1+T buffer, and then three times with 1× Optikinase buffer.
6. Phosphorylation of the 5′end of the DNA strands are done with 10 μl of γ-P^{32} ATP and 20 U of Optikinase™ enzyme in 30 μl for 45 min.
7. Beads are washed three times with TE buffer, pH 8.0, containing 0.1% SDS and three times with 1× D1+T buffer before resuspending them in 100 μl of 1× EcoR*I* buffer.
8. A small aliquot (0.5 μl) is removed for checking ligation and kinasing efficiency of DNA strands and is analyzed on a 6.5% denaturing PAGE containing 8 M urea.
9. The beads are divided into two equal halves; to one half 100 U of EcoR*I* is added and to the other half 100 U of EcoR*I* and 2 U of USER™ enzyme are added and incubated for 1 h at 37°C. EcoR*I* restriction enzyme releases probes from the

beads and USER™ enzyme creates a two nucleotide gap in the probe (see Note 2). Half μl of sample is removed from each reaction for analysis.

10. Probes are cleaned with a PCR purification kit and eluted in 50 μl of elution buffer.
11. DNA probes are concentrated by centrifugal evaporation under vacuum and reconstituted into nucleosomes with WT *X. laevis* octamer (see Subheading 3.1.1).
12. Probes taken out at different stages are denatured in formamide dye and separated on 6.5% denaturing gel containing 8 M urea.
13. The gel is dried and phosphorimaged. Final probes are also checked on a 4% native polyacrylamide gel (see Note 3).

3.1.4. Hydroxyl Radical Footprinting of Nucleosomes Bound to ISW2

1. Hydroxyl radical DNA footprinting is carried out with DNA, nucleosomes, ISW2-nucleosome complex, and nucleosomes remodeled by ISW2. The binding and sliding reactions for ISW2 contain 5% glycerol, and therefore glycerol is removed using spin columns containing S-200 Sephacryl resin (see Note 4).
2. Cleavage reaction is carried out at room temperature for 2 min by mixing the following components in order: 290 μM ferrous ammonium sulfate, 290 μM EDTA, 0.17% hydrogen peroxide, and 5.7 mM sodium ascorbate. This reaction should be done under reduced light condition as sodium ascorbate solution is light sensitive.
3. For rapid DNA footprinting of nucleosomes remodeled with ISW2 for different lengths of time, the cleavage conditions are slightly altered with ferrous ammonium sulfate, EDTA, hydrogen peroxide, and sodium ascorbate being added to final concentration of 1.9 mM, 2 mM, 0.14%, and 5.3 mM, respectively, and cleavage being carried out for 30 s (14).
4. The generation of hydroxyl radicals is terminated by addition of either termination mix containing 3.7 M ammonium acetate, 3.7 M thiourea, and 7.4 mM EDTA or glycerol to 10% final concentration in the reaction.
5. Water is added to increase the reaction volume to 200 μl.
6. Proteins are removed from the reaction by treating with 1:1 phenol/chloroform mixture and with equal volume of chloroform.
7. For efficient DNA precipitation, 500 ng of sheared salmon sperm DNA, 1/10 volume of 10 M lithium chloride, 5 μl of 1 M magnesium chloride, and 500 μl of 100% ethanol are added.
8. Precipitation is carried out at –20°C for 12–16 h.
9. Precipitated DNA is obtained by spinning the samples at 19,000 × *g* for 30 min at 4°C followed by washing the DNA pellet twice with 70% ethanol.

10. The pellet is dried and suspended in 8 μl of formamide dye (3).
11. Cleaved DNAs are separated on 6.5% denaturing polyacrylamide gel containing 8 M urea, 1× TBE buffer, and 40% acrylamide along with the sequencing ladder.
12. The polymerized gel is pre-run in 1× TBE buffer at 35 W for 1 h before loading the samples. The DNA sample in formamide dye is heated at 90°C for 2 min and 3 μl of the sample is immediately loaded on to the gel.
13. The gel is run for about 2 h and until the bromophenol blue dye front is 1 cm from the bottom of the gel.
14. The gel is dried and phosphorimaged for 1–4 days depending on the signal intensity and scanned with a Fuji FLA-5500 or comparable system.
15. To trap remodeling intermediates of ISW2 complex, DNA footprinting is carried out with nucleosomes containing DNA gaps at nucleotides −24/−25. DNA footprinting of ISW2–nucleosome is also done in the presence and absence of 400 μM ATP, γ-SATP, and ADP. Briefly, nucleosomes are incubated with ATP and its analogs with saturating amounts of ISW2 for 30 min at 30°C and then treated with hydroxyl radicals as described above.

3.1.5. Analysis and Interpretation of DNA Footprinting Data

1. The scanned image is exported to Image Quant™ software (GE Healthcare) for each sample and the pixel information is exported to Microsoft Excel™ to generate overlays.
2. Total lane signal for each sample is normalized and DNA footprint overlays are plotted for nucleosome alone and ISW2 bound to nucleosomes to determine the footprint of ISW2 on nucleosomes (3, 14). The nucleosomal DNA footprint clearly shows helical periodicity with peaks and troughs.
3. The identity of each peak is obtained by overlapping with the sequencing ladder (Fig. 2a). ISW2 generates footprints on nucleosomes at three distinct locations: 10–20 base pairs from the dyad axis, at the entry site of the nucleosomes, and at extranucleosomal DNA (3).

3.2. Mapping DNA–Protein Interactions by DNA Cross-Linking

DNA–protein interactions can also be studied by site-directed DNA cross-linking (15). Site-specific DNA cross-linking shows which subunit of a large multisubunit complex associates with different parts of the DNA either bound within the nucleosome or the extranucleosomal DNA. We have used two forms of photoreactive DNA (1): (1) 4-azidophenyl bromide (APB) conjugated to a phosphorothioate (S-P probes) moiety that is incorporated at specific position in the phosphate backbone of DNA and (2) an aryl azide attached to the base (i.e., AB probes) of deoxyuridine (AB-dUTP), deoxycytidine (AB-dCTP), or deoxyadenosine

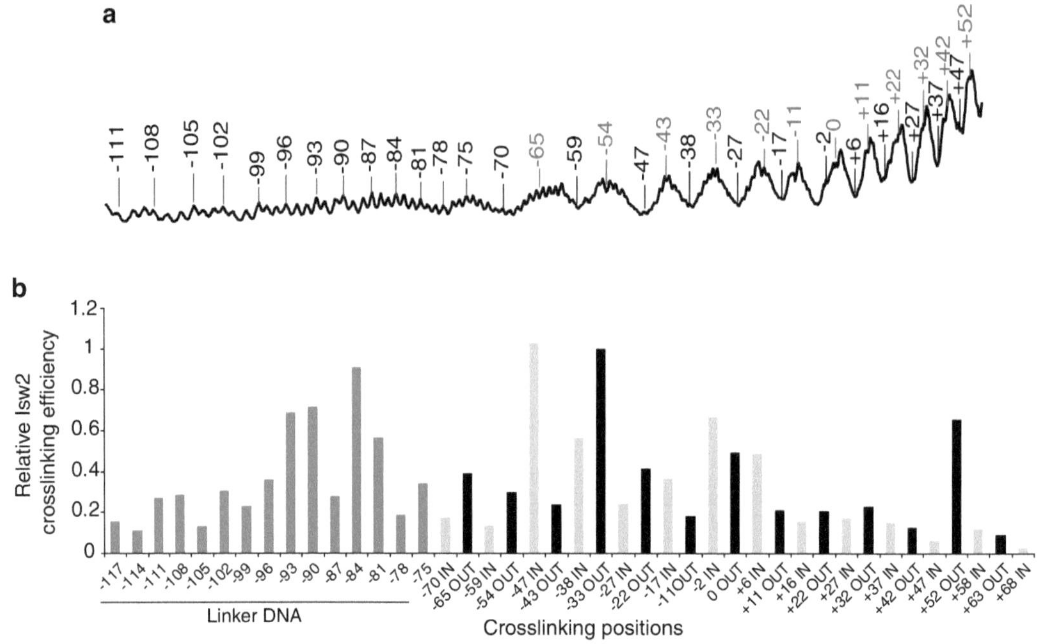

Fig. 2. Hydroxyl radical DNA footprinting and site-specific DNA cross-linking profiles. (**a**) End-positioned nucleosomes (70N0) are subjected to hydroxyl radical DNA footprinting. Cleavage of DNA on the nucleosome shows helical pattern of the DNA as the extent of cleavage reduces, where DNA is protected by the histone octamer. (**b**) The efficiency of Isw2 cross-linking with phosphorothioate-modified, 62N0 end-positioned nucleosomes is displayed. "In" and "out" positions represent the orientation of the photoreactive probe facing toward or away from the histone octamer, respectively. Each position in extranucleosomal (*dark grey bars*) and nucleosomal DNA (*light grey* and) is color coded as indicated.

(AB-dATP) (6, 15–19). The advantage of using two kinds of modifications on the DNA probes are: (1) phosphorothioate modification probes both the major and minor groove of DNA and (2) nucleotide modification exclusively scans the major groove of DNA (20).

A combination of DNA footprinting and site-specific DNA cross-linking gives a comprehensive view of DNA–protein interactions (3, 4). We have used DNA footprinting and DNA cross-linking techniques to determine the overall conformation and subunit interactions of ISW2, ISW1a, and SWI/SNF complexes bound to nucleosomes (3, 4, 21) and the progressive changes in these interactions associated with nucleosome remodeling (14). Remodeling intermediates of ISW2 complex were also trapped using nucleosomes having DNA gaps close to the internal site on the DNA that are important for ISW2 binding and therefore blocks DNA translocation by the remodeler (14). In this section, we discuss DNA cross-linking using azido-benzoyl- and phosphorothioate-based DNA modification, probe preparation, nucleosome assembly, and nucleosome binding and sliding with ISW2, DNA photoaffinity cross-linking, and analysis of cross-linking results.

3.2.1. Chemical Modification of Oligonucleotides for Site-Specific DNA Photoaffinity Cross-Linking

1. Commercially synthesized oligonucleotides are designed with a phosphorothioate group incorporated between the third and fourth nucleotides from the 5′ end of the oligo (see Note 6) (15, 20).
2. Commercially synthesized phosphorothioate oligonucleotides (IDT DNA) are resuspended in 100 mM TEAB, pH 8.0, to a concentration of 200 pmol/μl and 25 μl (5 nmol) of the oligonucleotide is mixed with 25 μl of 100 mM APB in DMF.
3. The coupling reaction of APB with the phosphorothioate is at room temp for 1 h under reduced lighting. A small amount (25 μl aliquot) of DMF is added to the modification reaction if any precipitate appears.
4. Next, samples are diluted to 300 μl by 100 mM TEAB to stop the reaction.
5. Modified oligonucleotides are extracted three times with isobutanol that has been saturated with TE. The oligonucleotide remains in the lower aqueous layer and APB partitions into the organic layer and is removed.
6. After extraction with isobutanol, oligonucleotides are extracted three times with ethyl ether that is saturated with TE.
7. The samples are incubated at 37°C for 10–15 min until residual ethyl ether evaporates.
8. Oligonucleotides are concentrated by centrifugal evaporation under vacuum (Labconco) for 2 h and dried samples are resuspend in 1 ml of water followed by drying.
9. The modified oligonucleotides are dissolved in 50 μl of 100 mM TEAB to a final concentration of ~100 pmol/μl and stored at −20°C.
10. Modification efficiency is determined by radiolabeling the modified and unmodified (control) oligonucleotides and separating them on a 10% gel with 8 M urea. Modified oligonucleotides have slower mobility as compared to unmodified oligonucleotides. After conjugation, 24 pmol of oligonucleotides are phosphorylated with γ-P^{32} ATP with T4 polynucleotide kinase (Optikinase™) for 45 min at 37°C followed by inactivation of the enzyme by heating at 65°C for 20 min. The unmodified oligonucleotide (1–2 pmol) is also phosphorylated for comparison.
11. The extent of modification is found by quantifying the intensities of modified and unmodified oligonucleotides. Greater than 95% modification of oligonucleotides is generally achieved using this method.

3.2.2. DNA Probe Preparation for DNA Containing Modified Nucleotide Bases (AB Probes)

1. The first or target oligonucleotide (40 pmol) is annealed to 4 pmol of immobilized single-stranded DNA template. Annealing is done as described in Subheading 3.1.3.
2. Oligonucleotides are extended using 1.5 μl of 100 μM specific photoreactive nucleotide and 2 μl of radioactive α-P^{32} deoxynucleotide. The particular photoreactive and radioactive nucleotide incorporated is dependent on the site being modified and the underlying sequence of the DNA template.
3. To create a 2 nucleotide gap at nt −24/−25, a −31/−9 oligonucleotide is used as a downstream oligonucleotide. The numbers in this case indicate the start and end positions of the oligonucleotide with reference to the dyad axis when the DNA is assembled into nucleosomes. This oligonucleotide has 2′ deoxyuridine incorporated at nt −24/−25. An upstream oligonucleotide −105/−82 is also used and both oligonucleotides are extended and ligated.
4. Probes are processed as described in Subheading 3.1.3 to create probes with and without 2 nucleotide gaps.

3.2.3. DNA Probes for Site-Specific DNA Photoaffinity Cross-Linking with Modified Phosphorothioates (SP Probes)

1. Radiolabeled photoreactive oligonucleotides (24 pmol) and an oligonucleotide upstream to the modified oligonucleotide (40 pmol) with an Xba*I* restriction enzyme site are annealed to the single-stranded immobilized DNA at 37°C for 30 min with intermittent mixing. Intermittent mixing with 30 s pause and 30 s mixing at 1,200 rpm is carried out to prevent magnetic beads from settling.
2. The beads are washed with 100 μl of 1× D1+T buffer twice to remove excess oligonucleotides and then resuspended in 20 μl of 1× D1+T buffer.
3. Annealed oligonucleotides are extended with T4 DNA polymerase (6 U) and ligated with T4 DNA ligase (50 U) under conditions described in Subheading 3.1.3.
4. Probes are released from the beads by digesting it with Xba*I* at 37°C for 30 min.
5. The digested probes are extracted with 1:1 phenol chloroform and precipitated with sodium acetate and ethanol.
6. Probes are analyzed on 6.5% urea and 4% native PAGE and the amount of radiolabeled probes are equalized upon nucleosome reconstitution. In this manner, 42 different site-specific photoreactive probes are synthesized within extranucleosomal DNA (3 bp apart) and nucleosomal DNA (5 bp apart) (Fig. 2b).

3.2.4. Nucleosome Reconstitution and Binding and Sliding Assays with ISW2

1. Nucleosomes are assembled with WT recombinant *Xenopus* histone octamer as described in Subheading 3.1.1.
2. Binding reaction with end-positioned nucleosomes are carried out with 2:1 molar ratio of ISW2 and nucleosomes in 60–70 mM

NaCl, 30 mM Na-Hepes, pH 7.8, 5 mM $MgCl_2$, 0.1 mg/ml BSA, and 5% glycerol in 25 μl reaction volume for 30 min at 30°C (see Note 13).

3. Two μl of binding reaction is mixed with 2 μl of 10% glycerol before and after treatment with hydroxyl radical and loaded on 4% native polyacrylamide gel to ensure that nucleosomes remain bound to ISW2 during the cleavage reaction. This control is critical for DNA footprinting assays.
4. Nucleosome sliding assays are carried out in 25 μl volume, similar to binding assays, except that 90 μM ATP is added to initiate remodeling (14). Prior to ATP addition, the temperature of the reaction is lowered to 18°C to slow down the remodeling reaction. Remodeling is stopped at different times by addition of 1 μl of a mixture of sheared salmon sperm DNA (10 mg/ml) and γ-SATP (20 mM).
5. Bound and remodeled nucleosomes are treated with hydroxyl radical as described in Subheading 3.1.4.
6. Similar binding and sliding reactions are carried out with photoreactive nucleosomes (−/+ gap at −24/−25 positions) under reduced lighting conditions.

3.2.5. DNA Photoaffinity Labeling SP or AB Probes

1. Nucleosome–ISW2 bound complex with different site-specific photoreactive groups is UV irradiated (310 nm, 2.65 mW/cm^2, at a distance of 8 cm) for 3 min.
2. DNA is digested with 4.6 U of DNase I at 30°C for 15 min (see Note 5).
3. DNaseI is terminated by SDS (final concentration of 0.4%) and samples are heated at 90°C for 3 min. Heating samples in the presence of SDS releases DNA from the histone octamer.
4. Samples are cooled to room temperature, supplemented with 1 mM zinc acetate, and further digested with 20 U of S1 nuclease at 30°C for 15 min (see Note 5).
5. S1 nuclease is stopped by adding 0.5 μl of 0.5 M Tris base and SDS gel loading buffer.
6. The samples are denatured for 3 min at 90°C and loaded on 4–20% SDS-PAGE.
7. Gels are stained with Coomassie brilliant blue R250 (0.1% in 50% methanol and 7% glacial acetic acid) and dried.
8. The locations of the stained ISW2 subunits are marked with radioactive ink and cross-linked ISW2 subunits are visualized by autoradiography.

3.2.6. Analysis and Interpretation of DNA Photoaffinity Cross-Linking Data

Cross-linking efficiency of Itc1, Isw2, and Dpb4 subunits at different DNA positions is determined and then normalized to the cross-linking efficiency of Isw2 cross-linked to base pair −33. Base pair −33 was chosen because it was one of the positions in DNA that cross-linked most efficiently.

1. In order to compare many samples either at different DNA positions or under different conditions, typically a reference standard sample is always included. In our case, we have used the level of Isw2 cross-linked at base pair −33 as our reference. Figure 2b shows the cross-linking profile of the Isw2 catalytic subunit with extranucleosomal (dark grey) and nucleosomal DNA (light grey and black). In this manner, we can compare the differences in cross-linking efficiency between subunits and at different positions and how these may change when remodeling is initiated. This approach is used for both type of DNA cross-linking (SP or AB).
2. This approach has been successfully used on nucleosomes with gaps at nucleotides −24/−25 that are used to trap remodeling intermediates and probe the conformation changes associated with these intermediates in ISW2 remodeling (14).

3.3. Mapping Protein Interactions with Nucleosomes by Histone Site-Directed Cross-Linking and Label Transfer

3.3.1. DNA for Nucleosome Reconstitution

1. A 276 bp DNA with 601 NPS is amplified by large-scale PCR (10 ml) from p159-1-G-27 plasmid with one of the two primers containing 5′end biotin (4). The 601 NPS is flanked by 69 (biotinylated) and 59 bp (with Gal4-binding site) linker DNA. The large-scale (10 ml) PCR reaction is performed by scaling up 100 μl test PCR reaction. A 100 μl reaction mix is distributed in 0.5 ml PCR tubes and amplified using the Eppendorf Mastercycler Gradient PCR device. The PCR product is pooled, concentrated with Microcon YM-50 filters, and washed three times with TE buffer.
2. The DNA is recovered in about 300 μl of TE and extracted by phenol/chloroform, and ethanol precipitated.

3.3.2. Nucleosomes with Site-Specific Cysteines

1. Cysteine mutant histones are generated by site-directed mutagenesis of the *X. laevis* core histone proteins and the original cysteine at residue 110 of H3 is mutated to alanine. Surface-exposed residues (Fig. 3a) are selected based on the crystal structure of the nucleosome (22) and is mutated to cysteine.
2. Histone octamers are refolded under reducing conditions with recombinant H2A, H2B, H3, and H4 in which one of the four histones contains a single cysteine and purified by size exclusion chromatography (12) (see Note 8).
3. Nucleosomes with a unique cysteine residue on the surface are reconstituted on 276 bp DNA with 601 NPS by salt dilution approach as in Subheading 3.1.1. A concentration of 1 mM BME is maintained during the whole nucleosome reconstitution processes.
4. Remove the reducing reagent (BME) in the nucleosome samples by dialysis against 1 L nucleosome storage buffer twice, each

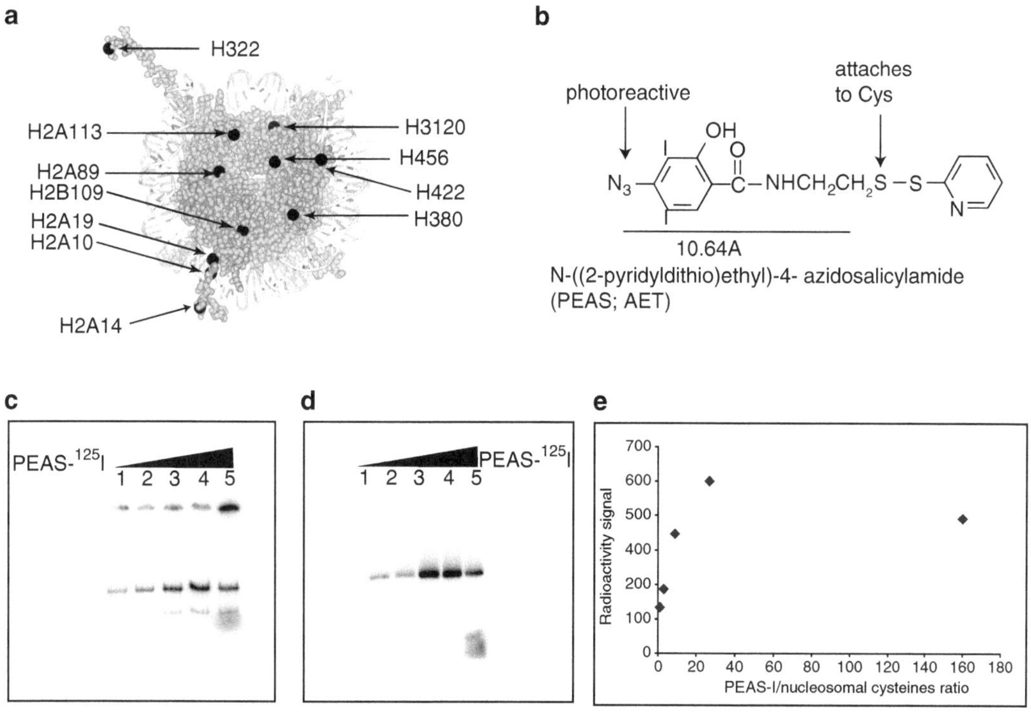

Fig. 3. Modification of nucleosomes at specific cysteines with PEAS. (**a**) The sites, where cysteine is engineered on the surface of the nucleosome, are displayed in *black* for both core domains and histone tails. (**b**) The chemical structure of PEAS or S-[2-(4-azidosalicylamido) ethylthio]-2-thiopyridine is shown. (**c**) Optimal modification conditions for cysteine-containing nucleosome with ^{125}I-PEAS. Nucleosome with cysteine residue at position 80 of H3 was modified with different concentrations of ^{125}I-PEAS. The molar ratio of ^{125}I-PEAS to cysteine residue in nucleosome was 1:1 (*lane 1*), 3:1 (*lane 2*), 9:1 (*lane 3*), 27:1 (*lane 4*), and 160:1 (*lane 5*). Modified nucleosomes were analyzed by 4% native-PAGE and phosphorimaging. (**d**) Samples in (**c**) were analyzed by SDS-PAGE (16%) under nonreducing condition. (**e**) Quantitative analysis of nucleosome modification. The relative nucleosome signal in (**c**) was plotted versus the ^{125}I-PEAS:cysteine ratio.

time for 3 h. Alternatively, Sephadex G-25 spin column equilibrated with nucleosome storage buffer can be used to remove reducing reagents from nucleosome samples immediately before modification with PEAS or ^{125}I-PEAS (23) (see Note 9).

3.3.3. Coupling of ^{125}I-PEAS to Cysteine-Containing Nucleosomes

We have used two kinds of approaches to modify the cysteine residue on the surface of nucleosomes with iodinated PEAS (^{125}I-PEAS) reagent (23) (see Notes 7 and 13).

1. PEAS reagent is iodinated with ^{125}I and then the iodinated PEAS is conjugated to the cysteine on the nucleosome surface.
2. The PEAS reagent is conjugated to the nucleosome, followed by iodination of the PEAS-modified nucleosomes. The conjugation of the PEAS reagent to cysteine thiol group is mediated by disulphide bond exchange (Fig. 3b).

Iodination of PEAS Reagent and Conjugation to Cysteine-Mutant Nucleosomes

In this approach, the cross-linking reagent PEAS is iodinated first and then used for labeling cysteine-mutant nucleosomes (23) (see Note 13).

1. In a typical iodination experiment, 2–5 mCi of $Na^{125}I$ (17 Ci/mg, PerkinElmer) and 2 nmol of PEAS are mixed in the IODO-GEN Pre-Coated Iodination Tube (Pierce) and the volume is adjusted to 90 μl with 100 mM sodium phosphate, pH 7.4, and incubated at 25°C for 1 min (10) (see Notes 10–12).
2. The reaction is stopped by transferring the sample to a tube with 1 μl of 2.5 mM tyrosine (27 μM final) and 1 μl of 80 mM methionine (final 0.87 mM) to quench free $^{125}I^{+}$.
3. Immediately, iodinated PEAS (^{125}I-PEAS) is used to modify cysteine-mutant nucleosomes. Add 10–15 μl (20 μM) of ^{125}I-PEAS to 50 μl (0.4 μM) cysteine-mutant nucleosomes or control nucleosome sample lacking cysteine residue and incubate on ice for 30 min. The optimal concentration of ^{125}I-PEAS can be determined by titrating different concentration of ^{125}I-PEAS against constant concentration of cysteine-mutant nucleosomes (Fig. 3c–e).
4. The excess free ^{125}I-PEAS and ^{125}I are removed by Sephadex G-25 spin column equilibrated with nucleosome storage buffer in an iodination safe hood. The modified nucleosomes are dialyzed further against 1 L nucleosome storage buffer three times with each dialysis lasting at least 3 h.
5. Analyze 4 μl of the modified nucleosomes on 4% native PAGE. Examine the stability and concentration of the modified nucleosomes by SYBR Gold or ethidium bromide staining (Fig. 4a). The concentration of the final modified nucleosome can be quantified by comparing with the original unmodified nucleosome samples. The gel is dried and analyzed for modification by phosphorimaging.
6. The specificity and efficiency of modification are examined by 16% SDS-PAGE under nonreducing conditions. Typically, add 2.5 μl of 4× SDS-loading buffer (without reducing reagent) to 7.5 μl of nucleosome and heat the samples for 3 min at 90°C. Run the samples on 16% SDS-PAGE and stain with Coomassie blue or SYPRO Ruby. After picture is taken, dry the gel, analyze for specificity (for cysteine) and level of modification (accessibility of cysteine residue) (Fig. 4b). As shown in Fig. 4b, only histones with cysteine residue are labeled with ^{125}I-PEAS and histones lacking cysteine are not labeled showing that the modification of nucleosome with ^{125}I-PEAS is cysteine side chain (thiol group) specific.

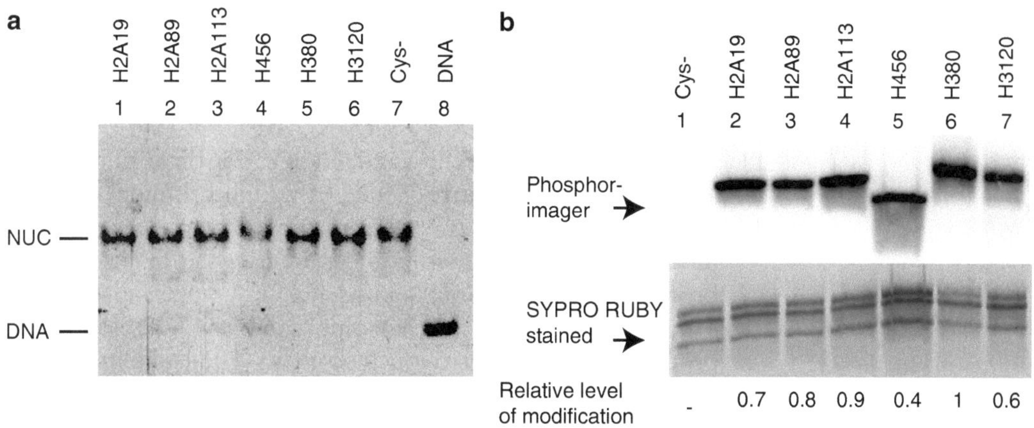

Fig. 4. Analysis of PEAS reagent-modified nucleosomes. (**a**) Cysteine-containing octamers were reconstituted into nucleosomes and modified with PEAS. Nucleosomes were analyzed by 4% native-PAGE and stained with SYPR Gold. The positions of the cysteine are indicated in terms of residue and histone. (**b**) The PEAS reagent specifically conjugated to cysteine-containing nucleosomes. ^{125}I-PEAS-modified nucleosomes were analyzed by SDS-PAGE under nonreducing condition and imaged using a phosphorimager (*top panel*). *Bottom panel* shows SYPRO Ruby staining of same gel. The relative level of modification is shown below the gel.

Indirect Iodination

In this approach, cysteine-containing nucleosomes are modified first with PEAS reagent followed by iodination of the PEAS-modified nucleosomes with ^{125}I (24, 25) (see Note 13).

1. Remove the reducing reagent from nucleosome samples by dialysis or spin column (see Note 8) as described in the preceding section.
2. Add 1 μl of 3 nmol/μl PEAS (in DMSO) to 50 μl (20 pmol, 0.4 μM) of nucleosomes and incubate for 1 h at room temperature.
3. Excess PEAS reagent is removed by dialysis (three times each for 3 h) against 1 L nucleosome storage buffer at 4°C in dark room.
4. Wet the IODO-GEN Pre-Coated Iodination Tube with 1 ml of 100 mM sodium phosphate buffer, pH 7.4, for 10–20 s and remove the buffer.
5. Add 90 μl of 100 mM phosphate buffer, pH 7.4, to the precoated tube. Add buffer directly to the bottom of the precoated tubes.
6. Add 1–2 mCi ^{125}I, mix, and incubate for 6 min at room temperature.
7. Transfer the oxidized ^{125}I ($^{125}I^{+}$) into 1.5 ml Eppendorf tube.
8. Add 10 μl of the activated $^{125}I^{+}$ (100–200 μCi) to 20 pmol PEAS-conjugated nucleosomes (0.3 μM) and incubate at room temperature for 10 min.
9. Remove the free ^{125}I by Sephadex G-25 spin column equilibrated with nucleosome storage buffer in iodination safe hood.

10. The excess free ^{125}I is further removed by dialysis against 1 L nucleosome storage buffer at 4°C in dark twice for at least 3 h each. The waste generated from spin column and dialysis is treated as radioactive ^{125}I waste (see Notes 10–12).
11. The concentration, stability, and modification of the nucleosome are assessed by 4% native-PAGE, ethidium bromide, or SYBR Gold staining and by phosphorimaging. The concentration of modified nucleosome samples can be assessed from ethidium bromide or SYBR Gold-stained gel by comparing with the original unmodified nucleosome samples. The iodination process can destabilize nucleosomes. The stability of the ^{125}I-PEAS-modified nucleosomes can be checked by assessing the enrichment of free DNA. Due to iodination of surface accessible tyrosine, in this iodination approach, it is difficult to assess the level of modification (see Note 13).

3.3.4. Photoaffinity Cross-Linking and Label Transfer

Chromatin remodelers SWI/SNF, ISW1a, and ISW1a (FLAG tagged in one of their subunits) are purified from *Saccharomyces cerevisiae* by affinity purification using anti-FLAG antibody M2 agarose resin (4). The optimal buffer conditions for nucleosome binding of these chromatin-remodeling enzymes are slightly different. Here, we describe the nucleosome binding and photoaffinity cross-linking of SWI/SNF. The modified nucleosomes are used directly or after being immobilized on streptavidin-coated magnetic beads (M280).

In-Solution Binding and Photoaffinity Cross-Linking

1. In a typical 25 μl reaction, yeast SWI/SNF (200 fmol) is incubated with ^{125}I-PEAS-modified nucleosome (200 fmol) in the presence or absence of Gal4-VP16 (320 fmol) at 30°C for 30 min. To make SWI/SNF binding to nucleosome Gal4-VP16 dependent, competitor DNA (50 ng sheared salmon sperm DNA) is included in the reaction (4). The concentration of competitor DNA required for SWI/SNF binding to nucleosomes plus Gal4-VP16 is determined empirically.
2. Load 4 μl of the reaction on a 4% native-PAGE to check for SWI/SNF binding to nucleosomes.
3. Samples are irradiated for 3 min to cross-link SWI/SNF to the ^{125}I-PEAS-modified nucleosomes.
4. Add 2.3 μl of 1 M DTT to 21 μl of irradiated samples and incubate at 37°C for 30 min. This step breaks the disulfide bridge between the cysteine side chain of the histone and the cross-linked SWI/SNF subunit(s). As a result, the ^{125}I-PEAS label is transferred to the cross-linked protein in close proximity to the cysteine-^{125}I-PEAS site.
5. Add 8 μl of 4× SDS-loading buffer and heat the samples for 3 min at 90°C.

6. Load the samples on to a 4–12% Bis–Tris or 4–20% Tris–glycine SDS-PAGE. Load uncross-liked SWI/SNF sample as a control. The 4–12% Bis–Tris SDS-PAGE gives better resolution for proteins under 50 kDa while the 4–20% Tris–glycine SDS-PAGE has better resolution for larger proteins (>100 kDa).
7. Fix the gel with 50% methanol, 7% glacial acetic acid for 30 min, and stain the gel with Coomassie blue.
8. Wash the gel with wash buffer (10% methanol, 7% glacial acetic acid) for 30 min and keep it in 10% glycerol for at least 1 h, dry, and visualize labeled proteins by phosphorimaging.

Photocross-Linking with Immobilized Nucleosomes

Alternatively, cross-linking is done using immobilized nucleosomes and it is helpful when labeling efficiency is low. The SWI/SNF–nucleosome reaction and cross-linking can be done in large volume, and finally the immobilized nucleosome/SWI/SNF is resuspended in small volume for analysis on SDS-PAGE.

1. Wash 20 μl (200 μg) of beads with 40 μl of 1×BW buffer, five times.
2. Wash the beads twice with 40 μl of nucleosome storage buffer.
3. Remove buffer and add 5 pmol of ^{125}I-PEAS-modified nucleosome (biotin-tagged DNA).
4. Incubate for 45 min at room temperature with gentle mixing.
5. Remove the supernatant (unbound fraction) and wash beads twice with 40 μl of nucleosome storage buffer. To assess the immobilization efficiency, run the unbound fraction and input sample on 4% native-PAGE, stain the gel with ethidium bromide or SYBR Gold, and quantify the nucleosome bands. The stability of the immobilized nucleosome can also be checked by 4% native-PAGE analysis of the nucleosome samples released from the beads with restriction enzyme digestion in linker DNA connecting the nucleosome to the bead.
6. In a typical reaction, immobilized nucleosomes (~100–300 fmol) are incubated with SWI/SNF under the conditions described in the preceding section.
7. Samples are cross-linked by UV irradiation for 3 min.
8. Unbound, free SWI/SNF is removed from the beads and the bound cross-linked SWI/SNF is resuspended in buffer with 100 mM DTT, 400 mM NaCl, and 10 mM Na-HEPES, pH 7.8, to break the disulfide bond connecting cross-linked nucleosomes to SWI/SNF and reverse SWI/SNF binding to the immobilized nucleosomes.
9. The samples are analyzed by 4–12% Bis–Tris or 4–20% Tris–glycine SDS-PAGE and phosphorimaging to identify the radiolabeled subunits as described in preceding section (Fig. 5).

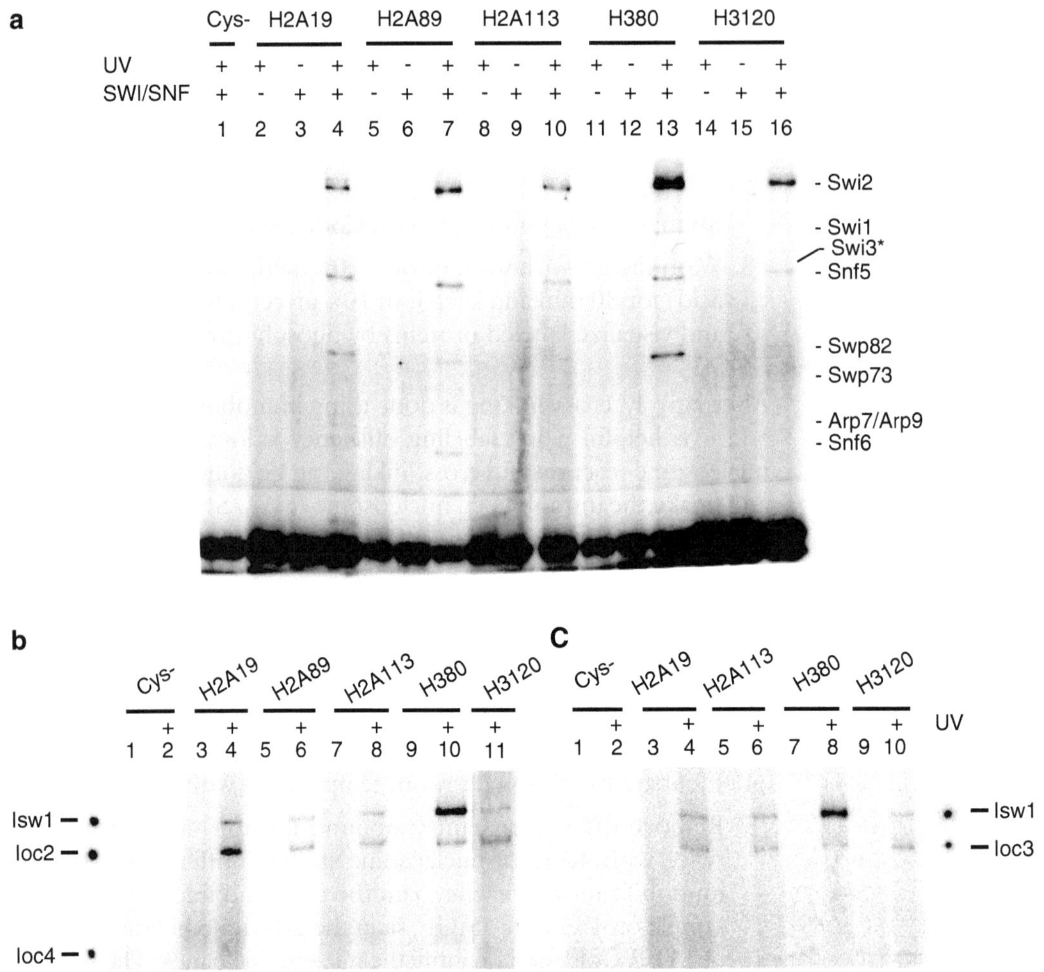

Fig. 5. Site-directed histone cross-linking of SWI/SNF and ISW2. (**a**) Photocross-linking of SWI/SNF bound to ^{125}I-PEAS nucleosomes. Nucleosomes were modified and labeled with PEAS-^{125}I reagent by indirect iodination. SWI/SNF was bound to immobilized nucleosomes. After photocross-linking with UV irradiation, the radiolabel was transferred by disulfide reduction and the samples were analyzed by 4–12% Bis–Tris SDS-PAGE and phosphorimaging. Location of modification sites is indicated above the lanes and refers to the particular histone (H2A or H3) and the residue changed to cysteine. Samples with SWI/SNF added or UV irradiated are indicated. The mobility of each SWI/SNF subunit is indicated and the *asterisk* (*) indicates that Swi3 was tagged with three epitopes (HA–V5–His6) at its C terminus. ISW1b (**b**) and ISW1a (**c**) were photocross-linked and were modified as described in (**a**).

3.4. Mapping SWI/SNF Interactions with Nucleosomes by Site-Directed Proteolysis Using FeBABE

In this approach, FeBABE, a FeEDTA derivative, is attached to either DNA or histones and used as a chemical protease to map the regions of chromatin-remodeling proteins that interact with DNA or nucleosomes (7, 9). FeBABE is attached to phosphorothioates that are incorporated into DNA or to thiol groups on the surface of nucleosomes. In the presence of ascorbate and hydrogen peroxide, the FeBABE attached to DNA or nucleosomes generates hydroxyl radicals locally and cleaves nearby peptide bonds of interacting proteins. The interacting regions of a protein are determined by identifying the cleavage sites by Western blot analysis using antibodies against the C- or N-terminal epitope of the protein.

3.4.1. Conjugation of FeBABE to DNA

1. Phosphorothioate-DNA is synthesized by PCR and replacing one of the deoxyribonucleotides in the reaction with deoxyribonucleotide (8). In the PCR product, the nonbridging oxygen of the phosphate backbone is replaced with sulfur. The primers are designed such that one 5′-end is biotin labeled.
2. DNA sample is concentrated and purified with Microcon-50 filter.
3. Phosphorothioate-DNA is modified by incubating with 10–20 folds excess FeBABE in conjugation buffer at 50°C for 3 h. The FeBABE is prepared immediately before use by dissolving with conjugation buffer (see Note 16).
4. Excess FeBABE is removed with Microcon-50 filters.

3.4.2. Conjugation of FeBABE to Nucleosomes

1. Nucleosomes are reconstituted with recombinant *X. laevis* histones and 276 bp DNA containing the 601 NPS as described in Subheading 3.1.1. The DNA has biotin incorporated at one 5′ end and the 601 NPS has 69 and 60 bp of flanking linker DNA.
2. To conjugate FeBABE to the solvent-accessible lysine residues in the nucleosome surface, add 0.85 mM FeBABE and 0.425 mM 2-IT to 0.4 μM nucleosomes and incubate at 37°C for 1 h in conjugation buffer (see Notes 16 and 17). The control nucleosomes are incubated with FeBABE alone. The 2-IT reacts with primary amines (–NH2) to introduce sulfhydryl (–SH) groups, which serve as conjugation sites for FeBABE.
3. Excess FeBABE and 2-IT are removed using a Sephadex G-25 spin column equilibrated with conjugation buffer.

3.4.3. In Vitro Translation of Truncated Snf2 with His_6–HA Tags

1. The DNA template for in vitro transcription and translation of N-terminally truncated Snf2–His_6–HA is prepared by PCR using Snf2–His_6–HA DNA construct cloned into pET-21b+ as template. The forward primers to generate DNA encoding for N-terminally truncated 55, 76, 98, 118, and 139 consist of a T7 promoter sequence. The reverse primer is the same for all Snf2–His_6–HA polypeptides, and it consists of poly (dA) sequence to improve translation efficiency.
2. Thaw the TNT T7 Quick Master Mix on ice.
3. Add 2–5 μl of PCR product and 2 μl methionine ^{35}S (1,175 Ci/mmol or 43.48 TBq/mmol) to the TNT T7 Quick Master Mix, mix gently, and incubate at 30°C for 90 min.
4. The synthesized protein products are analyzed on SDS-PAGE gel, dried, and visualized by phosphorimaging. The synthesized

protein products are further examined by Western blot analysis using antibodies against the C-terminus HA tag.

5. The truncated Snf2 polypeptides are used as molecular size standards to calculate the FeBABE cleavage site (10).

3.4.4. Mapping Interaction of Snf2 with DNA and Nucleosome by FeBABE Mediated Cleavage

1. Immobilization of FeBABE-modified DNA or nucleosome to streptavidin-coated magnetic beads (M-280) is done as indicated in Subheading 3.1.2.
2. Add ~6 nM SWI/SNF to the immobilized 4 nM DNA or nucleosome in the presence or absence of 6.4 nM Gal4-VP16 in SWI/SNF binding buffer.
3. Remove unbound Gal4-VP16 and SWI/SNF and resuspend the beads (immobilized SWI/SNF/Gal4-VP16 bound to DNA or to nucleosomes) with 10 μl of cleavage buffer (see Notes 14 and 15).
4. Hydroxyl radical cleavage is initiated by the addition of 1.25 μl of 50 mM sodium ascorbate and 1.25 μl of 0.48% H_2O_2 to final concentrations of 5 mM and 0.05%, respectively (see Note 15).
5. The reaction is stopped after 30–120 s by adding an equal volume of 2× SDS-PAGE loading buffer containing 40% glycerol.
6. Samples are analyzed immediately by SDS-PAGE or stored at −70°C for later analysis. Anti-HA antibody against the C-terminal HA tag of Snf2 is used in the detection of the cleavage Snf2 products by Western blotting.
7. The cleavage sites of Snf2 are determined using truncated fragments of the same protein as molecular size standards. The relative gel mobility (Rf) of the molecular size standards (Snf2 polypeptides) is determined by the distance (in cm) migrated by the polypeptide divided by the length (in cm) of the gel. A curve of molecular weight ($\log_{10}$ MWt) of the Snf2 size standards as a function of the distance of migration is used to determine the molecular weight of FeBABE cleavage products.

4. Notes

1. From our experience, the source of these nucleases can make a critical difference. Some of the nucleases obtained from other vendors appear to have significant protease contamination that does interfere with identifying the cross-linked target protein.
2. At this point, only lower strand is labeled and the upper strand label remains with the bead after Eco*RI* digestion.
3. Photoaffinity probes are checked on 4% native-PAGE to detect the presence of any secondary structure, for example strand

invasion due to the presence of residual Klenow during probe synthesis process. Strand invasion could happen during extension reaction, where residual Klenow displaces downstream annealed oligonucleotides, resulting in the formation of "fork" that migrates slower.

4. ISW2–nucleosome interactions are stable at low glycerol (0.4–0.8%) conditions. Reactions can be assembled at low glycerol concentration avoiding the need to remove glycerol later for hydroxyl radical footprinting assay.
5. Since incubation time is very short for DNaseI and S1 nuclease, it is important to incubate all the samples for the same time. Therefore, while handling multiple samples, it is critical to put the first sample in the incubator nuclease digestion followed by next sample with a 10–15 s time interval. In this same manner, the reaction is stopped such that all the samples get an equal incubation time.
6. Modification of the phosphorothioate destabilizes the DNA duplex. If the modification is at the 5′ end of the oligonucleotide, then it will destabilize duplex DNA making it difficult to ligate.
7. Due to photosensitivity of PEAS, the experiments are done with indirect lighting using 40 W incandescent lamps, and the samples are covered by foil.
8. Keeping histones, histone octamers, and nucleosomes under reducing conditions during purification, refolding, nucleosome reconstitution, and storage is critical for preventing oxidation of the cysteine thiol group.
9. Spin column and dialysis are used to remove reducing reagents and excess free PEAS, ^{125}I-PEAS, or ^{125}I. We use spin column for quick buffer change to remove reducing reagent and free radioactive or cross-linking reagents. Extensive dialysis of the samples against large volume of buffer (1 l) three times, each for at least 3 h, is more efficient to exchange buffer and remove the indicated reagents.
10. Iodination is done in a proper hood for volatile ^{125}I that captures any free iodine. Urine samples are taken from those who perform the iodination, generally within 6–24 h after the iodination reaction, to determine possible inhalation or other contamination of ^{125}I.
11. Lead acrylic shielding should be used for storage and handling of ^{125}I-labeled materials.
12. The ^{125}I waste products must be stored and disposed according to established guidelines.
13. Two iodination approaches are used to modify cysteine-mutant nucleosomes with iodinated PEAS reagent. The first approach,

which involves iodination of the PEAS reagent and quenching the activated free $^{125}I^{+}$ before nucleosome modification, prevents the iodination of surface-exposed tyrosine and allows assessing the efficiency and specificity of nucleosome modification with PEAS reagent. On the other hand, the second approach in which the cysteine-mutant nucleosomes are first modified with cold PEAS and followed by indirect iodination requires a lower amount of radioactivity.

14. MOPS and HEPES buffers are preferred for hydroxyl radical cleavage reactions.
15. A 10% glycerol is included in the buffer to scavenge diffusible hydroxyl radical during cleavage reaction.
16. Make fresh 2-IT, FeBABE, H_2O_2, and sodium ascorbate solutions immediately before use.
17. Conjugation of FeBABE to lysine through 2-IT generates a library of modified nucleosomes (9). The mapping of proteins with nucleosomes modified in this approach identifies the region(s) of proteins that interacts with the surface of nucleosomes, but it does not specify the nucleosome sites.

References

1. Clapier, C. R. and Cairns, B. R. (2009) The biology of chromatin remodeling complexes. *Annu Rev Biochem.* **78**, 273–304.
2. Gangaraju, V. K. and Bartholomew, B. (2007) Mechanisms of ATP dependent chromatin remodeling. *Mutat Res.* **618**, 3–17.
3. Kagalwala, M. N., Glaus, B. J., Dang, W., Zofall, M. and Bartholomew, B. (2004) Topography of the ISW2-nucleosome complex: insights into nucleosome spacing and chromatin remodeling. *EMBO J.* **23**, 2092–2104.
4. Dechassa, M. L., Sabri, A., Pondugula, S., Kassabov, S. R., Chatterjee, N., Kladde, M. P. and Bartholomew, B. (2010) SWI/SNF has intrinsic nucleosome disassembly activity that is dependent on adjacent nucleosomes. *Mol Cell.* **38**, 590–602.
5. Mizuguchi, G., Shen, X., Landry, J., Wu, W. H., Sen, S. and Wu, C. (2004) ATP-driven exchange of histone H2AZ variant catalyzed by SWR1 chromatin remodeling complex. *Science.* **303**, 343–348.
6. Sengupta, S. M., VanKanegan, M., Persinger, J., Logie, C., Cairns, B. R., Peterson, C. L. and Bartholomew, B. (2001) The interactions of yeast SWI/SNF and RSC with the nucleosome before and after chromatin remodeling. *J Biol Chem.* **276**, 12636–12644.
7. Datwyler, S. A. and Meares, C. F. (2001) Artificial iron-dependent proteases. *Met Ions Biol Syst.* **38**, 213–254.
8. Schmidt, B. D. and Meares, C. F. (2002) Proteolytic DNA for mapping protein-DNA interactions. *Biochemistry.* **41**, 4186–4192.
9. Traviglia, S. L., Datwyler, S. A. and Meares, C. F. (1999) Mapping protein-protein interactions with a library of tethered cutting reagents: the binding site of sigma 70 on Escherichia coli RNA polymerase. *Biochemistry.* **38**, 4259–4265.
10. Chen, H. T. and Hahn, S. (2003) Binding of TFIIB to RNA polymerase II: Mapping the binding site for the TFIIB zinc ribbon domain within the preinitiation complex. *Mol Cell.* **12**, 437–447.
11. Lowary, P. T. and Widom, J. (1998) New DNA sequence rules for high affinity binding to histone octamer and sequence-directed nucleosome positioning. *J Mol Biol.* **276**, 19–42.
12. Luger, K., Rechsteiner, T. J. and Richmond, T. J. (1999) Expression and purification of recombinant histones and nucleosome reconstitution. *Methods Mol Biol.* **119**, 1–16.
13. Shen, F., Triezenberg, S. J., Hensley, P., Porter, D. and Knutson, J. R. (1996) Critical amino acids in the transcriptional activation domain of the herpesvirus protein VP16 are solvent-exposed in highly mobile protein segments. An intrinsic fluorescence study. *J Biol Chem.* **271**, 4819–4826.
14. Gangaraju, V. K., Prasad, P., Srour, A., Kagalwala, M. N. and Bartholomew, B. (2009) Conformational

changes associated with template commitment in ATP-dependent chromatin remodeling by ISW2. *Mol Cell.* **35**, 58–69.

15. Persinger, J. and Bartholomew, B. (2009) Site-directed DNA crosslinking of large multisubunit protein-DNA complexes. *Methods Mol Biol.* **543**, 453–474.
16. Bartholomew, B., Kassavetis, G. A., Braun, B. R. and Geiduschek, E. P. (1990) The subunit structure of Saccharomyces cerevisiae transcription factor IIIC probed with a novel photocrosslinking reagent. *EMBO J.* **9**, 2197–2205.
17. Tate, J. J., Persinger, J. and Bartholomew, B. (1998) Survey of four different photoreactive moieties for DNA photoaffinity labeling of yeast RNA polymerase III transcription complexes. *Nucleic Acids Res.* **26**, 1421–1426.
18. Lannutti, B. J., Persinger, J. and Bartholomew, B. (1996) Probing the protein-DNA contacts of a yeast RNA polymerase III transcription complex in a crude extract: solid phase synthesis of DNA photoaffinity probes containing a novel photoreactive deoxycytidine analog. *Biochemistry.* **35**, 9821–9831.
19. Zofall, M. and Bartholomew, B. (2000) Two novel dATP analogs for DNA photoaffinity labeling. *Nucleic Acids Res.* **28**, 4382–4390.
20. Persinger, J. and Bartholomew, B. (1996) Mapping the contacts of yeast TFIIIB and RNA polymerase III at various distances from the major groove of DNA by DNA photoaffinity labeling. *J Biol Chem.* **271**, 33039–33046.
21. Gangaraju, V. K. and Bartholomew, B. (2007) Dependency of ISW1a chromatin remodeling on extranucleosomal DNA. *Mol Cell Biol.* **27**, 3217–3225.
22. Luger, K., Mader, A. W., Richmond, R. K., Sargent, D. F. and Richmond, T. J. (1997) Crystal structure of the nucleosome core particle at 2.8 A resolution. *Nature.* **389**, 251–260.
23. Ebright, Y. W., Chen, Y., Kim, Y. and Ebright, R. H. (1996) S-[2-(4-azidosalicylamido) ethylthio)-2-thiopyridine: radioiodinatable, cleavable, photoactivatible cross-linking agent. *Bioconjug Chem.* **7**, 380–384.
24. Chizzonite, R., Truitt, T., Desai, B. B., Nunes, P., Podlaski, F. J., Stern, A. S. and Gately, M. K. (1992) IL-12 receptor. I. Characterization of the receptor on phytohemagglutinin-activated human lymphoblasts. *J Immunol.* **148**, 3117–3124.
25. Chizzonite, R., Truitt, T., Podlaski, F. J., Wolitzky, A. G., Quinn, P. M., Nunes, P., Stern, A. S. and Gately, M. K. (1991) IL-12: monoclonal antibodies specific for the 40-kDa subunit block receptor binding and biologic activity on activated human lymphoblasts. *J Immunol.* **147**, 1548–1556.

Chapter 27

Evaluation of Histone-Modifying Enzymes in Stem Cell Populations

Leanne Stalker and Christopher Wynder

Abstract

The histone demethylases are a relatively novel family of histone-modifying enzymes. Their gene expression suggests that each of the subfamily members may have a discrete role in cell function. The KDM5 family of H3K4 histone demethylases has four members. Each family member has a distinct cellular role, including KDM5a, which is a tumor suppressor (Christensen et al. Cell 128: 1063–1076, 2007); KDM5b, which is an oncogene (Dey et al. Mol Cell Biol 17: 5312–5327, 2008); and KDM5c (Iwase et al. Cell 128: 1077–1088, 2007), which is expressed in terminally differentiated populations. To properly analyze how these enzymes are regulated, we interrogate their bioactivity in ES cells (ESCs) and during neural differentiation of ESCs. We evaluate the bioactivity from both affinity-purified complexes and reconstituted complexes and directly in the cell. These assays have allowed us to define a set of factors that regulate the KDM5 family of histone demethylases.

Key words: ES cells, Gene regulation, Histone demethylase, Neural differentiation, Lysine demethylation, Enzyme kinetics

1. Introduction

The KDM5 family of histone demethylases are developmentally regulated, but appear to have a conserved mechanism of regulation (1, 5, 6, 7). This regulation is primarily at the level of their enzymatic activity. Although multiple KDM5s are expressed at a given time point, their in vivo activity appears to be tightly regulated. Unlike the KDM1 family of H3K4 demethylases, purification of the KDM5-associated complexes through KDM5 does not yield an enzyme complex competent to demethylate H3K4 in its native conformation, the nucleosome. None of the KDM5 family members are capable of nucleosomal demethylation on their own. All of these

Ales Vancura (ed.), *Transcriptional Regulation: Methods and Protocols*, Methods in Molecular Biology, vol. 809, DOI 10.1007/978-1-61779-376-9_27, © Springer Science+Business Media, LLC 2012

enzymes are capable of demethylating a histone substrate and can cause alterations to H3K4 methylation when overexpressed in a variety of cell types confirming that the enzymes are in fact H3K4 demethylases. The lack of nucleosomal activity suggests that these proteins are tightly regulated and that the necessary components for full activity do not stably associate with the enzyme. Since ES cells (ESCs) can differentiate to a variety of cell types, we can use these as a source of cells. We focus here on protocols to define the activity of KDM5b which is highly expressed in ESCs and is downregulated during differentiation. Since we are primarily interested in neural development, we have also provided the protocols to interrogate these enzymes during neural differentiation. The enzymatic assays and fluorescent in vivo HDM assay, however, are readily applicable to differentiation to any cell type.

2. Materials

2.1. Reagents for Mouse ES Cell Culture Reagents (see Note 1)

1. Dulbecco's phosphate-buffered saline (PBS) without calcium and magnesium (Gibco-BRL).
2. β-mercaptoethanol stock solution: Add 70 μl of β-mercaptoethanol (Sigma) to 20 ml of distilled, deionized water (Gibco-BRL). Filter sterilize and store at 4°C for up to 2 weeks.
3. ESC medium: 1× DMEM medium (Invitrogen) supplemented with 2 mM glutamine (Gibco-BRL), 1 mM sodium pyruvate (Gibco-BRL), 1× nonessential amino acids (NEAAs), 10% (v/v) fetal bovine serum (characterized, Hyclone), a 1:1,000 dilution of β-mercaptoethanol stock solution, and 500–1,000 units per ml of leukocyte inhibitory factor (LIF; Chemicon) or 100 ng/ml recombinant LIF. Our laboratory makes our own LIF by expression of a GST-tagged LIF construct in bacteria by standard protocols (see Note 2).
4. Freezing medium: 10% dimethyl sulphoxide (DMSO) (tissue-culture grade, Sigma) in ESC medium (above). Filter sterilize. This solution can be keep for up to 1 week.
5. Gelatin-coated tissue culture plates: Make a sterile 0.1% gelatin solution in PBS by filter sterilization. Store at 4°C for up to 1 month. This solution is added to tissue culture plates for at least 5 min. Then, solution is removed and plates are allowed to dry for at least 4 h. Plates can be stored at room temperature in a sealed bag for up to 2 months.
6. Geneticin/G418 (Gibco-BRL): Dissolve powder in PBS to make a 125 mg/ml stock solution (active concentration). Filter sterilize and store at −20°C. The concentration of G418 should be titrated to determine the minimum concentration that kills nontransfected ES cells in 5 days. This must be done empirically.

7. Neurosphere media: 1% FBS, 1× B-27, 1× sodium pyruvate, 1× NEAAs, 10 ng/ml FGF2 in 1× DMEM.
8. Neural differentiation media: 5% FBS, 1× B-27, 1× sodium pyruvate, 1× NEAAs in 1× DMEM.

2.2. Poly-Lysine-Coated Tissue Culture Surfaces for Neural Differentiation

2.2.1. Part 1: To Coat Coverslips

1. Prepare a stock solution by preparing 25 mg/ml poly-D-lysine in water and filter sterilize through a 0.22-μm filter. Store in 100-μl aliquots at −20°C.
2. When ready to use, dilute one aliquot in 40 ml of water to prepare 13 μg/ml working solution.
3. Sterilize coverslips (use glass coverslips only) by autoclaving prior to coating. Dip coverslips in the working solution.
4. Incubate slides after dipping, for 15 min to several hours in a humidified 37°C, 5% CO_2 incubator. Allow surface to dry.

2.2.2. Part 2: To Coat Culture Dishes or 8-Well Chamber Slides (Uses Different Concentration of Working Stock)

1. Prepare a stock solution by dissolving 100 mg poly-D-lysine in 100 ml water and filter sterilize through a 0.22-μm filter. Store in 5-ml aliquots at −20°C.
2. When ready to use, dilute 1 part stock solution with 9 parts water to prepare 100 μg/ml working solution.
3. Fill tissue culture dishes or slide wells with the working solution and incubate for 1 h in a humidified 37°C, 5% CO_2 incubator.
4. Remove solution by vacuum aspiration and allow surface to dry.
5. Store coated tissue culture dishes or coverslips for up to 3 months at 4°C. Use diluted solutions, only once, but unused diluted aliquots can be stored for up to 3 months at 4°C.

2.3. Buffers for Nuclear Extract Protocol

1. Buffer A: 10 mM Tris–HCl, pH 7.9, 1.5 mM $MgCl_2$, 10 mM KCl, 0.5 mM DTT, 0.2 mM PMSF. DTT and PMSF should be added fresh at the time of use.
2. Buffer B (10×): 0.3 M Tris–HCl, pH 7.9, 1.4 M KCl, 30 mM $MgCl_2$.
3. Buffer C: 20 mM Tris–HCl, pH 7.9, 25% glycerol, 0.42 M NaCl, 1.5 mM $MgCl_2$, 0.2 mM EDTA, 0.5 mM PMSF, 0.5 mM DTT.
4. Buffer E: 50 mM Tris–HCl, pH 7.9, 25% glycerol, 0.5 mM EDTA, 5 mM $MgCl_2$, 0.5 mM DTT, 0.2 mM PMSF.

2.4. Basic Protein Wash Buffers (BC Buffers)

1. BC50: 20 mM Tris–HCl, pH 7.6, 500 mM EDTA, 10 mM β-mercaptoethanol, 10% glycerol, 50 mM potassium chloride.
2. BC250: 20 mM Tris–HCl, pH 7.6, 500 mM EDTA, 10 mM β-mercaptoethanol, 10% glycerol, 250 mM potassium chloride.
3. BC500: 20 mM Tris–HCl, pH 7.6, 500 mM EDTA, 10 mM β-mercaptoethanol, 10% glycerol, 500 mM potassium chloride.

2.5. Buffer for Histone Demethylase Assays

1. JHDM assay buffer: 50 mM HEPES, pH 8.0, 100 mM NH_4SO_4, 1 mM α-ketoglutarate, 2 mM ascorbate, 5% glycerol, and 0.2 mM PMSF (2).

2.6. Nucleosome Preparation Specific Buffers (3)

1. NP digestion buffer: 0.32 M sucrose, 50 mM Tris–HCl, pH 7.6, 4 mM $MgCl_2$, 1 mM $CaCl_2$, 0.1 mM PMSF, 5 mM sodium butyrate.
2. NP cell lysis buffer: 1.0 mM Tris–HCl, pH 7.6, 0.2 mM Na_2EDTA, 0.2 mM PMSF, 5 mM sodium butyrate.

3. Methods

The KDM5 proteins do not inherently interact with their cofactors. This represents a problem for understanding the interactions of the KDM5 family and, thus, for advancing the knowledge of the field. It does, however, provide an excellent opportunity to understand cellular conditions that allow for the cofactors to be present at sufficient levels to allow for the purification of an active nucleosomal demethylase. We have used this to tease out the cofactors required for KDM5b nucleosomal 3meH3K4 demethylase activity. These methods can also be used to test the relative activity of these enzymes in response to cell signaling.

We use feeder-independent ESC lines derived from the 129/Ola strain of mice. These lines were obtained from Baygenomics, which is now part of the Mouse Mutant Repository (http://www.mmrrc.org). We have found that feeder-free ESCs are preferable since there are significant differences in chromatin signature, and therefore the regulation of the epigenetic regulators, between the feeders (mitotically inactive mouse embryonic fibroblasts) and ESCs. We have found that it is not possible to tease these two populations apart in a manner that allows for in-depth biochemical analysis. We have also not been able to tease out the differences between these two populations in mixed cultures. Any ESCs can be adapted to feeder-free conditions; however, great care must be taken to ensure that the ESC lines maintain the ability to differentiate (see Note 1).

Since we use ESCs as a source of potential neural progenitors, we routinely test for the ability to differentiate toward the neural lineage. We ensure that by RT-PCR, stem cell markers, Oct4 and Nanog, are decreasing and neural lineage markers, neural filament and NeuroD2, are increasing. The advantage of the neural differentiation is the clear and obvious phenotype of the terminally differentiated cells. For researchers interested in understanding other processes, a set of criteria should be established and routinely tested on each batch of ESCs. We maintain at least five separate vials of each passage that has met our criteria.

3.1. Growing Feeder-Independent Embryonic Stem Cells

3.1.1. Thawing ES Cells

ESCs are frozen in medium containing 10% DMSO. Since DMSO can induce the differentiation of ESCs, we advise thawing the cells late in the day and changing the medium the following morning to minimize the effects of residual DMSO.

1. Coat a 10-cm^2 tissue culture dish with 0.1% gelatin. This can be done in advance and dishes can be stored at room temperature for up to 1 month.
2. Thaw ESCs (approximately 10^7 cells, equivalent to one confluent cm^2 tissue culture dish) in a 37°C water bath and dilute into 10 ml of prewarmed ESC medium.
3. Pellet the cells by spinning for 3 min at 140×*g* in a benchtop clinical centrifuge.
4. Aspirate off medium and gently resuspend cells in 5 ml of prewarmed medium.
5. Pellet the cells by spinning for 3 min at 140×*g* in a benchtop clinical centrifuge.
6. Aspirate off medium and gently resuspend cells in 10 ml of prewarmed medium.
7. Transfer cell suspension to a 10-cm^2 dish and grow at 37°C in a humidified 5% CO_2 incubator (see Note 2).
8. Change medium the following day to remove dead cells and residual DMSO.

3.1.2. Passage and Expansion of ES Cell Cultures

ESCs are routinely passaged every 2 days (see Note 2). In our experience, feeder-independent ESCs grow rapidly and quickly acidify the medium, turning it yellow (if phenol red-containing media is being used). Allowing the cells to acidify the medium (by passaging the cells at too high or too low a dilution) causes the cells to undergo crisis, triggering excess differentiation and cell death, after which their pluripotency cannot be guaranteed. Plating cells at too low a density, insufficient dispersion of cells during passage, or uneven plating can cause similar problems, as the cells form large clumps before reaching confluence and the cells within these clumps differentiate or die.

1. From thawed 10-cm^2 dish, allow to grow for approximately 2 days. A minimum time of 36 h should be required for mouse ESCs. The optimal time has to be derived empirically within each laboratory. Do not allow the cells to grow to greater than 80% confluency.
2. For a 70–80% confluent 10-cm^2 dish of cells, aspirate medium off and wash with 5–10 ml of prewarmed PBS, pipetting it away from the cells. Rock dish gently once and aspirate medium. Repeat (see Note 3).

3. Cover cells with 1 ml of 1× TrypLE solution (see Note 4) and return to 37°C incubator for 1–2 min or until cells are uniformly dispersed into small clumps.
4. Add 50 ml of media. Pipette gently to ensure a single-cell suspension. To split for maintenance, we usually take 200 μl of TrypLE solution, pipetting it to ensure single-cell density and add it to 10 ml of medium in a new gelatinized 10-cm^2 dish. If you are maintaining only one plate, take 200 μl of cells in TrypLE and add to 10 ml of media.
5. Add 10 ml of cells to each of five new gelatinized 10-cm^2 dishes.

3.2. Generation of Neurospheres from ESCs (see Note 5)

Allow mESC culture to reach 90–100% confluency (approximately 4 days after passage).

1. Change media to neurosphere media (Subheading 2.1, step 7). Leave for 48 h. (FGF2 can be omitted for straight differentiation experiments; it is only necessary if you want to passage neurospheres) (Day 0).
2. IMPORTANT: Do not touch nonadherent spheres for ANY reason for 48 h. The formation of spheres is very delicate, and constant perturbation of the dish alters their properties.
3. Scrape off cells and transfer to nonadherent plate (petri dish OR we also use ultralow-binding 96-well plates). The optimal density is to split a single 10-cm dish of mESCs into roughly two 10-cm petri dishes or 6-well dishes. Leave for 48 h; at this time, spheres should be easily seen. For neural differentiation, proceed to next section (omit step 5) (Day 2).
4. Change media by centrifugation (5 min in clinical-type centrifuge at 40 × *g*). Resuspend pellet in neurosphere media and replate. Cells can be passed as neurosphere cultures (go to step 5) or differentiated (next section, omit step 5).
5. Depending on the number and overall health of spheres, they can be split 1:2 or 1:1. Do not dissociate enzymatically, perform a mechanical disruption to form smaller spheres, and go to step 1. For neural stem cell cultures, please refer to Rietze and Reynolds (4).

3.3. Differentiation of Neurospheres

1. Take neurospheres and change media into neural differentiation media (Subheading 2.1, item 8) by centrifugation (5 min in clinical-type centrifuge at 40 × *g*).
2. Aspirate media and add neural differentiation media.
3. The spheres should be passed so that the density is no more than 5 spheres per 200 μl. One petri dish yields about four to five 96-well plates or three plates of either 6-well or 10-cm plates.

4. Important: Precoat tissue culture plates with either 1% gelatin OR poly-D-lysine (Subheading 2.2). This can be done up to a month in advance and stored at room temperature (remember to use sterilized gelatin by filtration), but be sure to coat them AT LEAST 4 h prior to use.
5. Spheres attach to matrix after about 24 h (Day 4). Do not touch them during the first day.
6. The first neurons migrate out at approximately day 7; under proper conditions, these neural cultures can be maintained for an additional week.

3.4. Nuclear Extract Preparation

This protocol yields three fractions: a cytoplasmic, nucleoplasm (soluble nuclear), and an insoluble protein/lipid rafts. The majority of the histone-modifying enzymes can be found in the soluble nuclear fraction. We often check the cytoplasmic as a control fraction. This is particularly important for measuring relative enzyme activity during differentiation using the abNDM assay (Subheading 3.8) (see Note 6).

3.4.1. First Step

1. Remove media and wash mESCs once with PBS to wash off dead cells and media. For neurospheres, collect spheres in media in a 15-ml falcon tube and proceed to step 3.
2. Trypsinize with 1 ml of TrypsinLE for 1–3 min, checking occasionally by tapping dish. Stop when cells are completely dissociated from plate.
3. Spin cells at 1,560 × *g* for 10 min at 4°C to pellet cells, remove TrypLE, and add PBS.
4. Transfer cells and PBS to 15-ml falcon tube.
5. Spin cells at 1,560 × *g* for 10 min at 4°C to pellet cells, and remove PBS.
6. Measure packed cell volume.
7. Add 2 volumes of buffer A (Subheading 2.3), e.g., if the top of the cell pellet reaches the 0.5-ml mark; then, this is 1 volume and 1 ml (2 volumes) would be added.
8. Resuspend by gentle pipetting and incubate for 5–10 min at 4°C.
9. Spin cells at 1,560 × *g* for 10 min at 4°C to pellet cells.
10. Remove supernatant with pipet and save both the supernatant, which is the cytoplasmic proteins and the pellet, which represents the nucleus and membrane fractions.

3.4.2. Supernatant (Cytoplasm)

1. Take the supernatant from step 10 above and add buffer B (Subheading 2.3; this is a 10× solution) to this to a final concentration of 1×. For example, if you have 1 ml of cytoplasm, you add 110 μl of buffer B to give a final volume of 1,110 ml of S100 with 1× buffer B.

2. Mix well by hand, turning the tube over three to five times.
3. Spin at 30,000 × *g* for 60 min (for volumes less than 500 μl, a microcentrifuge can be used at maximum rpm for 60 min).
4. Dialyze for 1–2 h against 20 volumes of JHDM buffer (Subheading 2.5) – this is specific to the use in histone demethylation assays. PBS or BC50 (Subheading 2.4) can be substituted.
5. Spin at 12,000 × *g* for 20 min (if you are using a microcentrifuge, increase this to 60 min).
6. Keep supernatant and discard any precipitate.
7. Add Protease inhibitor tablets (Roche) to the supernatant and store at −80°C.

3.4.3. (The Nucleus) Pellet

1. To the pellet from step 10 in the first step (Subheading 3.4.1), add 1 volume of buffer C (Subheading 2.3) and pipet vigorously to resuspend.
2. Rotate at 4°C for 30 min.
3. Spin at 12,000 × *g* for 30 min.
4. Remove supernatant (nucleoplasm) and save it together with the (nuclear) pellet.

3.4.4. Nucleoplasm (Soluble Nuclear Proteins Including Transcription Factors, Some Histones)

1. Dialyze for 1–2 h against 50 volumes of JHDM – this fraction is typically used for immunopurification of histone-modifying enzymes to be used in subsequent assays.
2. Spin at 12,000 × *g* for 20 min.
3. Add Protease inhibitor tablets and store at −80°C or use in assays below.

3.4.5. Nuclear Pellet (Nucleosomes, Some Chromatin Proteins) (see Note 7)

1. For mass spectrometry and western blot protein identification, the nuclear pellet can be solubilized by the addition of two volumes of 1 M guanidine hydrochloride. Proceed to step 5.
2. For immunoprecipitation or intact nucleosomes, follow steps below.
3. Resuspend pellet in one volume of buffer E.
4. Sonicate for 12 min (30 s on/30 s off) with a Diagenode bioruptor. For a stick sonicator, the settings must be determined empirically. This step is to break up the chromatin into small di- and tri-nucleosomes and to disrupt lipid membranes to release any bound proteins.
5. Spin at 12,000 × *g* for 20 min.
6. Keep supernatant and store at −80°C. Discard pellet, which is largely lipid and unsuitable for many of the assays.

3.5. Immunoprecipitation

This protocol can be used with either tagged cell lines or endogenous protein using traditional immunoprecipitation. The resulting complexes can be used in the two NDM assays. IP conditions should be confirmed by western blot. If the conditions below do not yield active complex, the high-stringency buffers can be replaced with PBS and low-percentage detergent (e.g., Triton X-100). We usually use a β-actin IP as a negative control. For affinity tag purifications, an unrelated tagged protein is the best control. The protein concentration in the nucleoplasm should be measured. We use BCA (Biorad), but any method can be used.

3.5.1. Affinity Tag IP

We have successfully used FLAG, V5, and 6Xhis tags on both enzymes and associated proteins.

1. Wash beads in BC50 (Subheading 2.4) or PBS and resuspend to make a 50% slurry (e.g., 100 μl of beads in 200 μl of PBS/BC50).
2. Add 5 μl of Flag, V5, or Nickel beads (or 10 μl of 50% slurry)/1 mg of protein.
3. Incubate with agitation for 1–3 h at 4°C.
4. Add to gravity flow column.
5. Wash with three bead volumes of BC500 + NP40 (0.01%) or BC250 + NP40 (0.01%).
6. Wash with two bead volumes of BC50.
7. Elute five times with one bead volume of 1 mg/ml Flag peptide dissolved in BC50 for flag-tagged proteins. Elute with 0.1 M glycine, pH 1.8, followed by neutralization with 2 volumes of Tris–HCl, pH 7.9, for V5-tagged proteins and standard imidizole elution for Nickel column. For Nickel columns, the manufacturer's guidelines should be followed as different beads have different binding capacity and strength. The second elution is usually the most concentrated. This can easily be tested by western blot for flag tag, V5 tag, or His tag.
8. Dialyze into JHDM buffer and store at −80°C.

3.5.2. Immunopurification of Endogenous Proteins

1. Add 3 μl of purified monoclonal or polyclonal antibody (for unpurified serum, we typically add 5 μl; for hybridomas, the binding must be tested empirically for each hybridoma) to 100 μl of nuclear extract. Incubate overnight at 4°C with agitation.
2. Similar to affinity beads, protein A–agarose beads are washed and resuspended in BC50 or PBS (these can be stored for long term).
3. Add 45 μl of 50% slurry of protein A beads to sample containing the antibody and incubate at 4°C with agitation for 2 h.

4. Remove samples, spin down at 3,000 × *g* for 3 min in a microcentrifuge, and remove supernatant.
5. Wash one time with 1 ml of BC250 + NP40 at 4°C with agitation for 2 min, and spin at 2,000 × *g* for 2 min in a microcentrifuge.
6. Wash one time with BC250 at 4°C with agitation for 2 min, and spin at 2,000 × *g* for 2 min in a microcentrifuge.
7. Elute by adding 0.1 M glycine, pH 1.8, and incubating with agitation for 2 min at room temperature. The elutions are as follows: first elution 50 μl, then second elution 20 μl, and a final elution in 50 μl of SDS loading buffer.
8. Add Tris–HCl, pH 7.9, to all samples to correct for pH.
9. Dialyze in JHDM and store at –80°C.

3.6. Chromatin Preparation from Cultured Cells

This protocol is used to make nucleosomes that can be used to test the nucleosome demethylase activity of the purified complexes. It can also be used to test if the protein of interest binds to nucleosomes by performing an IP from the nucleosomes. This has been adapted from Rampalli et al (3).

1. Cultured cells (any cell can be used; we use mESCs). Harvest cells: Centrifuge at 7,000 × *g*, 10 min, 4°C, and wash the cell pellet 3× in ice-cold 1 × phosphate-buttered saline (PBS)/5 mM sodium butyrate. It is essential that sodium butyrate is present in all solutions throughout chromatin isolation when using antibodies to acetylated histones to prevent deacetylation (see Note 8). Resuspend cell pellet in 1 volume of PBS and add an equal volume of 1.0% v/v Tween 20/1× PBS (e.g., to 100 μl of pellet, add 100 μl of PBS and 100 μl of 1% Tween20). Add PMSF to achieve a final concentration of 0.5 mM. Leave stirring gently on ice for 1 h on a rocker at the lowest setting.
2. Transfer cell lysate to an all-glass homogenizer (Dounce) and homogenize 7-ml aliquots with seven strokes using an "A" or "tight" pestle. Check that the nuclei have been released by phase-contrast microscopy; intact cells should have the central dark region of the nucleus surrounded by a halo, which is the less-dense cytoplasm. You may have to increase or decrease this homogenization step to maximize the yield of nuclei depending on cell line.
3. Gently transfer to 15-ml falcon tube and centrifuge at 7,700 × *g* for 20 min at 4°C.
4. Resuspend pelleted nuclei in 1 volume of 25% [w/v] sucrose in PBS.
5. Carefully add an underlay with 0.5 volume of 50% [w/v] sucrose in PBS. This is best achieved by slowly adding 50% sucrose solution to the side of the falcon tube and allowing it

to run down the side of the tube. Gently place in the centrifuge and centrifuge at 9,100 × *g* for 25 min at 4°C.

6. Discard supernatant, wash nuclei pellet in 5 ml of 25% sucrose, and centrifuge at 9,100 × *g* for 25 min at 4°C.
7. Resuspend nuclei pellet in 5 ml of digestion buffer (Subheading 2.6) and check absorbance ratios at 260 and 280 nm for a diluted sample of the nuclei suspension; calculate the approximate DNA concentration from the A260 reading (the ratio of A260/A280 should be about 1.1). Centrifuge at 10,000 × *g* for 10 min at 4°C and resuspend the nuclei pellet at 0.5 mg/ml in 1.7-ml Eppendorf tube(s). Do not freeze; proceed directly to micrococcal nuclease digestion.

3.7. Micrococcal Nuclease Digestion

This step reduces the nuclei into small polynucleosome arrays of different sizes. For most reactions, this oligonucleosome mix is a sufficient substrate for NDM assays.

1. Add 50 units of micrococcal nuclease per 0.5 mg DNA to the resuspended pellet from step 7 in Subheading 3.6 in a reaction volume of 1.0 ml. The nuclease comes as a powder from Pharmacia; dissolve it in ddH_2O as per vendor's protocol and store as small aliquots at −20°C. Aliquots may be refrozen and reused once. This step needs to be carefully controlled, especially in the initial preparations (see Note 9).
2. Perform micrococcal nuclease digestion at 37°C for 5 min.
3. Stop reaction by addition of 0.2 M EDTA to a final concentration of 5 mM.
4. Place all samples on ice for 5 min; centrifuge at 13,000 × *g* in microcentrifuge.
5. Remove the supernatant (this is called the S1 fraction; total volume is 1.0 ml); store at 4°C.
6. Resuspend the pellet in 1.0 ml of lysis buffer (Subheading 2.6) and dialyze overnight against 2 l of the same buffer.
7. After overnight dialysis, spin samples at 4° for 10 min at 2,000 × *g* in microcentrifuge.
8. Remove and keep the supernatant (called the S2 fraction; total volume is about 1.2 ml after dialysis); store at 4°C.
9. Check A260/A280 in all samples; the ratios for S1 and S2 fractions should be approx. 1.7 and 1.5, respectively. Use A260 as a rough approximation of protein concentration. As a substrate for histone demethylase assays, these fractions can be combined if the ratios for both are between 1.5 and 1.7.
10. Dilute samples down to approximately 1 μg per μl using JHDM buffer. Aliquot into 100-μl samples and freeze at −80°C. Samples can be freeze/thawed twice without any loss of nucleosome integrity.

3.8. Absorbance-Based Histone Demethylase Assay

Change in absorbance at 220 nm can be used to monitor the change in methyl groups (A220 being nonspecific for changes in small carbon groups, often used to assess purity for organic chemicals) in solution. This is not optimal in complex mixtures due to the carbons that are given off due to protein degradation. For this reason, assays involving peptide substrates and immunopurified complexes are not compatible with this assay. During optimization with either recombinant KDM5b or KDM4c, changes in A220 required the presence of both histones and enzyme. The monitoring of the A260 provides a control against protein degradation. Any samples with changes in A260 over time should be removed from analysis.

1. Immunopurified complexes and the nucleosomes are combined in JHDM buffer in a 96-well plate on ice. We use 15 μg of nucleosomes per reaction (15 μl). Depending on the concentration of the enzyme in the complex, between 20 and 35 μl of the IP is used. The final volume of 50 μl is reached by adding 1×JHDM. For very-low-concentration enzymes, a mix of JHDM and nucleosomes in 20-μl total volume can be added directly to beads. The sample is then overlaid with mineral oil to minimize sample loss.
2. The 96-well plate is then added to a plate reader with temperature control (we use a BMG Omega Fluorstar) and monitored at 7-min intervals at both A220 and A260 for 4 h. Samples are incubated for additional 5 h for a total incubation of 8 h (no fold increase in absorbance was seen after 2 h). The additional incubation is only required if the samples will be retested in the qNDM assay although there is no discernable difference in activity between hours 4 and 8. We know that for the qNDM assay 4 h is not sufficient to see reproducible and quantitative results.
3. If no additional tests are to be done, the mix can be discarded or used for western blot analysis to determine the levels of enzyme present in each sample. The reaction controls should include a mineral oil-only well, mineral plus JHDM buffer, IP in JHDM without nucleosomes, and nucleosome in JHDM. The average absorbance of these four controls should be used as the baseline value. We subtract this baseline from each time point prior to analysis.
4. Samples that show any change in A260 (suggesting protein degradation) should be removed from analysis. Seven-minute intervals were empirically chosen based on the parameters of the Flurostar and the minimal number of data points required for detailed analysis. The intervals chosen should be tested empirically based on the plate reader to be used. We express the values as changes versus time zero. The kinetics of different

conditions (e.g., Wnt3a treatment versus untreated) can then be measured and compared.

5. We use a minimum of six separate readings (three technical [replicates on the same plate] and three biological [samples from different preps]). This assay does not give an accurate measure of the level of demethylation seen, but is useful for comparative analysis based on changes, i.e., ESCs versus neurospheres. For detailed quantitation of the enzyme kinetics, the qNDM assay should be used.

3.9. Quantative Histone Demethylase Assay

50-μl samples are incubated overnight at 37°C in the plate reader in a 96-well plate or in individual 1.5-ml Eppendorf tubes in JHDM buffer plus either histones (Sigma) or purified nucleosomes.

3.9.1. qHDM Setup

1. 50-μl samples are diluted to 200 μl with 150 μl of PBS prior to analysis. This is then split into 100-μl aliquots per well. This allows both panH3 control and the experiment to be done from the same sample, alternating one row for the histone methylation and one row for pan-H3. The histone methylation and panH3 for each sample should be assayed on same plate for maximal control. We use panH3 as our loading control for all demethylation assays. To date, none of the other pan histone antibodies are as reliable and consistent as panH3. A variety of companies make reliable histone H3 antibodies, including Millipore, Abcam, Bethyl Labs, and Active Motif.
2. Incubate samples overnight at room temperature to cross-link them to maleic anhydride amine binding plates (Pierce). Plates are then quenched with blocking buffer (PBS–Tween 20, PBST20, [1%] plus 2% BSA) for 1 h at RT.
3. Plates are then washed in PBST20 and then blocked in PBST20 + 2% BSA for 1 h at RT.
4. Plates are then incubated with either a 1:2,000 dilution of rabbit anti-"histone methylation" antibody or rabbit pan-H3 antibody for 1 h at room temperature; this can also be done overnight at 4°C.
5. Plates are then washed at least five times with PBST20.
6. Secondary antibody is then added. We use 1:10,000 dilution of Alexa647, Alexa350, or Alexa488 conjugate anti-rabbit antibody. This must be determined empirically for your plate reader. Wash in PBST20 at least three times and add PBS to well for analysis.

3.9.2. qHDM Analysis

1. Using BMG Flurostar Omega microplate reader, fluorescence intensity is measured. Quantities are normalized using a duplicate standard curve of histone concentrations in μg (5, 2, 1, 0.1, 0.01, 0.001, 0.0001, and 0.00001), including a no-histone blank control (JHDM alone).

2. The RFU value for each well is then given a value in mg based on the standard curve. The BMG software does this automatically, but this can also be done in excel by graphing the standard curve and using the slope equation to determine the values.
3. Values (including standard curve) are normalized based against two blanks: JHDM buffer alone and protein (either recombinant or β-actin IP) alone. This provides a baseline value and represents "0," i.e., any RFU equal to this has no histones and is likely an empty well.
4. Bulk histones or nucleosomes are incubated in buffer alone and used as negative control. For immunoprecipitated samples, actin IP plus histones or nucleosomes serve as the negative control.
5. These histone methylation samples are then normalized to pan-H3 to yield a 3meH3K4 per unit pan-H3 value. This number is then divided by histone/nucleosome-only samples. All numbers are expressed as a fold change vs. histone/nucleosome-only sample.

3.10. In Vivo Assay for Changes in Histone Modifications

This assay can be modified to any cell type that grows adherently. This assay also works with mESCs. We have shown the neurosphere step as an example of our experimental procedure. This can also be combined with transfection-based experiments, but the transfection must have a fluorescent marker so that transfection efficiency can be monitored.

1. Neurospheres are plated down onto gelatin (or poly-D-lysine)-coated 96-well plates (clear-bottom, "optical tissue culture-treated plates"). For mESCs, the plating density must be carefully checked to limit the differentiation.
2. Cells are allowed to differentiate for 48 h.
3. Cells are then fixed in 1% paraformaldehyde at RT for 20 min.
4. Plates are then washed 1× with PBS.
5. Plates are then blocked using PBS + Triton (0.05%) with 4% FBS for 1 h at RT. Please note: Triton is a variable; this particular concentration works for histone antigens in cells that grow in multicellular colonies, such as neurospheres or ESCs.
6. Plates are then incubated with primary antibody at the same concentration as would be used to visualize cells at low magnification on coverslips, 96-well, or 10-cm plates (e.g., 3meH3K4 is 1:200 dilution).
7. Wash three to five times in PBS + T.
8. Fluorescent secondary antibody is then used. This concentration should be tested for the machine that is being used. We utilize

a BMG Fluorstar, which has very good fluorescent range and we use a 1:5,000 dilution as a starting point.

9. Wash three to five times in PBS+T. If possible, DAPI should be added during the first wash. This allows you to normalize the cell numbers per well. If you are using a transfection with a fluorescent protein, it is unnecessary to add DAPI.
10. PBS is then added to each well for plate reading. At no time should the cells be allowed to dehydrate. We use well scanning mode on our machine.
11. On the plate, there should be a NO CELLS blank, a NO SECONDARY BLANK, and a NO PRIMARY blank. We use the average of these three blanks to set our background.

4. Notes

1. These techniques as described are for mouse only; where we have tried these techniques from other stem cell populations, it has been noted.
2. For the Baygenomics ESCs, use 6% CO_2 as per vendor and phenol red-free media for two reasons: to reduce the activation of estrogen receptor and for easier live cell imaging using fluorescent proteins.
3. This protocol can be used with human ESCs and induced pluripotent cells (IPSCs).
4. It should take 2 days maximum to reach 70–80% confluency. If the cells have not reached this level of confluency, discard them and start with a new batch. If cells reach 80% confluency prior to 36 h, discard as well, as this is a sign of transformation.
5. 1× trypsin can be used but requires serum inactivation.
6. All steps, including centrifugation, should be performed on ice or in a cold room.
7. While this fraction does contain chromatin and chromatin-binding proteins, it also contains membrane lipids and membrane-bound proteins. Care must be taken to completely separate the lipid-bound pellet and soluble chromatin after treatment with either guanidine hydrochloride or buffer E.
8. The sodium butyrate can be omitted for histone demethylase assays.
9. High concentrations of micrococcal nuclease may overdigest the chromatin, leading to subnucleosomal particles. You should aim to obtain a long–medium oligonucleosome ladder. If pure mononucleosome preparations are required, carry out a linear sucrose gradient (5–20%); this increases resolution.

References

1. Benevolenskaya, E. V. (2007) Histone H3K4 demethylases are essential in development and differentiation. *Biochem Cell Biol* **85**, 435–43.
2. Lee, M. G., Norman, J., Shilatifard, A., and Shiekhattar, R. (2007) Physical and functional association of a trimethyl H3K4 demethylase and Ring6a/MBLR, a polycomb-like protein. *Cell* **128**, 877–87.
3. Rampalli, S., Li, L., Mak, E., Ge, K., Brand, M., Tapscott, S. J., and Dilworth, F. J. (2007) p38 MAPK signaling regulates recruitment of Ash2L-containing methyltransferase complexes to specific genes during differentiation. *Nat Struct Mol Biol* **14**, 1150–6.
4. Rietze, R. L., and Reynolds, B. A. (2006) Neural stem cell isolation and characterization. *Methods Enzymol* **419**, 3–23.
5. Christensen, J., Agger, K., Cloos, P. A., Pasini, D., Rose, S., Sennels, L., Rappsilber, J., Hansen, K. H., Salcini, A. E., and Helin, K. (2007) RBP2 belongs to a family of demethylases, specific for tri-and dimethylated lysine 4 on histone 3. *Cell* **128**, 1063–76.
6. Dey, B. K., Stalker, L., Schnerch, A., Bhatia, M., Taylor-Papidimtriou, J., and Wynder, C. (2008) The histone demethylase KDM5b/JARID1b plays a role in cell fate decisions by blocking terminal differentiation. *Mol Cell Biol.* **17**, 5312–27.
7. Iwase, S., Lan, F., Bayliss, P., de la Torre-Ubieta, L., Huarte, M., Qi, H. H., Whetstine, J. R., Bonni, A., Roberts, T. M., and Shi, Y. (2007) The X-linked mental retardation gene SMCX/JARID1C defines a family of histone H3 lysine 4 demethylases. *Cell* **128**, 1077–88.

Chapter 28

Purification of Multiprotein Histone Acetyltransferase Complexes

Yuan-Liang Wang, Francesco Faiola, and Ernest Martinez

Abstract

The reversible acetylation of specific lysine residues on core histones regulates gene transcription in eukaryotes. Since the discovery of GCN5 as the first transcription-regulating histone acetyltransferase (HAT), a variety of HATs have now been identified and shown to acetylate different sites on histones as well as on non-histone proteins, including transcription regulators. In general, purified recombinant HATs expressed in bacteria or in insect cells are able to acetylate free histones and sometimes other substrates in vitro. However, such activity is often restricted to certain substrates and/or is very weak on physiological substrates, such as nucleosomes. Moreover, it does not reflect the actual scenario inside the cell, where HATs generally associate with other proteins to form stable multisubunit complexes. Importantly, these peripheral proteins significantly influence the functions of the catalytic HAT subunit by regulating its intrinsic catalytic activity and/or by modulating its target substrate selectivity. In this chapter, we describe detailed methods for the rapid (two step) and efficient purification of large, multiprotein HAT complexes from nuclear extracts of mammalian epitope-tagged cell lines, including protocols for the generation and large-scale suspension culture of these cell lines. These methods have been used to purify and characterize different human GCN5 HAT complexes that retain activity toward their physiological substrates in vitro.

Key words: Transcription, Histone acetyltransferase, Protein complexes, GCN5, Affinity purification, FLAG tag, S-Sepharose, Cell line

1. Introduction

RNA polymerase II-mediated transcription in eukaryotes is a tightly regulated multistep process involving many types of regulatory proteins, such as transcriptional activators and repressors, different cofactors, and general transcription factors (1). The picture is further complicated as nuclear genomic DNA is packed into chromatin by both histone and non-histone proteins, rendering DNA

Ales Vancura (ed.), *Transcriptional Regulation: Methods and Protocols*, Methods in Molecular Biology, vol. 809,
DOI 10.1007/978-1-61779-376-9_28,

inaccessible to transcription regulators (2). To overcome this barrier, eukaryotes have evolved to use either ATP-dependent enzymes to move/evict the nucleosomes or chromatin-modifying enzymes to modify DNA or histone proteins (3). Indeed, posttranslational modifications of nucleosomal histones have been correlated with changes in chromatin structure and transcription regulation (4, 5). Although several different histone modifications have now been identified and shown to play a role in transcription, lysine acetylation was the first to be correlated with transcription activation (6) and is the best characterized.

Mechanistically, histone tail acetylation was first shown to facilitate binding of transcriptional activators to nucleosomal DNA (7), consistent with findings that active chromosomal domains are generally associated with hyperacetylated histones (8) while inactive or heterochromatin domains are associated with hypoacetylated histones (9, 10). However, a more direct link between histone acetylation and transcription came originally from the identification of the first histone acetyltransferase (HAT) Gcn5 and the first histone deacetylase (HDAC) Rpd3, two factors that were previously genetically defined as a coactivator and a corepressor of gene transcription, respectively (11, 12). Thus, the steady-state acetylation level of histone proteins is accomplished by an intricate balance between HATs and HDACs that is important for proper cellular function (13, 14). Acetylation affects high-order folding of chromatin fibers, loosens the contacts between the DNA and the nucleosomes, and alters the interactions between histones and non-histone proteins. Acetylated lysine residues on histone tails are also "marks" that are recognized by bromodomain-containing proteins, which include components of the transcription machinery and ATP-dependent nucleosome-remodeling enzymes that displace modified nucleosomes from promoters (15, 16). Not surprisingly, histone acetylation has been connected with changes in chromatin that occur not only during transcription, but also during DNA replication and repair in vivo (17–20). Although histone proteins are the primary targets of HAT activity, HAT enzymes also acetylate a growing number of non-histone substrates, including many transcription regulators (21, 22). In this respect, a direct proof that histone tail acetylation (rather than other non-histone substrate acetylation) is causally linked to gene activation was provided only relatively recently. This was achieved via reconstitution of transcription activation on chromatin in vitro with purified components, including a recombinant coactivator-HAT (p300) and recombinant nucleosomal templates bearing substitutions of the acetylated lysine residues on the tails of core histones (23, 24).

Based on their catalytic domains, HATs have been grouped into several families, including the Gcn5-related *N*-acetyltransferases (GNATs) family and the MYST (MOZ, Ybf2/Sas3, Sas2, and Tip60)-related family (25). As the prototypical HAT, Gcn5 has been the focus of intense study over the last decade. Although purified, recombinant Gcn5 displays HAT activity on free histones, it fails to acetylate the more physiological, nucleosomal histone substrates in vitro. This led to the discovery that Gcn5 exists as multisubunit complexes inside the cell – e.g., SAGA/STAGA [Spt3-Taf-Ada-Gcn5 Acetylase] complexes that (1) can acetylate nucleosomes, (2) are recruited to promoters by DNA-binding activators, and (3) have HAT activity-dependent transcription coactivator functions on nucleosomal genes/promoters in vitro and in vivo (26–33). Importantly, the functions and specificity of the GCN5 HAT – and most other HATs – depend largely on the context of other subunits within those complexes (15, 26). Moreover, HAT-associated subunits within these complexes have additional roles in transcription coactivation, as shown, for instance, for specific GCN5-associated subunits within SAGA/STAGA complexes, which directly interact with components of the general/basal transcription machinery (34) or have additional catalytic activities targeting non-histone proteins (35).

Several multiprotein HAT complexes have now been identified and characterized in different model systems and shown to be evolutionarily conserved from yeast to human (15). However, depending on the organism and the experimental design, different laboratories utilize different strategies to purify these multiprotein complexes. A successful purification scheme to purify multisubunit HAT complexes generally contains at least one highly specific and stringent affinity separation step that retains only the target protein and the stably associated subunits, but not proteins that interact only weakly or nonspecifically. Among all the purification strategies, immunoprecipitation/co-immunoprecipitation (IP/co-IP) with specific antibodies and affinity chromatography using epitope tags are the most commonly used methods (31, 33, 36, 37). In some cases, protein complexes can also be purified via their affinity for specific peptides/substrates, such as histone tails. For example, peptide pull-down assay has been shown to purify histone methyltransferase complexes (38). Conventional biochemical purification (gel filtration and ion-exchange chromatography) is, per se, generally not the method of choice, although it can be used as one additional step to complement the affinity step (as described below). A summary of these individual methods with their strengths and weaknesses is given in Table 1.

IP/co-IP experiments are most frequently used to test endogenous association of two or more proteins. Therefore, they are also suitable techniques to capture endogenous protein complexes. Typically, an antibody recognizing a particular component of the complex is incubated with cell lysates followed by immobilization on protein A or protein G resins. The bound protein complex is then washed extensively to remove the contaminants for downstream applications. However, there are some limitations with these methods.

Table 1
Commonly used strategies for protein complex purification

Approach	Strength	Weakness
Anti-epitope tag immunoaffinity	1. Highly specific monoclonal antibodies against tag commercially available 2. Standardized and reproducible assays 3. No disruption of protein complexes 4. Resistant to stringent washing conditions 5. Gentle elution of active complexes by competition with epitope peptides	1. Cloning required 2. Overexpression of ectopic-tagged subunit 3. Potential unnatural interactions introduced by tag 4. Tag needs to be exposed in the complex 5. Residual antibody cross-reactivity
Regular IP/Co-IP	1. No cloning required 2. Endogenous factors – No pleiotropic effect from overexpression	1. Specific antibody required 2. Possible disruption of complexes depending on location of the antigenic loci recognized by the antibody. 3. Difficult to eluate the protein complex without denaturing agents 4. Antibody cross-reactivity
Peptide affinity pull down	1. No cloning required 2. Endogenous factors – No pleiotropic effect from overexpression	1. Not generically applicable 2. Requires knowledge of target substrates/peptides 3. Complex–peptide interaction may be weak 4. Nonspecific peptide interactions possible
Ion-exchange and gel filtration chromatography	1. No cloning required 2. Endogenous factors – No pleiotropic effect from overexpression or tags 3. Purification based on different physical/chemical properties of distinct complexes	1. Slow/long procedures 2. Heterogeneity of complexes complicates purification 3. Complex subunits often dissociate during multiple steps – loss of activity 4. Difficult to achieve high purity

First, the antibody of interest may not be commercially available or may not work in IP. Second, the antibody has to be highly specific with minimal cross-reactivity against other nonspecific proteins, which might give rise to false-positive results. Therefore, re-IP against another component of the complex is sometimes performed to validate the interaction.

Another powerful method is immunoaffinity purification with specific monoclonal antibodies recognizing an epitope tag carried by one of the subunits of the complex (39, 40). Typically, the protein of interest (bait) is fused to an epitope tag (e.g., FLAG, HA) and then overexpressed in the cell. To retrieve the bait protein and its cognate protein complexes, an antibody directed against the tag is used. As a result, many different subunits of the same complex can be tagged and purified by the same procedure. A variety of expression vectors containing the tags and the corresponding antibody-immobilized resins are commercially available. The complexes can be further eluted (in an active form) from the antibody resin by competition with an excess of epitope peptides.

Often a highly pure protein complex is derived from more than one purification step. This is because no matter how stringent the purification condition is, it always carries trace amount of contaminants that are specific to that particular method. A second purification helps remove the contaminants and also concentrates the sample for downstream applications. Besides the aforementioned antibody-based purification methods, conventional size-exclusion and ion-exchange chromatography are also used to fractionate protein complexes and are most useful in conjunction with an affinity method (31, 33, 39–41).

Technically, there are a number of parameters involving protein complex purification. These include binding affinities between components of the complex, cellular expression level of each subunit, wash stringency, etc. Therefore, it is important to validate the complex of interest by a combination of methods. In this chapter, we describe a straightforward, two-step purification method that has been successful in purifying different human GCN5 HAT-containing complexes (STAGA and ATAC) for their subunit characterization and for analysis of their catalytic activities on physiological, nucleosomal histone and non-histone substrates (31, 33, 42).

2. Materials

2.1. Cell Culture

2.1.1. Equipment and Supplies

1. Tissue culture incubators (both a humidified incubator with 5% CO_2 and a nonhumidified/non-CO_2, "reach-in" incubator for large spinner flasks).
2. Carbon dioxide (CO_2) gas.
3. Dulbecco's modified Eagle's medium (DMEM; Cellgro).
4. Minimum essential medium (S-MEM; Gibco).
5. Penicillin–streptomycin solution (Cellgro).
6. Fetal bovine serum (FBS).
7. Bovine calf serum (BCS).
8. Trypsin-ethylenediaminetetraacetic acid (trypsin-EDTA) solution.
9. Phosphate-buffer saline (PBS).
10. Plastic tissue culture dish.

11. LipoD293™ DNA In Vitro Transfection Reagent (SignaGen Laboratories).
12. Geneticin/G418 (Gibco).
13. Conical tubes (15 and 50 ml).
14. pIRES vector (Clontech).
15. FLAG antibody (Sigma).
16. Spinner flasks and magnetic stirrers for cell suspension culture (many different types/providers work well).
17. Trypan blue stain 0.4% (Gibco).
18. Dounce homogenizer.

2.1.2. Buffers

1. Lysis buffer: 50 mM HEPES, pH 7.9, 250 mM NaCl, 0.1% (v/v) IGEPAL CA-630, 0.2 mM EDTA, 0.2 mM PMSF, 2 mM β-mercaptoethanol.

2.2. Nuclear Extracts' Preparation from Mammalian Cells

2.2.1. Buffers

1. Hypotonic buffer: 10 mM Tris–HCl, pH 7.9, 10 mM KCl, 1.5 mM $MgCl_2$. Autoclave and store at 4°C. To complete the buffer before use, take the required amount and add to final concentration 10 mM β-mercaptoethanol and 0.2 mM PMSF.
2. 2× BC-O: 40 mM Tris–HCl, pH 7.9, 40% (v/v) glycerol, 0.4 mM EDTA. Autoclave and store at 4°C. This is used to make all 1× BC buffers.
3. BC-500: 20 mM Tris–HCl, pH 7.9, 20% (v/v) glycerol, 0.2 mM EDTA, 500 mM KCl. To complete the buffer before use, take the required amount and add to final concentration 10 mM β-mercaptoethanol and 0.2 mM PMSF. (Working BC buffers are made fresh by diluting 2× BC-O with autoclaved deionized water and appropriate salt).

2.3. Purification of Multisubunit HAT Complexes

2.3.1. Buffers

1. BC-330: 20 mM Tris–HCl, pH 7.9, 20% glycerol, 0.2 mM EDTA, 330 mM KCl. To complete the buffer before use, take the required amount and add to final concentration 10 mM β-mercaptoethanol, 0.2 mM PMSF, and 0.05% (v/v) IGEPAL (CA-630).
2. BC-100: 20 mM Tris–HCl, pH 7.9, 20% glycerol, 0.2 mM EDTA, 100 mM KCl. To complete the buffer before use, take the required amount and add to final concentration 10 mM β-mercaptoethanol, 0.2 mM PMSF, and 0.05% IGEPAL (CA-630).
3. BC-400: 20 mM Tris–HCl, pH 7.9, 20% glycerol, 0.2 mM EDTA, 400 mM NaCl. To complete the buffer before use, take the required amount and add to final concentration 10 mM β-mercaptoethanol, 0.2 mM PMSF, and 0.05% IGEPAL (CA-630).

4. 1× SDS loading buffer: 60 mM Tris–HCl, pH 6.8, 10% glycerol, 2% (w/v) SDS, 0.1% (w/v) bromophenol blue.

2.3.2. Other Materials and Reagents

1. Thin, round gel-loading tips.
2. Protein A–Agarose (Pierce).
3. Anti-FLAG® M2 Affinity Gel/Resin (Sigma).
4. FLAG® peptide (Sigma).
5. S-Sepharose Fast Flow (Pharmacia).

3. Methods

3.1. Cell Culture and Cell Line Generation

1. Culture HEK293 cells (or other mammalian cell line of interest) in a 6-cm culture dish in DMEM supplemented with 10% FBS and penicillin–streptomycin solution at 37°C with 5% CO_2.
2. When the cells reach ~95% confluence, decant the medium from the dish and wash the cells three times with PBS. After the last wash, decant the PBS completely and add 0.3 ml trypsin-EDTA into the dish. Leave the cells in the incubator for 1 min to allow detaching (from the dish and from each other). Resuspend the cells in 6 ml prewarmed DMEM supplemented with 10% FBS and penicillin–streptomycin.
3. Distribute the cells into two 6-cm culture dishes evenly (each with a total volume ~3 ml). Wait overnight for the cells to attach (see Note 1).
4. Transfect one dish with 1 μg pFH-GeneX-IRESneo vector containing your gene X of interest with a FLAG (F) and/or HA (H) epitopes in frame at the N or C terminus (31). Gene X can be the catalytic HAT subunit (e.g., GCN5) or another subunit of the multisubunit HAT complex. Transfect the other dish with control empty vector (pFH-IRESneo) and handle the control cells in parallel afterward (see Note 2).
5. Change medium 24 h after transfection.
6. 24 h later, trypsinize the cells from the dish and transfer them into a 50-ml conical tube containing 45 ml prewarmed DMEM supplemented with 10% FBS and penicillin–streptomycin. Gently resuspend the cells by pipetting up and down. Distribute the cells evenly into two 15-cm dishes (~23 ml per dish).
7. 24 h later, add geneticin/G418 to a concentration of 500 μg/ml to select for stably transfected clones. Select for 2–3 weeks by changing medium every other day to refresh the selection drug and remove the dead cells.
8. Pick single geneticin/G418-resistent colony and subculture in 3.5-cm culture dish for expansion. Choose several colonies.

9. Duplicate each clone into two dishes. Use one dish to check for the expression of your FLAG-tagged protein and the other dish for expansion.
10. To check for expression, when the cells reach 95–100% confluence, decant the medium and wash the cells three times with 1–2 ml of PBS. After the last wash, use a pipettman to remove all the remaining PBS completely.
11. Add 150 μl of ice-cold lysis buffer to the dish and keep it on ice for 10 min.
12. Use a policeman to scrap the cells off the dish and pipet the lysate into a microcentrifuge tube.
13. Spin the lysate on a countertop microcentrifuge at 13,000 rpm/10 min/4°C. Take an aliquot of the supernatant and check for positive clones by SDS-PAGE and western blotting with the FLAG antibody.
14. Choose a cell clone that expresses your FLAG-tagged protein and adapt the cells from monolayer to suspension. To adapt the cells, trypsinize the cells from the dish and resuspend them gently in SMEM supplemented with 20% FBS in a spinner flask (see Note 3). The density of suspended cells should be $3–4 \times 10^5$ cells/ml (see Note 4).
15. 24 h later, check the viability of the cells by staining an aliquot (2–3 ml) of the cells with trypan blue and examine under a microscope (see Note 5).
16. When the cells reach $7–8 \times 10^5$ cells/ml, remove half of the culture medium (i.e., half of the cells) and replace with the same amount of fresh SMEM plus 10% FBS (see Note 6).
17. Repeat steps 13 and 14 by replacing the medium with different amounts of serum. In repeat 1, replace the medium with SMEM plus 5% FBS and 5% BCS; in repeat 2, replace with SMEM plus 20% BCS; in repeat 3, replace with SMEM plus 10% BCS; in repeat 4, replace with SMEM plus 5% BCS. The cells are ready for expansion after they are adapted to 5% BCS.
18. Expand the cells to the desired amount (~16 l) for nuclear extracts (NE)'s preparation. Grow the cells to a density of 1×10^6 cells/ml before harvest (see Note 7).

3.2. Preparation of Nuclear Extracts

All procedures should be performed on ice or in the cold room and finished within the same day.

1. Pour the suspended cells into 1-l bottles. Use a Beckman Coulter J6-HC centrifuge to spin down the cells at 3,000 rpm/10 min/4°C with a JS-4.2 rotor and swinging buckets. Decant the medium carefully. Repeat the process until all the cells in the spinner are collected (see Note 8).

2. Transfer cells into six 50-ml conical tubes. Use a Beckman Coulter J6-HC centrifuge to spin down the cells at 2,000 rpm/5 min/4°C (JS-4.2 rotor with swinging buckets).
3. Remove the medium and resuspend the cells gently with ice-cold PBS. Transfer the cell pellet into four 50-ml conical tubes.
4. Spin down the cells at 2,000 rpm/3 min/4°C, as above.
5. Remove the supernatant and wash the cells again with ice-cold PBS. Transfer the cell pellet into two 50-ml conical tubes.
6. Spin down the cells at 2,000 rpm/3 min/4°C.
7. Measure packed cell volume (PCV).
8. Remove the supernatant and resuspend the cells with 30–40 ml of ice-cold, complete hypotonic buffer by gently inverting the cells until completely resuspended.
9. Spin down the cells at 2,000 rpm/3 min/4°C. Use a pipette to remove the supernatant carefully (see Note 9).
10. Add 1/3 PCV ice-cold, complete hypotonic buffer and resuspend cells carefully by gently inverting the tube.
11. Swell cells on ice for 10 min.
12. Pour the cells into a 40 ml ice-cold Dounce homogenizer.
13. Homogenize the cells on ice using a "B" pestle. To homogenize the cells, slowly and carefully twist the pestle up and down for 15 strokes (see Note 10).
14. Use a microscope to check the cell breakage and release of nuclei by staining ~5 μl of the homogenate with 10–20 μl of trypan blue on a petri dish (see Note 11).
15. If there is less than 90% cell breakage, give two more strokes and check again. Repeat this process until more than 90% of the cells have released their nuclei (see Note 12).
16. Pour the nuclei from the Dounce homogenizer into a 50-ml conical tube and spin the nuclei at 3,500 rpm/20 min/4°C (Beckman JS-4.2 rotor with swinging buckets); measure nuclear pellet volume (NPV) (see Note 13).
17. Transfer the supernatant (cytoplasmic fraction) to another 50-ml conical tube and measure the volume (cytoplasmic fraction volume (CFV) (see Note 14).
18. Add 1/4 CFV of 100% glycerol, 1/29 CFV of 3 M KCl, and 1/1,000 CFV of 250 mM EDTA to the cytoplasmic fraction. Gently mix by inverting the tube. Keep it on ice.
19. Extract the nuclei by adding 2 NPV of ice-cold, complete BC-500 to the nuclear pellet (NP). Resuspend the pellet by pipetting up and down several times with a pipette aid (see Note 15).
20. Rotate for 1 h at 4°C (see Note 16).

21. Transfer the nuclei/nuclear extracts mixture and the cytoplasmic fraction into different centrifuge tubes for Sorvall SS-34 rotor.
22. Use a Sorvall RC-5B centrifuge (or equivalent) to spin at 14,000 rpm/30 min/4°C.
23. Transfer the supernatant from the nuclei/nuclear extracts mixture tube into a 50-ml conical tube and label nuclear extracts.
24. Scrap off the nuclear pellet into a 50-ml conical tube and label nuclear pellet.
25. Transfer the supernatant from the cytoplasmic fraction tube into a 50-ml conical tube and label cytoplasmic extracts (CEs). Discard the pellet.
26. Keep a small aliquot of NE, NP, and CE.
27. Snap freeze all these fractions by liquid nitrogen and store at –80°C (see Note 17).

3.3. Purification of FLAG-Tagged Complexes from Nuclear Extracts

3.3.1. First-Step Purification with Anti-FLAG® M2 Affinity Resin

1. Thaw nuclear extracts on ice (from both control parental cells and cell line expressing your FLAG-tagged protein).
2. Transfer 15 ml nuclear extracts into a centrifuge tube for Sorvall SS-34 rotor. Add 10% (v/v) of NP-40 (IGEPAL CA-630) to a final concentration of 0.05%. Mix gently by inverting.
3. Spin at 14,000 rpm/20 min/4°C.
4. Transfer ~14 ml of nuclear extracts into a 15-ml conical tube. Avoid lipid on the top and pellet at the bottom.
5. Aliquot 30 μl M2 resin (60 μl 50% slurry) and protein A resin for each 14 ml nuclear extracts (see Note 18).
6. Wash the resin three times with ice-cold/complete BC-330. After last wash, keep the resin as 50% slurry in BC-330 (see Note 19).
7. Preclear nuclear extracts by adding 30 μl of preequilibrated protein A resin into each tube. Rotate for 30 min/4°C.
8. Use a Beckman Coulter J6-HC centrifuge to spin down the resin at 2,000 rpm/1 min/4°C (JS-4.2 rotor with swinging buckets).
9. Transfer the nuclear extracts into a new 15-ml conical tube. Avoid taking the protein A resin at the bottom.
10. Add 30 μl of preequilibrated M2 resin into the precleared nuclear extracts (see Note 20).
11. Rotate for 3 h/4°C (see Note 21).
12. Spin down the resin at 2,000 rpm/1 min/4°C.
13. Transfer the supernatant (unbound) into a new 15-ml conical tube and label M2-1×. Snap freeze in liquid nitrogen and store at –80°C (see Note 22).
14. Wash the resin with 15 ml of ice-cold/complete BC-330 by gently inverting the tube.

15. Spin down the resin at 2,000 rpm/1 min/4°C. Carefully remove the wash without disturbing the resin.
16. Repeat the wash step twice.
17. After the third wash, add 1 ml of ice-cold/complete BC-330 to the 15-ml conical tube. Gently pipet up and down to resuspend the resin.
18. Transfer the resin into a microcentrifuge tube.
19. Use a countertop microcentrifuge (Eppendorf or equivalent) to spin down the resin at 2,000 rpm/1 min/4°C. Carefully remove the wash without disturbing the resin.
20. Repeat the wash step twice.
21. Adjust salt concentration to 100 mM by washing the resin twice with 1 ml ice-cold/complete BC-100.
22. Remove the final wash completely (see Note 23).
23. To elute, add 35 μl of 0.3 mg/ml FLAG peptide solution to the resin and shake it at 900 rpm/30 min/20°C (see Note 24).
24. Spin down the resin at 3,000 rpm/10 s/4°C.
25. Use a gel-loading tip to transfer the eluent into a new microcentrifuge tube and label "Eluate 1." Be careful not to take the resin. Keep eluate 1 on ice.
26. Repeat the elution step twice. Label the eluates as "Eluate 2" and "Eluate 3."
27. Snap freeze all the eluates in liquid nitrogen and store at −80°C (see Note 25).
28. Analyze 5–8 μl of each eluate (including the control eluates of mock purifications from parental untagged cells) by SDS-PAGE and silver staining.

3.3.2. Second-Step Concentration and Purification on S-Sepharose Resin

Generally, single-step FLAG/M2-purified HAT complexes are suitable for downstream applications, such as acetylation of histone and non-histone substrates. However, except for the first eluates of the initial M2 affinity round of purification, subsequent eluates may be too diluted; in addition, proteins that bind nonspecifically to the anti-FLAG/M2 resin (e.g., Hsp70) are still present. These contaminants can be detected by SDS-PAGE and silver staining in the control lane containing the mock purification sample from parental cells that do not express any tagged protein (step 28 above). A simple way to concentrate the pool of diluted complexes and to remove most of the contaminants (including the excess FLAG peptide) is to batch adsorb FLAG-tagged complexes onto a small amount of the negatively charged S-Sepharose (or SP-Sepharose) ion-exchange resin. Most contaminants do not bind tightly to S-Sepharose as illustrated in Fig. 1, which shows a representative S-Sepharose purification of GCN5 HAT-containing ATAC

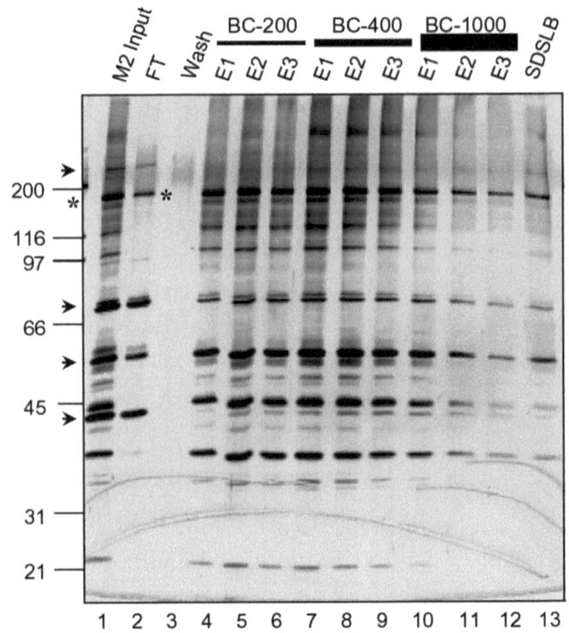

Fig. 1. S-Sepharose fractionation of anti-FLAG/M2-purified ATAC complexes. Shown is a silver-stained, SDS-PAGE gel containing different S-Sepharose elution fractions. M2-purified, FLAG-tagged ATAC complexes (M2 input, *lane 1*) were adsorbed onto S-Sepharose resin in BC-60, washed with BC-100 (Wash, *lane 3*), and eluted successively with three resin volumes (E1–E3) of the indicated BC buffers containing from 200 mM NaCl (BC-200) to 1,000 mM NaCl (BC-1000) and finally with SDS loading buffer (SDSLB). *FT* unbound "flow through" fraction. Proteins that bind nonspecifically to M2 agarose (*arrowheads*) and contaminate the M2-purified complexes (M2 input) do not bind to S-Sepharose (i.e., present in FT fraction, *lane 2*). Excess, "free," FLAG-tagged YEATS2 subunit (*asterisks*) does not bind to S-Sepharose either (*lane 2*).

complexes isolated from cells expressing the FLAG-tagged YEATS2 subunit (33).

1. Thaw M2-purified FLAG-tagged HAT complexes on ice (process the control/mock-purified eluates in parallel).
2. Pool M2 eluates together (~1 ml total) into one microcentrifuge tube (see Note 26).
3. Spin on a countertop centrifuge at 13,000 rpm/10 min/4°C.
4. Transfer 800 μl of the supernatant into a new microcentrifuge tube (see Note 27).
5. Adjust final salt concentration to 60 mM KCl by adding 533 μl of ice-cold/complete 1× BC-0 to a total of 1,333 μl. Gently mix by inverting the tube.
6. Keep 20 μl as "input."
7. Aliquot 20 μl of S-Sepharose resin (40 μl of 50% slurry) into a microcentrifuge tube and wash three times with ice-cold/complete BC-60. After last wash, keep the resin as 50% slurry in BC-60.

8. Add 20 μl of preequilibrated S-Sepharose resin (40 μl slurry) to the pooled M2 eluates.
9. Rotate for 3 h/4°C.
10. Spin down the resin at 2,000 rpm/1 min/4°C.
11. Transfer the supernatant into a new microcentrifuge tube and label "S-Sepharose flow through."
12. Wash the resin with 1 ml of ice-cold/complete BC-100 by gently inverting the tube.
13. Spin down the resin at 2,000 rpm/1 min/4°C. Keep the wash and label "wash 1."
14. Repeat the wash step three times and label "wash 2, wash 3, wash 4."
15. Remove the final wash completely.
16. To elute, add 25 μl of ice-cold/complete BC-400 to the resin (see Note 28). Gently flick the microcentrifuge tube several times with a finger.
17. Spin down the resin at 3,000 rpm/10 s/4°C.
18. Use a gel-loading tip to transfer the eluate into a new microcentrifuge tube and label "S-Sepharose eluate 1."
19. Repeat the elution step up to four times and label accordingly.
20. Add 20 μl of 1× SDS loading buffer to the resin.
21. Analyze 2–4 μl of each eluate by SDS-PAGE and silver staining. Snap freeze the remaining eluates in liquid nitrogen and store at −80°C.

4. Notes

1. Mammalian cells double approximately every 24 h. Therefore, by the next day, the cells should reach ~95% confluence again.
2. It is important to select appropriate transfection reagent and perform transfection according to manufacturer's protocol. To transfect HEK293 cells, we use LipoD293™ at 1:3 ratio (DNA [μg]:transfection reagent [μl]). To obtain good transfection efficiency and low cytotoxicity, transfection conditions should be tested by varying DNA and transfection reagent ratios. Optimal DNA (μg):transfection reagent (μl) ratio generally lies between 1:0.5–1:5. It may be optimal to transfect the cells again the next day if transfection efficiency is not good.
3. It is important not to use geneticin/G418 during the adaptation process; otherwise, the cells may die.
4. Depending on the cell type, one 10-cm dish generally equals to $1-2 \times 10^7$ cells. Assuming that one 10-cm dish contains

1×10^7 cells, three 10-cm dishes would be 3×10^7 cells. Therefore, cells from three confluent 10-cm dishes can be resuspended into 100 ml of SMEM plus serum at the density of 3×10^5 cells/ml.

5. Trypan blue goes inside the dead cells and stains them blue. On the contrary, the live ones are not stained and therefore have a shiny contour.
6. The cell density should be reduced in half but no less than 3×10^5 cells/ml. The cells do not grow if density drops below 3×10^5 cells/ml.
7. Depending on the cell type, serum concentration, and medium, the density of cells may vary. 16 l of HEK293 cells at a density of 1×10^6 cells/ml typically generate about 30–40 ml nuclear extracts.
8. Do not decant the medium completely, as this may result in cell loss. Leave approximately 200 ml of medium each time.
9. The cells swell and increase volume after adding hypotonic buffer.
10. The time to complete 1 stroke should be ~30 s. Homogenizing the cells too quickly may shear the nuclei. In addition, if the pestle is not twisted well, the cell membrane may not break.
11. Intact cells appear large and shiny while released nuclei are intensely stained by Trypan blue and tend to aggregate.
12. Avoid excessive homogenization, which leads to nuclei breakage and release of DNA/chromatin.
13. Typically, NPV is 15–18 ml for 16 l of HEK293 cell culture.
14. Typically, CFV is 10–12 ml for 16 l of HEK293 cell culture.
15. Nuclear pellet is very dense and viscous. Pipetting up and down for several minutes may be required. It is important to dissociate the clumps as much as possible to increase the surface area for optimal nuclear extraction.
16. Final salt/KCl concentration should be ~333 mM, which should be verified with a conductivity meter.
17. The protein concentration of the nuclear extracts and the cytoplasmic extracts should be 7–8 mg/ml by using the Bradford protein quantification method with BSA as a standard.
18. Optimal resin/NE volume ratio is ~1/500 for our FLAG-tagged GCN5 complex cell lines, but varies depending on specific tagged proteins and complexes and must be determined empirically. Avoid the use of too much resin, which increases the "sticking" of nonspecific contaminants.
19. 330 mM is the salt concentration of the nuclear extracts.
20. Resin may stick to the wall of the tip and microcentrifuge tube. Use the nuclear extracts to pipet up and down several times to wash the remaining resin into the 15-ml conical tube.

21. If incubation time is too short, FLAG-tagged complexes may not bind efficiently to the M2 resin; if incubation time is too long, nonspecific binding and protein degradation may occur.
22. The unbound fraction may be used up to three times to purify the remaining FLAG-tagged complexes. However, the yield decreases proportionally.
23. To completely remove the final wash, place a gel-loading tip against the bottom of the tube and carefully remove all remaining buffer. A small amount of resin may stick to the tip. Carefully circle the tip against the wall of the microcentrifuge tube to collect the tip-bound resin.
24. Dilute FLAG peptide in ice-cold/complete BC-100 at a concentration of 0.3 mg/ml.
25. Optional: Repeat the purification (steps 10–27) with the unbound protein extracts (step 13) up to twice more.
26. It is important to check each elution of FLAG-purified complexes by silver staining to decide which eluates are worth pooling.
27. Minor M2 resin carryover may be at the bottom. Be careful not to take the resin.
28. BC-400 contains 400 mM of NaCl instead of KCl. This is because loading 400 mM KCl into an SDS-PAGE gel may cause precipitation in the well.

Acknowledgments

The authors would like to thank Dr. Jennifer Liu for advice. This work was supported by grants R01CA100464 and MCB0448488 from NIH and NSF, respectively.

References

1. Roeder, R. G. (2005) Transcriptional regulation and the role of diverse coactivators in animal cells. *FEBS Lett* *579*, 909–915.
2. Kornberg, R. D., and Lorch, Y. (1999) Twenty-five years of the nucleosome, fundamental particle of the eukaryote chromosome. *Cell* ***98***, 285–294.
3. Berger, S. L. (2007) The complex language of chromatin regulation during transcription. *Nature* ***447***, 407–412.
4. Kingston, R. E., Bunker, C. A., and Imbalzano, A. N. (1996) Repression and activation by multiprotein complexes that alter chromatin structure. *Genes & Dev* ***10***, 905–920.
5. Eberharter, A., and Becker, P. B. (2002) Histone acetylation: a switch between repressive and permissive chromatin. Second in review series on chromatin dynamics, *EMBO Rep* *3*, 224–229.
6. Allfrey, V. G., Faulkner, R., and Mirsky, A. E. (1964) Acetylation and methylation of histones and their possible role in the regulation of RNA synthesis. *Proc Natl Acad Sci USA.* ***51***, 786–794.
7. Lee, D. Y., Hayes, J. J., Pruss, D., and Wolffe, A. P. (1993) A positive role for histone acetylation in transcription factor access to nucleosomal DNA. *Cell* *72*, 73–84.

8. Hebbes, T. R., Clayton, A. L., Thorne, A. W., and Crane-Robinson, C. (1994) Core histone hyperacetylation co-maps with generalized DNase I sensitivity in the chicken beta-globin chromosomal domain. *EMBO J 13*, 1823–1830.
9. Braunstein, M., Rose, A. B., Holmes, S. G., Allis, C. D., and Broach, J. R. (1993) Transcriptional silencing in yeast is associated with reduced nucleosome acetylation. *Genes & Dev 7*, 592–604.
10. Jeppesen, P., and Turner, B. M. (1993) The inactive X chromosome in female mammals is distinguished by a lack of histone H4 acetylation, a cytogenetic marker for gene expression. *Cell 74*, 281–289.
11. Brownell, J. E., Zhou, J., Ranalli, T., Kobayashi, R., Edmondson, D. G., Roth, S. Y., and Allis, C. D. (1996) Tetrahymena histone acetyltransferase A: a homolog to yeast Gcn5p linking histone acetylation to gene activation. *Cell 84*, 843–851.
12. Taunton, J., Hassig, C. A., and Schreiber, S. L. (1996) A mammalian histone deacetylase related to the yeast transcriptional regulator Rpd3p. *Science 272*, 408–411.
13. Cheung, W. L., Briggs, S. D., and Allis, C. D. (2000) Acetylation and chromosomal functions. *Curr Opin Cell Biol 12*, 326–333.
14. Narlikar, G. J., Fan, H. Y., and Kingston, R. E. (2002) Cooperation between complexes that regulate chromatin structure and transcription. *Cell 108*, 475–487.
15. Lee, K. K., and Workman, J. L. (2007) Histone acetyltransferase complexes: one size doesn't fit all. *Nat Rev 8*, 284–295.
16. Weake, V. M., Swanson, S. K., Mushegian, A., Florens, L., Washburn, M. P., Abmayr, S. M., and Workman, J. L. (2009) A novel histone fold domain-containing protein that replaces TAF6 in Drosophila SAGA is required for SAGA-dependent gene expression. *Genes & Dev 23*, 2818–2823.
17. Tse, C., Sera, T., Wolffe, A. P., and Hansen, J. C. (1998) Disruption of higher-order folding by core histone acetylation dramatically enhances transcription of nucleosomal arrays by RNA polymerase III. *Mol Cell Biol 18*, 4629–4638.
18. Carruthers, L. M., and Hansen, J. C. (2000) The core histone N termini function independently of linker histones during chromatin condensation. *J Biol Chem 275*, 37285–37290.
19. Wolffe, A. P., and Hansen, J. C. (2001) Nuclear visions: functional flexibility from structural instability. *Cell 104*, 631–634.
20. Shogren-Knaak, M., Ishii, H., Sun, J. M., Pazin, M. J., Davie, J. R., and Peterson, C. L. (2006) Histone H4-K16 acetylation controls chromatin structure and protein interactions. *Science 311*, 844–847.
21. Yang, X. J. (2004) The diverse superfamily of lysine acetyltransferases and their roles in leukemia and other diseases. *Nucleic Acids Res 32*, 959–976.
22. Berndsen, C. E., and Denu, J. M. (2008) Catalysis and substrate selection by histone/protein lysine acetyltransferases. *Curr Opin Struct Biol 18*, 682–689.
23. An, W., Palhan, V. B., Karymov, M. A., Leuba, S. H., and Roeder, R. G. (2002) Selective requirements for histone H3 and H4 N termini in p300-dependent transcriptional activation from chromatin. *Mol Cell 9*, 811–821.
24. An, W., Kim, J., and Roeder, R. G. (2004) Ordered cooperative functions of PRMT1, p300, and CARM1 in transcriptional activation by p53. *Cell 117*, 735–748.
25. Dekker, F. J., and Haisma, H. J. (2009) Histone acetyl transferases as emerging drug targets. *Drug Discov Today 14*, 942–948.
26. Grant, P. A., Eberharter, A., John, S., Cook, R. G., Turner, B. M., and Workman, J. L. (1999) Expanded lysine acetylation specificity of Gcn5 in native complexes. *J Biol Chem 274*, 5895–5900.
27. Grant, P. A., Duggan, L., Cote, J., Roberts, S. M., Brownell, J. E., Candau, R., Ohba, R., Owen-Hughes, T., Allis, C. D., Winston, F., Berger, S. L., and Workman, J. L. (1997) Yeast Gcn5 functions in two multisubunit complexes to acetylate nucleosomal histones: characterization of an Ada complex and the SAGA (Spt/Ada) complex. *Genes & Dev 11*, 1640–1650.
28. Candau, R., Zhou, J. X., Allis, C. D., and Berger, S. L. (1997) Histone acetyltransferase activity and interaction with ADA2 are critical for GCN5 function in vivo. *EMBO J 16*, 555–565.
29. Grant, P. A., Schieltz, D., Pray-Grant, M. G., Steger, D. J., Reese, J. C., Yates, J. R., 3 rd, and Workman, J. L. (1998) A subset of TAF(II)s are integral components of the SAGA complex required for nucleosome acetylation and transcriptional stimulation. *Cell 94*, 45–53.
30. Martinez, E., Kundu, T. K., Fu, J., and Roeder, R. G. (1998) A human SPT3-TAFII31-GCN5-L acetylase complex distinct from transcription factor IID. *J Biol Chem 273*, 23781–23785.
31. Martinez, E., Palhan, V. B., Tjernberg, A., Lymar, E. S., Gamper, A. M., Kundu, T. K., Chait, B. T., and Roeder, R. G. (2001) Human STAGA complex is a chromatin-acetylating transcription coactivator that interacts with pre-mRNA splicing and DNA damage-binding factors in vivo. *Mol Cell Biol 21*, 6782–6795.

32. Ogryzko, V. V., Kotani, T., Zhang, X., Schiltz, R. L., Howard, T., Yang, X. J., Howard, B. H., Qin, J., and Nakatani, Y. (1998) Histone-like TAFs within the PCAF histone acetylase complex. *Cell 94*, 35–44.
33. Wang, Y. L., Faiola, F., Xu, M., Pan, S., and Martinez, E. (2008) Human ATAC Is a GCN5/PCAF-containing acetylase complex with a novel NC2-like histone fold module that interacts with the TATA-binding protein. *J Biol Chem 283*, 33808–33815.
34. Liu, X., Vorontchikhina, M., Wang, Y. L., Faiola, F., and Martinez, E. (2008) STAGA recruits Mediator to the MYC oncoprotein to stimulate transcription and cell proliferation. *Mol Cell Biol 28*, 108–121.
35. Atanassov, B. S., Evrard, Y. A., Multani, A. S., Zhang, Z., Tora, L., Devys, D., Chang, S., and Dent, S. Y. (2009) Gcn5 and SAGA regulate shelterin protein turnover and telomere maintenance. *Mol Cell 35*, 352–364.
36. Nagy, Z., Riss, A., Romier, C., le Guezennec, X., Dongre, A. R., Orpinell, M., Han, J., Stunnenberg, H., and Tora, L. (2009) The human SPT20-containing SAGA complex plays a direct role in the regulation of endoplasmic reticulum stress-induced genes,. *Mol Cell Biol 29*, 1649–1660.
37. Li, S., and Shogren-Knaak, M. A. (2009) The Gcn5 bromodomain of the SAGA complex facilitates cooperative and cross-tail acetylation of nucleosomes. *J Biol Chen 284*, 9411–9417.
38. Wysocka, J., Swigut, T., Milne, T. A., Dou, Y., Zhang, X., Burlingame, A. L., Roeder, R. G., Brivanlou, A. H., and Allis, C. D. (2005) WDR5 associates with histone H3 methylated at K4 and is essential for H3 K4 methylation and vertebrate development. *Cell 121*, 859–872.
39. Zhou, Q., Lieberman, P. M., Boyer, T. G., and Berk, A. J. (1992) Holo-TFIID supports transcriptional stimulation by diverse activators and from a TATA-less promoter. *Genes & Dev 6*, 1964–1974.
40. Chiang, C. M., Ge, H., Wang, Z., Hoffmann, A., and Roeder, R. G. (1993) Unique TATA-binding protein-containing complexes and cofactors involved in transcription by RNA polymerases II and III. *EMBO J 12*, 2749–2762.
41. Guelman, S., Suganuma, T., Florens, L., Swanson, S. K., Kiesecker, C. L., Kusch, T., Anderson, S., Yates, J. R., 3 rd, Washburn, M. P., Abmayr, S. M., and Workman, J. L. (2006) Host cell factor and an uncharacterized SANT domain protein are stable components of ATAC, a novel dAda2A/dGcn5-containing histone acetyltransferase complex in Drosophila. *Mol Cell Biol 26*, 871–882.
42. Faiola, F., Liu, X., Lo, S., Pan, S., Zhang, K., Lymar, E., Farina, A., and Martinez, E. (2005) Dual regulation of c-Myc by p300 via acetylation-dependent control of Myc protein turnover and coactivation of Myc-induced transcription. *Mol Cell Biol 25*, 10220–10234.

Chapter 29

Reconstitution of Active and Stoichiometric Multisubunit Lysine Acetyltransferase Complexes in Insect Cells

Kezhi Yan, Chao-Jung Wu, Nadine Pelletier, and Xiang-Jiao Yang

Abstract

Protein lysine acetyltransferases (KATs) catalyze acetylation of the ε-amino group on a specific lysine residue, and this posttranslational modification is important for regulating the function and activities of thousands of proteins in diverse organisms from bacteria to humans. Interestingly, many known KATs exist in multisubunit complexes and complex formation is important for their proper structure, function, and regulation. Thus, it is necessary to reconstitute enzymatically active complexes for studying the relationship between subunits and determining structures of the complexes. Due to inherent limitations of bacterial and mammalian expression systems, baculovirus-mediated protein expression in insect cells has proven useful for assembling such multisubunit complexes. Related to this, we have adopted such an approach for reconstituting active tetrameric complexes of *mo*nocytic leukemia *z*inc (MOZ, finger protein, recently renamed MYST3 or KAT6A) and *MOZ*-*r*elated *f*actor (MORF, also known as MYST4 or KAT6B), two KATs directly linked to development of leukemia and self-renewal of stem cells. Herein, we use these complexes as examples to describe the related procedures. Similar methods have been used for reconstituting active complexes of histone deacetylases, lysine demethylases, and ubiquitin ligases, so this simple approach can be adapted for molecular dissection of various multisubunit complexes.

Key words: Lysine acetyltransferase, Complex reconstitution, Baculovirus, Protein expression, Affinity purification

1. Introduction

Lysine acetylation is a reversible posttranslational modification process that transfers the acetyl moiety from acetyl coenzyme A to the ε-amino group of a lysine residue. This modification was initially discovered in the early 1960s (1, 2) and is now known to play a major role in regulating various protein functions in diverse organisms (3–5). Recent proteomic studies indicate that acetylation

Ales Vancura (ed.), *Transcriptional Regulation: Methods and Protocols*, Methods in Molecular Biology, vol. 809,
DOI 10.1007/978-1-61779-376-9_29,

occurs not only on lysine residues of many nuclear and cytoplasmic proteins crucial in numerous eukaryotic cellular processes (6–8), but also on specific lysine residues of various bacterial proteins important for metabolism (9). In eukaryotic cells, lysine acetylation is catalyzed by lysine acetyltransferases (KATs) and reversed by deacetylases. While similar enzymatic reactions occur in bacteria, nonenzymatic acetylation involving high-energy intermediates, such as acetyl-phosphate or acetyl-AMP, also appears to be a viable mechanism (10). Since yeast *h*istone *a*cetyl*t*ransferase 1 (Hat1) and *Tetrahymena* *g*eneral *c*ontrol *n*onderepressible 5 (Gcn5) were identified as the first KATs in the mid-1990s (11, 12), over a dozen of eukaryotic proteins have been found to possess KAT activity (13–15). According to their sequence similarity, these proteins have been divided into different families. Two striking features about known KATs are that a majority of them are transcriptional coactivators and that many of them form chromatographically stable multisubunit complexes.

Related to the latter feature, KAT activity is often dependent on proper formation of complexes, which may contain one or two KAT catalytic subunits and additional noncatalytic subunits that play a regulatory role or serve as adaptors to substrates and binding partners of the enzymatic complexes. For example, yeast Hat1 associates with Hat2, a noncatalytic subunit with sequence similarity to the mammalian histone-binding protein RbAp46/48 (Rb-associated protein of 46 or 48 kDa) (16). Hat2 is required for Hat1 to bind and acetylate histone H4. This complex is typically cytoplasmic, but relocates to the nucleus upon association with the histone chaperon Hat1-interacting factor 1 (Hif1) (17, 18). These results indicate that interaction with different partners regulates both enzymatic activity and subcellular distribution of Hat1.

More interestingly, yeast Gcn5 is the catalytic subunit of four distinct multisubunit assemblies, including Spt-Ada-Gcn5-acetyltransferase (SAGA) and alteration/deficiency in activation 2 (ADA2) complexes (15, 19, 20). The former even contains ubiquitin-specific protease 8 (Ubp8), a deubiquitinating enzyme that removes ubiquitin from monoubiquitinated histone H2B. Furthermore, *Drosophila* Gcn5 interacts another KAT called Ada2a-containing complex component 2 (Atac2) and thus forms a unique complex that contains two acetyltransferases, Gcn5 and Atac2 (21–23). In mammals, there are two proteins, GCN5 and PCAF, homologous to yeast Gcn5, and both form complexes similar to those in yeast and *Drosophila*. Among these Gcn5 and related complexes, subunit interaction is important for activity regulation and substrate specificity determination. One challenging issue common to molecular characterization of different KAT complexes is to reconstitute them for dissecting how subunit interaction controls acetyltransferase activity and substrate specificity. In addition, reconstituted complexes are invaluable for structural analysis by X-ray diffraction,

NMR, and electron microscopy, as well as for drug screening through conventional and high-throughput approaches.

The simplest approach for complex reconstitution is to express different subunits in bacteria separately and then mix the proteins for subsequent complex purification, but this approach often fails to yield functional complexes due to peptide folding problems, especially when subunit interaction surface is extensive. To promote proper folding, recombinant subunits can be partially denatured and then renatured for refolding and complex formation in vitro. Alternatively, the subunits are coexpressed for complex formation in bacteria, such as *E. coli*. This can be achieved through cotransformation of different bacterial expression vectors containing compatible replication origins and distinct antibiotic markers (24). Polycistronic expression vectors able to express more than one protein per vector have also been developed, e.g., pST44 (25) and the Duet series of plasmids from Novagen. In addition to bacterial hosts, eukaryotic systems, such as yeast, rabbit reticulocyte lysate, and mammalian and insect cells, can be used for subunit coexpression and complex assembly. Compared to bacteria, one obvious advantage of these eukaryotic systems is better protein folding. In addition, disulfide bond formation and other posttranslational modifications occur in insect and mammalian cells. More importantly, it is often difficult to express large mammalian proteins in bacteria, with the yield inversely correlated to protein size for those larger than 50 kDa. Disadvantages of rabbit reticulocyte lysates and mammalian cells are higher cost and the difficulty to achieve high-expression yield. The insect cell system involves baculovirus-mediated expression and gives much higher expression levels than different mammalian systems. Moreover, unlike different viruses employed in the mammalian systems, baculovirus posts much smaller safety concerns to humans. Another advantage is that it is feasible to express large KATs using the baculovirus system. For example, p300, CBP, and MORF are larger than 200 kDa, and their active forms have been successfully expressed in insect cells (26, 27). Considering different factors altogether, it is reasonable to conclude that the baculovirus-mediated expression system is advantageous for complex assembly and characterization of mammalian KAT complexes, esp for those containing multiple and large subunits.

We have utilized insect cells to reconstitute and characterize complexes of mammalian MOZ finger protein, recently renamed MYST3, and MORF, also known as MYST4, two paralogous KATs that have been directly linked to acute myeloid leukemia, self-renewal of stem cells, and other developmental processes (28–30). Both were identified as the catalytic subunits of tetrameric complexes containing inhibitor of growth 5 (ING5), orthlog of Esa1-associated factor 6 (EAF6), and bromodomain PHD finger protein 1, 2, or 3 (BRPF1/2/3) (31–34). Through analyzing different deletion mutants of MOZ, MORF, BRPF1, and ING5, we have mapped regions important for tetrameric complex and found that the acetyltransferase domain of MOZ or MORF is sufficient for tetrameric complex formation (35). Here, we use MOZ/MORF complexes as examples to describe the experimental procedures for KAT complex assembly and characterization. Similar approaches have been successfully used for reconstituting and analyzing complexes of other chromatin-modifying enzymes, including histone deacetylases, demethylases, and ubiquitin ligases (for more details, see Note 1), so the simple procedures described herein should be easily adaptable for studying various multisubunit complexes.

2. Materials

2.1. Maintenance of Sf9 Insect Cells

1. *Spodoptera frugiperda* 9 (Sf9, see Note 2) insect cells (BD Biosciences, Cat. 551407 or Invitrogen Cat. 11496–015).
2. Incubator with a refrigerating system (such as Sanyo MIR153), with one or two magnetic stirrers (Bellco, multi-stir 2, Cat. 7785-D1502, or multi-stir 4, Cat. 7760–06005; see Note 3) put inside.
3. Complete Grace's medium: 88% (v/v) Grace's Insect Medium (Invitrogen, Cat. 11605), 10% (v/v) fetal bovine serum (FBS), 1% (v/v) Penicillin/Streptomycin (Invitrogen, Cat. 15140), and 1% (v/v) Pluronic F-68 (100×, GIBCO®, Cat. 24040). Typically, we just add 50 ml FBS, 5 ml Penicillin/Streptomycin, and 5 ml Pluronic F-68 to a 500-ml bottle of plain Grace's medium to prepare the complete medium, which can be kept at 4°C for a few months.
4. T-75 cell culture flasks (BD Biosciences, Cat. 353136).
5. Spinner flasks, 250 ml (Bellco Glass, Cat. 1965–00250) and 1 l (Bellco Glass, Cat. 1965–01000). see Note 4 for a proper cleaning procedure.
6. Conical tubes, 50 ml.
7. Hemacytometer.
8. Freezing medium: 60% (v/v) Grace's insect medium (Invitrogen, Cat. 11605), 30% (v/v) FBS (Invitrogen, Cat. 12483), and 10% (v/v) dimethyl sulfoxide (DMSO). This medium is freshly prepared just prior to use.
9. Cryovial, 1.2 ml (Sarstedt, CryoPure, Cat. 72.377.992; see Note 5).
10. Cryo 1°C freezing container (Nalgene, Cat. 5100–0001), containing 250 ml isopropanol. The container should be at room temperature prior to use.

2.2. Baculovirus Production in Sf9 Insect Cells

1. Sf9 (see Note 2) insect cells (BD Biosciences, Cat. 551407 or Invitrogen Cat. 11496–015).
2. Baculovirus shuttle vector, pAcSG2 (BD Biosciences, Cat. 554769), and its derivatives f-pAcSG2 and pNS24 (Fig. 1) for expression of Flag- and HA-tagged fusion proteins, respectively.
3. Baculovirus shuttle vectors for expression of Flag-H810 (HAT domain of MOZ), Flag-H361 (HAT domain of MORF), HA-BRPF1, HA-ING5, and HA-EAF6: these vectors were derived from f-pAcSG2 and pNS24, and have been described elsewhere (35).
4. 6-Well cell culture plate (BD Biosciences, Cat. 353502).
5. Cell culture dishes (Corning®, 10 cm, Cat. 430293 and 15 cm, Cat. 430599).
6. Linearized baculovirus DNA (BD Biosciences, Cat. 554739).
7. Cellfectin® II reagent (Invitrogen, Cat. 10362–100).
8. Cell scraper 2-posit blade 25 (Sarstedt, Cat. 83.1830).
9. Polystyrene round-bottom tubes (Falcon®, Cat. 352058).

2.3. Production of Multisubunit Complexes

1. Buffer B: 20 mM Tris–HCl, pH 8.0, 10% (w/v) glycerol, 5 mM $MgCl_2$, 0.15 M KCl, 0.1% Nonidet P-40 (NP-40), and protease inhibitors [1 mM phenylmethylsulfonyl fluoride (PMSF), 0.3 mM aprotinin, 1.2 mM leupeptin, and 1.5 mM

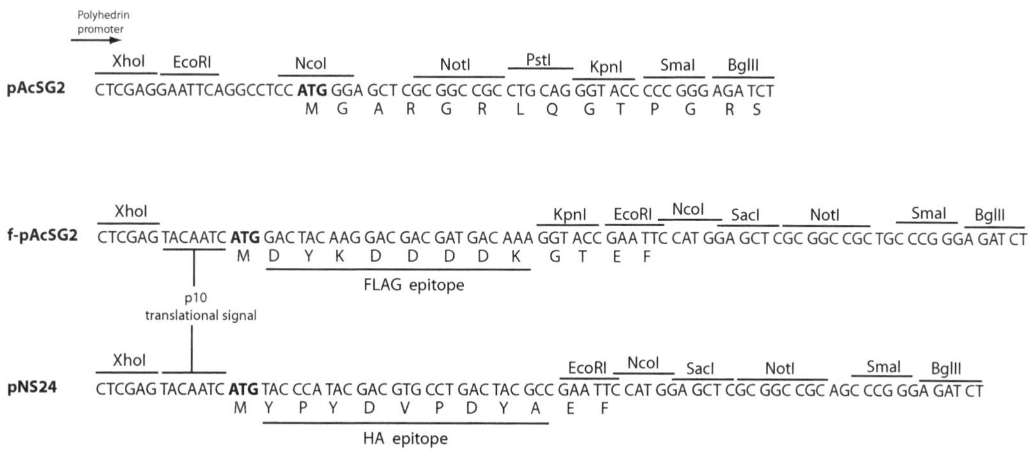

Fig. 1. Polylinker regions of baculovirus shuttle vectors pAcSG2, f-pAcSG2, and pNS24. The latter two were derived from pAcSG2 via insertion of complementary oligonucleotides. The arrow denotes transcription direction of the polyhyderin promoter upstream from the polylinker regions of these three vectors. For efficient translation of mRNA transcribed from the promoter, the sequence TACAATTC was included in the oligonucleotides so that this sequence is between the XhoI restriction site and the ATG translational initiation codon (*bold*). The corresponding RNA sequence is present immediately upstream from the translational initiation codon of mRNA for an abundant insect protein and is thus expected to serve as an efficient translational signal in insect cells. Also illustrated are positions and sequences of the FLAG and HA epitopes. It is important to note that although only Flag and HAT tags are described herein, other affinity tags have also been successfully used. For example, the 6xHis tag and related affinity matrix are more economical, esp when a large amount of proteins is needed (e.g., for structural studies).

pepstatin]. We use freshly prepared buffers: For this, a 2× stock containing the first three components is prepared in autoclaved water and kept at 4°C for fresh preparation of a 50- or 100-ml aliquot on ice just prior to use.

2. Buffer K [20 mM sodium phosphate, pH 7.0, 0.15 M KCl, 30 mM sodium pyrophosphate, 0.1% NP-40, 5 mM EDTA, 10 mM NaF, 0.1 mM Na_3VO_4, 2 μg/ml pepstatin A, 2 μg/ml aprotinin, 2 μg/ml leupeptin, and 1 mM PMSF].
3. Buffer N, same as for buffer K, except that NaCl is used instead of KCl and NP-40 concentration is 0.15%.
4. M2 agarose (anti-Flag® M2 Affinity Gel, Sigma, Cat. A2220, kept at −20°C): Pretreated with 0.1 M glycine–HCl, pH 2.5, for 2–5 min, neutralized with 0.1 M Tris–HCl, pH 7.5–8.0, washed with PBS, and pre-equilibrated with buffer B, K, or N for affinity purification.
5. Flag peptide (DYKDDDDK) (purchased from Sigma, Cat. F3290, or custom synthesized). The peptide solution (4 mg/ml) is prepared by dissolving lyophilized peptide powder in 0.1 M Tris–HCl, pH 8.0, yielding final pH 7.0–8.0. Small aliquots (50–100 μl) are kept frozen at −20°C.

2.4. Characterizing Multisubunit Complexes

1. Colloidal "blue silver" staining solution: 0.12% (w/v) Coomassie blue G-250, 10% (w/v) ammonium sulfate, 10% (v/v) orthophosphoric acid, 20% (v/v) methanol. This solution is freshly prepared as described in reference (36).
2. [^{14}C] acetyl-CoA, 50 μg/ml (55 mCi/mmol; Amersham Biosciences, Cat. CFA729).
3. 5× buffer A: 250 mM Tris–HCl, pH 8.0, 50% glycerol, 5 mM DTT, 0.5 mM EDTA, 5 mM PMSF, and 50 mM sodium butyrate. The isotope is secured and stored at −20°C.
4. Free histones or nucleosomes were prepared as described in reference (36, 37).
5. 3× SDS-PAGE sample buffer: 63 mM Tris–Cl, pH 6.8, 10% (v/v) glycerol, 2% SDS, 2.5% (v/v) β-mercaptoethanol, and 0.25% (w/v) bromophenol blue (stored as 1-ml aliquots at −20°C).
6. Phosphor imager (Amersham Bioscience, Typhoon 8600).

3. Methods

As reported previously, the HAT domains of MOZ and MORF are sufficient for tetrameric complex formation with BRPF1, ING5, and EAF6 (35). To assemble the HAT domain complex, we first prepared shuttle vectors for expressing the Flag-tagged mutant

H810 (MOZ HAT domain) and HA-tagged BRPF1, ING5, and EAF6. As illustrated in Fig. 2, upon coinfection of Sf9 cells with these viruses, the tetrameric complex was assembled in Sf9 cells; extracts were then prepared for affinity purification on anti-Flag affinity gel (or M2 agarose) and after extensive washing, the complex was eluted with the Flag peptide for subsequent determination of subunit stoichiometry by Coomassie blue staining and further analysis of acetyltransferase activity.

In what follows, we divide related procedures into four sections, with the first being about how to maintain Sf9 cell culture, which includes growing up the cells from a frozen stock and freezing them for long-term storage in liquid nitrogen tanks (Subheading 3.1). We then show how to produce baculoviruses, including construction of baculovirus shuttle vectors, production of recombinant baculovirus in Sf9 insect cells, and propagation of baculovirus (Subheading 3.2). Afterward, we present methods for assembling

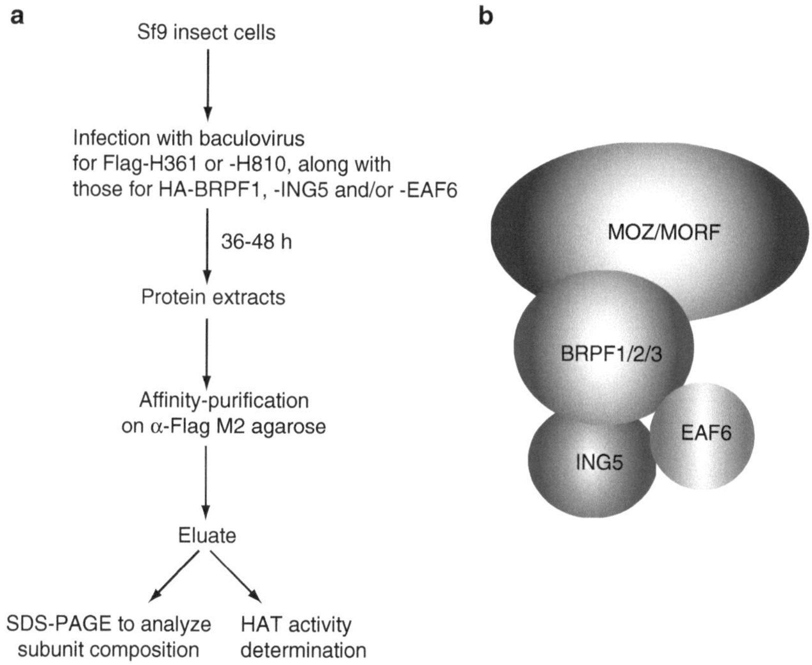

Fig. 2. Reconstitution of MOZ/MORF complexes in Sf9 cells. (**a**) Flow chart for assembling and purifying complexes. Sf9 insect cells are used as the host for coinfection of different combinations of recombinant baculoviruses to assemble multisubunit complexes. Upon coinfection, Flag-tagged H361 (deletion mutant corresponding to the MYST domain from residues 361 to 716 of MORF) or -H810 (mutant corresponding to the MYST domain from residues 507 to 810 of MOZ) forms a complex with HA-tagged BRPF1, which in turn recruits HA-tagged ING5 and EAF6. The complex is affinity purified on M2 agarose and eluted for further analysis of complex composition and determination of acetyltransferase activity. In addition to the HAT domains, we have tried full-length MOZ and MORF proteins. Although they also form stoichiometric complexes with BRPF1, ING5, and EAF6, the expression levels are much lower and much more difficult to prepare homogenous complexes (data not shown). (**b**) Cartoon showing the composition of tetrameric MOZ/MORF complexes. Although Flag/HA tags are described here, other affinity tags can also be used (see discussion in the legend to Fig. 1).

complexes by baculovirus coinfection of Sf9 insect cells and affinity purifying multisubunit complexes (Subheading 3.3). In the last section, we briefly discuss how to analyze affinity-purified complexes (Subheading 3.4).

3.1. Maintaining Sf9 Cell Culture

3.1.1. Starting and Maintaining Adherent Sf9 Cell Culture

1. A frozen Sf9 insect cell stock in a cryovial is taken from a liquid nitrogen tank for thawing in a 37°C water bath. An appropriate way is to hold the vial in an upright position and submerge the vial till the water bath level is close to the top of the frozen cell zone but well below the cap. It is a good idea to shake the vial gently, but under no circumstances the vial is inverted. Such precaution is to minimize potential contamination from the water bath and/or the cap area. Normally, thawing should be done within 1–2 min until ~80% of the frozen material becomes liquid.
2. Once the stock is ~80% liquid (i.e., there is only a small piece of ice left), spray the vial with 75% ethanol to decontaminate the vial surface.
3. Transfer the decontaminated vial to a tissue culture hood and use a sterile 2-ml pipette to transfer the cell suspension into a sterile 15-ml Falcon tube containing ~10 ml of complete Grace's medium (see Note 6). Make sure that the pipette does not touch the opening part of the vial, which could be contaminated from the cap.
4. Centrifuge the Falcon tube at 2,000 × *g* for 5 min and vacuum aspirate the supernatant in the hood to remove medium containing DMSO that is used to protect cells during freezing in many frozen stocks but is problematic for cell growth (see Note 3).
5. Add 2–5 ml of complete Grace's medium, and pipette up and down for ~5 times so that single-cell suspension is achieved.
6. Transfer the cell suspension into a sterile T-75 cell culture flask containing 10 ml complete Grace's medium.
7. Tightly cap the flask and wrap the neck and cap areas with aluminum foil.
8. Incubate the flask in a refrigerated 27°C incubator (see Note 4) and check cell growth every 2 days. Sf9 cells grow as semiattached adherent culture, so it is normal that there are some floating cells in the flask. Viable and healthy cells should be round and shining when examined under a regular phase-contrast microscope.
9. Split the cells when the culture is 80–90% confluent or change the medium whichever comes first.
10. For cell passage, spray the flask with 75% ethanol and hit or tap the flask from its side a few times so that a majority of cells become detached.

11. Transfer the flask to a cell culture hood, pipette the medium inside the flask, and use the stream to wash down the cells. Pipette up and down a few times and if some cells are still attached, use the stream to detach >95% cells. One-tenth of the cell suspension can be used to start a new T-75 flask of adherent culture, and the entire suspension can be used to start 50–100 ml suspension culture in a 250-ml spinner flask.

3.1.2. Maintaining Sf9 Suspension Culture

If large-scale expression is needed, suspension culture in a spinner flask is used, which is much more expensive than adherent culture, so it is important to use it when large-scale expression is needed. Do not use it when only small-scale expression is needed or at the initial stage of baculovirus generation.

1. Take an aliquot from step 11 above and count cells using a hemacytometer.
2. Transfer the cell suspension into a sterile 250-ml spinner flask through one side opening (see Note 7, sterilizing by flame) and add an appropriate amount of complete Grace's medium so that the cell density is around 0.2×10^6 cells/ml and the total volume is between 40 and 100 ml.
3. Stir the spinner flask at 75 rpm in an incubator set at 27°C (see Note 8).
4. Check the cell density every 2 days so that the suspension culture density does not exceed 2×10^6 cells/ml, which takes about 4 days with the doubling time of 22–24 h.
5. To split the culture, vacuum aspirate 90% of the culture, add back a similar amount of fresh medium, and repeat step 1–5.

3.1.3. Freezing Sf9 Cells for Long-Term Storage

Sf9 insect cells stocks are frozen at a density of 1×10^7 cells/ml in freezing medium for storage in liquid nitrogen tanks.

1. Remove 50 ml of mid-log phase suspension culture from the spinner flask and transfer the culture to a 50-ml conical tube.
2. A drop of culture is counted for cell density determination with a hemacytometer. A sterile glass Pastuer pipette can be used to transfer the culture from the flask to the hemacytometer. Alternatively, 10 μl of suspension is applied to the hemacytometer by a P20 Glison pipetman.
3. Cells culture is centrifuged at $2{,}000 \times g$ for 5 min at room temperature for subsequent vacuum aspiration of the supernatant in a tissue culture hood.
4. Add a corresponding volume of freshly prepared freezing medium to adjust the density to 1×10^7 cells/ml according to the original density counted at step 2.
5. Resuspend the pellet gently by using a sterile pipette.
6. Set up and label a desired number of 1.2-ml cryovials.

7. Aliquot 0.5–1 ml of the cell suspension into one cryovial.
8. Put the cryovials in a cryo-freezing container containing 250 ml of isopropanol and put the container in a −80°C freezer. After 4–16 h, transfer the cryovials to a liquid nitrogen tank for long-term storage.

3.2. Generating Recombinant Baculovirus

3.2.1. Construction of Baculovirus Shuttle Vectors

The baculovirus shuttle vectors for expressing Flag-tagged H361 (mutant corresponding to the MYST domain from amino acids 361 to 716 of MORF) and H810 (mutant corresponding to the MYST domain from amino acids 507 to 810 of MOZ) were generated from f-pAcSG2 transfer vector (Fig. 1) (35). To assemble active tetrameric complexes, additional baculovirus shuttle vectors were constructed by insertion of the coding sequences for BRPF1, ING5, and EAF6 proteins into pNS24, a derivative of pAcSG2 used for the expression of HA-tagged proteins (Fig. 1). As standard recombinant DNA techniques were used for preparing these constructs, there is no need to present the details here.

Baculovirus shuttle vectors were then individually transfected into Sf9 cells along with linearized BaculoGold viral DNA, which contains a deletion and does not produce viable virus unless there is homologous recombination with a proper baculoviral shuttle vector. The culture medium of transfected cells contains recombinant baculovirus and is thus collected for further propagation to obtain high-titer viral stocks (see Note 9). In addition to the BaculoGold system used here, the Bac-to-Bac system (formerly Gibco, now Invitrogen) can be used to prepare baculoviruses. Each system requires its own shuttle vectors and the vectors used herein are not compatible with the Bac-to-Bac system.

3.2.2. Generating Recombinant Baculovirus

1. Sf9 insect cells are seeded onto a 6-well plate, at 0.8×10^6 cells per well, yielding 90–95% confluence. Alternatively, 0.5×10^6 cells are seeded per well on the day before transfection (see Note 10).
2. Allow at least 30 min for cells to settle down and attach to the bottom of the plate.
3. Prepare two sterile 5-ml polystyrene, round-bottom tubes (Falcon 352058; unlike other turbid ones, these clear polystyrene tubes bind Cellfectin® minimally). Mix following components in Tube #1: 2 μg baculovirus shuttle vector (see Note 11), 0.5 μl linearized baculovirus DNA (25 ng), 100 μl plain Grace's medium.
4. Add the following to Tube #2: 100 μl plain Grace's medium, 6 μl Cellfectin® II Reagent.
5. Transfer the mixture from Tube #2 to Tube #1 and gently tap the tube several times to mix its content.
6. Incubate the mixture at room temperature for 30 min.

7. Aspirate the medium from the 6-well plate and add 2 ml plain Grace's medium. Aspirate the medium and add fresh 0.4 ml plain Grace's medium.
8. Add 0.4 ml plain Grace's medium to Tube #1 (step 5) and shake the tube to mix the content well.
9. Add the mixture in a dropwise manner onto one well of cells.
10. Incubate the cells at 27°C for 5 h.
11. Aspirate the transfection mixture and add 3 ml of complete Grace's medium.
12. Carefully seal the plate by aluminum foil or parafilm (avoid excessive evaporation over time).
13. Incubate cells at 27°C for 4 days.
14. On day 3, prepare a T-75 flask containing about 5×10^6 cells in 10 ml of complete Grace's medium and incubate it at 27°C.
15. Transfer the viral supernatant from one well (the conditioned culture medium from step 13) to the above flask and add 3 ml fresh complete medium back to the well.
16. Incubate at 27°C for 3 days and transfer the supernatant from the same well along with 4 ml of fresh complete Grace's medium into the T-75 flask from step 15. From steps above, it is clear that multiple viruses can be generated at the same time. For example, one 6-well plate in steps above can be used for generating 6 different viruses; if proper care is taken (e.g., different glass Pastuer pipettes for vacuum aspiration from individual wells), cross contamination between wells is minimal.
17. Discard the 6-well plate and incubate the flask at 27°C for 4 more days.
18. Transfer the culture medium from the flask to a 50-ml Falcon tube, centrifuge at $2,000 \times g$ for 5 min, and transfer the supernatant to a sterile tube. This supernatant is considered as the first generation of recombinant baculovirus or P1 viral stock. The viral stock can be kept dark at 4°C for short-term storage (up to a few months). For long-term storage, keep it as frozen aliquots at −80°C.

3.2.3. Propagating Recombinant Baculovirus

1. Seed Sf9 cells to two 15-cm cell culture dishes at $1.0–1.5 \times 10^7$ cells and with 25 ml complete Grace's medium per dish.
2. Incubate the dishes at 27°C for 1 h and add 1–2 ml P1 viral stock to each dish.
3. Incubate the cells at 27°C for 5–7 days until most cells become floating and granular with irregular shape (instead of being round and shining), which are characteristic signs of infection.
4. Collect the culture media from both dishes and transfer it to a 50-ml tube for centrifugation at $2,000 \times g$ for 5 min.

5. Transfer the supernatant to a new 50-ml conical tube and keep it dark at 4°C.

This is considered as the second generation or P2 viral stock and can be used for protein expression. In some cases, an additional round of amplification is required to produce a P3 viral stock for a high-titer viral stock and a good-expression yield. This can be achieved by replacing the P1 with P2 stock in step 2 and repeating steps 1–5 above.

3.3. Assembling and Purifying Tetrameric MOZ/MORF Complexes

For reconstitution of active MOZ/MORF complexes, baculovirus for Flag-H361 or Flag-H810 was used to infect Sf9 cells along with baculoviruses for HA-tagged BRPF1, ING5, and/or EAF6. After infection, protein extracts are prepared for affinity purification on M2 agarose. HA-tagged proteins that interact with Flag-H361 or Flag-H810 (35) remain bound to M2 agarose (Fig. 2), whereas free HA-tagged proteins are washed away. To obtain stoichiometric complexes, it is important that the HA-tagged proteins are expressed in higher levels than the Flag-tagged proteins. Thus, it is necessary to adjust the relative ratio of different viruses used for coinfection (see Note 12).

1. Seed 8×10^6 cells in 10 ml complete Grace's medium in a 10-cm cell culture dish to achieve 90–95% confluence.
2. Incubate at 27°C for 1 h and add 1 ml Flag-H361 or -H810 baculovirus along with appropriate amounts of baculoviruses for HA-BRPF1, -ING5, and/or -EAF6 (see Note 12). Typically, P2 or P3 stocks are used.
3. Incubate at 27°C for 36–48 h.
4. Scrape down the cells from each dish and transfer the culture to a 15-ml conical tube.
5. Harvest the cells by centrifugation at $2{,}000 \times g$ at 4°C for 5 min.
6. Discard the supernatant and gently resuspend the cell pellet in 10 ml PBS.
7. Repeat steps 5–6.
8. Resuspend the resulting pellet in 0.5 ml cold buffer B, K, or N (see Note 13) on ice.
9. Transfer the suspension to a 1.5-ml Eppendorf tube on ice. *This and all the following purification steps should be carried out either in a cold room or on ice.*
10. Sonicate the mixture on ice for 15 s using a microtip linked to a Virsonic 100 sonicator (setting 5; VirTis).
11. Rotate the tube on a rotator at 4°C or a cold room for 30 min.
12. Centrifuge at $14{,}500 \times g$ at 4°C for 3–5 min.
13. Mix 450 μl of the soluble supernatant to 20 μl pretreated M2 agarose (see Note 14) in a 0.5-ml Eppendorf tube.

14. Rotate the Eppendorf tube on a rotator for 2–4 h at 4°C.
15. Centrifuge the tube at 2,000 × g for 30 s at 4°C.
16. Vacuum aspirate the supernatant and add 300 μl buffer B to M2 agarose.
17. Repeat steps 15–16 three more times.
18. Repeat step 15 and vacuum aspirate the supernatant.
19. Add 30 μl buffer B supplemented with 1.5 μl Flag peptide (4 mg/ml, pH 7.5).
20. Rotate the Eppendorf tube on a rotator for 1–2 h at 4°C.
21. Centrifuge the tube at 2,000 × g for 30 s at 4°C.
22. Carefully pipette out 25 μl and save it as the eluate, which can be flash frozen in 5–10-μl aliquots on dry ice for storage at –80°C.

3.4. Characterizing Multisubunit MOZ/MORF Complexes

3.4.1. Analysis of Purity and Subunit Composition of Purified Complexes

An aliquot of the eluate (2–5 μl) is subject to separation by 10% SDS-PAGE. Then, the gel is stained by colloidal "blue silver" staining solution as described in reference (36). On the same gel, different amounts of BSA are loaded for estimating the concentration of the purified protein.

For example, by using different combinations of P3 baculoviral stock of Flag-H810 or -H361 along with P3 baculoviral stocks of HA-BRPF1, -ING5, and -EAF6 for coinfection of Sf9 cells, we reconstituted tetrameric complexes and their subcomplexes containing two or three subunits (Fig. 3a, b). Colloidal "blue silver" staining showed that HA-BRPF1 copurified with Flag-H361 to an almost stoichiometric level (Fig. 3a, lanes 1 to 2). However, HA-ING5 did not efficiently copurify with Flag-H361 unless HA-BRPF1 was present (Fig. 3a, lanes 3 and 5), indicating that interaction of ING5 with MORF is mediated by BRPF1. HA-EAF6 copurified with Flag-H361 only when both HA-ING5 and HA-BRPF1 were present (Fig. 3a, lanes 4 to 6), supporting that BRPF1 serves as a scaffold. Similar results were obtained with Flag-tagged H810 (Fig. 3b). As insect cells contain many KATs and some of them may bind to M2 agarose nonspecifically, proper negative controls are needed to address this. For this, we infected Sf9 cells with baculoviruses for HA-tagged BRPF1, ING5, and EAF6 and prepared extracts for affinity purification as other samples. Importantly, no detectable complexes or enzymatic activities were found in the negative controls (Fig. 3a, b, lanes 7). Therefore, the reconstitution experiments indicate that BRPF1 plays a scaffolding role in the formation of MOZ and MORF complexes (35).

3.4.2. KAT Activity Determination

KAT activity of the reconstituted complexes was determined according to two published procedures (36). One of them is known as the P81 filter paper assay, which is quick and quantitative and is very easy to perform. The other has been referred to as the gel

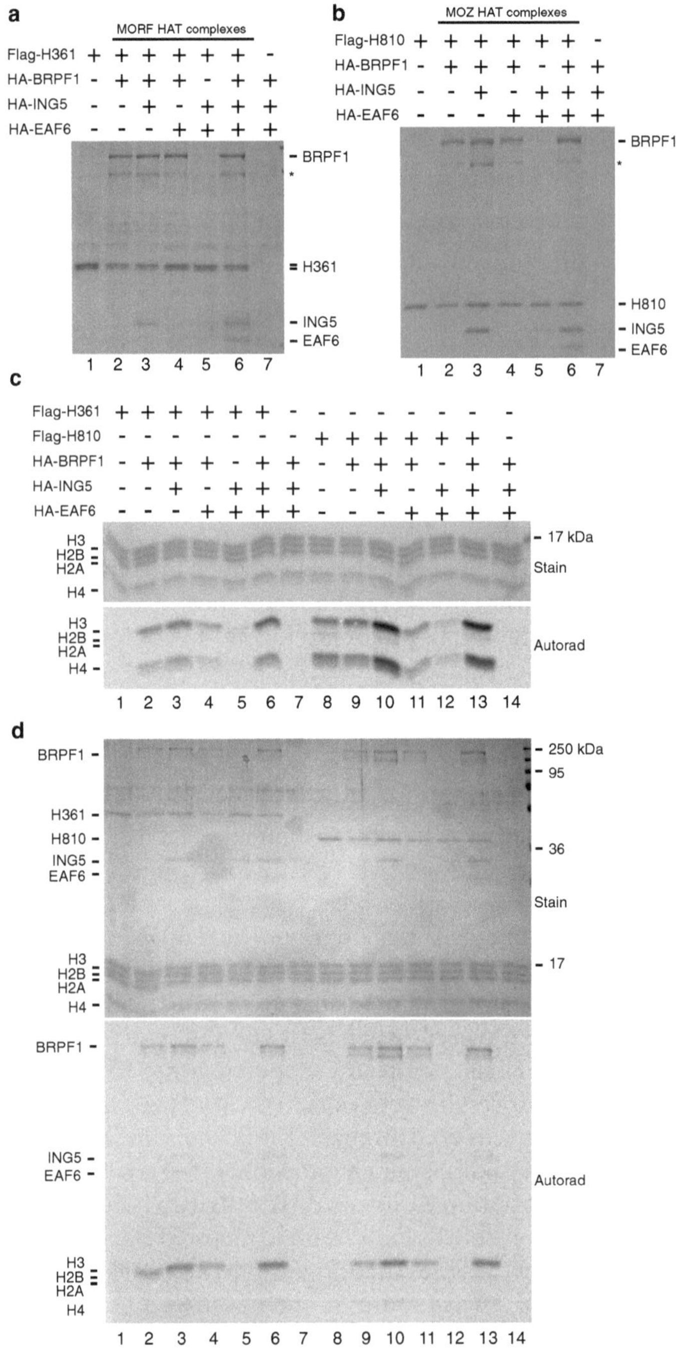

Fig. 3. Purification and analysis of MOZ/MORF complexes. (**a**, **b**) Sf9 cells were infected with baculovirus for Flag-H361 (**a**) or -H810 (**b**) along with baculoviruses for HA-tagged BRPF1, ING5, and/or EAF6 as indicated. Extracts were prepared in buffer B for affinity purification on M2 agarose, and bound proteins were eluted with Flag peptide. Eluted proteins were resolved by SDS-PAGE and detected by colloidal blue-derived "blue silver" staining. (**c**, **d**) KAT activity determination. Complexes reconstituted as in (**a**, **b**) were incubated with HeLa histone octamers (**c**) or oligonucleosomes (**d**) in the presence of [^{14}C]acetyl-CoA. Reaction mixtures were resolved by SDS-PAGE (15%) for "blue sliver" staining (*top*) and dried for autoradiography (*bottom*). Note that ~50 times of the enzymes were used for assays with nucleosomal substrates as compared to that for assays with histone octamers. This figure was adapted from ref. 35, with permission.

assay, which yields detailed information as to which proteins are acetylated and is briefly described as follows.

1. Prepare a 15 μl acetylation reaction mix in 0.5-ml sterile Eppendorf tube on ice: 0.75 μl free histones (2 μg/μl) or nucleosomes (1 μg/μl), 3.0 μl 5× buffer A, 0.2 μl [^{14}C] acetyl-CoA, 10.0 μl autoclaved water, 1 μl KAT (25–200 ng/μl, diluted in an appropriate buffer).
2. Incubate at 30°C for 10 or 20 min. This is a fixed-time enzymatic assay, so it is important to try two concentration of each enzyme and gain information as to whether the assay is within the linear range of the enzymatic reaction. In addition, the length of incubation or reaction time should be strictly controlled so that different samples are comparable.
3. Stop the reaction with 7.5 μl of the 3× SDS-PAGE sample buffer.
4. Boil the protein sample in a water bath for 5 min for subsequent separation by SDS-PAGE. It is important to use 15% gels for histone and nucleosome substrates; lower percentage cannot yield good separation of the four core histones, which are 10–15 kDa in size.
5. Stain the gel with colloidal "blue sliver" staining solution and dry the gel for phosphoimaging. This procedure has been used to analyze MOZ and MORF complexes assembled in Sf9 cells (Fig. 3c, d). In the presence of BRPF1, KAT activity of the MORF MYST domain (H361) toward free histones H3 and H4 increased drastically (Fig. 3c, compare lanes 1 and 5 with lane 2). ING5 also stimulated the KAT activity, but EAF6 had minimal effects (Fig. 3c, lanes 2 and 3 with 4 and 6, respectively). With both free and nucleosomal histones, the MOZ MYST domain (H810) exhibited similar responses to BRPF1 and ING5 (Fig. 3c, lanes 8 to 14). These results indicate that BRPF1 and ING5 activate the MYST domains to acetylate nucleosomal histone H3 and free histones H3 and H4. The assays also revealed that the MYST domain of MOZ or MORF acetylates BRPF1 (Fig. 3d). ING5 appeared to stimulate the acetylation (Fig. 3d, compare lane 3 with lanes 2 and 4; and also lane 10 with lanes 9 and 11). No activity was detected in mock purification from extracts of Sf9 cells infected with BRPF1, ING5, and EAF6 baculoviruses (Fig. 3c, d, lanes 7). Together, these observations indicate that subunit interaction regulates KAT activity of MOZ and MORF.

4. Notes

1. The baculovirus expression system has also been used for analyzing complexes of other chromatin-modifying enzymes, including histone deacetylases, lysine demethylases, and ubiquitin E3 ligase. The deacetylase complex *nu*cleosome-*re*modeling histone *d*eacetylase (NuRD) was one of the first complexes reconstituted in such a manner (38). Its core subunits include *h*istone *deac*etylase 1 (HDAC1), HDAC2, and RbAp46/48. By coinfection of Sf9 cells viruses for these subunits, an active core complex was obtained. Coexpression of other subunits of the complex demonstrated that *met*astasis *a*ssociated 1 family, member 2 (MTA2) promotes complex assembly and enhances deacetylase activity (38). Mammalian *p*olycomb *r*epressive *c*omplex 1 (PRC1) is another complex that has been reconstituted in insect cells (39). This complex plays an important role in histone ubiquitination and Hox gene silencing, and contains five components, B lymphoma Mo-MLV insertion region 1 (Bmi1), Ring1A, Ring1B, polycomb 3 (Pc3), and polyhomeotic-like protein 2 (Ph2), with Ring1B being the catalytic subunit. By coexpression of Bmi1, Ring1A, and Ring1B in insect cells, it was shown that Bmi1 stimulates ubiquitin ligase activity of Ring1B and that incorporation of Ring1A into the Ring1B/Bmi1 subcomplex augments E3 ligase activity (39). Furthermore, lysine demethylase complexes have been reconstituted and studied similarly (40).
2. The Sf9 cell line is a subclone of Sf21 and was selected for fast growth and high cell density. Sf21 itself was originally derived from pupal ovarian tissue of Fall armyworm *Spodoptera frugiperda*. Sf9 cells are highly susceptible to infection with *Autographa california* nuclear polyhedrosis (AcNP) baculovirus and are commonly used to produce and propagate recombinant baculoviral stocks for expression of recombinant proteins.
3. As Sf9 cells grow well at 27°C (just a few degrees above typical air-conditioned room temperature), an incubator with a refrigerating system, such as SANYO MIR-153 and equivalents, offers an ideal growth environment. Different from mammalian cells, no CO_2 is required. Incubators can have certain humidity supported by a beaker of water. An easy way is to put a water-containing 1-liter beaker inside the incubator. On the other hand, Sf9 insect cells grow well as semiattached adherent culture in regular plastic tissue culture dishes/flasks or as suspension culture in glass spinner flasks even when the containers are tightly capped, so the humidity is not an issue.

4. To wash spinner flasks, we do not use any regular detergents as a trace amount might cause problems to proper growth of insect cells. Typically, we hand wash a new or used flask and rinse it with deionized or distilled water extensively. The flask is then filled with double-distilled or NanoPure Water to about one-third of the full level. With the cap of one side arm loosely attached and covered with aluminum foil but the top and the other side opening tightly capped, the flask is autoclaved for 30 min, which not only sterilizes but also cleans up the flask further. An autoclaved flask can be stored at room temperature for a few weeks. To prepare it for suspension insect cell culture, the flask is transferred to a sterile hood for pouring out or vacuum aspirating the water. If a burner is available, the flask outsurface and the sidearm opening from which to aspirate the water and put in culture media are briefly flamed to minimize contamination.
5. We found that liquid nitrogen accidentally leaked into cryovials of some brands after long-term storage in liquid nitrogen tank (storage for several years, presumably due to aging and decay of plastic and rubber materials). This not only caused contamination, but also led to explosion of some vials when they were brought to room temperature or 37°C for thawing. Thus, it is important to use a good brand of vials and to wear protective device for protection of face, especially eyes, when frozen vials are thawed.
6. As cell suspension freezes, ice crystals form, which can puncture the plasma membrane and cause cell death. DMSO protects cells by partially solubilizing the membrane so that it is less prone to puncture and by interrupting the lattice of the ice so that fewer crystals form. But DMSO is toxic to cells as it partially solubilizes the membrane, so when to thaw cells, it is operative to dilute and remove it as soon as possible.
7. To minimize contamination, we intentionally use only one sidearm of an autoclaved flask. Only one side of the spinner flask has the volume scale printed by the manufacturer, so it is a good idea to select one sidearm of spinner flask to take out and put in cells and culture media according to handedness of conductor so that the scale faces the conductor.
8. Before placing a spinner flask back to incubator, the sidearm openings are wrapped with aluminum foil. Cells should be examined every 3–4 days for density and healthy state.
9. Cross contamination between different viruses occurs very easily, so it is important to take high standard of aseptic operation measures. This tends to be more problematic and wastes more time than regular bacterial or yeast contamination as it takes lots of time and energy to figure out when confusing results are obtained due to cross contamination.

10. Once a shuttle vector is made, it takes one day to prepare transfection-ready DNA. It is better to initiate the insect cell transfection procedure early in the morning because the incubation time is 5 h. It takes 2 weeks to obtain virus from the second propagation, 2 days to infect and express the proteins, 2–3 days to immunoprecipitate and analyze proteins to be expressed, and a few days to obtain KAT assay results if gel assays are used. In addition to gel assays described here, KAT activity can be quickly and quantitatively determined by filter assays in just half a day (36).
11. Transfection of Sf9 cells requires high quality of plasmid DNA. Because it takes a long time and many steps to confirm whether a baculovirus is produced and a recombinant protein is expressed, it is very important to make sure that the plasmid is correct by multiple restriction digestion to check both the inert and backbone, and to use high quality of plasmid DNA for transfection. We often use the QIAGEN Plasmid Mini Kit (Tip-20, QIAGEN, Cat. 12123) for plasmid DNA preparation. The 260/280 OD ratio should be close to 1.8.
12. To obtain stoichiometric complexes, it is important that HA-tagged proteins are expressed in higher levels than Flag-tagged proteins. Thus, it is necessary to adjust the relative ratio of different viruses used for coinfection. In theory, it is necessary to use plaque assays to determining the titer of each viral stock and adjust the relative amounts accordingly. Two drawbacks are that plaque assays are lengthy and not so easy to execute and that baculoviruses for different proteins give different expression yields even the same amount of virus is used for infection. For example, ING5 and EAF6 viruses yield much higher expression levels than the BRPF1 virus. Therefore, a practical way is to take a trial-and-error approach to adjust the relative ratio of viruses used for coinfection.
13. Buffers are chosen according to the stability and other properties of complexes to be reconstituted. Buffers K and N are more stringent than buffer B and, thus, yield complexes with better purity. We have used all three buffers for affinity purifying MOZ and MORF complexes. As insect cells contain orthologs of many mammalian KATs and/or their noncatalytic subunits, it is important to have proper controls to ensure that detected activity is not due to association of insect proteins.
14. While the tube is rotating in **step 11**, 20–30 μl of M2 agarose is put into a new 0.5-ml Eppendorf tube. Add 400 μl buffer B, K, or N to the tube and mix it well. Centrifuge at 2,000 × *g* for 10 s, aspirate the supernatant, and M2 agarose resin is then ready for use.

Acknowledgments

We acknowledge that the f-pAcSG2 plasmid was initially constructed when X.J.Y. was a postdoctoral fellow in Dr. Yoshihiro Nakatani's laboratory at National Institutes of Health (NIH), the USA. This work was supported by operating grants from Canadian Cancer Society and Canadian Institutes of Health Research (to X.J.Y.).

References

1. Gershey, E.L., *et al.* (1968) Chemical studies of histone acetylation. The occurrence of epsilon-N-acetyllysine in the f2a1 histone. *J Biol Chem* 243, 5018–5022
2. Vidali, G., *et al.* (1968) Chemical studies of histone acetylation. The distribution of epsilon-N-acetyllysine in calf thymus histones. *J Biol Chem* 243, 6361–6366
3. Kouzarides, T. (2000) Acetylation: a regulatory modification to rival phosphorylation? *EMBO J* 19, 1176–1179
4. Glozak, M.A., *et al.* (2005) Acetylation and deacetylation of non-histone proteins. *Gene* 363, 15–23
5. Kim, G.W., and Yang, X.J. (2010) Comprehensive lysine acetylomes emerging from bacteria to humans. *Trends Biochem Sci* 36, 211–220
6. Kim, S.C., *et al.* (2006) Substrate and functional diversity of lysine acetylation revealed by a proteomics survey. *Mol Cell* 23, 607–618
7. Choudhary, C., *et al.* (2009) Lysine acetylation targets protein complexes and co-regulates major cellular functions. *Science* 325, 834–840
8. Zhao, S., *et al.* (2010) Regulation of cellular metabolism by protein lysine acetylation. *Science* 327, 1000–1004
9. Wang, Q., *et al.* (2010) Acetylation of metabolic enzymes coordinates carbon source utilization and metabolic flux. *Science* 327, 1004–1007
10. Ramakrishnan, R., *et al.* (1998) Acetylation at Lys-92 enhances signaling by the chemotaxis response regulator protein CheY. *Proc Natl Acad Sci U S A* 95, 4918–4923
11. Kleff, S., *et al.* (1995) Identification of a gene encoding a yeast histone H4 acetyltransferase. *J Biol Chem* 270, 24674–24677
12. Brownell, J.E., *et al.* (1996) Tetrahymena histone acetyltransferase A: a homolog to yeast Gcn5p linking histone acetylation to gene activation. *Cell* 84, 843–851
13. Roth, S.Y., *et al.* (2001) Histone acetyltransferases. *Annu Rev Biochem* 70, 81–120
14. Yang, X.J. (2004) The diverse superfamily of lysine acetyltransferases and their roles in leukemia and other diseases. *Nucleic Acids Res* 32, 959–976
15. Lee, K.K., and Workman, J.L. (2007) Histone acetyltransferase complexes: one size doesn't fit all. *Nat Rev Mol Cell Biol* 8, 284–295
16. Parthun, M.R. (2007) Hat1: the emerging cellular roles of a type B histone acetyltransferase. *Oncogene* 26, 5319–5328
17. Poveda, A., *et al.* (2004) Hif1 is a component of yeast histone acetyltransferase B, a complex mainly localized in the nucleus. *J Biol Chem* 279, 16033–16043
18. Ai, X., and Parthun, M.R. (2004) The nuclear Hat1p/Hat2p complex: a molecular link between type B histone acetyltransferases and chromatin assembly. *Mol Cell* 14, 195–205
19. Nagy, Z., and Tora, L. (2007) Distinct GCN5/PCAF-containing complexes function as co-activators and are involved in transcription factor and global histone acetylation. *Oncogene* 26, 5341–5357
20. Baker, S.P., and Grant, P.A. (2007) The SAGA continues: expanding the cellular role of a transcriptional co-activator complex. *Oncogene* 26, 5329–5340
21. Suganuma, T., *et al.* (2008) ATAC is a double histone acetyltransferase complex that stimulates nucleosome sliding. *Nat Struct Mol Biol* 15, 364–372
22. Wang, Y.L., *et al.* (2008) Human ATAC Is a GCN5/PCAF-containing acetylase complex with a novel NC2-like histone fold module that interacts with the TATA-binding protein. *J Biol Chem* 283, 33808–33815
23. Guelman, S., *et al.* (2009) The double-histone-acetyltransferase complex ATAC is essential for mammalian development. *Mol Cell Biol* 29, 1176–1188

24. Samara, N.L., *et al.* (2010) Structural insights into the assembly and function of the SAGA deubiquitinating module. *Science* 328, 1025–1029
25. Barrios, A., *et al.* (2007) Expression and purification of recombinant yeast Ada2/Ada3/Gcn5 and Piccolo NuA4 histone acetyltransferase complexes. *Methods* 41, 271–277
26. Santoso, B., and Kadonaga, J.T. (2006) Reconstitution of chromatin transcription with purified components reveals a chromatin-specific repressive activity of p300. *Nat Struct Mol Biol* 13, 131–139
27. Champagne, N., *et al.* (1999) Identification of a human histone acetyltransferase related to monocytic leukemia zinc finger protein. *J Biol Chem* 274, 28528–28536
28. Yang, X.J., and Ullah, M. (2007) MOZ and MORF, two large MYSTic HATs in normal and cancer stem cells. *Oncogene* 26, 5408–5419
29. Voss, A.K., and Thomas, T. (2009) MYST family histone acetyltransferases take center stage in stem cells and development. *Bioessays* 31, 1050–1061
30. Katsumoto, T., *et al.* (2006) MOZ is essential for maintenance of hematopoietic stem cells. *Genes Dev* 20, 1321–1330
31. Doyon, Y., *et al.* (2006) ING tumor suppressor proteins are critical regulators of chromatin acetylation required for genome expression and perpetuation. *Mol Cell* 21, 51–64
32. Vezzoli, A., *et al.* (2010) Molecular basis of histone H3K36me3 recognition by the PWWP domain of Brpf1. *Nat Struct Mol Biol* 17, 617–619
33. Hibiya, K., *et al.* (2009) Brpf1, a subunit of the MOZ histone acetyl transferase complex, maintains expression of anterior and posterior Hox genes for proper patterning of craniofacial and caudal skeletons. *Dev Biol* 329, 176–190
34. Laue, K., *et al.* (2008) The multidomain protein Brpf1 binds histones and is required for Hox gene expression and segmental identity. *Development* 135, 1935–1946
35. Ullah, M., *et al.* (2008) Molecular architecture of quartet MOZ/MORF histone acetyltransferase complexes. *Mol Cell Biol* 28, 6828–6843
36. Pelletier, N., *et al.* (2008) Analysis of protein lysine acetylation in vitro and in vivo. *Curr Protoc Protein Sci* Chapter 14, Unit 14 11
37. Cote, J., *et al.* (1995) Basic analysis of transcription factor binding to nucleosomes. *Methods Mol. Genet.* 6, 108–128
38. Zhang, Y., *et al.* (1999) Analysis of the NuRD subunits reveals a histone deacetylase core complex and a connection with DNA methylation. *Genes Dev* 13, 1924–1935
39. Cao, R., *et al.* (2005) Role of Bmi-1 and Ring1A in H2A ubiquitylation and Hox gene silencing. *Mol Cell* 20, 845–854
40. Shi, Y.J., *et al.* (2005) Regulation of LSD1 histone demethylase activity by its associated factors. *Mol Cell* 19, 857–864

Chapter 30

Affinity Purification of MLL3/MLL4 Histone H3K4 Methyltransferase Complex

Young-Wook Cho, SunHwa Hong, and Kai Ge

Abstract

Methylation on histone H3 lysine 4 (H3K4) correlates with actively transcribed genes. In mammalian cells, there exist multiple Set1-like histone H3K4 methyltransferase complexes, which have overlapping but distinct subunit compositions. Developing methods to isolate each of these histone H3K4 methyltransferase complexes would help understand the molecular mechanisms by which histone H3K4 methylation regulates mammalian gene expression. In this chapter, we provide a one-step affinity purification protocol on isolation of the MLL3/MLL4 histone H3K4 methyltransferase complex using FLAG-tagged PA1, a unique subunit of the MLL3/MLL4 complex.

Key words: Histone H3K4 methyltransferase complex, MLL3, MLL4, PA1

1. Introduction

Histone lysine methylation plays essential roles in chromatin dynamics, transcription, and DNA repair. Histone lysine methylation is dynamically regulated by site-specific methyltransferases and demethylases. In yeast, a single Set1 complex, also known as COMPASS, is responsible for all methylations on histone H3K4 (1–3). In mammalian cells, at least six Set1-like histone methyltransferase (HMT) complexes with robust H3K4 methyltransferase activities have been isolated (4). Each of these complexes contains one SET domain-containing homolog of yeast Set1, such as Set1A (also known as Setd1a, KMT2F) (5, 6), Set1B (also known as Setd1b, KMT2G) (7), MLL1 (mixed-lineage leukemia 1, also known as MLL, HRX, ALL1, KMT2A) (8–10), MLL2 (mixed-lineage leukemia 2, also known as TRX2, MLL4, KMT2B) (10, 11),

Ales Vancura (ed.), *Transcriptional Regulation: Methods and Protocols*, Methods in Molecular Biology, vol. 809,
DOI 10.1007/978-1-61779-376-9_30,

	Human Set1-like Complexes			
Yeast Set1 Complex	Set1A/B Complex	MLL1/MLL2 Complex	MLL3/MLL4 Complex	
Set1	Set1A / B	MLL1 / MLL2	MLL3 / MLL4	Enzymatic subunit
Bre2	ASH2L	ASH2L	ASH2L	Common subunits form a subcomplex
Swd1	RbBP5	RbBP5	RbBP5	
Swd3	WDR5	WDR5	WDR5	
Sdc1	DPY30	DPY30	DPY30	
Swd2	WDR82	-	-	Distinct subunits
Spp1	CXXC1	-	-	
		Menin	PTIP & PA1	
		HCF1	NCOA6	
			UTX	

Fig. 1. Subunit compositions of the yeast Set1 and human Set1-like histone H3K4 methyltransferase complexes.

MLL3 (mixed-lineage leukemia 3, also known as KMT2C), and MLL4 (mixed-lineage leukemia 4, also known as ALR, MLL2, KMT2D) (4, 12–14), which carries the enzymatic activity for the associated complex. Based on the homologies in both protein sequences and domain structures, the six Set1-like HMTs fall into three subgroups, Set1A and Set1B, MLL1 and MLL2, and MLL3 and MLL4. ASH2L, RbBP5, WDR5, and DPY30, which are homologs of yeast Set1/COMPASS complex subunits Bre2, Swd1, Swd3, and Sdc1, respectively, form a 4-subunit subcomplex that is not only shared by all mammalian Set1-like HMT complexes, but also critical for the H3K4 methyltransferase activities of these complexes (4, 8). In addition, each of these complexes contains distinct but overlapping subunits (Fig. 1) (10, 11). For example, WDR82 and CXXC1, which are homologs of the Swd2 and Spp1 subunits of yeast Set1/COMPASS complex, selectively associate with Set1A/B complexes (5). Menin, a protein with no homology with any of the yeast Set1/COMPASS complex components, selectively associates with MLL1 and MLL2 complexes (10, 11).

In cells, PTIP and a novel protein PA1 are both unique subunits of the MLL3/MLL4 histone H3K4 methyltransferase complex that contains the enzymatic subunits MLL3 and MLL4, and the histone H3K27 demethylase UTX (4, 15). Methylation on H3K4 is an activating epigenetic mark while methylation on H3K27 is a repressive one. The finding that H3K4 methyltransferases MLL3/MLL4 physically associate with H3K27 demethylase UTX suggests that by adding an activating epigenetic mark and removing a repressive one, the MLL3/MLL4 complex may use two distinct histone-modifying activities to synergistically activate target gene expression. Recent evidence also suggests that MLL3 and MLL4 may exist in the HMT complex in a mutually exclusive manner (16). Here, we describe the one-step isolation of the MLL3/MLL4 histone H3K4 methyltransferase complex from nuclear extracts prepared from a HeLaS cell line stably expressing FLAG-tagged PA1(4).

2. Materials

2.1. Generation of Stable Cell Lines Using Retrovirus Vector

All cells are routinely cultured in Dulbecco's modified Eagle's medium (DMEM) supplemented with 10% fetal bovine serum (FBS).

1. Phoenix-Ampho retrovirus packaging cell line (ATCC product # SD 3443) and HeLaS cell line.
2. GenJet™ in vitro DNA transfection reagent (SignaGen Laboratories).
3. Bovine serum (for large-scale cell culture only).
4. Retroviral infection medium (RIM) (DMEM + 10% heat-inactivated FBS).
5. Sterile 0.45-μm syringe filter (Millipore).
6. Polybrene (Sigma H9268).
7. G418 (Invitrogen).
8. 1× PBS.
9. Anti-FLAG M2 antibody (Sigma F3165).

2.2. Isolation of MLL3/MLL4 Complex by Anti-FLAG Antibody Immunoprecipitation

1. Retrovirus plasmid pWZLneo-F-PA1 expressing FLAG-tagged PA1 (available from the authors upon request).
2. HeLaS cells expressing FLAG-tagged PA1 (available from the authors upon request).
3. 15- and 50-ml conical tubes.
4. 200-μl and 1-ml wide-orifice tips (see Note 1).
5. Mouse IgG-agarose (Sigma A0919).
6. M2 agarose (anti-FLAG antibody conjugated to agarose, Sigma A2220).
7. 5 mg/ml FLAG Peptide (Sigma F3290).
8. Buffer A: 20 mM HEPES, pH 7.9, 180 mM KCl, 0.2 mM EGTA, 1.5 mM $MgCl_2$, 20% (v/v) glycerol, 0.1% (v/v) Nonidet P-40.
9. Elution buffer: 20 mM HEPES, pH 7.9, 180 mM KCl, 0.2 mM EGTA, 1.5 mM $MgCl_2$.

10. Ultrafree-MC centrifugal filter units with microporous membrane 0.45 µm (Millipore).
11. Vivaspin 500 centrifugal filter units, MWCF 10 kDa (Fisher NC9924359).
12. 4–15% SDS-PAGE gel (Bio-Rad).
13. Protease inhibitors: Aprotinin, leupeptin, and pepstatin (Roche).
14. Dithiothreitol (DTT, Bio-Rad, stock solution 1 M).
15. PMSF (Sigma, stock solution 200 mM).

3. Methods

PTIP and PA1 both associate with the MLL3/MLL4 histone H3K4 methyltransferase complex (4). However, ectopically expressed FLAG-tagged PTIP associates not only with the MLL3/MLL4 complex, but also with proteins involved in DNA damage response and repair. In contrast, FLAG-tagged PA1 selectively associates with the MLL3/MLL4 complex (4). Using retrovirus-mediated gene transfer, a HeLaS cell line stably expressing FLAG-tagged, full-length, human PA1 (F-PA1) is established. The MLL3/MLL4 complex can be purified in one step using anti-FLAG antibody immunoprecipitation from nuclear extracts prepared from this cell line.

3.1. Generation of Stable Cell Lines Using Retrovirus Vector

Day 1:

1. In the afternoon, plate 1×10^6 Phoenix-Ampho amphotropic retrovirus packaging cells in 4 ml of culture medium in a 6-cm dish. Incubate cells in a 37°C incubator for 24 h.

Day 2:

2. In the afternoon, dilute 2.5 µg pWZLneo-F-PA1 plasmid in 100 µl of plain DMEM without serum and antibiotics. Dilute 7.5 µl of GenJet™ transfection reagent in 100 µl of plain DMEM. Mix the two solutions and incubate at room temperature for 15 min. Add the 200 µl of mixture to the dish of cells from step 1. Incubate cells in a 37°C incubator for 2 days.

Day 4:

3. In the afternoon, replace the medium with 4 ml of RIM. Plate 3.5×10^5 HeLaS cells in another 6-cm dish.

Day 5:

4. In the morning, collect the 4 ml of retrovirus-containing supernatant, filter through 0.45-µm syringe filter, dilute with an equal volume of fresh RIM, and add 8 µg/ml polybrene.

5. Replace the supernatant of the HeLaS cells with 3–4 ml of the diluted virus-containing supernatant. Incubate cells at 37°C for 2 days.

Day 7:

6. Trypsinze the infected HeLaS cells in the 6-cm dish, transfer 1/5, 1/25, 1/125, and 1/625, respectively, into four 15-cm dishes (i.e., 1:35, 1:175, 1:875, and 1:4,375 dilutions). Add 1 mg/ml G418. Every 3 days, change to fresh medium supplemented with 1 mg/ml G418.

Days 21–30:

7. 2 weeks since the splitting of the infected HeLaS cells, pick 24 well-isolated single colonies using 200-μl pipette tips (see Note 2). Transfer each colony into one well of a 24-well plate filled with 1 ml of culture medium containing 0.5 mg/ml G418. Pipette up and down with 1-ml tips to disperse the cells.
8. Choose 12–18 fast-growing colonies in the 24-well plate. When cells reach confluency, trypsinize and split cells in one well of the 24-well plate into two wells of 6-well plates (one for freezing down and the other for western blotting). Add 0.5 mg/ml G418.
9. After cells reach confluency in 6-well plates, trypsinize cells from one well and freeze down in −80°C freezer (see Note 3). Collect cells from the second well and prepare whole cell extracts for western blot analysis using anti-FLAG M2 antibody. The extracts from the parental HeLaS cells serve as the negative control in western blot. PA1 runs at ~42 kDa on SDS-PAGE gels.

3.2. Isolation of MLL3/MLL4 Complex by Anti-FLAG Antibody Immunoprecipitation

After identifying HeLaS cell lines expressing FLAG-tagged PA1 (HeLaS/F-PA1) by western blot, the HeLaS/F-PA1 cells are cultured till confluence in 100 × 15-cm dishes in DMEM containing 8% bovine serum, 2% FBS, and 0.1 mg/ml G418. Cells are collected, and nuclear extracts (N.E.) are prepared exactly as described (17).

Next, we describe one-step purification of the MLL3/MLL4 complex by immunoprecipitation with anti-FLAG antibody conjugated to agarose (M2 agarose) from N.E. prepared from HeLaS/F-PA1 cells. All steps described below are performed in cold room or on ice. All buffers are freshly supplemented with 1 mM DTT and protease inhibitors 0.5 mM PMSF, 1 μg/ml aprotinin, 2 μg/ml leupeptin, and 0.7 μg/ml pepstatin.

1. Dilute 100 mg of N.E. protein in buffer A to get a final protein concentration of 2–3 mg/ml. Centrifuge at 25,000 × *g* for 30 min at 4°C to remove denatured proteins and cell debris. Transfer the supernatant to a 50-ml conical tube.

2. To remove proteins that nonspecifically bind to M2 agarose from the diluted N.E., the supernatant from step 1 is precleared by adding 0.2 ml of mouse IgG–agarose that has been washed twice with 1 ml of buffer A. Rotate the tube for 2 h.
3. Centrifuge at 1,500 × *g* for 5 min in a swinging bucket rotor, and transfer the supernatant into a new 50-ml conical tube (see Note 4).
4. Repeat steps 2 and 3 once.
5. While doing steps 2–4, equilibrate 0.2 ml of anti-FLAG M2 agarose by washing three times with 1 ml of buffer A for 5 min.
6. Incubate precleared N.E. from step 4 with 0.2 ml of anti-FLAG M2 agarose for overnight immunoprecipitation. Rotate the tube in the cold room.
7. The next morning, centrifuge the tube at 1,500 × *g* for 10 min in a swinging bucket rotor. Transfer the supernatant to a new tube. Keep the agarose beads.
8. Wash the agarose beads by adding 40 ml of buffer A, rotate for 30 min, spin the tube at 1,500 × *g* for 5 min, and remove the supernatant.
9. Add 10 ml of buffer A to the agarose beads, resuspend, and transfer to a 15-ml conical tube. Rotate for 5 min. Centrifuge the tube at 1,500 × *g* for 2 min and remove the supernatant. Repeat the wash three more times.
10. Add 1 ml of buffer A to the agarose beads, resuspend, and transfer to a 1.5-ml tube. Rotate for 2 min. Spin at 1,500 × *g* in a microcentrifuge for 2 min and remove the supernatant. Repeat the wash three more times. After the final wash, spin down the agarose beads and completely remove the supernatant with a 27-G needle.
11. To elute proteins bound to the M2 agarose, dilute 5 mg/ml FLAG peptide solution 20-fold in 1 ml of elution buffer to get a final concentration of 0.25 mg/ml.
12. Elution: Add 0.5 ml of diluted FLAG peptide to the agarose beads after the final wash at step 10. Rotate for 30 min, spin at 1,500 × *g* in a microcentrifuge for 2 min, and transfer the supernatant to a new 1.5-ml tube. Add 0.5 ml of diluted FLAG peptide to repeat the elution once.
13. Combine the first and second eluates and filter through an Ultrafree-MC centrifugal filter unit by a quick spin to completely remove agarose beads.
14. Concentrate with a Vivaspin 500 (10-kDa cutoff) by centrifuging at 15,000 × *g* in a microcentrifuge for 4× 5 min until the volume is less than 100 μl.

15. Run the concentrated sample on a 4–15% SDS-PAGE gel and analyze by western blot or mass spectrometry. Alternatively, the purified MLL3/MLL4 complex is subjected to HMT assays.

4. Notes

1. Wide-orifice tips have a larger opening to allow the pipetting of viscous solutions. Wide-orifice tips can be homemade by simply cutting off the end of regulator tips with a razor blade. Use wide-orifice tips to pipette antibody-conjugated agarose.
2. To pick single HeLaS cell colonies, first identify well-isolated single colonies under the microscope and draw circles around the single colonies by labeling at the bottom of the 15-cm dishes. Prepare a 24-well cell culture plate filled with 1 ml of culture medium per well. Remove the majority of the culture medium from the 15-cm dishes. In the tissue culture hood, quickly pick single HeLaS cell colonies using 200-μl pipette tips in a way similar to picking bacteria colonies from agar plates.
3. One day before freezing down cells, remove G418-containing medium and change to fresh culture medium without G418.
4. After centrifugation, the pellet is loose in 50-ml conical tubes. Do not disturb the pellet when transferring the supernatant.

References

1. Roguev, A., Schaft, D., Shevchenko, A., Pijnappel, W. W., Wilm, M., Aasland, R. & Stewart, A. F. (2001). The *Saccharomyces cerevisiae* Set1 complex includes an Ash2 homologue and methylates histone 3 lysine 4. *EMBO J* **20**, 7137–48.
2. Briggs, S. D., Bryk, M., Strahl, B. D., Cheung, W. L., Davie, J. K., Dent, S. Y., Winston, F. & Allis, C. D. (2001). Histone H3 lysine 4 methylation is mediated by Set1 and required for cell growth and rDNA silencing in *Saccharomyces cerevisiae*. *Genes Dev* **15**, 3286–95.
3. Miller, T., Krogan, N. J., Dover, J., Erdjument-Bromage, H., Tempst, P., Johnston, M., Greenblatt, J. F. & Shilatifard, A. (2001). COMPASS: a complex of proteins associated with a trithorax-related SET domain protein. *Proc Natl Acad Sci USA* **98**, 12902–7.
4. Cho, Y.-W., Hong, T., Hong, S., Guo, H., Yu, H., Kim, D., Guszczynski, T., Dressler, G. R., Copeland, T. D., Kalkum, M. & Ge, K. (2007). PTIP Associates with MLL3- and MLL4-containing Histone H3 Lysine 4 Methyltransferase Complex. *J. Biol. Chem.* **282**, 20395–20406.
5. Lee, J. H. & Skalnik, D. G. (2005). CpG-binding protein (CXXC finger protein 1) is a component of the mammalian Set1 histone H3-Lys4 methyltransferase complex, the analogue of the yeast Set1/COMPASS complex. *J Biol Chem* **280**, 41725–31.
6. Wysocka, J., Myers, M. P., Laherty, C. D., Eisenman, R. N. & Herr, W. (2003). Human Sin3 deacetylase and trithorax-related Set1/Ash2 histone H3-K4 methyltransferase are tethered together selectively by the cell-proliferation factor HCF-1. *Genes Dev* **17**, 896–911.
7. Lee, J. H., Tate, C. M., You, J. S. & Skalnik, D. G. (2007). Identification and characterization of the human Set1B histone H3-Lys4

methyltransferase complex. *J Biol Chem* **282**, 13419–28.

8. Dou, Y., Milne, T. A., Ruthenburg, A. J., Lee, S., Lee, J. W., Verdine, G. L., Allis, C. D. & Roeder, R. G. (2006). Regulation of MLL1 H3K4 methyltransferase activity by its core components. *Nat Struct Mol Biol* **13**, 713–9.
9. Milne, T. A., Briggs, S. D., Brock, H. W., Martin, M. E., Gibbs, D., Allis, C. D. & Hess, J. L. (2002). MLL targets SET domain methyltransferase activity to Hox gene promoters. *Mol Cell* **10**, 1107–17.
10. Yokoyama, A., Wang, Z., Wysocka, J., Sanyal, M., Aufiero, D. J., Kitabayashi, I., Herr, W. & Cleary, M. L. (2004). Leukemia proto-oncoprotein MLL forms a SET1-like histone methyltransferase complex with menin to regulate Hox gene expression. *Mol Cell Biol* **24**, 5639–49.
11. Hughes, C. M., Rozenblatt-Rosen, O., Milne, T. A., Copeland, T. D., Levine, S. S., Lee, J. C., Hayes, D. N., Shanmugam, K. S., Bhattacharjee, A., Biondi, C. A., Kay, G. F., Hayward, N. K., Hess, J. L. & Meyerson, M. (2004). Menin associates with a trithorax family histone methyltransferase complex and with the hoxc8 locus. *Mol Cell* **13**, 587–97.
12. Issaeva, I., Zonis, Y., Rozovskaia, T., Orlovsky, K., Croce, C. M., Nakamura, T., Mazo, A., Eisenbach, L. & Canaani, E. (2007). Knockdown of ALR (MLL2) reveals ALR target genes and leads to alterations in cell adhesion and growth. *Mol Cell Biol* **27**, 1889–903.
13. Patel, S. R., Kim, D., Levitan, I. & Dressler, G. R. (2007). The BRCT-Domain Containing Protein PTIP Links PAX2 to a Histone H3, Lysine 4 Methyltransferase Complex. *Developmental Cell* **13**, 580.
14. Goo, Y. H., Sohn, Y. C., Kim, D. H., Kim, S. W., Kang, M. J., Jung, D. J., Kwak, E., Barlev, N. A., Berger, S. L., Chow, V. T., Roeder, R. G., Azorsa, D. O., Meltzer, P. S., Suh, P. G., Song, E. J., Lee, K. J., Lee, Y. C. & Lee, J. W. (2003). Activating signal cointegrator 2 belongs to a novel steady-state complex that contains a subset of trithorax group proteins. *Mol Cell Biol* **23**, 140–9.
15. Hong, S., Cho, Y.-W., Yu, L.-R., Yu, H., Veenstra, T. D. & Ge, K. (2007). Identification of JmjC domain-containing UTX and JMJD3 as histone H3 lysine 27 demethylases. *Proc Natl Acad Sci USA* **104**, 18439–18444.
16. Lee, S., Lee, D.-K., Dou, Y., Lee, J., Lee, B., Kwak, E., Kong, Y.-Y., Lee, S.-K., Roeder, R. G. & Lee, J. W. (2006). Coactivator as a target gene specificity determinant for histone H3 lysine 4 methyltransferases. *Proc Natl Acad Sci U S A* **103**, 15392–15397.
17. Abmayr, S. M., Yao, T., Parmely, T. & Workman, J. L. (2006). Preparation of nuclear and cytoplasmic extracts from mammalian cells. *Curr Protoc Mol Biol* **Chapter 12**, Unit 12 1.

Chapter 31

Methods for Analyzing Histone Citrullination in Chromatin Structure and Gene Regulation

Pingxin Li, Jing Hu, and Yanming Wang

Abstract

Histone posttranslational modifications play significant roles in regulating chromatin structure and gene expression. One of the histone modifications, histone citrullination, is catalyzed by an enzyme called peptidylarginine deiminase 4 (PAD4, also called PADI4), which converts both histone arginine (Arg) and mono-methyl arginine residues to citrulline. Recent studies have found that histone citrullination counteracts the effect of histone arginine methylation and functions as a repressive marker to turn off gene expression. Here, we describe assays to study histone citrullination by PAD4 in vitro and in vivo. We also describe approaches to measure histone citrullination levels at gene promoters using chromatin immunoprecipitation assay and analyze the effects of PAD4 inhibitor on cell cycle and apoptosis by flow cytometry. These methods would be useful techniques to study this unique histone modification.

Key words: Histone citrullination, Histone arginine methylation, Peptidylarginine deiminase 4, Nucleosome, Chromatin immunoprecipitation

1. Introduction

In eukaryotic cells, 147 bp of DNA is wrapped around a histone octamer to form the nucleosome core particle structure, which is the fundamental repeating unit of chromatin to store the genetic information (1–3). Each histone octamer is composed of a tetramer of two H3–H4 dimers and two individual H2A–H2B dimers (2, 4). Upon binding of the linker histone H1, nucleosomes are further packed into higher order chromatin structures (1, 5). It is believed that chromatin condensation and nucleosomal structures restrain the access of multiple DNA-and histone-binding factors to their cognate sites (6, 7). Therefore, in order to efficiently carry out DNA-involved metabolic events, such as transcription, DNA damage

Ales Vancura (ed.), *Transcriptional Regulation: Methods and Protocols*, Methods in Molecular Biology, vol. 809,
DOI 10.1007/978-1-61779-376-9_31, © Springer Science+Business Media, LLC 2012

repair, DNA recombination, and chromosome rearrangements, the regulation of histone–DNA interaction and the condensation/decondensation of chromatin structure become very essential (8–10). So far, several mechanisms have been discovered to regulate chromatin structure, including chromatin remodeling, histone modifications, and histone variant incorporation (8, 9, 11).

Nucleosomal histones contain a middle globular domain through which they interact with each other to form the histone octamer and a flexible N- or C-terminal tail that protrudes from the surface of the nucleosome to interact with other protein factors or neighbor nucleosomes (1). These unstructured histone tails are rich in basic residues and subjected to various kinds of posttranslational modifications, including methylation, acetylation, phosphorylation, and citrullination (12–15).

Arginine (Arg) methylation is a common modification on histone tails, which is catalyzed by one class of methylating enzymes named protein arginine methyltransferases (PRMTs). Arg methylation is involved in multiple cellular processes, including protein transportation, signal transduction, and transcriptional regulation (16, 17). Histones are common substrates for PRMTs, and histone Arg methylation catalyzed by different PRMTs could either activate or repress transcription (18–20).

Histone modification enzymes usually work in pairs to counteract each other's effect, such as histone acetyltransferases (HATs) and histone deacetylases (HDACs), kinases and phosphatases (13, 21). For a quite long time, Arg methylation was considered stable modifications because the turnover of methylation in chromatin is relatively slow (22). Recently, JmjC domain-containing 6 (JMJD6) protein has been suggested as a histone H3R2 and H4R3 demethylase (23). However, another report described JMJD6 as a lysyl hydroxylase that functions in the regulation of RNA splicing instead of Arg demethylation (24). In addition, PAD4 was identified in searching for enzymes that can revert Arg methylation. In particular, PAD4 was found to be able to convert monomethyl-Arg to citrulline (Cit) through a biochemical reaction termed demethylimination (15, 25).

PAD4 is one of the five isotypes of PAD family proteins. Except for PAD6, PAD1–4 are able to catalyze the conversion of protein-bound Arg to citrulline in a calcium-dependent manner (26). Citrulline is a nonconventional amino acid, which can be produced only by posttranslational modifications since there is no citrullinyl-tRNA available. PAD4 is detected mainly in white blood cells, such as granulocytes and monocytes, and is the only PAD enzyme localized in nucleus. Its targets include nuclear proteins, such as histones and nucleophosmin (27, 28).

Citrullination neutralizes the net positive charge on arginine residue, which as a result affects the intramolecular and intermolecular interactions (29). Numerous studies reveal that citrullination plays a significant role in the structures and functions of the target proteins, and is related to the etiology of several autoimmnune diseases, including rheumatoid arthritis (RA) related to PAD4. PAD4 attracts more and more attention recently because of its ability to use histones as enzymatic substrate to regulate transcription (30). PAD4 deiminates histones H2A, H3, and H4 at their N-terminal tails (28). Several studies have shown that PAD4 could serve as a transcriptional corepressor to inhibit transcription under the control of estrogen-responsive pS2 promoter by converting methyl Arg to Cit (15, 25) and PAD4 is also overexpressed in various cancer tissues (31, 32). We found that PAD4 expresses at comparable levels in several human cancer cell lines (Fig. 1a). The PAD4 inhibitor, Cl-amidine (Fig. 1b), was generated recently (33). Cl-amidine is structurally similar to the PAD4 substrate peptidylarginine (Fig. 1c), and inhibits PAD4 by covalent modification of a cysteine residue (Cys645) at the active site of the enzyme (33).

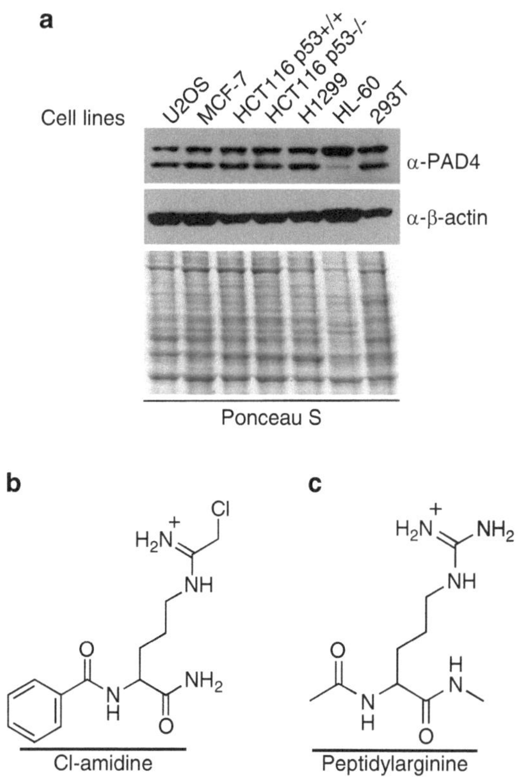

Fig. 1. PAD4 protein levels in different human cancer cell lines and PAD4 inhibitor Cl-amidine. (**a**) Western blot detected high and comparable levels of PAD4 protein expressed in different human cancer cell lines. (**b**, **c**) The structure of Cl-amidine (**b**) is similar to the PAD4 substrate peptidylarginine (**c**).

In order to study the mechanisms of carcinogenesis induced by PAD4 deregulation, we investigated the role of PAD4 and PAD4 catalyzed histone citrullination in the transcriptional repression of p53 target genes. Using one of the p53 target genes p21/CIP1/WAF1 as a model gene, we found that PAD4 is recruited to the p21 gene promoter in a p53-dependent manner. PAD4 association and the RNA Pol II activity at the p21 promoter are dynamically regulated after UV irradiation (34). We also found that PAD4 and histone citrullination coordinate with HDAC2, a histone lysine deacetylase, in repressing tumor-suppressor gene expression. PAD4 inhibitor Cl-amidine and HDAC inhibitor SAHA induced p53 target gene expression and inhibited cancer cell growth additively in a p53-dependent manner (35). In addition, we found that histone citrullination induces chromatin decondensation in both human leukemia HL-60 cells and mouse neutrophils (36). In the following sections, we describe the experimental procedures for analyzing in vitro and in vivo histone citrullination. We also describe the procedures to study in vivo PAD4 association and histone arginine modifications at individual gene promoter, as well as the effects of PAD4 inhibitor on cell cycle and apoptosis using flow cytometry method.

2. Materials

2.1. In Vitro Histone Citrullination Assay

1. Substrates: Recombinant histone H3, and mononucleosome prepared from Hela cells (37).
2. Reaction buffer: Prepare 10× PAD buffer from stock solutions right before use: 500 mM Tris–HCl, pH 7.6, 40 mM $CaCl_2$, 40 mM DTT, 10 mM PMSF.
3. Other reagents: Recombinant PAD4 protein purified from *E. coli* (34), anti-H3Cit antibody (Abcam, ab5103), anti-H3 antibody (Abcam, ab1791).
4. Equipment: SDS-PAGE apparatus, Semi-dry transfer unit (GE healthcare Life Sciences) and power supply.

2.2. In Vivo Histone Citrullination Assay

1. Cells: Human leukemia HL-60 cells and mouse neutrophils. HL-60 cells are cultured in RPMI 1640 Medium (Invitrogen) supplemented with 10% fetal bovine serum and 1% penicillin–streptomycin in a 37°C incubator with 5% CO_2. Mouse neutrophils are purified from mouse peripheral blood.
2. Buffers:

   Locke's solution: 10 mM HEPES–HCl, pH 7.3, 150 mM NaCl, 5 mM KCl, 2 mM $CaCl_2$, 0.18% D-Glucose.

   IP buffer: 20 mM Tris–HCl, pH 8.0, 150 mM NaCl, 10 mM EDTA, 0.2% Triton X-100, 0.2% NP-40. Add protease inhibitors before use: 1 mM PMSF, 10 nM aprotinin, 10 μM leupeptin, 1 μg/ml pepstatin A.

Fixation solution: 1× PBS supplemented with 3.7% paraformaldehyde (prepared from 5× PFA: 4 ml ddH_2O, 0.74 g paraformaldehyde, 28 μl 1 M KOH, heat to 65°C to dissolve), 0.1% Triton X-100, and 0.2% NP-40.

Mounting solution: 2% *N*-proply gallate in 10 mM Tris–HCl, pH 9.0, and 80% glycerol.

3. Other reagents: Anti-H3Cit antibody (Abcam, ab5103), anti-H3 antibody (Abcam, ab1791), goat anti rabbit Cy3 antibody, DNA dye Hoechst (1 mg/ml), Histopaque-1119 (Sigma, 11191), Histopaque-1077 (Sigma, 10771), lipopolysaccharides (Sigma, L2143), H_2O_2 (Sigma, H3410), phorbol myristate acetate (Sigma, P1585), and PAD4 inhibitor Cl-amidine (Custom made).
4. Equipment: SDS-PAGE apparatus, semi-dry transfer unit (GE Healthcare Life Sciences), power supply, Bioruptor (Diagenode Inc.), and fluorescence microscope (Zeiss).

2.3. Chromatin Immunoprecipitation Assay to Measure Histone Citrullination at Gene Promoters

1. Cells: U2OS cells are cultured in Dulbecco Modified Eagle Medium-High Glucose (Invitrogen) supplemented with 10% fetal bovine serum and 1% penicillin–streptomycin in a 37°C incubator with 5% CO_2.
2. Buffers:

Protease inhibitor cocktail: 1 mM PMSF, 10 nM aprotinin, 10 μM leupeptin, 1 μg/ml pepstatin A.

SDS-lysis buffer: 100 mM NaCl, 50 mM Tris–HCl, pH 8.0, 5 mM EDTA, 0.4% SDS, supplemented with protease inhibitor cocktail (final concentrations: 1 mM PMSF, 10 nM aprotinin, 10 μM leupeptin, 1 μg/ml pepstatin A).

HB buffer: 10 mM Tris–HCl, pH 7.3, 10 mM KCl, 1.5 mM $MgCl_2$, 1% Triton X-100, 1 mM DTT, supplemented with protease inhibitor cocktail (final concentrations: 1 mM PMSF, 10 nM aprotinin, 10 μM leupeptin, 1 μg/ml pepstatin A).

NP-S buffer: 10 mM Tris–HCl, pH 7.5, 50 mM NaCl, 5 mM $MgCl_2$, 1 mM $CaCl_2$, 0.5 mM DTT, 0.1 mM PMSF.

IP buffer: 100 mM NaCl, 67 mM Tris–HCl, pH 8.0, 5 mM EDTA, 0.2% SDS, 1.6% Triton X-100, supplemented with protease inhibitor cocktail (final concentrations: 1 mM PMSF, 10 nM aprotinin, 10 μM leupeptin, 1 μg/ml pepstatin A).

Mixed Micelle buffer: 150 mM NaCl, 20 mM Tris–HCl, pH 8.0, 5 mM EDTA, 5.2% sucrose, 1% Triton X-100, 0.2% SDS, supplemented with protease inhibitor cocktail (final concentrations: 1 mM PMSF, 10 nM aprotinin, 10 μM leupeptin, 1 μg/ml pepstatin A).

Buffer 500: 10 mM Tris–HCl, pH 8.0, 50 mM HEPES, pH 8.0, 500 mM NaCl, 1% Triton X-100, 0.1% sodium deoxycholate, 1 mM EDTA, supplemented with protease inhibitor cocktail (final concentrations: 1 mM PMSF, 10 nM aprotinin, 10 μM leupeptin, 1 μg/ml pepstatin A).

LiCl/detergent buffer: 10 mM Tris–HCl, pH 8.0, 250 mM LiCl, 0.5% NP-40, 0.5% sodium deoxycholate, 10 mM EDTA, supplemented with protease inhibitor cocktail (final concentrations: 1 mM PMSF, 10 nM aprotinin, 10 μM leupeptin, 1 μg/ml pepstatin A).

TE buffer: 10 mM Tris–HCl, pH 8.0, 1 mM EDTA.

Bicarbonate/SDS buffer: 100 mM $NaHCO_3$, 1% SDS. Prepare right before use.

3. Other reagents: Anti-PAD4 antibody (custom made), anti-H3Cit antibody (Abcam, ab5103), anti-H3R17Me antibody (Abcam, ab8284), Micrococcal nuclease (Sigma, N5386), protein A/salmon sperm DNA agarose (Millipore, 16-157), PCR purification Kit (Qiagen, 28104), 2× SYBR Green master mixture (Applied Biosystems, 4309155), and PAD4 inhibitor Cl-amidine (custom made).
4. Equipment: Bioruptor (Diagenode Inc.), real-time PCR machine (Applied Biosystems), StepOne software (Applied Biosystems).

2.4. Analyze Cell Cycle, Cell Death, Mitochondria Membrane Potential, and Reactive Oxygen Species After PAD4 Inhibitor Treatment by Flow Cytometry

1. Cells: U2OS cells are cultured in Dulbecco Modified Eagle Medium-High Glucose (Invitrogen, 11965) supplemented with 10% fetal bovine serum and 1% penicillin–streptomycin in a 37°C incubator with 5% CO_2.
2. Buffers: Propidium iodide (PI)/Triton X-100 staining solution: For 10 ml of 0.1% Triton X-100 in PBS, add 200 μl of 1 mg/ml PI (Invitrogen, P3566) and 2 mg DNase-free RNase A (Qiagen, 1007885). Prepare freshly before use.
3. Other reagents: Annexin V-FITC Apoptosis Detection Kit (Abcam, ab14085), Rhodamine 123 (Rh123, Sigma, 83702), 2′,7′-Dichlorofluorescein (DCF, Sigma, D6665), H_2O_2 (Sigma, H3410), and PAD4 inhibitor Cl-amidine (Custom made).
4. Equipment: FC500 flow cytometer (Beckmann), CXP software (Beckmann), WinMDI 2.9 software.

3. Methods

3.1. In Vitro Histone Citrullination Assay

1. Assemble 20-μl reaction system as follows: 2 μl 10× PAD buffer, 1 μg PAD4 protein, 2 μg substrate (histone or nucleosome), add ddH_2O to 20 μl.
2. Incubate the reaction at 37°C for 2 h.

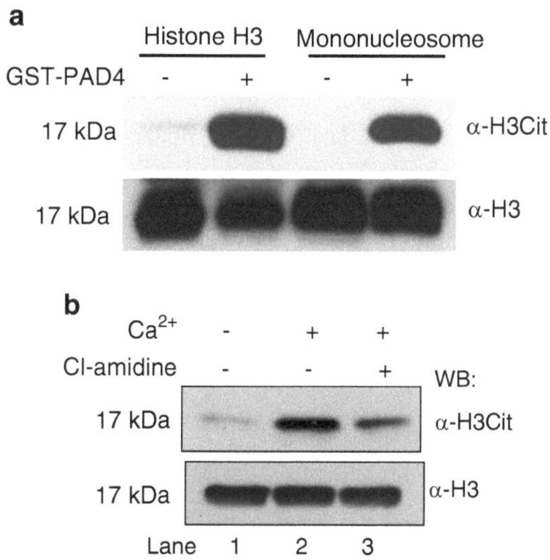

Fig. 2. In vitro and in vivo histone citrullination assay. (**a**) Western blotting of histone citrullination of histone H3 or mononucleosome before and after PAD4 treatment. General H3 antibody was used to show the amount of H3 in each lane. (**b**) In vivo histone citrullination assay. Western blot assays of histone H3 citrullination. Histone H3 blot serves as a loading control.

3. Resolve the proteins in a 13% SDS-PAGE and transfer proteins to a 0.1-μm nitrocellulose membrane.
4. Perform a standard Western blotting assay with the anti-H3Cit antibody and the anti-H3 antibody (Fig. 2a).

3.2. In Vivo Histone Citrullination Assay

1. In vivo histone citrullination assay with HL-60 cells
   (a) HL-60 cell treatment: To differentiate HL-60 along the granulocytic pathway, cells are first treated with 1.25% DMSO for 3 days at a starting cell density of 4×10^5 per ml. About 5×10^6 differentiated HL-60 cells are used for each following experiment. Treat cells with 4 μM calcium ionophore in Locke's solution for 30 min. For Cl-amidine treatment, first add 200 μM Cl-amidine to cells and incubate for 15 min before adding calcium ionophore. Split sample into two parts for Western blotting and immunostaining assays.
   (b) Detect histone citrullination by Western blotting
      - Collect cells by centrifugation at 1,000 rpm for 5 min. Wash once with cold PBS. Resuspend cells in IP buffer at 1×10^7 cell/ml concentration. Sonicate samples in Bioruptor for 5 min (30 s on, 30 s off) at high energy level. Spin down at $16{,}100 \times g$ for 5 min and collect supernatant as total protein extract.
      - Resolve the proteins in a 13% SDS-PAGE and transfer proteins to a 0.1-μm nitrocellulose membrane.

- Perform a standard Western blotting assay with the anti-H3Cit antibody and the anti-H3 antibody (Fig. 2b).

(c) Detect histone citrullination by immunostaining

- Cells are seeded on coverslip by spinning to coverslip at 800×*g* for 10 min. To enhance cell adhesion, coverslips are pretreated with 50–100 μl 0.01% poly-L-lysine, air dry, rinse with ddH_2O, and air dry again for 30–60 min.
- Fix cells with the fixation solution for 10–15 min at room temperature.
- Wash the cells with PBST for three times, 10 min each.
- Block the cells with 2% BSA in PBST for at least 30 min at room temperature.
- Stain the cells in a dark humid chamber with anti-H3Cit antibody in PBST with 2% BSA and 5% normal goat serum overnight at 4°C.
- Next day, wash the cells with PBST for three times, 10 min each.
- Incubate the cells in a dark humid chamber with goat anti-rabbit Cy3 antibody in PBST with 2% BSA and 5% normal goat serum for 2 h at room temperature.
- Wash the cells with PBST for three times, 10 min each and PBS for 5 min.
- Stain the cells briefly with DNA dye Hoechst at 1 μg/ml dissolved in PBS.
- Wash the cells with PBS for 5 min and ddH_2O for 5 min.
- Mount the coverslips to slides with the mounting solution, air dry briefly, and seal the edge with nail polish.
- Images are captured using a fluorescence microscope.

2. In vivo histone citrullination assay with mouse neutrophils

(a) Mouse neutrophil purification

- Collect mice peripheral blood from mandibular vein using 22-G needles to EDTA-coated tubes (see Note 1). Collect a total of ~2 ml blood for each purification from several mice.
- Prepare a Histopaque density gradient by carefully layering 3 ml of Histopaque-1077 on top of 3 ml Histopaque-1119 in a 15-ml tube.
- Load the blood on top of the gradient and centrifuge at 700×*g* for 30 min at room temperature (see Note 2).

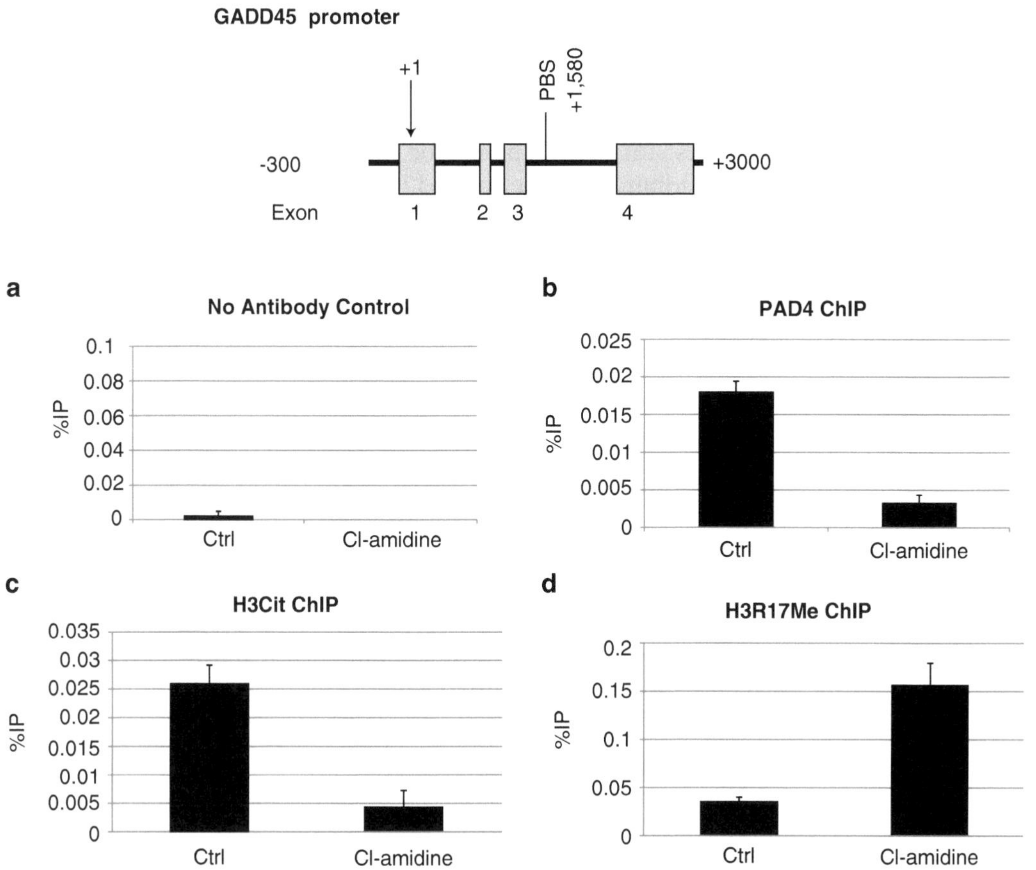

Fig. 3. In vivo PAD4 association and histone arginine modifications at the GADD45 promoter before and after Cl-amidine treatment. The exon and intron structure of GADD45 gene is illustrated. The p53 binding site (PBS) is indicated. No antibody control was shown in (**a**). ChIP experiments were performed to analyze the level of PAD4 (**b**), H3 citrullination (H3Cit) (**c**), or H3 Arg17 methylation (H3R17Me) (**d**) in U2OS cells before and after treatment with the PAD4 inhibitor Cl-amidine at the p53 binding site of GADD45 promoter. ChIP signals were calculated as percentages of the input (% IP).

- Collect the middle layer between Histopaque-1077 and Histopaque-1119 that contains neutrophils (see Note 3). Pellet cells by spinning at 450 ×*g* for 5 min at 4°C, and remove supernatant.
- To remove contaminating erythrocytes, resuspend the cell pellet in 1 ml of ice-cold 0.2% NaCl solution for 30 s. Immediately restore the isotonicity by adding 1 ml of ice-cold 1.6% NaCl solution. Spin down cells at 450 ×*g* for 5 min, and remove supernatant. Repeat this step a couple of times to remove most erythrocytes (see Note 4).
- Finally, resuspend neutrophils in ice-cold PBS containing 10 mM D-Glucose.

(b) Mouse neutrophil treatment with chemokines

Purified mouse neutrophils were stimulated with 1 μg/ml lipopolysaccharides, 100 μM H_2O_2, and 25 nM phorbol myristate acetate, respectively, in the presence of 2 mM calcium at 37°C and 5% CO_2 in PBS supplemented with 10 mM D-Glucose for 3 h. Split samples into two parts for Western blotting similar as in Subheading 3.2, step 1b, and immunostaining assays similar as in Subheading 3.2, step 1c.

3.3. Measure Histone Citrullination at Gene Promoter by Chromatin Immunoprecipitation Assay

1. U2OS cell treatment: Start with one 10-cm plate of cells with 30–40% confluency for each treatment. Treat one plate of cells with 200 μM Cl-amidine for 72 h.
2. After treatment, shear DNA into short fragments by sonication or micrococcal nuclease digestion (see Note 5).
   (a) Add 37% formaldehyde to cells to a final concentration of 1% to cross-link cells, and agitate gently for 10 min.
   (b) Add 2 M glycine to a final concentration of 0.125 M to stop cross-linking, and agitate gently for 5 min.
   (c) Wash three times in cold PBS with protease inhibitor cocktail. Collect cells by spinning down at 450×*g* for 5 min at 4°C. Discard supernatant.
   - Shear DNA by sonication to produce DNA fragments between 500 and 1,000 bp.
     - Resuspend cells with 8 ml SDS-lysis buffer.
     - Rotate for 10 min at 4°C, and warm up at room temperature until SDS precipitate disappears. Spin down at 1,500 rpm for 6 min at 4°C. Remove supernatant.
     - Resuspend the pellet with 2 ml ice-cold IP buffer.
     - Leave in ice for 10 min, and invert occasionally.
     - Sonicate the sample for five rounds (30 s on, 30 s off) in Bioruptor at high energy level (see Note 6). Spin down at 13,200 rpm for 15 min at 4°C. Collect supernatant. Proceed to next step after checking the size of sonicated DNA (see Note 7).
   - Shear DNA by micrococcal nuclease digestion to produce mononucleosome.
     - Resuspend cells in 1 ml HB buffer, put on ice for 10 min, and invert from time to time.
     - Homogenize cells in a Dounce homogenizer (Wheaton) for 15 strokes with a tight pestle. Spin down at 1,800×*g* for 10 min to pellet nuclei.

- Resuspend cells in NP-S buffer at 1×10^7 cell/ml concentration, aliquot into 500 µl/sample, and add $CaCl_2$ to reach the final concentration of 5 mM.
- Add 10 U micrococcal nuclease to cells, and place at 37°C for 15 min for digestion (see Note 6).
- Add 10 mM EDTA to stop the digestion.
- Add SDS to a final concentration of 1%, and put on ice for 10 min.
- Sonicate for 30 s at high energy level. Spin down at 13,200 rpm for 15 min at 4°C. Collect supernatant. Proceed to next step if the micrococcal nuclease digestion is successful (see Note 7).

3. Chromatin immunuprecipitation assay to measure histone citrullination level at gene promoter
   (a) Wash protein A/salmon sperm DNA (ssDNA) agarose three times with 1 ml IP buffer.
   (b) Add protein A/ssDNA agarose to sheared sample, and use 50 µl agarose per 1 ml sample. Preclear sample for 2 h at 4°C with continuous rotation.
   (c) Spin down at 3,000 rpm for 2 min at 4°C. Collect supernatant and spin again at 13,200 rpm for 15 min at 4°C to discard possible aggregates.
   (d) Distribute sample to 1.5-ml tubes. Add about 2 µg of each anti-PAD4 antibody, anti-H3Cit antibody, and anti-H3R17Me antibody to each sample, aiming for about $1–5 \times 10^5$ cells per antibody. Bring up total volume to 500 µl, and rotate overnight at 4°C. Set up one tube without antibody that represents the background and the source for the total input.
   (e) Next day, spin down the IP samples at 13,200 rpm for 15 min at 4°C. Collect supernatant and discard possible aggregates.
   (f) Collect 30 µl from the "no antibody" sample and keep on ice; this is processed at reverse cross-linking step as the "total input" sample.
   (g) Add about 20 µl washed protein A agarose/ssDNA to each sample, and rotate for 1.5 h at 4°C.
   (h) Let the agarose settle down by sitting at 4°C for 20 min. Discard supernatant.
   (i) Wash the agarose in the order of: three washes with mixed Micelle buffer, two washes with buffer 500, two washes with LiCl/detergent buffer, two washes with TE buffer. Use 1 ml buffer and rotate for 5 min at 4°C for every wash.

Spin down the agarose at 3,000 rpm for 1 min at 4°C. Discard supernatant.

(j) Resuspend the agarose in 170 μl bicarbonate/SDS buffer. Add 120 μl to "total input" sample.

(k) Vortex at low speed for 15 min at room temperature. Spin down at 3,000 rpm for 2 min and collect 150 μl supernatant.

(l) Resuspend the agarose in 150 μl bicarbonate/SDS buffer and repeat step (k).

(m) Incubate the samples at 65°C for at least 5 h to reverse cross-linking.

(n) Spin down briefly to collect drops under the lid. Incubate the samples at room temperature for 10 min to cool down.

(o) Add 3 μl of 10 mg/ml protease K and 1 μl of 10 mg/ml RNase A and incubate at 37°C for 2.5 h.

(p) Add 30 μl of 3 M NaAc (pH 5.0) to each sample and mix. Purify DNA with the PCR purification kit.

(q) Measure the enrichment of each protein or histone modifications at the GADD45 promoter by real-time PCR (see Note 8). Data is recorded and analyzed using StepOne Software v2.1. ChIP efficiencies of each antibody are calculated by the Microsoft Excel program (see Note 9) (Fig. 3).

3.4. Analyze Cell Cycle, Cell Death, Mitochondria Membrane Potential, and Reactive Oxygen Species After PAD4 Inhibitor Treatment by Flow Cytometry

1. U2OS cell treatment. Start with three 10-cm dishes of cells with 70–80% confluence for each assay. Treat one dish of cells with PAD4 inhibitor at appropriate dilution for 12 h. Use one dish of cells without any treatment as negative control. For cell cycle and cell death assays, treat one dish of cells with 4 μM of doxorubicin for 12 h as positive control; for mitochondria membrane potential and ROS generation analysis, treat one dish of cells with 100 μM of H_2O_2 for 15 min as positive control (see Note 10).
2. Cell cycle assay

   (a) After treatment, harvest cells by trypsin with centrifugation at 1,000 rpm for 5 min. Wash 1×10^6–1×10^7 cells once with cold PBS and resuspend thoroughly in 0.5 ml of PBS. Transfer the cell suspension into 4.5 ml of 70% ethanol for fixation. Keep cells on ice for at least 2 h.

   (b) Centrifuge the ethanol-suspended cells at 1,000 rpm for 5 min. Decant ethanol and wash once with cold PBS. Suspend the cell pellet in 1 ml of PI/Triton X-100 staining solution with RNase A. Keep for 15 min at 37°C or 30 min at room temperature.

(c) Measure cell fluorescence in the flow cytometer using 488-nm excitation and 610-nm emission.

3. Apoptosis assay: After treatment, collect 1×10^5 cells and incubate with Annexin V-FITC and PI provided by Apoptosis detection kit for 5 min in dark at room temperature. Analyze cell fluorescence in the flow cytometer using 488-nm excitation. Detect Annexin V-FITC staining at the emission of 525 nm and PI staining at the emission of 610 nm.
4. Analyze mitochondria membrane potential by flow cytometry: After treatment, collect 1×10^6 cells and wash once with prewarmed PBS. Incubate cell pellet with 500 μl of Rh123 (1 μg/ml in PBS) for 10 min in dark at 37°C. Wash once with PBS and keep on ice until the next step. Analyze cell fluorescence by flow cytometer using 488-nm excitation and 525-nm emission and analyze with WinMDI 2.9 software.
5. Measure ROS by flow cytometry: After treatment, collect 1×10^6 cells and resuspend in 1 ml of prewarmed DMEM. Incubate with 10 μM of DCF (10 mM in DMSO as working solution) for 30 min in dark at 37°C. Wash once with DMEM and keep on ice until the next step. Assess cell fluorescence by flow cytometer using 488-nm excitation and 525-nm emission and analyze with WinMDI 2.9 software.

4. Notes

1. Process the sample quickly. The peripheral blood coagulates in a very short time. Shake the EDTA-coated tubes gently while collection to mix the blood and EDTA to prevent coagulation.
2. Gently load the blood drop by drop on top of the Histopaque gradient; do not disturb the surface. Decelerate slowly after the centrifugation.
3. Most erythrocytes are at the bottom of the tube, and macrophages and lymphocytes are at the top of the gradient. Take the middle layer of neutrophils with an 18-G needle and syringe very gently.
4. Incubation with 0.2% NaCl too long can cause the burst of neutrophils as well, restore the isotonicity no longer than 30 s. From this step on, always keep the neutrophils on ice before future experiment to ensure their viabilities.
5. We have used two methods to shear chromatin in our ChIP experiments. Micrococcal nuclease can digest DNA to mononucleosome length, thus giving a better ChIP resolution, especially for histone modification ChIP. However, some proteins

bind to nucleosome-free region that would be digested away by micrococcal nuclease; therefore, DNA fragments sheared by sonication would be better to detect the association of these proteins with chromatin. The experiment shown here (Fig. 3) used sonication to shear chromatin.

6. It is recommended to optimize the sonication conditions according to individual sonicator and sample in order to produce DNA fragments with the proper size. The digestion efficacy of micrococcal nuclease varies with different cells; it is also important to first optimize the conditions of micrococcal nuclease used in every experiment with different samples by adjusting nuclease amount or digestion time. The efficiency of micrococcal nuclease may also vary from batch to batch.
7. To check the size of sonicated DNA:
   (a) Take a small aliquot from each sample, and incubate the sample at 65°C for 5 h to reverse cross-linking.
   (b) Digest with 10 μl of 10 mg/ml RNase A for 1 h and then 10 μl of 10 mg/ml protease K for 2 h at 37°C.
   (c) Purify DNA by phenol–chloroform extraction and ethanol precipitation. Add 10–20 μg glycogen before precipitation as carrier if necessary.
   (d) Spin down at 13,200 rpm for 10 min at 4°C. Wash twice with 70% cold ethanol. Air dry at 37°C and resuspend the pellet in 20 μl ddH_2O.
   (e) Resolve samples on agarose gel electrophoresis to check DNA size. A smear of DNA between 500 and 1,000 bp after sonication or most DNA fragments is around 150 bp after micrococcal nuclease digestion is good for proceeding to the next step.
8. Real-time PCR reaction:
   (a) Assemble 12 μl real-time PCR reaction in a 96-well plate:

       6 μl 2× Power SYBR Green PCR Master Mix

       3 μl 1 μM forward and reverse primer mixture

       3 μl DNA
   (b) PCR is performed in real-time PCR machine using the below program:

       95°C for 10 min

       40 cycles of:

       95°C for 15 s

       55°C for 30 s

       70°C for 15 s (collect data at this step)

Add melt curve after the PCR program to check primer quality.

GADD45 promoter primer sequences: Forward 5′-actttc agccgagatgtgct-3′; Reverse 5′-gagtagctgggctgactgct-3′.

9. After export threshold PCR cycle numbers from real-time PCR reaction. ChIP efficiency of each antibody was analyzed using the Microsoft Excel program by calculating percentage of IP compared with total input.
10. It is recommended to add H_2O_2 after the staining step instead of adding at beginning for better results.

Acknowledgment

The authors would like to thank the Flow Cytometry Facility at the Pennsylvania State University for its technical help. Research in the Wang laboratory is supported by NIH grant R01 CA136856.

References

1. Kornberg, R. D., and Lorch, Y. (1999) Twenty-five years of the nucleosome, fundamental particle of the eukaryote chromosome. *Cell* **98**, 285–94.
2. Luger, K., Mader, A. W., Richmond, R. K., Sargent, D. F., and Richmond, T. J. (1997) Crystal structure of the nucleosome core particle at 2.8 A resolution. *Nature* **389**, 251–60.
3. Richmond, T. J., and Davey, C. A. (2003) The structure of DNA in the nucleosome core. *Nature* **423**, 145–50.
4. Davey, C. A., Sargent, D. F., Luger, K., Maeder, A. W., and Richmond, T. J. (2002) Solvent mediated interactions in the structure of the nucleosome core particle at 1.9 a resolution. *J Mol Biol* **319**, 1097–113.
5. Ramakrishnan, V. (1997) Histone H1 and chromatin higher-order structure. *Crit Rev Eukaryot Gene Expr* 7, 215–30.
6. Narlikar, G. J., Fan, H. Y., and Kingston, R. E. (2002) Cooperation between complexes that regulate chromatin structure and transcription. *Cell* **108**, 475–87.
7. Campos, E. I., and Reinberg, D. (2009) Histones: annotating chromatin. *Annu Rev Genet* **43**, 559–99.
8. Li, B., Carey, M., and Workman, J. L. (2007) The role of chromatin during transcription. *Cell* **128**, 707–19.
9. Kouzarides, T. (2007) Chromatin modifications and their function. *Cell* **128**, 693–705.
10. Groth, A., Rocha, W., Verreault, A., and Almouzni, G. (2007) Chromatin challenges during DNA replication and repair. *Cell* **128**, 721–33.
11. Jenuwein, T., and Allis, C. D. (2001) Translating the histone code. *Science* **293**, 1074–80.
12. Kouzarides, T. (2002) Histone methylation in transcriptional control. *Curr Opin Genet Dev* **12**, 198–209.
13. Kuo, M. H., and Allis, C. D. (1998) Roles of histone acetyltransferases and deacetylases in gene regulation. *Bioessays* **20**, 615–26.
14. Nowak, S. J., and Corces, V. G. (2004) Phosphorylation of histone H3: a balancing act between chromosome condensation and transcriptional activation. *Trends Genet* **20**, 214–20.
15. Wang, Y., Wysocka, J., Sayegh, J., Lee, Y. H., Perlin, J. R., Leonelli, L., Sonbuchner, L. S., McDonald, C. H., Cook, R. G., Dou, Y., Roeder, R. G., Clarke, S., Stallcup, M. R., Allis, C. D., and Coonrod, S. A. (2004) Human PAD4 regulates histone arginine methylation levels via demethylimination. *Science* **306**, 279–83.
16. McBride, A. E., and Silver, P. A. (2001) State of the arg: protein methylation at arginine comes of age. *Cell* **106**, 5–8.

17. Bedford, M. T., and Clarke, S. G. (2009) Protein arginine methylation in mammals: who, what, and why. *Mol Cell* **33**, 1–13.
18. Chen, D., Ma, H., Hong, H., Koh, S. S., Huang, S. M., Schurter, B. T., Aswad, D. W., and Stallcup, M. R. (1999) Regulation of transcription by a protein methyltransferase. *Science* **284**, 2174–7.
19. Wang, H., Huang, Z. Q., Xia, L., Feng, Q., Erdjument-Bromage, H., Strahl, B. D., Briggs, S. D., Allis, C. D., Wong, J., Tempst, P., and Zhang, Y. (2001) Methylation of histone H4 at arginine 3 facilitating transcriptional activation by nuclear hormone receptor. *Science* **293**, 853–7.
20. Pal, S., Vishwanath, S. N., Erdjument-Bromage, H., Tempst, P., and Sif, S. (2004) Human SWI/SNF-associated PRMT5 methylates histone H3 arginine 8 and negatively regulates expression of ST7 and NM23 tumor suppressor genes. *Mol Cell Biol* **24**, 9630–45.
21. Nowak, S. J., Pai, C. Y., and Corces, V. G. (2003) Protein phosphatase 2A activity affects histone H3 phosphorylation and transcription in Drosophila melanogaster. *Mol Cell Biol* **23**, 6129–38.
22. Bannister, A. J., Schneider, R., and Kouzarides, T. (2002) Histone methylation: dynamic or static? *Cell* **109**, 801–6.
23. Chang, B., Chen, Y., Zhao, Y., and Bruick, R. K. (2007) JMJD6 is a histone arginine demethylase. *Science* **318**, 444–7.
24. Webby, C. J., Wolf, A., Gromak, N., Dreger, M., Kramer, H., Kessler, B., Nielsen, M. L., Schmitz, C., Butler, D. S., Yates, J. R., 3rd, Delahunty, C. M., Hahn, P., Lengeling, A., Mann, M., Proudfoot, N. J., Schofield, C. J., and Bottger, A. (2009) Jmjd6 catalyses lysylhydroxylation of U2AF65, a protein associated with RNA splicing. *Science* **325**, 90–3.
25. Cuthbert, G. L., Daujat, S., Snowden, A. W., Erdjument-Bromage, H., Hagiwara, T., Yamada, M., Schneider, R., Gregory, P. D., Tempst, P., Bannister, A. J., and Kouzarides, T. (2004) Histone deimination antagonizes arginine methylation. *Cell* **118**, 545–53.
26. Vossenaar, E. R., Zendman, A. J., van Venrooij, W. J., and Pruijn, G. J. (2003) PAD, a growing family of citrullinating enzymes: genes, features and involvement in disease. *Bioessays* **25**, 1106–18.
27. Nakashima, K., Hagiwara, T., Ishigami, A., Nagata, S., Asaga, H., Kuramoto, M., Senshu, T., and Yamada, M. (1999) Molecular characterization of peptidylarginine deiminase in HL-60 cells induced by retinoic acid and 1alpha,25-dihydroxyvitamin D(3). *J Biol Chem* **274**, 27786–92.
28. Hagiwara, T., Nakashima, K., Hirano, H., Senshu, T., and Yamada, M. (2002) Deimination of arginine residues in nucleophosmin/B23 and histones in HL-60 granulocytes. *Biochem Biophys Res Commun* **290**, 979–83.
29. Tarcsa, E., Marekov, L. N., Mei, G., Melino, G., Lee, S. C., and Steinert, P. M. (1996) Protein unfolding by peptidylarginine deiminase. Substrate specificity and structural relationships of the natural substrates trichohyalin and filaggrin. *J Biol Chem* **271**, 30709–16.
30. Thompson, P. R., and Fast, W. (2006) Histone citrullination by protein arginine deiminase: is arginine methylation a green light or a roadblock? *ACS Chem Biol* **1**, 433–41.
31. Chang, X., and Han, J. (2006) Expression of peptidylarginine deiminase type 4 (PAD4) in various tumors. *Mol Carcinog* **45**, 183–96.
32. Chang, X., Han, J., Pang, L., Zhao, Y., Yang, Y., and Shen, Z. (2009) Increased PADI4 expression in blood and tissues of patients with malignant tumors. *BMC Cancer* **9**, 40.
33. Luo, Y., Arita, K., Bhatia, M., Knuckley, B., Lee, Y. H., Stallcup, M. R., Sato, M., and Thompson, P. R. (2006) Inhibitors and inactivators of protein arginine deiminase 4: functional and structural characterization. *Biochemistry* **45**, 11727–36.
34. Li, P., Yao, H., Zhang, Z., Li, M., Luo, Y., Thompson, P. R., Gilmour, D. S., and Wang, Y. (2008) Regulation of p53 target gene expression by peptidylarginine deiminase 4. *Mol Cell Biol* **28**, 4745–58.
35. Li, P., Wang, D., Yao, H., Doret, P., Hao, G., Shen, Q., Qiu, H., Zhang, X., Wang, Y., and Chen, G. Coordination of PAD4 and HDAC2 in the regulation of p53-target gene expression. *Oncogene* **29**, 3153–62.
36. Wang, Y., Li, M., Stadler, S., Correll, S., Li, P., Wang, D., Hayama, R., Leonelli, L., Han, H., Grigoryev, S. A., Allis, C. D., and Coonrod, S. A. (2009) Histone hypercitrullination mediates chromatin decondensation and neutrophil extracellular trap formation. *J Cell Biol* **184**, 205–13.
37. Fang, J., Wang, H., and Zhang, Y. (2004) Purification of histone methyltransferases from HeLa cells. *Methods Enzymol* **377**, 213–26.

Part IV

RNA Synthesis and Regulation

Chapter 32

Analysis of mRNA Abundance and Stability by Ribonuclease Protection Assay

Cristina Romero-López, Alicia Barroso-delJesus, Pablo Menendez, and Alfredo Berzal-Herranz

Abstract

Gene expression is a multi-step process, which proceeds from DNA through RNA to protein. The tight regulation of this process is essential for overall cellular integrity and physiological homeostasis. Regulation of the messenger RNA (mRNA) levels has emerged as a crucial event in the modulation of the expression of genetic information. The mechanisms by which this process occurs have been extensively studied and begin to be much better understood. They involve a network of complex pathways that use intrinsic features of the target mRNA, like stability, to control its relative abundance in the cytoplasm. Thus, the analysis of the mRNA stability and abundance is essential to properly undertake gene expression studies. This chapter describes the ribonuclease protection assay, a widely accepted approach to evaluate the quality and amount of a target mRNA. This technique displays a higher sensitivity than classical Northern blot analysis and may be used either individually or in combination with other quantitative methods, such as quantitative reverse-transcription PCR, as complementary procedures rendering more complete and reliable information on gene expression.

Key words: RNA stability, Ribonuclease protection assay, Gene expression analysis, RNA quality, RNA abundance

1. Introduction

Gene expression control is a key point for cell development and survival across all living kingdoms, from bacteria to mammals. Since RNA molecules are the central effectors of the genetic information flow, the precise regulation of their relative abundance and stability is essential for the proper synthesis, assembly, and localization of subcellular structures (1, 2). Regulation of the RNA levels is a multi-step process that requires an accurate coordination

Ales Vancura (ed.), *Transcriptional Regulation: Methods and Protocols*, Methods in Molecular Biology, vol. 809, DOI 10.1007/978-1-61779-376-9_32,

between all the cellular processes. In eukaryotes, it begins in the nucleus, where protein factors specifically bind to DNA sequences activating or inhibiting the RNA synthesis. Nascent transcripts recruit diverse proteins that mediate multiple RNA modifications, such as splicing, 5′-capping, 3′-polyadenylation, and editing. Mature molecules are subsequently exported to the cytoplasm and directed to macromolecular complexes, where they play essential roles in protein synthesis (rRNAs, mRNAs; see ref. 1) and mRNA post-transcriptional regulation (miRNAs; see refs. 3, 4), among others. RNA molecules are eventually eliminated following diverse degradation pathways (5–7). In consequence, gene expression studies inevitably require the evaluation of the quality, stability, and relative abundance of the intracellular RNA transcripts.

This chapter focuses on a classical strategy for the analysis of specific transcripts, the ribonuclease protection assay (RPA). It constitutes a very reliable and sensitive method for detection, quantification, and analysis of RNA species in complex mixtures of total cellular RNA (8–11). RPA is based on the ribonucleases (RNases)' ability to specifically discriminate between double- and single-stranded RNA templates. In this approach, RNA transcripts are hybridized in solution with a radioactive-labelled RNA probe that is complementary to the molecule under investigation (Fig. 1). This interaction is performed under highly stringent conditions that are usually assessed by adding formamide to the hybridization buffer, which considerably reduces non-specific interactions. After hybridization, subsequent nuclease treatment removes non-interacting molecules. Several RNases specific for single-stranded RNA, such as RNase T1 or RNase A, can be used either individually or in combination. Thus, non-hybridized RNA molecules or mismatched RNA duplexes are degraded while intact, fully matched probe:transcript duplexes can be recovered by ethanol precipitation and resolved in denaturing polyacrylamide gels. With this technique, the size, amount, and integrity of the desired RNA can be interrogated by subsequent exposure to autoradiography film or phosphoimaging (Fig. 2). Ideally, when this procedure is carried out with a molar excess of the probe over the target RNA, the intensity of the signal from the protected fragment is proportional to the amount of complementary RNA in the sample.

RPA is a widely accepted technique, which complements the information obtained from quantitative reverse-transcription PCR (qRT-PCR), with the advantage that it does not require a reverse transcription step. In addition, it provides additional data that cannot be obtained from qRT-PCR, like the size and integrity of a specific transcript. It may also substitute the Northern blotting analysis, in which the sensitivity, specificity, and resolution limitations are often compromised by the inability to load large RNA amounts in the gel, inefficient transfer from gel to the blot membrane, and cross or non-specific hybridization between probes and

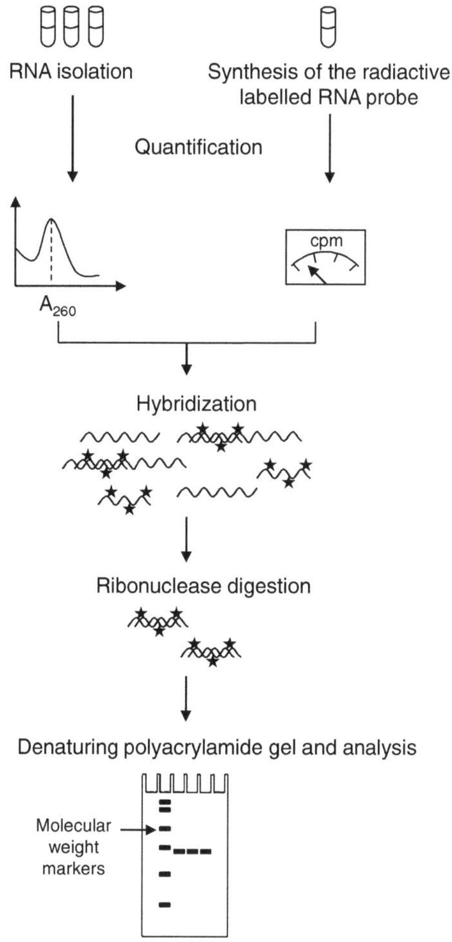

Fig. 1. Cartoon describing the ribonuclease protection assay (RPA). Purified RNA samples are hybridized in solution with a molar excess of a radioactive-labelled RNA probe to generate RNA duplexes. Unpaired RNA molecules are digested by the ribonuclease cocktail and protected fragments are resolved in denaturing polyacrylamide gels. These are subsequently exposed to evaluate the size and amount of the RNA under investigation.

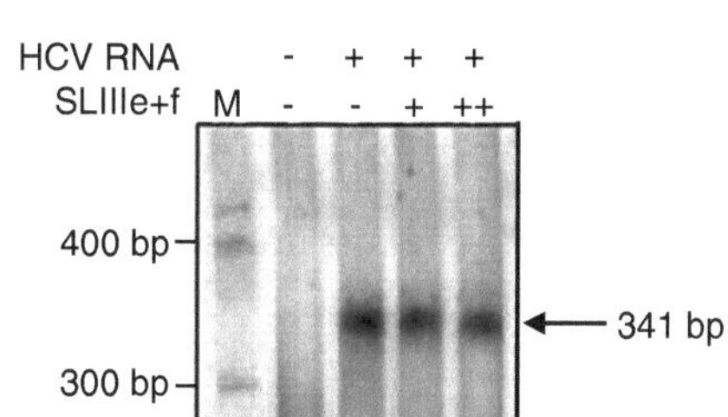

Fig. 2. Representative ribonuclease probing assay (RPA) for relative quantification of viral RNA. A human-derived hepatoma cell line containing a replicon system for the hepatitis C virus (HCV) was transfected with different amounts of the HCV RNA domain SLIIIe+f. Total cellular RNA was hybridized overnight with 10,000 cpm of a sense RNA probe for the 5′ end of the HCV genome. After ribonuclease digestion, products were resolved on 10% denaturing polyacrylamide gels. *Lane M* denotes molecular weight marker. Figure is a kind gift of Dr. Saumitra Das (18) by permission of Oxford University Press.

target RNA molecules. In addition, the hybridization in solution in the RPA favours the detection of rare transcripts. Furthermore, RPA replaces other classical, nuclease-based methods, such as the S1 nuclease analysis, which usually employs DNA probes that are synthesized as double-stranded molecules. This entails that reconstitution of probe duplexes is probable, therefore reducing the sensitivity. Hence, careful purification of the probe strand is a requisite for a proper analysis.

RPA assays are well-suited for mapping positions of external and internal junctions in RNA, such as transcription initiation and termination sites and intron/exon boundaries (12–14). With a proper probe design, RPA reveals the presence of sense and antisense transcripts from gene units (15) and allows for the estimation of genetic diversity in complex viral populations (16). It has been successfully used in the evaluation of the stability of therapeutic ribonucleoprotein complexes (17) and to quantify viral RNA replication levels (Fig. 2; ref. 18). RPA also allows to discriminate between closely related targets, like members of a multigene family, by using probes designed to span the differing regions (19). These reactions can also be performed as multiplex assays to simultaneously analyze several transcripts (20), as long as the protected fragments exhibit different sizes. Moreover, its high sensitivity and specificity have been exploited for detection and quantification of microRNAs (miRNAs; see ref. 21). Recently, modifications in the RPA technique have been reported to provide exclusive data that cannot be supplied by other methodologies. For instance, a sequential RPA/primer extension analysis has been described to distinguish among paralogous miRNAs that differ from each other by only one nucleotide located at their 5′ end (22).

In summary, RPA is a reproducible and independent method that provides significant and valuable information in gene expression studies. In this chapter, we describe in detail the analysis of specific eukaryotic mRNAs by RPA analysis.

2. Materials

All reagents and solutions, as well as plastic and glassware, must be free of RNases. Gloves must be worn at all the times.

2.1. RNA Isolation

1. Denaturing solution (see Note 1): 4 M guanidinium thiocyanate, 25 mM sodium citrate, pH 7.0, 0.5% (w/v) *N*-laurosylsarcosine (Sarkosyl), 0.1 M 2-mercaptoethanol.
2. 2 M Sodium acetate, pH 4.0.
3. Water-saturated phenol.
4. Chloroform:isoamyl alcohol (49:1, v/v) or bromochloropropane.

5. Isopropanol.
6. 80% ethanol.
7. RNase-free, sterile, distilled water.
8. Glass-Teflon homogenizer.

2.2. Probe Synthesis

1. DNA templates (plasmid DNA or PCR products).
2. Restriction enzymes to digest the plasmid templates to define the proper 3′ end of the RNA.
3. NTP mix stock solution: 10 mM adenosine 5′ triphosphate (ATP), cytosine 5′ triphosphate (CTP), and guanosine 5′ triphosphate (GTP), 1 mM uridine 5′ triphosphate (UTP).
4. Transcription buffer made according to the manufacturer's instructions. A common recipe is 40 mM Tris–HCl, pH 8.0, 6 mM $MgCl_2$, 1 mM spermidine, 4 mM NaCl, 10 mM dithiothreitol, 0.01% Triton X-100.
5. [α-^{32}P-UTP], 10 mCi/ml, 800 Ci/mmol. It is recommended to be used within 14 days of the labelling date to ensure high specific activity. It is a hazardous material, so appropriate protection is required when handling.
6. Placental ribonuclease inhibitor (40 U/μl).
7. T7 RNA polymerase.
8. RNase-free DNase I (1–10 U/μl; e.g. RQ1 DNase from Promega).
9. 40% polyacrylamide stock solution (19:1 acrylamide:bis-acrylamide).
10. 10× TBE buffer: 0.89 M Tris base, 0.89 M boric acid, 25 mM ethylenediamine-tetraacetic acid (EDTA).
11. 10% (w/v) ammonium persulphate (APS).
12. *N,N,N,N*-tetramethyl-ethylene diamine (TEMED).
13. Formamide loading buffer: 97% formamide deionized ultra-pure grade, 17 mM EDTA, 0.025% xylene cyanol, 0.025% bromophenol blue.
14. RNA elution buffer: 0.5 M ammonium acetate, 0.1% sodium dodecyl sulphate (SDS), 1 mM EDTA.
15. Buffered saturated phenol.
16. Chloroform:isopropyl alcohol (24:1).
17. 3 M sodium acetate, pH 5.2 and pH 6.
18. 100% ethanol.
19. RNase-free, sterile, distilled water.
20. Sterile razor blades.

2.3. Ribonuclease Protection Assay

1. Hybridization buffer: 80% deionized formamide, 40 mM PIPES, pH 6.4, 400 mM NaCl, 1 mM EDTA.
2. Ribonuclease digestion buffer: 300 mM NaCl, 10 mM Tris–HCl, pH 7.4, 5 mM EDTA, 10 U RNase T1, 40 μg/ml RNase A.
3. 20% (w/v) SDS.
4. Proteinase K, 20 mg/ml.
5. tRNA, 10 μg.
6. Buffered saturated phenol.
7. Chloroform:isopropyl alcohol (24:1).
8. 3 M sodium acetate, pH 5.2 and pH 6.
9. 100% ethanol.
10. Formamide loading buffer.
11. 40% polyacrylamide stock solution (19:1 acrylamide:bis-acrylamide).
12. TBE buffer.
13. 10% (w/v) APS.
14. TEMED.
15. RNase-free, sterile, distilled water.

3. Methods

3.1. Total Cellular RNA Extraction

The isolation of clean, intact RNA is essential for the RPA. Procedures developed for extracting RNA from cultured cells involve complete cell lysis that liberates all the nucleic acids from different organelles. This involves that DNA and RNA are co-purified in the initial steps. Additional treatments are then required to remove the DNA. Other difficulty in RNA isolation is the contamination with intra- and extracellular RNases, which are very stable and remain active under very suboptimal conditions. Hence, cell lysis in a chemical environment that inactivates nucleases is indispensable. Subsequent treatments allow for the selective purification of RNA from other macromolecular complexes.

There are several methods for preparing RNA from eukaryotic cells. Most of them employ common laboratory reagents, but commercial kits are also available. Here, we present a widely used protocol for total RNA extraction from adherent and suspension culture cells, the so-called guanidinium thiocyanate–phenol–chloroform extraction method (23). It is based on the potent protein-denaturing properties of the guanidinium thiocyanate. This chaotropic compound had been traditionally used as RNase inhibitor by other authors in RNA extraction procedures (24, 25), but these required several ultracentrifugation steps to completely remove genomic DNA. This was notably time consuming and limited the number of samples to be processed. The discovery that only total RNA remained soluble in the acidic aqueous phase after treatment with acid guanidinium thiocyanate–phenol–chloroform prompted the optimization of a faster purification protocol. The guanidinium thiocyanate–phenol–chloroform extraction method renders high yield of RNA from a small amount of start sample (tissue, cell culture), retrieves a good representation of the intracellular RNA pool, and can be easily scaled up. The development of a single-step protocol was an important advance in the field and led to the marketing of kits that reduce hazardous handling and time for reagents' preparation (RNAzol from MRC, TRI Reagent from Sigma–Aldrich, or Trizol from Invitrogen-Life Technologies). The standard protocol for total RNA extraction is detailed below.

1. Homogenization:

   For tissues: Add 1 ml of denaturing solution to 100 mg of fresh tissue. Homogenize in a glass-Teflon homogenizer with a few strokes.

   For culture cells: Discard cell culture medium either by centrifugation (in case of cells growing in suspension) or by removing it from adherent cells. Wash cells twice in PBS. Then, add 1 ml of denaturing solution per 10^7 cells and mix by pipetting, which favours cell lysis and DNA fragmentation.

   Samples must not remain in denaturing buffer for more than 30 min.

2. Transfer the homogenate to a polypropylene tube.
3. Add 0.1 volumes of 2 M sodium acetate, pH 4.0, and mix thoroughly by inverting the tube.
4. One volume of water-saturated phenol is then supplied and mixed by inversion (see Note 2).
5. Extraction is completed by the addition of 0.2 volumes of chloroform:isoamyl alcohol (49:1) solution or bromochloropropane and vigorous shaking during 15 s. This aids the dissociation of the nucleoprotein complexes.
6. Preserve the samples at room temperature for 3 min.
7. Centrifuge for 20 min at 10,000 × *g*, 4°C.
8. The aqueous upper phase mostly containing the cellular RNA pool is carefully transferred to a new tube. RNA is then precipitated by adding one volume of 100% isopropanol and incubating at room temperature for 10 min (see Note 3). Centrifuge the samples at 10,000 × *g* for 20 min to pellet total RNA. It should appear as a gel-like precipitate.

9. Resuspend the precipitate in 1 ml of ethanol 80% and vortex it to wash away the residual guanidinium. Incubate for 10 min at room temperature.
10. Centrifuge for 10 min at 10,000 × *g*, 4°C. Discard the supernatant and air dry the pellet.
11. Dissolve the RNA pellet in hybridization buffer by briefly vortexing or heating at 60°C for 5 min.
12. Spectrophotometric readings at wavelengths of 260 and 280 nm (A_{260}, A_{280}), respectively, are required to analyze the RNA amount and the quality of the preparation. Measurements at 260 nm allow for the calculation of nucleic acid concentration, whereas the ratio A_{260}/A_{280} estimates the contamination with phenol or proteins. This value should be between 1.8 and 2 in pure preparations.

3.2. Synthesis and Purification of the RNA Probes

A DNA construct must be designed and produced for each specific RNA probe. The minimal requirement is a double-stranded promoter sequence upstream of the template for the desired RNA. The most common approach is the use of plasmid vectors carrying the T7 RNA polymerase promoter upstream of a multicloning site, where the specific template sequences can be inserted. Alternatively, the use of PCR products as DNA templates carrying the T7 promoter sequence at their 5′ end is also possible (see Note 4). Recommended probe size is 200–400 nt, though shorter probes may be used for abundant targets. Nevertheless, probe size is determined by the molecular feature to be sought. In any case, a careful removal of the DNA template must be accomplished to avoid the formation of heteroduplexes DNA template:RNA probe, which could promote misleading results. Probe gel purification is generally recommended, since incomplete transcripts may increase background levels and lost of sensitivity.

1. DNA template preparation:

   For plasmid constructs: Digest the plasmid containing the probe coding sequence with a restriction enzyme that cuts at the 3′ end of the probe coding sequence. This provides a proper template for the RNA synthesis (see Note 5).

   For PCR templates: Amplify the sequence of interest by PCR using a 5′ primer containing the T7 RNA polymerase promoter sequence.
2. Check the quality of the template by agarose electrophoresis.
3. Extract DNA by adding an equal volume of buffered-saturated phenol. Mix well to form an emulsion and centrifuge for 2 min at 12,000 × *g*.
4. Transfer the aqueous upper phase to a new tube and add one volume of chloroform:isoamyl alcohol (24:1). Mix by vortexing and centrifuge again.

5. Keep the upper phase and add 0.1 volume of 3 M sodium acetate, pH 6, along with three volumes of cold (−20°C) absolute ethanol.
6. Incubate at −20°C for 30 min and centrifuge for 30 min at 12,000 × *g*, 4°C.
7. Discard the supernatant and wash the pellet with 300 μl of 70% ethanol.
8. Centrifuge for 5 min at 12,000 × *g*, 4°C.
9. Remove the supernatant and vacuum dry briefly the DNA pellet.
10. Dissolve the template in RNase-free, sterile, distilled water.
11. Set up 50 μl transcription reactions in transcription buffer 1× containing 1 μg of purified DNA template, 5 μCi of [α-^{32}P] UTP (3,000 Ci/mmol), 1 mM GTP, 1 mM ATP, 1 mM CTP, 0.1 mM UTP (to achieve high specific activity of the RNA probe; see Note 6), 0.5 U/μl of placental RNase inhibitor, and 20 μg/ml of purified T7 RNA polymerase.
12. Incubate the reaction mix at 37°C for 2 h.
13. Stop the reaction by adding 1 U of RNase-free DNase and incubate at 37°C for 15 min.
14. An equal volume of formamide loading buffer (50 μl) is added and samples are heated at 95°C in a dry bath before loading directly on preparative denaturing polyacrylamide–7 M urea gels. The gel concentration should be chosen according to the size of the RNA molecule to be purified (3.5–20%). Electrophoresis should be performed under denaturing conditions in 0.5× TBE.
15. Cover the gel with a saran wrap and place it on an exposure cassette with a sheet of X-ray film. Expose it in a darkroom (see Note 7). The acrylamide slice containing the full-length probe is excised using a razor blade.
16. The gel slices are then soaked in 300 μl of RNA elution buffer and incubated overnight at 4°C.
17. Purify the RNA by consecutive phenol and chloroform:isoamyl alcohol extraction as indicated in steps 3–5.
18. Precipitate the RNA by adding 0.1 volume of 3 M sodium acetate, pH 5.2, and three volumes of cold (−20°C) absolute ethanol.
19. Pellet the RNA as noted in steps 6–9.
20. Dissolve the labelled RNA in hybridization buffer.
21. Check the specific activity of the probe in a scintillation counter. The product should be stored at −20°C until use.

3.3. Ribonuclease Protection Assay

1. Mix 10 μg of total cellular RNA with 10^5 cpm of RNA probe in 30 μl of hybridization buffer (see Note 8). Include a negative control with tRNA instead of total RNA. This provides information about the background hybridization and the yield of the RNase digestion reaction (see Note 9).
2. Denature RNA by heating at 85°C for 5 min.
3. Immediately transfer the mix to the hybridization temperature (see Note 10).
4. Anneal overnight.
5. Add 300 μl of RNase digestion buffer and incubate at 30°C for 30–60 min.
6. Stop the reaction by the addition of 1% SDS and 10 mg/ml of proteinase K. Incubate for 30 min at 37°C.
7. Undigested RNA is extracted with phenol:chloroform:isoamyl alcohol and precipitated with ethanol as described above.
8. Dry pellets are resuspended in formamide loading buffer and resolved by denaturing polyacrylamide–urea gels.

3.4. Analysis

Gels are exposed to autoradiography film or phosphoimaging and subsequently analyzed and quantified by densitometric or fluorescence scanning.

One of the main RPA applications is the absolute quantification of a specific RNA. For that purpose, a standard curve must be established using known amounts of the in vitro-transcribed target hybridized with an excess of the probe. After RNase treatment, products are resolved in denaturing polyacrylamide gels. The intensity of the signal is compared to that generated by the experimental samples and used to determine the absolute amount of target RNA.

4. Notes

1. Prepare a 7 M guanidinium thiocyanate stock solution in RNase-free, sterile, distilled water by heating at 65°C. Then, add sodium citrate, pH 7.0, to reach a final concentration of 45 mM and 26.4 ml of 10% (w/v) Sarkosyl (*N*-laurosylsarcosine). The stock solution can be stored for up to 3 months at room temperature. To prepare the working denaturing solution, just add 2-mercaptoethanol (0.5% final concentration). This solution can be stored up for 1 month at room temperature. To minimize handling of guanidinium thiocyanate, dissolve it directly in the manufacturer's bottle. The 2-mercaptoethanol should be handled under a fume hood.

2. Preservation of the acidic pH is a key factor in the success of the RNA isolation method. A low pH value ensures the proper separation of RNA from DNA and proteins. For this reason, the use of water-saturated phenol is recommended instead of buffered-saturated phenol.
3. RNA precipitation can be improved by increasing the incubation time up to 1 h. In this case, samples should be maintained at –20°C. This is specially recommended when working with low amount of starting material.
4. T7 RNA polymerase efficiently synthesizes RNA molecules in vitro with minimal requirements. These include the 17 bp double-stranded promoter sequence followed by GGG and a single-stranded template.

   5′-**TAATACGACTCACTATA**↓GGG

   3′-**ATTATGCTGAGTGATAT** CCCXXXX…XX-5′

   Structure of the partially double-stranded DNA template.

   T7 promoter sequence is shown in bold; ↓ indicates the transcription initiation site.
5. The enzyme may cut the plasmid in multiple positions, as long as it does not cut into the promoter or the probe sequence. Complete digestion of the plasmid template should be checked by agarose electrophoresis. The presence of non-digested molecules in the reaction results in run-on transcripts that use up nucleotides and reduce the yield of the probe synthesis.
6. This is a general protocol that must be optimized for each specific probe. The nucleotide concentration is especially relevant: the lower amount of nucleotide, the higher proportion of incomplete transcripts. For short RNA probes (less than 400 nt), 1 mM of each non-labelled NTP generally renders high transcription yields. Long probes require specific transcription buffers with higher amounts of nucleotides and divalent cations. In many cases, the addition of non-radioactive limiting nucleotides may be necessary to preserve the nucleotide concentration high enough to produce significant amounts of full-length transcript. This leads to a decreased specific activity that is compensated by the increased sensitivity of the longer probe. Alternatively, commercial kits are currently available that produce high amounts of long RNA molecules (up to 3,000 nt).
7. The position of the film on the gel is recorded using a marker, with non-symmetric lines drawn across all edges of the film to ensure its correct repositioning on top of the gel after development. Alternatively, a variety of phosphorescent labels or markers available from different commercial sources can be used to guarantee the correct alignment of the film on the gel. A 2–5-min

exposure should be enough for efficient transcription when using nucleotides with high specific activity.

8. For total cellular RNA, 10 μg is enough to detect most of RNA molecules. However, this amount can be increased for low abundant species.
9. Background hybridization is commonly due to incomplete digestion of the DNA template after the probe transcription reaction. Residual DNA template molecules can efficiently interact with the probe, leading to heteroduplexes DNA:RNA that are not cleaved by RNases. Background can be also caused by the presence of trace amounts of non-digested single-stranded probe. In both cases, this generates smears that appear in all the reactions. A tRNA negative control allows for the evaluation of the background.
10. There is an optimal annealing temperature for each probe, which depends not only on its sequence but also on its secondary structure. It is, therefore, advisable to experimentally test the optimal temperature by using a range from 30 to 60°C.

Acknowledgements

A.B.-H's group is funded by grant BFU2009-08137 from the Spanish Ministry of Science and Innovation, grant CTS-5077 from the *Junta de Andalucía*, and by FEDER funds from the EU. P.M.'s group is supported by the Andalusian Health Department, Andalusian Innovation and Science Department (P08-CTS-3678 to P.M), the FIS (PI070026), the MICINN (PLE-2009-0111), and the Marie Curie (PIIF-GA-2009-236430). We are indebted to Dr. Saumitra Das for his kind gift of Fig. 2.

References

1. Moore, M. J. (2005) From birth to death: the complex lives of eukaryotic mRNAs. *Science* **309**, 1514–1518.
2. Halbeisen, R. E., Galgano, A., Scherrer, T., and Gerber, A. P. (2008) Post-transcriptional gene regulation: from genome-wide studies to principles. *Cell Mol Life Sci* **65**, 798–813.
3. Eulalio, A., Huntzinger, E., and Izaurralde, E. (2008) Getting to the root of miRNA-mediated gene silencing. *Cell* **132**, 9–14.
4. Chekulaeva, M., and Filipowicz, W. (2009) Mechanisms of miRNA-mediated post-transcriptional regulation in animal cells. *Curr Opin Cell Biol* **21**, 452–460.
5. Fritz, D. T., Bergman, N., Kilpatrick, W. J., Wilusz, C. J., and Wilusz, J. (2004) Messenger RNA decay in mammalian cells: the exonuclease perspective. *Cell Biochem Biophys* **41**, 265–278.
6. Parker, R., and Song, H. (2004) The enzymes and control of eukaryotic mRNA turnover. *Nat Struct Mol Biol* **11**, 121–127.
7. Schmid, M., and Jensen, T. H. (2008) The exosome: a multipurpose RNA-decay machine. *Trends Biochem Sci* **33**, 501–510.
8. Zinn, K., DiMaio, D., and Maniatis, T. (1983) Identification of two distinct regulatory regions adjacent to the human beta-interferon gene. *Cell* **34**, 865–879.

9. Melton, D. A., Krieg, P. A., Rebagliati, M. R., Maniatis, T., Zinn, K., and Green, M. R. (1984) Efficient in vitro synthesis of biologically active RNA and RNA hybridization probes from plasmids containing a bacteriophage SP6 promoter. *Nucleic Acids Res* **12**, 7035–7056.
10. Winter, E., Yamamoto, F., Almoguera, C., and Perucho, M. (1985) A method to detect and characterize point mutations in transcribed genes: amplification and overexpression of the mutant c-Ki-ras allele in human tumor cells. *Proc Natl Acad Sci USA* **82**, 7575–7579.
11. Lee, J. J., and Costlow, N. A. (1987) A molecular titration assay to measure transcript prevalence levels. *Methods Enzymol* **152**, 633–648.
12. Burczynski, M. E., Lin, H. K., and Penning, T. M. (1999) Isoform-specific induction of a human aldo-keto reductase by polycyclic aromatic hydrocarbons (PAHs), electrophiles, and oxidative stress: implications for the alternative pathway of PAH activation catalyzed by human dihydrodiol dehydrogenase. *Cancer Res* **59**, 607–614.
13. Benkusky, N. A., Fergus, D. J., Zucchero, T. M., and England, S. K. (2000) Regulation of the Ca^{2+}-sensitive domains of the maxi-K channel in the mouse myometrium during gestation. *J Biol Chem* **275**, 27712–27719.
14. Stabell, F. B., Tourasse, N. J., Ravnum, S., and Kolsto, A. B. (2007) Group II intron in Bacillus cereus has an unusual 3′ extension and splices 56 nucleotides downstream of the predicted site. *Nucleic Acids Res* **35**, 1612–1623.
15. Chen, G. L., and Miller, G. M. (2009) 5 -Untranslated region of the tryptophan hydroxylase-2 gene harbors an asymmetric bidirectional promoter but not internal ribosome entry site in vitro. *Gene* **435**, 53–62.
16. Cabrera, O., Roossinck, M. J., and Scholthof, K. B. (2000) Genetic Diversity of Panicum mosaic virus Satellite RNAs in St. Augustinegrass. *Phytopathology* **90**, 977–980.
17. Kim, J., Lee, S. H., Choe, J., and Park, T. G. (2009) Intracellular small interfering RNA delivery using genetically engineered double-stranded RNA binding protein domain. *J Gene Med* **11**, 804–812.
18. Ray, P. S., and Das, S. (2004) Inhibition of hepatitis C virus IRES-mediated translation by small RNAs analogous to stem-loop structures of the 5′-untranslated region. *Nucleic Acids Res* **32**, 1678–1687.
19. Brown, P. C., Thorgeirsson, S. S., and Silverman, J. A. (1993) Cloning and regulation of the rat mdr2 gene. *Nucleic Acids Res* **21**, 3885–3891.
20. Cloutier, N., Gravel, A., and Flamand, L. (2004) Multiplex detection and quantitation of latent and lytic transcripts of human herpesvirus-8 using RNase Protection Assay. *J Virol Methods* **122**, 1–7.
21. Lee, Y., Jeon, K., Lee, J. T., Kim, S., and Kim, V. N. (2002) MicroRNA maturation: stepwise processing and subcellular localization. *EMBO J* **21**, 4663–4670.
22. Zeiner, G. M., and Boothroyd, J. C. Use of two novel approaches to discriminate between closely related host microRNAs that are manipulated by Toxoplasma gondii during infection. *RNA* **16**, 1268–1274.
23. Chomczynski, P., and Sacchi, N. (1987) Single-step method of RNA isolation by acid guanidinium thiocyanate-phenol-chloroform extraction. *Anal Biochem* **162**, 156–159.
24. Chirgwin, J. M., Przybyla, A. E., MacDonald, R. J., and Rutter, W. J. (1979) Isolation of biologically active ribonucleic acid from sources enriched in ribonuclease. *Biochemistry* **18**, 5294–5299.
25. Katoch, V. M., and Cox, R. A. (1986) Stepwise isolation of RNA and DNA from mycobacteria. *Int J Lepr Other Mycobact Dis* **54**, 409–415.

Chapter 33

Array-Based Nuclear Run-On Analysis

Jinshui Fan, Yu-Chi Chen, Tonya Watkins, Chi V. Dang, Myriam Gorospe, and Chris Cheadle

Abstract

There is extensive evidence that posttranscriptional mechanisms of gene regulation, such as mRNA turnover, critically affect the patterns of expressed mRNAs. Conventional microarray analysis measures steady-state messenger RNA (mRNA) levels, which represents the dynamic balance between new transcription and mRNA degradation. Accordingly, only de novo transcription can accurately reflect the temporal and spatial events of transcriptional regulation. In this chapter, we describe a recently reported method to study transcription systematically. It involves the genome-wide labeling of nascent transcripts using nonradioactive modified nucleotides, their isolation for amplification, and their hybridization and analysis using commercial microarrays.

Key words: Nascent RNA, Nuclear run-on, Biotin, Microarray, Posttranscriptional regulation

1. Introduction

Biological processes are critically controlled by changes in gene expression. Many such changes are initiated through the altered binding of transcription factors (TFs) to specific DNA *cis*-elements that control the initiation of transcription and by-factors that affect the elongation and termination of transcription. Together, the temporal and spatial regulation of transcription is the primary determinant of both the fate and function of the cell. Virtually all microarray studies to-date have measured changes in the levels of steady-state mRNA by harvesting total cellular RNA and using it to generate mRNA-specific probes through a variety of strategies. However, an increasing appreciation for the scope and importance of posttranscriptional regulatory mechanisms such as changes in mRNA stability (1–4) has highlighted the need for a more precise

Ales Vancura (ed.), *Transcriptional Regulation: Methods and Protocols*, Methods in Molecular Biology, vol. 809,
DOI 10.1007/978-1-61779-376-9_33,

and direct measurement of TF-controlled transcription. Conventional microarray analysis cannot be used toward this end, as it employs steady-state/total RNA as the starting material to study a process that represents a changing balance between the rates of mRNA synthesis and decay. This approach almost certainly masks true transcriptional events, since mRNA turnover contributes measurably (sometimes exclusively) to changes in the abundance of expressed mRNAs.

In recent years, powerful array-based methods such as ChIP on Chip (chromatin immunoprecipitation using specific TFs cross-linked to genomic DNA binding sites followed by array interrogation of the captured nucleic acid sequences) have emerged to catalog and quantify the binding of TFs to DNA (5–7). These strategies are capable of identifying a large majority of TF-binding sites at the genomic level, including previously unknown regulatory regions. However, studies using these high-throughput ChIP approaches reveal that TFs often bind to gene loci whose transcription is not altered (8, 9). In addition, their application is limited by (1) the requirement of prior knowledge about the involvement of a particular TF in a specific biological event, (2) the inherent difficulty in identifying functionally coordinated TFs by this method, and (3) insufficient information about the functional consequences of the TF upon gene transcription (i.e., whether the TF–DNA binding activates, represses, or leaves unchanged the rate of transcription) (7, 10). Attempts to link the ChIP–Chip studies to actual changes in mRNA expression profiles are limited by the confounding contributions of transcription and decay rates, as described above. Thus, a reliable genome-wide method for detecting nascent mRNA transcripts is useful for the interpretation ChIP–Chip data by properly identifying changes in gene expression which appear to be directly influenced by TF binding.

Classical nuclear run-on (NRO) techniques have been used extensively for many years on a single-gene basis to directly measure nascent gene transcription rates. When used in conjunction with RNA polymerase inhibitors, these techniques have also been used to calculate mRNA decay rates. Because single-gene NRO methods both are quite laborious and require considerable skill, we first developed a simplified medium-throughput NRO assay. Based on the hybridization of metabolically radiolabeled nascent RNAs with early generation nylon membrane microarrays, this assay permitted the simultaneous measurement of nascent transcription for thousands of genes per sample per experiment (3). We extended the use of this technique to map nascent gene transcription profiles in human cervical cancer cells following stress stimulation (10). Using bioinformatic approaches to locate the possible regulatory *cis*-elements (promoters) of genes that were coordinately transcribed, we identified possible TFs involved; by subsequent RNAi and ChIP-PCR analyses, we validated the interactions between the predicted TFs and corresponding *cis*-elements (10). This strategy is advantageous because (1) it can potentially identify coordinately acting TFs, (2) it does not require prior knowledge of the involvement of TFs, and (3) the results directly reflect changes in nascent transcription as the functional consequences of the TF–DNA interactions.

However, the radioactive/nylon NRO arrays were largely limited by (1) the requirement of large amounts of radioactive material (typically 0.5 mCi [α^{33}P]-UTP per reaction) and large numbers of cells (typically >50 million per sample) and (2) technically demanding procedures and potential artifacts from hybridization of labeled antisense transcripts to double-stranded spotted cDNAs. To circumvent these issues, we have established an efficient genome-wide, nonradioactive labeled NRO array, which is compatible with most commercially available oligomer arrays (11).

2. Materials

Precautions should be taken to minimize RNase contamination throughout all of the steps (see Notes 1 and 2).

2.1. Preparation of Nuclei from Culture Cells

1. Phosphate-buffered saline (PBS), which can be prepared from 10× stock and precooled at 4°C.
2. DEPC-treated water, molecular biology grade.
3. Cell lysis buffer: 20 mM Tris–HCl, pH 7.5, 5 mM $MgCl_2$, 20 mM NaCl, 100 mM sucrose, 0.25% NP-40 (Sigma), prepared in nuclease-free water (see Note 3).
4. 2× Nuclei resuspension buffer: 50 mM Tris–HCl, pH 8.0, 5 mM $MgCl_2$, 0.1 mM EDTA, 45% glycerol, prepared in nuclease-free water (see Note 4).
5. Qiagen RNeasy Mini Kit.

2.2. Nascent RNA Labeling and RNA Purification

1. 2× NRO reaction buffer: 300 mM KCl, 10 mM $MgCl_2$, 2 mM each of rATP, rCTP, rGTP (Promega), 1 mM Bio-16-UTP (Enzo Life Sciences), 800 U/ml RNaseOUT (Invitrogen).
2. rUTP (Promega).
3. DNase I, RNase-free (Roche Applied Sciences).
4. Proteinase K, RNA Grade (Invitrogen).
5. Qiagen RNeasy® Mini Kit.
6. Ethanol, absolute.
7. TURBO DNase™ (Ambion).
8. (Optional) RNeasy MinElute Cleanup Kit (Qiagen).

2.3. Selection of Biotin-Labeled RNA, cDNA Synthesis, and cRNA Amplification

1. 1× PBS.
2. BeadWash Solution A: 100 mM NaOH, 50 mM NaCl, prepared with nuclease-free water.
3. Bead Wash Solution B: 100 mM NaCl, prepared with nuclease-free water.
4. Dynabeads® kilobaseBINDER™ Kit (Invitrogen/Dynal).
5. RNaseOUT (Invitrogen).
6. DEPC-treated water, molecular biology grade.
7. T7N6 Primer: 5′-GTA ATA CGA CTC ACT ATA GGG CTN NNN NN-3′ (Integrated DNA Technologies).
8. Illumina® TotalPrep RNA Amplification Kit (Ambion).

2.4. Affymetrix Exon Array

1. GeneChip® Whole Transcript (WT) cDNA Synthesis and Amplification Kit (Affymetrix).
2. GeneChip® WT Terminal Labeling Kit (Affymetrix).
3. GeneChip® Hybridization, Wash, and Stain Kit (HWS kit) (Affymetrix).
4. GeneChip® Human Exon 1.0 ST Array (Affymetrix).

2.5. Equipment

1. Low-speed table-top centrifuge.
2. Microcentrifuge.
3. Dynal MPC®-S Magnet Separator.
4. Dynal® RKDynal Sample Mixer.
5. Eppendorf Thermomixer® R Dry Block Heating and Cooling Shaker.
6. Hybridization Oven.
7. Nanodrop ND1000 Spectrophotometer.
8. Illumina Hybridization and Scanner System.
9. Affymetrix Hybridization and Wash Station and GeneChip® Scanner.

3. Methods

The method described here has been successfully applied to both human and mouse cell lines including human Jurkat T cells, human A549 lung adenocarcinoma epithelial cells, human P493-6 B lymphocytes, human H9 stem cells, and mouse embryonic fibroblasts (see Note 5). The major steps of this method are outlined in Fig. 1. Biotin UTP labeling of all of the nascent transcription allows for the potential labeling of nascent nonprotein-coding transcripts (antisense RNA, primary microRNA, etc.), as well as nascent pre-mRNA (see Note 6).

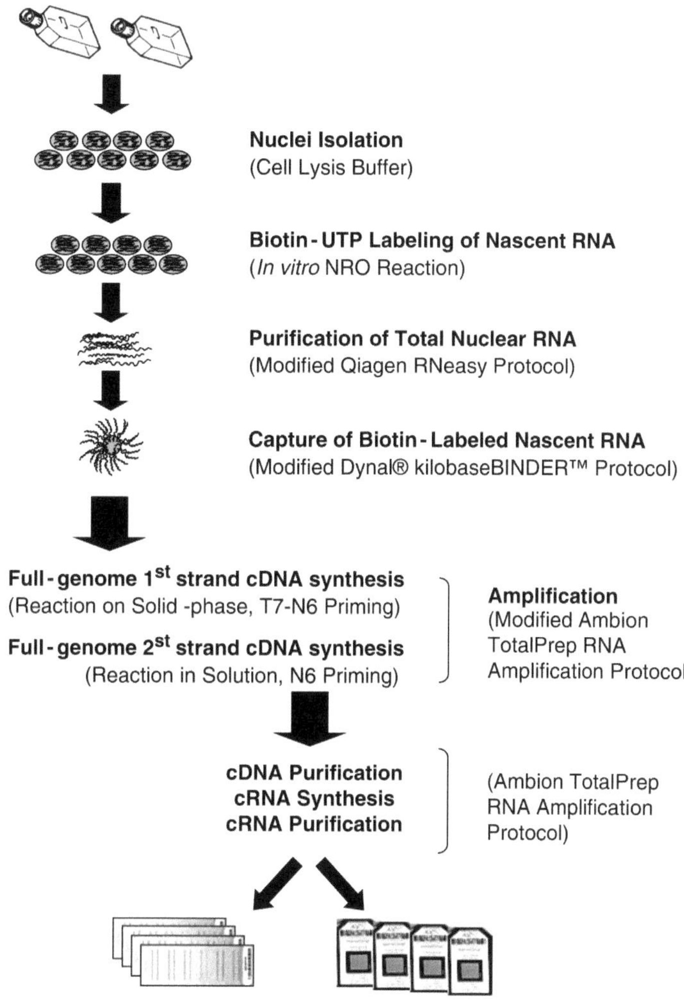

Fig. 1. Schematic of the protocol.

3.1. Preparation of Nuclei from Culture Cells

1. For adherent cultured cells, collect all cells of each condition into a 50 ml conical tube after trypsinization; for suspension cultured cells, directly collect cells into a 50 ml conical tube.
2. Pellet cells by spinning for 5 min at $216\times g$ (~1,000 rpm, tabletop centrifuge).
3. Carefully remove and discard the supernatant; wash cells with 40 ml of precooled 1× PBS and spin as in previous step.
4. Carefully remove and discard the supernatant; add 10 ml of precooled 1× PBS, resuspend cells by pipetting up and down several times or by inverting the tube several times.

5. Aliquot ~1 ml (~10%) of the cell suspension into microcentrifuge 1.5 ml tubes and spin 3–5 min at 2,000 × *g* in a microcentrifuge. These cells will be used for total RNA isolation and conventional microarrays (see Note 7).
6. For the remaining 90% of cells, spin again for 5 min at 216 × *g* in a table-top centrifuge; decant supernatant carefully, and proceed to isolate nuclei as follows.
7. Add 10 ml of cold Cell Lysis Buffer to the above cell pellet; mix by pipetting up and down several times or by inverting the tube several times; let tubes sit on ice for 6 min and invert tubes several times.
8. Spin 5 min. at 216 × *g* (~1,000 rpm, table top centrifuge).
9. Aspirate carefully (so as not to disturb the nuclei pellet) and discard the supernatant. Invert the 50 ml tube (with nuclei pellet at bottom) upside down on a piece of clean wipe paper on the laboratory bench to drain out excess solution inside the tube (see Note 8).
10. Place all tubes on ice horizontally to avoid collecting excess solution; resuspend each nuclei pellet with 100 µl of Nuclei Resuspension Buffer; transfer nuclei suspensions into appropriately labeled 1.5 ml microcentrifuge tubes and keep on ice until all of samples have been processed.

3.2. NRO Reaction for Nascent RNA Labeling and Purification of Total Nuclear RNA

1. Add equal volume of 2× NRO reaction buffer (usually the nuclei suspension reaches ~120 µl), mix well by inverting the tubes several times.
2. Incubate the NRO labeling reactions at 30°C for 30 min with constant mixing in an oven (see Note 9).
3. (Optional) Add cold rUTP to 1 mM (2 µl of 100 mM of stock per 200 µl of NRO reaction); continue the incubation for an additional 5–10 min.
4. Remove NRO reaction tubes from 30°C incubation and reset temperature at 37°C. Add 200 U of DNase I (10 U/µl, Roche Applied Sciences) to each reaction and incubate for 20 min at 37°C.
5. Add 400 U of Proteinase K (20 mg/ml, Ambion, premixed with 10% SDS at 3:1) and incubate 15 min at 37°C.
6. Adjust the NRO Reaction volume to 600 µl by adding RLT Buffer (Qiagen RNeasy Mini Kit); add equal volume (600 µl) absolute ethanol and mix well.
7. Load onto RNeasy column and follow RNeasy Mini Kit instructions thereafter.
8. Elute total nuclear RNA with 100 µl of nuclease-free water.

9. Measure RNA concentration using a Nanodrop ND1000 Spectrophotometer.
10. (Optional) Aliquot 10 μg of total nuclear RNA, treat with Ambion TURBO DNase™, and clean it using Qiagen RNeasy MinElute Cleanup Kit.

3.3. Preparation of Dynabeads and Immobilization of Biotin-Labeled Nascent RNA

1. Aliquot total amount of kilobaseBINDER Binding Solution (60 μl per sample, Dynabeads® kilobaseBINDER Kit) in a 1.5-ml microcentrifuge tube; add RNaseOUT at 5 μl per 100 μl Binding Solution; put on ice for later use.
2. Aliquot total amount of Dynabeads needed (30 μl per sample, Dynabeads® kilobaseBINDER™ Kit) in a 1.5 ml microcentrifuge tube, remove solution on the Magnetic Separator Stand.
3. Wash beads sequentially with two volumes of the total bead pellet volume using the following buffer: 1× PBS (cell culture grade), Bead Wash Solution A, and Bead Wash Solution B. For each wash, let beads settle completely on the Magnetic Separator Stand before discarding the supernatant.
4. Wash beads once with one volume of the above-described kilobaseBINDER Binding Solution (see Note 10).
5. Resuspend beads in equal volume (total bead amount) of above prepared kilobaseBINDER Binding Solution; put on ice for later use.
6. Denature 10 μg (in 30 μl nuclease-free water) purified biotinylated nuclear run-on RNA at 68°C for 3 min; pulse-spin to collect samples and immediately place on ice (see Note 11).
7. Dispense 30 μl of prepared beads into each RNA tube, pipette up and down to mix; mount onto Dynal® RKDynal Sample Mixer and set rotation speed at 8–10 rpm; incubate for 3 h at room temperature (see Note 12).

3.4. Bead Wash and T7N6 Primer Annealing for cDNA Synthesis

1. Prepare T7N6 Primer at 3 mM in nuclease-free water; put on ice for later use.
2. Carefully add 600 μl of nuclease-free water to each bead-RNA binding tube (Subheading 3.3); close tubes and invert several times to mix, spin 5–10 s at 2,000 × *g* in a microcentrifuge; place tubes on the Magnetic Separator Stand and remove the supernatant.
3. Thoroughly wash beads with nuclease-free water as follows: once with 600 μl, twice with 400 μl, and twice with 200 μl; each time allow the beads to settle down completely on the Magnetic Separator Stand (see Note 13).
4. After removing water from last wash, immediately resuspend beads in 24 μl of 3 mM T7N6 primer and place on ice (see Note 14).

5. Denature the T7N6 primer and beads-NRO RNA mix at 68°C for 3 min in a Thermomixer with constant shaking (1,000 rpm); immediately afterward, transfer tubes onto Sample Mixer and incubate at room temperature for 5–10 min with constant rotation (8–10 rpm).

3.5. Solid-Phase First-Strand and Solution-Based Second-Strand cDNA Synthesis

1. Prepare Master Mix for first-strand cDNA Synthesis at 16 μl for each sample (4 μl of 10× first-strand Buffer, 8 μl of dNTP solution, 2 μl of RNase Inhibitor, and 2 μl of ArrayScript, Illumina® TotalPrep RNA Amplification Kit).
2. Remove bead tubes from Sample Mixer, spin 5–10 s at 2,000 × *g* in a microcentrifuge to collect all beads and solutions at the bottom of the tubes.
3. Add 16 μl of the above prepared Master Mix for first-strand synthesis into each beads-NRO RNA tube and pipette up and down to mix well (now the total volume is 40 μl per reaction). Incubate the solid-phase first-strand cDNA Synthesis reaction for 2 h at the preset 42°C oven with constant rotation (8–10 rpm) (see Note 15).
4. Prepare Master Mix for RNase H digestion at 20 μl per reaction: 10 μl of nuclease-free water, 6 μl of 10× second-Strand Buffer, 2 μl of Invitrogen Random Primer N_6 (Invitrogen), and 2 μl of RNase H, Illumina® TotalPrep RNA Amplification Kit.
5. Remove all samples from the 42°C oven, spin 5–10 s at 2,000 × *g* in a microcentrifuge to collect all beads and solutions; at the same time, reset oven temperature at 37°C.
6. Add 20 μl of the Master Mix for RNase H digestion prepared above into each tube and pipette up and down to mix well (now the total volume reaches 60 μl per reaction). Incubate 30 min at 37°C oven with constant rotation (8–10 rpm).
7. Prepare Master Mix for second-strand cDNA Synthesis at 40 μl for each sample (30 μl of nuclease-free water, 4 μl of 10× second-strand cDNA Buffer, 4 μl of dNTP solution, and 2 μl of DNA Polymerase, Illumina® TotalPrep RNA Amplification Kit).
8. Remove all samples from 37°C oven, spin 5–10 s at 2,000 × *g* in a microcentrifuge to collect all beads and solutions.
9. Incubate tubes in 68°C for 3 min in a Thermomixer with vigorous mixing (above 1,000 rpm); immediately spin 5–10 s at 2,000 × *g* in a microcentrifuge to collect all beads and solutions and place tubes in the Magnet Separator Stand; remove solution fraction (will contain the first-strand cDNA) into appropriately labeled 1.5 ml microcentrifuge tubes and place on ice until all of the samples are collected (see Note 16).

10. Dispense 40 μl of above prepared Master Mix for second-strand cDNA Synthesis to each first-strand cDNA tube, pipette up and down to mix well (now the total volume is 100 μl per reaction); incubate 2 h in a Thermomixer at 16°C with constant shaking (650–850 rpm).

3.6. cDNA Purification, cRNA Synthesis, cRNA Purification, and Illumina Array Hybridization, Wash, Scan and Data Analysis

1. For cDNA purification and cRNA synthesis, follow instructions of Illumina® TotalPrep RNA Amplification Kit (see Note 17).
2. For cRNA purification, directly add 350 μl of cRNA Binding Buffer and 250 μl of absolute ethanol into the cRNA synthesis reaction tube, and follow instructions of Illumina® TotalPrep RNA Amplification Kit.
3. We recommend loading 5–10 μg of total nuclear run-on cRNA on Illumina BeadChips for hybridization at 58°C for 14–16 h. For array wash, scan and data analysis, follow instructions as for conventional Illumina arrays.

3.7. Application to Other Microarray Platforms

The biotin-labeled NRO cRNA prepared here can be applied to other microarray platforms with minor protocol modifications. Here, we describe our application to Affymetrix Exon Arrays (see Note 18).

1. First-strand cDNA Synthesis: Aliquot 10 μg NRO cRNA and follow instructions of GeneChip Whole Transcript (WT) cDNA Synthesis and Amplification Kit.
2. Cleanup Single-strand DNA (ssDNA): follow instructions of GeneChip WT cDNA Synthesis Kit.
3. Fragmentation of ssDNA: follow instructions of GeneChip® WT Terminal Labeling Kit.
4. Labeling of fragmented ssDNA: follow instructions of GeneChip® WT Terminal Labeling Kit.
5. Microarray hybridization and wash: follow instructions of GeneChip Hybridization, Wash and Stain Kit (HWS Kit).

4. Notes

1. For general precautions when working with RNA, all instruments that touch sample tubes should be kept nuclease-free. We typically wipe laboratory bench and centrifuge (including rotor and tube slots) and any other equipment in direct contact with tubes with RNaseZap® (Ambion, cat. # 9780).
2. We suggest purchasing solutions which are certified DNase-free and RNase-free from the manufacturers. In our experience, RNase-free solutions from Quality Biologicals perform well.

3. We typically prepare 10% NP-40 as regular stock and add the needed amount when preparing solutions. Cell Lysis Buffer can be stored at 4°C for long-term use (more than 6 months).
4. The 2× Nuclei Resuspension Buffer can be dispensed into 1 ml aliquots and stored at −80°C for extended periods of time (more than 6 months).
5. To get consistent adequate labeling, we typically use $20–30 \times 10^6$ of either adherent or suspension cells for isolation of nuclei for each condition (untreated and treated cells); we use trypsinization to collect adherent cells. For comparison between nascent transcription and steady-state RNA expression profiles, we recommend aliquoting a small portion of cells (such as 1/10) for total RNA isolation and conventional microarray analysis.
6. We have successfully detected nascent primary microRNA transcripts using real-time PCR with the double-stranded cDNAs (before cRNA synthesis) as the template. Modified procedures of the cRNA synthesis step (no addition of biotin-modified ribonucleotides) followed by labeling of first-strand cDNA (using the NRO cRNA as prepared above as template) with biotin-modified dCTP can be applied to detect antisense transcripts.
7. Cell pellet can be stored at −80°C for total RNA isolation at later times. We typically use Qiagen RNeasy Mini Kit with DNase I treatment to isolate total RNA for conventional microarray.
8. Watch out for the nuclei pellet sliding down the wall of the conical tube!
9. We recommend carrying out the reaction in a hybridization oven. First, set the oven temperature at 30°C; then fix the microcentrifuge 1.5 ml tubes on the rotor with rubber bands; remount the rotor and keep rotation at 8–10 rpm.
10. The kilobaseBINDER Binding Solution is viscous and it takes longer for the beads to settle onto the wall of the tube on the Magnet Separator Stand. After placing the tubes on the Magnetic Separator Stand and solution, we recommend pipetting up and down several times while the solution turns clear to prevent the loss of beads retained in the tips.
11. We typically apply 10 μg of the total NRO RNA for bead capture although it can go down to 5 μg depending on the labeling efficiency.
12. As mentioned above, the kilobaseBINDER Binding Solution is viscous, thus be careful to not make too many bubbles when pipette mixing the beads with RNA samples. On the other hand, a few unavoidable bubbles are actually helpful in maintaining a homogenous mixing state in the Dynal® RKDynal

Sample Mixer. We typically use two ways of binding depending on the time frame: one is incubating 3 h at room temperature and then forward to the next step immediately; the other is incubating 2 h at room temperature, and then transfer the Sample Mixer to the 4°C cold room to incubate over night and begin subsequent steps the following day.

13. We recommend resuspending the beads at each wash by flipping or inverting the tubes. For the last 200 μl wash, we typically remove 150 μl and flip the remainder to mix beads, spin 2,000 × *g* in a microcentrifuge, and finally remove the remaining 50 μl of water on the Magnetic Separator Stand. In addition, to prevent the beads from over drying during handling, we suggest handling no more than four samples for each wash, and no more than two samples when removing 150 μl from the last 200 μl wash.
14. We recommend adding 24 μl of 3 mM T7N6 primer solution right above the beads on the wall of each tube, tightening the tube, gently flipping to resuspend all beads in solution, then putting on ice until all samples are finished.
15. We recommend placing the tubes immediately on the rotor in the 42°C oven to keep constant rotation after mixing each sample tube. To further prevent beads from settling in the tube, after mounting all tubes on the rotor, one measure that can be taken is to set the rotor a little unevenly for a slightly asymmetric rotation.
16. We recommend handling no more than two samples at the 68°C 3-min step. After putting two tubes on the Magnetic Separator Stand, load the next two samples for the 68°C 3-min step while waiting for the beads of the first two samples to settle on the Magnetic Separator Stands for separating the first-strand *cDNA in solution.*
17. For consistent results, we recommend using Zymo DNA Clean & Concentrator™-5 Kit (Zymo Research cat. #D4014) for cDNA purification.
18. The protocols for application of NRO cRNA to other Affymetrix Arrays are not described in detail here. The Affy exon array data highlighted here demonstrates that the ANRO array method described above can effectively detect full-length, nascent transcripts as exemplified by the detection of the nascent transcript of PRKDC gene (Fig. 2), which is composed of 91 exons and spans 190 kb of genomic DNA. In our experience, the Affymetrix Exon Arrays generated far fewer statistically significant regulated genes as compared to the Illumina BeadChips, although there is a high correlation between the genes detected by both platforms.

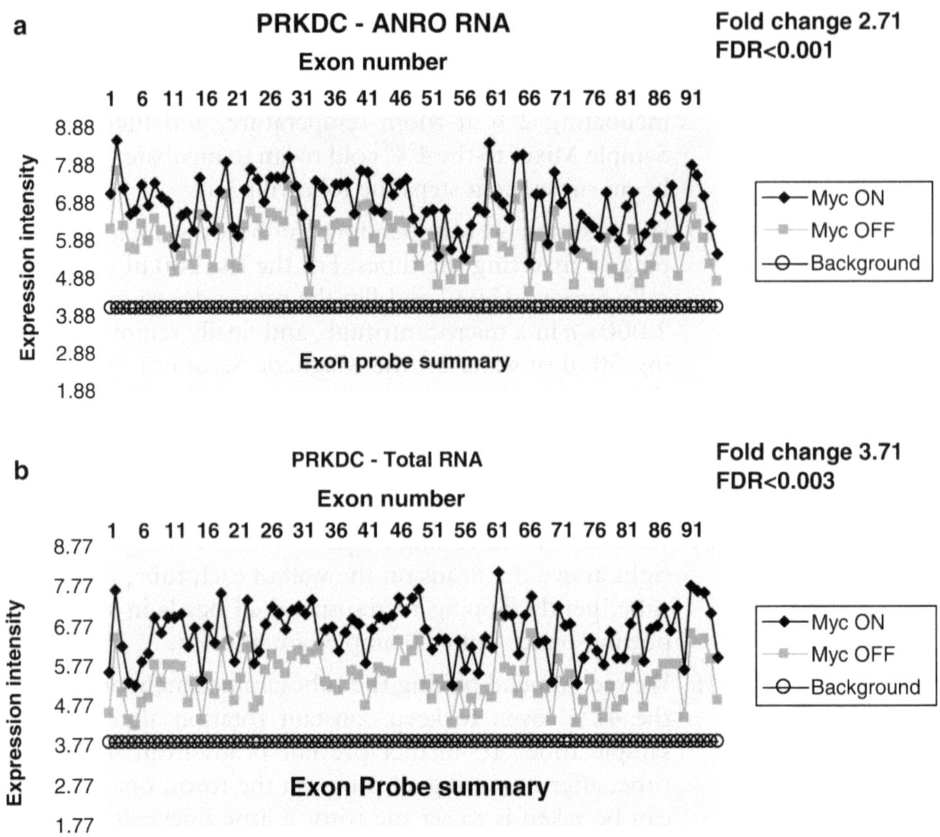

Fig 2. Detection of full-length nascent transcripts by the ANRO method. Affymetrix exon arrays were used to detect the 91 exons of the PRKDC gene spanning 190 kb of genomic DNA. As the PRKDC gene transcription is under control by the Myc promoter, PRKDC transcription rates and mRNA abundance were compared under conditions of high Myc expression (Myc ON) and no Myc expression (Myc OFF). *Graphs* represent the expression intensity of each exon using nascent pre-mRNA prepared by the ANRO method (**a**) and using total RNA prepared using conventional methods (**b**). As shown, all exons are represented in ANRO arrays, indicating that the nascent PRKDC pre-mRNA is labeled evenly throughout the entire transcript (**a**). Moreover, the label incorporation in the newly transcribed PRKDC pre-mRNA (**a**) mirrors the abundance of each exon in the mature PRKDC mRNA (**b**). When comparing Myc ON vs. Myc OFF, PRKDC pre-mRNA levels are induced by 2.71-fold (**a**), PRKDC mRNA levels by 3.71-fold. *FDR* false discovery rate.

References

1. Raghavan A, Ogilvie RL, Reilly C, Abelson ML, Raghavan S, Vasdewani J, Krathwohl M, Bohjanen PR (2002) Genome-wide analysis of mRNA decay in resting and activated primary human T lymphocytes. *Nucleic Acid Res* 30: 5529–5538
2. Wang Y, Liu CL, Storey JD, Tibshirani RJ, Herschlag D, Brown PO (2002) Precision and Functional Specificity in mRNA Decay. *Proc Natl Acad Sci USA* 99: 5860–5865
3. Fan J, Yang X, Wang W, Wood WH III, Becker KG, Gorospe M (2002) Global Analysis of Stress-regulated mRNA Turnover Using cDNA Arrays. *Proc Natl Acad Sci USA* 99: 10611–10616
4. Cheadle C, Fan J, Cho-Chung YS, et al. (2005) Control of gene expression during T cell activation: alternate regulation of mRNA transcription and mRNA stability. *BMC Genomics* 6: 75

5. Wei CL, Wu Q, Vega VB, et al. (2006) A global map of p53 transcription-factor binding sites in the human genome. *Cell* 124: 207–219
6. Hallikas O, Palin K, Sinjushina N, Rautiainen R, Partanen J, Ukkonen E, Taipale J (2006) Genome-wide prediction of mammalian enhancers based on analysis of transcription-factor binding affinity. *Cell* 124: 47–59
7. Weinmann AS (2004) Novel ChIP-based strategies to uncover transcription factor target genes in the immune system. *Nature Rev Immunol* 4: 381–386
8. Zeller KI, Zhao X, Lee CW, et al. (2006) Global mapping of c-Myc binding sites and target gene networks in human B cells. *Proc Natl Acad Sci USA* 103: 17834–17839
9. Kim J, Chu J, Shen X, Wang J, Orkin SH (2008). An extended transcriptional network for pluripotency of embryonic stem cells. *Cell* 132: 1049–1061
10. Fan J, Zhan M, Shen J, Martindale JL, Yang X, Kawai T, Gorospe M (2006) En masse nascent transcription analysis to elucidate regulatory transcription factors. *Nucleic Acids Res* 34: 1492–1500
11. Fan J, Zeller K, Chen YC, Watkins T, Barnes KC, Becker KG, Dang CV, Cheadle C (2010) Time-Dependent c-Myc Transactomes Mapped by Array-Based Nuclear Run-On Reveal Transcriptional Modules in Human B Cells. *PLoS ONE* 5: e9691

Chapter 34

In Vivo Run-On Assays to Monitor Nascent Precursor RNA Transcripts

Piergiorgio Percipalle and Emilie Louvet

Abstract

Biochemical methods have provided mechanistic insights into the different transcription phases during which the RNA polymerase is assembled at gene promoter and becomes engaged in the elongation of nascent transcripts. Evidence that transcription takes place in specific regions of the nucleus has fuelled the need to develop assays that can be performed in living cells and provide information on the location of the specific foci, where transcription takes place. In this chapter, we describe a method that is based on the incorporation of a fluorine-conjugated uridine analogue, incorporation that can be monitored by immunofluorescence and light microscopy using specific fluorochrome-conjugated monoclonal antibodies. This assay allows direct monitoring of active transcription foci in living cells. When coupled to suitable software, the method outlined here also provides a semiquantitative approach to measure the number of active transcription foci that correlate with the proliferation state of the cell. Therefore, the assay we present here is a sensitive analytical tool to monitor the topology of transcription foci in the eukaryotic cell nucleus and to gain insight into transcription rates.

Key words: Nascent RNA transcripts, Pre-mRNA, transcription inhibitors, Immunofluorescence, Transcription foci, Confocal microscopy, Image processing and analysis

1. Introduction

Gene transcription is highly conserved from prokaryotes to eukaryotes and represents the first step in the flow of gene expression. Mechanistically, transcription occurs in multiple phases. Following initial assembly at gene promoter, the RNA polymerase machinery becomes engaged in the elongation of nascent RNA transcripts (1). Commitment of eukaryotic polymerases into the elongation phase is governed by multiple factors. In the case of protein-coding genes, an essential part is played by the carboxy-terminal domain

Ales Vancura (ed.), *Transcriptional Regulation: Methods and Protocols*, Methods in Molecular Biology, vol. 809,
DOI 10.1007/978-1-61779-376-9_34, © Springer Science+Business Media, LLC 2012

(CTD) of the RNA polymerase II (pol II) large subunit. The CTD becomes phosphorylated at different serine residues within its conserved heptapeptide repeats (2, 3). These phosphorylation events contribute to the establishment of a very specific molecular platform that mediates recruitment of several elongation factors to active gene. Transcription is terminated when the polymerase encounters terminator sequences located towards the 3′ end of the gene (4). RNA polymerase I (pol I)- and RNA polymerase III-mediated transcription elongations are less understood. In any case, all three eukaryotic polymerases require a profound interplay with chromatin structure for efficient elongation of precursor RNA molecules.

The above mechanisms are essential for basal gene expression and play an important role in differentiation. Yet we lack a clear picture of how and where transcription takes place in vivo. The fact that eukaryotic RNA polymerases are recruited to the promoters of genes when they become active suggests that transcription sites form de novo on individual genes. It is also believed that nascent transcription is compartmented within limited numbers of discrete foci (5, 6) containing high concentration of active RNA polymerases. The importance of compartments in transcription is especially evident for the non-coding rRNA genes. In this case, transcription occurs in nucleoli, at the boundary between dense fibrillar centres and fibrillar component (7). Interestingly, during mitosis, certain subunits of the RNA polymerase I machinery remain associated with inactive nucleolar organizer regions (NORs) after nucleoli have been disassembled (8), indicating that certain polymerase subunits associate with both active and inactive genes.

In this complex scenario, it is important to be able to monitor nascent RNA transcripts. In vitro transcription assays have contributed to our understanding of the different transcription phases and their crosstalk with the chromatin template (9, 10). However, these studies do not allow us to visualize transcripts in living cells and above all they do not provide topological information on the transcription foci within the nucleoplasmic volume. To address these questions, we and others have used the bromine-conjugated UTP analogue known as BrUTP in run-on transcription assays performed on living cells (11), a method originally devised by Moore and Ringertz using radioactively labelled UTP (12). In these assays, BrUTP is allowed to be incorporated into nascent RNA molecules. The incorporated BrUTP can be subsequently visualized by indirect immunofluorescence using commercially available fluorochrome-conjugated monoclonal antibodies in combination with light microscopy. In these assays, the use of buffers with different ionic strengths allows BrUTP incorporation in nucleoplasmic or nucleolar foci, providing the opportunity to monitor global transcription or just pol I transcription, respectively. However, BrUTP is not cell permeable and therefore cells subjected to run-on assays must be permeabilized as condition for BrUTP incorporation. Recent developments have led to the use of fluorine-conjugated UTP (FUrd) analogues. Here, the advantage is that FUrd is readily incorporated by living cells and does not require cell permeabilization with detergents. FUrd is incorporated into nucleolar foci much more rapidly in comparison to nucleoplasmic foci (13, 14), reflecting the fact that rRNA comprises more than 80% of cellular RNA. This, in combination with the use of fluorochrome-conjugated secondary antibodies and confocal microscopy, provides a highly reliable methodology to visualize discrete and active transcription sites in living cells.

Here, we describe the steps to perform in vivo transcription assays using FUrd incorporation. In our applications, the number of transcription foci corresponds to the immunolocalization of FUrd incorporation sites. Therefore, the method also provides the bases for a semiquantitative analysis of pol I and pol II transcription in living cells. Finally, since FUrd is preferentially incorporated into nascent transcripts, this method is highly suited to monitor the transcription elongation phase in living cells.

2. Materials

2.1. Cell Culture

1. Dulbecco's modified Eagle's medium (DMEM), 10% fetal calf serum (FCS), penicillin (0.01 mg/ml), streptomycin (0.01 mg/ml).
2. Trypsin (2.5 mg/ml), cell culture grade. Aliquot and store at −20°C.
3. Low-passage HeLa cells.
4. Water-jacketed incubator or equivalent tissue cell culture incubator with maintained standard temperature, humidity, and CO_2 level.
5. Sterile 10× phosphate-buffered saline (PBS). Dilute 1:10 with sterile water when required.
6. Sterile glass coverslips, 13 mm diameter No. 1 (VWR).
7. 24- or 12-well plates

2.2. Transcription Inhibition

1. 10 mg/ml actinomycin D stock solution in 70% ethanol. Keep aliquots at −20°C in the dark (Fig. 1).
2. 1 mg/ml 5,6-dichlorobenzimidazole 1-β-D-ribofuranoside, DRB (Sigma) stock solution in 70% ethanol (Fig. 2).

2.3. Run-On Transcription Assay on Living Cells

1. 50–80%-confluent HeLa cells grown on coverslips.
2. 200 mM FUrd (Sigma) stock solution in dimethyl sulfoxide (DMSO). Keep aliquots at −20°C.

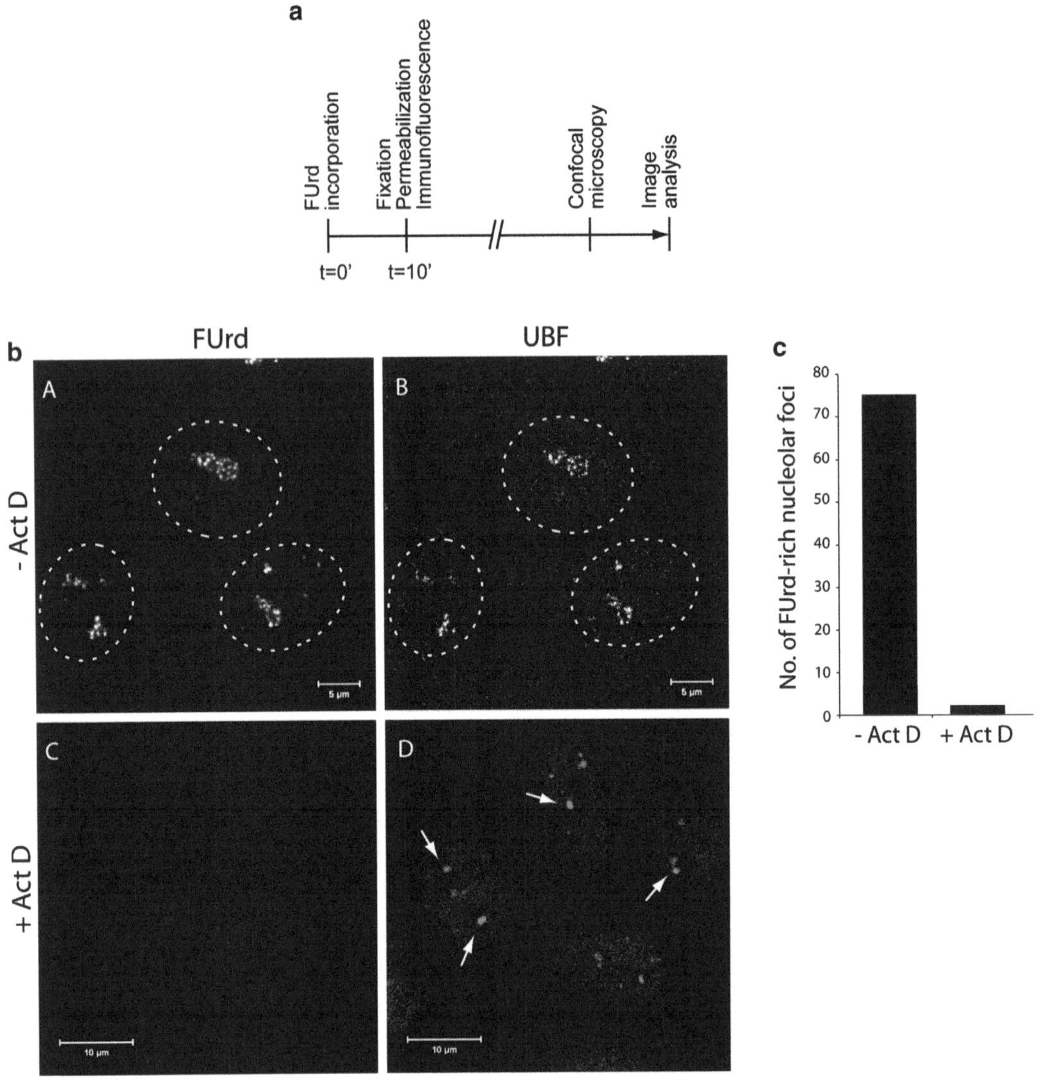

Fig. 1. Monitoring pol I transcription foci in living cells. (**a**) Experimental set-up. (**b**) Run-on assay performed on living HeLa cells: where indicated cells are treated with 0.04 μg/ml actinomycin D (Act D) for a selective pol I transcriptional block. In untreated and actinomycin D-treated cells, transcription is monitored by following the FUrd incorporated into nascent rRNA molecules during the time course of the experiment. Incorporated FUrd was visualized by immunostaining and confocal microscopy. Panels **a** and **b**, dotted lines, identify examples of nuclei, where nucleolar FUrd incorporation and UBF labelling are detected (scale bar, 5 μm). Panels **c** and **d**, *arrows*, point towards examples of UBF-rich nucleolar caps which result from the actinomycin D block of pol I transcription (scale bar, 10 μm). (**c**) Quantification of FUrd foci was by measurement on randomly selected nucleolar regions using the ImageJ software. The results are displayed as bar diagrams.

3. DMEM supplemented with FCS and antibiotics, pre-warmed at 37°C.
4. 10× PBS buffer. Dilute 1:10 with bi-distilled water. Keep at 4°C.

2.4. Immuno-fluorescence

1. Glass slides (Menzel) for mounting coverslips.
2. 10× PBS buffer. Dilute 1:10 with bi-distilled water. Keep at room temperature (RT).

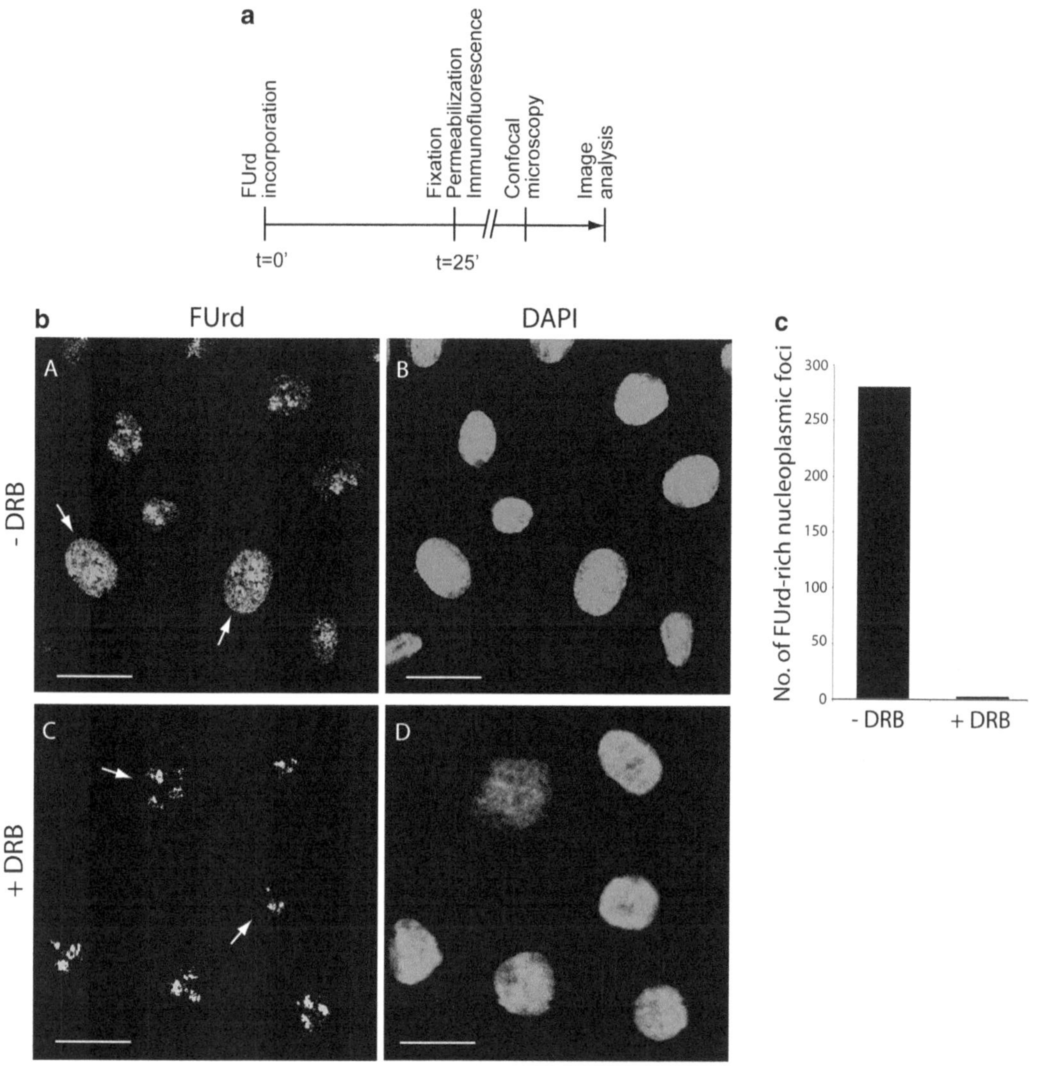

Fig. 2. Run-on assay to reveal global nucleoplasmic transcription foci in living cells. (**a**) Experimental set-up. (**b**) Run-on assay on living HeLa cells: where indicated that the experiment is performed after a specific block of the elongating pol II machinery with 75 μM DRB. FUrd incorporated in elongating transcripts was monitored by immunostaining and confocal microscopy (scale bar, 10 μm). In support of the specificity of the reaction, FUrd incorporation in nucleoli is hardly affected in the presence of the pol II-specific drug DRB. Panels **a**, **c**, *arrows*, point towards examples of nucleoplasmic FUrd incorporation. (**c**) Quantification of nucleoplasmic FUrd foci was by measurement of randomly selected nucleoplasmic regions by ImageJ. The data set is displayed as bar diagram.

3. 37% formaldehyde stock solution (Sigma). For cell fixation, dilute 1:10 with 1× PBS. Protect from light and keep at RT (see Note 1).
4. Permeabilization solution: 0.5% (v/v) Triton X100 (Sigma) diluted in 1× PBS.
5. Blocking solution: 0.5% milk powder dissolved in PBS. Keep refrigerated at 4°C.

6. Primary and secondary antibodies dilution buffer: 0.5% milk in PBS.
7. Primary antibodies: Mouse monoclonal antibody against BrdU (B2531, Sigma) and rabbit polyclonal antibody against the Upstream Binding Factor, UBF (sc-9131, Santa Cruz Biotechnology) (see Note 2).
8. Secondary antibodies: Alexa Fluor® 568 goat anti-mouse antibodies (A11004, Invitrogen) and Alexa Fluor® 488 donkey anti-rabbit antibodies (A21206, Invitrogen).
9. Nuclear stain: 300 nM 4,6-diamino-2-phenylindole (DAPI) in water. Store at 4°C in the dark (see Note 3).
10. Mounting medium: Mowiol (Mowiol 4.88, Calbiochem). Prepare stock solutions and store at −20°C (see Note 4).

2.5. Confocal Microscopy, Data Acquisition, and Image Processing Software

1. LSM 510 meta confocal microscope equipped with data acquisition system (Zeiss) (see Note 5).
2. ImageJ, free software downloadable at http://www.rsbweb.nih.gov/ij/ (see Note 6).

3. Methods

As a general guideline, laboratory work benches must be clean and protected from dust. Through the entire image processing and analysis work, data acquisition should be performed as soon as possible after preparation of the biological sample for microscopy. For image processing and analysis, always keep raw image data and work on copies.

3.1. Run-On Transcription Assay

1. Grow HeLa cells on coverslips in 12- or 24-well plates up to 50–80% confluence.
2. Remove medium. For DRB treatment, add fresh DMEM medium containing 75 μM DRB. Incubate for 1 h. For actinomycin D treatment, add fresh DMEM containing 0.04 μg/ml actinomycin D. Incubate for 2 h. In both cases, perform incubation at 37°C (see Note 7).
3. Wash cells with fresh DMEM medium pre-warmed at 37°C.
4. Incubate untreated, DRB-treated, and actinomycin D-treated cells with pre-warmed DMEM medium containing 2 mM FUrd (see Note 8). Perform incubation for up to 10 min to reveal pol I transcription foci (Fig. 1). Perform incubation for up to 20 min to reveal nucleoplasmic transcription foci. In all cases, perform incubation at 37°C (Fig. 2).

5. To stop the run-on assay, remove the FUrd-containing DMEM medium and wash cells with cold PBS.
6. Keep cells in PBS and proceed to the immunofluorescence protocol.

3.2. Immuno-fluorescence

1. Fix cells for 10 min with a solution containing 3.7% formaldehyde in PBS (see Note 9).
2. Wash cells 3× for 5 min with PBS.
3. Permeabilize the cells with a solution containing 0.5% Triton X100 in PBS for 5 min at RT.
4. Wash cells 3× for 5 min with PBS.
5. Block for 1 h at RT with a solution containing 0.5% milk dissolved in PBS. Alternatively, blocking can be done overnight at 4°C.
6. Wash cells 3× for 10 min.
7. Incubate cells with solutions of primary antibodies against BrdU and UBF supplemented with 0.5% milk in PBS for 1 h at RT.
8. Wash cells 3× for 10 min.
9. Incubate cells with Alexa488- and Alexa568-conjugated mouse and rabbit secondary antibodies diluted according to the manufacturer's instruction leaflet in the same solution as the primary antibodies.
10. Stain cells with 300 nM DAPI for 3 min.
11. Mount coverslips on glass slides using Mowiol (see Note 4). For mounting, place a drop of Mowiol on coverslip. Next, carefully take coverslips with tweezers and place upside down (cells facing down towards glass slide) on the Mowiol drop pre-spotted on the slide. Leave in the dark overnight to allow Mowiol to harden before oil immersion lenses are used.
12. After Mowiol polymerization, proceed to confocal microscopy.

3.3. Fluorescence Confocal Microscopy and Data Acquisition

1. Use an LSM 510 meta confocal microscope and a 63× numerical aperture (NA) 1.3 or 1.4 oil immersion objective (see Note 10).
2. To visualize DAPI staining, use a 405-nm laser diode. The use of other excitation wavelengths depends on the fluorochromes conjugated to the secondary antibodies. To visualize the Alexa488-conjugated secondary antibody, use an excitation wavelength at 488 nm produced by the argon laser in the LSM system. To reveal the Alexa568-conjugated secondary antibody, use the excitation wavelength at 543 nm produced by the helium–neon laser in the LSM system (see Note 11).

3. Collect a number of images to allow enough data for statistical analysis and image processing (Figs. 1 and 2).
4. Save images as tagged image file format (tiff).

3.4. Image Processing for Quantification of Transcription Foci

On a general basis, there is no strict definition of image processing. Moreover, the difference between image processing and image analysis (see Subheading 3.5) is sometime not easy to establish. Briefly, in image processing, the input is an image and the output of the processed image is an image. For example, an 8-bit, grey-level picture is processed into a binarized picture (22).

One should keep in mind the reason why image processing is necessary. Is it for image visualization? Is it to facilitate the analysis? What kind of analysis does one want to perform? Processing of the data should be done by keeping in mind the final goal. In our application, the final goal is to have an image with objects segmented from the background (Fig. 3).

1. For image processing, open tiff images using the ImageJ software (see Note 12).
2. Convert RGB images to greyscale and 8-bit mode (see Note 13).
3. Correct noise if necessary (see Notes 14 and 15).
4. If necessary, improve brightness and contrast so that objects are clearly distinguishable by applying *Image > Adjust > Brightness/Contrast*.
5. Binarize the picture either by *Process > Binary > Make Binary* or by *Image > Adjust > Threshold* (see Notes 16–18).

3.5. Image Analysis on Transcription Foci

The overall objective of image analysis is to extract meaningful information. In the specific example presented here, the input is an image containing transcription foci and the output generated is a table of numbers. For image analysis with the ImageJ software, we use the command "Analyze particles", which counts and measures objects in binary or thresholded images by scanning the image until it finds the edge of an object (see http://www.rsb.info.nih.gov/ij/docs/menus/analyze).

1. Analyze binary picture (or a thresholded picture) from above using the submenu command *Analyze > Analyze Particles.* A dialogue box opens which enables configuration of the particle analyzer (see Note 19).
2. Set the *Size* parameter to ignore particles outside the selected range. For pol I transcription foci, set lower value to 10 pixel^2; for pol II transcription foci, set lower value to 0 or 5 pixel^2 (see Note 20).
3. Set the *Circularity* parameter to ignore particles outside the range specified.

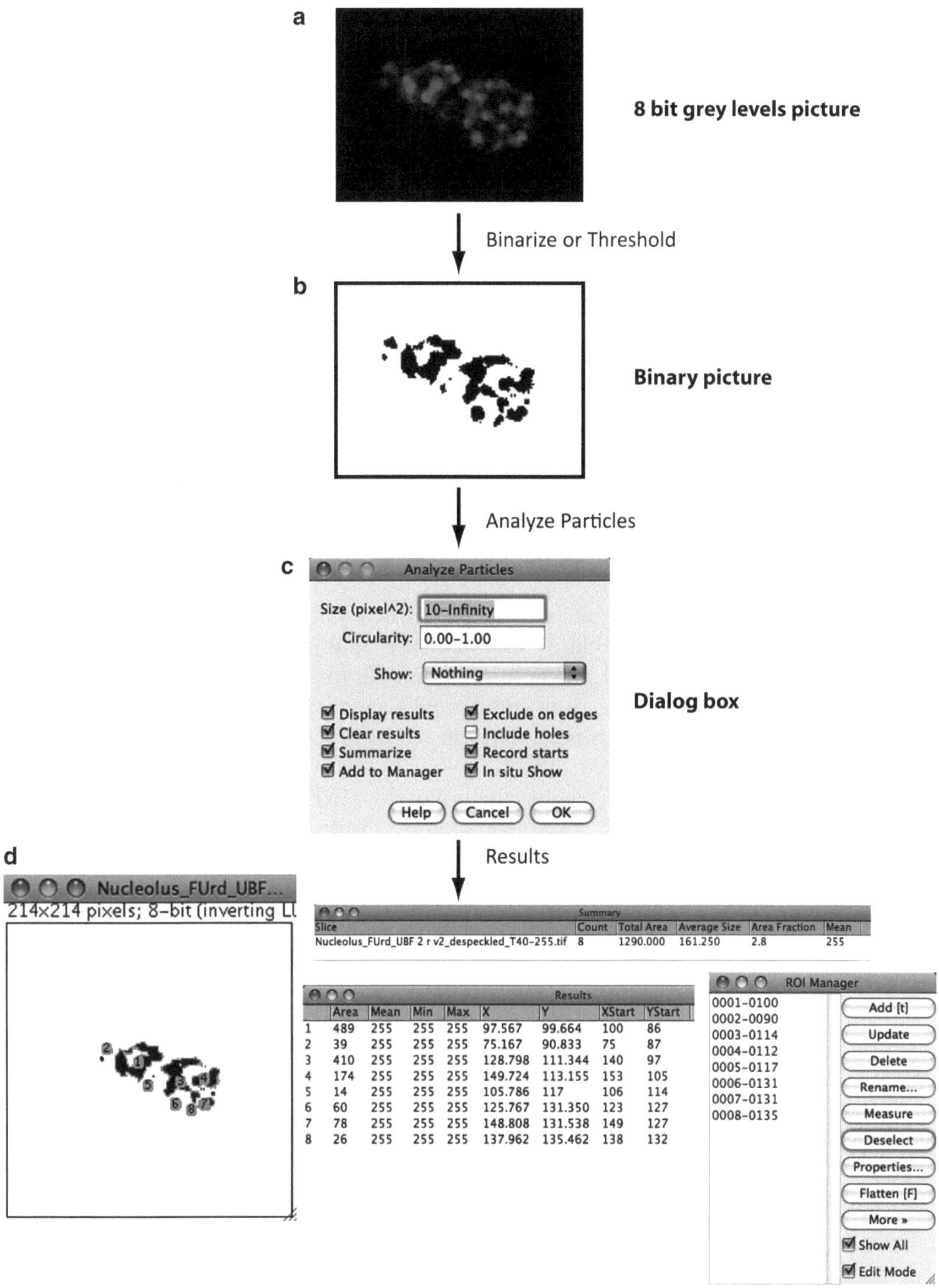

	Area	Mean	Min	Max	X	Y	XStart	YStart
1	489	255	255	255	97.567	99.664	100	86
2	39	255	255	255	75.167	90.833	75	87
3	410	255	255	255	128.798	111.344	140	97
4	174	255	255	255	149.724	113.155	153	105
5	14	255	255	255	105.786	117	106	114
6	60	255	255	255	125.767	131.350	123	127
7	78	255	255	255	148.808	131.538	149	127
8	26	255	255	255	137.962	135.462	138	132

Fig. 3. A workflow for the quantification of active transcription foci. (**a**) 8-bit, *grey*-level picture of pol I transcription foci in one nucleolus and (**b**) corresponding binary picture. (**c**) Dialogue box appearing after selection of the particle analyzer. (**d**) An example of windows that appear once the calculations have been done. The binary picture appears with outlined particles/transcription foci with a number corresponding to each one. One *Summary* table informing about the total count and average values. The Results table provides information for each individual particle.

4. Select the command *Show* to visualize particles. Select the function *Nothing* to attribute a number to each particle and outline them with a blue line (grey in Fig. 3) (see Note 21).
5. Chose the command *Display results* to show measurements of each particle in the "Results" window (Fig. 3). The option *Clear results* is used to erase any previous measurements.
6. Display particles count, total particles area, average particle size, and area fraction in a separate window using the command *Summarize*.
7. Add the measured particles in the *R*egion *O*f *I*nterest (*ROI*) *Manager* using the command *Add to manager* (see Note 22).
8. Ignore particles on edges (i.e. borders of the image or selection) using *Exclude on edges* (see Note 23). Follow the link http://www.rsb.info.nih.gov/ij/docs/menus/analyze for further details on the parameters and functions discussed above.
9. The outlines of the counted particles corresponding to individual transcription foci appear on the binary pictures. Save as tiff file. The results are displayed in three tables: the "Results" that can be saved as a Microsoft Excel file (other options are available in the menu), a "Summary" table that can also be saved in Microsoft Excel file format, and the "ROI manager" (the option *More* provides several possibilities: one could save it with the extension *.roi* or as *RoiSet.zip*).

3.6. Data Representation

Open the results table using the Microsoft Excel software. Evaluate statistical parameters and display results as bar diagrams (histograms).

4. Notes

1. Formaldehyde is light sensitive and must be kept in the dark.
2. The commercially available mouse monoclonal antibody against bromine-conjugated deoxyuridine (BrdU) is known to cross-react with both BrUTP and FUrd.
3. DAPI is a cell-permeable fluorescent stain that binds strongly to DNA and is extensively used to label nuclei. DAPI binds also to RNA, but it is not as strongly fluorescent. The DAPI solution is light sensitive and must be kept in the dark. There are other alternatives to label the cell nucleus. The Hoechst stains are used to label DNA and are commonly used in fluorescence microscopy to visualize nuclei and mitochondria. Hoechst 33258 and the more lipophilic Hoechst 33342 are commonly used. The Hoechst stains bind to DNA and impair

DNA replication during cell division. These dyes are potentially mutagenic and carcinogenic; therefore, it is important to be careful in their handling and disposal. Propidium iodide (or PI) is an intercalating agent staining DNA. PI is used both for flow cytometry and microscopy to visualize the nucleus and other DNA-containing organelles. It also binds to RNA, and therefore nuclease treatment is required to distinguish between RNA and DNA staining. PI is not membrane permeable.

4. Mowiol (Mowiol 4.88, Calbiochem, catalogue number 475904) is a solution of polyvinyl alcohol, which normally hardens overnight after slide preparation and does not require the coverslips to be sealed with nail polish (15). For preparation of Mowiol, place 6 g glycerol in a 50-ml centrifuge tube, add 2.4 g Mowiol, and stir to mix. While stirring, add 6 ml distilled water and leave for 2 h at RT. Add 12 ml 2 M Tris–HCL, pH 8.5, and sodium azide (NaN_3) to a final concentration of 0.02% (optional). Incubate the tube in a hot water bath (50–60°C) for 10 min to dissolve the Mowiol. This can be repeated over several hours if necessary. Centrifuge at 5,000 × *g* for 15 min to remove any particulate. For storage, keep Mowiol as a 1-ml aliquot at −20°C. Before use, warm tubes to room temperature. Opened tubes can be stored at 4°C for approximately 1 month or at −20°C for longer storage. Discard if any crystalline material is seen in the tube or on the slides.
5. A wide-field fluorescence microscope equipped with a CCD camera and data acquisition software can be used as an alternative to the confocal microscope.
6. ImageJ was and is continuously being developed by Wayne Rasband, the US National Institutes of Health (NIH). It offers a set of ready-made tools for viewing and interactive manipulation of images. It can also be extended easily by writing new software components in Java (16). We routinely use ImageJ because it is user-friendly software. No license is required and it runs on Linux, Mac OS X, as well as Windows. Other software can be used as alternatives, including Volocity (Perkin-Elmer), which is user friendly and can also be used for advanced applications or Matlab (MathWorks), coupled to a toolbox like DIPimage (license free) downloadable on http://www.diplib.org/ (created by Cris Luengo at the Centre for Image Analysis, Uppsala University, Sweden).
7. DRB specifically inhibits pol II-mediated transcription elongation by inhibiting kinases that phosphorylate Ser2 within the heptapeptide repeats of the pol II large subunit CTD (17–20). Actonomycin D is a general transcription inhibitor. When used at low concentration, it selectively inhibits pol I transcription (9, 11). When used at higher concentration, it irreversibly blocks global transcription. Actinomycin D is a potent mutagen

and should be handled with extreme care using gloves and laboratory coat.

8. FUrd is incorporated into all nascent RNA transcripts. Since pol I transcription occurs at a very fast rate and the total rRNA synthesized abundantly exceeds the other RNA species, a short incubation time is appropriate to detect pol I transcription sites. A longer incubation time with FUrd is necessary to reveal also pol II transcription foci. Indeed, after a 20-min incubation, immunodetection of incorporated FUrd shows a large number of transcription foci dispersed in the nucleoplasmic volume and saturated nucleoli (which are easily identified).
9. Alternatively, cells can be kept in 3.7% formaldehyde at 4°C overnight.
10. The 63× objective is the most commonly used when analyzing the cell nucleus by confocal microscopy. Image brightness decreases concomitantly with image magnification. Therefore, for the same NA, the 63× objective should be preferred to the 100× objective (for further details, visit http://www.microscopyu.com, section on the properties of microscopes objectives).
11. Make sure that your confocal system allows UV excitation. Alexa conjugates are recommended in this application since they are considerably more stable (less bleaching under the laser beam) than the standard fluorescein or rhodamine fluorochromes. For emitted light, make sure that you have selected the proper filters. For the data acquisition parameters, a lot depends on the microscope system utilized in the specific application (wide field or confocal). However, the goal remains the same. The signal-to-noise ratio should be optimized, meaning that the background should be minimized and the intensity of the objects of interest should be as high as possible without reaching saturation. This is achieved by fine-tuning the gain/offset ratio (21). For clarity and to optimize the outcome of the image processing step, we recommend to select fields where objects are well-separated. During confocal microscopy, always avoid direct eye contact with the laser beams.
12. When processing images, we recommend saving them at each step in a dedicated folder corresponding to each specific stage.
13. In the ImageJ software, it is possible to check the type of image before processing it using the submenu command *Image > Type*.
14. When pictures are acquired on a confocal microscope, the background is most often corrected (because of the offset correction). Despite this, pictures may present a "salt-and-pepper" noise often due to a low signal. The salt and pepper noise is an

impulsive noise which can be corrected by a non-linear filter, such as a median filter (22). The median filter has a smoothing effect and often preserves edges. Use the commands Despeckle (*Process > Noise > Despeckle*) or Median Filter (*Process > Filter > Median*). Visit the link http://www.rsb.info.nih.gov/ij/docs/menus/process for further details. The salt-and-pepper noise must be removed because it may be mistakenly considered as source of genuine particles/transcription foci by the software.

15. Pictures obtained from wide-field microscopes need to be corrected for background more often than confocal microscopes. For smooth continue background the background correction can be done either by using the "rolling ball" algorithm or by mathematically subtracting in the following submenus *Process > Substract background* or *Process > Math > Substract*, respectively. Details for the rolling ball algorithm are given on http://www.rsb.info.nih.gov/ij/docs/menus/process. In case of salt and pepper noise utilize Despeckle or Median Filter.

16. The process that leads to a binary picture generates a picture characterized by two grey levels, white and black, corresponding to background and objects, respectively (Fig. 3). In the ImageJ software, after binarizing a picture, the background is white with 0 value and objects are black with value 255. Be aware that this can differ depending on the software that is utilized.

17. Thresholding a picture allows setting lower and upper threshold values, thus segmenting the image into features of interest, i.e. objects to be counted and background (*see* http://www.rsb.info.nih.gov/ij/docs/menus/image). The resulting picture is a binary picture. The thresholding is useful if the objects of interest can be separated from the background and from other objects which are not relevant for the analysis according to their intensities. For the pol I transcription foci, our threshold values were between 40 and 255. For the pol II transcription foci, we used threshold values between 45 and 95. The upper value was set to 95 to discard nucleoli. Segmentation of two objects with similar or overlapping intensities may be difficult to achieve.

18. For those objects which are not well-separated, the ImageJ software provides the option to proceed to an opening or watershed. Follow *Process > Binary > Open* or *Process > Binary > Watershed*. Keep in mind that segmenting objects is a difficult task and remains one of the main limitations when counting objects. Therefore, preparation of good biological samples and acquisition of clean pictures considerably facilitate the segmentation step.

19. In this case, we proceed to the particle counting of the entire picture, but it is possible to proceed to the counting of a selection.
20. We selected a lower value of 10 for pol I transcription foci counting to avoid artefactual counting of background particles. For pol II transcription foci, we selected a lower value of 0 or 5 because the pol II transcription foci were smaller and we did not want to lose some in the counting.
21. The alternative functions *Outlines* or *Mask* can be very convenient for further analysis. For details concerning the *Show* options, we recommend to visit the documents menu at http://www.rsb.info.nih.gov/ij/docs/menus/analyze.
22. This function is very convenient since the ROI Manager allows many options, including drawing, measuring, or being saved.
23. It is possible to include interior holes when counting particles using the function *Include holes.* This option was not used when counting the transcription foci in the present application.

Acknowledgements

Our work is supported by grants from the Swedish Research Council and Cancerfonden to PP. EL is supported by a postdoctoral fellowship from the Wenner-Gren Foundation, Stockholm, Sweden.

References

1. Alberts B, Johnson A, Lewis J et al (2008) Molecular Biology of the Cell. Garland Science, New York
2. Eissenberg J C, Shilatifard A (2006) Leaving a mark: the many footprints of the elongating RNA polymerase II. *Curr Opin Genet Dev* 16:184–190
3. Hirose Y, Ohkuma Y (2007) Phosphorylation of the C-terminal Domain of RNA Polymerase II Plays Central Roles in the Integrated Events of Eucaryotic Gene Expression. *J Biochem* 141:601–608
4. Rondon A G, Mischo H E, Proudfoot N J (2008) Terminating transcription in yeast: whether to be a 'nerd' or a 'rat'. *Nat Struct Mol Biol* 15:775–776
5. Jackson D A, Hassan A B, Errington R J et al (1993) Visualization of focal sites of transcription within human nuclei. *EMBO J.* 12: 1059–1065
6. Wansink D G, Schul W, van der Kraan I et al (1993) Fluorescent labelling of nascent RNA reveals transcription by RNA polymerase II in domains scattered throughout the nucleus. *J Cell Biol* 122:283–293
7. Boisvert F M, van Koningsbruggen S, Navascue´s J et al (2007) The multifunctional nucleolus. Nat Rev Mol Cell Biol 8:574–585
8. Sirri V, Urcuqui-Inchima S, Roussel P et al (2008). Nucleolus: The fascinating nuclear body. *Histochem Cell Biol* 129:13–31
9. Percipalle P, Fomproix N, Cavellan E et al (2006) The chromatin remodelling complex WSTF-SNF2h interacts with nuclear myosin 1 and has a role in RNA polymerase I transcription. *EMBO Rep* 7:525–530

10. Längst G, Blank T A, Becker P B et al (1997) RNA polymerase I transcription on nucleosomal templates: TTF-I induces chromatin remodeling and relieves transcriptional repression. *EMBO J* 16:760–768
11. Fomproix N, Percipalle P (2004) An actin-myosin complex on actively transcribing genes. *Exp Cell Res* 294:140–148
12. Moore G P, Ringertz N R (1973) Localization of DNA-dependent RNA polymerase activities in fixed human fibroblasts by autoradiography. *Exp Cell Res* 76:223–228
13. Mitchell J A, Fraser P (2008) Transcription factories are nuclear subcompartments that remain in the absence of transcription. *Genes Dev* 22:20–25
14. Kruhlak M, Crouch E E, Orlov M et al (2007) The ATM repair pathway inhibits RNA polymerase I transcription in response to chromosome breaks. *Nature* 447:730–734
15. Percipalle P, Obrdlik A (2009) Analysis of nascent RNA transcripts by chromatin RNA immunoprecipitations. *Meth Mol Biol* 567:215–235
16. Burger W, Burge M J (2008) Digital Image Processing. Springer, New York
17. Fraser N W, Sehgal P B, Darnell J E (1978) DRB-induced premature termination of late adenovirus transcription. *Nature* 272:590–593
18. Marshall N F, Price D H (1995) Purification of P-TEFb, a transcription factor required for the transition into productive elongation. *J Biol Chem* 270:12335–12338
19. Marshall N F, Peng J, Xie Z et al (1996) Control of RNA polymerase II elongation potential by a novel carboxyl-terminal domain kinase. *J Biol Chem* 271:27176–27183
20. Gribnau J, Diderich K, Pruzina S et al (2000) Intergenic transcription and developmental remodeling of chromatin subdomains in the human β-globin locus. *Mol Cell* 5:377–386
21. Louvet E, Tramier M, Angelier N et al (2008) Time-Lapse Microscopy and Fluorescence Resonance Energy Transfer to Analyse the Dynamics and Interactions of Nucleolar Proteins in Living Cells. *Meth Mol Biol* 463:123–135
22. Gonzales R C, Woods R E (2008) Digital Image Processing. Pearson Education, London

Chapter 35

Genome Wide Full-Length Transcript Analysis Using 5′ and 3′ Paired-End-Tag Next Generation Sequencing (RNA-PET)

Xiaoan Ruan and Yijun Ruan

Abstract

RNA-PET is a paired end tag (PET) sequencing method for full-length mRNA transcripts analysis using the next generation sequencer platforms such as Illumina GA and SOLiD. Unlike RNA-Seq method that sequences randomly sheared shotgun RNA short fragments, RNA-PET captures and sequences the 5′ and 3′ end tags of full-length cDNA fragments of all expressed genes in a biological sample. When mapped to reference genome, RNA-PET sequences can demarcate the boundaries of transcription units genome-wide, in addition to its ability to quantify the transcription level of each expression genes. Furthermore, the unique feature of RNA-PET is to identify fusion transcripts. Therefore, RNA-PET has been regarded as the best PET for genome annotation (1). Here in this chapter, we describe the details of the RNA-PET protocol and discuss the critical issues.

Key words: NGS, PET sequencing, Full length mRNA, Gene expression, Illumina GA, SOLiD, Transcriptome, RNA boundary, Splice variants, Fusion transcripts, EcoP15I

1. Introduction

Genomics holds much promise for huge improvements in human healthcare, and the next-generation sequencing technologies are becoming a driving force that penetrates the entire field of genomic science. As the current sequencing technologies are limited by short sequencing reads, an important part of the sequencing strategy is to use the PET sequencing approaches to analyze nucleic acid templates such as RNA for transcriptome (1) and DNA for genome (2).

To fully understand gene transcription regulation in the whole genome content, it is important to define where precisely gene transcription starts and terminates. To obtain such information, we developed an efficient strategy to demarcate the boundaries of

Ales Vancura (ed.), *Transcriptional Regulation: Methods and Protocols*, Methods in Molecular Biology, vol. 809,
DOI 10.1007/978-1-61779-376-9_35, © Springer Science+Business Media, LLC 2012

transcription units for the whole genome (1). The core concept is to obtain only linked 5′ and 3′ short tag sequences for each transcript, map these terminal "signatures" to the genome and thereby infer the complete transcription units by the genome sequence encompassed between these 5′ and 3′ signatures. As an intermediate step, we first developed the 5′ and 3′ LongSAGE protocols to capture the 5′ and 3′ tag sequences separately of expressed genes (3). With this ability, we then combined these two separate protocols into one for extracting the paired end 5′ and 3′ tags for sequencing and mapping analysis (4).

In the early version of the PET sequencing for full-length mRNA transcripts, short tag fragments of 20 bp (5′) and 20 bp (3′) were extracted through bacterial cloning-based process, and the paired ditag fragments were concatenated into longer DNA fragments for Snager capillary sequencing (AB3730xl). Later, we adapted the Roche 454 GS-FLX pyrosequencer for such analysis (5). However, the bacterial-based cloning method for PET extraction was a long and laborious process, and the 20-bp tag information is a limitation for high mapping specificity. We now further improved the PET analysis protocol by developing a complete in vitro cloning-free protocol and adapting the enzyme EcoP5I for extracting longer PET fragments (27–27 bp).

This new version of PET analysis for full-length mRNA (we called RNA-PET) consists of six major steps (Fig. 1): (1) Capture and synthesize full-length cDNAs from mRNA transcripts using Cap-Trapper approach and a specifically designed GsuI-dT (16) oligo for reverse transcription (RT). (2) Ligate the captured full-length cDNAs with specific DNA linkers. (3) Circularize the linker-ligated full-length cDNAs and excise 27 bp of terminal tags from each of the 3′ and 5′ end. (4) Isolate and purify the PETs and ligate with next generation sequencer adaptors. (5) Paired End (PE) sequencing the captured ends with either Illumina GA or SOLiD v4, and mapping both sequence tags back to the reference genome. (6) Cluster and annotate the tags onto existing reference database and visualize the annotated sequence data on genome browser, and produce digital gene expressions as well as detect unusual gene transcripts caused by a variety of genome rearrangements or abnormal transcription events.

To start RNA-PET, total RNA sample is used as starting material and poly (T) Oligos are used to enrich and purify mRNAs. Approximately 1–5 μg poly (A) mRNA is used in the RNA-PET library construction. The Cap-Trapper approach (6–8) is combined with GsuI-poly(T) Oligos to capture full-length cDNAs. After full-length cDNAs are obtained, the full-length cDNA are methylated to block EcoP15I recognition site at the fifth residue A. The cDNA is then ligated to the specifically designed linker sequences and circularized at larger ligation volume (~0.1 ng DNA/μl). Un-circularized molecules are further cleaved out by using a plasmid-save treatment and the remaining circular cDNAs are gone through EcoP15I

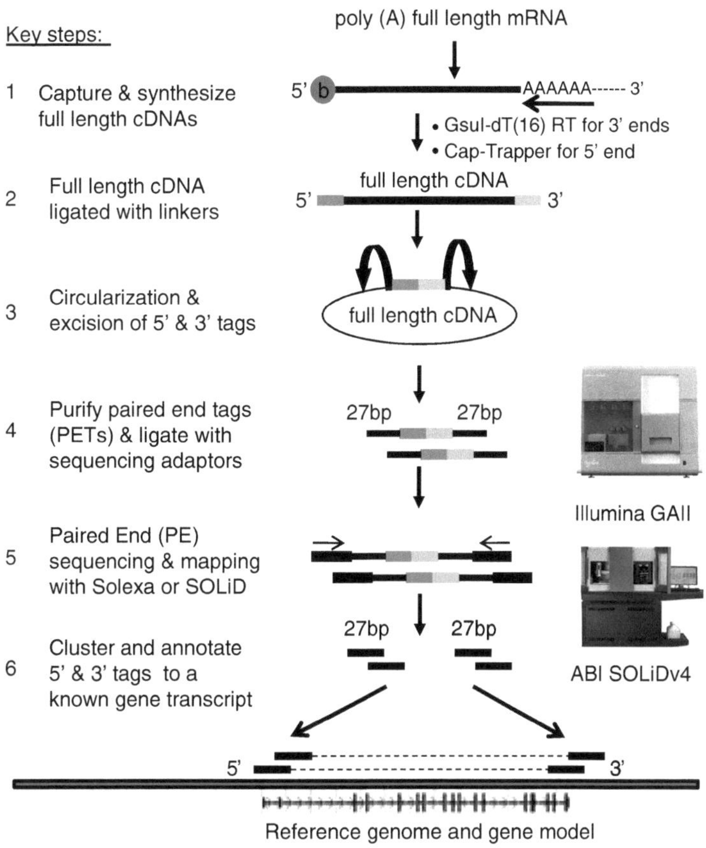

Fig. 1. Schematic overview of the RNA-PET analysis for full-length whole transcriptome. Six sections of steps are involved in the RNA-PET analysis as illustrated and include full-length cDNA capturing, PET excision, sequencing, mapping, and annotation analysis.

digestion to release the PETs at 27 bp from each end. The resulting PETs are modified with specific paired end (PE) sequencing adaptors compatible to either Illumina GAII or SOLiD4. After ligated to the sequencing adaptors, the PET template are further PCR amplified and sequenced through PE sequencing format. Approximately 20–30 million sequence PETs are generated through HT sequencing. After filter out redundant and noise tags, the unique PETs proceed to mapping on the specific reference genome. A approximate 90% of PETs, mapped to known transcripts or splicing variants, are named as *concordant* PETs, however, a small portion of the misaligned PETs (named as *discordant* PETs) are mapped either to wrong orientations in the same chromosome, or to different strands, or on different chromosomes. For *concordant* PETs, the digital expression level can be easily obtained from mapped sequence counts. Even though a majority of the *discordant* PETs are derived from noise ligations, a collection of these PETs serves as a valuable pool for identification of many novel transcripts and its associated splice variants.

It should be noted that as RNA-PET only characterizes the 5′ and 3′ ends of transcripts whereas RNA-seq is robust for tagging internal exons but poor on transcript terminal regions, the combination of RNA-PET and RNA-seq should be viewed as the ultimate solution for comprehensive transcriptome characterization.

2. Materials

2.1. Poly (A) mRNA and GsuI-dT (16) Oligo for RT

1. 1–5 μg of poly (A) mRNA isolated from total RNA using Miltenyi Biotec μMACs mRNA Isolation Kit.
2. GsuI-dT(16)oligo:5′-GAGCTAGTTCTGGAGTTTTTTTTT TTTTTTTVN-3′ in 1 μg/μl, stored in −20°C.
3. DNA low-bind tube used throughout DNA or RNA procedures (Eppendorf, see Note 1).
4. 3 M sodium acetate (NaOAc), pH 5.5.
5. Isopropanol.
6. 75% Ethanol (EtOH, see Note 2).
7. RNasin-Plus RNase inhibitor (Promega).
8. 2× GC-I Buffer (Takara).
9. dNTP mix (with 5-Me-dCTP in replace of dCTP): 10 mM dATP, 10 mM dTTP, 10 mM dGTP, 5 mM 5-Me-dCTP in 10 mM Tris–HCl, pH 8.0.
10. 4.9 M D-sorbitol (Sigma).
11. Superscript II and III reverse transcriptases (Invitrogen).
12. Saturated trehalose (RNase-free) (Sigma): Heat water in a 1.7-ml tube to 42°C in a heat block. Slowly add trehalose powder to the tube and dissolve it by vortex. Maintain temperature at around 42°C and continue add trehalose until saturation is reached. Upon cooling down the solution to room temperature, trehalose will form crystals and saturated solution is obtained (see Note 3). Aliquot the solution and store in −20°C.
13. Proteinase K 20 mg/ml (Ambion).
14. Phenol:Chloroform:IAA solution, 25:24:1, pH 6.6 (Ambion).
15. Water (nuclease-free).

2.2. Reagents for Oxidation of Diol Structures and Biotinylation of 5′ End

1. Sodium periodate ($NaIO_4$) (Sigma).
2. 1.1 M NaOAc, pH 4.5: Diluted from 3 M NaOAc, and adjusted to pH 4.5.
3. 10% SDS solution.
4. 5 M NaCl.

5. 10 mM biotin hydrazide (long arm) (Vector Laboratories).
6. 1 M NaOAc, pH 6.1: Diluted from 3 M NaOAc and adjusted to pH 6.1.

2.3. RNase One Selection

1. 5 M NaCl.
2. 1 M NaOAc, pH 6.1.
3. 10× RNase ONE buffer (Promega).
4. RNase ONE Ribonuclease 10 U/μl (Promega).
5. Yeast tRNA 10 mg/ml (Ambion).

2.4. Prepare Dynalbead M-280 Streptavidin

1. Dynal MPC-S (now replaced by DynalMag-2) (Invitrogen).
2. Dynalbead M-280 Streptavidin (Invitrogen).
3. 1× Binding buffer (BB): 2 M NaCl, 50 mM EDTA, pH 8.0.
4. 1× BB + Yeast tRNA: 2 M NaCl, 50 mM EDTA, pH 8.0, Yeast tRNA 0.25 μg/μl.
5. Eppendorf thermomixer.

2.5. Binding of Full-Length Biotinylated (−)DNA/RNA Heteroduplex

1. Intelli Mixer (Elmi).
2. Dynal MPC-S (now replaced by DynalMag-2) (Invitrogen).
3. 1× BB: 2 M NaCl, 50 mM EDTA, pH 8.0.
4. 1× Blocking buffer: 0.4% SDS, 50 μg/ml Yeast tRNA.
5. 1× Washing buffer: 10 mM Tris–HCl, pH 7.5, 0.2 mM EDTA, 10 mM NaCl, 20% glycerol, 40 μg/ml Yeast tRNA.

2.6. Hydrolytic Degradation of Bound RNA to Release Full Length (−) cDNA Strand

1. Alkaline hydrolysis buffer: 50 mM NaOH, 5 mM EDTA, pH 8.0. Prepare fresh each time.
2. 2 ml MaXtract High Density Tube (Qiagen).
3. Phenol:Chloroform:IAA, 25:24:1, pH 7.9 (Ambion).
4. Buffer EB (Qiagen).
5. Isopropanol precipitation: 3 M NaOAc, pH 5.5, isopropanol.

2.7. Synthesis of Double-Stranded Full-Length cDNA

1. 1× Tris–NaCl–EDTA (TNE) buffer: 10 mM Tris–HCl, pH 8.0, 50 mM NaCl, 0.1 mM EDTA. Sterilize the buffer by syringe filter or autoclave.
2. DNA linkers for Cap-Trapper of 5′-end of the full-length cDNAs, and double-stranded cDNA synthesis. Linker E-E2-GsuI-N5 and linker E-E2-GsuI-N6 (see Table 1 for sequence detail and Subheading 3.9.1 for oligo annealing).
3. Oligos are synthesized by HPLC purification from Integrated DNA Technologies (IDT). Oligos are annealed to dsDNA and stored in aliquots in −20°C at 0.4 μg/μl for use.
4. 4–20% TBE gels (Invitrogen).

Table 1
Sequences for DNA linkers and adptors used in RNA–PET construction

Oligos and DNA adaptors	Sequences	Length (nt)
Linkers for capture of 3′-end of full length poly A mRNA and RT synthesis:		
GsuI-dT (16) oligo	5′-GAGCTAGTTCTGGAGTTTTTTTTTTTTTTTTVN-3′	33
Linkers for Cap-Trapper of 5′-end of full length cDNA and second strand cDNA synthesis:		
Linker E-E2-GsuI-N5 (top)	5′-CTACCTGGAGAACATGAGGCAGCCAGGNNNNN-3′	32
Linker E-E2-GsuI-N5 (bottom)	5′-Phos-CTGGCTGCCTCATGTTCTCCAGGTAG-3′	26
Linker E-E2-GsuI-N6 (top)	5′-CTACCTGGAGAACATGAGGCAGCCAGNNNNNN-3′	32
Linker E-E2-GsuI-N6 (bottom)	5′-Phos-CTGGCTGCCTCATGTTCTCCAGGTAG-3′	26
Linkers for 5′ and 3′ paired end tag (PET) capture:		
Illumina E-E2-h-AGG linker (top)	5′-GGCCGCGA*T*ATCGGACAGCAG-3′ (Internal Biotin at the ninth d*T*)	21
Illumina E-E2-h-AGG linker (bottom)	5′-Phos-GCTGTCCGATATCGC-3′	15
Illumina E-E2-t-AGG linker (top)	5′-Phos-CTGCTGTCCGATATCGC-3′	17
Illumina E-E2-t-AGG linker (bottom)	5′-GGCCGCGA*T*ATCGGACAGCAGTT-3′ (Internal Biotin at the ninth d*T*)	23
SOLiD linker E5v4 (top)	5′-CCGCC*T*TGGCCGTACAGCAG-3′ (Internal Biotin at the sixth d*T*)	20
SOLiD linker E5v4 (bottom)	5′-Phos-GCTGTACGGCCAAG-3′	14
SOLiD linker E3v3 (top)	5′-GCGGA*T*GTACGGTACAGCAGTT-3′ (Internal Biotin at the sixth d*T*)	22
SOLiD linker E3v3 (bottom)	5′-Phos-CTGCTGTACCGTACAT-3′	16
Sequencing adaptors and PCR amplification primers for Solexa GA, or SOLiD V4:		
Illumina 454-adaptor E/A (top)	5′-CCATCTCATCCCTGCGTGTCCCATCTGTTCCCTCCCTGTCTCAG-3′	44

Illumina 454-adaptor E/A (bottom)	5-NNCTGAGACAGGGAGGGAACAGATGGGACACGCAGGG-ATGAGATGG-3′	46
Illumina 454-adaptor E/B (top)	5′-GACTCTGTGCGTTGTCCCCTATCCGTTCCGTGTGTCCC-CTATCC-3′	44
Illumina 454-adaptor E/B (bottom)	5′-GGATAGGGGACACACGGAACGGATAGGGGACAACGCAC-AGAGTCNN-3′	46
SOLiD P1 adaptor (dsDNA bought from ABI)	5-CCACTACGCCTCCGCTTTCCTCTCTATGGGCAGTCGGTGAT-3′ 3′-TTGGTGATGCGGAGGCGAAAGGAGAGATACCCGTCAGCCACTA-5′	41/43
SOLiD P2 adaptor (dsDNA bought from ABI)	5′-AGAGAATGAGGAACCCGGGGCAGTT-3′ 3′-TCTCTTACTCCTTGGGCCCCGTC-5′	25/23
RNA-PET library PCR primers:		
Illumina-454 PCR primer-1	5′-AATGATACGGCGACCACCGAGATCTACACCCTATCCCC-TGTGTGCCTTG-3′	49
Illumina-454 PCR primer-2	5′-CAAGCAGAAGACGGCATACGAGATCGGTCCATCTCAT-CCCTGCGTGTC-3′	48
SOLiD library PCR primer-1 (bought from ABI)	5′-CCACTACGCCTCCGCTTTCCTCTCTATG-3′	28
SOLiD library PCR primer-2 (bought from ABI)	5′-CTGCCCCGGGTTCCTCATTCT-3′	21

5. Takara Solution I (Takara).
6. Takara Solution II (Takara).
7. 10× ExTaq buffer (Takara).
8. Takara ExTaq (Takara).
9. dNTP mix at 2.5 mM each (Takara).
10. Proteinase K at 20 mg/ml (Ambion).
11. 2 ml MaXtract High Density Tube (Qiagen).
12. Phenol:Chloroform:IAA, 25:24:1, pH 7.9 (Ambion).
13. Glycoblue, 15 mg/ml (Ambion).
14. Isopropanol precipitation: 3 M NaOAc, pH 5.5, isopropanol.
15. 70% EtOH in nuclease-free water.
16. GsuI, 5 U/μl (Fermentas).
17. cDNA size fractionation columns (Invitrogen).
18. 6× loading dye (Fermentas).
19. TEN buffer: 10 mM Tris–HCl, pH 8.0, 0.1 mM EDTA, pH 8.0, 25 mM NaCl.
20. Molecular probes SYBR Green, 10,000× in DMSO (Invitrogen).
21. Gel staining buffer: 1× TBE, Molecular probes SYBR Green 1×.
22. Gel handler, gel support (Sigma).
23. Molecular probes Quant-iT PicoGreen dsDNA reagent (Invitrogen).
24. Dynalbead M-280 Streptavidin beads (Invitrogen).
25. Calf thymus DNA (Sigma).
26. S-adenosylmethionine (SAM), 0.5 mM (NEB).

2.8. Double Strand cDNA Methylation by EcoP15I Enzyme (Recognition:CAGCAG)

1. EcoP15I enzyme (NEB).
2. 10× Buffer 3 (NEB).
3. 100× BSA (NEB).
4. SAM, 32 mM (NEB).
5. Isopropanol precipitation: 3 M NaOAc, pH 5.5, isopropanol.

2.9. Linker Ligation for 5′ and 3′ Paired End Tag Capture of ds Full-Length cDNAs

1. 5× T4 DNA ligase buffer with PEG (Invitrogen).
2. T4 DNA ligase, 30 U/μl (Fermentas).
3. Solexa E-E2-h-AGG linker and Solexa E-E2-t-AGG linker each at 200 ng/μl (see Table 1 for sequence detail and Subheading 3.9.1 for oligo annealing).
4. SOLiD linker E5v4 and SOLiD linker E3v3 each at 200 ng/μl (see Table 1 for sequence detail and Subheading 3.9.1 for oligo annealing).
5. 2 ml MaXtract High Density Tube (Qiagen).

6. Phenol:Chloroform:IAA, 25:24:1, pH 7.9 (Ambion).
7. Glycoblue, 15 mg/ml (Ambion).
8. Isopropanol precipitation: 3 M NaOAc, pH 5.5, isopropanol.

2.10. Add Phosphate Group at 5′-Ends of Linker-Ligated DNA Fragments

1. 10× T4 DNA ligase (NEB).
2. T4 DNA polynucleotide kinase, 10 U/μl (NEB).
3. Fermentas T4 DNA ligase (30 U/μl).
4. Isopropanol precipitation: 3 M NaOAc, pH 5.5, Isopropanol.

2.11. Circularization of Linker-Ligated cDNA Fragments

1. 10× T4 DNA ligase (NEB).
2. T4 DNA ligase, 30 U/μl (Fermentas).
3. 15 ml MaXtract High Density Tube (Qiagen).
4. Phenol:Chloroform:IAA, 25:24:1, pH 7.9 (Ambion).
5. Glycoblue, 15 mg/ml (Ambion).
6. Isopropanol precipitation: 3 M NaOAc, pH 5.5, isopropanol.
7. Buffer EB (Qiagen).

2.12. Nick Translation Repair of DNA

1. 10× *Escherichia coli* DNA ligation buffer (Qiagen).
2. 10 mM dNTP mix (Eppendorf).
3. *E. coli* ligase (NEB).
4. *E. coli* DNA polymerase I (NEB).
5. 2 ml MaXtract High Density Tube (Qiagen).
6. Phenol:Chloroform:IAA, 25:24:1, pH 7.9 (Ambion).
7. Glycoblue, 15 mg/ml (Ambion).
8. Isopropanol precipitation: 3 M NaOAc, pH 5.5, isopropanol.

2.13. Plasmid-Safe Treatment to Cleave Linear DNA Molecules

1. Plasmid-safe reaction: 25 mM ATP, 10× reaction buffer, Plasmid-safe DNase, 10 U/μl (Epicentre).
2. 2 ml MaXtract High Density Tube (Qiagen).
3. Phenol:Chloroform:IAA, 25:24:1, pH 7.9.
4. Glycoblue, 15 mg/ml.
5. Isopropanol precipitation: 3 M NaOAc, pH 5.5, isopropanol.

2.14. EcoP15I Digestion to Release 5′ and 3′ Paired End Tags from Full Length cDNAs

1. 10× Buffer 3 (NEB).
2. 100× BSA (NEB).
3. 10 mM Sinefungin (Calbiochem).
4. Adenosine 5′-triphosphate, (ATP) (NEB).
5. EcoP15I enzyme 10 U/μl (NEB).

2.15. Binding of 5′ and 3′ Pair End Tags to Dynalbeads M280 Streptavidin

1. Intelli Mixer (Elmi).
2. Dynal MPC-S (now DynalMag-2): Magnetic stand (Invitrogen).
3. Dynalbeads M-280 Streptavidin (Invitrogen).
4. 2× B&W buffer: 10 mM Tris–HCl, pH 7.5, 1 mM EDTA, 2 M NaCl.
5. 1× B&W buffer: 5 mM Tris–HCl, pH 7.5, 0.5 mM EDTA, 1 M NaCl.

2.16. Ligation of Sequencing Adaptors to 5′ and 3′ PETs

1. Solexa 454-adaptor E/A and Solexa 454-adaptor E/B each at 200 ng/μl.
2. SOLiD P1 and SOLiD P2 adaptor each at 200 ng/μl (see Table 1 for sequence detail).
3. T4 DNA ligase (30 U/μl) (Fermentas).
4. 10× T4 DNA ligase buffer (NEB).
5. Intelli Mixer (Elmi).
6. 1× B&W buffer: 5 mM Tris–HCl, pH 7.5, 0.5 mM EDTA, 1 M NaCl.

2.17. Nick Translation for DNA Repair

1. 10× Buffer 2 (Qiagen).
2. 10 mM dNTP mix (Eppendorf).
3. *E. coli* DNA polymerase I (NEB).
4. Nuclease-free water (Ambion).
5. Buffer EB (Qiagen).
6. Intelli Mixer (Elmi).
7. 1× B&W buffer: 5 mM Tris–HCl, pH 7.5, 0.5 mM EDTA, 1 M NaCl.

2.18. PCR Amplification for PET Sequencing Templates

1. 2× Phusion High-Fidelity PCR Master Mix with HF Buffer (Finnzymes).
2. Molecular probes SYBR Green 1, 10,000× in DMSO (Invitrogen).
3. Gel staining buffer: 1× TBE, Molecular probes SYBR Green 1.
4. Gel handler gel support (Sigma).
5. Solexa-454 PCR primer-1 and Solexa-454 PCR primer-2, each at 25 μM.
6. SOLiD library PCR primer-1 and SOLiD library PCR primer-2, each at 2 μM (see Table 1 for sequence detail).
7. 4–20% TBE gels (Invitrogen).
8. 25-bp DNA ladder at 1 μg/μl (Invitrogen).
9. 1× TBE buffer (First Base Inc.) diluted from 10× stock.

2.19. Scale-Up PCR Amplification for Sequencing Templates

1. 2× Phusion High-Fidelity PCR Master Mix with HF Buffer (Finnzymes).
2. Molecular probes SYBR Green 1, 10,000× in DMSO (Invitrogen).
3. Gel staining buffer: 1× TBE, Molecular probes SYBR Green 1.
4. Gel handler gel support (Sigma).
5. Solexa-454 PCR primer – 1 and 2.
6. SOLiD library PCR primer – 1 and 2.
7. 6% TBE gel with 5-wells (Invitrogen).
8. 25-bp DNA ladder at 1 μg/μl (Invitrogen).
9. 1× TBE buffer (First Base Inc.) diluted from 10× stock.
10. QIAquick PCR Purification Kit (Qiagen).

2.20. Purification of PCR Templates Using Gel-Crush Elution Method

1. Spin-X centrifuge tube filters, CA membrane, 0.22 pore size (Costar).
2. TE buffer, pH 8.0 (Ambion).
3. 21-G needle (Becton Dickinson).
4. 0.6-ml microtube (Axygen).
5. 1.5-ml screw cap tube (Axygen).
6. Glycoblue, 15 mg/ml (Ambion).
7. 70% Ethanol in nuclease-free water.
8. Nuclease-free water (Ambion).
9. Isopropanol precipitation: 3 M NaOAc, pH 5.5, isopropanol.

3. Methods

3.1. Mix Poly (A) mRNA and GsuI-dT (16) Oligo

1. Mix following reagents in order in a 0.2-ml PCR tube on ice (see Notes 4–6):

Poly (A) mRNA (1–5 μg)	10 μl containg 1 μl poly (A) mRNA
GsuI-dT oligo (1 μg/μl)	0.7 μg oligo
3 M NaOAc, pH 5.5	1/10 volume
Isopropanol	Equal volume

2. Keep tube at –80°C for 30 min.
3. Centrifuge at 24,000×*g* for 30 min at 4°C in a microcentrifuge.
4. Wash 2× with 500 μl of cold 75% EtOH.

5. Air dry pellet, resuspend pellet in 19-μl of nuclease-free water, and add 1 μl of RNase-Plus inhibitor to the mixture and transfer solution into a 0.2-ml PCR tube.

3.2. Set Up Reverse Transcription (RT)

1. Heat the reaction mixture at 65°C for 10 min and cool to 37°C for 1 min, then hold at 42°C on a thermal cycler while waiting for the other components to be prepared.
2. Set up the RT mix on ice in a 0.2-ml thin-walled PCR tube:

2× GC-I buffer	75 μl
RNasin-Plus RNase inhibitor	1 μl
10 mM dNTP (with 5-Me-dCTP in place of dCTP)	4 μl
4.9 M sorbitol	26 μl
Superscript II reverse transcriptase	8 μl
Superscript III reverse transcriptase	4 μl

3. Put 10 μl of saturated trehalose into another 0.2-ml PCR tube and leave warming at 42°C in a thermal cycler.
4. When the oligo-dT/mRNA annealing step is complete, place the RT mix into the thermal cycler that was preset at 42°C for at least 2 min.
5. Mix the warm trehalose together with the RT mix (volume = 128 μl), and quickly transfer the entire reaction mix into the tube containing the annealed oligo/mRNA (volume now = 148 μl), and immediately start the incubation:

42°C for 50 min
50°C for 25 min
55°C for 25 min
Hold at 4°C

6. Add 2 μl of proteinase K to digest all enzymes by incubation at 45°C for 15 min.
7. Transfer the solution into a lo-bind tube and add equal volume of Phenol:Chloroform:IAA, 25:24:1, pH 6.6 directly into the tube.
8. Mix the organic and aqueous phases thoroughly for about 1 min.
9. Centrifuge at 24,000×*g* for 3 min in a microcentrifuge to separate the phases.
10. Remove the upper, nucleic acid-containing phase by carefully pipetting into a new tube.
11. Repeat the extraction with another 150 μl of nuclease-free water.
12. Combine two aqueous phases together.

13. Isopropanol precipitates the RNA/DNA heteroduplex.
14. Resuspend the pellet in 44.5 μl of nuclease-free water.

3.3. Oxidation of Diol Structures

1. Prepare the following stocks fresh each time, using 1.7-ml tubes:

10 mM biotin hydrazide (long arm)
200 mM $NaIO_4$

2. In a 1.7 ml siliconized tube, add the following and mix well:

RNA/DNA heteroduplex	44.5 μl
1.1 M NaOAc, pH 4.5	3 μl
Fresh 200 mM $NaIO_4$	2.5 μl

3. Incubate reaction mix on ice for 45 min in the dark.
4. Isopropanol precipitates the DNA/RNA heteroduplex.
5. Resuspend the pellet in 50 μl of nuclease-free water.

3.4. Biotinylation of 5′-End of mRNA

1. To the 50 μl oxidized (−) cDNA/RNA heteroduplex, add the following:

1 M NaOAc pH 6.1	5 μl
10% SDS	5 μl
Fresh-made 10 mM biotin hydrazide	150 μl

2. Incubate at room temperature (RT) for overnight and keep in the dark.
3. Isopropanol precipitates the 210 μl biotinylated (−) cDNA/RNA heteroduplex.
4. Resuspend the pellet in 170 μl of nuclease-free water.
5. Preset the Eppendorf shaking incubator to cool down to 4°C.

3.5. RNase ONE Selection

1. Set up the following in a 1.7-ml lo-bind tube (see Note 7):

Biotinylated (−) DNA/RNA sample:	170 μl
10× RNase ONE buffer	20 μl
RNase ONE Ribonuclease (10 U/μl)	xx μl
Nuclease-free water	Add up to 200 μl

2. Incubate at 37°C for 30 min (see Note 8).
3. Quench the reaction by adding:

10 mg/ml Yeast tRNA	4 μl
5 M NaCl	50 μl

4. Leave the RNase ONE-treated sample on ice.

3.6. Preparation of Dynalbeads M-280 Streptavidin

1. Use the magnetic stand and lo-binding tubes for all steps involving M-280 beads. Use 200 μl of the Dynalbeads M-280 Streptavidin suspension per RNA sample (see Note 9).
2. Wash Dynalbead to remove preservatives (see Note 10).
3. Wash Dynalbeads 3× with 200 μl of 1× Binding buffer at room temperature (RT).
4. Add 200 μl of 1× BB + Yeast tRNA to the beads and incubate at 4°C for 30 min using Eppendorf shaker at 800 rpm (pre-cooled to 4°C, see Subheading 3.5, step 5).
5. Wash beads 3× with 1× BB at RT.

3.7. Binding of Full-Length Biotinylated (–) DNA/RNA Heteroduplex

1. Remove the supernatant (S/N) from beads and add the chilled RNase ONE-treated sample to the beads.
2. Rotate 30 min at RT on an Intelli-Mixer for binding to occur (Program F8: U = 50, u = 60, 30 rpm) (see Notes 11 and 12).
3. Wash the heteroduplex-bound beads at RT as follows:

2× with 200 μl of 1× BB
1× with 200 μl of 1× Block
1× with 200 μl of 1× Wash
1× with 200 μl of 50 μg/ml Yeast tRNA.

3.8. Hydrolytic Degradation of Bound RNA to Release Full-Length (–) cDNA Strand

1. Prepare alkaline hydrolysis buffer (see Note 13).
2. Prepare a tube containing 150 μl of 1 M Tris–HCl, pH 7.5 (for neutralization).
3. Remove S/N (50 μg/ml Yeast tRNA) from the beads and add 50 μl of alkaline hydrolysis buffer.
4. Shake the mixture at 65°C for 10 min using the Eppendorf shaker at 2,400 × *g*.
5. Collect the S/N containing full-length (–) cDNA into the tube containing 150 μl of 1 M Tris–HCl, pH 7.5, for neutralization.
6. Repeat the hydrolysis and collection steps twice, collecting all fractions into the same tube to a final volume of 300 μl.
7. Immediately before use, pellet MaXtract High Density Tube by centrifugation at 19,300 × *g* for 30 s in a microcentrifuge (see Note 14).
8. Transfer the DNA into the MaXtract tube and directly add equal volume of Phenol:Chloroform:IAA, 25:24:1, pH 7.9, to the MaXtract tube.
9. Mix the organic and aqueous phases thoroughly for about 1 min.
10. Centrifuge at 19,300 × *g* for 3 min in a microcentrifuge to separate the phases.

11. Remove the upper, nucleic acid-containing phase by carefully pipetting into a new tube.
12. Isopropanol precipitates the DNA and resuspend the pellet in 5 μl of Buffer EB.

3.9. Synthesis of Double-Stranded cDNA

3.9.1. Annealing Oligos for Double Strand DNA Linkers (see Note 15)

1. Thaw single strand oligos at room temperature for 15 min.
2. Spin at maximum speed at 4°C for 1 min to collect any dislodged oligos.
3. Add 1× TNE buffer to thawed oligos to make it 100 μM.
4. Vortex for approximately 1 min to resuspend oligos and brief spin down to collect the resuspended oligos to the bottom of tubes.
5. Perform Nanodrop and use OD constant value to calculate concentration. Check if the measured concentration by Nanodrop falls within the expected range.
6. In a 0.2-ml PCR tube, mix together the Oligos (see Table 1 for sequences), e.g.,

Oligonucleotide A (top strand) (100 μM)	40 μl
Oligonucleotide B (bottom strand) (100 μM)	40 μl

7. Heat 95°C for 10 min, then, turn off the program with lid closed, and let the tube cool down to room temperature slowly to allow Oligos annealed together. It may take 90 min to complete the naturally cool-down process. For long term storage, keep the annealed Oligos (or called dsDNA linkers or adaptors) in −80°C.
8. Keep all annealed dsDNA linkers or adaptors on ice before use.
9. Measure the DNA concentration by Nanodrop and dilute it to a concentration of 200 ng/μl with 1× TNE.
10. Run 200 ng each of the single strand oligos together with 200 ng of annealed dsDNA linkers on the 4–20% PAGE gel to ensure the annealing result satisfactory.

3.9.2. 5′-End Cap-Trapper Ligation and Second Strand cDNA Synthesis

1. Set following reagents on ice in a 1.7-ml lo-bind tube:

Full-length single strand (−) cDNA	5 μl
0.4 μg/μl linker E-E2-GsuI-N5	4 μl
0.4 μg/μl linker E-E2-GsuI-N6	1 μl
Takara Solution II (A3101-1) (see Note 16)	10 μl
Takara Solution I (ligase, A201-1)	20 μl

2. Mix by flicking and short spin at 4°C.
3. Incubate at 16°C for overnight to allow degenerate oligo to anneal and ligate.

3.9.3. Primer Extension for Second Strand cDNA Synthesis

1. Set the following on ice in a 0.2-ml thin-walled PCR tube:

Overnight ligation mix	40 μl
Nuclease-free water	20 μl
10× ExTaq buffer with Mg^{2+}	8 μl
2.5 mM dNTP	8 μl
ExTaq polymerase	4 μl

2. Incubate the PCR reaction in the warmed up thermal cycler at 65°C for 5 min, 68°C for 30 min, 72°C for 10 min, and hold at 4°C.
3. Add 2 μl of proteinase K and mix by pipetting up and down and incubate at 45°C for 15 min to digest the enzyme.
4. Immediately before use, pellet MaXtract High Density Tube by centrifugation at 23,000×*g* for 30 s in a microcentrifuge.
5. Transfer the DNA into the MaXtract tube and adjust the reaction volume to 200 μl.
6. Purify DNA by Phenol:Chloroform:IAA as described in Subheading 3.8.
7. Precipitate the DNA with isopropanol and resuspend in 66.8 μl of nuclease-free water.

3.9.4. GsuI Digestion to Remove Poly (A) Tail and Produce 3′ Terminal Ends

1. Freshly dilute 32 mM SAM to 0.5 mM.
2. Set the following on ice in a 1.7-ml lo-bind tube:

Full-length double-stranded cDNA	66.8 μl
10× buffer TANGO with BSA (Fermentas)	8.6 μl
0.5 mM SAM (1 μl SAM + 63 μl dH_2O)	8.6 μl
GsuI (5U/μl, Fermentas, #ER0462)	2 μl

3. Mix by flicking and brief spin at 4°C.
4. Incubate at 30°C overnight.
5. Inactivate GsuI at 65°C for 20 min and transfer sample to ice.

3.9.5. Isolate Full-Length cDNAs by Size-Fractionation and Separation from Linkers

1. Prepare size fractionation column according to manufacturer's instructions while sample is held at 65°C for 20 min. Equilibrate column to room temperature before use (see Note 17). Remove top cap first, then bottom and then allow liquid to drain off completely. Add 0.8 ml TEN buffer and allow it to drain off. Repeat the washes three times (see Note 18).
2. Label twenty 1.7-ml tubes for fractionation purpose.
3. Add 2 μl of 6× loading dye to the cDNA sample and keep it on ice.

4. Transfer the overnight digested mixture into the prepared column and collect the entire flow out into the first collection tube.
5. Add 100 μl of TEN buffer and collect the entire flow through into the second collection tube.
6. Add another 100 μl of TEN buffer and start to collect cDNA by single drop per tube into the third tube and so on until the 20th tube.
7. During the collection process allow each drain-off completely before adding the next 100 μl of buffer. It may need adding a total seven times at each with 100 μl of TEN buffer.
8. After finishing the size fractionation, the collected tubes from 9 through 18 will be selected to run on a 4–20% TBE gel at 200 V for 15 min to assess cDNA collection results. Usually the tubes from 3 to 13 will be pooled and purified as full-length cDNA going to next step (see Note 19). The tubes beyond 14–20 will be discarded as they usually contain linker DNA added.
9. The quantity of the purified full-length cDNA is measured using Quant-iT PicoGreen method following manufacturer's instruction.

3.10. Methylation of Full-Length cDNA Using Enzyme EcoP15I

1. Set the following reaction mix on ice in a 1.7-ml lo-bind tube:

Full-length double strand cDNA	Top up to 100 μl with dH2O
10× Buffer 3	10 μl
100× BSA	1 μl
32 mM SAM (1 μl SAM + 63 μl dH2O)	25 μl
EcoP15I enzyme (see Note 20)	10 U/μg cDNA

2. Incubate reaction mix at 37°C for overnight.
3. Immediately before use, pellet MaXtract High Density Tube by centrifugation at 14,000 rpm for 30 s in a microcentrifuge.
4. Transfer DNA into MaXtract tube, adjust the reaction volume to 200 μl with water.
5. Purify DNA by Phenol:Chloroform:IAA as described in Subheading 3.8.
6. Precipitate with isopropanol and resuspend cDNA in 50 μl of water.

3.11. Ligation of Sequencing Linkers to Full-Length cDNAs

1. Estimate the amount of linkers (h/or t) to be used with a formula below:

$$[(\text{ng cDNA} \times 200 \times 20\text{bp}) / 2,500\text{ bp}] / [200\text{ng}/\mu\text{l linkers}] = \mu\text{l of linkers to be used}$$

(see Note 21).

2. Add the proper linkers and reagents on ice in order and incubate at 16°C for overnight.

Sequencing linker-1 (200 ng/μl) (see Note 15)	x μl (see formula above)
Sequencing linker-2 (200 ng/μl)	x μl (see formula above)
Full-length cDNA	50 μl
5× T4 DNA ligase buffer + PEG	40 μl
Nuclease-free water	Top up to 200 μl
T4 DNA ligase (see Note 22)	1 μl

3. Prepare MaXtract High Density Tube as previous procedures.
4. Transfer the DNA into the MaXtract tube.
5. Purify DNA by Phenol:Chloroform:IAA as described in Subheading 3.8.
6. Precipitate with isopropanol and resuspend DNA in 44 μl of water.

3.12. Addition of Phosphate Group to the 5′-Ends of Linker-Ligated Full-Length cDNAs

1. Set the following on ice in a 1.7-ml tube:

Linker-ligated full-length cDNAs	44 μl
10× T4 DNA ligase buffer	5 μl
T4 DNA polynucleotide kinase	1 μl (final conc. ~0.2 U/μl)

2. Incubate at 37°C for 30 min.

3.13. Circularization of Linker-Ligated cDNA by Ligation in a 5-ml Volume

1. Prepare the following enzyme reaction mix (5 ml) on ice into a 15-ml tube:

Nuclease-free water	4,425 μl
10× T4 DNA ligase buffer	495 μl
T4 DNA ligase	30 μl (final conc. ~0.18 U/μl)
Linker-ligated cDNA mix (from Subheading 3.12)	50 μl

   Transfer the premix (50 μl) described in Subheading 3.12 into the above ligase mix (15 μl tube).

2. Incubate the ligation at 16°C for overnight (see Note 23).
3. Purify DNA by Phenol:Chloroform:IAA as described in Subheading 3.8.
4. Precipitate the DNA with isopropanol (see Note 24) and resuspend it in 78 μl of Buffer EB.

3.14. DNA Nick Repair

1. Set the following reagents on ice in a 1.7-ml tube:

DNA in Buffer EB	78 μl
10× *E. coli* DNA ligation buffer	10 μl
10 mM dNTP	2 μl
E. coli DNA ligase	2 μl
E. coli DNA polymerase I	8 μl

2. Incubate at 16°C for 2 h.
3. Immediately before use, pellet a 2-ml MaXtract High Density Tube by centrifugation at 13,200 rpm for 1 min in a microcentrifuge.
4. Transfer DNA nick repair reaction mix into the MaXtract tube.
5. Purify DNAby Phenol:Chloroform:IAA as described in Subheading 3.8.
6. Precipitate DNA with isopropanol and resuspend in 84 μl of water.

3.15. Plasmid-Safe Treatment to Cleave Remaining Linear DNA Molecules

1. Set the following on ice in a 1.7-ml lo-bind tube:

Circularized DNA	84 μl
25 mM ATP (Epicentre)	4 μl
10× reaction buffer (Epicentre)	10 μl

2. Incubate at 37°C, 40 min (see Note 25).
3. Prepare the 2 ml MaXtract High Density Tube as described above.
4. Transfer the DNA into the MaXtract tube; adjust reaction volume to 200 μl.
5. Purify DNA by Phenol:Chloroform:IAA as described in Subheading 3.8.
6. Precipitate DNA with isopropanol and resuspend in 50 μl of water.

3.16. EcoP15I Digestion to Release 5′ and 3′ Tags from Circularized Full-Length cDNA

1. Set the following on ice in a 1.7-ml lo-bind tube:

Circularized full-length cDNA	50 μl
10× Buffer 3	10 μl
100× BSA	1 μl
10 mM Sinefugin	1 μl
10× ATP (NEB) (see Note 26):	2 μl
EcoP15I (NEB)	10 U/μg DNA
Nuclease-free water	Adjust volume to 100 μl

2. Incubate at 37°C, 2 h.

3.17. Binding of EcoP15I Digested DNA Tags to Dynalbeads M-280 Streptavidin

1. Swirl bottle of Dynalbeads M-280 streptavidin suspension thoroughly.
2. Transfer 50 μl of Dynalbeads M-280 Streptavidin suspension to a 1.7-ml tube.
3. Using the MPC, wash beads with 150 μl of 2× B&W buffer by pipetting up and down.
4. Resuspend beads in 100 μl of 2× B&W buffer.
5. Add 100 μl of EcoP15I-digested DNA to the resuspended Dynalbeads, mix well.
6. Incubate at room temperature with rotation on the Intelli-Mixer (Program F8, 30 rpm) for 30 min. During the incubation, biotinylated linkers associated with the captured DNA tags are to be bound and remained on the beads.
7. With the help of MPC, the reaction beads are washed twice with 150 μl of 1× B&W buffer by pipetting up and down, which will remove those DNA fragments with no linker attached.

3.18. Ligation of (Solexa or SOLiD) Sequencing Adaptors to DNA Template for HT Sequencing

1. For Solexa adaptor ligation, set the following on ice in a 1.7-ml tube:

Nuclease-free water	36 μl
Solexa 454-adaptor E/A (200 ng/μl)	4 μl
Solexa 454-adaptor E/B (200 ng/μl)	4 μl
10× T4 DNA ligase buffer	5 μl

2. For SOLiD adaptor ligation, set the following on ice in a 1.7-ml tube:

Nuclease-free water	36 μl
SOLiD P1 Adaptor (200 ng/μl)	4 μl
SOLiD P2 Adaptor (200 ng/μl)	4 μl
10× T4 DNA ligase buffer	5 μl

   Since the SOLiD sequencing adaptors (P1 and P2) are blunt-end, therefore, the EcoP15I released tags (5′-overhang) have to be blunt-end first by a "filled-in" reaction step before adaptor ligation (see Note 27 for detail).

3. Resuspend tag-bound beads with the above ligation mix based on specific sequencing platform chosen.
4. Add 1 μl of T4 DNA ligase to the beads suspension to make a final concentration at 0.6 U/μl to ligate adaptors to the captured PETs.
5. Incubate at RT for overnight with rotation on an Intelli-Mixer (Program F8, 30 rpm, U = 50, u = 60)
6. Wash the beads twice with 150 μl of 1× B&W buffer and go for next Nick Repair.

3.19. Nick Translation Repair

1. Set the following reagents on ice in a 1.7-ml tube:

Nuclease-free water	38.5 μl
10× Buffer 2	5 μl
10 mM dNTP	2.5 μl (final conc. 500 μM)
E. coli DNA polymerase I	4 μl

2. Resuspend Dynalbeads in the above reaction mix.
3. Incubate at RT with rotation for 2 h on an Intelli-Mixer (F8, 30 rpm).
4. Wash the beads twice with 150 μl of 1× B&W buffer using the MPC.
5. Resuspend the Dynalbeads in 50 μl of Buffer EB.

3.20. PCR Amplification to Assess Captured 5′ and 3′ cDNA Tags

1. Set the following on ice in 0.2-ml thin-walled PCR tube:

Nuclease-free water	21 μl
Dynalbeads suspension	2 μl
Solexa or SOLiD PCR primer-1	1 μl
Solexa or SOLiD PCR primer-2	1 μl
2× Phusion Master Mix with enzyme	25 μl

PCR cycling conditions:

98°C	30 s
98°C	10 s
65°C	30 s
72°C	30 s
72°C	5 min
4 C	hold

Repeat 20 times from step "98°C/10 s" to "step 72°C/30 s."

2. After PCR reaction is done, take 25 μl of PCR products and run on a 10-well 4–20% TBE gel at 200 V for 45 min. Stain the gel for 10 min in SYBR-TBE buffer before taking picture.
3. Load 500 ng of a 25-bp DNA ladder side by side for size determination.
4. As shown in the gel picture (Fig. 2) below, an expected 237 bp (Solexa) and 154 bp (SOLiD) bands are observed which are consistent with the structures of the captured cDNA tags associated with platform-specific linkers and adaptors.

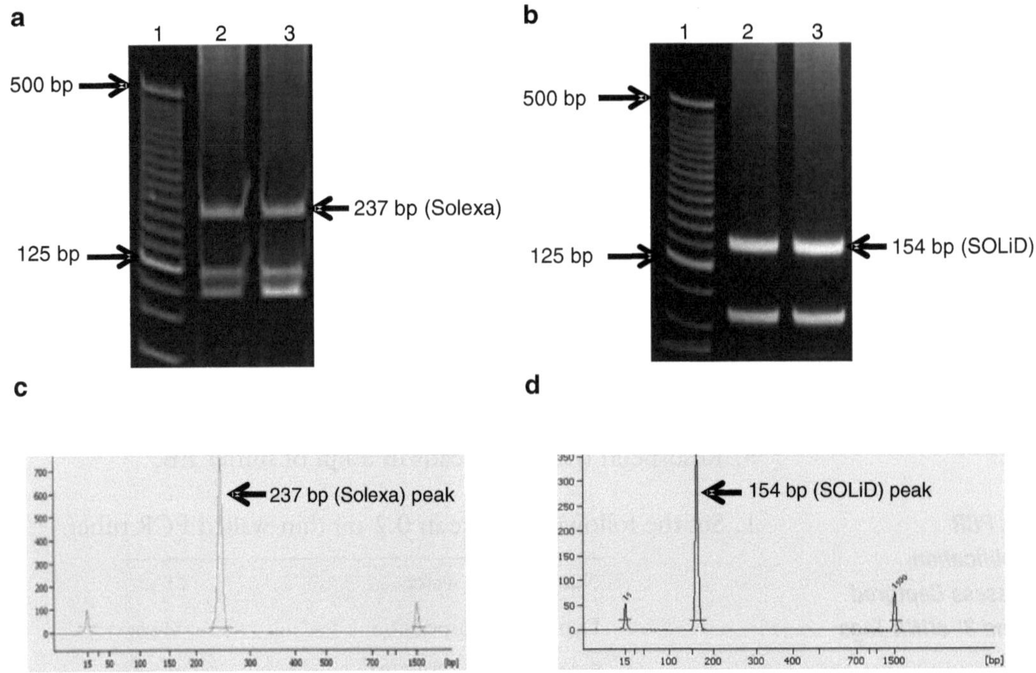

Fig. 2. RNA-PET sequence template QC assessments. Examples of RNA-PET libraries constructed for Illumina GAII (**a**) and ABI SOLiD4 (**b**) sequencing methods. The libraries were purified as specific PCR fragments (237 bp for Illumina GAII 154 bp for SOLiD), which are composed of 5′ and 3′ end tags, a linker sequence and two sequencer-specific adaptors ligated at each end. After gel purification, the library materials were further analyzed by Agilent Bioanalyzer profiles, showing unique DNA fragment peak at 237 bp for Illumina GAII (**c**) and 154 bp for SOLiD (**d**), respectively.

3.21. PCR Scale-Up for Preparation of Sequencing Templates

1. Set PCR scale-up reactions using all or at least half amount of available Dynalbeads as amplification template and collect expected PCR fragments separated on 6% TBE gel.
2. Depending on the number of PCR reactions set up, the PCR products need to be concentrated before loading into the gel.
3. A 25-bp DNA ladder is critical to be loaded side by side in the gel for size reference when harvesting the desired band.

3.22. Purify PCR Fragment from 6% TBE gel Using Gel-Crush Method

1. PCR fragment of the interest is carefully excised and collected into several 0.6-ml microtubes that have been pierced at the bottom with a 21-G needle. Two or more gel slices can be put into each 0.6-ml microtube according to the size of the gel slices. The pierced tube is placed inside a 1.5-ml screw-cap microtube and centrifuged at 13,200 rpm for 5 min in a microcentrifuge. The gel slices are thus conveniently shredded and collected in the bottom of each 1.5-ml tube.
2. Add 400 μl of TE buffer to each 1.5-ml screw-cap tube, stir the gel pieces with the pipette tip to ensure gel pieces are immersed in the TE buffer.

3. Transfer the 1.5-ml screw-cap tubes containing shredded to −80°C freezer for 1–2 h; then bring tubes back to a 37°C incubator for an overnight incubation. The DNA from the shredded gel will come out from the gel into TE buffer during the incubation.
4. After overnight incubation, transfer gel pieces together with the buffer to the filter cup of a Spin-X column and centrifuge at 13,200 rpm, 4°C for 10 min in a microcentrifuge.
5. After centrifugation, add 200 μl of TE to each filter cup and stir to loosen the gel pie with a pipette tip. Centrifuge again at 13,200 rpm for 10 min to recover remaining DNA.
6. Transfer the filter-through liquid to a new tube. Precipitate the DNA with isopropanol and resuspend in 20 μl of TE buffer.
7. Take 1 μl of DNA template to perform QC check with Agilent 2100 Bioanalyzer using DNA-1000 kit according to manufacturer's instruction. The Agilent profile should show a clean and strong DNA fragment peak at expected size and no noise background detected in the sample.
8. Sequence the RNA-PET library templates by the paired end (PE) sequencing method according to manufacturer's (Illumina GA or SOLiD v4) guidelines and instructions.

3.23. PET Sequencing and Mapping

3.23.1. Paired End Tag Sequencing with Illumina (Solexa) GA and SOLiD v4

1. Paired end (PE) sequencing run-type is performed with the respective library made using either Illumina (Solexa) GA or SOLiD v4 following manufacturer's guidelines and instructions. For Illumina GA, the PE sequencing generates two paired end reads with each at 36 bp length. SOLiD v4 PE sequencing generates two paired end reads at each 35 bp long.
2. In Solexa sequences, each 36 bp read consists of a 27-bp tag plus 9-bp linker sequence, which is part of the linker immediately close to the tag. For SOLiD PE sequence, each 35 bp read contains a 27-bp tag plus 8-bp linker sequence that is close to the tag. The PET structure is simply illustrated as 27 bp tag + 9/or 8 bp – linker – 9/or 8 bp + 27 bp tag.
3. In our experience, one lane of a Solexa flow cell (FC), after filter out the noise reads, can usually produce over 20–30 millions of pass filter PETs, or PETs. For SOLiD run, a spot equivalent to 1/8 of a slide can also produce approximately 30 millions of usable paired end reads.
4. Since the sequencing adaptors can be ligated to either end of a given transcript PET, to differentiate which end tag is come from 3′ or from 5′ end of a transcript, a specific signature sequence (*AA*CTGCTG), characteristic by a double *AA* residue in the beginning of the signature, and is also the two end residues at the 3′-end of a 27-bp tag, services as an indicator for identifying the 3′-end tag first.

5. To begin the PET analysis, the signature sequence is searched out first from both paired end sequence reads. As long as the *AA*CTGCTG (5→3) sequence is identified from one end of a PET, this tag is considered as the 3′-end tag and the other end is defined as the 5′-end tag. There is a small portion of the PETs which do not have any signature sequence being identified, or in rarely cases, the signature sequence could appear on both paired ends; these PETs are discarded from future analysis.

3.23.2. Mapping of the Paired End Tags

1. After PET orientation (5′>3′) is identified, regular mapping of 5′ and 3′ tags to reference genome is performed through Solexa (ELANE) or SOLiD (BioScope) analysis pipeline, followed by further analysis approaches specifically designed for RNA-PET.
2. In Solexa and SOLiD mappings a seed of 25 bp of tag is used and a maximum 2 bp mismatches are allowed within the seed sequence for each tag mapping.
3. For those PETs which have uniquely mapped both to the 5′ and 3′ ends of the reference genome are classified as uniquely mapped PETs.
4. Approximately 90% of PETs which are mapped on the same chromosome, in the same strand, and in the same correct orientation to the known transcripts or known transcript variants, are defined as *concordant* PETs.
5. Whereas, a small portion ~10% of PETs, which are mapped incorrectly to the reference genome, are referred as *discordant* PETs, that makes a class of the PETs which are mapped either in the wrong orientations in the same strand (e.g., 3′-end tag mapped before the 5′-end tag), or the paired tags mapped on two different strands, or two different chromosomes, e.g., one end mapped on chromosome 3, another end mapped on chromosome 8.
6. A majority of *discordant* PETs are derived from ligation noise; however, this class of the PETs serves as a valuable pool to identify those novel, fusion or transcriptional variants which might be caused by a variety of genome rearrangements such as deletions, inversions, insertions, tandem repeats, and translocations; or from transcriptional variations of *trans*-splicing or mutational events etc.

3.23.3. PET Clustering and Genome Annotation

1. After *concordant* PETs are identified, they are clustered each other with nearby PETs based on searching a 100-bp distance as a window to extend each PET at 5′- and 3′-end, respectively. Specifically, the mapping location of the 5′ and 3′ tag of a given PET is extended in a 100-bp wide window from both directions.
2. If the 5′ and 3′ tags of a second PET mapped within the 5′ and 3′ search window of the first PET then the two PETs are clustered

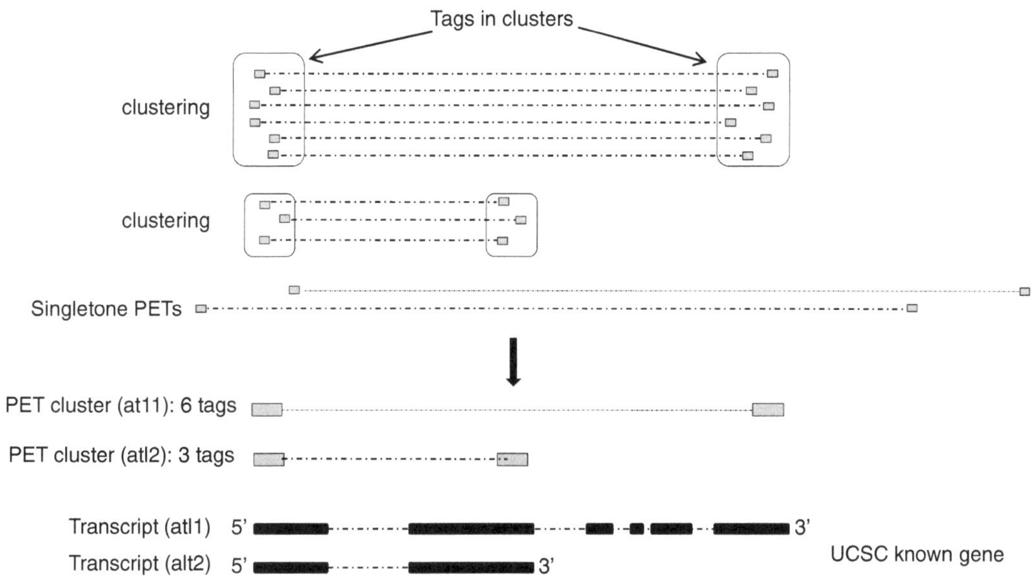

Fig. 3. Scheme of RNA-PET sequence clustering and annotation to reference genome. After mapping to reference genome, tag sequences aligned in defined span range are cluster together to represent molecules derived from the same transcripts. The illustrated example showing 6 PET sequences clustered together to represent one transcript of the gene encoded at this location, and three other PET sequences cluster together tp represent a short alternative of transcript of this gene. Singletons are discarded from further analysis.

and the search windows are re-adjusted and continue the expansion to acquire new PETs. This process is dynamic and iterative, and continues until no new PETs can be found within the allowed window distance.

3. At the end of clustering process, most related *concordant* PETs are clustered each other and those PETs which do not fall into this search window are classified as singletons (Fig. 3), and filtered out from the dataset and not go for further analysis. Furthermore, any PETs to be clustered, 5′-end of tags should be within 100-bp each other, and the same criteria applies to the 3′-end tags.
4. The following (Fig. 3) diagram illustrates the mapping and annotation concepts.
5. Figure 4 shows the sequence tags mapped on the RNA-PET browser from two libraries derived from breast cancer cell line samples MCF7 with Control vs. Estrogen diol treatment.

Two RNA-PET library datasets are shown in Fig. 4. PET sequence tags mapped on the browser are from two breast cancer cell samples i.e., MCF7-Control vs. Estrogen diol treated. As the concordant PET data shown on the browser, expression levels of two libraries

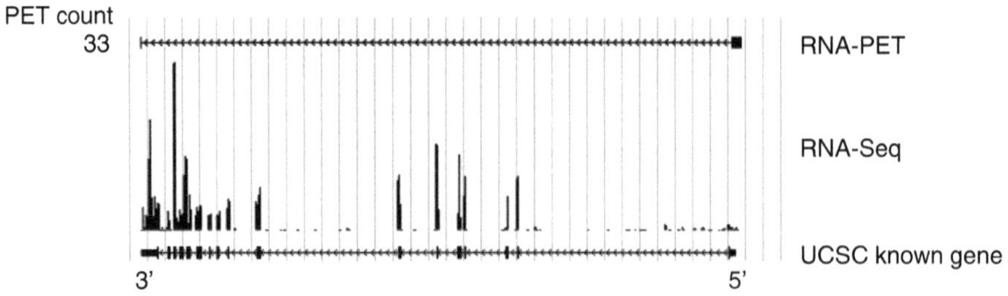

Fig. 4. An Example of RNA-PET data visualized in genome browser. RNA-PET sequence tags mapped in genome browser are from breast cancer cell MCF7. The PET count reflects the expression level and the tag mapping locations demarcate the boundaries of the transcription unit. RNA-Seq data from the same cells showed the marks of each exons. The combination of RNA-PET and RNA-Seq data provides a comprehensive view of transcript structure and expression abundance.

are clearly represented by the PET counts. In addition, novel splice transcript can be seen from the mapped PETs.

4. Notes

1. For all steps involving single-stranded DNA or RNA, it is preferable to use "lo-bind" microfuge tubes to avoid loss of nucleic acid.
2. Commercially available nuclease-free water (instead of DEPC-treated water) was used for all RNA-containing enzymatic reactions to avoid possible inhibition of enzymatic reactions by residual DEPC or ethanol.
3. Because the solubility of trehalose increases with temperature, it is important to maintain the liquid temperature at or around 42°C (in which temperature trehalose is used in the protocol) when making trehalose solution.
4. For all steps involving RNA manipulations, ensure RNase-free conditions are maintained, including all reaction buffers.
5. This step is necessary when combined volume of polyA mRNA and GsuI-dT oligo exceeds 9 μl.
6. Do not use glycogen at any stage where it is not specifically mentioned as glycogen will interfere with the cap-trapper selection process.
7. Use 2.5 U of RNase ONE (Promega) per μg of starting polyA mRNA.
8. During RNase One digestion, proceed to Subheading 3.6 to prepare the Dynalbeads.
9. Before using the Dynalbeads, resuspend the beads in suspension by strong shaking.
10. Cautions in Dynalbeads washing procedures:

(a) After transferring the required amount beads (e.g., 200 μl) to a fresh lo-bind tube, the tube should be placed on a MPC stand (magnet) for at least 1–2 min.

(b) Remove wash supernatant only when the tube is positioned on a MPC. Never remove S/N while the tube is taken off from the MPC.

(c) Add washing buffer along the inside wall of a tube.

(d) Resuspend and mix the beads only when the tube is taken off the MPC.

11. As Dynabeads are heavy and can easily settle down to the bottom of the tube, the immobilization of (−) cDNA/RNA heteroduplex onto the beads should be done on an Intellimixer with constant rotations.
12. The incubation time of immobilization can be increased if the samples are diluted.
13. Always use freshly prepared alkaline hydrolysis buffer.
14. RNase-free condition is not necessary from this step onward.
15. DNA linkers should be annealed beforehand and stored in aliquots at −20°C. Always keep the annealed linkers on ice and avoid warming them when thawing. This precaution could prevent possible denaturation and subsequent complications. Centrifugation of the annealed adaptors is also recommended in cold (4°C) condition.
16. Ensure that the cDNA and linkers are well mixed before adding solution II as the latter contains PEG, which could lead to precipitation of glycoblue.
17. It is important to use only columns that do not show any visible bubbles trapped within the matrix as these tubes are likely faulty in our experiences. Also, the chances of bubble formation can be reduced by allowing the columns to equilibrate to room temperature.
18. It is preferred that each draining of the cDNA columns does not take longer than 25 min.
19. Avoid collecting any fractionation tubes showing presence of small molecular weight linker DNA bands as the presence of linkers in following reactions will cause problems in quenching enzymatic reactions.
20. Use about 10 U of EcoP15I (NEB) per μg of starting double-stranded cDNA.
21. It is assumed that the average length of cDNA is 2.5 kbp.
22. Add T4 DNA ligase to the reaction at the last and keep the reaction in cold at all times.
23. Circularization can be performed for as long as >24 h at 16°C.

24. Precipitation can also be done in multiple 1.7-ml tubes.
25. The maximum incubation time for plasmid-safe treatment can be 2 h.
26. The ATP that is supplied with EcoP15I (NEB, B6101S) is at 10×, thus 20 μl is needed to obtain a 2× final concentration while the ATP sold separately (NEB, P0756) is usually at 100×, and hence 2 μl is used instead in the protocol.
27. The "filled-in" blunt-end reaction is conducted using Endit kit (Epicentre Inc.) following manufacturer's instruction. After "filled-in" reaction, the beads are washed following the steps described in Subheading 3.17 (7) and are ready for next blunt-end ligation for SOLiD adaptors.

Acknowledgments

The authors would like to thank GOH, Yufen; Ho, Andrea; QUEK, Kelly, KHNG, Alexis, CHOY, Yeen Hui, Atif, Shahab; ONG, WaiLoon; POH, WanTing; Lavanya Veeravalli; Thoreau, Herve; Ong, ChinThing; Dawn, Sum; Leong, SeeTing and all other members of the GTB research and sequence group. This work was supported by NIH ENCODE grant and A-STAR, Singapore.

References

1. Peters, B. A and Velculescu, V. E. (2005). Transcriptome PETs: A genome's best friends. *Nature Methods* 2, 93–94.
2. Fullwood, M. J., CL. Wei, E.T., Liu, and Y. Ruan (2009) Next-Generation DNA Sequencing of Paired End diTags for Transcriptome and Genome Analysis (Review). *Genome Research* 19, 521–532.
3. Wei, C.L., Ng Patrick W.P., Chiu, K.P., Wong, C.K., Ang, C.C., Lipovich L., Liu, E.T., and Y. Ruan (2004). 5′ Long serial analysis of gene expression (LongSAGE) and 3′ LongSAGE for transcriptome characterization and genome annotation. *PNAS*, USA. 101, 11701–11706.
4. Ng P, Wei CL, Sung WK, Chiu KP, Lipovich L, Ang CC, Gupta S, Shahab A, Ridwan A, Wong CH, et al. (2005) Gene identification signature (GIS) analysis for transcriptome characterization and genome annotation. *Nature Methods*. 2, 105–111.
5. Ng P, Tan JJ, Ooi HS, Lee YL, Chiu KP, Fullwood MJ, Srinivasan KG, Perbost C, Du L, Sung WK, et al. (2006) Multiplex sequencing of paired-end ditags (MS-PET): a strategy for the ultra-high-throughput analysis of transcriptomes and genomes. *Nucleic Acids Res.* 34, e84.
6. Carninci, P., Kvam, C., Kitamura, A., Ohsumi, T., Okazaki, Y., Itoh, M., Kamiya, M., Shibata, K., Sasaki, N., Izawa, M., Muramatsu, M., Hayashizaki, Y., Schneider, C. (1996) High-efficiency full-length cDNA cloning by biotinylated CAP trapper. *Genomics*. 37, 327–336.
7. Carninci, P., Westover, A., Nishiyama, Y., Ohsumi, T., Itoh, M., Nagaoka, S., Sasaki, N., Okazaki, Y., Muramatsu, M., Schneider, C., Hayashizaki, Y. (1997) High efficiency selection of full-length cDNA by improved biotinylated cap trapper. *DNA Res.* 4, 61–66.
8. Carninci P, Hayashizaki Y. (1999) High-efficiency full-length cDNA cloning. *Methods Enzymol.* 303, 19–44.

Chapter 36

Analysis of Co-transcriptional RNA Processing by RNA-ChIP Assay

Danielle Bittencourt and Didier Auboeuf

Abstract

It was initially assumed that RNA biogenesis and processing were two independent processes with transcripts undergoing splicing only after being completely synthesized and released from the DNA template. However, transcription and splicing are tightly linked and increasing evidence shows that nascent transcripts can undergo splicing in the vicinity of chromatin while still attached to the RNA polymerase II (RNAPII) transcriptional machinery. These co-transcriptionally spliced RNA molecules are very labile due to dynamic processing and represent a minor subpopulation among total cellular RNA species. Thus, it is difficult to isolate these RNAs in order to study the dynamics and mechanisms of co-transcriptional RNA splicing. To overcome this problem, the RNA-chromatin immunoprecipitation (ChIP) assay, adapted from classical ChIP, allows to co-purify and isolate nascent RNAs after immunoprecipitation of RNAPII. Thanks to this technique, we have shown that co-transcriptional RNA splicing occurs with distinct efficiencies for different genes and different exons of a given transcript and can represent a rate-limiting step in the biological response of messenger RNA synthesis to extracellular stimuli and drug treatments.

Key words: RNA, Cross link, Immunoprecipitation, RNA polymerase II, Splicing, Nascent transcript, Co-transcriptional

1. Introduction

It is common in the transcription field to find publications, where the levels of mature messenger RNA (mRNA) are used as a measure of transcription rate. However, mRNA levels are a result of multiple steps including not only transcription, but also transcript 5′ end capping, splicing (i.e. intron removal), 3′ end cleavage, and polyadenylation (1, 2). Moreover, steady-state mRNA levels are biased by differences in the degree of RNA stability for different

Ales Vancura (ed.), *Transcriptional Regulation: Methods and Protocols,* Methods in Molecular Biology, vol. 809,
DOI 10.1007/978-1-61779-376-9_36, © Springer Science+Business Media, LLC 2012

mRNA molecules. Thus, in order to directly measure transcription rate, in vitro assays, such as nuclear run on, are popularly used. However, a more physiologically relevant indicator of whether drug treatments or extracellular stimuli have a direct impact on gene transcription is the measure of nascent RNA levels in vivo. Moreover, since RNA splicing has been shown to occur co-transcriptionally (2–6), it is important to consider the dynamics of co-transcriptional RNA processing in order to understand the biological outcome of a given transcriptional stimuli. Indeed, transcription and splicing are tightly linked and increasing evidence show that nascent transcripts can undergo splicing in the vicinity of chromatin while still attached to the RNA polymerase II (RNAPII) transcriptional machinery. As these co-transcriptionally spliced RNA molecules are very labile due to dynamic processing and represent a minor subpopulation among total cellular RNA species, it is difficult to isolate these RNAs in order to study the dynamics and mechanisms of co-transcriptional RNA splicing. In this context, the RNA-chromatin immunoprecipitation (ChIP) assay (7, 8), adapted from classical ChIP assay and that allows to co-purify and isolate nascent RNAs after immunoprecipitation (IP) of RNAPII, provides a powerful tool for analyzing co-transcriptional splicing.

First, covalent bonds between RNAPII and associated nascent transcripts are created by treating cells with a chemical cross-linking agent (formaldehyde) (see Fig. 1a). Cross-linked cells are then lysed and the cell lysate is sonicated to solubilize chromatin and release RNAPII protein epitopes that might be "trapped" or embedded in chromatin environment. Importantly, sonication does not result in the fragmentation of nascent transcripts (see Fig. 1b, c). The sonicated cell lysate is next pre-cleared by incubation with agarose beads in order to eliminate non-specific interactions between proteins and nucleic acids with the agarose bead surface, thus reducing background noise. Pre-cleared sonicated cell lysate is next incubated with anti-RNAPII antibody-bound beads in order to recover the RNAPII-nascent transcript immmunocomplex. Subsequent washing of the bead-bound immunocomplex allows removal of weak non-specific interactions, thus increasing signal:noise ratio for the immunoprecipitations. Heat-induced reverse cross linking then elutes the immunocomplex. RNA is then purified from the eluate and the latter is extensively treated with DNAse I to remove DNA contaminants (see Fig. 2a). Finally, RT-qPCR analysis using primers amplifying pre-mRNA or partially spliced pre-mRNA intermediates is used to measure the rate of pre-mRNA synthesis and co-transcriptional splicing, respectively, compared to the total amount of these RNAs in the nucleus (nuclear RNA extracts are used as inputs). It is, thus, possible to determine for a given gene product which exons are spliced co-transcriptionally and in which order (see Fig. 2b).

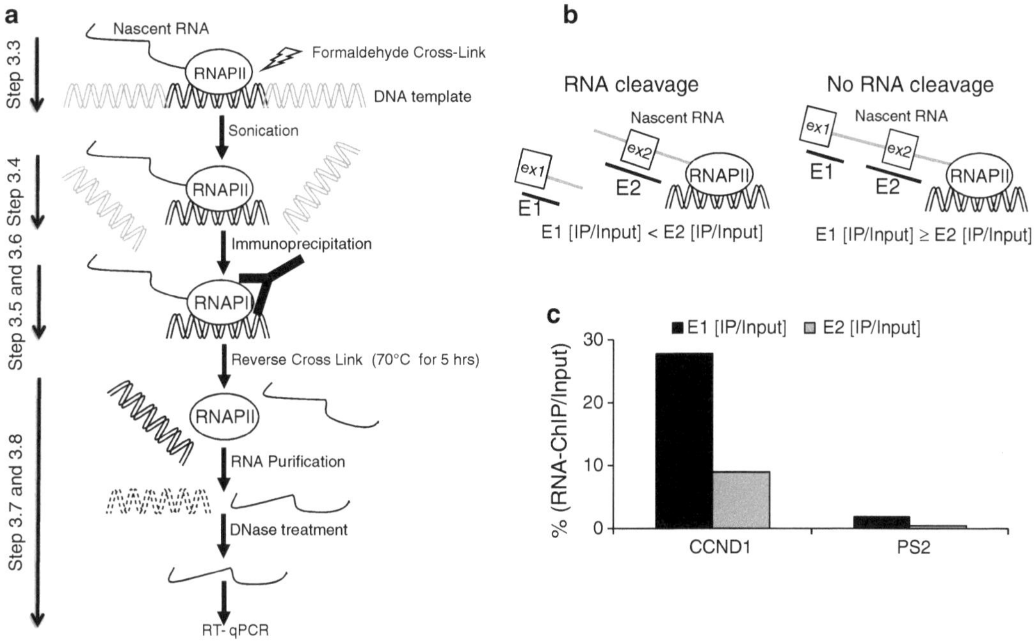

Fig. 1. RNA-ChIP assay. (**a**) Schematic representation of the individual steps (from top to bottom) of the RNA-ChIP assay. Oval, double helix, curvy line, and black forkhead represent RNA polymerase II (RNAPII), DNA, nascent transcript, and anti-RNAPII antibody, respectively. (**b**) Transcript level was measured in nuclear RNA (Input) or RNA co-immunoprecipitated with an anti-RNAPII antibody (IP) by qPCR with sets of PCR primers amplifying the Exon1–Intron1 boundary (PCR amplicon E1) or primers located in the introns flanking Exon2 (PCR amplicon E2). If sonication results in cleavage of nascent transcripts (left panel), E1 enrichment in IP compared to Input will be weaker than for E2, since E1 has a greater probability of being separated from RNAPII due to its location further apart from the RNAPII. In contrast, if sonication does not result in cleavage of nascent transcripts (right panel), E1 enrichment in IP compared to input will be either similar or greater than for E2 as E1 is synthesized before E2. (**c**) RNAPII-associated RNAs are not cleaved by sonication: ratio between IP and input for the E1 amplicon (*black bars*) and E2 amplicon (*grey bars*) regions for CCND1 (*left*) and PS2 (*right*) transcripts. As this ratio is superior for E1 amplicon when compared to E2 amplicon for both PS2 and CCND1 transcripts, we conclude that the integrity of RNAPII-associated transcripts is not significantly affected by sonication.

2. Materials

RNAse/DNAse-free filter tips have to be used throughout the RNA-ChIP assay.

2.1. Preparation of Nuclear RNA Extracts (INPUT)

1. One confluent 10-cm dish (~$2-3\times10^6$ cells).
2. PBS (1×).
3. Cell lifter (Corning).
4. Cell lysis buffer A: 10 mM Tris–HCl, pH 7.5, 140 mM NaCl, 1.5 mM $MgCl_2$, 10 mM EDTA, 0.5% NP-40, 100 U/ml RNaseOUT (Invitrogen) (see Note 1).
5. Trizol reagent (Invitrogen).
6. 20-gauge needle (0.9-mm diameter).

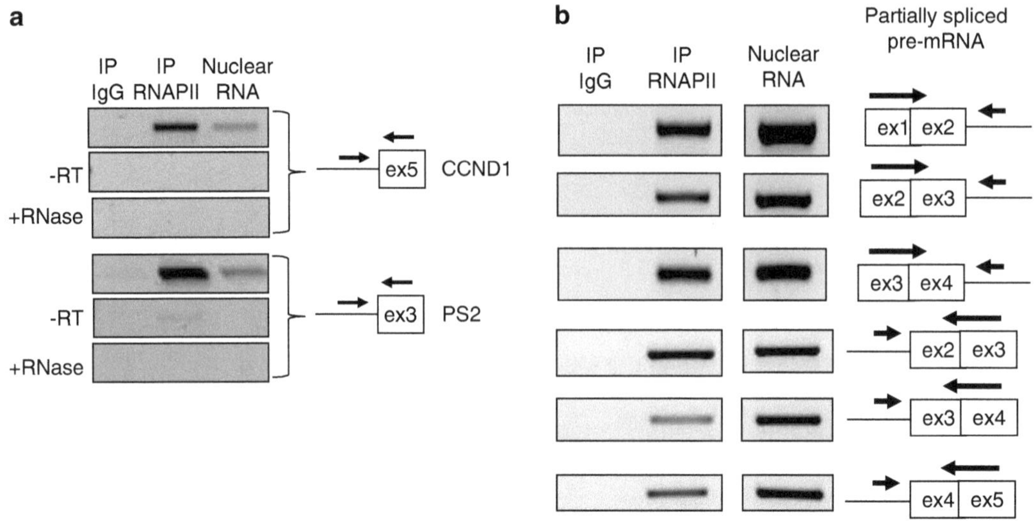

Fig. 2. Co-immunoprecipitation of partially spliced pre-mRNAs. (**a**) PCR amplification of Cyclin D1 and pS2 nascent transcripts co-immunoprecipitated with an anti-RNA polymerase II antibody (IP RNAPII) or with a control IgG (IP IgG). Top slot shows RT-PCR amplification products. Lack of PCR signals in the absence of RT reaction (–RT; middle slot) or when RNAse is added prior to the RT reaction (+RNase, bottom slot) demonstrates that PCR signals reflect co-immunoprecipitated RNAs and are not due to amplification of DNA contaminants. *Arrows*, *boxes*,and *lines* represent primers, exons, and introns, respectively. Numbers reflect exon position from the 5′ to the 3′ gene end. Nuclear RNA extracts were used as an input control for RT-PCR. (**b**) Primers spanning exon–exon junctions or located in intronic regions of the Cyclin D1 nascent transcript were used for RT-PCR analysis of RNAs co-immunoprecipitated with an anti-RNA polymerase II antibody or with a control IgG. *Arrows*, *boxes*, and *lines* represent primers, exons, and introns, respectively. Numbers reflect exon position from the 5′ to the 3′ gene end (ex1, ex2, and so on). Nuclear RNA extracts were used as an input control for RT-PCR. The specificity of the primers must be validated as shown in Fig. 3.

2.2. Preparation of Agarose Beads

1. Protein A/G PLUS-Agarose (Santa Cruz).
2. RIPA buffer: 50 mM Tris–HCl, pH 7.5, 150 mM KCl, 1% NP-40, 50 mM NaF, 1% (v/v) Protease Inhibitor Cocktail (Sigma P8340), 1% (v/v) Phosphatase Inhibitor Cocktail I (Sigma P2850), 1% (v/v) Phosphatase Inhibitor Cocktail II (Sigma P5726) (see Note 1).
3. Salmon sperm DNA.
4. Yeast tRNA (Invitrogen 15401-011).
5. RNAseOUT recombinant ribonuclease inhibitor (Invitrogen 10777-019).
6. Anti-RNA polymerase II antibody, clone CTD4H8 (Upstate 05-623).
7. Normal mouse IgG.

2.3. Cross Linking of Cells

1. One confluent 10-cm dish of cells (~$2-3\times10^6$ cells) is required per immunoprecipitation reaction.
2. Formaldehyde solution, 37%.
3. Glycine, pH 7–7.5.
4. Cell lifter (Corning).

2.4. Preparation of Sonicated Cell Lysate

1. Lysis buffer B: 50 mM Tris–HCl, pH 7.5, 150 mM KCl, 5 mM EDTA, 1% NP-40, 0.1% SDS, 0.5% sodium deoxycholate, 50 mM NaF, 5% (v/v) Protease Inhibitor Cocktail (Sigma P8340), 1% (v/v) Phosphatase Inhibitor Cocktail I (Sigma P2850), 1% (v/v) Phosphatase Inhibitor Cocktail II (Sigma P5726), 100 U/ml RNaseOUT (Invitrogen) (see Note 1).
2. Sonicator device or sonicating water bath.
3. RIPA buffer.

2.5. Preclearing and Immunoprecipitations

1. Normal mouse IgG.
2. RNAseOUT Recombinant Ribonuclease Inhibitor (Invitrogen 10777-019).

2.6. Wash, Elution, and Reverse Cross Link

1. RIPA wash buffer: 50 mM Tris–HCl, pH 7.5, 150 mM KCl, 1% NP-40, 0.25% sodium deoxycholate (see Note 1).
2. TE 1×: 10 mM Tris–HCl, pH 7.5, 1 mM EDTA (see Note 1).
3. RIPA elution buffer: 50 mM Tris–HCl, pH 7.5, 5 mM EDTA, 10 mM DTT, 1% SDS, 400 U/ml RNaseOUT (Invitrogen) (see Note 1).
4. Cap lock for 1.0/2.0 ml microcentrifuge tubes (STARLAB I1415-1508).

2.7. Purification of RNA Samples and Reverse Transcription

1. Molecular biology-grade chloroform.
2. Glycoblue™ (Ambion AM9516).
3. Molecular biology-grade isopropanol.
4. 70% EtOH (EtOH diluted in UltraPure DNAse, RNAse-free distilled water).
5. UltraPure DNAse, RNAse-free distilled water.
6. DNA-free™ (Ambion).
7. Random primers (Invitrogen).
8. dNTP (Invitrogen).
9. SuperScript II reverse transcriptase (Invitrogen).
10. GoTaq Green Master Mix (Promega).
11. Master Sybr Green I (Roche) for Roche LightCycler.

3. Methods

For the RNA-ChIP assay, one confluent 10-cm dish is required (1) per immunoprecipitation reaction; (2) per input (one input per condition; i.e. treated and untreated); and (3) per chromatin size check post sonication (one per condition; i.e. treated and untreated). The experiment is performed over two consecutive days (time needed to seed and grow cells is not included). The first day consists of Subheadings 3.1 through 3.5. The second day consists of Subheading 3.6. RT-qPCR analysis (Subheadings 3.7 and 3.8) can subsequently be performed upon convenience. Alternatively, preparation of nuclear RNA extracts and cross link of cells (Subheadings 3.1 and 3.3) can be done in advance and samples stored at −80°C for up to 2 months. In this case, the first day of the experiment begins with Subheading 3.2 and then proceed directly to Subheading 3.4.

3.1. Preparation of Nuclear RNA Extracts (INPUT)

1. One confluent 10-cm dish (~$2-3\times10^6$ cells) is required for preparation of nuclear RNA extracts (see Note 2).
2. Discard cell medium and rinse cells twice with 10 ml of 1× PBS.
3. Harvest cells by mechanically scraping in 1 ml of ice-cold 1× PBS and transfer to a 1.5 ml microcentrifuge tube.
4. Centrifuge in a microcentrifuge for 3 min at 800×*g* at 4°C.
5. Discard supernatant and wash pellet with 1 ml of 1× PBS.
6. Centrifuge for 5 min at 3,000 rpm at 4°C in a microcentrifuge, and discard supernatant.
7. Carefully (slowly, without making bubbles) resuspend cell pellets in 200 μl of cold cell lysis buffer A, and then transfer to a clean 1.5 ml microcentrifuge tube.
8. Incubate for 5 min on ice with occasional homogenization by inverting tube head over end.
9. Centrifuge for 3 min at 3,000 rpm at 4°C in a microcentrifuge.
10. Recover the supernatant in a clean microcentrifuge tube (this is the cytoplasmic cell fraction) and set aside pellets on ice (these are the nuclei) (see Note 3).
11. Centrifuge the supernatant for 1 min at 15,700×*g* at 4°C in a microcentrifuge to pellet cell debris and contaminating nuclei.
12. Recover supernatant (cytoplasmic fraction) in a clean microcentrifuge tube and add 1 ml of Trizol reagent. Vortex and store at −80°C.
13. Briefly resuspend pellets (nuclei) in 200 μl of cold cell lysis buffer A to remove cytoplasmic contamination from the nuclear fraction.
14. Centrifuge for 3 min at 3,000 rpm at 4°C in a microcentrifuge.
15. Discard all supernatant and resuspend pellet (nuclear fraction) in 200 μl of cold cell lysis buffer A (see Note 4).
16. Add 1 ml of Trizol reagent to nuclear RNA fraction and pass the lysate at least five times through a 20-gauge needle (0.9-mm diameter).
17. Vortex vigorously and store at −80°C.

3.2. Preparation of Agarose Beads

1. 40 μl of stacked bead volume are required per immunoprecipitation reaction (see Note 5). Transfer total amount of beads necessary for the experiment to a 1.5 ml tube (see Notes 6 and 7).
2. Centrifuge tube for 1 min at 3,300 × *g* in a table-top microcentrifuge to pellet the beads (see Note 8).
3. Discard supernatant (see Note 9).
4. Add 1 ml of RIPA buffer (without protease and phosphatase inhibitors) to beads (see Note 10).
5. Homogenize by tilting tube head over end several times until all beads are dislodged from the bottom of tube.
6. Centrifuge for 1 min at 600 × *g* in a table-top microcentrifuge.
7. Leave the tube on a rack in vertical position at room temperature for 1–2 min to allow bead pellet to settle.
8. Discard supernatant (see Note 9).
9. Repeat steps 4 through 8 four times (five times in total).
10. Add the equivalent of one stacked bead volume of RIPA buffer to bead pellet to obtain 50% bead slurry (see Note 10).
11. Add 40 μg of sheared salmon sperm DNA, 200 μg of yeast tRNA, and 25 μl of RNaseOUT per ml of 50% bead slurry and incubate for 1 h at room temperature on a rocker platform or rotation device (see Note 11).
12. Set aside half of the volume of 50% saturated bead slurry at 4°C (these are used for pre-clearing in Subheading 3.5).
13. Dispatch the remaining 50% saturated bead slurry into 40 μl batches per 1.5 ml microcentrifuge tube (see Note 7). One batch is necessary per immunoprecipitation reaction.
14. Add 160 μl of RIPA buffer to each batch (200 μl final volume).
15. Add 3 μg of anti-RNAPII (clone CTD4H8) antibody per batch (for specific immunoprecipitation reaction) or 3 μg of normal mouse IgG per batch (for control non-specific immunoprecipitation reaction).
16. Incubate all batches at 4°C on a rocker platform or rotation device for at least 4 h (these are used later on for immunoprecipitation in Subheading 3.5).

3.3. Cross Linking of Cells

1. One confluent 10-cm dish ($\sim 2-3 \times 10^6$ cells) is required per immunoprecipitation reaction (see Note 2).
2. Add formaldehyde directly to cell media (1% final concentration) and incubate for 10 min at room temperature on a rocker platform.

3. Stop cross-link reaction by adding glycine, pH 7.5 (0.25 M final concentration), to cell media and incubate for 5 min at room temperature on a rocker platform.
4. Discard media and rinse cells twice with 10 ml of ice-cold PBS.
5. Harvest cells by mechanical scraping in 1 ml of ice-cold PBS and transfer to a 1.5 ml microcentrifuge tube.
6. Centrifuge for 4 min at 2,300 × *g* at 4°C in a microcentrifuge, and discard supernatant.
7. Add 1 ml of PBS and centrifuge for 4 min at 5,000 rpm at 4°C in a microcentrifuge.
8. Discard all supernatant, either proceed to sonication step or fast freeze the pellet in liquid nitrogen, and then store at −80°C until usage.

3.4. Preparation of Sonicated Cell Lysate

1. On ice, resuspend each cell pellet in 200 μl of lysis buffer B by pipetting up and down (see Note 12).
2. Sonicate samples to shear DNA (see Note 13).
3. Add 300 μl of RIPA buffer to 200 μl of sonicated cell lysate (500 μl final volume).
4. Centrifuge for 10 min at full speed and 4°C in a microcentrifuge to pellet cell debris.
5. Recover supernatants in a clean microcentrifuge tube and discard pellet.

3.5. Pre-clearing and Immunoprecipitations

1. Add 3 μg of normal mouse IgG to each 500 μl supernatant.
2. Add 40 μl of 50% saturated bead slurry (from step 12 of Subheading 3.2) and incubate for 1 h at 4°C on a rocker platform or rotation device.
3. Centrifuge for 1 min at 2,600 rpm at 4°C in a microcentrifuge, and place the tube on ice in vertical position for 1–2 min to allow the bead pellet to settle.
4. Recover supernatants and discard bead pellet (see Note 9).
5. Pool together all similar pre-cleared lysates and then re-dispatch equal volumes of pre-cleared lysates to the tubes containing 50% saturated bead slurry pre-incubated with desired antibody or control immunoglobulins (from step 16 of Subheading 3.2).
6. Add 2.5 μl of RNAseOUT to each tube and incubate overnight at 4°C on a rocker platform or rotation device.

3.6. Wash, Elution, and Reverse Cross Link

1. Centrifuge tubes for 1 min at 3,000 rpm at 4°C in a microcentrifuge, and place the tube on ice in vertical position for 1–2 min to allow bead pellet to settle.
2. Discard supernatant (see Note 9).

3. Add 1 ml of RIPA wash buffer to each tube (see Note 10) and homogenize by inverting tube head over end several times until all beads are dislodged from the bottom of tube.
4. Incubate for 5 min at room temperature on a rocker platform or rotation device.
5. Incubate for 5 min on ice.
6. Repeat steps 1 through 5 five more times (six times total).
7. Repeat steps 1 through 5 two times using 1× TE instead of RIPA wash buffer.
8. Discard all supernatant (see Note 9) and add 100 μl of RIPA elution buffer to each bead pellet (see Note 10).
9. Reverse cross-link samples by incubating at 70°C for 5 h with occasional homogenization by flicking tubes (see Note 14).
10. Chill samples on ice and microcentrifuge briefly to pellet any condensation formed on the lid of the tube.
11. Add 500 μl of Trizol reagent to each sample, vortex, and proceed to purification or store at −80°C.

3.7. Purification of RNA Samples and Reverse Transcription

3.7.1. Purification of RNA

1. If samples were previously frozen in Trizol reagent, thaw at 37°C for 5 min (see Note 15).
2. Add 200 μl of chloroform per 1 ml of Trizol sample and vortex vigorously for 10 s.
3. Centrifuge at 9,300 × *g* for 10 min at 4°C in a microcentrifuge, and recover the aqueous phase (upper phase) in a clean microcentrifuge tube (see Note 16).
4. Add 1 μl of glycoblue (see Note 17).
5. Add 500 μl of isopropanol per 1 ml of (initial) Trizol sample and vortex vigorously.
6. Incubate at room temperature for 10 min.
7. Centrifuge at 13,000 rpm for 10 min at 4°C in a microcentrifuge.
8. Discard supernatant and add 1 ml of 70% EtOH to the pellet.
9. Invert tube head over end several times until pellet is dislodged from the bottom of the tube.
10. Centrifuge at 13,000 rpm for 10 min at 4°C.
11. Discard supernatant and air dry pellets by leaving microcentrifuge tube open under a hood for 5–10 min to remove all traces of EtOH.
12. Add 24 μl of nuclease-free water per RNA pellet (see Note 18).
13. Freeze RNA at −80°C for 10–15 min, then thaw, and vortex vigorously (see Note 19).

3.7.2. DNase I Treatment of Purified RNA

1. On ice, in a microcentrifuge tube, mix 21.5 μl of purified RNA from immunoprecipitation, 2.5 μl of 10× DNAse buffer, and 1 μl DNAse I (25 μl final reaction volume). For input samples, use only 0.75 μg of RNA for DNAse treatment and bring the volume up to 10.25 μl with nuclease-free water. Add to input sample 1.25 μl of 10× DNAse buffer and 1 μl of DNAse I (12.5 μl final reaction volume) (see Note 15).
2. Incubate at 37°C for 30 min.
3. For input samples, proceed directly to step 5.
4. For immunoprecipitation samples, add 1 μl of DNase I and incubate again at 37°C for 30 min. This is important to remove any traces of DNA contamination since the amount of RNA in immunoprecipitation samples is very low and thus any background coming from DNA contamination may interfere with accurate quantification by RT-qPCR (see Fig. 2a).
5. Add 2.5 μl of DNAse inactivation reagent.
6. Incubate for 2 min at room temperature, flicking the tubes occasionally to homogenize.
7. Centrifuge at 10,000 rpm for 2 min in a microcentrifuge.
8. Recover 20 μl of the supernatant for immunoprecipitation (10 μl of the supernatant for input) being very careful not to pipette the DNase inactivation reagent. Transfer to a clean microcentrifuge tube.

3.7.3. Reverse Transcription of DNase I-Treated Samples

1. In a tube, mix together by pipetting up and down 2 μl of random primers (100 ng/μl), 2 μl of dNTP mix (10 mM each), and 20 μl of DNase I-treated RNA sample from immunoprecipitation. For input RNA, mix together by pipetting up and down 1 μl of random primers (100 ng/μl), 1 μl of dNTP mix (10 mM each), and 10 μl of DNase I-treated RNA sample (see Note 15).
2. Incubate at 65°C for 5 min and then quickly chill on ice.
3. Centrifuge briefly in a microcentrifuge to pellet any condensation that may have formed on the lid of the tube.
4. For immunoprecipitation samples: Add 8 μl of 5× first-strand buffer, 4 μl of 0.1 M DTT, and 2 μl of RNaseOUT to each tube and mix by pipetting gently up and down. For input samples: Add 4 μl of 5× first-strand buffer, 2 μl of 0.1 M DTT, and 1 μl of RNaseOUT to each tube and mix by pipetting gently up and down.
5. Incubate at room temperature for 2 min.
6. For immunoprecipitation samples: Add 0.5 μl of SuperScript II reverse transcriptase to each tube and mix by pipetting gently up and down. For input samples: Add 1 μl of SuperScript II reverse transcriptase to each tube and mix by pipetting gently up and down.

7. Incubate at room temperature for 10 min.
8. Incubate at 42°C for 50 min.
9. Inactivate the reaction by heating at 70°C for 15 min.
10. Store the cDNA at –20°C for PCR/qPCR.

3.8. Quantitative Real-Time PCR

3.8.1. PCR Primer Design

For amplification of pre-mRNA, at least one PCR primer (forward or reverse) should be located in an intron. For amplification of partially spliced RNA intermediates, one PCR primer should span the exon–exon junction resulting from intron removal and the other PCR primer should be located in the intron adjacent to the exon–exon junction. When designing primers spanning exon–exon junctions, it is critical that a minimum of 2 and a maximum of 4 nucleotides of the 3′ primer end be complementary to the sequence of the exon located closest to the adjacent intron. Indeed, primers containing several nucleotides of the 3′ end located in the exon closest to the adjacent intron may be false prime and amplify the more abundant pre-mRNA in addition to partially spliced RNA variants (see Fig. 3).

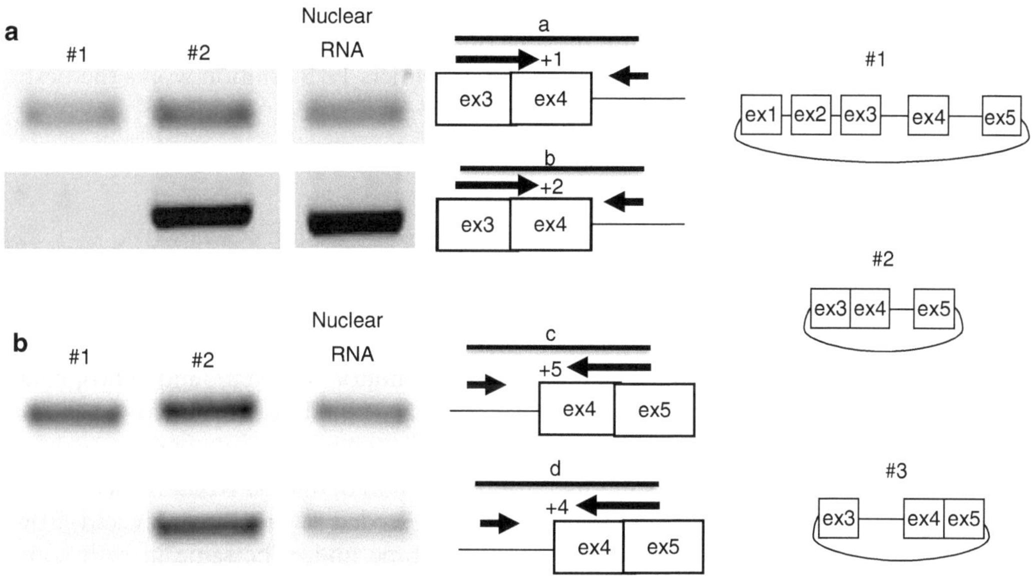

Fig. 3. PCR primer design for the amplification of partially spliced pre-mRNA. Specificity of primers (*arrows*) designed to amplify cyclin D1 partially spliced pre-mRNA was tested by performing classical PCR using as a substrate nuclear RNA or plasmids containing the sequences of the corresponding cyclin D1 pre-mRNA (#1) or cyclin D1 exon–exon junctions (#2 and #3). (**a**) Forward primers spanning exon3–exon4 junction used in top (a) and bottom panels (b) were similar, except that either the first (+1) or first two (+2) bases of the 3′ primer end were located in exon 4 (ex4). Reverse primer located in intron 4 was identical in top and bottom panels. PCR products were obtained with primer set (a) using plasmids corresponding to both partially spliced pre-mRNA (#2) and unspliced pre-mRNA (#1) due to false priming to the intron 3–exon 4 boundary. A PCR product was obtained using primer set (b) when using only the plasmid corresponding to partially spliced pre-mRNA (#2). (**b**) Reverse primers spanning exon4–exon5 junction used in top (c) and bottom (d) panels were similar, except that either the first five (+5) or first four (+4) bases of the 3′ primer end were located in exon 4 (ex4). Forward primer located in intron 3 was identical in top and bottom panels. PCR products were obtained with primer set (c) using plasmids corresponding to both partially spliced pre-mRNA (#3) and unspliced pre-mRNA (#1) due to priming to exon 4. A PCR product was obtained using primer set (b) when using only the plasmid corresponding to partially spliced pre-mRNA (#3).

Importantly, in the case of the genes we have studied, both ordered and non-ordered partially spliced variants were observed since the order of intron splicing does not necessarily obey the sequential order of transcription (9) (see Fig. 2b).

3.8.2. PCR Reaction

For classic PCR reactions, 2.5 μl of reverse transcription product (non-diluted cDNA) are used per 25 μl PCR reaction. Typically, 35–38 PCR cycles are necessary to amplify pre-mRNA, depending on the efficiency of the PCR primer set. Forty cycles are used to amplify partially spliced pre-mRNA. For quantitative real-time PCR, reverse transcription products of IP and INPUT samples are diluted 1:15 in nuclease-free water and 5 μl of this diluted cDNA sample is used per 20 μl of real-time PCR reaction and 40 cycles of amplification are carried out. For some primers designed to amplify partially spliced pre-mRNAs, it may be necessary to decrease the dilution factor of cDNA in order to quantify products by qPCR. This may be either because these splice intermediates are very scarce or because of weak efficiency of the primer set. However, the less the RNA is diluted, the more RT buffer is present in your sample and this may inhibit the efficiency of your qPCR reaction. Thus, dilution must be optimized per primer set, although we have found that for most of our primers 1:15 dilution works the best.

4. Notes

1. All buffers are freshly made using molecular biology-grade reagents and UltraPure DNAse, RNAse-free, distilled water (GIBCO). All buffers are used ice cold, except when specified. NaF, DTT, Protease Inhibitor Cocktail and Phosphatase Inhibitor Cocktails, and RNase inhibitor are added to solutions immediately before usage.
2. Cell plates used for preparation of nuclear RNA extracts and for immunoprecipitations are seeded on the same day and grown for the same amount of time under the same growth conditions. It is recommended to seed an extra 10-cm dish in case any technical problems arise during the preparation of nuclear RNA extracts (such as rupture of the nuclear membrane) making it necessary to re-commence nuclear RNA extract preparation from fresh cells. Nuclear extract preparation must be optimized for each cell line; see ref. 10 for more information.
3. At this stage, pellets (nuclei) should have reduced size and appear whiter in colour.
4. Pellet should easily re-suspend in cell lysis buffer A. If pellet is viscous and does not re-suspend (filamentous chromatin sticks

to the microtip), this means that the nuclei are not intact and have been lysed. If this occurs, make extract again from fresh cells using shorter incubations in cell lysis buffer A and handling the pellets very gently.

5. It is strongly recommended to overestimate the amount of beads needed to be prepared for the experiment in order to be sure to have enough bead slurry for all your immunoprecipitations as some beads are lost due to sticking of beads to pipette tip. Typically, an extra 40–80 μl of stacked bead volume is prepared.
6. Protein A/G PLUS-Agarose (Santa Cruz sc-2003) is provided commercially as 0.5 ml agarose in 2 ml final volume; in order to pipette an accurate amount of beads, it is critical beforehand to homogenize by inverting the tube several times until all beads are dislodged from the bottom of the tube.
7. When possible, it is advised to use a 1 ml pipette tip to aspirate beads as the diameter of the pipette tip is wide enough to allow the beads to pass through the tip. If using a 200 μl pipette tip, cut the end of the tip with a sterile cutter or scissors before aspirating bead slurry.
8. Avoid centrifuging beads at a speed higher than 6,000 rpm to prevent deformation of the agarose bead surface.
9. To avoid aspirating beads, aspirate supernatant very slowly and keep the pipette tip as close to the surface of the liquid as possible when aspirating. Remove supernatant until only a meniscus remains on top of the bead pellet. Do not attempt to further recover supernatant as this may result in aspiration of beads.
10. Gently add the buffer along the side of the tube avoiding contact of the tip with the bead pellet.
11. Saturation of the agarose bead surface with sheared salmon sperm and yeast tRNA allows to reduce background nonspecific binding of DNA and RNA to bead surface.
12. Lysis buffer is chilled on ice only moments before usage to avoid SDS precipitation. No chunks of cells should remain after re-suspension of pellet in lysis buffer.
13. It is possible to pool similar samples for sonication in order to sonicate in a greater volume and afterwards re-dispatch the sonicated cell lysate in 200 μl batches. Before performing RNA-ChIP experiment, sonication conditions should be optimized to generate shearing of DNA into fragments of ~ 500 bp. To check for DNA fragment size under different sonication conditions, phenol/chloroform purify sonicated DNA and migrate between 5 and 15 μg of DNA sample on a large,

low-percentage agarose (0.75–1%) gel for several hours at low voltage. A DNA smear around 500 pb (<1 kb and >100 bp) should be observed. Note that cell type and sonication volume affect the size of DNA fragments. It is recommended to anticipate using one 10-cm dish to check the DNA fragment size for each RNA-ChIP experiment as variability in DNA fragmentation can affect the reproducibility of the results of this assay. Sonication conditions must be optimized for each cell line.

14. When incubating samples at 70°C, it is strongly recommended to seal tubes with cap locks to prevent tubes from opening, thus causing evaporation of samples.
15. It is advised to process RNA from input and immunoprecipitation samples simultaneously. It is best to process RNAs from immunoprecipitation in a first moment and next process input samples in a second moment in order to avoid contamination of IP by input samples during the purification process.
16. Following centrifugation, the mixture separates into a lower red phenol–chloroform phase, a white interphase, and a colourless upper aqueous phase. To collect the aqueous phase, aspirate very slowly and keep the pipette tip as close to the surface of the liquid as possible when aspirating. Collect aqueous phase until only a meniscus of aqueous phase remains over the white interphase. Do not attempt to further recover aqueous phase as this may result in aspiration of the interphase.
17. Glycoblue facilitates precipitation of the RNA and, in addition, has a blue colour to facilitate visualization of the RNA pellet. This may be especially helpful in the immunoprecipitation samples as they contain a very small amount of RNA resulting in tiny pellets.
18. Pellets are very small and it is recommended to avoid contact of pipette tip with pellet upon adding water. In addition, it is recommended to avoid re-suspending the pellet by pipetting up and down. These precautions are to avoid loss of the pellet by it sticking to the pipette tip.
19. Freeze–thaw of the pellet followed by vortexing leads to its re-suspension in water.

Acknowledgements

This work was supported by the EURASNET European Network of Excellence and ANR.

References

1. Bentley, D. L. (2005) Rules of engagement: co-transcriptional recruitment of pre-mRNA processing factors. *Curr Opin Cell Biol* **17**, 251–256.
2. Moore M. J., Proudfoot, N. J. (2009) Pre-mRNA processing reaches back to transcription and ahead to translation. Cell **136**, 688–700.
3. Neugebauer, K. M. (2002) On the importance of being co-transcriptional. *J Cell Sci* **115**, 3865–3871.
4. Pandya-Jones, A. Black, D. L. (2009) Co-transcriptional splicing of constitutive and alternative exons. *RNA* **15**, 1896–1908.
5. Wada, Y., Ohta, Y., Xu, M., Tsutsumi, S., Minami, T., Inoue, K., Komura, D., Kitakami, J., Oshida, N., Papantonis, A. *et al.* (2009) A wave of nascent transcription on activated human genes. *Proc Natl Acad Sci USA* **106**, 18357–18361.
6. Beyer, A. L. Osheim, Y. N. (1988) Splice site selection, rate of splicing, and alternative splicing on nascent transcripts. *Genes Dev* **2**, 754–765.
7. Bittencourt, D., Dutertre, M., Sanchez, G., Barbier, J., Gratadou, L. and Auboeuf, D. (2008) Cotranscriptional splicing potentiates the mRNA production from a subset of estradiol-stimulated genes. *Mol Cell Biol* **28**, 5811–5824.
8. Kaneko, S. Manley, J. L. (2005) The mammalian RNA polymerase II C-terminal domain interacts with RNA to suppress transcription-coupled 3' end formation. *Mol Cell* **20**, 91–103.
9. Wetterberg, I., Bauren, G. Wieslander, L. (1996) The intranuclear site of excision of each intron in Balbiani ring 3 pre-mRNA is influenced by the time remaining to transcription termination and different excision efficiencies for the various introns. *RNA* **2**, 641–651.
10. Barbier, J., Dutertre, M., Bittencourt, D., Sanchez, G., Gratadou, L., de la Grange, P. and Auboeuf, D. (2007) Regulation of H-ras splice variant expression by cross talk between the p53 and nonsense-mediated mRNA decay pathways. *Mol Cell Biol* **27**, 7315–7333.

Chapter 37

Quantitative Analysis of Transcription Elongation by RNA Polymerase I In Vitro

David Alan Schneider

Abstract

The elongation step in transcription has gained attention for its roles in regulation of eukaryotic gene expression and for its influence on RNA processing. Sophisticated genetic analyses have identified factors and/or conditions that may affect transcription elongation rate or processivity; however, differentiation of direct and indirect effects on transcription is difficult using in vivo strategies. Therefore, effective, reproducible in vitro assays have been developed to test whether a given factor or condition can have a direct effect on the kinetics of transcription elongation. We have adapted a fully reconstituted transcription system for RNA polymerase I (Pol I) for kinetic analysis of transcription elongation rate in vitro. The assay described here has proven to be effective in the characterization of defects or enhancement of wild-type transcription elongation by RNA Pol I. Since transcription elongation by RNA Pol I has only recently gained significant attention, this assay will be a valuable resource for years to come.

Key words: Transcription, rRNA, Ribosome, Elongation rate, Pause site, RNA polymerase I

1. Introduction

Transcription of ribosomal RNA constitutes more than 60% of total transcription in proliferating eukaryotic cells, and is a target for regulation of ribosome synthesis (1, 2). Due to the intimate link between rRNA synthesis rate and cell proliferation, the molecular mechanisms that regulate polymerase I (Pol I) transcription are directly relevant to our understanding of important human diseases like cancer (3). However, the mechanisms by which rRNA synthesis is controlled have not yet been fully described. Recent studies have identified factors that influence transcription elongation by RNA Pol I to have a role in the regulation of rRNA synthesis rates (4, 5).

Ales Vancura (ed.), *Transcriptional Regulation: Methods and Protocols,* Methods in Molecular Biology, vol. 809,
DOI 10.1007/978-1-61779-376-9_37, © Springer Science+Business Media, LLC 2012

Thus, a reliable assay for RNA Pol I transcription elongation is essential. We have adapted a previously described multiround transcription elongation system for analysis of transcription elongation rate as well as intrinsic pausing and arrest (5–7).

When developing this system, every effort was made to preserve the context of transcription elongation encountered in vivo. Methods exist for formation of transcription elongation complexes in vitro (e.g., scaffold templates); however, we have chosen to use promoter-dependent transcription initiation and nucleotide limitation to synchronize transcription elongation complexes. As with all experiments in vitro, this assay is subject to the criticism that it does not fully recapitulate in vivo conditions. Nevertheless, we have minimized the potential for artifacts induced by the assay and we have purified all of the factors essential for transcription by RNA Pol I. Furthermore, we have demonstrated that this assay is effective in quantification of both positive and negative effects on transcription elongation by Pol I. Thus, the assay described here is the most robust assay for measurement of direct effects of a factor or condition on the transcription elongation by Pol I to date.

2. Materials

2.1. Cell Growth, Harvest, and Lysis

1. Luria–Bertani broth (Fisher Scientific; New Jersey) supplemented with 100 μg/ml of ampicillin.
2. Isopropyl β-D-1-thiogalactopyranoside (IPTG).
3. YEPD medium: 1% (w/v) Yeast extract (Becton, Dickinson and Company, BD, Maryland), 2% (w/v) peptone, 2% (w/v) dextrose.
4. Breakage buffer: 500 mM KCl, 50 mM Tris–HCl, pH 7.6, 20% (v/v) glycerol, 5 mM $MgCl_2$, 5 mM imidazole, 1 mM PMSF, 0.1% (v/v) Tween 20.

2.2. Factor Purification

1. Nickel Sepharose 6 Fast Flow (GE Healthcare Bio-Sciences, GE; Uppsala).
2. Monoclonal anti-HA agarose conjugate, clone HA-7 (Sigma-Aldrich, Saint Louis).
3. HA peptide (GenScript, New Jersey).
4. Nickel wash buffer: 200 mM KCl, 50 mM Tris–HCl, pH 7.6, 5 mM $MgCl_2$, 10 mM imidazole, 0.1% Tween 20, 20% glycerol.
5. Nickel elution buffer: 200 mM KCl, 50 mM Tris–HCl, pH 7.6, 5 mM $MgCl_2$, 250 mM imidazole, 0.1% Tween 20, 20% glycerol

6. Columns: HiTrap Heparin HP, HiTrap Q FF, MonoQ 5/50 GL, MonoS 5/50 GL (GE).
7. Standard gradient buffer: 200 mM (A) or 1 M (B) KCl, 20 mM Tris–HCl, pH 7.8, 20% glycerol, 0.1 mM EDTA (see Note 1).
8. Upstream activating factor (UAF) extraction buffer: 200 mM Tris acetate, pH 8.0, 400 mM ammonium sulfate, 10% glycerol, 0.1% Tween 20, 20 mM imidazole.
9. AKTA Purifier, UPC-10 with fraction collector (GE) or suitable similar instrument for liquid chromatography.

2.3. DNA Template Preparation

1. Plasmid pNOY746: 35S rDNA with promoter, all C residues in nontemplate strand mutated to G between +1 and +56, relative to transcription start site.
2. Oligonucleotides (Integrated DNA technologies, Iowa): –247 oligo: 5′-gttgtaaaacgacggccagtgc-3′ and +763 oligo: 5′-cttaga-cat-gcatggcttaatc-3′.
3. PCR purification kit (Qiagen, Maryland).

2.4. Transcription Assays

1. 2.5× transcription buffer: 50 mM Tris acetate, pH 7.9, 250 mM potassium glutamate, 20 mM magnesium acetate, 5 mM dithiothreitol, RNasin Plus (Promega), 0.5 mg/ml acetylated bovine serum albumin.
2. Ultrapure NTPs.
3. α^{32}P GTP. Radioactive materials must be handled with extreme caution using proper personal protection and disposal.
4. 1 mg/ml heparin, sodium salt.
5. Phenol, 3 mg/ml (w/v) glycogen, 1 M ammonium acetate in 100% ethanol.
6. 10× TBE buffer: 108 g/l Tris base, 55 g/l boric acid, 7.44 g/l ethylenediaminetetraacetic acid (EDTA).
7. 90% formamide loading dye: 90% deionized formamide, 1× TBE, 0.05% bromophenol blue, 0.05% xylene cyanol.

2.5. Gel Electrophoresis and Quantification

1. 40% (w/v) acrylamide/bis solution (29:1).
2. Urea.
3. *N*,*N*,*N*,*N*′-Tetramethylethylenediamine (TEMED).
4. 10% (w/v) ammonium persulfate.
5. 16.5 × 28-cm vertical gel plates with 0.7-mm spacers and 20-well comb.
6. Storm phosphorimager with 8 × 10-in. phosphor screens or equivalent imaging instrumentation.

3. Methods

Detailed analysis of transcription elongation in vitro requires highly purified, active protein samples and properly prepared template DNA and reagents. Since this assay uses promoter-dependent transcription initiation and a modified rDNA template, it closely mimics the in vivo state of transcription elongation (as closely as possible for an in vitro assay). As a consequence, all of the transcription initiation factors must be purified as well as the polymerase. Thus, most of the effort in executing this experiment is invested in protein purification. There are other protocols for analysis of short RNA synthesis from scaffold DNA templates using only the polymerase (8), but those methods are not described here.

The transcription elongation assay for RNA Pol I is an adaptation of a previously published assay for multiround transcription. Two previous papers describe the reconstituted transcription assay in detail and can be seen for additional reference (6, 9). To aid reproducibility, the entire procedure is described here. Some steps in the method described here are redundant with those published previously.

Since quantification of transcription elongation requires single-round transcription, the signal-to-noise ratio of these experiments is lower than for more robust multiround assays. Therefore, it is essential that extreme care be devoted to purification of the factors required, since minor contamination can lead to RNA degradation or nonspecific transcription, both of which increase the background in the assay. The following protocol is subdivided by purification of each factor and by the assay description itself.

3.1. Preparation of Cell Extracts

1. Rrn3p and Tata-binding protein (TBP) are expressed in *Escherichia coli* (BL21 DE3) from pET plasmid derivatives (His-tagged Rrn3p, pNOY3162 (6), and His-tagged TBP, 6HisT-pET11 (10)).
2. *E. coli* cells carrying the appropriate plasmid are grown in Luria–Bertani medium, supplemented with 100 μg/ml ampicillin. Cultures are grown with aeration at 25°C to an $A_{600} = 0.5$, and protein expression is induced with 0.5 mM IPTG.
3. After >12 h of induction, cells are harvested by centrifugation (10,000 × *g* for 30 min), washed once in breakage buffer, spun down again, and frozen at −80°C.
4. To purify Pol I, core factor, and UAF, strains expressing 6-his, 3-hemagglutinnin (HA)-tagged proteins are used (strains: NOY760 = tagged A135 to purify Pol I; NOY797 = tagged Rrn7p to purify core factor; and NOY798 = tagged Rrn5p to purify UAF).

5. Yeast cultures are grown with aeration in YEPD medium at 30°C to an A_{600} = 0.8.
6. Cells are harvested by centrifugation (6,000 × *g* for 20 min), washed with breakage buffer, spun down again, and the pellets are frozen at −80°C.

3.2. Rrn3p Purification

1. Frozen cell pellets are thawed on ice, suspended in threefold excess (v/w) breakage buffer, and disrupted using a French press (8,000 psi, internal pressure) (see Note 2).
2. After lysis, the extract is cleared by centrifugation (30,000 × *g* for 30 min).
3. The supernatant is mixed with Ni-Sepharose FF that has been prewashed twice with breakage buffer. This mixture is incubated at 4°C with gentle agitation for 2 h. Approximately 1 ml of settled resin is used per 2 g of wet cell mass.
4. Resin is separated from "flow through" by gentle centrifugation (500 × *g* for 5 min). Resin is resuspended in equal volume of breakage buffer and poured onto an empty column containing fresh, prewashed Ni-Sepharose FF (10% of volume used in binding) (see Note 3).
5. Using either gravity flow or pump-assisted flow, resin is washed with five-column volumes of Ni wash buffer.
6. Protein is eluted with three-column volumes of Ni elution buffer.
7. Eluted protein is loaded onto a Q-Sepharose column using an AKTA Purifier UPC-10 (or equivalent instrument for liquid chromatography) (see Note 4).
8. Sample is loaded and flow through is collected in 100% gradient buffer A (200 mM KCl).
9. After stable baseline is achieved, a 50-ml gradient from 200 mM to 1 M KCl is applied, collecting 1-ml fractions. The flow rate is 0.5 ml/min. Rrn3 elutes at ~400 mM KCl (see Note 5).
10. Peak fractions are pooled and rebound in batch to Ni-Sepharose FF. Resin volume is reduced to 20% of previous Ni purification.
11. Ni-binding, washing, and elution are repeated as described above.
12. Eluted protein is applied to a MonoQ column and eluted as described for the Q-Sepharose column. Pure Rrn3p elutes at ~400 mM KCl.

3.3. Tata-Binding Protein Purification

1. Cells are thawed and broken and first Ni column is performed as for Rrn3p.
2. Ni eluate is loaded onto an HiTrap heparin column in gradient buffer A, and flow through is collected.

3. After achieving a stable baseline, a 180-ml gradient from 0 to 60% gradient B is executed, collecting 1-ml fractions. His-TBP elutes at ~430 mM KCl.
4. Peak fractions are pooled and dialyzed into 100 mM KCl-modified gradient buffer.
5. Sample is then loaded onto a MonoS column and after a stable baseline is achieved, an 80-ml gradient from 100 mM to 360 mM KCl is run, collecting 1-ml fractions. TBP elutes at ~200 mM KCl.

3.4. RNA Polymerase I Purification

1. Cell pellets are thawed on ice as for *E. coli* cells (see Note 6).
2. Cells are resuspended in twofold excess (v/w) breakage buffer and disrupted in a French press (25,000 psi internal pressure) (see Note 2).
3. Extract is cleared by centrifugation (30,000 × *g* for 30 min).
4. Supernatant is mixed with prewashed Ni-Sepharose FF (~1 ml settled resin per 2 g of wet cell paste) and incubated at 4°C for 2 h with gentle agitation.
5. Resin is separated from the flow through by gentle centrifugation (500 × *g* for 5 min) and then poured into a column containing prewashed Ni resin, ~10% of resin volume used in batch binding.
6. Using gravity or pump-assisted flow, column is washed with five-column volumes of Ni wash buffer and proteins are eluted with three-column volumes of Ni elution buffer.
7. Eluted material is loaded onto an HiTrap heparin column in 100% gradient buffer A, and flow through is collected. After reaching a stable baseline, a 50-ml gradient from 200 mM to 1 M KCl is run, collecting 1-ml fractions. Pol I elutes at ~375 mM KCl.
8. Peak fractions are pooled and diluted with an equal volume of "no-salt" gradient buffer. This sample is then loaded onto a MonoQ column. After baseline is reached in 160 mM buffer A, a 50-ml gradient from 0 to 50% buffer B (500 mM KCl) is run, collecting 1-ml fractions. Pol I elutes at ~330 mM KCl (see Note 7).

3.5. Core Factor Purification

1. Cell pellets are thawed, broken, and fractionated with Ni-Sepharose and HiTrap heparin as described for Pol I. Core factor elutes from heparin at ~350 mM KCl.
2. Peak fractions are pooled and mixed with 1 ml of anti-HA affinity resin. The sample is incubated with gentle agitation at 4°C for 2 h (see Note 8).
3. The mixture is then applied to an empty column and washed with greater than five-column volumes of gradient buffer A.

4. Before elution, a HiTrap Q column is attached downstream of the column holding the anti-HA resin.
5. Gradient buffer A plus 1 mg/ml HA peptide is then recirculated over the two columns for greater than 2 h (see Note 9).
6. The Q column is detached and loaded in the AKTA purifier. A 50-ml gradient from 0 to 60% buffer B is executed, collecting 1-ml fractions. Core factor elutes at ~350 mM KCl.

3.6. UAF Purification

1. Cells are thawed, broken, and fractionated with Ni-Sepharose as described for Pol I, except that UAF extraction buffer is used to break and wash the beads.
2. After washing the Ni resin, an HiTrap heparin column (that has been previously washed into the elution buffer) is attached downstream of the nickel column.
3. Approximately five-column volumes of modified gradient buffer with 450 mM KCl and 250 mM imidazole are passed over the resin to elute UAF from the Ni-Sepharose onto the heparin resin.
4. The heparin column is then attached to an FPLC and eluted with a 50-min gradient from 450 mM KCl to 1 M KCl (flow rate 0.5 ml/min). UAF elutes at ~650 mM KCl.
5. The peak fractions are pooled and mixed with sufficient anti-HA immune affinity resin and incubated at 4°C with gentle agitation (see Note 8).
6. The beads are centrifuged (2 min at 1,000 × *g*) to pellet the beads and the supernatant is removed and saved. Five resin volumes of 400 mM KCl gradient buffer is added to the beads; they are gently mixed and spun down again at 1,000 × *g*.
7. This wash procedure is repeated three more times.
8. The beads are then resuspended and transferred to a small empty column.
9. An HiTrap heparin column is attached downstream of the column, and 400 mM gradient buffer + 0.5 mg/ml HA peptide is recirculated over the anti-HA beads and through the heparin column. Eluted UAF remains attached to the heparin, but HA peptide flows through.
10. After overnight recirculation (at 4°C), the heparin column is detached and mounted onto an FPLC. Fractionation is achieved with a 50-min gradient from 400 mM to 1 M KCl. One-milliliter fractions are collected. Pure UAF elutes at approximately 650 mM KCl.

3.7. DNA Template Preparation

1. Since only 6 C residues are present in the first 55 nucleotides of the transcribed rRNA, we chose to mutate those Cs to Gs. These mutations conserved the G/C content of the transcript,

but permitted transcription initiation and promoter clearance in the absence of CTP. This modified rDNA template is encoded in pNOY746 (7).

2. Using pNOY746 as a template and –247 and +763 oligonucleotides (sequences provided above), we use standard PCR conditions to amplify the C-less promoter region.
3. The resulting 1,010 bp PCR product is purified using the Qiagen PCR purification kit (see Note 10).
4. After elution from the kit, DNA concentration is measured spectrophotometrically at 260-nm wavelength.

3.8. Transcription Elongation Assay

1. Having purified all of the necessary components, the transcription elongation rate assay is performed at 23°C using all RNase-free reagents.
2. All nonradioactive reagents are mixed first. These steps can be performed without shielding.
3. For each condition to be tested, a large master mix is prepared (see Note 11). [For example, if a factor's effect on transcription elongation rate is to be tested, then two mixes are prepared, one with the appropriate amount of factor, and one with an equal volume of buffer only or heat-inactivated factor.]
4. Each reaction equals 20 μl; 17 μl of master mix; 1 μl of NTP mix (ATP, GTP, UTP, and α^{32}P GTP); 1 μl of heparin (1 mg/ml); and 1 μl of CTP.
5. Recipe per reaction: 1× transcription buffer, 6 ng DNA template, 0.2 μl of active UAF preparation, 50 nM TBP, 0.2 μl of active core factor preparation, 20 nM Pol I/Rrn3 complex (see Note 12), plus any additional factors or variables.
6. After preparing the master mix, NTP mix is added. The concentration of NTPs is exceptionally important for the elongation rate in vitro (Fig. 1) and can be modified to optimize the reaction for probing faster or slower effects. Since each sample is ultimately 20 μl, we make 20× stocks of everything, and our standard final NTP concentrations are 100 μM ATP, 100 μM UTP, 10 μM GTP, and 10 μCi α^{32}P GTP. To a 11× master mix, we add 11 μl of our 20× stock (see Note 13).
7. After 5 min, we add 11 μl of 1 mg/ml heparin (50 μg/ml final concentration). Heparin prevents reinitiation and disrupts initiation complexes that have not cleared the promoter.
8. After 4 min, we remove a 20-μl sample and add it to a tube containing 60 μl of acid phenol and 50 μl of 3 mg/ml glycogen. This sample is "stopped" and stable for later processing.
9. After 1 more minute (at $t=0$), 10 μl of 20× CTP is added to a final concentration of 100 μM.

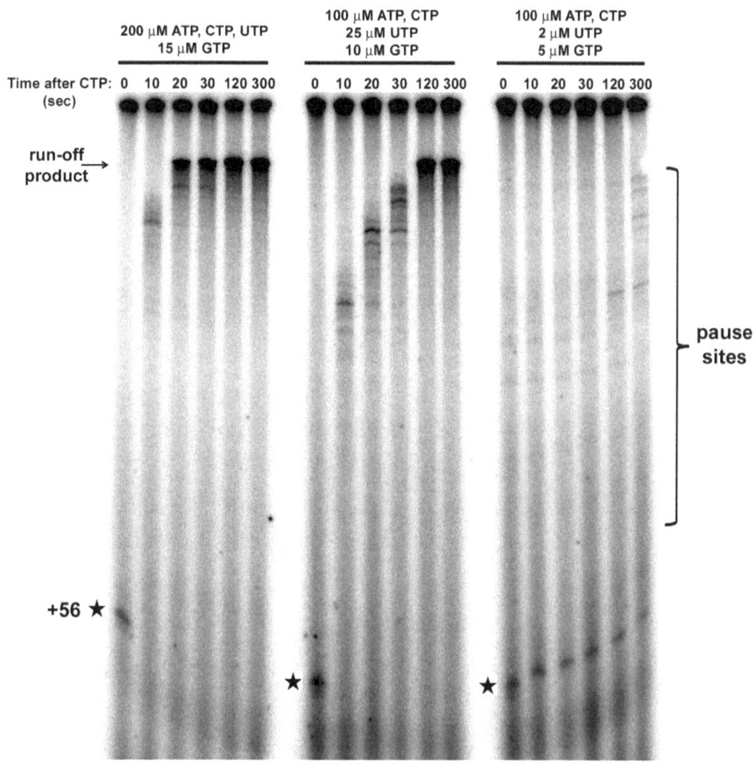

Fig. 1. Nucleoside triphosphate concentration dependence of RNA polymerase I transcription elongation rate in vitro. Elongation assays were performed as described herein with three different NTP concentrations (indicated on *top* of gel). Samples were collected at identical time points under all three conditions, purified, and run on an 8% denaturing polyacrylamide gel. The gel was dried and analyzed by phosphorimaging. The +56 CTP arrest (*stars*), intrinsic pause sites, and runoff product positions are indicated.

10. As a function of time, 20-µl samples are collected and added directly to tubes containing phenol and glycogen (as in step 8).
11. Typical time points = 0, 30 s, 60 s, 2 min, 3 min, 4 min, 5 min, 6 min, 8 min, and 10 min. These vary based on your reaction conditions and desired focus of the experiment.
12. After completion of the time course, each sample is mixed again by pipetting vigorously up and down.
13. Samples are centrifuged at 17,000 × *g* for 10 min.
14. Aqueous phase is removed to a fresh tube containing 330 µl of 1 M ammonium acetate in 95% ethanol (see Note 14).
15. Samples are incubated on crushed dry ice for 15 min.
16. Samples are spun at 17,000 × *g* for 10 min.
17. Ethanol is carefully and completely removed.
18. 20 µl of formamide loading dye is added to each dried pellet, and the samples are heated to 95°C for 10 min prior to electrophoresis.

3.9. Electrophoresis and Quantification

1. The gel contains 1× TBE, 8% polyacrylamide (29:1, acrylamide to bis), and 7 M urea. 50 ml of this solution is dissolved with stirring at room temperature. The urea requires >15 min to dissolve.
2. The spacers and plates are assembled and clipped with medium binder clips in place and a small amount of Vaseline is used at each of the two bottom corners of the spacers.
3. When mix is dissolved, 250 μl of 10% APS and 50 μl of TEMED are added and stirred for ~1 min.
4. This mix is then poured into the plates avoiding the introduction of any bubbles.
5. Once the mix is poured into the plates and bubbles are removed (typically by "vigorously" tapping the glass), the comb is inserted, being careful to avoid bubbles in the wells.
6. In ~30 min, the gel is sufficiently solidified to be wrapped on top (over the comb) with cellophane.
7. Gel is kept at room temperature overnight.
8. The next day, the gel is pre-run for 45 min at 700 V in 1× TBE.
9. During the pre-run, RNA samples are incubated at 95°C for 10 min.
10. 10 μl of each RNA sample is loaded (directly from the heat block) into the gel (see Note 15).
11. Once loaded, the gel is run at 700 V until the dark blue band (bromophenol blue) reaches the bottom of the gel (see Note 16).
12. The gel is removed and dried onto a double layer of Whatman 3-mm paper under a sheet of cellophane at 80°C under vacuum for 1 h.
13. The dried gel is exposed ~16 h to a phosphorimager cassette (depending on the age of the ^{32}P and the strength of the transcription).
14. After generation of the image with a phosphorimager, the amount of runoff product is quantified with the imager's packaged software (see Note 17).
15. If runoff product accumulation is plotted as a function of time, a sigmoidal plot is typically generated.
16. There are several defensible methods for calculating elongation rates. One easy and reproducible method is to divide the length (in nucleotides) of the transcript (downstream of the +55 pause) by the time (in seconds) at which the sigmoidal curve crosses 100% of runoff accumulation. This yields an elongation rate in nucleotides/second.
17. In addition to quantification of the net elongation rate in your sample, visual analysis of your gel yields qualitative detection of intrinsic pause/arrest sites (Fig. 1).

4. Notes

1. We use the same base buffer for all gradient elutions. In some cases, as noted in the method, we use a different starting and finishing salt concentration. This can be achieved by making fresh buffers with different KCl concentrations or by using the gradient mixer in the FPLC to adjust the starting and final concentrations of KCl. Either method works. We also make a stock of gradient buffer with no salt that is used to dilute salty fractions for additional fractionation.
2. After one pass through the press, extracts tend to be viscous. A second pass reduces viscosity making subsequent steps more efficient. Also note that PMSF is added to a final concentration of 1 mM just before cell disruption.
3. We use the XK16 empty column (GE) since it can accommodate variable bed heights and has proper fittings for use with the AKTA Purifier FPLC. Any manufacturer of empty columns can be used.
4. HiTrap columns are sold in 1- or 5-ml sizes. The binding capacity of these resins is high, so the 1 ml variety is usually sufficient, but if a very large preparation is performed, one can use the 5-ml cartridges.
5. Concentrations for elution are estimates from the resulting chromatograph. Slight variations in pH or buffer preparation can affect retention time of a given protein on the column; thus, elution should be confirmed by Western Blot using commercially available antibodies.
6. Larger mass of cells is required for factors purified from yeast than from *E. coli*. Typical preparative mass of cells used for Pol I is ~100 g of wet cell mass versus ~20 g for proteins produced in *E. coli*.
7. Pol I is relatively pure after MonoQ; however, WT Pol I also binds to MonoS and elutes with a standard gradient. However, purification through MonoQ is standard and sufficient for enzymology.
8. The anti-HA resin has a very high binding capacity, but for retention on small disposable columns for recirculation, usually >500 μl of settled resin is used.
9. Recirculation of HA peptide for elution onto downstream columns is generally performed at 4°C using a peristaltic pump overnight for highest yield.
10. Qiagen has begun including pH indicators in their DNA purification kits. Kits without this indicator are available and must be used for efficient transcription in vitro.

11. Each reaction is arbitrarily 20 μl; thus, for ten time points, we typically make a 11× (220 μl) mix.
12. Molar quantities are provided for all proteins, but core factor and UAF. Since these complexes are purified in very low quantities, accurate quantification of the samples is not feasible. Thus, the activity of each preparation is determined empirically by titration into transcription assays. The volumes described are representative for fractions purified as described herein. Pol I and a slight molar excess of Rrn3p are incubated together at room temperature for 4 h to preform the initiation-competent complex. Without preincubation, transcription initiation efficiency is reduced.
13. After addition of the NTPs, everything is timed. It is useful to write out the exact time at which every supplement or sample is taken and then keep a timer running in plain sight throughout the experiment.
14. For quantification, it is important that the volume of aqueous phase removed be identical between samples. Thus, it is useful to set the pipette to a low volume (e.g., 60 μl) to keep error low.
15. Urea gels accumulate debris in the wells rapidly which affects the RNA migration. Use a fine needle and 1× TBE to "spray" each well and remove debris prior to loading.
16. Radioactive GTP has migrated into the bottom of buffer tank, so great care should be taken in removing and disposing of this buffer after running the gel.
17. We have used GE or Bio-Rad imagers as well as ImageQuant or Quantity One software. There is very little difference in data quality; it is just a matter of becoming familiar with one program/instrument or the other.

Acknowledgments

The author wishes to thank Professor Masayasu Nomura for his pioneering development of the reconstituted transcription assay for RNA Pol I and for guidance in the adaptation of the system to its current form. This work is supported by the National Institutes of Health grant #GM84946.

References

1. Nomura, M., Nogi, Y., Oakes, M. (2004) Transcription of rDNA in the Yeast *Saccharomyces cerevisiae*. In: Olson, M. O. J. (ed) The Nucleolus, Kluwer Academic / Plenum Publishers, London, pp. 128–153.
2. Warner, J. R. (1999) The economics of ribosome biosynthesis in yeast. *Trends Biochem Sci* **24**, 437–440.
3. Drygin, D., Rice, W. G., Grummt, I. (2010) The RNA polymerase I transcription machinery: an

emerging target for the treatment of cancer. *Annu Rev Pharmacol* **50**, 131–156.

4. Stefanovsky, V., *et al.* (2006) Growth factor signaling regulates elongation of RNA polymerase I transcription in mammals via UBF phosphorylation and r-chromatin remodeling. *Mol Cell* **21**, 629–639.
5. Zhang, Y., Smith, A. D. t., Renfrow, M. B., Schneider, D. A. (2010) The RNA polymerase-associated factor 1 complex (Paf1C) directly increases the elongation rate of RNA polymerase I and is required for efficient regulation of rRNA synthesis. *J Biol Chem* **285**, 14152–14159.
6. Keener, J., Josaitis, C. A., Dodd, J. A., Nomura, M. (1998) Reconstitution of yeast RNA polymerase I transcription in vitro from purified components. TATA-binding protein is not required for basal transcription. *J Biol Chem* **273**, 33795–33802.
7. Schneider, D. A., *et al.* (2007) Transcription elongation by RNA polymerase I is linked to efficient rRNA processing and ribosome assembly. *Mol Cell* **26**, 217–229.
8. Kuhn, C. D., *et al.* (2007) Functional architecture of RNA polymerase I. *Cell* **131**, 1260–1272.
9. Tongaonkar, P., Dodd, J. A., Nomura, M. (2003) Purification and assay of upstream activation factor, core factor, Rrn3p, and yeast RNA polymerase I. *Methods Enzymol* **370**, 109–120.
10. Hoffmann, A., Roeder, R. G. (1991) Purification of his-tagged proteins in non-denaturing conditions suggests a convenient method for protein interaction studies. *Nucleic Acids Res* **19**, 6337–6338.

Chapter 38

Detection and Characterization of Transcription Termination

Ghada Ghazal, Jules Gagnon, and Sherif Abou Elela

Abstract

In most eukaryotes, the generation of the 3′ end and transcription termination are initiated by cleavage of the pre-mRNA upstream of the polyadenylation site. This cleavage initiates 5′–3′ degradation of the 3′ end cleavage product by the exoribonuclease Rat1p leading to the dissociation of the RNA polymerase II (RNAPII) complex. The Rat1p-dependent transcription termination was also shown to be initiated by a polyadenylation-independent cleavage performed by the double-stranded RNA-specific ribonuclease (RNase) III (Rnt1p) suggesting that the majority of transcription termination events are RNase dependent. Therefore, it became essential for future studies on transcription termination to carefully consider both the nature of the RNase-dependent RNA transcripts and the association pattern of the RNAPII with the transcriptional unit. Here, we present methods allowing the evaluation of the impact of yeast RNases on the 3′ end formation and their contribution to transcription termination. Northern blot analysis of transcripts generated downstream of known genes in the absence of RNases identifies potential transcription termination sites while chromatin immunoprecipitation of RNAPII differentiates between termination- and transcription-independent processing events.

Key words: RNase III, Rat1p, Rnt1p, Termination, ChIP, Northern blot, Ribonucleases

1. Introduction

Traditionally, transcription termination and formation of the 3′ end were considered two separate and sequential processes. However, recent studies are indicating that these two events are tightly linked and the interference with one may impair the other (1, 2). In most eukaryotes, the generation of the 3′ end and transcription termination are initiated by cleavage of the pre-mRNA 20–30 nt upstream of the polyadenylation site. This endonucleolytic cleavage occurs within a consensus sequence of AAUAAA by a multisubunit cleavage/ polyadenylation specificity factor (CPSF) (3). This endonucleolytic cleavage triggers transcription termination by giving access to the

Ales Vancura (ed.), *Transcriptional Regulation: Methods and Protocols*, Methods in Molecular Biology, vol. 809, DOI 10.1007/978-1-61779-376-9_38, © Springer Science+Business Media, LLC 2012

5′–3′ exoribonuclease Rat1p leading to the dissociation of the elongation complex (4). This mechanism of transcription termination was nicknamed the "torpedo model" and it is conserved between human and yeast (5, 6). However, many RNA polymerase II (RNAPII)-transcribed RNAs are not polyadenylated, necessitating the existence of a polyadenylation-independent mechanism of transcription termination. In the case of short noncoding RNAs like snoRNA, it is believed that transcription termination may occur through a polyadenylation-independent mechanism (7, 8). In some cases, it appears that termination can occur without endoribonucleolytic cleavage (9–12). Instead, the binding of the Nrd1/Nab3 changes the conformation of the transcription complex leading to transcription termination of short transcripts. Recently, it was shown in yeast that termination of long, noncoding RNA occurs through a modified torpedo like termination (13, 14). In this case, cleavage of the nascent transcript is performed by the double-stranded RNA (dsRNA)-specific ribonuclease (RNase) III (Rnt1p) independently of the polyadenylation signal. Cleavage by Rnt1p generates an entry site for the 5′–3′ exoribonuclease Rat1p that leads to transcription termination. In all these different modes of transcription termination, the nature of the RNA transcript and factors that affect RNA stability play an important role in determining the location and nature of the transcript 3′ end. Here, we present a comprehensive method for the detection of transcription termination that takes in account the activity of endoribonucleases and their impact on 3′ end formation.

2. Materials

2.1. Cell Culture

1. YEPD media: 10 g yeast extract, 20 g Bacto-peptone, 0.1 g adenine in 900 mL double-distilled water (ddH_2O) and autoclaved. The dextrose is prepared separately (20 g in 100 mL ddH_2O) and added to the medium after autoclaving.

2.2. RNA Extraction

1. 0.2% (v/v) diethyl pyrocarbonate treated, autoclaved ddH_2O (see Note 1); stored at 4°C.
2. LETS buffer: 0.01 M tris(hydroxymethyl)aminomethane (Tris–HCl), pH 7.5, 0.1 M lithium chloride (LiCl), 0.01 M disodium ethylenediaminetetraacetate (EDTA), pH 8.0, 0.2% (w/v) sodium dodecyl sulfate (SDS). The buffer is made in a final volume of 500 mL of DEPC-treated ddH_2O.
3. Sterile, acid-washed glass beads with a diameter 425–600 μm (see Note 2).
4. Phenol (see Note 3) equilibrated to pH 7.5–8.0 with LETS buffer.
5. Chloroform/isoamyl alcohol solution (24:1 (v/v)).

6. Salted ethanol: Ethanol 95% containing 2% (w/v) potassium acetate (KOAc).

2.3. Northern Blots

1. MOPS buffer 10×: 0.4 M 3-(*N*-morpholino)propanesulfonic acid (MOPS), 0.1 M sodium acetate (NaOAc), 0.01 M EDTA. The buffer is prepared in a volume of 1 L using ddH_2O and filtered on Whatman number 1 paper. The buffer can be stored in the dark at 4°C for no more than 3 months.
2. RNA loading dye: 0.05% (w/v) xylene cyanol, 0.05% (w/v) bromophenol blue, 94% formamide (see Note 4).
3. SSPE buffer 20×: 3.6 M sodium chloride (NaCl), 0.2 M sodium phosphate monobasic (NaH_2PO_4), 0.02 M EDTA. The solution is prepared in a volume of 1 L using ddH_2O. The pH is adjusted to 7.7 using sodium hydroxide (NaOH), filtered using Whatman number 1 paper and stored at room temperature for up to 1 year.
4. The transfer apparatus is assembled from a glass container (e.g., Pyrex), a thick sponge larger than the gel, Whatman 3MM paper, paper towels, four plastic or acetate strips, a level, a glass plate, and a weight (flat book or flat metal weight of about 1 kg).
5. Methylene blue stain: The solution is prepared in a volume of 1 L using ddH_2O, 11.5 mL glacial acetic acid, 27.2 g NaOAc, 2.0 g methylene blue. The solution should be filtered using Whatman number 1 paper and can be stored for up to 6 months at room temperature.
6. SET buffer 10×: 1.5 M NaCl, 300 mM Tris-HCl, 20 mM EDTA. The buffer is made in a volume of 1 L using ddH_2O, adjusted to pH 8.0 using NaOH and filtered using Whatman number 1 paper. The buffer can be stored at room temperature for up to a year.
7. Denhardt's buffer 100×: 2% (w/v) bovine serum albumin (BSA), 2% (w/v) Ficoll 400, 2% (w/v) polyvinylpyrolidone. The solution is prepared in 100-mL volume using ddH_2O, filtered on Whatman number 1 paper, and stored at −20°C for up to 3 months.
8. Hybridization solution: 4× SET buffer, 0.25 mg/mL Herring sperm DNA, 0.1% (w/v) SDS, 10× Denhardt's. The solution is freshly prepared in a volume of 100 mL in ddH_2O.

2.4. Chromatin Extraction and Immunoprecipitation

1. Cross-linking stop buffer: 2.5 M glycine is prepared in a 500-mL volume using sterile ddH_2O, sterilized by filtration, and can be stored at room temperature for a month.
2. HBS buffer: 50 mM 4-(2-hydroxyethyl)-1-piperazineethanesulfonic acid (HEPES), pH 7.5, 140 mM NaCl. The buffer is made in a volume of 500 mL using ddH_2O, filtered, and stored at 4°C.

3. ChIP lysis buffer: 50 mM HEPES, pH 7.5, 140 mM sodium chloride, 1 mM EDTA, 1% (v/v) Igepal CA-630, 0.1% (v/v) sodium deoxycholate. The buffer is freshly prepared in a 1-L volume using sterile ddH_2O and 1 mM phenylmethyl sulfonyl fluoride (PMSF) is normally added just before use.
4. TE buffer: 20 mM Tris, pH 7.5, 0.1 mM EDTA, pH 8.0, prepared in 100 mL with ddH_2O.
5. Elution buffer: Add 1% (w/v) SDS to TE buffer. Prepare freshly.
6. T/L buffer: 20 mM Tris, pH 7.5, 250 mM LiCl, 1 mM EDTA, pH 8.0, complete with ddH_2O to 100 mL. Just before starting the washes, add 0.15% (v/v) Igepal CA-630 and 0.15% (v/v) sodium deoxycholate.

3. Methods

The products of the RNA cleavage events leading to transcription termination are inherently labile and very difficult to detect under normal conditions. To detect nascent RNA transcripts and RNA-carrying sequence corresponding to the termination site, the RNA extraction and sample handling should be carried out at 4°C using RNase-free labware and solutions. Rapid extraction and analysis of the samples greatly contribute to the accuracy and reproducibility of the results. To distinguish RNA transcripts corresponding to transcription termination events from those produced by subsequent cleavage or processing, the pattern of RNAPII association near the RNA mature 3′ end of the RNA should be analyzed. By comparing the RNAPII association pattern (15) and the RNA transcript in the presence and absence of different RNases, we could reconstruct the events leading to transcription termination and formation of RNA 3′ end.

3.1. Cell Culture

1. Wild-type cells and cells carrying deletion in the genes coding for the dsRNA-specific RNase III (Rnt1p), the 5′–3′ cytoplasmic exoribonuclease Xrn1p, or the 3′–5′ nuclear exoribonuclease Rrp6p are grown in 50 mL of YEPD in a 250-mL flask at 26°C with shaking to an OD_{600} between 0.6 and 0.8. Cells carrying a temperature-sensitive allele of the nuclear 5′–3′ exoribonuclease Rat1p alone or in combination with deletion of the gene coding for Xrn1p are grown similarly in 50-mL cultures to an OD_{600} of 0.3, and then shifted for 4 h to 37°C.
2. The doubling time of the different strains is verified every 2 h by following the increase in optical density, and strains that grow 20% slower or faster than the preestablished rate (~2 h for the wild type, ~2 h for *rrp6Δ* (13), ~2 h for *xrn1Δ* (13), ~2.5 h for

rat1-1 (13), ~3.5 h for *rat1-1 xrn1*Δ (13), and ~6.5 h for *rnt1*Δ (16)) are rejected. The absence of contamination with bacteria or other fungi is verified by examining samples of the culture using light microscopy before harvesting for extraction.

3.2. RNA Extraction for Northern Blot Analysis of Transcription Termination

1. Harvest the cells grown as described in Subheading 3.1 in a 50-mL Oakridge tube by spinning at 5,000 × *g* for 3 min and remove the media.
2. Resuspend the cell pellet in 20 mL cold, sterile ddH_2O and respin the cells at 5,000 x*g* for 3 min. Repeat this step two times.
3. Resuspend the clean cell pellet in 5 mL cold, sterile ddH_2O, transfer the suspension to a 13-mL round-bottom polypropylene tube, collect the cells by spinning at 5,000 x*g* for 3 min, discard the water, and place the tube on ice.
4. Resuspend the cells in 300 μL LETS buffer, and add 300 μL phenol and 0.75 mL of glass beads.
5. Vortex the tube for 30 s, let it rest on ice for 30 s, and repeat for seven to ten times or until more than 80% of the cells are broken (see Note 5).
6. The liquid containing the cell lysate is retrieved using a 1-mL pipette and transferred to a clean 1.5-mL microtube. Wash the glass beads with 200 μL of LETS buffer, vortex for 15 s, and add it to the rest of the lysate.
7. Spin the cell lysate at 8,000 x*g* in a microfuge for 3 min.
8. Transfer the aqueous layer to a clean microtube, add 0.5 volumes of phenol and chloroform/isoamyl alcohol, vortex for 15 s, and spin the tube once more at 8,000 x*g*. Repeat the process twice or until no opaque material remains in the phenol interface.
9. Transfer the clean supernatant to a clean microtube and precipitate the RNA using 3 volumes of salted ethanol (stored at −20°C). Place the tube on dry ice for 10 min.
10. Pellet the RNA by spinning the tube at 8,000 x*g* for 10 min.
11. Wash the pellet twice with 95% ethanol and dry the pellet in a speedvac.
12. Resuspend the RNA on ice with 60 μL DEPC-treated water.
13. Dose on a spectrophotometer at 260 and 280 nm (1 μL RNA in 999 μL water). A value of 1 OD_{260} = 40 μg/μL (see Note 6).
14. Evaluate the quality of the RNA by loading 2 μg total RNA on a small, denaturating 1% agarose gel (see Subheading 3.4) and calculate the 25S/18S ratio following methylene blue staining. Reject RNA samples with ratios lower than 1.5.

3.3. Design of Northern Blot Probes

1. Retrieve the genomic sequence corresponding to the gene of interest from the *Saccharomyces* Genome Database (SGD).
2. Determine the position of the transcripts associated with the gene of interest using data obtained from the whole genome tiling array (17).
3. Select a target sequence unique to the extended form of each transcript and a sequence common to all transcripts (e.g., ORF sequence).
4. Use Primer3 (18) to design pairs of PCR primers 20–32-nt long within the region of interest (for example, see Fig. 3.1). Successful primer pairs should exhibit a melting temperature around 60°C (calculated using recommended parameters (19)) and generate PCR products 150–400-bp long (20). Primers exhibiting less than three mismatches or score over 90 with WU-BLAST2 (21) are excluded.
5. Assess the specificity of the predicted PCR product using WU-BLAST2. PCR products with target expectation value below 0.01 are excluded.

3.4. RNA Separation and Hybridization

1. Prepare a 1.2% agarose gel by dissolving 1.8 g of agarose in 135 mL of sterile ddH_2O and boiling it in a microwave oven.
2. Let the solution cool down to about 60°C at room temperature.
3. Add 15 mL of 10× MOPS buffer and stir gently.
4. Add 1.2 mL formaldehyde and stir gently. This step should be carried under a certified chemical hood.
5. Pour the solution in a thoroughly cleaned electrophoresis apparatus (see Note 7).
6. Allow gel to set on a leveled surface. When gel hardens, remove the comb and place the gel in the tank.
7. Add 1× MOPS to cover the gel.
8. Prepare the RNA sample by mixing 10–20 μg of RNA (dissolved in a maximum of 4 μL of DEPC-treated sterile ddH_2O) with 1.6 μL of 10× MOPS buffer, 3 μL formaldehyde, and 7 μL formamide.
9. Heat the RNA sample for 5 min at 55°C.
10. Add 5 μL of loading dye, mix, spin in a microfuge for 10 s, and load into the gel.
11. Run the gel at 50 V for 10–20 min and then 5 V/cm of gel for the desired length depending on the size of the RNA targeted. Normally, the gel is run until the bromophenol blue in the loading dye reaches 5 cm from the bottom of a 16-cm-long denaturing agarose gel.
12. Once the gel run is complete, the gel is removed from the electrophoresis apparatus, placed in glass dish, and rinsed with ddH_2O to remove the MOPS.

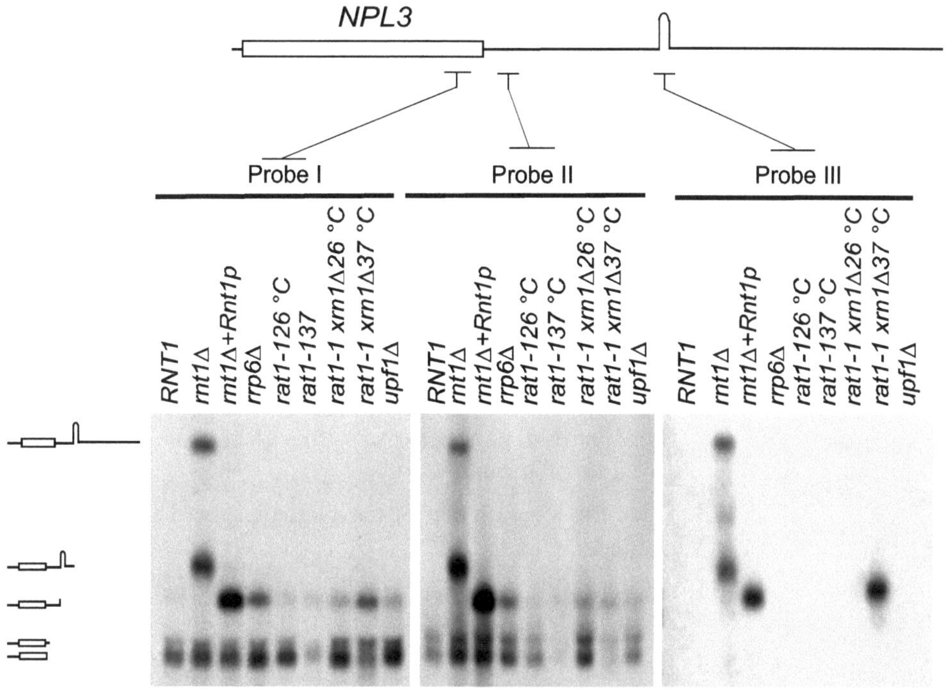

Fig. 1. Example of Northern blot used to detect ribonuclease-dependent transcriptional read-through. RNA was extracted from cells lacking the dsRNA-specific endoribonuclease Rnt1p and the 3′–5′ nuclear exoribonuclease Rrp6p (*rrp6Δ*), cells expressing a temperature-sensitive allele of the 5′–3′ nuclear exoribonuclease *RAT1* grown at permissive (*rat1-1* 26°C) or restrictive conditions (*rat1-1* 37°C), *rat1-1* cells lacking 5′–3′ cytoplasmic exoribonuclease *XRN1* grown at the permissive (*rat1-1 xrn1Δ* 26°C) or restrictive (*rat1-1 xrn1Δ* 37°C) temperature, and cells lacking the nonsense-mediated decay ribonuclease Upf1p (*upf1Δ*). RNA extracted from *rnt1Δ* cell was incubated with recombinant Rnt1p enzyme as a marker for Rnt1p cleavage product size. The different RNAs were visualized by probes complementary to different regions in or downstream of the *NPL3* ORF.

13. Replace the water by 500 mL of a 50 mM sodium hydroxide solution and soak the gel for 10 min with gentle shaking to fragment the RNA and facilitate the transfer of larger RNA species (see Note 8).
14. Rinse the gel with ddH_2O, then replace the water with 10× SSPE, and shake lightly for 10 min. At the 10-min mark, discard the SSPE, add new buffer, and shake for another 10 min. This newly added 10× SSPE can be used later for the transfer.
15. While the gel soaks in SSPE, cut a corner of the nylon membrane to mark its orientation and wet it for 5 min in ddH_2O, and then transfer it to a clean dish with 10× SSPE until needed.
16. Assemble the RNA transfer setup by placing the sponge in a Pyrex dish and soaking it with 10× SSPE buffer.
17. Cut six Whatman 3MM papers to fit the sponge size, wet them, one at a time, with 10× SSPE, and place them on the gel one by one while taking care not to introduce air bubbles between

the different layers. If bubbles are introduced, remove them by rolling a glass pipette over the paper stack.

18. Place the gel on the Whatman paper and remove any air bubble by rolling over with a glass pipette.
19. Cover all exposed area of the Whatman paper with plastic strips to prevent buffer from passing outside the gel surface.
20. Place the wet nylon membrane on the gel and remove any air bubbles.
21. Place six layers of Whatman papers wetted in 10× SSPE on top of the nylon membrane and remove any air bubbles.
22. Place a stack of paper towels about 5–7-cm high on top of the Whatman papers, cover with a glass plate, and place on top a weight of about 1-2 kg.
23. Place the level on top of the glass plate and adjust the setup if needed.
24. Add enough 10× SSPE for the dish to have a 3-cm-thick layer of buffer.
25. Let the RNA transfer for 12–16 h.
26. Once transfer is complete, remove the nylon membrane and wrap it with Saran wrap. Cross-link the RNA to the membrane by irradiating each side with a 254-nm lamp to 0.5 J/cm^2 (see Note 9).
27. After cross-linking, stain the membrane with the methylene blue solution for 5 min, destain with ddH_2O until the ribosomal RNA appears with minimum background, scan the stained membrane, and keep it for your record (see Note 10).
28. Store at 4°C, wrapped in a Saran wrap, until you are ready for hybridization.
29. Place the membrane in a hybridization tube and prehybridize in 50 mL of hybridization solution at 55°C for 2 h.
30. Discard the solution used for prehybridization and replace it with 50 mL of fresh hybridization buffer.
31. Place 10–20 μL of randomly primed probe (>10^6 cpm) in a microtube containing 200 μL of hybridization buffer and heat for 2 min at 95°C.
32. Add the denatured probe to the hybridization tube taking care not to pour the probe directly on the membrane.
33. Leave the hybridization tube rotating at 55°C for 12–16 h.
34. Discard the hybridization solution and rinse the membrane with 4× SET–0.1% SDS buffer twice.
35. Wash the membrane twice for 20 min in 4× SET–0.1% SDS and once for 20 min with 3× SET–0.1% SDS at 55°C.

36. Wrap the membrane in Saran wrap and check for background at the instant imager (PerkinElmer, Whatham, MA) or a Geiger counter (VWR CANLAB, Mississauga, Canada). If needed, further wash for another 20 min with 3× SET–0.1% SDS. The membrane is then exposed to film depending on the signal strength. A typical example is shown in Fig. 1.

3.5. Preparation of Soluble Chromatin

1. Grow cells in 50 mL YEPD as indicated in Subheading 3.1 until cell density reaches an OD_{600} of 0.8–1.
2. Cross-link the cells by adding formaldehyde to the medium to a final concentration of 1% and shake at room temperature for 5 min.
3. At the 5-min mark, stop the reaction by adding cross-linking stop buffer to a final concentration of 125 mM glycine and continue to shake for another 5 min.
4. Transfer the cells to prechilled, 50-mL Oakridge tubes placed on ice, spin the cell at 4°C for 5 min at 1,200 x*g*, and discard the supernatant.
5. Wash the cells with 30 mL of ice-cold HBS, spin at 1,200 x*g* for 5 min, and discard the supernatant.
6. Wash the cells with 30 mL ice-cold ChIP lysis buffer, spin at 1,200 ×*g* for 5 min, and discard the supernatant.
7. Resuspend the cells in 400 μL ChIP lysis-buffer, transfer the suspension to a prechilled, 0.6-mL microtube, and freeze the sample by placing the tube on dry ice. They can be stored at −80°C at this point.
8. When ready to extract chromatin, thaw the cells on ice and add glass beads to about two-thirds of the liquid volume.
9. Vortex the sample at 4°C (three times for 15 min, each time with intermittent incubation on ice).
10. Once vortexing is complete, pierce a hole in the bottom of the tube with a hot 21-gauge needle, place it on 1.5-mL microtube, and gently spin for 1 min at 137 x*g* at 4°C.
11. Respin the liquid collected in the 1.5-mL microtube at 23,000 x*g* at 4°C.
12. Aspirate the supernatant using a 1-mL pipette without touching the pellet.
13. Resuspend the pellet in 500 μL ChIP lysis buffer using a pipette tip while leaving the tube on ice.
14. Sonicate the cell suspension using 3 series of 15 pulses at a power output of about 30 W. The pulse time and power should be adjusted based on the type of cell used and sonicator to obtain final DNA fragments size of about 0.5–1.0 kb.

15. Spin the lysate for 3 min at 1,200 x*g* at 4°C and transfer the supernatant to a 1.5-mL microtube while paying attention not to disturb the pellet.
16. Repeat this step until no more pellet is detected after spinning.
17. Resuspend in 500 μL ChIP lysis buffer; samples can be stored at –80°C ready for immunoprecipitation.
18. Take 30 μL of the soluble chromatin extract and transfer it to a 1.5-mL microfuge tube.
19. Add 170 μL elution buffer and incubate in a block heater at 65°C overnight.

The amount of chromatin extracts should be compared and equalized by southern blot analysis performed using "The Convertible" system available from Biometra, Goettingen, Germany.

3.6. Chromatin Immunoprecipitation

1. Add 2 μg of the antibodies of choice to 500 μL of the soluble chromatin sample.
2. Shake the mix for 2 h at 4°C using a rotary shaker.
3. Transfer 30 μL to a 1.5-mL microtube containing 170 μL of elution buffer and incubate in a block heater at 65°C overnight to use as control (input DNA).
4. Add 50 μL of protein G Sepharose beads prewashed in 1 mL ChIP lysis buffer to the rest of the chromatin sample and shake for 1 h at 4°C using a rotary shaker.
5. Collect the beads by spinning the mix at 1,200 x*g* in a microfuge at room temperature.
6. Aspirate the supernatant and wash the beads once with 1.5 mL ChIP lysis buffer.
7. Wash the beads twice with elution buffer for 5 min at room temperature and remove the supernatant after spinning at 1,200 x*g*.
8. Wash once with T/L buffer for 5 min at room temperature and spin at 1,200 x*g* for 1 min.
9. Wash twice with TE buffer for 5 min at room temperature and spin at 1,200 x*g* for 1 min.
10. To elute the immunoprecipitate from the beads, add 100 μL elution buffer, vortex, heat at 65°C for 2 min, then vortex again, and incubate the tube at 65°C for an additional 3 min.
11. Spin at 8,800×*g* for 1 min at room temperature. Transfer 100 μL of the eluate to a new microtube.
12. Add 100 μL elution buffer to the beads and repeat steps 10 and 11. Pool the two eluates.

13. Take 15 μL of the pooled eluates and transfer to a new Eppendorf tube. Add 15 μL 2× Laemmli loading buffer (22) and freeze the sample at −80°C. This is the immunoprecipitated protein sample to be analyzed by western blotting (23).
14. The remaining eluate goes into a block heater at 65°C overnight for cross-link reversal to be used as the eluted sample for the PCR reaction.

3.7. Purification and Analysis of Immunoprecipitated DNA

1. Add 200 μL of TE to all samples (input DNAs and eluates) after overnight cross-link reversal.
2. Add RNase A to 0.2 mg/mL final (8 μL of 10 mg/mL) and incubate for 1 h at 37°C.
3. Add 3.5 μL of TE, 5 μL of 10 mg/mL proteinase K, and 1.5 μL of 20 mg/mL glycogen per sample and incubate for 6 h to overnight at 37°C.
4. Extract with 400 μL phenol/chloroform/isoamyl alcohol, and transfer 300 μL of the aqueous phase to a new tube.
5. Add 150 μL of TE to the tube containing phenol/chloroform/isoamyl alcohol, re-extract the organic phase, take 200 μL of the aqueous phase, and pool the extractions.
6. Extract once with 500 μL chloroform/isoamyl alcohol.
7. Precipitate the DNA with 0.3 M sodium acetate (50 μL of 3 M, pH 5.2) and 2 volumes (1 mL) of 95% ethanol overnight at −20°C.
8. Spin for 10 min at 2,300 x*g* at 4°C. Resuspend the "input DNA" samples in 40 μL of TE and the "eluted DNA" samples in 20 μL of TE.
9. Immunoprecipitated (eluted) and input DNA can now be analyzed by real-time PCR.

3.8. Real-Time PCR Probe Design

1. Retrieve the genomic sequence for the region of interest from the SGD (24).
2. Retrieve the genomic sequence for an untranscribed intergenic region located far from any gene. That region should not have any expression on the whole genome tiling array (17). Regions near the telomeric repeats can be used.
3. Select regions along the extended transcript starting from 500 nt upstream of the 5′ end to 500 nt downstream of the 3′ end. On average, design one primer pair every 500 nt. Also design one primer pair for the control intergenic region.
4. Use Primer3 to design pairs of primers capable of generating PCR products ranging from 50 to 300 bp (optimally, 200 bp). Ideal primer pairs should have a melting temperature of 60°C

(58–62°C), a melting temperature difference of less than 3°C, and a GC content of 50% (30–80%). The generated product should have a melting temperature of 82°C (minimum 75°C) (20).

5. Using WU-BLAST2, ensure that the primer pairs cannot generate off-target amplification products.

3.9. Real-Time PCR

1. Prepare the 10 μL PCR reactions in a 96-well microplate by mixing 1 μL of input DNA (prepared in Subheading 3.6) or eluant DNA (prepared in Subheading 3.7) with 1 μM final concentration of each primer (designed in Subheading 3.8) and 5 μL FastStart Universal SYBR green Master mix.
2. Set the 7500 ABI apparatus to the following cycling: 10 min at 95°C; 50 cycles: 15 s at 95°C, 30 s at 60°C, 30 s at 72°C; and melting curve: 15 s at 95°C, 60 s at 60°C, 1°C/min temperature gradient, 15 s at 95°C.

3.10. Calculation and Interpretation of Real-Time PCR Data

1. Determine the relative precipitations (fold differences) using the following formula: $\Delta\Delta Ct = 2 - (\Delta Ct\ IP - \Delta Ct\ background)$, where Ct stands for cycle threshold, the Ct IP is the cycle number for immunoprecipitate, and Ct background is the cycle number for external control (untranscribed intergenic region).
2. Evaluate the background of ΔCt IP and the ΔCt background with these equations: $\Delta Ct\ IP = Ct\ IP - Ct\ Input\ DNA$ and the $\Delta Ct\ background = Ct\ background - Ct\ input\ DNA$, respectively (for examples of the results, see Fig. 2). In order to determine the site of termination, the RNAPII pattern (Fig. 2) should be compared with the size of the 3′ end RNA fragment produced from the same gene as observed by Northern blot (Fig. 1). Deletion of endoribonucleases that affect the termination directly, like Rnt1p in the shown example, should lead to an increase in the RNAPII association downstream of the gene's mature sequence and the production of extended species. Exoribonucleases, like Rat1p, should also change the RNAPII association pattern, but only leads to the detection of a short 3′ end fragment that is normally degraded after endoribonucleolytic cleavage by the polyadenylation complexes or other endoribonucleases. In many cases, the deletion of both nuclear (Rat1p) and cytoplasmic (Xrn1p) exoribonucleases is needed to detect the fragment generated by the termination-dependent end endoribonucleolytic activity because RNA fragments that are not degraded in the nucleus are transported and degraded in the cytoplasm (Fig. 1 and (13)).

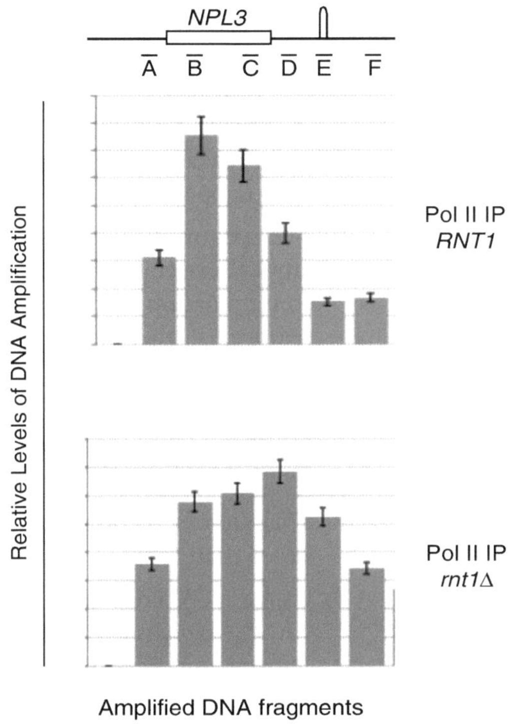

Fig. 2. Chromatin immunopreciptiation of RNAPII as an indicator of transcription termination. Chromatin immunoprecipitations were performed using antibodies against the RNAPII protein subunit Rpb1p in the presence (*top panel*) or absence (*bottom panel*) of *RNT1*. The precipitated DNA was amplified by real-time PCR using primers specific to different regions upstream, within or downstream of *NPL3*. A total of two biological and three technical replicates were used to calculate the relative levels of DNA precipitated, and the average values are indicated. A primer pair amplifying a known, untranscribed region of chromosome V was used as negative control (Ctl). Standard deviations between replicate experiments were ±0.05.

4. Notes

1. DEPC is added directly to the water, and mixed thoroughly by shaking. DEPC-treated water is autoclaved and incubated in oven at 95 C overnight.
2. Glass beads can be bought acid washed or unwashed. Unwashed or previously used beads are cleaned by acid wash (10 M HCl), successive ddH_2O rinse, 95% ethanol wash, and baked overnight in a 90°C oven.
3. Calibrated liquid phenol is prepared by mixing 500 mL phenol with 500 mL LETS buffer (pH 11), 0.5 g of 8-hydroxyquinoline (SDS), and 500 mL of LETS buffer adjusted to pH 11. The pH of the final phenol phase should be ~8.

4. Xylene cyanol 5% (w/v) and bromophenol blue 1% (w/v) solutions are prepared in DEPC-treated ddH_2O. The solutions are filtered before storage. We routinely store the solution in aliquots of 10 mL.
5. Cell breakage is verified by looking at the samples using a light microscope. Lysed cells have darker and less-defined contours.
6. The spectrophotometer 260/280 ratio of nucleic acid should range between 1.9 and 2.1. Lower ratio indicates protein contamination and higher ratio may result from incomplete removal of phenol. Repeating the phenol/chloroform extraction may improve the nucleic acid quality.
7. Electrophoresis apparatus used for preparing gels destined for Northern blot analysis should be clean and free from ethidium bromide.
8. This step is recommended when analyzing RNA transcripts larger than 3 kb.
9. Using a UV cross-linker model HL-2000 Hybrylinker (VWR CANLAB, Mississauga, Canada), 0.5 J/cm^2 corresponds to about 2 min of exposure.
10. A real-size printed copy of the membrane scan serves as a marker for the rRNA position.

Acknowledgments

This work was supported by grants from the Natural Sciences and Engineering Research Council of Canada (NSERC) and the Canadian Institute of Health Research (CIHR).

References

1. Prescott EM, Osheim YN, Jones HS, et al (2004) Transcriptional termination by RNA polymerase I requires the small subunit Rpa12p. Proc Natl Acad Sci USA 101:6068–73
2. Cui M, Allen MA, Larsen A, Macmorris M, Han M, Blumenthal T (2008) Genes involved in pre-mRNA 3′-end formation and transcription termination revealed by a lin-15 operon Muv suppressor screen. Proc Natl Acad Sci USA 105:16665–70
3. Zarudnaya MI, Kolomiets IM, Hovorun DM (2002) What nuclease cleaves pre-mRNA in the process of polyadenylation? IUBMB Life 54:27–31
4. Tollervey D (2004) Molecular biology: termination by torpedo. Nature 432:456–7
5. Luo W, Bentley D. (2004) A ribonucleolytic rat torpedoes RNA polymerase II. Cell 119: 911–4
6. West S, Gromak N, Proudfoot NJ (2004) Human 5′→3′ exonuclease Xrn2 promotes transcription termination at co-transcriptional cleavage sites. Nature 432:522–5
7. Kim M, Vasiljeva L, Rando OJ, Zhelkovsky A, Moore C, Buratowski S (2006) Distinct pathways for snoRNA and mRNA termination. Mol Cell 24:723–34
8. Carroll KL, Pradhan DA, Granek JA, Clarke ND, Corden JL (2004) Identification of cis elements directing termination of yeast non-polyadenylated snoRNA transcripts. Mol Cell Biol 24:6241–52

9. Osheim YN, French SL, Keck KM, et al (2004) Pre-18 S ribosomal RNA is structurally compacted into the SSU processome prior to being cleaved from nascent transcripts in Saccharomyces cerevisiae. Mol Cell 16:943–54
10. Steinmetz EJ, Warren CL, Kuehner JN, Panbehi B, Ansari AZ, Brow DA (2006) Genome-wide distribution of yeast RNA polymerase II and its control by Sen1 helicase. Mol Cell 24:735–46
11. Egloff S, O'Reilly D, Murphy S (2008) Expression of human snRNA genes from beginning to end. Biochem Soc Trans 36:590–4
12. Richard P, Manley JL (2009) Transcription termination by nuclear RNA polymerases. Genes Dev 23:1247–69
13. Ghazal G, Gagnon J, Jacques PE, Landry JR, Robert F, Elela SA (2009) Yeast RNase III triggers polyadenylation-independent transcription termination. Mol Cell 36:99–109
14. Rondon AG, Mischo HE, Kawauchi J, Proudfoot NJ (2009) Fail-safe transcriptional termination for protein-coding genes in S. cerevisiae. Mol Cell 36:88–98
15. Taggart AK, Teng SC, Zakian VA (2002) Est1p as a cell cycle-regulated activator of telomere-bound telomerase. Science 297:1023–6
16. Lamontagne B, Tremblay A, Abou Elela S (2000) The N-terminal domain that distinguishes yeast from bacterial RNase III contains a dimerization signal required for efficient double-stranded RNA cleavage. Mol Cell Biol 20:1104–15
17. David L, Huber W, Granovskaia M, et al (2006) A high-resolution map of transcription in the yeast genome. Proc Natl Acad Sci U S A 103:5320–5
18. Rozen S, Skaletsky H (2000) Primer3 on the WWW for general users and for biologist programmers. Methods Mol Biol 132:365–86
19. SantaLucia J, Jr (1998) A unified view of polymer, dumbbell, and oligonucleotide DNA nearest-neighbor thermodynamics. Proc Natl Acad Sci U S A 95:1460–5
20. Brosseau JP, Lucier JF, Lapointe E, et al (2010) High-throughput quantification of splicing isoforms. RNA 16:442–9
21. Lopez R, Silventoinen V, Robinson S, Kibria A, Gish W (2003) WU-Blast2 server at the European Bioinformatics Institute. Nucleic Acids Res 31:3795–8
22. Laemmli UK (1970) Cleavage of structural proteins during the assembly of the head of bacteriophage T4. Nature 227:680–5
23. Catala M, Tremblay M, Samson E, Conconi A, Abou Elela S (2008) Deletion of Rnt1p alters the proportion of open versus closed rRNA gene repeats in yeast. Mol Cell Biol 28:619–29
24. Cherry JM, Ball C, Weng S, et al (1997) Genetic and physical maps of Saccharomyces cerevisiae. Nature 387:67–73

Chapter 39

Promoter-Associated Noncoding RNA from the *CCND1* Promoter

Xiaoyuan Song*, Xiangting Wang*, Shigeki Arai, and Riki Kurokawa

Abstract

More than 90% of the human genome have been found to be transcribed and most of the transcripts are noncoding (nc) RNAs (Willingham et al., Science 309:1570–1573, 2005; ENCODE-consortium, Science 306:636–640, 2004; Carninci et al., Science 309:1559–1563, 2005; Bertone et al., Science 306:2242–2246, 2004). Studies on ncRNAs have been radically progressed mainly regarding microRNAs, piRNAs, siRNAs, and related small ncRNAs of which length are relatively short nucleotides (Fire et al., Nature 391:806–811, 1998; Filipowicz et al., Nat Rev Genet 9:102–114, 2008; Lau et al., Science 313:363–367, 2006; Brennecke et al., Science 322:1387–1392, 2008; Siomi and Siomi, Nature 457:396–404, 2009). These small RNAs play roles in regulation of translation and gene silencing while long ncRNAs with length more than 200 nucleotides have been emerging and turn out to be involved in regulation of transcription (Kapranov et al., Science 316:1484–1488, 2007; Ponting et al., Cell 136:629–641, 2009; Kurokawa et al., RNA Biol 6:233–236, 2009). Recently, we have identified novel, long ncRNAs bearing capability of repression of transcription (Wang et al., Nature 454:126–130, 2008).

RNA-binding protein, translocated in liposarcoma (TLS), binds CREB-binding protein CBP/adenovirus p300 and inhibits their histone acetyltransferase (HAT) activities (Wang et al., Nature 454:126–130, 2008). The HAT inhibitory activity of TLS requires specific binding of RNA. The systematic evolution of ligands by exponential enrichment experiments with randomized sequences revealed that TLS specifically recognizes RNA oligonucleotides containing GGUG as a consensus sequence although the GGUG sequence is not an absolute requirement for the TLS binding (Lerga et al., J Biol Chem 276:6807–6816, 2001). TLS is specifically recruited to the CBP/p300-associated binding sites of the cyclin D1 gene (*CCND1*) and the cyclin E1 gene (*CCNE1*) promoters (Wang et al., Nature 454:126–130, 2008; Impey et al., Cell 119:1041–1054, 2004). Our extensive exploration for naturally occurring RNA molecule that binds TLS has indicated that long ncRNAs (promoter-associated ncRNAs: pncRNAs) transcribed from the *CCND1* promoter bind TLS and inhibit the HAT activities on the sites to repress the transcription of the *CCND1* gene (Wang et al., Nature 454:126–130, 2008). We have optimized RT-PCR, chromatin immunoprecipitation, RNA immunoprecipitation, and RNA gel-shift assay in order to detect these pncRNAs. The methods that we have developed successfully identified these low-abundant, long ncRNAs and provide the data showing that the *CCND1* pncRNAs bind TLS and induce its HAT inhibitory activity to repress the transcription of *CCND1* gene upon genotoxic stress.

Key words: Promoter-associated noncoding RNA, Translocated in liposarcoma, RNA immunoprecipitation, Chromatin immunoprecipitation, Gel-shift assay, RNA oligonucleotide

Ales Vancura (ed.), *Transcriptional Regulation: Methods and Protocols,* Methods in Molecular Biology, vol. 809, DOI 10.1007/978-1-61779-376-9_39, © Springer Science+Business Media, LLC 2012

1. Introduction

Recently, we have identified novel, long ncRNAs transcribed from the CCND1 gene promoter (pncRNAs) (13). The CCND1 pncRNAs are expressed at a low level as a few copies per cell and bear capability of repression of transcription through an RNA-binding protein TLS, a HAT inhibitor. Histone posttranslational modifications play pivotal roles in positive and negative regulations of gene expression (1–12, 16). Histone acetyltransferase (HAT) is one of the major enzymes of transcription activation and was proposed to be regulated by a regulatory factor (15, 17, 18), most probably an inhibitor. Indeed, we have identified an RNA-binding protein, translocated in liposarcoma (TLS)/FUS, as an inhibitory molecule for CREB-binding protein (CBP)/p300-HAT (13). TLS is a unique enzyme inhibitor that specifically targets CBP/p300-HAT on the specific genes, *CCND1* and *CCNE1* (13).

TLS was initially identified as TLS–CHOP, a fusion protein arising from a chromosomal translocation presumed to be causative gene for myxoid liposarcoma (19). TLS has been shown to play roles in RNA processing, homologous DNA pairing, and transcription (19, 20). Deletion of the TLS gene in mice resulted in male infertility, chromosomal instability, and increased sensitivity to DNA-damaging agents (20). The data from TLS null mice showed that TLS possesses crucial functions in maintaining genomic integrity (20, 21), indicating that TLS has significant biological consequences.

Amyotrophic lateral sclerosis (ALS) is a late-onset, fatal neurodegenerative disease that causes degeneration of the motor neurons in the patient (22–24). The disorder causes muscle weakness and atrophy throughout the body by disruption of sending messages to muscle. Most of the patients die from respiratory failure usually within 3–6 years after the onset of the symptom. The superoxide dismutase gene was reported as a causative gene (25), although development of the therapeutics has not been fruitful. Indeed, no cure has yet been developed. Recently, TLS has also been identified as a causative gene for ALS (26, 27). More than a dozen mutations have been reported in amino acid sequence of TLS, including H517Q and R521G involved in ALS. These amino acid residues are located at the very C-terminus of TLS and can function to bind target RNAs. Involvement of the TLS RNA-binding domain mutations in causing ALS implies that RNA binding ability of TLS is essential to maintain normal functions of the motor neurons. Therefore, TLS and the (*CCND1*) promoter-associated ncRNAs (pncRNAs) are possible targets of therapeutics for ALS.

It has been shown by in vitro binding selection assays with randomized RNA oligonucleotide pool that RNA oligonucleotides containing GGUG sequence bind TLS (14). We have attempted to search naturally occurring RNA molecule that binds TLS and enhances its inhibition against the HAT activity, and identified novel, long noncoding (nc) RNAs (pncRNAs) transcribed from the *CCND1* gene promoter (13).

We developed high-sensitivity detection systems of reverse transcription (RT)-PCR and real-time PCR, chromatin immunoprecipitation (ChIP), RNA immunoprecipitation (RIP), and RNA gel-shift assays. The detailed procedures of these techniques are described in this chapter.

2. Materials

2.1. Cell Culture and Treatment

1. Dulbecco's modified Eagle's medium (DMEM; Gibco/BRL, Bethesda, MD) supplemented with 10% fetal bovine serum (FBS, HyClone, Ogden, UT).
2. Solutions of 0.25% (v/v) trypsin and 1 mM ethylenediamine tetraacetic acid (EDTA; Gibco/BRL).
3. Forskolin (Sigma, St. Louis, MO) is dissolved in ethanol at 10 mmol/mL. The cells are treated for 6 h at a final concentration of 10 μmol/mL.
4. Gamma-irradiation (IR): J.L. Shepherd & Associates, Model: Mark 1 Model 30 Irradiator; Source: 137 Cesium 8000 Curies (April 10, 1980).

2.2. Transfection

Lipofectamine 2000 (Invitrogen, Carlsbad, CA) is used for transfection.

2.3. siRNAs

siRNAs (Qiagen):

siA, 5′-GGCGCCUCAGGGAUGGCUU-3′.
siD, 5′-AAUUCAGUCCCAGGGCAAA-3′.
siE, 5′-GACCCGGAAUAUUAGUAAU-3′.
siC, 5′-GGCUAGAAGGACAAGAUGA-3′.
siF, 5′-GAGUGGGCGAGCCUCUUUA-3′.
si5′UTR, 5′-GGACUUUGCAACUUCAACA-3′.
siCCND1, SI02654547.
siCTL, 5′-AAUUCUCCGAACGUGUCAC-3′.
siTLS(mouse), 5′-CAGAGUUACAGUGGUUAUG-3′.
5′-UUCUCUGGGAAUCCUAUUA-3′.

2.4. Antibodies

1. Anti-TLS antibody (BD Biosciences).
2. Anti-CBP, anti-p300, anti-p/CAF, anti-TAFII68, and anti-EWS antibodies (Santa Cruz Biotechnology).
3. Anti-TIP60 and anti-acetylated histone H3 antibodies (Upstate Biotechnology).

2.5. RNA Extraction

1. TRIzol 12.5 (Invitrogen).
2. RNase-free DNase I (DNA-free; Applied Biosystems).

2.6. RT and Real-Time PCR

1. RT kit: SuperScript III First-Strand (Invitrogen).
2. Real-time PCR is performed with the Mx3000P (Stratagene, La Jolla, CA).

2.7. Fractionation (see Note 1)

1. Nuclease-free water.
2. Digitonin: 40 μg/mL in RSB-100 buffer (100 mM Tris–HCl, pH 7.4, 100 mM NaCl, 2.5 mM $MgCl_2$).
3. RSB-100T buffer: 0.5% Triton X-100 in RSB-100 buffer.
4. Microtip sonicator: Fisher Sonic Dismembrator, Model 300 (Fisher Scientific, Pittsburgh, PA).
5. 30% sucrose/RSB-100: One volume of 30% sucrose (w/v) in one volume of RSB-100.

2.8. Northern Blotting

1. 15% TBE–Urea PAGE gels (Invitrogen).
2. Hybond N + Nylon membranes (GE Healthcare).
3. 0.25× TBE running buffer: Made from 20× TBE running buffer stock (Invitrogen).
4. UV cross-linker (Stratalinker, Stratagene).
5. ExpressHyb hybridization solution (Clontech).
6. ULTRAhyb hybridization solution (Applied Biosystems/Ambion).
7. Rediprime II Random Prime Labelling system (Amersham/GE Healthcare).
8. Nick column (Amersham/GE Healthcare).
9. Riboprobe combination system (Promega).
10. X-ray film.
11. Chemiluminescent Nucleic Acid Detection Module (Pierce).

2.9. RNA Gel-Shift Assay

1. Synthetic RNA oligonucleotides (purified with HPLC) dissolved in TE buffer at 100 μM and stored at −80°C.
2. [γ^{32}P]ATP is stored at 4°C. Fresh lot is OK to use until 1 month.
3. T4 poly nucleotide kinase (PNK) is stored at −20°C.
4. Nucleotide purification columns, Qiaquick Nucleotide Removal kit (Qiagen).
5. Yeast transfer RNA (Sigma–Aldrich) dissolved in TE buffer at 10 mg/mL and stored at −20°C.
6. 40% acrylamide solution (19:1) is stored at 4°C.
7. Polyacrylamide gel plates (16 × 16 cm × 1 mm).
8. Buffer W: 100 mM Tris–HCl, pH 8.0, 150 mM NaCl, 1 mM EDTA.

9. TED buffer: 10 mM Tris–HCl, pH 7.5, 20 mM EDTA, 1 mM dithiothreitol, 5% (v/v) glycerol.
10. Loading buffer: 0.2% (w/v) bromophenol blue, 20% (v/v) glycerol.
11. Fix solution: 5% (v/v) methanol, 7.5% (v/v) acetic acid.

2.10. RIP (see Note 1)

1. Lysis buffer: NETN buffer (125 mM NaCl, 1 mM EDTA, 20 mM Tris–HCl, pH 8.1, 0.5% Nonidet NP40, 10% glycerol, protease inhibitor cocktail).
2. Microtip sonicator: Fisher Sonic Dismembrator, Model 300 (Fisher Scientific).
3. Protein A/G–Sepharose beads (Sigma–Aldrich).
4. Antibody: See Subheading 2.4.

2.11. ChIP

1. Fixation: 1% formaldehyde made from 36.5% (v/v) formaldehyde stock.
2. 2.5 M glycine.
3. Nuclei extraction:

   Buffer I: 0.25% Triton X-100, 10 mM EDTA, 0.5 mM EGTA, 10 mM HEPES, pH 6.5.

   Buffer II: 200 mM NaCl, 1 mM EDTA, 0.5 mM EGTA, 10 mM HEPES, pH 6.5.

   Lysis buffer: 1% SDS, 10 mM EDTA, 50 mM Tris–HCl, pH 8.1, protease inhibitor cocktail (complete protease inhibitor cocktail tablets, one tablet in 1 mL of buffer = 50×; Roche Applied Science).
4. Microtip sonicator: See Subheading 2.10.
5. Protein A/G–Sepharose beads.
6. Dilution buffer: 1% Triton X-100, 2 mM EDTA, 150 mM NaCl, 20 mM Tris–HCl, pH 8.1, protease inhibitor cocktail.
7. Antibodies: See Subheading 2.4.
8. Wash buffer:

   TSE I buffer: 0.1% SDS, 1% Triton X-100, 2 mM EDTA, 20 mM Tris–HCl, pH 8.1, 150 mM NaCl.

   TSE II buffer: 0.1% SDS, 1% Triton X-100, 2 mM EDTA, 20 mM Tris–HCl, pH 8.1, 500 mM NaCl.

   Buffer III: 0.25 M LiCl, 1% Nonidet NP40, 1 mM EDTA, 10 mM Tris–HCl, pH 8.1.

   TE buffer.
9. Elution buffer: 1% SDS, 0.1 M $NaHCO_3$.
10. Reverse cross-linking: 65°C water bath overnight.
11. ChIP DNA purification: QIAquick Spin Kit (Qiagen).

3. Methods

Starting materials are always crucial for following procedures. The condition of HeLa cells at the experiments described here is one of the most critical parameters. Empirically, fresh HeLa cells are prone to lose their regulatory functions after certain numbers of passages. Therefore, it is recommended to periodically prepare freshly thawed HeLa cells from frozen stocks and to keep the cells at an optimal condition for the experiments.

Every procedure using RNA is required to eliminate any contact with RNases that are extremely stable even after 20 min of boiling. Whenever living cells are disrupted, the extract must be placed into denaturing reagents, such as TRIzol. RNA is also labile against high pH, UV irradiation, and physical agitations, such as strong pipettings. Therefore, it is required to keep RNA sample away from these agents as well.

Northern blotting of the pncRNAs is highlighted in the chapter because we have successfully utilized it for detection of the pncRNAs. For the experiment, the quality of the RNA sample is one of the most crucial factors. Selection of the probe sequences and the hybridization conditions are also important. However, the condition of the Northern blotting is just one example for our application with the *CCND1* pncRNAs. Therefore, optimization for each specific ncRNA might be required.

The RNA gel-shift assay is a powerful tool to detect interaction between ncRNAs and its cognate binding proteins. Two important points are the preparations of the RNA-binding protein TLS and of ^{32}P-RNA probes. Concentration of the gel running buffer (TBE) is also critical for the interaction between TLS and the pncRNAs. Higher concentration reduces binding of TLS to the pncRNAs while lower concentration increases nonspecific bindings.

RIP and ChIP are powerful techniques to detect the TLS interaction with pncRNAs and target promoter sites, respectively. As described below, nonspecific binding to agarose beads is a tough business for these experiments. Especially, TLS is a sticky protein that binds to agarose beads. For reducing the background, the process of preclearing agarose beads is one of the solutions, although it is not always a perfect one. A negative control for this RIP is required to show the specific interaction between the RNA-binding protein and the ncRNA of your interest.

It is crucial to show that the *CCND1* pncRNAs are functionally required for TLS. Therefore, we knocked down the *CCND1* pncRNAs in HeLa cells by specific siRNAs and then performed RT-PCR and ChIP assays mentioned above.

Because of diversity of ncRNAs, it is hard to set a single standardized protocol that can be applied to all ncRNAs. The protocols present in this chapter could be a compass for voyage of discovery of ncRNAs, although the experimental conditions shown here need to be adjusted or optimized for particular ncRNAs of your interest.

3.1. Preparation of RNA Samples

3.1.1. Fractionation (see Note 2)

1. Subconfluenced HeLa cells are rinsed three times with PBS on plate. Cells are then collected with a cell scraper and pelleted with centrifugation at 5,000 × *g* for 1 min.
2. Resuspend cells with 1 mL of RSB-100/digitonin buffer containing protease inhibitor cocktail per 3×10^6 cells, incubate on ice for 5 min (see Note 3), and separate the soluble cytosolic fraction from the nuclear and digitonin-insoluble fraction by centrifugation at 2,000 × *g* for 8 min at 4°C. The supernatant fraction is collected as a cytosolic fraction.
3. The cell pellet is resuspended in RSB-100T buffer containing protease inhibitor cocktail and incubated on ice for 5 min. After centrifugation at 2,000 × *g* for 8 min, the supernatant is collected as a nuclear fraction.
4. The resulting cell pellet is resuspended in RSB-100T containing protease inhibitor cocktail and disrupted by two 5-s exposures to sonication on ice using a microtip sonicator, set as scale 3. Layer the sonicated material onto a 30% sucrose cushion in RSB-100 by gently adding the sonicated lysate on top of one volume of 30% sucrose/RSB-100. The soluble supernatant is collected as DNA-bound fraction by centrifugation at 4,000 × *g* for 15 min.

3.1.2. Preparation of RNA Samples

RNA is extracted with TRIzol (see Note 4) and treated with RNase-free DNase I.

3.2. Identification of the pncRNA

3.2.1. RT and Real-Time PCR

Normal reverse transcription experiments are performed using a random hexamer following manufacturer's protocol. Real-time PCR is performed on the 1:5 diluted cDNA. To detect the strandedness, we perform the RT using strand-specific primers, followed with real-time PCR. In an effort to normalize the RNA level used for strand-specific RT, an aliquot of RNA sample is used for RT using random hexamer primers followed with real-time PCR for housekeeping genes. Reaction without reverse transcriptase is performed as a no-RT control for all RT experiment.

3.2.2. Northern Blotting

Total RNA is extracted using TRIzol and treated with RNase-free DNase I. Small transcript Northern blotting is performed as follows.

1. 15% TBE–Urea PAGE gels are loaded with 25–30 μg of total RNA, and then run at 200 V for about 1 h until the bromophenol blue dye reaches the gel bottom.
2. The RNA is then electrotransferred to Hybond N+Nylon membranes for 3 h at 25 V in 0.25× TBE running buffer.

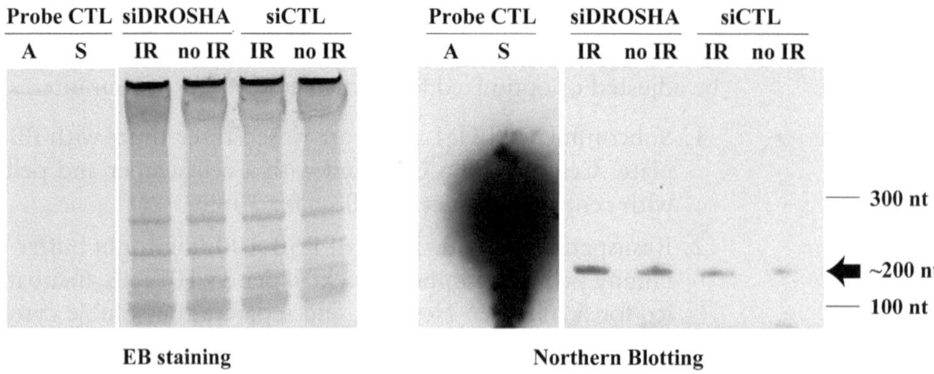

Fig. 1. Northern blotting with *CCND1* pncRNAs. The *CCND1* pncRNAs were analyzed on 15% TBE–Urea gel. After running the gel, the gel was stained with ethidium bromide. Then, the RNAs were transferred to a Hybond N+Nylon membrane and Northern blotting was performed. The probe was a biotin-labeled DNA in vitro transcript (~240 bp) of antisense of the region D (13). The analysis showed sense-strand transcript, ~200 bp, which was increased upon the siDrosha transfections.

3. Membranes are UV cross-linked and prehybridized in ExpressHyb hybridization solution (for dsDNA-probed Northern) or in ULTRAhyb hybridization solution (for strand-specific Northern, in vitro-transcribed probe) at 68°C for 1 h.
4. For dsDNA-probed Northern blotting, 25 ng of dsDNA probes are labeled with [α-^{32}P]-dCTP using Rediprime II Random Prime Labelling system and purified with Nick column. Hybridization is performed at 42°C overnight. For strand-specific Northern blotting, biotin-labeled probes are in vitro transcribed using Riboprobe combination system plus biotin-11-UTP and purified by $LiCl_2$ precipitation. 10 ng/ml probes are used, and hybridization is carried out at 68°C overnight in ULTRAhyb hybridization solution (see Note5).
5. Membranes are extensively washed in high-stringency wash buffer (0.1× SSC, 0.1% SDS) at 65°C and exposed to film (ds Northern) or at 68°C (strand-specific Northern, in vitro-transcribed probe). Probe and biotin signal are detected by chemiluminescent nucleic acid detection module.

Our data indicate that the *CCND1* pncRNA is DROSHA dependent (Fig. 1). Northern blotting shows that DROSHA siRNA enhances the level of *CCND1* pncRNA (Fig. 1).

3.3. pncRNA Interaction with TLS

3.3.1. Gel-Shift Assay

Preparation of strep-tagged protein for gel-shift assay:

1. TLS cDNA (NM_004960) is inserted into pASK-IBA5 vector (IBA, Göttingen, Germany) and expressed in *Escherichia coli*.
2. The bacterially expressed strep-tagged TLS protein is extracted with sonication in buffer W, followed by centrifugation at 14,000 ×*g* for 10 min at 4°C. The extracted protein is purified

with StrepTactin-sepharose (IBA) beads using elution buffer (buffer W plus 2.5 mM desthiobiotin).

3. The eluted protein sample is analyzed by 10% SDS-PAGE. Approximate amount of the eluted protein is estimated by a loading control of bovine serum albumin on the same gel.
4. The eluted protein sample is stored at 4°C until use.

Labeling of RNA oligonucleotide:

1. RNA oligonucleotides are labeled at the 5′-terminus by T4 PNK with [γ-^{32}P] ATP.
2. Excess amount of [γ-^{32}P] ATP and degraded RNA oligonucleotides are removed by a Qiaquick column.
3. Count cpm of the labeled RNA oligonucleotides with a Beckman LS-6500 liquid scintillation counter.

Gel-shift assay:

1. The ^{32}P-labeled RNA oligonucleotides (50,000 cpm) is incubated with approximately 50 ng of recombinant strep-TLS for 20 min at room temperature in TED buffer containing 0.5 μg/μl bovine serum albumin and 10 ng/μl yeast tRNA (see Note 6).
2. After addition of 2 μL of loading buffer, the sample is immediately analyzed by a 4% native PAGE gel in TBE. Typical condition is at room temperature and 150 V for 90 min.
3. The gel is soaked in a fix solution for 20 min. After drying, the gel is analyzed by autoradiograph or an imaging analyzer BAS-5000 (Fuji Film).

Our data show that TLS binds the GGUG RNA oligonucleotides in the gel-shift assay (Fig. 2) while it does not bind the CCUC RNA oligonucleotide that has inactive mutation of GGUG to CCUC.

3.3.2. RIP (see Note 7)

1. Before collecting the cells, prepare the conjugated antibody/protein A/G–sepharose beads as following. Thoroughly wash 45 μl of the protein A/G–sepharose beads six times with NETN buffer, and then add one volume of NETN buffer to resuspend the beads. The beads are incubated with 2–4 μg of RNase inhibitor-pretreated antibodies at 4°C for 2 h.
2. While waiting, rinse the monolayer cells on plate with PBS once and then collect cells in the proteinase inhibitors containing NETN buffer and incubate on ice for at least 15 min.
3. During waiting, pretreat the sonication tip by immersing it in NETN buffer with RNase inhibitors. The cell lysate is then sonicated on ice (10 s per sonication and 10 s in between sonications). After sonication, spin down the cell debris by

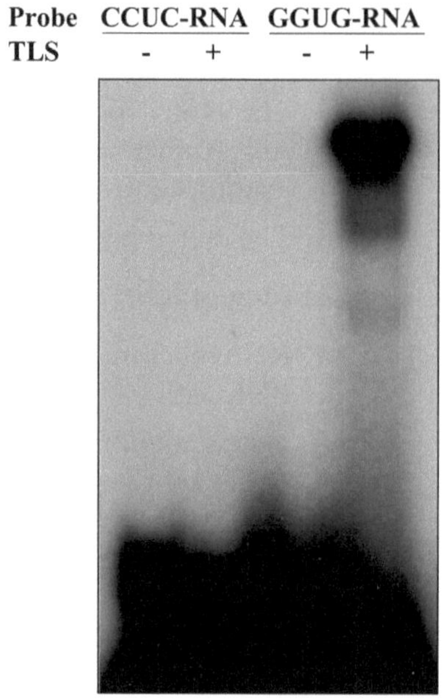

Fig. 2. RNA gel-shift assay. The RNA oligonucleotides shown below were loaded onto the gel and run as described in Subheading 3. The bacterially expressed TLS bound the GGUG-RNA oligonucleotides while it did not bind the CCUC-RNA oligonucleotides. GGUG-RNA oligonucleotides: UUGUAUUUUGAGCUAGUUUGGUGAU; CCUC-RNA oligonucleotides: UUGUAUUUUGAGCUAGUUUCCUCAU.

centrifugation at 14,000 × *g* for 10 min and transfer the supernatant into a fresh tube for the next step.

4. The soluble portion is then precleared by adding 45 μl of protein A/G–sepharose beads and incubated and rotated for 1 h at 4°C.
5. After preclearing, spin down the beads by centrifugation at 3,000 × *g* for 1 min, transfer the supernatant to a new Eppendorf tube, and add the conjugated antibody/protein A/G–sepharose beads from step 1 for a further incubation at 4°C overnight with rotation.
6. The next day, spin down the beads by centrifugation at 3,000 × *g* for 1 min and wash the beads at least six times for 10 min each at 4°C in NETN buffer.
7. The bound RNA is then eluted from the beads by directly adding TRIzol to the beads, followed by RNA extraction and RT–real-time PCR as described above.

3.3.3 ChIP (see Note 7)

1. Cells are cross-linked with 1% formaldehyde at room temperature for 10 min. The reaction is stopped with 125 mM glycine (final concentration) for 10 min at room temperature.

2. The cells are then sequentially washed in ice-cold buffer I and buffer II.
3. The cell pellets are resuspended in appropriate amount of lysis buffer (see Note 8) and sonicated as previously described in Subheading 3.3.2. The sonication times are determined by the resulting DNA sizes (see Note 9).
4. The soluble chromatin is precleared as previously mentioned in Subheading 3.3.2.
5. The precleared chromatin is then diluted in dilution buffer (1:10). Specific antibody (2–4 μg) is added to the aliquots of chromatin and incubated by rotating at 4°C overnight. 10% (v/v) of the chromatin is saved as input.
6. The next day, 45 μL of protein A/G–sepharose beads are added to the chromatin and rotated for 2 h at 4°C.
7. The beads with the chromatin are harvested by centrifugation at 3,000 × *g* at microcentrifuge and washed sequentially in TSEI, TSE II, and buffer III (see Note 10), and twice in TE.
8. The chromatin on the beads is reverse cross-linked at 65°C overnight, and proteins are digested by adding proteinase K.
9. The resulting DNA is purified with a QIAquick Spin kit.

3.3.4. siRNA and Transfection

All siRNAs are custom designed and ordered from Qiagen. A series of dose and time-curve experiments are performed to test the efficiency of siRNA. The best working conditions based on our experience are the following. HeLa cells are plated the day before transfection with a confluency around 50–60%. On the day of transfection, fresh medium is added on the plate and a final concentration of 20 nM of siRNA is transfected with Lipofectamine 2000. The cells are transferred to media without transfection complex 6 h after the transfection and cells are continually cultured for another 18 h. Cells are then treated and collected for desired experiments. Figure 3 shows a nuclear localization of a fluorescence-labeled siRNA after 18 h of transfection in HeLa cells.

4. Notes

1. All buffers are dissolved in nuclease-free water.
2. All fractionation steps are carried out on ice.
3. The lysis time is determined by testing the lysis efficiency with Trypan blue staining.
4. It is more convenient to extract RNA from the different fractions using TRIzol-LS (Invitrogen), which is for liquid samples.

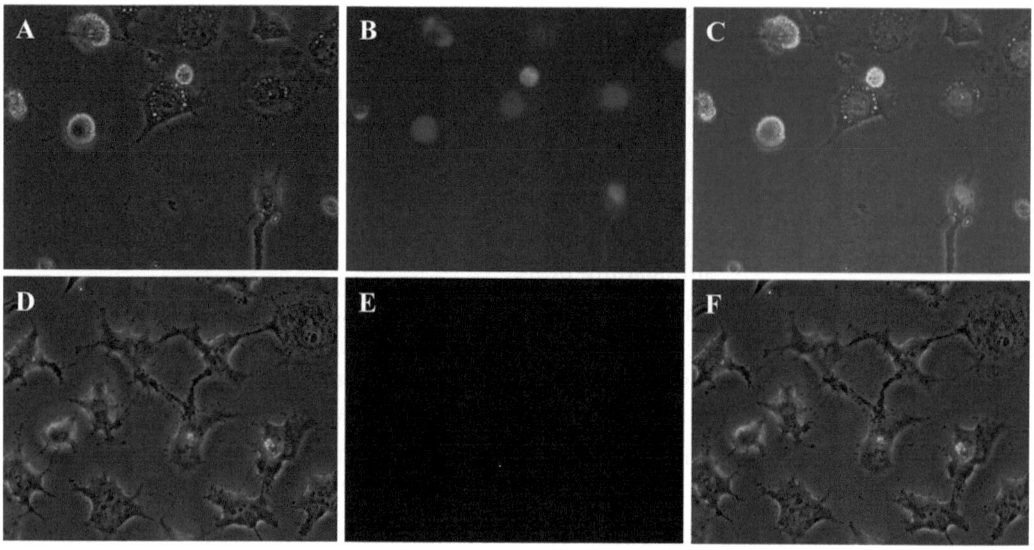

Fig. 3. The transfection of the siRNAs (Block-iT fluorescein-labeled dsRNA oligomer designed for use in RNAi analysis, Invitrogen) into HeLa cells. HeLa cells were transfected for 18 h as described in Subheading 3 (**A-C**). D-F were no transfection control. Phase-contrast microscopic images. (**B**, **E**) Fluorescence microscopic images. (**C**, **F**) The merged images of (**A**, **D**) and (**B**, **E**), respectively. The data show that the transfected siRNAs localized mainly in nuclei.

5. Customer-synthesized, 5′ biotin-labeled DNA oligonucleotide (25–49 bp) can also be used as biotin-labeled probe in Northern blotting. However, in that case, ULTRAhyb-Oligo hybridization solution (Ambion) should be used and hybridization performed at 42°C.
6. To prepare a long ncRNA probe, in vitro transcription system is more suitable. In this case, transcribed RNA is separated by agarose gel and purified from the gel.
7. It is critical to optimize the RIP and ChIP conditions for each antibody. The critical steps are fixation (ChIP only), sonication, and washes.
8. The amount of lysis buffer is determined by starting cell numbers and is suggested to optimize for the desired antibody. In our case, a near-confluence, 10-cm plate of HeLa cells was dissolved in 300 μl of lysis buffer for three individual ChIPs.
9. To determine the resulting DNA sizes, ~10 μl of the soluble chromatin fraction is transferred to a fresh tube and decrosslinked by boiling for 10 min followed by treating with RNase A for 2 min. Then, the DNA can be run on agarose gel to monitor the sonication efficiency (~200–500 bp).
10. Depending on different antibody signals, some of the washing buffers may need to be replaced with TE.

Acknowledgments

The authors thank Ms. R. Tanji for preparation of the manuscript, and Dr. C.K. Glass and Dr. M.G. Rosenfeld for critical discussion. This work was supported by Takeda Science Foundation, the Naito foundations, Astellas Foundation for Research on Metabolic Disorders Foundation, and also Grant-in-Aid for Scientific Research (B: nos22390057) and Grant-in-aid for "Support Project of Strategic Research Center in Private Universities" from the Ministry of Education, Culture, Sports, Science and Technology to Saitama Medical University Research Center for Genomic Medicine.

References

1. Willingham AT, Orth AP, Batalov S, Peters EC, Wen BG, Aza-Blanc P, Hogenesch JB, Schultz PG (2005) A strategy for probing the function of noncoding RNAs finds a repressor of NFAT. Science **309**,1570–1573.
2. ENCODE-consortium 2004 The ENCODE (ENCyclopedia Of DNA Elements) Project. *Science* **306**, 636–640.
3. Carninci P, Kasukawa T, Katayama S, Gough J, Frith MC, Maeda N, et al (2005) The transcriptional landscape of the mammalian genome. *Science* **309**, 1559–1563.
4. Bertone P, Stolc V, Royce TE, Rozowsky JS, Urban AE, Zhu X, Rinn JL, Tongprasit W, Samanta M, Weissman S, Gerstein M, Snyder M (2004) Global identification of human transcribed sequences with genome tiling arrays. *Science* **306**, 2242–2246.
5. Fire A, Xu S, Montgomery MK, Kostas SA, Driver SE, Mello CC (1998) Potent and specific genetic interference by double-stranded RNA in Caenorhabditis elegans. *Nature* **391**, 806–811.
6. Filipowicz W, Bhattacharyya SN, Sonenberg N (2008) Mechanisms of post-transcriptional regulation by microRNAs: are the answers in sight? *Nat Rev Genet* **9**, 102–114.
7. Lau NC, Seto AG, Kim J, Kuramochi-Miyagawa S, Nakano T, Bartel DP, Kingston RE (2006) Characterization of the piRNA complex from rat testes. *Science* 313,363–367.
8. Brennecke J, Malone CD, Aravin AA, Sachidanandam R, Stark A, Hannon GJ (2008) An epigenetic role for maternally inherited piRNAs in transposon silencing. *Science* **322**, 1387–1392.
9. Siomi H, Siomi MC (2009) On the road to reading the RNA-interference code. *Nature* 457, 396–404.
10. Kapranov P, Cheng J, Dike S, Nix DA, Duttagupta R, Willingham AT, Stadler PF, Hertel J, Hackermuller J, Hofacker IL, Bell I, Cheung E, Drenkow J, Dumais E, Patel S, Helt G, Ganesh M, Ghosh S, Piccolboni A, Sementchenko V, Tammana H, Gingeras TR (2007) RNA maps reveal new RNA classes and a possible function for pervasive transcription. *Science* **316**, 1484–1488.
11. Ponting CP, Oliver PL, Reik W (2009) Evolution and functions of long noncoding RNAs. *Cell* **136**, 629–641.
12. Kurokawa R, Rosenfeld MG, Glass CK (2009) Transcriptional regulation through noncoding RNAs and epigenetic modifications. *RNA Biol* **6**, 233–236.
13. Wang X, Arai S, Song X, Reichart D, Du K, Pascual G, Tempst P, Rosenfeld MG, Glass CK, Kurokawa R (2008) Induced ncRNAs allosterically modify RNA-binding proteins in cis to inhibit transcription. *Nature* **454**, 126–130.
14. Lerga A, Hallier M, Delva L, Orvain C, Gallais I, Marie J, Moreau-Gachelin F (2001) Identification of an RNA binding specificity for the potential splicing factor TLS. *J Biol Chem* **276**, 6807–6816.
15. Impey S, McCorkle SR, Cha-Molstad H, Dwyer JM, Yochum GS, Boss JM, McWeeney S, Dunn JJ, Mandel G, Goodman RH (2004) Defining the CREB regulon: a genome-wide analysis of transcription factor regulatory regions. *Cell* **119**, 1041–1054.

16. Glass CK, Rosenfeld MG (2000) The coregulator exchange in transcriptional functions of nuclear receptors. *Genes Dev* **14**, 121–141.
17. Murata T, Kurokawa R, Krones A, Tatsumi K, Ishii M, Taki T, Masuno M, Ohashi H, Yanagisawa M, Rosenfeld MG, Glass CK, Hayashi Y (2001) Defect of histone acetyltransferase activity of the nuclear transcriptional coactivator CBP in Rubinstein-Taybi syndrome. *Hum Mol Genet* **10**, 1071–1076.
18. Steffan JS, Bodai L, Pallos J, Poelman M, McCampbell A, Apostol BL, Kazantsev A, Schmidt E, Zhu YZ, Greenwald M, Kurokawa R, Housman DE, Jackson GR, Marsh JL, Thompson LM (2001) Histone deacetylase inhibitors arrest polyglutamine-dependent neurodegeneration in Drosophila. *Nature* **413**, 39–743.
19. Crozat A, Aman P, Mandahl N, Ron D (1993) Fusion of CHOP to a novel RNA-binding protein in human myxoid liposarcoma. *Nature* **363**, 640–644.
20. Kuroda M, Sok J, Webb L, Baechtold H, Urano F, Yin Y, Chung P, de Rooij DG, Akhmedov A, Ashley T, Ron D (2000) Male sterility and enhanced radiation sensitivity in TLS(–/–) mice. *EMBO* J **19**, 453–462.
21. Hicks GG, Singh N, Nashabi A, Mai S, Bozek G, Klewes L, Arapovic D, White EK, Koury MJ, Oltz EM, Van Kaer L, Ruley HE (2000) Fus deficiency in mice results in defective B-lymphocyte development and activation, high levels of chromosomal instability and perinatal death. *Nat Genet* **24**, 175–179.
22. Al-Chalabi A, Leigh PN (2000) Recent advances in amyotrophic lateral sclerosis. *Curr Opin Neurol* **13**, 397–405.
23. Boillee S, Vande Velde C, Cleveland DW (2006) ALS: a disease of motor neurons and their nonneuronal neighbors. *Neuron* **52**, 39–59.
24. Pasinelli P, Brown RH (2006) Molecular biology of amyotrophic lateral sclerosis: insights from genetics. *Nat Rev Neurosci* 7, 710–723.
25. Rosen DR (1993) Mutations in Cu/Zn superoxide dismutase gene are associated with familial amyotrophic lateral sclerosis. *Nature* **364**, 362.
26. Kwiatkowski TJ, Jr., Bosco DA, Leclerc AL, Tamrazian E, Vanderburg CR, Russ C, Davis A, Gilchrist J, Kasarskis EJ, Munsat T, Valdmanis P, Rouleau GA, Hosler BA, Cortelli P, de Jong PJ, Yoshinaga Y, Haines JL, Pericak-Vance MA, Yan J, Ticozzi N, Siddique T, McKenna-Yasek D, Sapp PC, Horvitz HR, Landers JE, Brown RH, Jr. (2009) Mutations in the FUS/TLS gene on chromosome 16 cause familial amyotrophic lateral sclerosis. *Science* **323**, 1205–1208.
27. Vance C, Rogelj B, Hortobagyi T, De Vos KJ, Nishimura AL, Sreedharan J, Hu X, Smith B, Ruddy D, Wright P, Ganesalingam J, Williams KL, Tripathi V, Al-Saraj S, Al-Chalabi A, Leigh PN, Blair IP, Nicholson G, de Belleroche J, Gallo JM, Miller CC, Shaw CE (2009) Mutations in FUS, an RNA processing protein, cause familial amyotrophic lateral sclerosis type 6. *Science* **323**, 1208–1211.

INDEX

Ales Vancura (ed.), *Transcriptional Regulation: Methods and Protocols*, Methods in Molecular Biology, vol. 809, DOI 10.1007/978-1-61779-376-9, © Springer Science+Business Media, LLC 2012

MIX
Papier aus verantwortungsvollen Quellen
Paper from responsible sources
FSC® C105338

If you have any concerns about our products,
you can contact us on
ProductSafety@springernature.com

In case Publisher is established outside the EU,
the EU authorized representative is:
Springer Nature Customer Service Center GmbH
Europaplatz 3, 69115 Heidelberg, Germany

Printed by Libri Plureos GmbH
in Hamburg, Germany